Marine Geology and Oceanography of the Pacific Manganese Nodule Province

MARINE SCIENCE

Coordinating Editor: Ronald J. Gibbs, *University of Delaware*

Volume 1 – *Physics of Sound in Marine Sediments*
Edited by Loyd Hampton

An Office of Naval Research symposium
Consulting Editors: Alexander Malahoff and Donald Heinrichs
Department of the Navy

Volume 2 – *Deep-Sea Sediments: Physical and Mechanical Properties*
Edited by Anton L. Inderbitzen

An Office of Naval Research symposium
Consulting Editors: Alexander Malahoff and Donald Heinrichs
Department of the Navy

Volume 3 – *Natural Gases in Marine Sediments*
Edited by Isaac R. Kaplan

An Office of Naval Research symposium
Consulting Editors: Alexander Malahoff and Donald Heinrichs
Department of the Navy

Volume 4 – *Suspended Solids in Water*
Edited by Ronald J. Gibbs

An Office of Naval Research symposium
Consulting Editors: Alexander Malahoff and Donald Heinrichs
Department of the Navy

Volume 5 – *Oceanic Sound Scattering Prediction*
Edited by Neil R. Andersen and Bernard J. Zahuranec

An Office of Naval Research symposium

Volume 6 – *The Fate of Fossil Fuel CO_2 in the Oceans*
Edited by Neil R. Andersen and Alexander Malahoff

An Office of Naval Research symposium

Volume 7 – *Transport Processes in Lakes and Oceans*
Edited by Ronald J. Gibbs

Consulting Editor: Richard P. Shaw

Volume 8 – *Ocean Wave Climate*
Edited by Marshall D. Earle and Alexander Malahoff

Volume 9 – *Marine Geology and Oceanography of the Pacific Manganese Nodule Province*
Edited by James L. Bischoff and David Z. Piper

A Continuation Order Plan is available for this series. A continuation order will bring delivery of each new volume immediately upon publication. Volumes are billed only upon actual shipment. For further information please contact the publisher.

Marine Geology and Oceanography of the Pacific Manganese Nodule Province

Edited by
James L. Bischoff
and
David Z. Piper
U. S. Geological Survey
Menlo Park, California

PLENUM PRESS • NEW YORK AND LONDON

Library of Congress Cataloging in Publication Data

Main entry under title:

Marine geology and oceanography of the Pacific manganese nodule province.

(Marine science; v. 9)
Includes index.
1. Submarine geology–Pacific Ocean. 2. Oceanography–Pacific Ocean. 3. Manganese nodules–Pacific Ocean. I. Bischoff, James L. II. Piper, David Z.
QE350.4.M37 551.4′65 79-12475
ISBN 0-306-40187-8

An appendix in the form of six microfiche cards,
which appears in the pocket of the inside back cover,
forms an integral part of this volume.

A Division of Plenum Publishing Corporation
227 West 17th Street, New York, N.Y. 10011

Printed in the United States of America

Preface

Deep-sea manganese nodules, once an obscure scientific curiosity, have, in the brief span of two decades, become a potential mineral resource of major importance. Nodules that cover the sea floor of the tropical North Pacific may represent a vast ore deposit of manganese, nickel, cobalt, and copper. Modern technology has apparently surmounted the incredible problem of recovering nodules in water depths of 5000 meters and the extraction of metals from the complex chemical nodule matrix is a reality. Both the recovery and the extraction appear to be economically feasible. Exploitation of this resource is, however, hindered more by the lack of an international legal structure allowing for recognition of mining sites and exploitation rights, than by any other factor.

Often, when a mineral deposit becomes identified as an exploitable resource, scientific study burgeons. Interest in the nature and genesis of the deposit increases and much is learned from large scale exploration. The case is self evident for petroleum and ore deposits on land. The study of manganese nodules is just now entering this phase. What was the esoteric field of a few scientists has become the subject of active exploration and research by most of the industrialized nations. Unfortunately for our general understanding of manganese nodules, exploration results remain largely proprietary. However, scientific study has greatly increased and the results are becoming widely available.

Early research focused on nodule distribution, morphology, mineralogy, and chemistry and, in so doing, defined questions concerning nodule genesis. World-wide exploration indicated that economic-grade nodules are primarily confined to the tropical North Pacific, from latitude 5° to 20°N and longitude 110° to 180° W. Are there properties unique to this area that control nodule genesis? Are these properties related to bottom currents, sediment geochemistry, surface primary productivity, or even to tectonic processes? Deep-sea nodules grow in diameter at rates estimated to be 1000 times slower than the accumulation rates of the sediments upon which they rest. How can these estimated growth rates be real if the nodules remain unburied? Do benthic biota move the nodules to the surface as they burrow for food and protection, or is the finer-grained sediment swept away by bottom currents? Because such questions cross disciplines, it was recognized that the environment of the nodule province should be studied as a complete system including physical and biological oceanography as well as marine geology.

The DOMES (Deep Ocean Mining Environmental Study) program provided the opportunity for just such a study. Phase I of DOMES was undertaken by the National Oceanic and Atmospheric Administration (NOAA) of the US Department of Commerce to provide environmental baseline information on three representative mining sites in the Pacific manganese nodule province. The purpose was to identify possible environmental harm that might occur from eventual commercial exploitation and to provide information useful to the development of environmental guidelines for controlling such exploitation. Three sites, each covering an area of approximately 200 km on a side and located at 8°25'N-150°47'W (Site A), 11°42'N-138°24'W (Site B), and 15°00'N-126°00'W (Site C), were chosen with advice from industry and the scientific community.

Twelve cruises of NOAA's research vessel OCEANOGRAPHER were carried out from August 1975 through November 1976, totaling approximately 240 ship days at the three sites. Scientists from U.S. academic and governmental institutions planned the cruises and gathered the data. Scientific disciplines represented were physical oceanography (studies of solar radiation and ocean currents), biological oceanography (studies of phytoplankton and benthic fauna), chemical oceanography (studies of nutrient chemistry and suspended matter), and marine geology (studies of sediment, nodules, acoustic stratigraphy).

DOMES was designed primarily as a data gathering effort and final data reports were submitted to NOAA by early 1978. The scientists involved recognized that the DOMES results are of great scientific value and represent perhaps the most complete and thorough study of a small part of the deep sea yet carried out. They decided that the results and scientific inferences thereof should be published together in this single volume rather than dispersed throughout the various scientific journals.

One of the unfortunate trends in the rapid expansion of the scientific literature is that less and less space is available for data presentation, and this unfortunately is occurring at a time when the rate of accumulation and the quality of data is improving. Although interpretations of data are made only transitory by the very nature of advancing science, the data upon which they are based are the raw materials of new ideas. In order to preserve the large DOMES data base, each contributor to this volume was encouraged to present his complete data in the microfiche appendix accompanying the book. We hope these data spawn new ideas.

Coincident with the DOMES investigation, the Pacific nodule province was studied by a number of institutions representing several countries. These include the U.S. National Science Foundation (I.D.O.E.-sea bed assessment program); the French Centre National pour l'Exploration des Oceans (Afernod program); P.P. Shirshov Institute of Oceanology of the U.S.S.R.; the Geological Survey of Japan; Institute of Oceanography of New Zealand, and the Bundesanstalt fur Geowissenschaften of West Germany. Scientists active in each of these programs were invited to contribute to this volume.

These coincident programs were primarily geological and include studies of nodule geochemistry, nodule growth rates, and the sea floor. The combination of these studies with the DOMES investigations presents a current "state of the art" report of international scientific effort in the region.

First drafts of all manuscripts were received between mid-June and late September, 1978. Each underwent rigorous and anonymous peer review by at least two reviewers. Those that were accepted were returned to the authors for appropriate modification and rewriting. Final versions of most manuscripts were received between October and November, 1978.

We thank the many peer reviewers who generously gave of their time to improve the manuscripts. Helpful cooperation, encouragement and permission to publish the DOMES material by Robert Burns of NOAA and manager of DOMES is gratefully acknowledged. Jeanne Henning, assisted on occasion by Terry Coit, cheerfully and competently shouldered much of the copy-editing, reference checking and final typing, for which we are grateful.

JAMES L. BISCHOFF
DAVID Z. PIPER

Menlo Park, California
April, 1979

Contents

I. DOMES

A. Water Column Studies

B. Biological Studies

III. APPENDIX

Appendices for the various reports in this volume appear in order on six microfiche cards in a pocket on the inside of the back cover.

THE SOLAR RADIATION ENVIRONMENT IN THE

EASTERN TROPICAL NORTH PACIFIC OCEAN

Guy A. Franceschini

Department of Meteorology
Texas A&M University
College Station, Texas 77843, U.S.A.

Abstract

Solar radiation measurements were made in the DOMES area of the eastern tropical Pacific Ocean, between 5° and 18° North and 126° and 151° West, during the late summer 1975 and winter of 1976, aboard the NOAA ship OSS OCEANOGRAPHER. These were of the net photon irradiance above the surface, and the downward photon irradiance to a depth of 150 m. Measurements were of the broad waveband between 400 and 700 nm, i.e., the photosynthetically active radiation.

Results show an average net irradiation of 39 (E/m^2-day) during the summer, and a 16% greater amount in winter, i.e, 45 (E/m^2-day), due to reduced cloudiness. During the same interval, biological primary productivity increased by 20%. Clouds were responsible also for: the unequal partitioning of radiation with respect to local apparent noon; a daily photon north-south minimum at 12°N; and an east-west maximum of daily averages in the central part of the region during both seasons.

Submarine measurements show the water to be a clean open-ocean type with marked, alternating, horizontal stratifications, fewer in number during the winter. Characteristic features include: low transmittance in a shallow layer near the surface; an increase of transmittance in the layer below to the mixed layer depth, MLD; a decrease of transmittance in the stable upper thermocline; and a general increase below. Several deviations existed. Diagnoses indicate clearer water toward the north, and toward the west away from Central America, in general. The MLD also increased toward the west. Axes of maximum values of phytoplankton parameters were generally found between the levels of 1 to 10% of near-surface values, and coincided with minimum-transmittance layers. In the summer, these biological maxima were always below the MLD (the 10-25% light level); in the winter they were above and below the deeper MLD (the 1-10% light level). In the latter instance, they could be seen in the photon signature only in the eastern transect. With the deeper winter MLD, the near-surface transmittance was higher than with the summer situation. This was not the case near 126°N.

Introduction

Solar radiation is the primary source of energy for all physical and biological systems. Hence, it was necessary to establish, as a baseline, the availability of this energy to the water mass of the DOMES region. In addition, since the planned mining operations would change the transparency of the water, at least initially, it was necessary to establish, also as a baseline, the optical properties of the water column. A major concern of the effort was with the marine biosphere. Therefore, a solar radiation baseline was needed for the study of primary productivity in the region (El-Sayed et al., 1979). Consequently, to meet these needs, measurements were made of the photosynthetically active radiation (PAR) contained within the visible waveband from 400 to 700 nm.

According to Jerlov (1976), the optical classification of near-surface waters in our region of interest is Type IB, clear open-ocean water. To the north and south of the region, the waters are somewhat more transparent, i.e., Types I and IA. In addition, the particle content in the upper 50 m is closely correlated with water motion (both horizontal currents and upwelling), and with the mixed layer depth. Consequently, the horizontal pattern of particle content is characterized by zonal bands. Jerlov shows an axis of maximum content traversing the region, approximately parallel to 9°N latitude, in which the content decreases westward.

Measurements

All solar radiation observations were made aboard the NOAA vessel OSS OCEANOGRAPHER during expeditions RP80C75 and RP80C76.

Net photon irradiance, PAR

Cosine-corrected quantum sensors (Li-Cor, Lambda Instrument Corp.) with filtered silicon photodiodes were used to determine this parameter. These sensors employ a visible bandpass interference filter to produce sharp cutoff at 400 nm and at 700 nm, and colored glass to obtain a desired spectral response (Biggs et al., 1971). Separate measurements were made of the downwelling and upwelling photon irradiances. For the former, the upfacing sensor was rigidly mounted on a bow pedestal approximately 1.7 m above the cowl on the starboard side; for the latter, the downfacing sensor was installed on the end of a bow boom at a distance of 10 m ahead of the vessel. Both exposures were selected to minimize the influence of the ship on the measurements. In order to monitor the integrity and reliability of the sensors, a precision spectral pyranometer (Eppley) was also installed on the bow pedestal to measure the incident, or downwelling, flux. Signals from the sensors, which were in continuous operation during daylight hours, were recorded on analogue or digital acquisition systems, as well as magnetic tape, when possible.

The net PAR, i.e., the photons made available to the water mass and the marine biosphere, is represented by the downwelling photon irradiance minus the upwelling, since the latter represents losses due to surface reflection and internal backscattering by the water. The manufacturer's calibration constants were accepted. Reliability of the sensors was verified by a post-expedition intercalibration with a secondary standard. Data were processed for 30-min intervals. These were combined to give daily integrated amounts, as well as integrated values for the biological *in situ* incubation periods. Data obtained with the pyranometer are not presented.

Ship shadows occasionally interfered with an otherwise clean record. The shadow signatures were easily recognized, and usually occurred in the early morning or late afternoon with an aft sun. In such cases, the data were rehabilitated by assuming a linear change during the brief shadow intervals.

Submarine light

Measurements of the vertical distribution of PAR within the water were made with quantum sensors similar to those used for the in-air observations. Due to recorder difficulties, only the downward photon irradiances were obtained. In addition, to establish the depths at which water samples were taken for primary productivity studies, a special photometer was employed (Carpenter and Jitts, 1973) as recommended by the SCOR Working Group 15 (1966). This gimbals-mounted cosine collector was a Schott-filtered selenium-barrier photocell which was sensitive to blue light at 475 nm with a 50 nm half-band width. Both underwater sensors, mounted on a common frame, were used for profiling to a depth of 150 m. The operation was performed, with the STD winch, off the sunny side of the vessel, either astern or off the port side, during mid- or late-morning preceding the *in situ* incubation experiments.

In order to allow for occasional changing cloud conditions, a deck cell was mounted on the bow pedestal. This sensor was similar to the underwater photometer, i.e., the blue-light sensor. In combination with the pyranometer, use of the deck cell made it possible to determine ratio corrections associated with fluctuating irradiances during a submarine cast so that gross transmittances of water layers could be calculated. As a first approximation, these occasional ratio corrections were also used to adjust the submarine quantum irradiances. Although the quality of PAR is influenced by cloudiness (Franceschini, 1971), such ratio corrections made it possible to adjust the submarine values relative to a fixed near-surface reference. Such adjustments were rarely necessary, however, since the down-trace and the up-trace of each cast furnished sufficient information from which a reliable profile could be established.

Signals from the submarine quantum sensor and its related pressure probe were recorded on the STD recorder (Leeds and Northrup Speedomax) available on the vessel. Quantum sensor outputs were corrected for the immersion effect according to the manufacturer's recommendation. The signals for both of the blue-light photometers, the deck-cell and its underwater counterpart, were recorded on a two-channel analogue recorder (Hewlett-Packard). Depth marks were entered manually on the analog trace.

The manufacturer's calibrations of the quantum sensors between expeditions were accepted as correct. Secondary cross-calibrations in-air before and after each expedition verified the reliability of the exceedingly stable sensors.

The depth at which the light was reduced to 1 percent of its near-surface value, as measured by the blue-light photometer, was used to define the lower limit of the euphotic zone. The near-surface light was taken as that measured at a depth of approximately 0.5 m.

Cloud data

Standard weather observations were made and recorded hourly by qualified members of the ship's scientific contingent. Sky conditions, critical to the solar radiation study, included types of clouds, and amounts in oktas. Daytime average values used in this study were based on these observations. Since solar radiation is influenced more by the clouds at low levels than those at middle and high levels (Franceschini, 1968; List, 1971), only those at low levels were used for determining the daily mean cloud values here presented. In addition, each observed hourly cloud amount, $C(i)$, was weighted according to the fractional part, $W(i)$, of undepleted daily irradiation, $q(o)$, that would be available during the associated hourly interval, i.e.,

$$W(i) = q(i)/q(o), \qquad (1)$$

where $q(i)$ is the undepleted irradiation available during the i-th hourly interval. In general, for the interval from 1 to 2,

$$q \propto \int_1^2 (\sin \phi \sin \delta + \cos \phi \cos \delta \cos H)dt, \qquad (2)$$

where: ϕ is the mid-latitude of a north-south transect; δ is the associated solar declination angle; and H is the solar hour angle which varies with time, t. Daily values of the weighted-mean, low-level cloudiness, C, were calculated from

$$C = \sum_{i=1}^{12} W(i)C(i)/100. \tag{3}$$

For this approximation, the weighting factors used for both expeditions were: 2, 5, 8, 10, 12, 13, 13, 12, 10, 8, 5 and 2. These were applied to the hourly values of C(i) from 07:00 to 18:00 local standard time.

Positions occupied

Stations occupied were primarily along three north-south transects each of which extended 3 degrees of latitude north and south of a central station. Transect A, T-A, was centered near 8°N and 151°W; T-B was centered near 12°N and 138°W; and T-C, near 15°N and 126°W. Location of an individual station along a transect is indicated by a numeric suffix showing its distance, in degrees of latitude, north (plus) or south (minus) of the central station, e.g., B+2 indicates the station is 2 deg north of the central station of transect B. (See Fig. 1, El-Sayed and Taguchi, this volume).

Units

Values of quantum irradiation are in units of Einsteins per square meter (E/m^2) during the interval of time considered, e.g., day or incubation period. Submarine quantum-irradiance values are given in units of ($\mu E/m^2 s$). One Einstein is defined as 6.023×10^{23} photons. Weighted-mean, low-level cloud cover, C, is given in oktas, i.e., eighths.

Results, Net Photon Irradiation

Legs 1 and 2, 29 Aug to 12 Oct 1975

Results for these legs are presented in Table 1. In general, the southward increase of radiation that might have been anticipated on the basis of astronomical considerations was not observed on any of the transects. The influence of clouds, a major controlling factor, has masked the dependence on mean solar altitude. For example, clear-sky daily irradiation should have been a maximum during the period of Transect C, and a minimum during Transect B. However, when transect-mean values are considered, a maximum was observed during T-B, a minimum during T-A, viz., 40.7 E/m^2-day and 37.4 E/m^2-day, respectively, with associated cloudiness values of 3.9 and 4.9 oktas, respectively. An anomalous minimum during each transect was found at 12°N latitude in association with excessive cloudiness. In addition, though not shown in Table 1, cloudiness was responsible for an unequal partitioning of solar energy with respect to local apparent noon, LAN.

TABLE 1

Net photon irradiation. PAR gained by the water mass during Legs 1 and 2 of DOMES/OCEO Expedition, 29 Aug.-12 Oct., 1975. Beginning times (LST) of in-situ incubation periods are given; all end at sunset. 1 Einstein = 6.023 x 10^{23} photons.

Date	Station	Cloudiness (oktas)	Day (E/m^2)	In-situ (E/m^2)	In-situ (LST)
Aug 29	C+3	8.0	22.9	5.8	13:49
30	C+2	6.0	34.5	19.0	12:43
31	C+1	2.0	49.7	25.9	12:42
Sep 1	C	1.0	52.9	52.9	06:00
2	C	4.9	39.9	-	-
3	C	5.8	30.6	-	-
4	C-1	2.2	46.8	21.0	12:26
5	C-2	3.9	40.3	18.7	12:26
Sep 6	C-3	7.1	26.4	16.5	12:37
Oct 12	B+3	3.6	41.9	-	-
11	B+2	-	41.9	-	-
10	B+1	2.7	50.1	28.2	11:34
9	B	6.5	23.7	-	-
7	B	3.3	45.3	23.2	12:12
6	B	3.2	41.0	-	-
5	B-1	3.9	41.6	26.3	10:59
4	B-2	4.3	40.4	27.4	10:58
Oct 3	B-3	5.4	-	-	-
Sep 22	A+3	8.0	21.8	21.8	05:46
23	A+2	6.0	30.8	17.2	11:44
24	A+1	4.0	42.9	23.3	11:56
25	A	4.2	41.6	41.6	05:45
26	A	2.3	48.0	-	-
27	A	2.9	42.7	-	-
28	A-1	7.4	-	-	-
29	A-2	6.3	32.5	16.2	11:53
Sep 30	A-3	5.3	38.7	17.5	11:59

Although the sample is small, i.e., 24 pairs for the three transects, the data show a good linear relationship between the daily values of net irradiation of quanta, Q, and the weighted-mean low-level cloudiness, C, as defined in (3). The least-square line of regression is

$$Q = a(1 - bC), \tag{4}$$

where: a is 59.1 and b is 0.077; and the correlation coefficient is -0.98. In (4), Q is expressed as (E/m^2-day), and C in oktas. Mean values for the three fall transects are: 38.7 E/m^2-day, and 4.4 oktas.

Based on our findings, the daily values of net photon irradiation were 95 percent of the downwelling global values. Thus, the upwelling losses amounted to 5 percent of the incident flux.

Leg 4, 18 Feb to 15 Mar 1976

Data for this leg are seen in Table 2. Unlike the fall/late-summer expedition, this case showed a southward increase of solar energy during each transect, i.e., the southern portion of each transect had larger photon irradiations than the northern half. However, the influence of cloudiness was again apparent. Based on astronomical factors, the maximum clear-sky daily value of the transect means would have been expected during T-A, and a minimum during T-C. The minimum, 41.6 E/m^2-day, was indeed observed during T-C, when mean cloudiness was 4.4 oktas, but the maximum, 47.2 E/m^2-day, occurred during T-B when average cloudiness was 2.9 oktas. An anomalous minimum during each transect was again observed at 12°N latitude, the zone of maximum cloudiness. As during the fall expedition: cloudiness was responsible for the unequal a.m.-p.m. partitioning of solar energy; and the daily values of net photon irradiation were found to be 95 percent of the downwelling values.

The small sample, viz., 20 pairs for this leg, also shows a good linear relationship between Q and C, as defined in (4). Related values of the regression coefficients are: a = 63.6, and b = 0.082. The correlation coefficient is -0.96. Mean values for the three transects are: 45.0 E/m^2-day, and 3.6 oktas.

Seasonal changes

The first expedition, Legs 1 and 2, was approximately centered on the autumnal equinox; the second cruise preceded the vernal equinox. Based on consideration of the local solar altitude, no significant change in clear-sky daily values of the transect means would have been expected for T-A and T-B, and an appreciably lower spring value for T-C. However, due to the lesser amounts of cloudiness in the late winter, all transect-mean irradiation values during Leg 4 were larger than during Legs 1 and 2. In particular, these increases in daily values were: T-A, 24 percent; T-B, 16 percent; and T-C, 9 percent. The associated percentage

TABLE 2

Net photon irradiation. PAR gained by the water mass during Leg 4 of DOMES/OCEO Expedition, 18 Feb.-15 Mar., 1976. Beginning times (LST) of in-situ incubation periods are given; all end at sunset except the one on 3 Mar., which ended at 13:44 LST. 1 Einstein = 6.023 x 10^{23} photons. (Estimated values, in parentheses, are 95% of incident fluxes.)

Date	Station	Cloudiness (oktas)	Day (E/m^2)	In-situ (E/m^2)	In-situ (LST)
Feb 18	C+3	7.7	(24.3)	(9.0)	13:16
19	C+1.5	5.4	(35.2)	(17.1)	12:35
20	C	5.0	(41.0)	-	-
21	C	3.7	(46.6)	(24.5)	12:32
22	C	3.2	(49.9)	(49.9)	07:35
23	C-1.5	2.8	(50.2)	(34.7)	11:33
Feb 24	C-3	3.2	(43.7)	(19.9)	12:45
Feb 26	12N/135W	-	39.6	-	-
Feb 27	12N/138W	-	46.2	-	-
Feb 28	B+3	2.5	50.5	22.9	12:37
29	B+1.5	3.1	44.7	20.6	12:50
Mar 1	B	4.1	38.3	-	-
2	B	3.5	41.5	25.9	12:35
3	B	2.6	52.5	27.8	06:09
4	B-1.5	2.0	55.6	20.0	13:22
Mar 5	B-3	2.7	47.5	27.4	12:27
Mar 6	8N/140W	-	33.6	-	-
15	18N/156W	-	11.5	-	-
Mar 14	14N/152W	-	18.7	-	-
Mar 13	A+3	7.8	22.0	11.7	12:15
12	A+1.5	3.1	46.2	17.7	12:30
11	A	1.9	53.2	29.9	11:51
10	A	1.6	54.1	-	-
9	A-1.5	2.4	52.3	28.2	12:06
Mar 8	A-3	3.3	50.2	-	-
Mar 7	7N/146W	-	32.3	-	-

decreases in cloudiness were: 31, 26 and 2, respectively. Such changes attest to the dominant role of cloudiness on the number of photons made available to the marine biomass for primary productivity, as well as to the physical system.

Average conditions

Due to the paucity of solar radiation observations in ocean areas, it is often necessary to rely on empirical relations when estimates of irradiation are required. Such estimates could be based on cloud information which is normally included in standard weather observations. Although few in number, shipboard weather observations are relatively more plentiful than those of radiation. Consequently, data for both expeditions were combined. Based on the 44 pairs of values, the line of regression, (4), is

$$Q = 61.1(1 - 0.079C), \tag{5}$$

and the correlation coefficent is -0.94. The coefficient, 61.1, may be interpreted as the best estimate of quantum-irradiation for a day with no low-level clouds in the DOMES area. In addition, for overcast conditions, 8 oktas of clouds at low levels, (5) gives an estimated irradiation of 22.5 E/m^2-day. Mean values for all data are: $Q = 41.6$ E/m^2-day, and $C = 4.0$ oktas of clouds at low levels. Although the data sample is small, these results may be useful for other tropical oceanic regions.

Results, Submarine Light

A portraiture of the results of submarine measurements is presented in three forms for each season and each transect: (a) tabular values of the downwelling photon irradiances; (b) transect diagnostic figures of these values expressed as a percentage of the near-surface submarine flux, viz., the irradiance at approximately 0.5 m; and (c) diagnostic figures of the transmittance of 10-m layers, i.e., the percentage of the incident downwelling irradiance which is transmitted by each 10-m layer. Included in (b) are the depths of the 1-percent level as determined by the blue-light photometer, as well as the mixed-layer depth, MLD. The latter is also included in the figures of 10-m layer transmittance. This gross optical property, the 10-m layer transmittance, is a measure of the clarity of the water; 100 minus this value expresses the percentage depletion of the downward stream of photons. The combined effects of absorption and scattering account for this reduction.

It may be well to indicate that the magnitude of a percentage value at the depth of the bottom of the n-th 10-m layer, $T(10n)$ as employed in (b), is related to the product of 10-m layer transmittances, $\tau(j)$ represented in (c), for all layers above that depth. The relation may be expressed as:

$$T(10n) = (1/100)^n \prod_{j=1}^{n} \tau(j), \tag{6}$$

where j = 1 for the top 10-m layer, and j = 15 for the 10-m layer with base at the 150-m depth. In addition, although the isopleths in presentations of the 10-m layer transmittance in (c) are labeled in percent, they may be interpreted in terms of a related gross attenuation coefficient, k(1/m). For the j-th layer, this downward quantum-irradiance coefficient is defined as:

$$k(j) = - [\ln\{\tau(j)/100\}]/10. \tag{7}$$

The equivalent values, according to (7), are:

τ(%)	10	20	30	40	50	60	70	80	90
k(1/m)	.230	.161	.120	.092	.069	.051	.036	.022	.011

Longitudinal sections, along 12°N latitude, of the percentage penetration of light as well as the gross transmittance are also presented. The seasonal changes of these optical parameters are included to complete the portraiture of the baseline milieu of spatial and temporal distributions of the submarine solar photon irradiances.

Transect A, Leg 2, 22 to 30 Sep 1975

Irradiance. The downward photon fluxes of PAR are given in Table 3. The maximum near-surface value occurred at A-2; the minimum at A-1. However, with depth these extrema shifted northward because of the variable turbidity of the water. As a result, at 150 m the maximum photon flux was found at the central A station, the minimum at A+1. Hence, subsurface irradiances cannot be based solely on near-surface measurements. Percentage penetrations are shown in Fig. 1. In the figure, MLD is indicated by X; the 1-percent blue-light level by a circle. It will be noted that these latter depths are greater, from 10 to 40 m, than the 1-percent depths of total photon fluxes. This is to be expected, since the water is more transparent to the higher energy photons. In general, the larger differences occurred on the less-cloudy days when a smaller fraction of the total energy was in the blue part of the spectrum (Franceschini, 1973). Large attenuations, in excess of 50 percent, were found in the upper 10 m, especially at the southern stations of the transect. This was due to the selective absorption of low energy photons associated with the longer wavelengths. In addition, 10 percent or more of the near-surface light penetrated to the MLD north of A-1, whereas only 1 percent reached the deeper MLD at the two southernmost stations. There was an apparent northward deeping of the isopleths, in general.

TABLE 3

Downward flux of photosynthetically active solar radiation (PAR) as a function of depth during DOMES/OCEO Expedition, Leg 2, Transect A, 22-30 Sept. 1975. Times are of the beginning of each cast in 150th-meridian standard times; units are $\mu E/m^2 s$ where 1 Einstein = 6.023×10^{23} photons.

Station Depth (m)	A+3	A+2	A+1	A	A	A	A-1	A-2	A-3
0	400	655	1747	515	1223	1878	286	1944	728
5	188	379	946	240	480	909	140	808	395
10	149	298	612	169	376	699	112	563	278
20	119	226	335	95.1	258	463	83.2	262	185
30	96.1	175	219	60.3	170	321	63.5	140	144
40	72.5	137	160	40.6	109	238	47.5	98.3	118
50	49.4	78.6	112	24.5	65.5	141	34.0	74.3	74.3
60	33.2	48.1	55.0	15.7	39.3	83.0	22.2	45.9	47.3
70	22.7	29.8	33.2	10.5	22.5	48.9	13.9	21.8	30.0
80	14.1	17.6	9.61	7.57	13.9	27.5	8.70	13.5	17.2
90	9.46	10.5	4.80	5.97	9.52	17.1	5.43	9.39	13.1
100	6.26	6.55	2.84	5.10	7.08	10.8	3.72	6.77	9.81
110	4.37	4.80	2.01	4.37	5.68	7.43	2.72	4.80	6.84
120	3.32	3.32	1.46	3.93	4.89	5.37	2.11	3.36	5.53
130	2.40	2.66	1.14	3.73	4.28	4.02	1.85	1.92	3.13
140	1.72	2.37	0.98	3.43	4.06	3.28	1.69	1.27	1.89
150	1.41	2.24	0.90	-	3.93	2.78	1.56	0.98	-
Time (LST)	11:20	09:22	10:46	11:50	13:23	10:25	09:47	09:40	9:49
Date (1975)	22 Sep	23 Sep	24 Sep	25 Sep	26 Sep	27 Sep	28 Sep	29 Sep	30 Sep

The depth of the euphotic zone and of the 1% level of PAR-photons, both of which are shown in Fig. 1, serve as useful parameters for the classification of waters for biological purposes. In an optical sense, each depth, Z, to which 1% (or 0.01) of the near-surface light penetrates, defines a diffuse attenuation coefficient, K, analogous to k in (7): $K = -[\ln(0.01)]/Z = 4.605/Z$. Equivalent values are:

Z(m)	50	70	90	110	130	150
K(1/m)	.092	.066	.051	.042	.035	.031

For the blue light, the euphotic-zone coefficient, K, ranges from 0.035 to 0.046 per m; for the PAR-photons, K varies from 0.042 to 0.066 per m for the related layer. The maximum value for the blue light occurs at 9°N latitude, A+1, where upwelling is suggested.

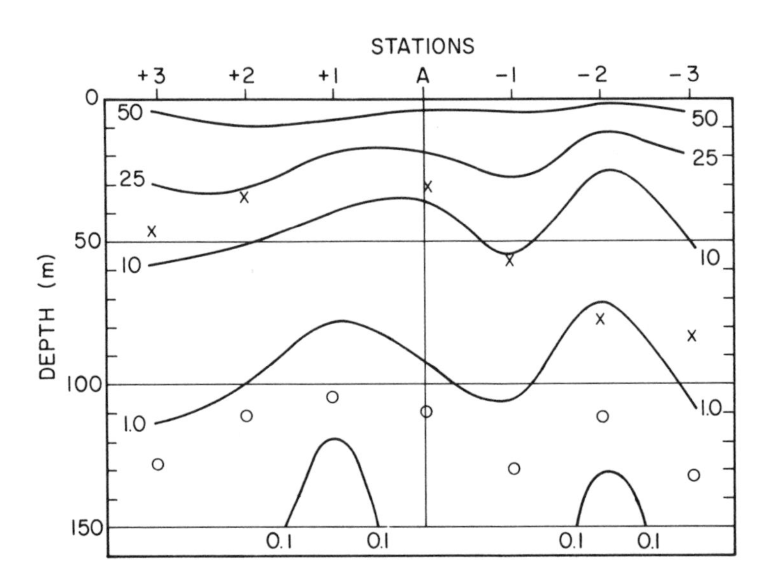

Fig. 1. Vertical distribution of submarine PAR-photon irradiance in percent of 0.5-m flux. x, MLD; o, 1% blue-light level. DOMES, Leg 2, 22-30 Sep 1975. (From Table 3).

<u>Transmittance of 10-m layers.</u> An analysis of this optical property is shown in Fig. 2. Small values near the surface gave way with depth to alternating high and low values in a stratified pattern. The axes of the quasi-horizontal layers of low transmittance, high attenuation, are indicated by the heavy dashed lines. Except at A-2 and A-3, the shallowest of these was in the stable thermocline below the MLD, indicated by x's. The marked features of this transect were the turbid waters between 50 and 100 m north of A, and the general increase of transmittance below 100 m. The turbid waters at 9°N (A+1), with a minimum transmit-

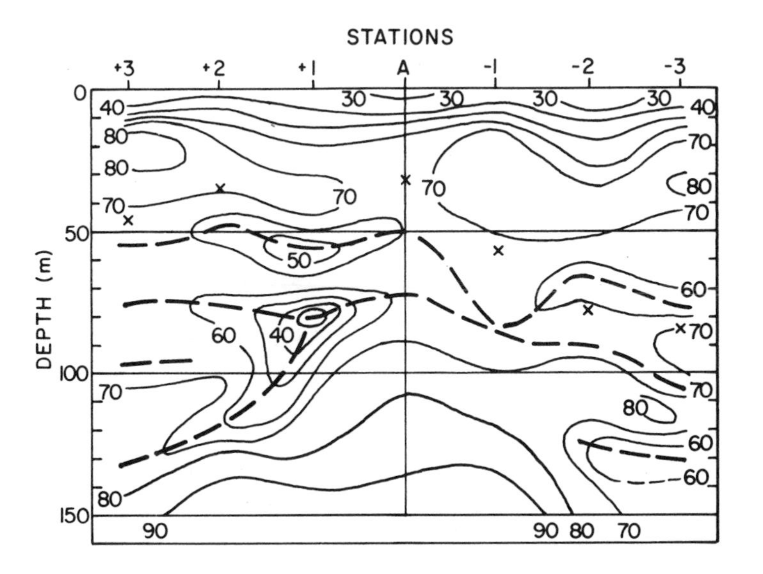

Fig. 2. Vertical distribution of submarine photon transmittance (%) for 10-m layers. Heavy dashed line, axis of minimum values; x, MLD. DOMES, Leg 2, 22-30 Sep 1975. (From Table 3).

tance of 30% and related 10-m layer gross attenuation coefficient of 0.120 per m at 80 m, may be associated with upwelling at the northern boundary of the Equatorial Countercurrent (Jerlov, 1976). There was excellent agreement between positions of the upper two turbid layers and the maximum concentration of chlorophyll a and phaeopigments, as well as maximum primary productivity (El-Sayed, Fig. 3 and 7, this volume). This is especially noteworthy at A-2 and A-3 where these features were found above the deeper MLD.

Transect B, Leg 2, 3-12 Oct 1975

Irradiance. Table 4 presents measured values of the downward photon flux. Here we see the near-surface maximum occurred at B+1; the minimum at B-3. As with T-A, these extrema were found at more northerly positions at the bottom of the runs, viz., at B+3 and B-2, respectively. Percentage penetrations are given in Fig. 3. Again, we note that large attenuations, in excess of 50 percent, occurred in the upper 10 m, especially in the southern sector, and that more than 10 percent of the near-surface flux penetrated the shallow MLD. The 1-percent blue-light levels were found again at depths greater, by 10 to 20 m, than the corresponding depths of the total photon flux. The gross attenuation coefficient, K, for the euphotic zone ranged from 0.044 to 0.061 per m, with an average of 0.053 per m; for the PAR-photons, the related K values were 0.054 to 0.079 per m, respectively, and 0.066 per m as an average. The maximum value of K, i.e., one of

TABLE 4

Downward flux of photosynthetically active solar radiation (PAR) as a function of depth during DOMES/OCEO Expedition, Leg 2, Transect B, 3-12 Oct. 1975. Times are of the beginning of each cast in 135th-meridian standard time; units are $\mu E/m^2 s$ where 1 Einstein = 6.023×10^{23} photons. (Values in parentheses refer to the level 5 m above the table depth)

Station Depth (m)	B+3	B+2	B+1	B	B	B	B-1	B-2	B-3
0	1529	1092	1835	1354	1136	1441	1143	1070	845
5	725	612	917	612	524	764	341	280	379
10	568	428	507	437	393	489	236	205	291
20	419	267	301	262	216	295	131	108	194
30	260	147	184	153	123	175	79.5	72.1	124
40	186	89.1	103	78.6	72.5	100	42.4	39.3	79.4
50	128	47.2	66.4	39.3	37.6	51.5	19.2	19.2	44.1
60	86.5	24.9	36.3	19.7	19.0	27.1	9.17	9.61	22.0
70	52.0	13.8	20.5	10.5	9.92	15.3	5.21	5.15	10.5
80	21.4	7.86	11.2	5.33	5.02	8.43	3.67	2.84	4.95
90	12.8	3.58	6.81	2.69	2.36	4.46	2.05	1.51	2.27
100	6.12	1.94	3.06	1.57	1.27	2.53	1.14	0.87	1.06
110	3.65	1.01	1.36	0.94	0.73	1.40	0.61	0.50	0.57
120	2.49	0.59	1.10	0.52	0.40	0.73	0.36	0.30	0.30
130	1.79	0.35	0.66	0.31	0.24	0.26	0.22	0.15	0.17
140	1.03	0.24	0.35	(0.26)	0.15	-	0.11	(0.11)	-
150	0.70	0.15	0.10	-	0.10	-	-	-	-
Time (LST)	09:48	09:45	09:36	09:29	14:48	09:50	09:12	09:36	09:21
Date (1975)	12 Oct	10 Oct	9 Oct	8 Oct	7 Oct	6 Oct	5 Oct	4 Oct	3 Oct

the shallowest euphotic zones of the entire area, namely 76 m, was found at B-2, near 10°N latitude. This characteristic feature may well be connected to the similar feature at A+1 near 9°N as shown in Fig. 1 and 2. As in T-A, there was a northward deepening of the isopleths.

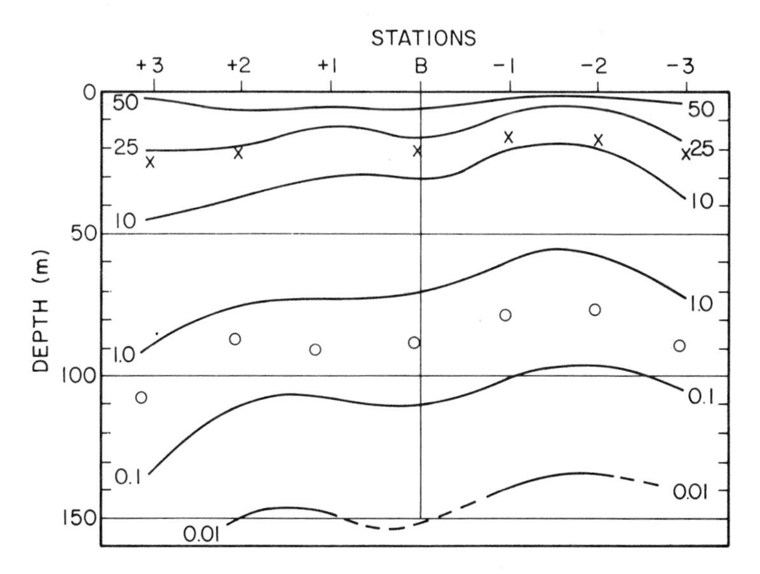

Fig. 3. Vertical distribution of submarine PAR-photon irradiance in percent of 0.5-m flux. x, MLD; o, 1% blue-light level. DOMES, Leg 2, 3-12 Oct 1975. (From Table 4)

Transmittance of 10-m layers. An analysis is presented in Fig. 4. As with T-A, low magnitudes near the surface gave way with depth to alternating high and low values, but in a more orderly and horizontally stratified pattern. Of the four well defined minimum axes, the shallowest was found again in the stable thermocline immediately below the MLD. However, it is along the axis immediately below this that we found excellent corresponsence with phytoplankton parameters, i.e., the axis of maximum primary productivity, chlorophyll a, and phaeopigments (El-Sayed, Fig. 3 and 7, this volume). Related transmittance values of less than 50%, somewhat more extensive south of Station B, correspond to a PAR-quantum absorption coefficient, k in (7), of greater than 0.070 per m, which is in good agreement with Jerlov (1976). This may be associated with upwelling near the northern boundary of the Equatorial Countercurrent. Comparison with Fig. 2 shows this feature is not as well marked as its counterpart at A+1. There was also good agreement with the increased concentration of phaeopigments below 110 m south of Station B. The less turbid waters observed at the greater depths in T-A were not apparent in this transect.

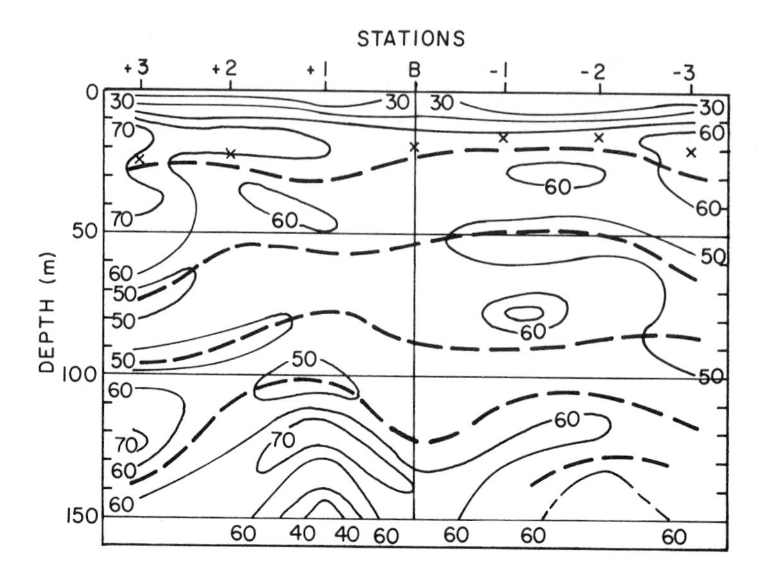

Fig. 4. Vertical distribution of submarine photon transmittance (%) for 10-m layers. Heavy dashed line, axis of minimum values; x, MLD. DOMES, Leg 2, 3-12 Oct 1975. (From Table 4).

Transect C, Leg 1, 29 Aug to 6 Sep 1975

Irradiance. Observed values of the downward photon irradiances are given in Table 5. The near-surface maximum occurred at C-2; the minimum, at C. At all depths below 10 m, the minimum was found at C-3, while the maximum had shifted to a more northerly position, C+1, below 30 m. An analysis of these variations will be discussed in the next section. Fig. 5 shows the vertical cross-sectional pattern of the percentage penetration. Except in the water of C+1, characteristic depletions in excess of 50 percent were the rule in the top 10-m layer. In general, more than 25 percent of the near-surface irradiance reached the shallow MLD, except at the extremities of the transect. Characteristic northward deepening of the isopleths was especially well marked from C-3 to C+1. In the remainder of this transect, it was masked by the anomolously clear water at C+1. The deepest euphotic zone of the DOMES region, viz., 141 m, was associated with this mass of clear water.

The 1-percent blue-light depth was 20 to 35 m greater than its total photon counterpart. Associated K values, for the 1%-layer, ranged from 0.033 to 0.055 per m for the blue light, and from 0.041 to 0.082 per m for the PAR-photons. Related averages were 0.039 and 0.051 per m, respectively. Although the euphotic zone was shallow at station C-3 (12°N), because of the zonal character of the current system in this tropical region, it may well have been shallower south of the transect at this longitude.

TABLE 5

Downward flux of photosynthetically active solar radiation (PAR) as a function of depth during DOMES/OCEO Expedition, Leg 1, Transect C, 29 Aug.-6 Sept. 1975. Times are of the beginning of each cast in 120th-meridian standard time; units are $\mu E/m^2 s$ where 1 Einstein = 6.023×10^{23} photons. (Values in parentheses refer to the level 5 m above the table depth).

Station Depth (m)	C+3	C+2	C+1	C	C	C	C-1	C-2	C-3
0	582	568	670	590	684	451	1573	1835	936
5	320	335	575	480	320	367	917	917	363
10	280	240	502	437	255	332	728	681	234
20	216	160	398	315	214	262	451	437	109
30	166	105	300	258	170	211	328	274	60.1
40	118	74.3	236	210	142	169	218	163	31.7
50	75.7	56.1	172	171	103	122	149	90.3	14.9
60	44.3	40.5	138	134	69.9	74.3	91.7	43.7	6.19
70	26.9	29.1	104	97.4	44.4	39.8	49.8	20.4	2.24
80	16.0	20.8	69.6	62.0	23.6	19.2	26.2	9.90	0.60
90	8.74	15.1	45.1	35.4	10.6	8.74	13.3	4.51	0.15
100	4.51	9.76	24.8	15.4	4.08	3.71	5.81	1.49	0.04
110	1.92	5.53	7.72	5.90	1.24	1.08	2.14	(0.44)	0.01
120	0.60	2.91	2.81	1.03	(0.98)	(0.41)	0.35	-	-
130	0.09	1.31	1.49	(0.44)	-	-	-	-	-
140	-	0.51	0.82	-	-	-	-	-	-
150	-	0.01	0.51	-	-	-	-	-	-
Time (LST)	10:42	10:35	10:00	13:00	10:55	13:40	10:30	11:00	10:30
Date (1975)	29 Aug	30 Aug	31 Aug	1 Sep	2 Sep	3 Sep	4 Sep	5 Sep	6 Sep

Presumably, T-C lies primarily within the North Equatorial Current. However, at its southern extremity, conditions strongly suggest the presence of upwelling associated with horizontal divergence near the boundary between the Equatorial Countercurrent and the North Equatorial Current. The zonal continuity of this feature is apparent in transects A, B and C.

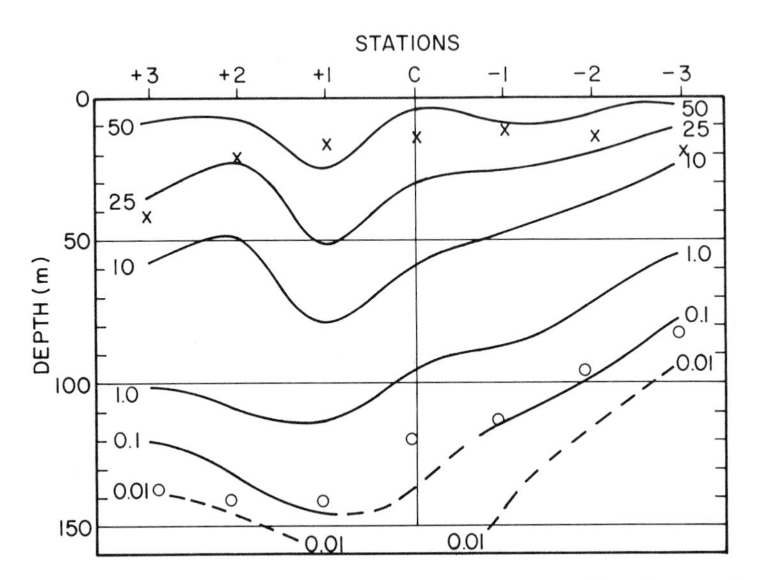

Fig. 5. Vertical distribution of submarine PAR-photon irradiance in percent of 0.5-m flux. x, MLD; o, 1% blue-light level. DOMES, Leg 1, 29 Aug - 6 Sep 1975. (From Table 5).

Transmittance of 10-m layers. Clarification of the anomalous character along this transect can be seen in Fig. 6. The highly transparent mass of water above 100 m, centered north of C, was completely surrounded by more turbid water on all sides in the cross section. This is in good agreement with particle-content conditions in the upper 50 m of the area as presented by Jerlov, Fig. 131 (1976). Associated with these centers of maximum photon-transmittance, the gross attenuation coefficient was 0.022 per m or less. Although this water mass, in an optical sense, dominated the pattern, stratification existed. A well marked minimum axis was situated immediately below the MLD. This corresponded to the axis of maximum primary productivity. At greater depths, there was good agreement between the position of the northward deepening min-axis and the phytoplankton concentration parameters (El-Sayed, Fig. 3 and 7 this volume). Compared to T-A and T-B this transect had the most turbid water below 100 m, and the lowest values of transmittance at C-3, i.e., less than 20%. The related 10-m layer attenuation coefficient was greater than 0.161 per m, a maximum for the expedition.

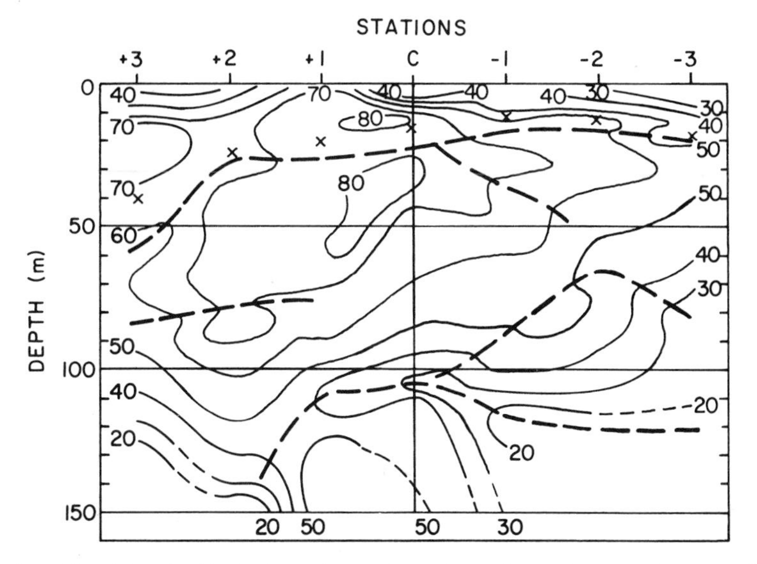

Fig. 6. Vertical distribution of submarine photon transmittance (%) for 10-m layers. Heavy dashed line, axis of minimum values; x, MLD. DOMES, Leg 1, 29 Aug - 6 Sep 1975. (From Table 5).

Longitudinal variations near 12°N, Legs 1 and 2

The west-east spatial distribution of the percentage levels between A+3, B and C-3 is presented in Fig. 7. The following features are apparent; more than 50 percent of the downward irradiance was attenuated in the top 10 m; the isopleths deepened from the turbid water in the east to the more transparent water in the west; more than 10 percent reached the westward-deepening MLD; and the 1-percent blue-light level exceeded its total photon counterpart by 10 to 30 m. The associated attenuation coefficient, K for the euphotic zone, ranged from 0.055 per m in the east to 0.037 per m in the less turbid water in the west. Corresponding coefficients for the PAR-photon 1%-layer ranged from 0.082 per m in the more turbid water in the east to 0.041 per m in the west. In terms of the 10-m layer transmittance of ambient light, shown in Fig. 8: the water became more transparent toward the west at all levels; the pattern was characteristically stratified; and the shallowest axis of minimum values was found in the stable layer below the MLD, with a zone of maximum values immediately above it and just below the less transparent 10-m surface layer. It was suggested that the zone of upwelling in the region lies south of this cross-section. The more turbid water in the east may suggest the zone is farther north, or better defined, in this sector.

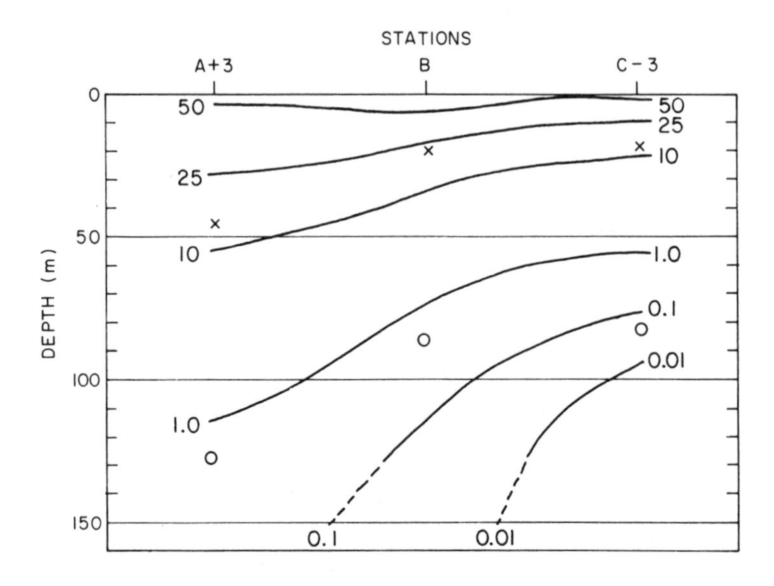

Fig. 7. Vertical distribution of submarine photon irradiance, in percent of the 0.5-m value, along 12°N latitude. x, MLD; o, 1% blue-light level. DOMES, late summer 1975.

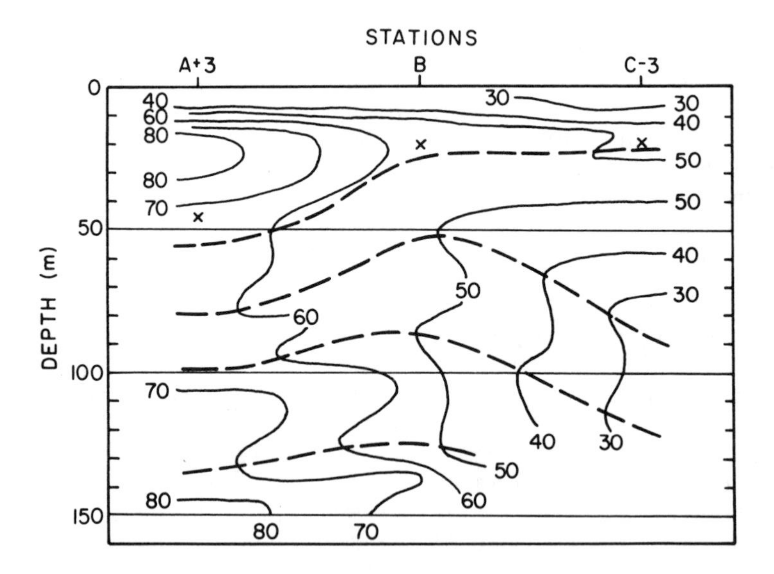

Fig. 8. Vertical distribution of submarine photon transmittance (%) for 10-m layers, along 12°N latitude. Heavy dashed line, axis of minimum values; x, MLD. DOMES, late summer 1975.

TABLE 6

Downward flux of photosynthetically active solar radiation (PAR) as a function of depth during DOMES/OCEO Expedition, Leg 4, Transect A, 8-13 Mar. 1976. Times are of the beginning of each cast in 150th-meridian standard time; units are $\mu E/m^2 s$ where 1 Einstein = 6.023×10^{23} photons.

Station Depth (m)	A+3	A+1.5	A	A-.25	A-1.5	A-3
0	275		2053	1872	711	1041
5	138		946	970	382	658
10	113		582	693	285	444
20	77.3		284	453	170	242
30	54.2		149	290	112	157
40	38.9		80.1	185	73.4	102
50	28.2		48.5	114	48.6	63.9
60	20.2		34.5	65.3	34.8	42.5
70	14.4		23.3	38.7	27.2	27.1
80	9.04		15.7	21.6	21.1	18.9
90	4.50		11.8	13.9	15.3	13.9
100	2.32		8.87	9.41	10.8	10.5
110	1.24		6.73	6.80	7.81	7.63
120	0.75		4.76	5.01	4.51	5.67
130	0.45		3.10	3.90	2.70	3.92
140	0.26		1.83	-	0.61	3.18
150	0.17		0.96	-	-	2.61
Time (LST)	10:27	-	09:08	11:16	10:58	10:27
Date (1976)	13 Mar	12 Mar	11 Mar	10 Mar	9 Mar	8 Mar

Transect A, Leg 4, 8-13 Mar 1976

Irradiance. Results are presented in Table 6. Due to equipment failure, data are not available for A+1.5. The maximum photon flux level near the surface was at A; the minimum at A+3. Below 110 m the maximum was found at A-3; the minimum remained at A+3. In terms of the percentage values, shown in Fig. 9, less than 50 percent was transmitted by the top 10-m layer. Only between 1 and 10 percent of the near-surface irradiance penetrated to the much depressed late winter MLD; and the blue-light photometer indicated 1-percent levels which were 15 to 40 m deeper than those of the total photon flux. Of the two southern stations, the smaller difference was found at A-1.5, where cloudier conditions had reduced the total flux and increased its fractional blue component. Isopleths deepened both north and south of the central station. This central upward bulging, seen also in Fig. 1, may indicate upwelling at the northern edge of the Equatorial Countercurrent. The related K-value is 0.049 per m for the euphotic zone.

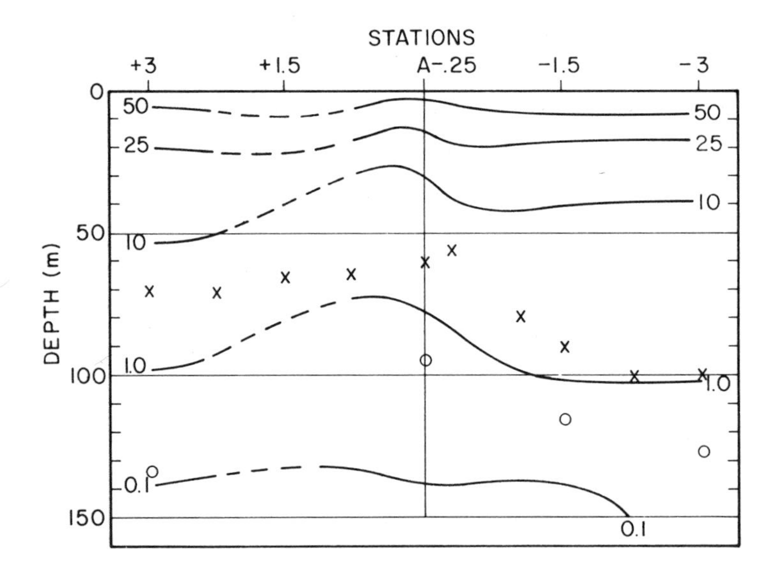

Fig. 9. Vertical distribution of submarine PAR-photon irradiance in percent of 0.5-m flux. x, MLD; o, 1% blue-light level. DOMES, Leg 4, 8-13 Mar 1976. (From Table 6).

Transmittance of 10-m layers. The pattern of this optical property, somewhat less ordered, is shown in Fig. 10. The typical signatures, however, are apparent. The more turbid surface waters gave way with depth to a more transparent layer the axis of which was above the deep MLD. Embedded in the thermocline, below the MLD, was the quasi-horizontal layer of more turbid water marked by high concentrations of chlorophyll *a* and phaeopigments (El-Sayed,

Fig. 4, this volume). In general, the water above 60 m and below 120 m, was more transparent to the photon flux in the northern section of this transect than in the southern parts. The reverse was true between 60 m and 120 m. Of particular importance, the layers of high productivity and maximum concentration of chlorophyll a and phaeopigments above the deep MLD were not apparent in the transmittance pattern (El-Sayed, Fig. 4 and 8, this volume). Near A-0.25, a turbid mass (k > 0.051 per m) appears to have penetrated the MLD at a depth of around 60 m, and to extend toward the surface north of A, in concert with the high values of phytoplankton parameters. The feature may be a consequence of the upwelling mentioned above. At depth, it may extend southward.

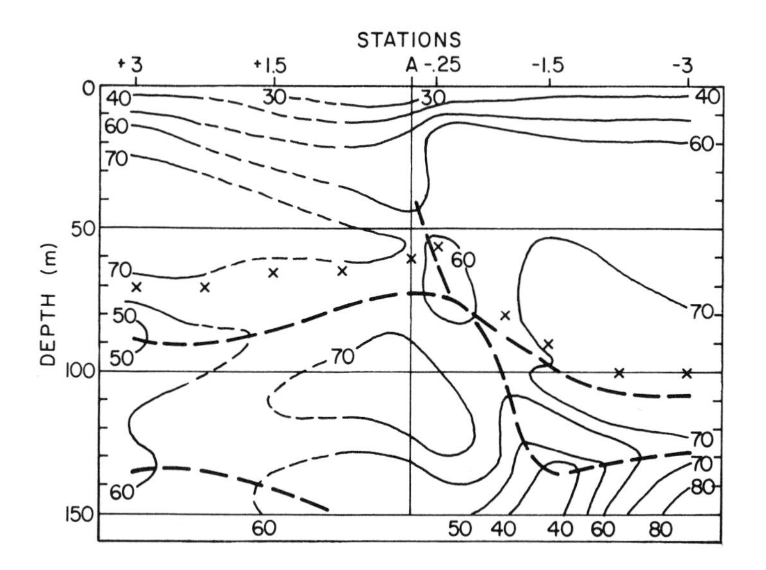

Fig. 10. Vertical distribution of submarine photon transmittance (%) for 10-m layers. Heavy dashed line, axis of minimum values; x, MLD. DOMES, Leg 4, 8-13 Mar 1976. (From Table 6).

Transect B, Leg 4, 28 Feb to 5 Mar 1976

Irradiance. The downward, submarine, photon irradiance of PAR is presented in Table 7. The vertical axis of the transect minimum remained at station B. However, the maximum, found at B-1.5 near the surface, was found at station B+3 immediately below the surface, and remained at this station to a depth of 80 m, much to the advantage of the local phytoplankton community. Below this level, the largest irradiances apparently shifted southward to B+1.5.

Percentage levels of downward irradiance are seen in Fig. 11. Less than 50 percent of the near-surface photons reached the 10-m level. Except near B+1.5, less than 10 percent penetrated to the

TABLE 7

Downward flux of photosynthetically active solar radiation (PAR) as a function of depth during DOMES/OCEO Expedition, Leg 4, Transect B, 28 Feb.-5 Mar. 1976. Times are of the beginning of each cast in 135th-meridian standard time; units are $\mu E/m^2s$ where 1 Einstein = 6.023×10^{23} photons.

Station Depth (m)	B+3	B+1.5	B	B-1.5	B-3
0	1318	818	238	1341	578
5	744	439	143	577	331
10	482	375	100	380	235
20	251	221	58.7	178	132
30	159	156	37.5	96.0	79.5
40	111	112	24.2	57.6	50.3
50	71.6	68.3	14.8	35.2	29.0
60	45.3	35.6	8.41	22.5	19.4
70	27.9	18.8	4.40	10.2	12.1
80	16.1	10.8	1.90	5.1	7.38
90	6.5	7.20	0.81	2.4	4.02
100	-	4.31	-	1.2	2.23
110	-	2.80	-	0.59	1.36
120	-	1.31	-	0.40	0.83
130	-	0.60	-	0.21	0.40
140	-	-	-	-	0.09
150	-	-	-	-	-
Time (LST)	11:00	11:11	09:46	10:32	10:38
Date (1976)	28 Feb	29 Feb	2 Mar	4 Mar	5 Mar

MLD. Here, the 1-percent blue-light level was observed 10 to 25 m deeper than that of the total photon flux. Again, the smaller differences were associated with cloudier skies when the total incident flux was decreased, but its fractional blue component was increased. Extremum values of the attenuation coefficient of the euphotic zone were 0.044 and 0.052 per m; for the 1%-layer of the total quanta, 0.055 to 0.069 per m. As in Fig. 3, the upward bulging of the isopleths near B-1.5, i.e., 10°N, suggests the upwelling of detrital material, i.e., nutrients, which gives rise to increased primary biological activity.

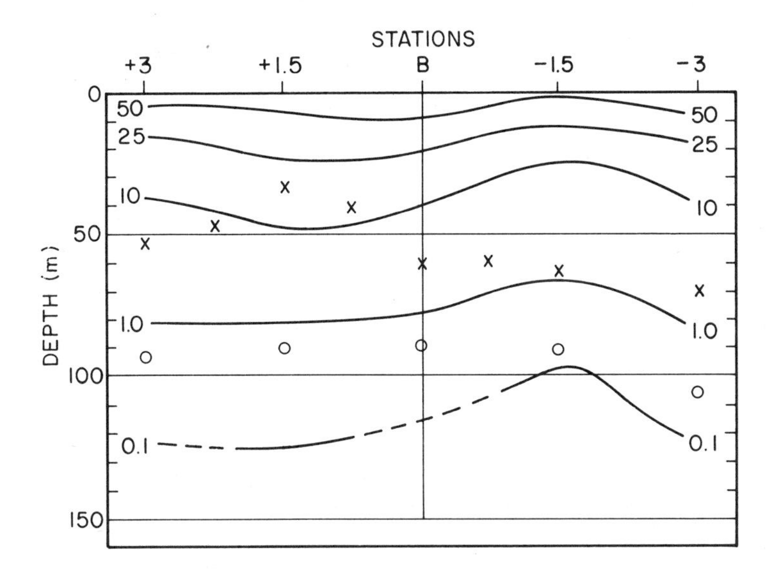

Fig. 11. Vertical distribution of submarine PAR-photon irradiance in percent of 0.5 m flux. x, MLD; o, 1% blue-light level. DOMES, Leg 4, 28 Feb - 5 Mar 1976. (From Table 7).

<u>Transmittance of 10-m layers.</u> As shown in Fig. 12, the near-surface water of low transmittance gave way to a zone of greater transparency near the base of the surface mixed layer, below which was found the axis of a more turbid stratum. There was good correspondence between the depth of the latter and that of the maxima of the phytoplankton concentrations (El-Sayed, Fig. 4 and 8, this volume). Good agreement was also found with the excessive concentrations of phaeopigments near 140 m at B-1.5. However, the primary productivity maxima within the mixed layer were not apparent in the transmittance pattern, thus duplicating the situation found with T-A of this leg. Above 100 m, the least transparent water, with $k > 0.069$ per m, was observed at B-1.5, the location of suspected upwelling. Values decreased at the greatest depths.

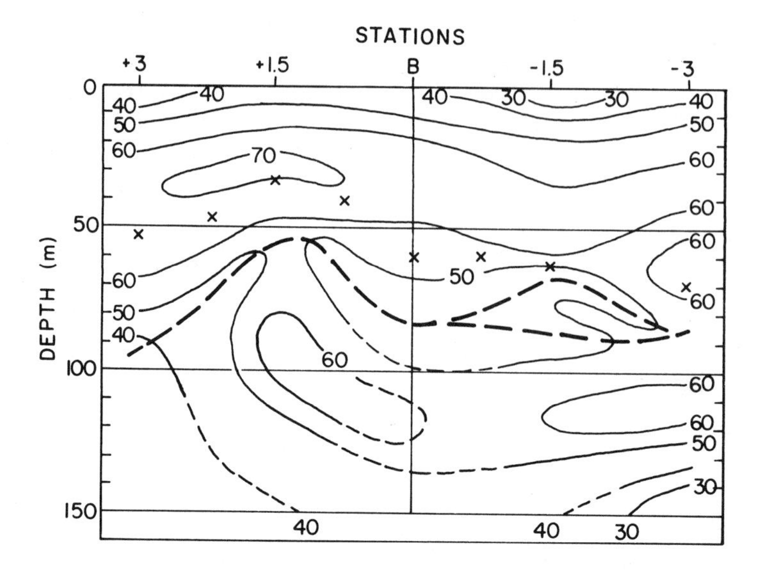

Fig. 12. Vertical distribution of submarine photon transmittance (%) for 10-m layers. Heavy dashed line, axis of minimum values; x, MLD. DOMES, Leg 4, 28 Feb - 5 Mar 1976. (From Table 7).

Transect C, Leg 4, 18-24 Feb 1976

Irradiance. Table 8 gives the downward flux of quanta. The maximum value, found at C near the surface, remains at that station to the 140-m level. However, the minimum value, in the surface layer at C+1.5, is found below 40 m at the southern end of the transect. The pattern of the related percentage values is shown in Fig. 13. Characteristic features are similar to those of the other winter transects, namely: more than 50 percent of the photons did not penetrate the top 10-m layer; less than 10 percent reached the large MLD; and the 1-percent blue-light level was 10 to 20 m deeper than that of the total quanta, with the smaller differences in the northern branch associated with increased cloudiness. The overall pattern resembles the one for this transect during the summer, as seen in Fig. 5, i.e., the isopleths reached their greatest depth in the clear water north of the central station. For this clear water in the euphotic zone, the gross attenuation coefficient is 0.040 per m, slightly greater than during the preceding summer. For the shallow euphotic zone at C-3 (12°N), Z = 78 m and K = 0.059 per m. Again, this latter feature very likely represents the consequence of upwelling of dendritus near the southern boundary of the North Equatorial Current. Zonal continuity is reflected in all transects.

Transmittance of 10-m layers. A portraiture is presented in Fig. 14. The characteristic features are similar to those of the other transects, except for one major difference. A prominent and

TABLE 8

Downward flux of photosynthetically **active** solar radiation (PAR) as a result of depth during DOMES/OCEO Expedition, Leg 4, Transect C, 18-24 Feb. 1976. Times are of the beginning of each cast in 120th-meridian standard time; units are $\mu E/m^2$ where 1 Einstein = 6.023×10^{23} photons.

Station Depth (m)	C+3	C+1.5	C	C	C-1.5	C-3
0	946	277	996	1660	1463	1420
5	456	134	481	801	655	910
10	340	100	358	597	415	459
20	173	80.8	239	342	190	192
30	108	60.6	163	225	107	102
40	72.2	44.4	105	142	56.6	54.0
50	45.5	30.3	69.6	91.7	30.8	29.0
60	29.1	20.7	53.7	59.0	16.2	14.7
70	18.3	13.1	36.4	33.6	7.90	7.13
80	12.8	9.09	23.1	18.3	4.08	3.44
90	8.47	5.56	11.7	9.61	2.40	2.10
100	5.85	3.54	7.21	4.80	1.50	1.40
110	3.63	1.92	4.80	2.49	0.98	1.02
120	2.49	1.26	3.19	1.27	0.68	0.79
130	1.69	0.94	2.39	0.68	0.47	0.65
140	1.12	0.64	1.85	0.38	0.36	0.56
150	0.75	0.49	-	0.17	0.29	0.50
Time (LST)	09:50	09:30	09:10	10:52	10:00	09:47
Date (1976)	18 Feb	19 Feb	21 Feb	22 Feb	23 Feb	24 Feb

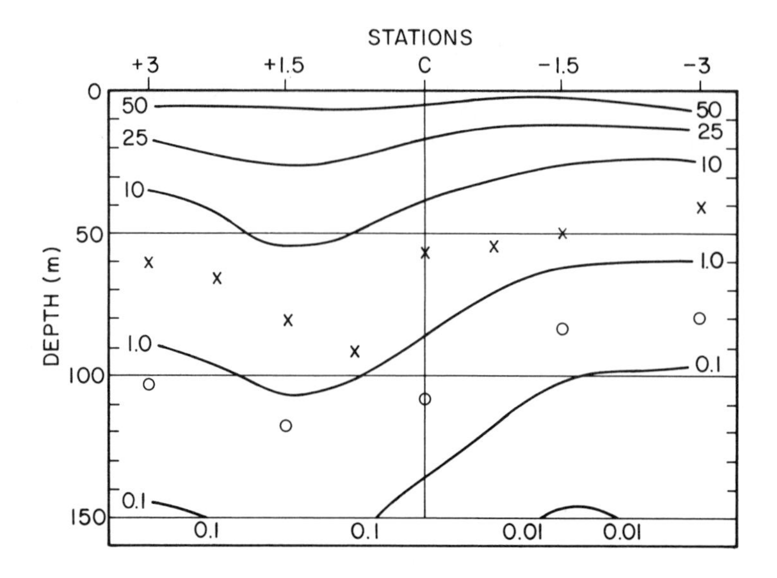

Fig. 13. Vertical distribution of submarine PAR-photon irradiance in percent of 0.5 m flux. x, MLD; o, 1% blue-light level. DOMES, Leg 4, 18-24 Feb 1976. (From Table 8).

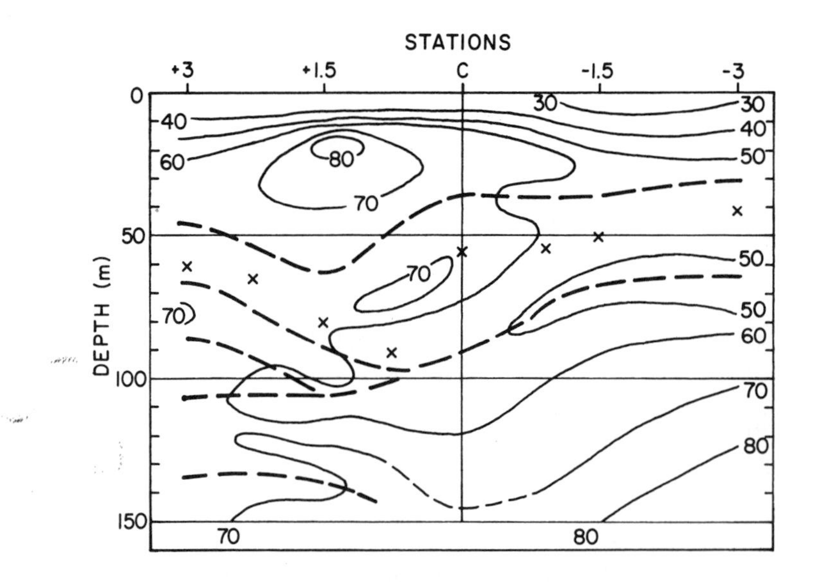

Fig. 14. Vertical distribution of submarine photon transmittance (%) for 10-m layers. Heavy dashed line, axis of minimum values; x, MLD. DOMES, Leg 4, 18-24 Feb 1976. (From Table 8).

continuous layer of water with low transmittance is found sandwiched between two less turbid layers, all above the MLD. Except for the southern extremity of T-A in summer (Fig. 2), and the central region of T-A during the winter (Fig. 10), this feature was not observed elsewhere, and is therefore considered to be an anomaly. However, when comparison is made with phytoplankton data, there is excellent agreement between the position of this turbid layer and the maxima in primary productivity and chlorophyll a concentration. The turbid layer in the thermocline also corresponds to maxima in concentration of chlorophyll and phaeopigments (El-Sayed, Fig. 4 and 8, this volume).

The excessively turbid water seen below 100 m in Fig. 6, was not present during this transect (Fig. 14). As with other features, it may have shifted to greater depths as well as southward. The appearance of less turbid water at C-3 below 100 m, a feature more typical of the western sector, suggests a much reduced dendrital upwelling. However, the clear water in the upper layers north of C appears to be a permanent feature. The 10-m layer transmittance of 80% corresponds to a value of 0.022 per m for the attenuation coefficient.

Longitudinal variations near 12°N, Leg 4

A portraiture of the west-east spatial distribution of the percentage levels between A+3, B and C-3 is given in Fig. 15.

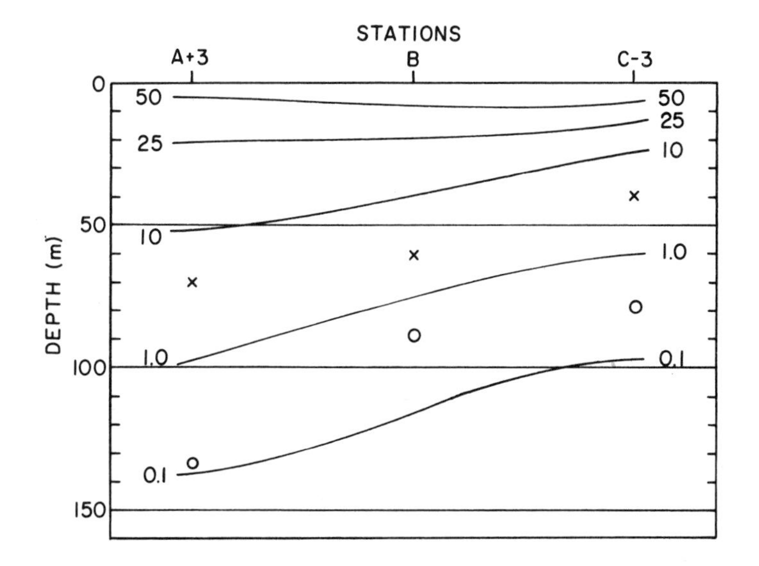

Fig. 15. Vertical distribution of submarine photon irradiance, in percent of the 0.5-m value, along 12°N latitude. x, MLD; o, 1% blue-light level. DOMES, late winter 1976.

The characteristic features noted are: more than half of the photons were attenuated in the top 10 m; the isopleths deepened westward, except in the top 10-m layer; less than 10 percent penetrated to the westward deepening MLD; and the 1-percent blue-light level exceeded that of the total quanta by 15 to 40 m. These latter differences are larger than those found during the late summer, as seen in Fig. 7, possibly due to the reduced cloudiness during the late winter. The attenuation coefficient, K, which characterizes the euphotic zone, ranged from 0.059 per m at C-3 to 0.034 per m in the clearer water to the west. Corresponding coefficients, which characterize the 1%-layer of the PAR-quanta, ranged from 0.077 per m in the turbid water in the east to 0.047 per m in the west. The related pattern of the 10-m layer transmittance of ambient light, with its characteristic features, is shown in Fig. 16. Above 80 m, the values increased toward the west; below, the water was more transparent at the eastern extremity. The major quasi-horizontal axis of minimum transmittance was found in the stable thermocline below the westward deepening MLD.

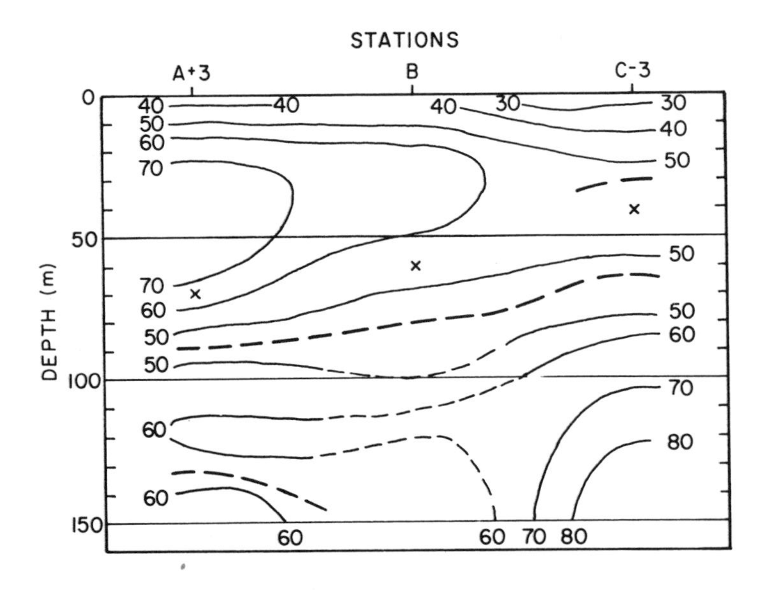

Fig. 16. Vertical distribution of submarine photon transmittance (%) for 10-m layers, along 12°N latitude. Heavy dashed line, axis of minimum values; x, MLD. DOMES, late winter 1976.

Seasonal changes

General. Since the values of percentages are small at greater depths, seasonal changes are consequently small. On the other hand, fractional changes would be excessively large, and quite possibly unreliable, because the initial measured values may involve large fractional errors. Consequently, the differences in percentages, from late summer to late winter, are treated. It must be remembered that the initial percentage value for a given level represents the total transmittance, T, of the entire layer above that level, and is equal to the product of the separate slab-transmittance values of the discrete layers above, as given by (6). Such total transmittance values represent optical signatures of the related suprajacent water mass, in much the same way that K optically characterizes the euphotic zone, i.e., the layer for which T = 1%.

Model considerations. As we have seen, there are certain characteristic features of the slab-transmittance pattern between the surface and the thermocline well below the MLD. These give guidance to what we might expect of temporal changes associated with vertical migrations of the MLD. We noted that the surface-layer transmittance was low. Below, to the MLD the waters were more transparent, and in the upper portions of the thermocline, just below the MLD, the turbidity increased, i.e., the transmittance decreased. Below this, the turbidity decreased. This distribution, in the absence of advection, may be readily explained. A large fraction of the low energy photons is absorbed by the top 10 m of the water. In addition, since the surface acts as an efficient collector and temporary repository of debris from the atmosphere, as well as from interior waters, the surface layer is more turbid. At levels below the surface layer, since they are removed from their intermediate source of debris, i.e., the surface, and since the available light is richer in more penetrating high energy photons, the waters are more transparent. If the water mass were not stably stratified, the transmittance would remain high or respond to local sources of material, either biogenic or anthropogenic. However, within the upper portion of the thermocline, immediately below the MLD, because of the great stability, particulate debris is temporarily restricted and attenuation is high. Since the motion in some sectors of the region is marked by horizontal divergence, the associated upwelling is a likely source of this nutrient laden debris which increases the potential for primary productivity. Below this layer, the stability is lower and the suspended matter is less restricted, hence transmittance is higher.

We have noted a moderate deepening of the MLD between the periods studied. Consequently, in view of the above, there are definite changes we should anticipate, in the absence of advective processes (and marked local sources). Relative to the late winter MLD, we would anticipate a decrease of transmittance in the subjacent layer, and an increase in the suprajacent layer which would be enhanced by the vertical spreading of the isopleths in the larger volume. In this connection, vertical mixing in the surface layer would produce an increase of transparency in the near-surface region, but a decrease immediately below it. Where such changes are not observed, other processes are involved, e.g., three dimensional advection of optically different water masses due to current shifts and/or vertical motion, and the local introduction or extinction (sources or sinks) of turbid matter. The latter may be biogenically related. Diagnostic discussions will be presented within the framework of this somewhat simplistic, empirical model.

Central stations. To get a gross view of temporal changes of the downward flux of PAR, comparisons were made of the initial and final submarine photon irradiances at central stations A, B and C. Values, in terms of percentage of the near-surface flux, are given in Table 9. A general decrease in total transmittance was observed at stations A and C, but an increase occurred at B. At the latter station, this increase was apparent over the shorter period from 11 September to 6 October, also. All appear cloud related.

Transect A. The seasonal change of percent irradiance, shown in Fig. 17, was marked by an increase in total transmittance of the water above 20 m, and at all levels at A-2. Except for a small region near 70 m at A+2, and the sector below 100 m at A+1, the remainder of the transect experienced a decrease in transparency relative to the photon flux. The pattern was derived from Fig. 1 and 9. Increases at A-2 were due to the initial presence of turbid water, seen in Fig. 1. The feature, possibly a signature of the current system, was not apparent in late winter. The speculative suggestion is that it translated southward.

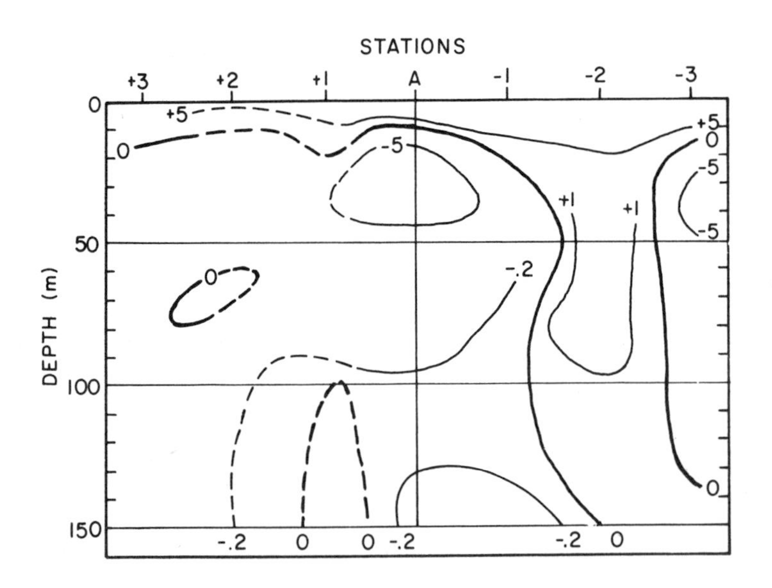

Fig. 17. Vertical distribution of the change of percent photon irradiance, from late summer 1975 to late winter 1976, along Transect A. (Derived from Fig. 1 and 9).

TABLE 9

A comparison of winter and summer values of the downward irradiance of photosynthetically active solar radiation (PAR) as a function of depth at Stations A, B, and C. Values are in percent of the 0.5-m submarine flux. Time, LST, of the beginning of each profile.

Station	A		B			C	
Leg	2	4	1	2	4	1	4
Depth (m)							
0	100	100	100	100	100	100	100
5	47	46	44	53	60	47	48
10	33	28	25	34	42	37	36
20	19	14	15	21	25	31	21
30	12	7.3	7.6	12	16	25	14
40	7.9	3.9	4.3	6.9	10	21	8.6
50	4.8	2.4	2.2	3.6	6.0	15	5.5
60	3.1	1.7	1.1	1.9	4.0	10	3.6
70	2.0	1.1	0.60	1.1	1.8	6.5	2.0
80	1.5	0.77	0.32	0.59	0.81	3.5	1.1
90	1.2	0.58	0.18	0.31	0.30	1.5	0.58
100	0.99	0.43	0.10	0.18	-	0.59	0.29
110	0.85	0.33	0.06	0.10	-	0.18	0.15
120	0.76	0.23	0.03	0.05	-	-	0.08
130	0.72	0.15	0.02	0.02	-	-	0.04
140	0.67	0.09	0.01	-	-	-	0.02
150	-	0.05	0.01	-	-	-	0.01
Time (LST)	11:50	09.08	15:11	09:50	09:42	10:55	10:52
Date	25 Sep 75	11 Mar 76	11 Sep 75	6 Oct 75	2 Mar 76	2 Sep 75	22 Feb 76

Transmittance change for 10-m layers is presented in Fig. 18. The pattern of change of slab or individual-layer transmittance is rather complex. The consistent, but speculative, dashed features at A+1.5 are the result of a lack of data during Leg. 4. However, in the cross section, which represents the graphical difference between Fig. 2 and 10, the marked changes are a consequence of the initially very turbid regions seen in Fig. 2. In general, the large decreases below 100 m, especially near A-1.5, were overlain by a broad band of rises which extended to near the 50-m level. Above this, the waters became less transparent, except for the layer near the surface. This near-surface decrease in attenuation extended to 30 m near A-2. Most of the observed changes are in keeping with anticipated results associated with the lowering of the MLD. Where these results were not observed, it is reasonable to assume that intrusions of different water masses occurred, possibly in association with vertical motion, e.g., a latitudinal shift of the equatorial currents. A slight southward shift is suggested. Hence, the rather large decreases in the upper 50-m layer north of A, the prominent rise center at 85 m at A+1 flanked by decreases south of A, and the prominent decrease below 100 m at A-1 may well represent such intrusions. The north-south tilt of this feature is consistent with discussion presented with Fig. 10.

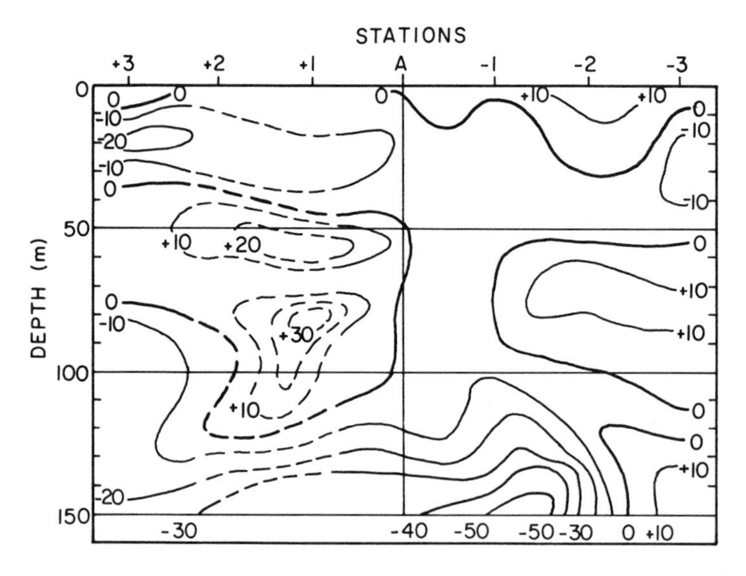

Fig. 18. Vertical distribution of the change of transmittance of 10-m layers, from late summer 1975 to late winter 1976, along Transect A. (Derived from Fig. 2 and 10).

Transect B. Changes of percent irradiance are presented in Fig. 19. Increases in total transmittance, i.e., decreases in attenuation, although generally small, dominated the region,

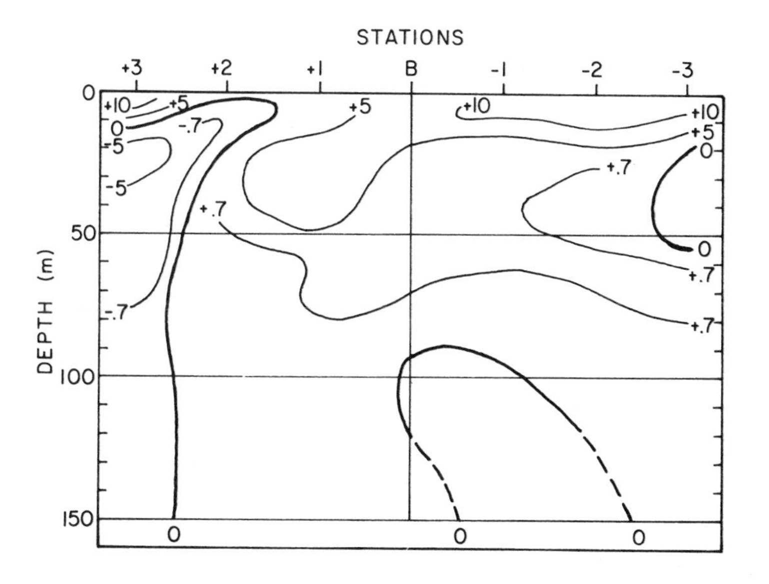

Fig. 19. Vertical distribution of the change of percent photon irradiance, from late summer 1975 to late winter 1976, along Transect B. (Derived from Fig. 3 and 11).

except: at the northern end of the transect, at B-1 below 100 m, and around 40 m at B-3. Consequently, a larger fraction of incident photons penetrated to greater depths in the late winter than during the late summer, except at B+3 where the reverse occurred. The pattern was obtained from Fig. 3 and 11.

The stratification seen in the slab-transmittance change pattern, Fig. 20, is in keeping with the layering shown in Fig. 4 and 12, from which the pattern was derived. Axes of the alternating layers of rises and falls have a wavelike character. The relatively large band of increases above 50 m, and the parallel band of decreases below it were apparently associated with the downward migration of the MLD, as seen in T-A, also. Other extrema may have been due to intrusions of different water masses, or a latitudinal shift of the upwelling feature, e.g., the large positive center near 100 m at B+1, which appears to be related to a similar feature along T-A and T-C.

Transect C. Changes of percent irradiance shown in Fig. 21 were based on Fig. 5 and 13. Large decreases in the depth of penetration of fractional amounts of downwelling photons characterize the changes in the region above 100 m and north of C-3. Since the MLD deepened, it is likely that these increases in attenuation, even near the surface, were associated with turbid water intrusions as suggested by the appearance of the biogenically related axis of low slab-transmittance values above the MLD seen in Fig. 14.

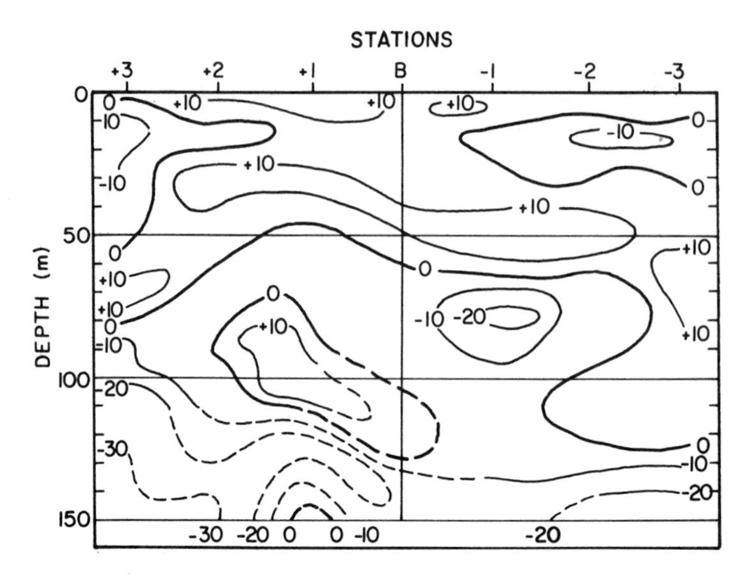

Fig. 20. Vertical distribution of the change of transmittance of 10-m layers, from late summer 1975 to late winter 1976, along Transect B. (Derived from Fig. 4 and 12).

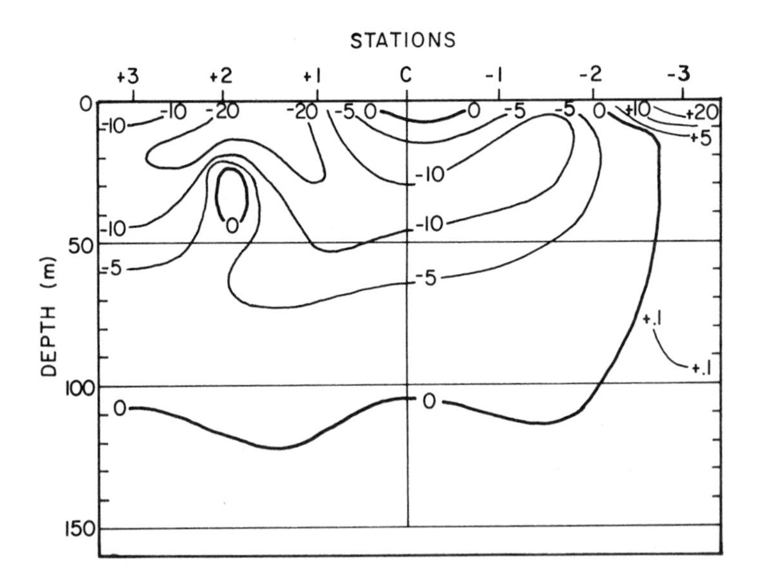

Fig. 21. Vertical distribution of the change of percent photon irradiance, from late summer 1975 to late winter 1976, along Transect C. (Derived from Fig. 5 and 13).

There appears to be a paradox in this situation. Consider C+1. Although the fraction of photons that penetrated to nearly 120 m was smaller in winter than in summer, a larger fraction reached the greater depths. A partial answer to this is that the surface-incident irradiances were smaller in the cloudier weather. The situation is clarified by consideration of changes in layer transmittance.

Transmittance changes for 10-m layers, presented in Fig. 22, were derived from Fig. 6 and 14. This portraiture represents the most dramatic changes that occurred in the DOMES region. Simplistic model changes, associated with the lowering of the MLD, were obscured. Large decreases in slab-transmittance dominated the surface layer, and extended from near C-2, along a northward arc, to a depth of 100 m at C+2. Excessive increases occurred in the deeper waters, and penetrated upward at C+3 to a depth of 40 m. From this point, and connected to it, a well defined region of increases developed. It was centered near 25 m at C+1.5. The large increases in transmittance below 100 m were responsible for the apparent paradox discussed above. It is likely that these well defined changes were due to differential advection associated with a changed current regime. A northward shift is suggested, at least in the upper layers. The changes associated with a deepened MLD are well masked because of the appearance of the more turbid layer, of biogenic origin, well above the thermocline, as seen in Fig. 14.

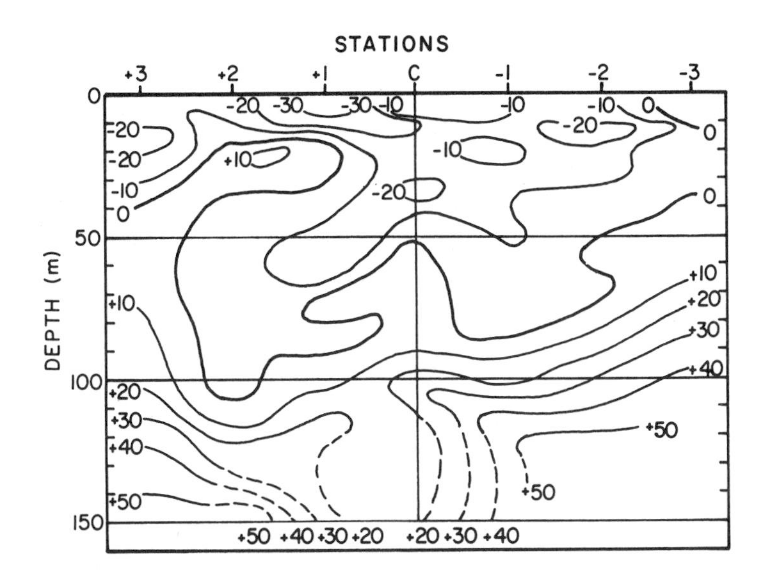

Fig. 22. Vertical distribution of the change of transmittance of 10-m layers, from late summer 1975 to late winter 1976, along Transect C. (Derived from Fig. 6 and 14).

Longitudinal changes near 12°N. The seasonal change of percentage penetration, the total transmittance, is shown in Fig. 23. Increases in the fractional photon penetration occurred in surface layers along the entire latitudinal zone, but were significantly greater in the eastern half. Below these layers the changes were small but well ordered, i.e., increases in the eastern half, and decreases in the western portions with a minimum near 70 m (compare with Fig. 17, 19 and 21).

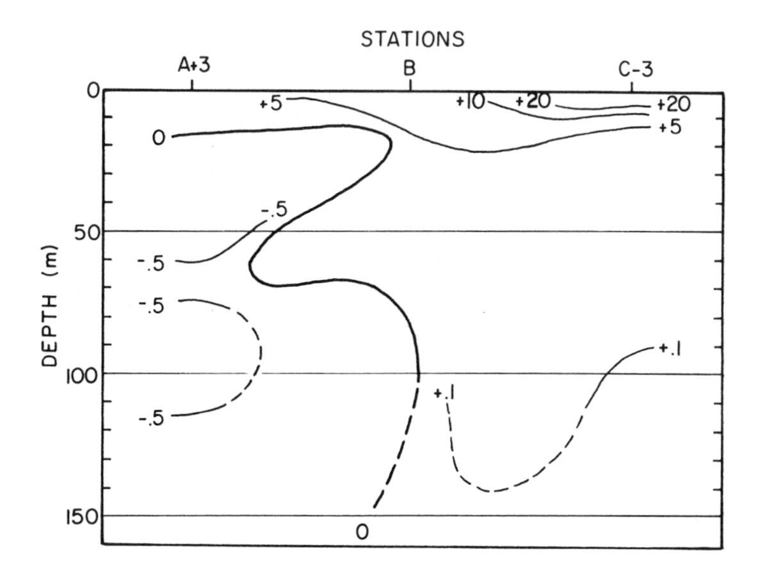

Fig. 23. Vertical distribution of the change of percent photon irradiance, from late summer 1975 to late winter 1976, along 12°N latitude. (Derived from Fig. 7 and 15).

Figure 24 shows the pattern of the changes which occurred in the slab-transmittance features. The changes were in phase with the MLD-lowering model, viz.: rises near the surface underlain by decreases, a zone of increases above the MLD, and decreases in the thermocline. The rather large falls above 40 m at A+3, and the excessive increases in the deeper eastern portion have been mentioned (compare with Fig. 18, 20 and 22). In this latter regard, the axis of these increases extends westward through B near 120 m, and is present farther north (B+1.5) above 100 m as shown in Fig. 20. Because of the zonal character of the surface currents in this sector of the tropical Pacific Ocean, it is tempting to make further speculations regarding longitudinal interconnections. However, since profiling was done only at numbered stations, the lack of data is limiting.

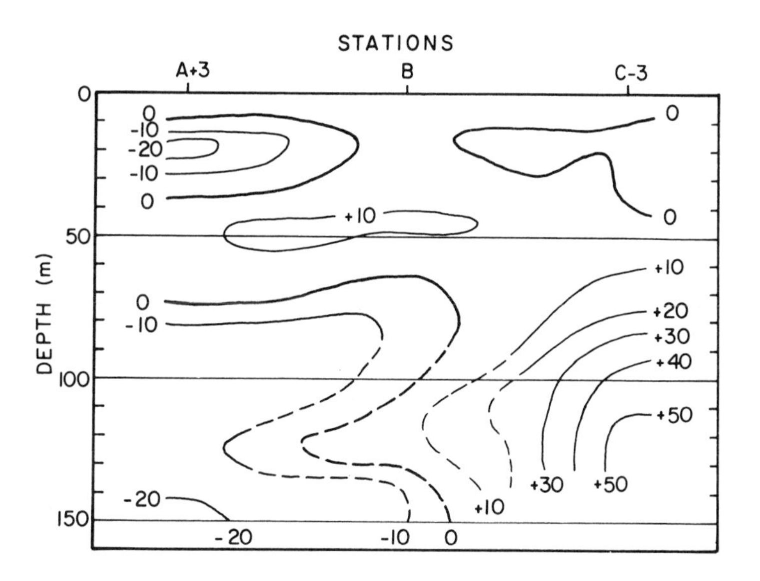

Fig. 24. Vertical distribution of the change of transmittance of 10-m layers, from late summer 1975 to late winter 1976, along 12°N latitude. (Derived from Fig. 8 and 16).

Summary and Conclusions

The net photon irradiation above the ocean surface represents the number of quanta made available to the biosphere and water mass per unit area and during a time interval. Since photosynthesis is a function of the PAR-photons, it is not surprising to find a close correlation between the seasonal changes of net irradiation and of primary productivity. What is relevant, however, is that the maximum of each occurred in winter when cloudiness, as defined for this study, was a minimum. This suggests a meteorological constraint or influence on the biology of the ocean in the DOMES area.

The close correlation between the vertical distributions of submarine PAR and phytoplankton parameters shows an interdependence between the two. Not only does primary productivity require photons, but in the process of using them, it also changes the radiation environment. As a result of this mutual interaction which involves absorption and scattering, maximum attenuation of photons often coincides with maximum concentration of chlorophyll. When such concentrations occur in stable layers, the attenuation may be enhanced by other dissolved and particulate materials. However, if these latter materials are concentrated above the phytoplankton maxima, e.g., at the base of a shallow MLD, they may reduce the PAR enough to restrict the productivity. On the other hand, if they are at a greater depth, this restriction would not happen. Indeed, associated backscattering could enhance the photosynthetic activity in the latter case.

Results of the *in situ* measurements of magnitudes and spatial variations found for this manganese-nodule province agree with those reported by Jerlov (1976). The water is characterized as a clear-open-ocean type. Depth of the euphotic zone compares with that in the Sargasso Sea, e.g., 140 m, at times. Ordinarily, the surface layer, which is somewhat less transmissive than the subjacent layer, is the feature which is optically classified. However, because the water is markedly stratified, with alternating high and low values of transmittance, the characterization of larger water masses is more difficult. Data suggest that the depth of the mixed layer must be considered in such classification. For example, vertical gradients of slab transmittance above the MLD vary inversely with the depth of the mixed layer. In addition, in the stable thermocline below the MLD, as in other stable layers, the slab-transmittance tends to be low.

Depth of the 1%-layer of PAR-quanta and of blue light (the euphotic zone) indicated a west-east axis of minimum values, smaller in the east, at about 10°N latitude during both periods studied. It is suggested that this axis coincides with the zone of upwelling between the North Equatorial Current and the Equatorial Countercurrent. A gross attenuation coefficient which optically characterizes the 1%-layer, has been suggested.

Temporal variations of optical properties may be anticipated, in part, from consideration of the diagnostic, qualitative model which has been suggested. For example: wind mixing will reduce vertical gradients in the surface layer; and a deepening of the MLD will depress the associated pattern features. Anomalous changes must be ascribed to three dimensional advective changes, or the presence of sources or sinks, viz., biogenic or anthropogenic.

With the introduction of particulate material associated with future mining operations, drastic increases in the attenuation of photons could be produced. The attendant inhibition of primary productivity could be damaging, although the related increases in nutrients might be potentially useful. Continuous monitoring of changes in the natural environment is needed during the planned mining operations.

Acknowledgments

My sincere thanks go to: B. Benson for dependable and patient clerical assistance; G. Chen and D. Franceschini for their tireless efforts with data reduction. The study was conducted in part under NOAA Contract No. 03-022-35115 (El-Sayed *et al*., 1977).

References

Biggs, W.W., Edison, A.R., Eastin, J.D., Brown, K.W., Maranville, J.W., and Clegg, M.D. (1971) Photosynthesis light sensor and meter, Ecology 52, 125-131.

Carpenter, D.J., and Jitts, H.R. (1973) A remote operating submarine irradiance meter, Deep Sea Res. 20, 859-865.

El-Sayed, S.Z., Franceschini, G.A., Fryxell, G.A., Gates, E.C., and Taguchi, S., (1977) Phytoplankton and primary productivity studies (in connection with the Deep Ocean Mining Environmental Study, DOMES), Final Report, NOAA Contract No. 03-6-022-35115, 1-197.

El-Sayed, S.Z., and Taguchi, S., (1979) Phytoplankton standing crop and primary productivity in the Tropical Pacific, (this volume).

Franceschini, G.A., (1968) The influence of clouds on solar radiation at sea, Deutsche Hydrographische Zeit 21(4), 162-168.

Franceschini, G.A., (1971) Observations of net solar radiation for oceanic biological study, Antarctic Jour. of U.S., VI (5), 157-158.

Franceschini, G.A., (1973) Solar radiation, Antarctic Jour. of U.S., VIII (3), 108-110.

Jerlov, N.G., (1976) "Marine Optics", Elsevier, New York, 231 p.

List, R.J., (1971) Smithsonian Meteorological Tables, Smithsonian Instit., Wash., D.C., 527 p.

Working Group 15 (SCOR, UNESCO, IAPO, 1966) Report of the second meeting of the joint group of experts on photosynthetic radiant energy, UNESCO Tech. Paper in Marine Science , No. 5.

OBSERVATIONS OF UPPER OCEAN CURRENTS AT DOMES SITES A,B, AND C IN THE TROPICAL CENTRAL NORTH PACIFIC OCEAN DURING 1975 AND 1976

David Halpern

NOAA Pacific Marine Environmental Laboratory

Seattle, Washington 98105

Abstract

The time and space (vertical) scales of the upper ocean current field are described from moored current measurements made at 20 m, 50 m, 100 m, 200 m, and 300 m depths at 15°02'N, 126°01'W (Site C) during September and October 1975, at 11°45'N, 138°23'W (Site B) during March and April 1976, and at 8°27'N, 150°45'W (Site A) during August, September and October 1976. During the periods when the measurements were made, Sites A and B were located within the eastward-flowing North Equatorial Countercurrent and Site C was within the westward-flowing North Equatorial Current. At Sites A and B the vector-mean eastward currents below the thermocline at 200 m and 300 m were approximately 10 cm sec^{-1} and 20 cm sec^{-1}, respectively, which were much larger than _a priori_ expected. At Site C the vector-mean current at 200 m and 300 m was about 5 cm sec^{-1} towards the east, indicating the presence of a subsurface current flowing in opposite direction to the surface flow which heretofore was not previously detected. At each site the northward direction of the vector-mean meridional current component in the mixed layer at 20 m depth was consistent with the direction of wind-generated Ekman transport resulting from the Northeast Tradewinds. The standard deviations were typically larger than the mean values indicating that the upper ocean currents were extremely time-dependent. Significant current fluctuations with inertial- and semidiurnal-periods were measured at each of the depths of the current meters. The amplitude of the inertial motion was typically 2 times larger than the semidiurnal tidal internal gravity wave motion; however, the inertial oscillations were largest in the near-surface mixed layer and the internal wave amplitude was ∿ 5 m or ∿ 1.5 cm sec^{-1} throughout the thermocline. Fluctuations with amplitudes of about 5 cm sec^{-1} to 10 cm sec^{-1} and time-scales greater than 20-days were measured beneath the thermocline, indicating that caution must be used in regard to the representativeness of month-long averaged currents.

The wind and mixed layer currents were coupled at time-scales greater than the local inertial-period. When the monthly averaged westward wind-stress of the Northeast Tradewinds increased, the near-surface eastward current speeds in the North Equatorial Countercurrent decreased and the speed of the westward-flowing North Equatorial Current increased, and vice-versa. Good agreement was found between the monthly averaged Ekman transport produced by the zonal wind-stress and the meridional transport computed from the near-surface current measurements. At the inertial-period the wind and the 20 m current fluctuations were statistically coherent (with 95% confidence) with approximately zero phase difference. The amplitude of the near-surface inertial motion was largest at Site A where the local inertial-period was approximately equivalent to the period of the easterly waves in the atmosphere between 5°N and 10°N, indicating the possibility of a resonant air-sea interaction process.

A comparison of the zonal components of baroclinic geostrophic currents computed from hydrographic measurements obtained coincidently with the moored current observations was made with low-pass filtered currents with the fluctuations at inertial-frequency and higher frequencies removed. Above 100 m the geostrophic and observed shears were significantly different because the uppermost 100 m contained the wind-driven currents. Between 100 m and 300 m the agreement between the geostrophic and measured shears was weak. At 300 m the observed currents were larger than the geostrophic currents by approximately 15 cm sec^{-1} indicating that the 500 m level-of-no-motion was not deep enough. However, the geostrophic current at 500 m relative to 1000 m was small (∿ 2 cm sec^{-1}). Calculations of the magnitudes of various terms in the equations of motion indicated the importance of the time-dependence of the long-period fluctuations in the dynamical balance of the North Equatorial Countercurrent.

The large time-variability of the current speed and direction, the large vertical shear, and the large current speeds beneath the thermocline will produce huge, variable and unpredictable current drag on the pipe and cable assembly connecting the nodule mining ship moving at slow speeds and the collector moving along bottom contours. The presence of internal waves, which have a maximum vertical velocity about 100 times larger than typical settling velocities for bottom sediments discharged at the surface during the mining operation, will cause in situ measurements of settling velocity and of the corresponding identity of the effluent to be spurious unless measurements are made for time intervals comparable to the semidiurnal-period.

Introduction

The circulation of the upper ocean within the 13 x 10^6 km^2 Deep Ocean Mining Environmental Studies (DOMES) area bounded by the 110°W and 180°W longitudes and the 5°N and 20°N latitudes (Fig. 1) is influenced by several large-scale current and wind patterns. The North Equatorial Current, which occurs north of ~ 10°N, flows westward in the same general direction as the

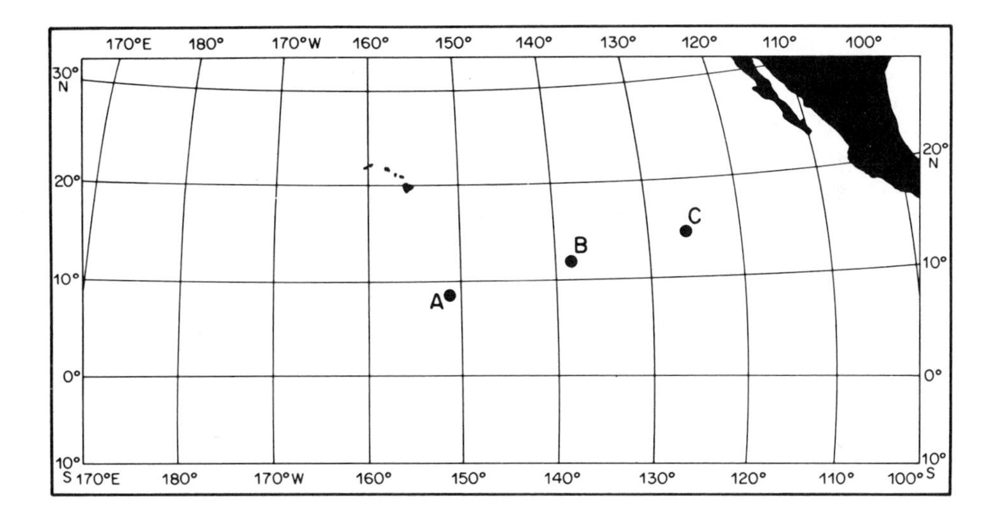

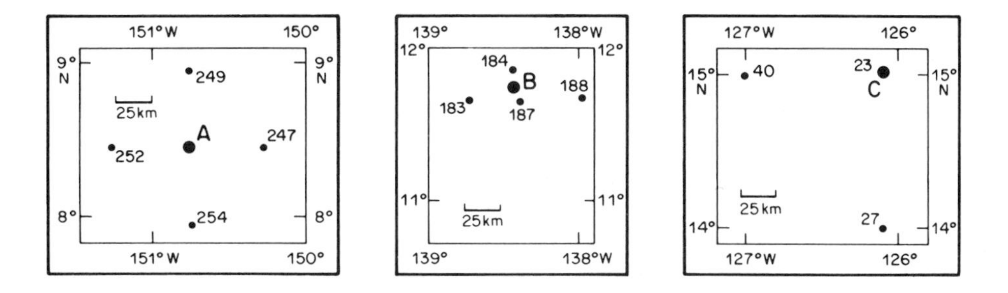

Fig. 1. (Upper) Locations of DOMES Sites A, B and C in the central tropical Pacific Ocean. (Lower) Locations of CTD measurements used in Fig. 4 and 11.

Northeast Tradewinds. Between ∿ 5°N and ∿ 10°N the North Equatorial Countercurrent flows eastward against the prevailing westward direction of the tradewinds. During the northern hemisphere winter the convergence zone of the Northeast and Southeast Tradewinds (i.e., the Intertropical Convergence Zone [ITCZ]) occurs at ∿ 2°N and during the northern hemisphere summer, when the center of the North Pacific High has moved ∿ 1500 km northwest from its wintertime position, the Northeast Tradewinds are weaker and the ITCZ shifts to ∿ 10°N. Because the large-scale current patterns are related to the winds (e.g., Ekman, 1905; Sverdrup, 1947), the upper ocean circulation in the central tropical Pacific Ocean undergoes considerable variation in response to the shifting of the major wind systems. In spring when the Northeast Trades are most vigorous, the intensities of the North Equatorial Countercurrent and the North Equatorial Current are weak, and in autumn when the Northeast Trades are weak and the ITCZ is near its northern position, the strengths of the North Equatorial Countercurrent and North Equatorial Current are maximum (Wyrtki, 1974; Meyers, 1975).

Virtually all our information about the structure of the upper ocean currents within the DOMES region (excluding the region adjacent to the Hawaiian Islands) has been obtained from (1) monthly charts of surface currents based on numerous observations of ships' drift (Wyrtki, 1965), (2) charts of dynamic height topography (Reid, 1961), and (3) water mass analysis and tracer techniques (Reid, 1965; Tsuchiya, 1968). To the author's knowledge, Wyrtki's (1967) 11-day direct current measurements at 6°25'N, 157°50'W were the longest measurements made in the upper 500 m of the DOMES region prior to the beginning of the DOMES Project in 1975. In fact, no direct current measurements longer than 4-days existed in the National Oceanographic Data Center archives prior to May 1976. Because interpretation of the DOMES biological and chemical measurements and because initialization and verification of dispersion models require knowledge of the vertical structure of the current at time-scales from hours to a month, direct observations of upper ocean currents were made at each of the DOMES sites. The locations of the sites were chosen by an interdisciplinary group of researchers including biological, chemical, geological and physical oceanographers, ocean-mining engineers and representatives of the ocean-mining industry. It was intended that the environmental conditions encountered at the sites would encompass nearly all of the environmental features of the DOMES region.

The limited resources prevented concurrent recording of currents at the 3 sites. Moored current measurements were made at 20 m, 50 m, 100 m, 200 m and 300 m depths at 15°02'N, 126°01'W (Site C) during September and October 1975, at 11°45'N, 138°23'W (Site B) during March and April 1976, and at 8°27'N, 150°45'W

(Site A) during August, September and October 1976. Moored wind measurements were made at Site A and upper ocean temperature observations were recorded at the 3 sites. The moored observations were recorded at 7.5-min intervals. During the times of deployment and/or recovery of the surface mooring, vertical profiles of temperature and salinity were recorded with a Plessey 9006 CTD (conductivity, temperature, depth) system in the vicinity of the mooring. Locations of the moorings and the CTD measurements used herein are shown in Fig. 1. The water depths at Sites A, B and C were approximately 5030 m, 4850 m and 4600 m, respectively. The depths and record-lengths of the moored current measurements are shown in Fig. 2. Failures

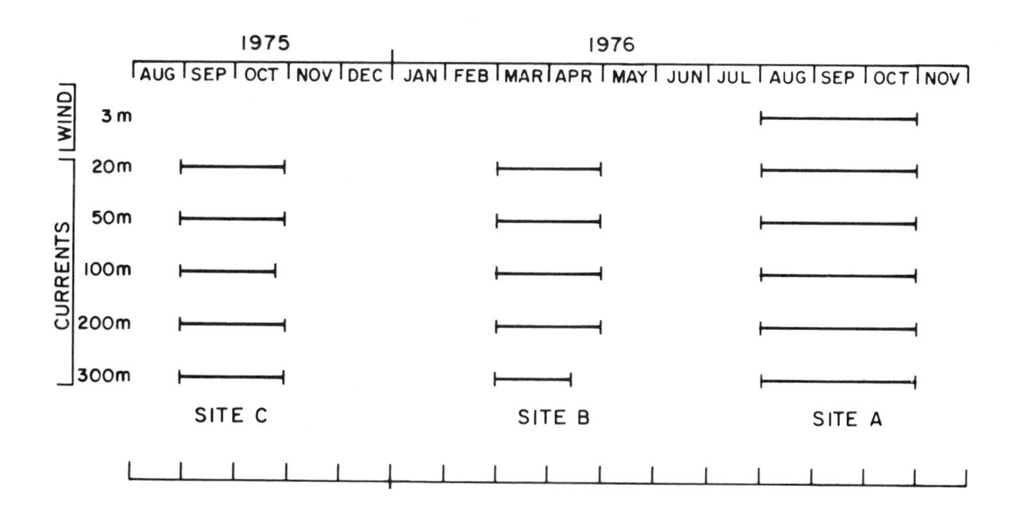

Fig. 2. Record-lengths of moored wind and current observations at Sites A, B and C.

to obtain moored wind observations at Sites B and C were due to an electronic malfunction of the wind recorder and the loss of the surface buoy, respectively. The moored current measurement system used in DOMES is shown in Fig. 3, and descriptions of the design and of the deployment and recovery techniques are given by Berteaux (1975). In this paper the time-scales and the vertical structure of the upper ocean current field at DOMES Sites A, B and C will be described.

Comparison of CTD measurements made at times of mooring deployment and recovery suggested that the vertical profiles of temperature and sigma-t shown in Fig. 4 were representative of the period when each mooring was on station. The mixed

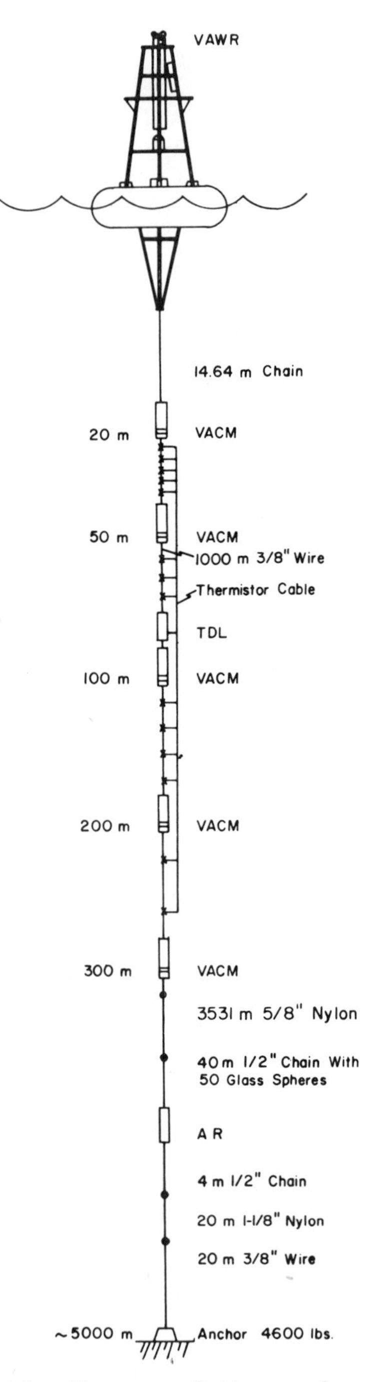

Fig. 3. Schematic diagram of the surface mooring used at DOMES Sites A, B, and C.

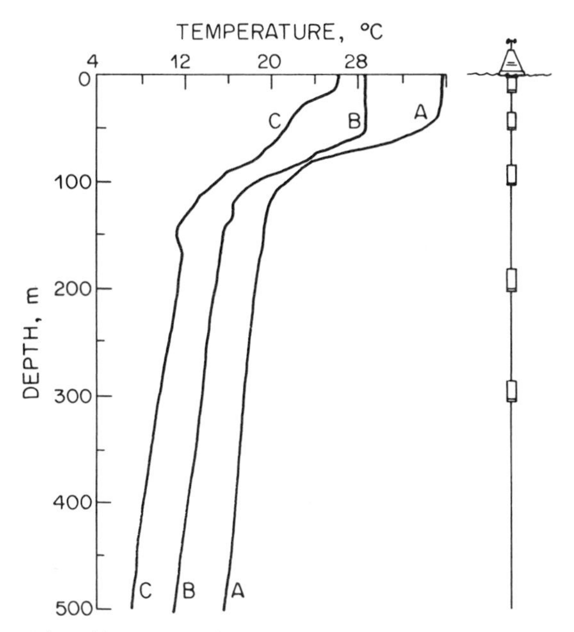

Fig. 4. Schematic diagram of depths of current measurements and vertical profiles of temperature measured at Site A (CTD 249; 29 Aug 76), at Site B (CTD 184; 6 Apr 76), and Site C (CTD 23; 3 Sep 75). Locations of CTD measurements are shown in Fig. 1. The Sites B and A profiles are offset from the Site C profile by 4°C and 8°C, respectively.

layer depths at Sites A, B and C were approximately 50 m, 70 m and 20 m, respectively. Thus, at Site A the 20 m vector-averaging current meter (VACM) was placed in the mixed layer, the 50 m VACM was at the base of the mixed layer or at the top of the thermocline, the 100 m VACM was near the base of the strong (0.30°C m^{-1}) thermocline, and the 200 m and 300 m VACMs were in the weakly stratified water beneath the thermocline. At Site B the 20 m and 50 m VACMs were placed in the mixed layer, the 100 m VACM was in the middle of the thermocline ($\sim$ 0.17°C m^{-1}) which extended to about 150 m, and the 200 m and 300 m VACMs were beneath the thermocline. At Site C where a series of thermoclines occurred the 20 m VACM was placed near the base of the mixed layer, the 50 m VACM was in a layer between two thermocline regions, the 100 m VACM was in the middle of the main thermocline ($\sim$ 0.15°C m^{-1}), the 200 m VACM was near the base of a temperature inversion layer of approximately 50 m thick and the 300 m VACM was in weakly stratified water. The depths of the current measurements relative to the upper ocean thermal structure are shown in Fig. 4.

Current Meter Instrumentation

To measure currents in the upper ocean, current meters must be placed near the surface and the presence of surface waves affects the quality of moored current measurements. The large amplitude high-frequency orbital wave motions are not easily recorded and are aliased into the low-frequency current variations. If the current meter is placed beneath a buoy which moves horizontally, the relative motion recorded by the instrument will not be equal to the absolute water velocity. If the surface float is displaced vertically by surface waves, the current meter is moved through a vertical shear and, in addition, axial and transverse vibrations of the mooring line produce spurious motions of the velocity sensors.

The VACM, which represented a significant advancement of oceanographic technology when it was developed in 1971 at the Woods Hole Oceanographic Institution, contains a magnetic compass and at the lower end of the instrument are a Savonius rotor of 10 cm diameter with 4 s-shaped rotors and a small (17 cm x 9 cm) vane, both turning on a vertical axis. Every eighth of a revolution of the rotor, vane and compass orientations are measured and internally combined into a discrete current direction which is converted internally to Cartesian components of velocity. For the duration of the sampling interval the east and north speed values are summed internally (McCullough, 1975). Such a sampling scheme has the effect of reducing contamination of the current spectrum by surface wave 'noise' (Halpern and Pillsbury, 1976).

Several _in situ_ intercomparison tests reviewed by Halpern (1979a) indicated that reliable VACM measurements could be made with $\sim$ 2 cm sec^{-1} accuracy for frequencies less than about 4 cph within the uppermost $\sim$ 100 m beneath surface-following buoys moored in deep and shallow waters. However, the VACM did not yield reliable data when placed beneath surface-following buoys at depths greater than a few hundred meters in a region of low current speeds (Gould _et al_., 1974) because the data were affected by surface wave energy transmitted to instruments at depth by motion of the mooring cable. Though laboratory experiments have demonstrated the amplification of rotor speeds by high frequency axial motions, the quantitative relationship between cable motion and the rotation of a rotor is unknown; _i.e._, it is not known what surface wave height will produce a spurious 1 cm sec^{-1} current by high frequency displacements of the mooring line and instruments. Answers to the question: To what depth can reliable current records be made beneath a surface-following buoy? are being sought. During the August-September 1977 Mixed Layer Experiment at 50°N, 145°W, current measurements were made with a VACM and a dual orthogonal vector-measuring current meter (VMCM) at a nominal depth of 93 m beneath a surface-following buoy in a region of low current speeds. Even during periods when significant surface wave heights reached 5 m, the VACM spectral levels exceeded those of the VMCM by a factor of 1.7 below 4 cph, suggesting that mooring motion changes the effective calibration constant of a VACM (Davis and Halpern, 1978). Thus, we believe that to a first approximation, the VACM speeds recorded at 200 m or 300 m depths during the DOMES Program were probably about 1.5 times larger than the speeds measured at 20 m or 50 m assuming uniform current speeds throughout the upper ocean.

Observations

Wind

Direct wind observations were made at Site A. At Sites B and C surface wind speed and direction values were obtained at 6-hour intervals from the Fleet Numerical Weather Central (FNWC), Monterey. We believe that the moored wind measurements at Site A were accurate to within 1 m sec^{-1} and 20° (Day, 1970; Halpern, 1974a). Comparison of FNWC-derived surface winds and buoy winds by Halpern and Reynolds (1973) at a midlatitude site, where the correspondence between atmospheric pressure gradients and wind vectors is expected to be better than at low latitudes, indicated that the FNWC-derived surface winds were representative of the buoy winds for time-scales longer than about 1 week.

The largest mean wind speeds ($\sim$ 9 m sec^{-1}) occurred at Site B (Table 1). The vector-mean wind speeds and directions were 1.9 m sec^{-1} from 078° True at Site A, 8.9 m sec^{-1} from 062° True at Site B, and 5.4 m sec^{-1} from 047° True at Site C. The averaged winds were representative of the Northeast Tradewinds. At Sites B and C the standard deviations were smaller than the mean values (Table 1) indicating that at these sites the winds were relatively steady in speed and direction. In contrast, the variability of the wind was relatively large at Site A. However, a time-series of 6-hour wind vectors computed from atmospheric pressure gradients averaged over hundreds of kilometers is expected to contain a smaller amount of time-variations than wind observations made at 7.5-min intervals.

Current

The east-west (u; positive eastward) and north-south (v; positive northward) components of the mean currents (Fig. 5) computed from each record contained several features consistent with the structure of tropical currents and several characteristics of the mean currents which were not *a priori* expected. At Sites A and B the eastward directions of the vector-mean

TABLE 1

Mean and standard deviation values of the speed (s), east-west (u; positive eastward) and north-south (v; positive northward) wind components. At Sites B and C the wind records consisted of 6-hourly speed and direction values produced by Fleet Numerical Weather Central, Monterey, from sea level atmospheric pressure. At Site A vector-averaged u and v wind components were recorded at 7.5-min intervals. Units are m sec^{-1}.

	Site A		Site B		Site C	
	Mean	Std. Dev.	Mean	Std. Dev.	Mean	Std. Dev.
s	4.4	2.1	9.1	2.2	6.1	2.1
u	-1.9	3.7	-7.9	1.8	-4.1	2.5
v	-0.4	2.6	-4.2	2.2	-3.9	1.8

zonal component of the 20 m depth currents were representative of the North Equatorial Countercurrent. However, the near-surface currents encountered at Site B were probably not representative of the climatological-mean circulation because Wyrtki's (1965) March and April monthly averaged surface current charts contained westward flow of 25 cm sec^{-1} at the location of Site B. The eastward directions of the mean zonal currents at Site B indicated that the northern limit of the North Equatorial Countercurrent extended northward of the latitude of Site B. In the meridional section of temperature made from 9°N to 14°30'N along 138°23'W during 28 February - 5 March (i.e., after the mooring was placed on station), the thermocline sloped upwards towards the north from 9°N to 13°30'N which, according to Wyrtki and Kendall (1967), indicated that 13°30'N was the northern limit of the North Equatorial Countercurrent. When a similar temperature section was made during 4-13 October 1975 the thermocline ridge was south of the latitude of Site B indicating the usual occurrence of the westward-flowing North Equatorial Current at the location of Site B. At Site C the -17 cm sec^{-1} mean westward speed at 20 m depth was in excellent agreement with Wyrtki's (1965) surface current charts; however, the 10 cm sec^{-1} mean northward speed at 20 m was not in agreement with Tsuchiya's (1974) surface geostrophic current relative to 500 db or with Wyrtki's (1965) monthly surface current charts. At Site C, located north of the North Equatorial Countercurrent, the eastward-flowing undercurrent observed beneath the shallow westward North Equatorial Current was not a priori expected (e.g., Reid, 1973; Tsuchiya, 1975). At each site the northward

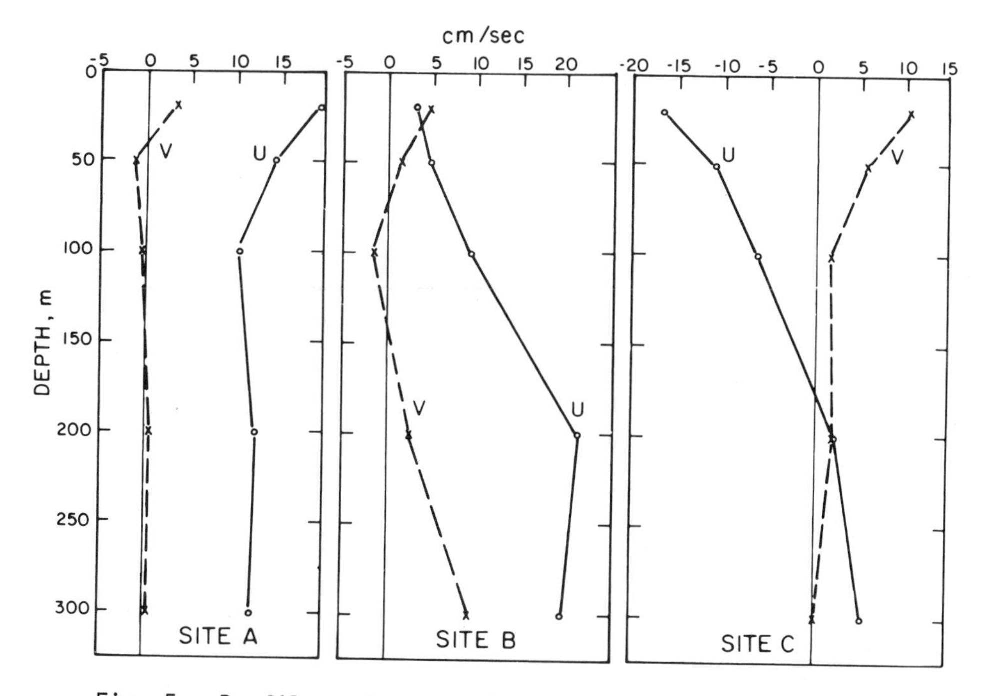

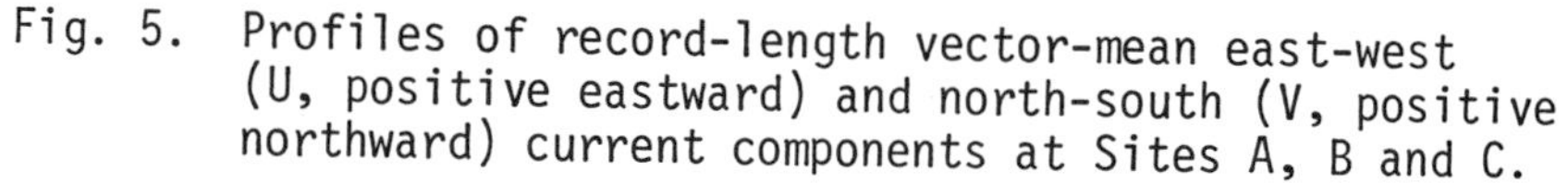
Fig. 5. Profiles of record-length vector-mean east-west (U, positive eastward) and north-south (V, positive northward) current components at Sites A, B and C.

direction of the vector-mean meridional component at 20 m was consistent with the northward direction of the wind-generated Ekman (1905) transport resulting from the Northeast Tradewinds. A similar persistent northward flow in the near-surface layer has recently been observed from trajectories of surface buoys drogued at about 30 m depth and tracked by satellite for long periods (Hansen _et al._, 1978). For the buoys deployed between 2°N and 12°N the northward speed averaged over many months was $\sim$ 3 cm sec^{-1} which was equivalent to the mean meridional current measured at 20 m depth at Sites A and B.

In contrast to the concept that each component of the tropical current system (except the Equatorial Undercurrent) is adequately represented by a zonal baroclinic geostrophic current decreasing in speed monotonically from beneath the mixed layer to about 300 m to 500 m depth with most of the shear occurring above 200 m (Knauss, 1961; Wyrtki, 1978a), at Site A the mean eastward speed between 100 m and 300 m was approximately uniform at 11 cm sec^{-1}, which was much larger than _a priori_ expected. However, this was not a persistent feature because the vector-mean zonal current speed measured at 200 m at 7°N, 150°W from 4 November 1977 to 23 March 1978 was ∿ 1 cm sec^{-1} (Halpern, 1978). Another unexpected feature was observed at Site B where the eastward speed increased from 3 cm sec^{-1} at 20 m depth to 20 cm sec^{-1} at 200 m and 300 m. The possibility that the data from the Site B VACMs were mislabelled so that the deeper records were from the shallow depths and vice-versa is negligible because the VACM temperature records were in excellent agreement with the CTD data.

The standard deviation is a measure of the total temporal variability. Most of the ratios of the standard deviations to the mean values (Table 2) were greater than unity, indicating

TABLE 2

Absolute values of the ratios of the standard deviation to mean values of the vector-averaged east (u) and north (v) current components computed for 88.7-days, 61.5-days and 60.5-days record-lengths at Site A, B and C, respectively. The vector-mean u and v components are shown in Figure 5. An asterisk indicates that the record-length was 43.3-days. R = Ratio.

	Site A		Site B		Site C	
	R (u)	R (v)	R (u)	R (v)	R (u)	R (v)
20 m	1.0	5.9	2.9	1.8	0.9	1.4
50 m	1.2	13.1	2.0	5.8	1.4	1.7
100 m	1.0	47.5	0.7	4.5	1.9	5.3
200 m	0.7	17.0	0.3	4.4	3.6	4.3
300 m	0.6	13.2	0.3*	1.2*	1.5	45.0

that the upper ocean currents were extremely time-dependent. The variability of the north-south component was larger than the east-west component because the mean directions of the currents were primarily zonal and the processes producing much of the time-variations (e.g., inertial-internal waves, eddies) have approximately equivalent Cartesian speeds.

Spectral estimates of the kinetic energy were computed by first estimating the spectra of the east-component series and the north-component series from the Cooley-Tukey Fourier transforms using the perfect Daniell frequency window of variable width. For each spectrum the sum over positive frequencies was equal to the total variance. At each frequency band the east and north spectral estimates were combined by taking half the sum of the two to form the horizontal kinetic energy spectrum. At all frequencies there was an approximate 1:1 correspondence between the shapes of the spectra (Fig. 6). At Site B for frequencies above 0.25 cycles per hour (cph) there was an energy

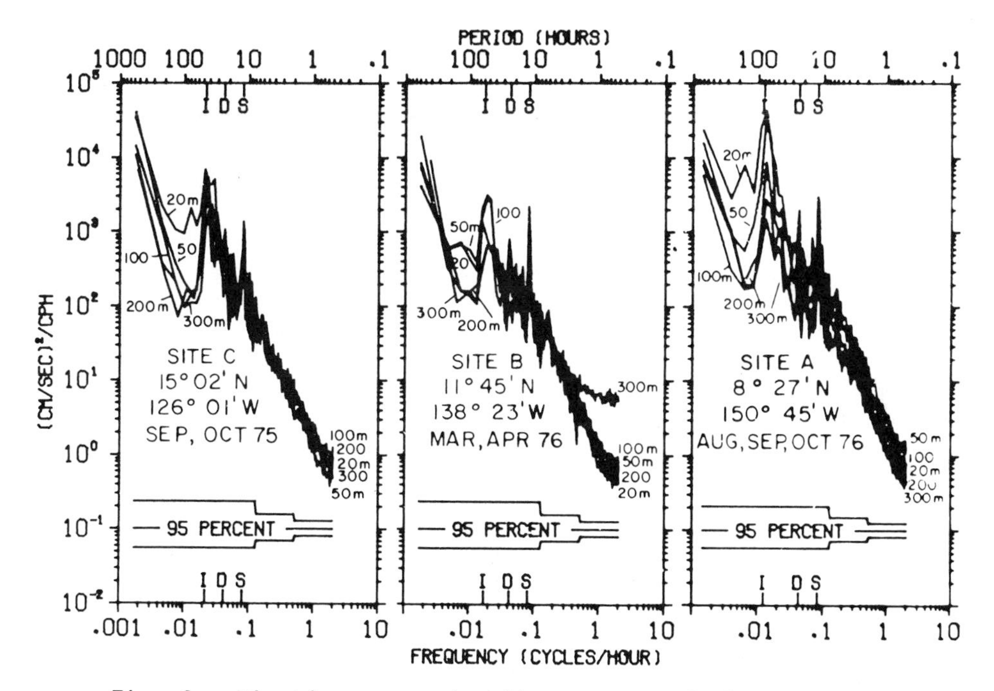

Fig. 6. Kinetic energy density spectra of the current measurements recorded at Sites A, B and C. The inertial, diurnal and semidiurnal frequencies are indicated by I, D and S. The "95 percent" represents the 95% confidence levels determined from the chi-square distribution and applies to each curve.

mismatch at the 95% confidence level between the 300 m VACM record and the other VACM records because each sample of the 300 m VACM record consisted of rotor count speeds and an instantaneous direction; at frequencies less than 0.1 cph the 300 m scalar speed and direction VACM record was approximately equivalent to the vector-averaged east and north VACM record. The spectral trough occurring at frequencies less than the inertial frequency was more pronounced below 100 m than nearer the surface, and the energy levels of the spectral trough at 100 m, 200 m and 300 m were about the same. Below the mixed layer the energy level of the trough decreased with increasing latitude, as suggested by Wunsch (1978), indicating that at lower latitudes a larger amount of mesoscale energy occurs between the low-frequency fluctuation of the large-scale circulation and the inertial-internal wave motions. At Sites A, B and C the average energy levels of the trough were 2.1×10^2 (cm sec^{-1})^2/cph, 1.5×10^2 (cm sec^{-1})^2/cph and 1.0×10^2 (cm sec^{-1})^2/cph, respectively. Further evidence of energetic motions filling up the mesoscale spectral trough was found from current measurements made by the author on the equator at 125°W for 98.5 days beneath the thermocline at 200 m depth; the kinetic energy density spectrum did not contain a significant spectral trough at frequencies less than the diurnal tidal frequency.

On the basis of the features of the kinetic energy density spectrum each current record was divided into three frequency bands. Fluctuations with frequencies less than the frequency (f_c) of the low-frequency spectral trough (i.e., $f_c < 7.5 \times 10^{-3}$ cph at Site A, $f_c < 1.25 \times 10^{-2}$ cph at Site B, $f_c < 1.67 \times 10^{-2}$ cph at Site C) were called low-frequency variations. At each site the intermediate frequency interval contained statistically significant (with 95% confidence) inertial- and semidiurnal-period oscillations. At Sites A and B the diurnal-period fluctuations were also statistically significant. High-frequency fluctuations corresponded to frequencies greater than 1.25×10^{-1} cph.

The intermediate- and high-frequency fluctuations were removed from the current records in the following way: for each u and v series, the Fourier coefficients were computed, the harmonics with frequencies greater than the low-frequency cutoff were subtracted, and the residual coefficients were retransformed. Time-series of current vectors ('sticks') in polar form were produced at 0000 GMT (Fig. 7). The low-frequency wind fluctuations, defined as daily vector-averaged wind vectors, are also shown in Fig. 7. The sample correlation coefficients between the u and v components of the 'sticks' are given in Table 3. The winds and the currents at 20 m depth were weakly correlated at the 95% confidence level. The low-frequency zonal current variations between adjacent depths were strongly correlated. The meridional components were also correlated between adjacent depths but the correlation coefficients were not as large as for the zonal currents. The highest correlations (>0.90) were found between the current records at 200 m and 300 m depth.

Significant current fluctuations with near-inertial and semidiurnal periods were measured at each of the depths of the current meters. At each depth the root-mean-square (r.m.s.) amplitude of the inertial-period motion was larger than the semidiurnal-period fluctuations (Table 4). The depth-averaged ± 1 standard deviation of the r.m.s. amplitudes of the semidiurnal-period motions at Sites A, B and C were 2.1 ± 0.3 cm sec^{-1}, 1.9 ± 0.6 cm sec^{-1} and 1.8 ± 0.6 cm sec^{-1}, respectively, indicating that the amplitudes of the tidal currents in the upper ocean were approximately the same throughout the region. Earle (1975) reported similar amplitude values at 1000 m depth near 38°N, 144°W. The 0.5 cm sec^{-1} r.m.s. amplitude of the semidiurnal tidal current computed from Hayes' (1977) 6-month current measurements recorded 30 m above the bottom at Site A presumably represented the barotropic component of the tidal current, indicating that the semidiurnal-period motion in the upper ocean was primarily internal gravity wave motion. The amplitude of the inertial currents decreased with depth; at each

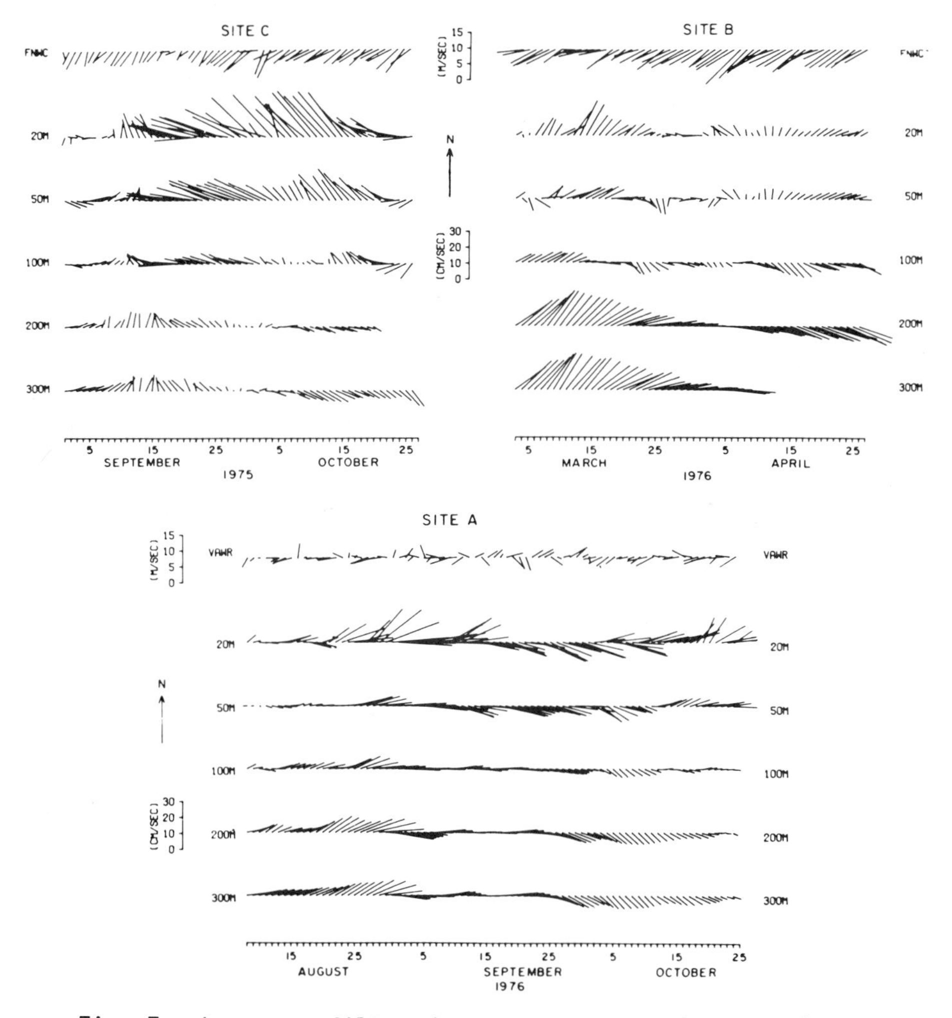

Fig. 7. Low-pass filtered current vectors ('sticks') of wind and current observations at Sites A, B and C.

TABLE 3

Sample correlation coefficients (r_u, r_v) between the (u, v) low-frequency current components shown in Fig. 7.

$$r = \frac{\Sigma\ (x_1 - \overline{x}_1)(x_2 - \overline{x}_2)}{(n - 1)\ S_1\ S_2}$$ where x_1 and x_2 represent the two low-frequency u (or v) time-series of n values, and S_1 and S_2 are the standard deviations of each time series with mean values $\overline{x}_1$ and $\overline{x}_2$. The number (n) of low-frequency current vectors computed at 0000 GMT at Sites A, B and C were 77, 54 and 55, respectively. The 95% confidence levels of r_u and r_v computed at Sites A, B and C are 0.22, 0.27 and 0.27, respectively. Blank spaces indicate coefficients were not significant. A single asterisk indicates that n = 36 and the 95% level was 0.33 and two asterisks mean that n = 49 and the 95% level was 0.28.

	Site A		Site B		Site C	
	r_u	r_v	r_u	r_v	r_u	r_v
Wind / 20 m	.23			.28	.36	.43
20 m /50 m	.75	.42	.97	.55	.81	.75
50 m/100 m	.69	.52	.56		.85	.64
100 m/200 m	.91	.77	.38	.78	.75**	
200 m/300 m	.90	.94	.92*	.97*	.95**	.95**

site it was at least 2 times larger in the mixed layer than below the thermocline at 300 m depth. In contrast, Bubnov _et al_. (1979) current measurements at 10°N, 23°30'W near the northern limit of the Atlantic North Equatorial Countercurrent showed constant kinetic energy at the inertial frequency between 20 m and 1000 m. The depth-averaged r.m.s. amplitudes of the inertial-period motions, which were 6.2 ± 4.2 cm sec^{-1} at Site A, 2.6 ± 1.1 cm sec^{-1} at Site B and 3.8 ± 1.3 cm sec^{-1} at Site C, were 2 to 3 times larger than the tidal currents. At the inertial period the energy densities of the u and v components were virtually equivalent and the ratios of the energy density of the clockwise- (_i.e._, when looking downward; _cum sole_) to counter-clockwise-rotating components were 50-100, indicating that the inertial-period motion measured at each site was nearly circularly polarized and rotated in the clockwise direction.

Spectral peaks nearly significant at the 95% confidence

TABLE 4

Root-mean-square (r.m.s.) amplitude of the inertial (I) and semidiurnal (S) period motions. The bandwidth corresponding to the inertial and semidiurnal estimates was approximately 1.6×10^{-3} cph. Units are cm sec^{-1}. An asterisk indicates that the record-length was 43.3 days.

	Site A		Site B		Site C	
	I	S	I	S	I	S
20 m	11.0	2.5	3.5	2.4	5.5	1.4
50 m	10.3	2.1	4.0	2.8	4.4	1.1
100 m	4.9	2.1	1.8	1.6	2.7	1.8
200 m	2.8	2.1	1.8	1.6	4.1	2.4
300 m	2.1	1.7	1.7*	1.3*	2.5	2.3

level occurred at the diurnal frequency at Sites A and B. Hayes (1977) also observed these prominent diurnal-period oscillations near the ocean bottom at Sites A and B. However, Earle (1975) did not observe diurnal tidal currents north of 30°N in the eastern central North Pacific because propagating diurnal-period internal wave motion cannot occur north of 30°N where the local inertial frequency is greater than the diurnal frequency.

Temperature

Temperature time-series measurements were made with VACMs and thermistor-chains at each DOMES site. Electronic problems plagued the thermistor-chains and the combination of the longest temperature records at the largest number of depths occurred at Site A where 17-day (2-19 August) records were obtained at 20 m, 25 m, 30 m, 45 m, 50 m, 60 m, 80 m, 90 m, 110 m, 115 m, 150 m, 220 m, 225 m and 300 m. Assuming a continuous, smooth temperature distribution with depth (*viz.*, without a layered structure), such as the type shown in Fig. 4, time-series of the depths of isotherms (Fig. 8) were produced by a linear interpolation between hourly averaged temperature measurements made at discrete depths. Figure 8 shows semidiurnal-period fluctuations of the isotherms which were in phase throughout the uppermost 300 m. The average values of the standard deviations of the vertical displacements of the 25°C, 20°C, 15°C, 14°C, 13°C and 12°C isotherms (*i.e.*, throughout the thermocline region) and of the 11°C and 10°C isotherms were 5.2 m and 13.5 m,

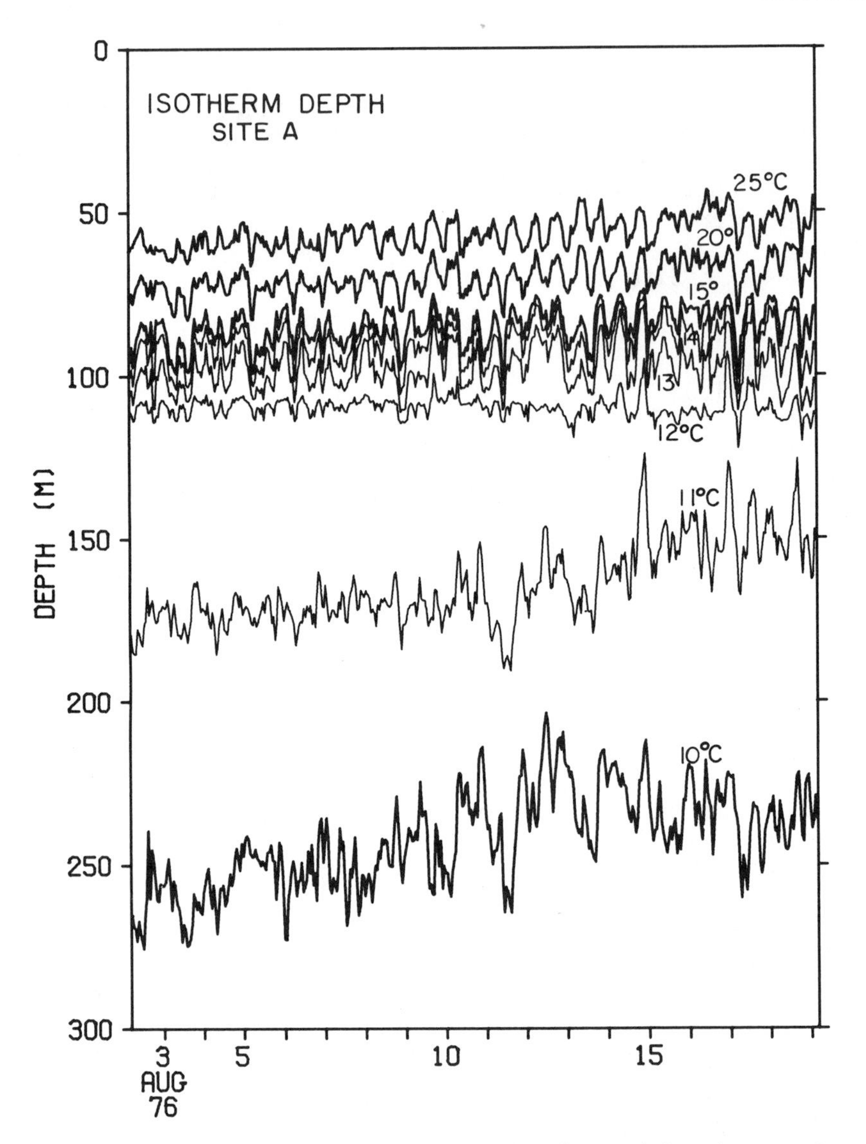

Fig. 8. Hourly values of depths of selected isotherms at Site A.

respectively. The average ± 1 standard deviation of the r.m.s. amplitudes of the semidiurnal-period fluctuations of the 8 isotherms shown in Fig. 8 was 4.9 m ± 1.3 m. These vertical displacements were much larger than mooring-induced vertical movement of the thermistors, which were less than 0.25 m within the thermocline region, determined from numerical model studies of the configuration of the mooring line.

Results

Monthly Current Variations

Large differences occurred between profiles of vector-mean monthly current components (Fig. 9), suggesting that caution be used in regard to the representativeness of month-long averaged currents and of the record-length mean currents. For example, at Site A the September and October u-component currents were different by $\sim$ 10 cm sec^{-1}, or $\sim$ 50%, throughout the 20 m to 300 m layer.

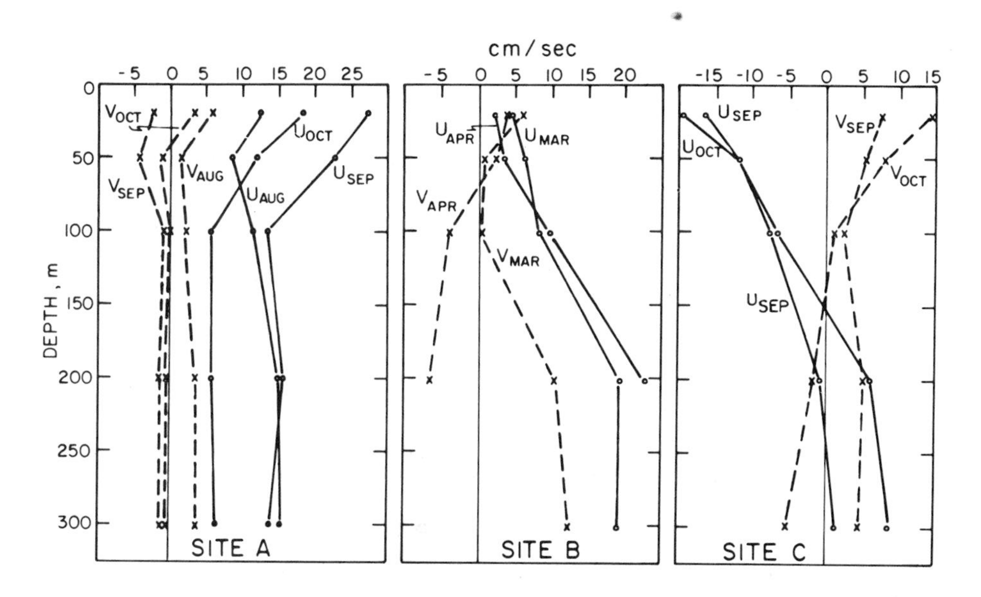

Fig. 9. Vector-mean monthly east-west (solid line) and north-south (dashed line) current components at Sites A, B and C.

The variations of the vector-mean monthly zonal components of the wind-stress, $\tau_o^x = \rho_a C_D U_W^2$ where ρ_a, C_D and U_W represent the density of air (1.25 g cm^{-3}), the dimensionless drag coefficient (2.5 x 10^{-3}) and the vector-mean zonal wind component, were related to the fluctuations of the monthly u-component currents at 20 m and 50 m. At Site A the near-surface currents increased from $\sim$ 10 cm sec^{-1} to $\sim$ 25 cm sec^{-1} from August to September when τ_o^x increased from -0.19 dynes cm^{-2} to

0.01 dynes cm^{-2}; during October when τ_o^x (-0.48 dynes cm^{-2}) was again towards the west, the eastward October near-surface current was ∿ 10 cm sec^{-1} less than in September. At Site B the increase in the westward τ_o^x from -1.76 dynes cm^{-2} in March to -2.05 dynes cm^{-2} in April was accompanied by a decrease in the eastward current from ∿ 5 cm sec^{-1}in March to ∿ 3 cm sec^{-1} in April. A similar result was observed at Site C where the westward τ_o^x increased from -0.3 dynes cm^{-2} to -1.05 dynes cm^{-2} and the westward current at 20 m increased from ∿ -16 cm sec^{-1} to ∿ -19 cm sec^{-1}. Comparison of the wind and near-surface current observations at Sites A and B, where stronger winds and smaller eastward currents occurred at Site B than at Site A, provided additional evidence of the relationship between the zonal wind-stress and the near-surface eastward current. The coincident variations of monthly averages of the near-surface eastward current at Sites A and B and the zonal wind-stress implied that the strength of the North Equatorial Countercurrent was stronger when it flowed through a region of minimum westward wind-stress, as suggested by Wyrtki (1974) in his study of climatological-mean data.

At each site the directions of the vector-mean monthly zonal wind-stress and the meridional component of the near-surface current were consistent with Ekman (1905) circulation. When the wind-stress was westward, the near-surface current was northward, and on one occasion (i.e., September at Site A) when the wind-stress was eastward, the near-surface current was southward. The average monthly Ekman transport (E.T.) per unit width of longitude, $\tau_o^x/\rho f$ where ρ and f represent the density of water and the Coriolis parameter, was about 1/3 larger than the average meridional transport (M.T.) computed from the mean monthly northward current within the mixed layer (Table 5). The surface current was defined by linear extrapolation. Because the mean direction of the v-component surface geostrophic current relative to 500 db is southward (Wyrtki, 1975), the difference might have been less than 1/3 if the geostrophic component of the total transport was removed. The good agreements found between the directions of the zonal wind-stress and the current measurements and between the transports computed from the wind-stress and from the current measurements indicated a significant interaction between the atmosphere and ocean.

Long-Period Fluctuations

The envelopes encompassing the endpoints of the low-frequency current vectors (Fig. 7) recorded below the thermocline at each

TABLE 5

Monthly average values of zonal wind-stress (τ_o^x; dyne cm^{-2}), Ekman transport (E.T.; 10^4 cm^2 sec^{-1}), meridional current (v_{ML}; cm sec^{-1}) in the mixed layer and meridional transport (M.T.; 10^4 cm^2 sec^{-1}) computed from the currents measured in the mixed layer. The mixed layer depth (M.L.D.; m) was determined from CTD profiles shown in Fig. 4.

	Site A		Site B		Site C	
	Aug	Oct	Mar	Apr	Sep	Oct
τ_o^x	-0.2	-0.5	-1.8	-2.1	-0.3	-1.1
E.T.	0.9	2.3	5.9	6.9	0.8	2.8
M.L.D.	50		70		20	
v_{ML}	6.0	3.5	5.0	2.0	8.0	16.5
M.T.	3.0	1.8	3.5	1.4	1.6	3.3

DOMES site suggested the occurrence of long-period fluctuations with time-scales greater than 20 days. Clearly, records of 60 to 90 days duration are too short to use spectral analysis to substantiate this result. Hand-drawn smoothed curves of the daily u and v low-frequency current components with zero mean are used to outline the long-period fluctuations (Fig. 10). The long-period time-scales of the u-component were 24-days at Site A, 24-days at Site B, and 30-days at Site C. For the v-component the 'half-period' time-scales were approximately 35-days at Site A, 43-days at Site B and 25-days at Site C. The long-period v-component fluctuations were reminiscent of lateral motion along a meridian of zonal currents containing meridional shear. For example, Wyrtki (1978b) has inferred north-south oscillations of the central Pacific North Equatorial Countercurrent with a time-scale of about 34 days. Because the time-scales of the long-period u-component were less than one-half the v-component, the long-period flow pattern was more complicated than a north-south meandering of a zonal current with a single high-velocity core (see, e.g., Düing et al., 1975).

Similar long-period fluctuations have been observed elsewhere in the tropical current system. Current fluctuations with a 45-day and a 60-day time-scale were found in Hayes' (1977) low-frequency current vectors (i.e., sticks) recorded 30 m above the bottom (i.e., at 4954 m depth) at Site A during May-November 1976. Current measurements made during the summer of 1974 at 5°N, 23°30'W in the Atlantic North Equatorial Countercurrent,

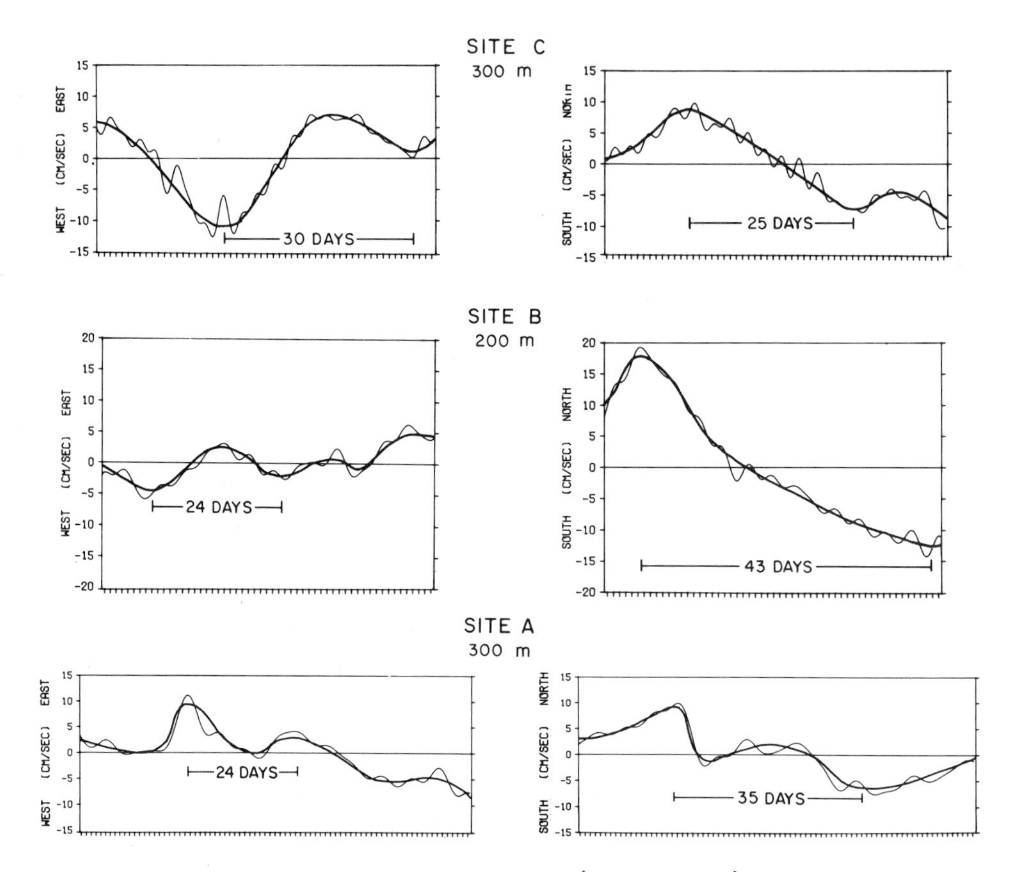

Fig. 10. Low-pass filtered east-west (right-side) and north-south (left-side) currents recorded at 300 m at Sites A and C and at 200 m at Site B. The hand-drawn thick line is the smoothed version of the low-frequency fluctuations.

described by Bubnov _et al_. (1979), contained a single wave of ∿ 30-day period which was interpreted to be a solitary Rossby wave. Ramp's (1976) 272-day current record obtained during 1975 at 100 m above the bottom in water 5870 m deep at ∿ 31°N, 158°W, approximately 1000 km north of Hawaii, contained a single occurrence of a well-defined fluctuation with a time-scale of about 60-days. Niiler (1978, personal communication) reported finding several 60-day fluctuations in year-long current measurements recorded at 15°N in the Atlantic North Equatorial Current. In addition to the occurrence of long-period fluctuations in current measurements, spectral estimates of long time-series of sea level measurements recorded in the central and western Pacific [_e.g._, at Hilo (20°N, 155°W), Christmas (2°N, 155°W), Kwajalein (9°N, 168°E)] contained a peak at periods of 50-60 days (Luther, 1978, personal communication). Although the sparse data seemed to confirm the existence of long-period fluctuations in tropical currents, the data are too meager to discriminate whether the fluctuations were planetary waves or eddies and whether the fluctuations were generated by atmospheric forcing, instability of the zonal currents (Philander, 1978a), or flow over rough terrain. Tropical wind oscillations with periods of ∿ 50-days (Madden and Julian, 1972) represent a possible generation mechanism.

Geostrophic Currents

Figure 11 contains estimates of the u- and v-components of the baroclinic geostrophic currents relative to 500 m computed from the CTD casts whose positions are shown in Fig. 1. The distances between the CTD stations at Sites A, B and C were about 100 km, 25 km, and 100 km, respectively. Anomalies of specific volume were computed from equations developed by Ekman and Knudsen (Sweers, 1971). Depth in meters and pressure in decibars were defined as equivalent and are used interchangeably. The trapezoidal method of integration was used to compute dynamic height anomalies.

Baroclinic geostrophic currents will contain errors due to the quasi-synoptic nature of the CTD measurements and to the internal variations of the temperature and salinity fields. From the temperature time-series measurements (see, _e.g._, Fig. 8), we expect the semidiurnal tidal frequency to be the dominant short-period fluctuation of the density field in this region. At Site B time-series of CTD casts were made at 1-hr intervals for 24-hr to estimate the time-variations of the dynamic height anomaly (ΔD) over a tidal period. The standard deviations from the mean geopotential anomalies were: 20/500 db - 0.009 dyn m; 50/500 db - 0.009 dyn m; 100/500 db - 0.006 dyn m; 200/500 db - 0.003 dyn m; and 300/500 db - 0.002 dyn m. All the deviations were appreciably greater than

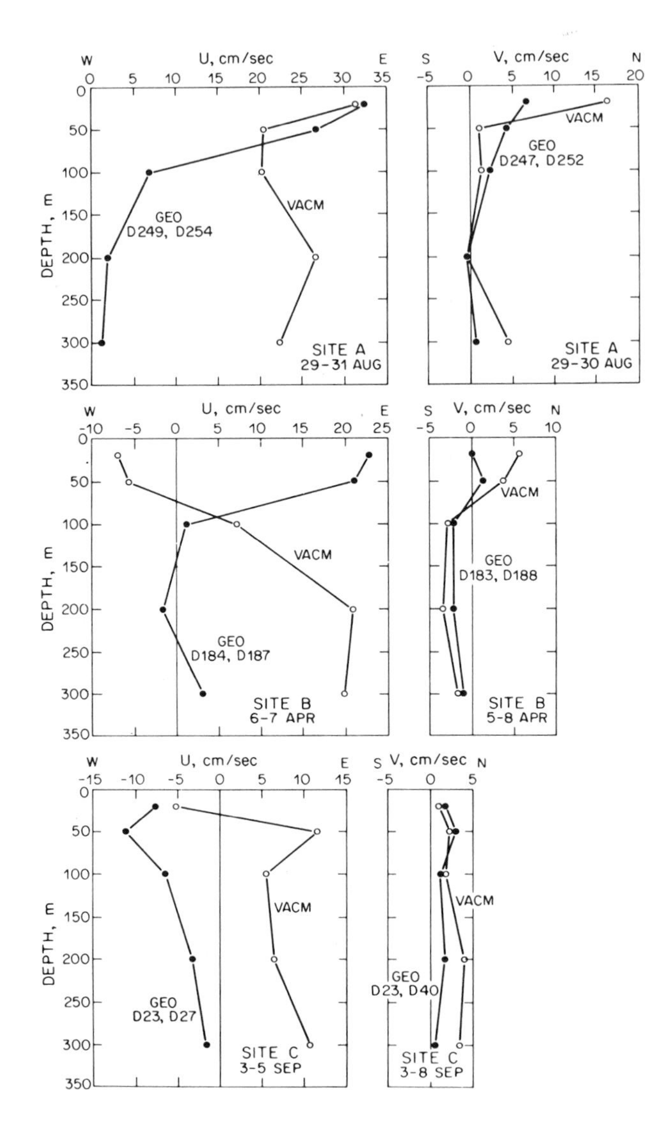

Fig. 11. (Right) Each panel (upper-Site A; middle-Site B; lower-Site C) contains the east-west (u; positive eastward) component of (1) the baroclinic geostrophic current relative to a 500 m level-of-no-motion (GEO curve; closed circles) computed from CTD data recorded near the mooring (Fig. 1), and (2) the low-pass filtered currents (VACM curve; open circles) averaged over the time interval of the CTD measurements. (Left) Same as right-side except for the north-south (v; positive northward) component.

that due to random measurement error, which should be less than 0.001 dyn m according to the method of Wooster and Taft (1958). Suitable CTD time-series measurements to compute variations over a tidal period were not made at Sites A and C. If we assume that the time-variations of ΔD were produced primarily by ubiquitous internal gravity wave motions, then the standard deviations of the ΔD field at Sites A and C would be similar to the Site B measurements. If, between the 2 CTD stations used to compute geostrophic currents, the dynamic height difference produced by the time-variations was equal to twice the standard deviation, then at Site A an upper limit of the uncertainties of the u-component geostrophic current at 20 m, 50 m, 100 m, 200 m and 300 m relative to 500 m, respectively, were 7.7 cm sec^{-1}, 7.7 cm sec^{-1}, 5.1 cm sec^{-1}, 2.6 cm sec^{-1} and 1.7 cm sec^{-1}. At Site B the corresponding uncertainties were 24 cm sec^{-1}, 24 cm sec^{-1}, 16 cm sec^{-1}, 8.1 cm sec^{-1} and 5.4 cm sec^{-1}, and at Site C the corresponding uncertainties were 4 cm sec^{-1}, 4 cm sec^{-1}, 2.9 cm sec^{-1}, 1.4 cm sec^{-1} and 1.0 cm sec^{-1}. Thus, the u-component geostrophic currents computed at Sites A and C were larger than the uncertainty of the geostrophic currents computed from short-period variations of the density field. In contrast, at Site B the uncertainty of the u-geostrophic current speed was greater than the measured geostrophic current. However, if the CTD station spacing at Site B had been larger, such as $\sim$ 100 km as at Sites A and C, then the computed uncertainties due to time-variations would probably have been about 75% smaller.

Vertical profiles of the time-averaged low-frequency current vectors (shown in Fig. 7) corresponding to the time-intervals when geostrophic currents were calculated are also shown in Fig. 11. The measured profiles shown in Fig. 11 did not persist for durations much longer than about 10-days because of the long-period fluctuations of the current field. Whether variations of the geostrophic current field also occurred with similar time-scales is unknown. The geostrophic and measured shears are given in Table 6. The averages of the absolute values of the differences of the u- and v-component shears were 2.18×10^{-3} sec^{-1} and 0.63×10^{-3} sec^{-1} respectively. Much better agreement occurred between the v-shears than the u-shears. Above 100 m depth the average difference between observed and geostrophic u-component shears was 3.72×10^{-3} sec^{-1}; below 100 m it was 0.70×10^{-3} sec^{-1} or about 5 times less than the corresponding value of the uppermost 100 m. We consider differences in shear greater than 0.5×10^{-3} sec^{-1} to be significant. If at 500 m the geostrophic current is zero, a linear geostrophic shear of 0.5×10^{-3} sec^{-1} corresponds to a surface current speed of 25 cm sec^{-1} and if the current profile is uniform over the 5-degree width of the North Equatorial Countercurrent, then the geostrophic transport of the North Equatorial Countercurrent

TABLE 6

East-west ($\Delta U/\Delta z$) and north-south ($\Delta V/\Delta z$) components of the shears computed from the baroclinic geostrophic current components relative to 500 m (GEO) and from the moored current measurements (VACM). The absolute value of the difference between the u-component (v-component) GEO and VACM shears is denoted $|DIFF(u)|$ ($|DIFF(v)|$). The interval between depths z_1 and z_2 is given by $z_1 - z_2$ where the depth z is positive upwards (i.e., $z_1 > z_2$). Units of $\left(\frac{\Delta U}{\Delta z}, \frac{\Delta V}{\Delta z}\right)$, $|DIFF|$ and $(z_1 - z_2)$ are 10^{-3} sec^{-1}, 10^{-3} sec^{-1} and m, respectively.

SITE A	$\Delta U/\Delta z$		$\|DIFF(u)\|$	$\Delta V/\Delta z$		$\|DIFF(v)\|$
$z_1 - z_2$	GEO	VACM		GEO	VACM	
20-50	1.90	3.68	1.78	0.73	5.00	4.27
50-100	4.24	0.05	4.19	0.42	-0.05	0.47
100-200	0.37	-0.64	1.01	0.26	0.16	0.10
200-300	0.08	0.44	0.36	-0.09	-0.46	0.37

SITE B	$\Delta U/\Delta z$		$\|DIFF(u)\|$	$\Delta V/\Delta z$		$\|DIFF(v)\|$
$z_1 - z_2$	GEO	VACM		GEO	VACM	
20-50	0.53	-0.47	1.00	-0.43	0.58	1.01
50-100	3.98	-2.52	6.50	0.68	1.32	0.64
100-200	0.30	-1.39	1.69	0	0.08	0.08
200-300	-0.46	0.11	0.57	-0.08	-0.19	0.11

SITE C	$\Delta U/\Delta z$		$\|DIFF(u)\|$	$\Delta V/\Delta z$		$\|DIFF(v)\|$
$z_1 - z_2$	GEO	VACM		GEO	VACM	
20-50	1.10	-5.60	6.70	-0.43	-0.40	0.03
50-100	-0.90	1.24	2.14	0.38	0.16	0.22
100-200	-0.35	-0.09	0.26	-0.06	-0.27	0.21
200-300	-0.14	-0.42	0.28	0.08	0.06	0.02

would be $\sim$ 35 Sv (1 Sv = 1 x 10^{12} cm^3 sec^{-1}) which lies within the range of geostrophic transports determined by Knauss (1961) and Wyrtki (1978a). Thus, above 100 m depth the geostrophic and measured shears were significantly different, which is not too surprising because the uppermost 100 m contained the wind-driven currents. In agreement with our result, Knauss (1961) reported that the flow in the mixed layer in the North Equatorial Countercurrent between 108°W and 120°W was a combination of geostrophic current plus Ekman wind drift. Between 100 m and 300 m depths a weak agreement was found between the geostrophic and observed shears indicating that further comparisons are warranted. A more extensive comparison of observed current and geostrophic current shears using airborne-XBT and CTD measurements made 50 km apart along 150°W at 20 intervals during 4 months and moored current measurements made simultaneously at 6°N, 7°N and 8°N along 150°W is in preparation.

Geostrophic current speeds are relative to a level-of-no-motion and a change of the value of the depth of zero current speed produces a corresponding adjustment of the current speeds throughout the layer. Below 100 m depth the agreement between geostrophic and measured u-component shears seemed good enough to examine the validity of our assumed 500 m level-of-no-motion. At 300 m depth the u geostrophic current was smaller than the measured current by 20 cm sec^{-1}, 17 cm sec^{-1} and 13 cm sec^{-1}, respectively, at Sites A, B and C. These large differences suggest that a level-of-no-motion of 500 m was inappropriate and, perhaps, not deep enough. The east-west component of the geostrophic current at 500 m relative to 1000 m computed from several CTD casts made to 1000 m depth was 2-3 cm sec^{-1}, indicating that the geostrophic component of tropical currents was concentrated within the upper ocean. Wyrtki (1975) found the dynamic topography at 500 db relative to 1000 db to be essentially flat between 15°N and 15°S and, furthermore, Wyrtki (1978a) indicated that there was essentially no meridional pressure gradient beneath the North Equatorial Current and Countercurrent below 300 m and, therefore, 300 m depth was an acceptable level-of-no-motion.

Although we a priori did not expect an excellent correspondence between comparisons of geostrophic and observed currents involving only 1 sample at each site, the lack of agreement found between the geostrophically-inferred currents and the low-frequency measured currents is perplexing. In other oceanographic regions, such as in the California Current over the continental shelf (Smith, 1974) and continental slope (Halpern et al., 1978) and at intermediate depths in a mid-ocean location in the Sargasso Sea (Bryden, 1977), the geostrophic and observed current shears were in good agreement. Where agreements were found the current speeds were much less than those observed in the DOMES region. Geostrophic currents represent a balance between the acceleration due to the horizontal pressure gradient and the Coriolis acceleration and, perhaps, the poor agreement we found indicated that the magnitudes of other terms in the equation of motion were comparable to the Coriolis acceleration. For example, at Site A at 300 m depth the time-rate of change of the low-frequency u-component (Fig. 9), $\frac{\partial u}{\partial t}$, during the time of the CTD measurements was approximately 4.5 cm sec^{-1} over 3 days or $\sim 1.7 \times 10^{-6}$ cm sec^{-2}, which was about 1/3 of the Coriolis acceleration, $fv[(2.14 \times 10^{-5}$ rad $sec^{-1})$ $(2.5$ cm $sec^{-1}) = 5.3 \times 10^{-6}$ cm $sec^{-2}]$; $\frac{\partial v}{\partial t}$ was about 1% of fu. In contrast, at Sites B and C the two ratios $\frac{\partial u}{\partial t}/fv$ and $\frac{\partial v}{\partial t}/fu$ were about 0.01. In the mixed layer, we showed that the currents contained a substantial wind-driven (ageostrophic) component. Thus, based upon our Site A observations, the dynamical balance of the low-frequency current variations in the upper ocean of the North Equatorial Countercurrent seemed to be

$$\frac{\partial u}{\partial t} - fv = - \frac{1}{\rho} \frac{\partial p}{\partial x} + \frac{\partial}{\partial z} A_V \frac{\partial u}{\partial z}$$

$$fu = - \frac{1}{\rho} \frac{\partial p}{\partial y} + \frac{\partial}{\partial z} A_V \frac{\partial v}{\partial z}$$

where $\frac{1}{\rho} \frac{\partial p}{\partial x}$ and A_V represent the acceleration due to the horizontal pressure gradient in the zonal direction and the vertical eddy viscosity coefficient. The depth, z, is zero at the sea surface and positive upwards. The DOMES data were insufficient to estimate the field acceleration and lateral friction terms and Kendall (1970) has indicated that field acceleration (e.g., $u \frac{\partial u}{\partial x}$) and lateral friction terms were also important in the x-equation of motion. An understanding of the relative role of importance of these terms is required before accurate models can be made of the coupled tropical ocean-atmosphere circulation system.

Low-Frequency Transport Variations

The low-frequency variations of the eastward transport per unit meridional width within the uppermost 300 m (Fig. 12), $\int_{-300\,m}^{0} u\,dz$ was computed from the Site A current vectors shown in Fig. 7. The surface current was equal to the current at 20 m depth and the trapezoidal method of integration was used. Assuming uniform currents across a 5-degree wide North Equatorial Countercurrent (Knauss, 1961), the mean ± 1 standard deviation of the transport of the Countercurrent was 23 Sv ± 8 Sv. The computed Countercurrent transport ranged from a high of 40 Sv to a low of 10 Sv. An understanding of these time-variations and the spatial variations along the east-west direction (Wyrtki, 1966; Kendall, 1970) is one of the most challenging problems in oceanography today.

Inertial-Period Motion

Significant current fluctuations with near-inertial periods were measured at each of the depths of the current meters and the largest amplitudes occurred at Site A. Estimates of the temporal variation of the Site A inertial-period motions (Fig. 13) were computed from the periodogram. Each estimate represented an average value over a time interval equal to 2 inertial periods;

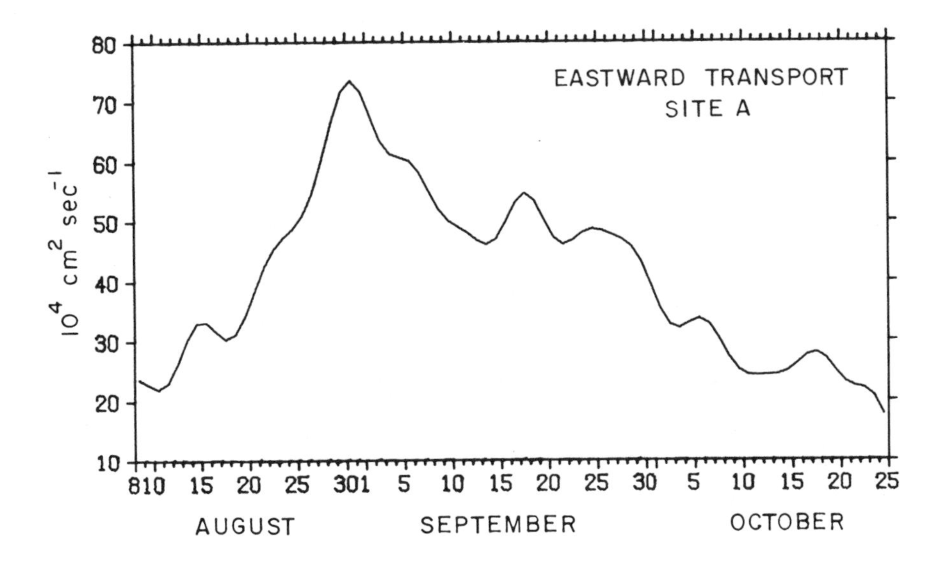

Fig. 12. Low-frequency variations of the eastward transport of the upper 300 m per unit meridional width at Site A.

each interval overlapped the previous one by one inertial period. Fig. 13 shows that the amplitudes of the inertial-period motions in the wind and 20 m current records were large at the beginning of the records but that the times of occurrence of large inertial-period motion were not the same at 20 m and 50 m which, perhaps, can be explained if the near-surface inertial currents were wind-generated and if the 50 m depth current meter was located below the mixed layer (see, e.g., Halpern, 1974b). At Site A the wind fluctuations at 2- to 6-day periods rotated in the clockwise direction, similar to the sense of rotation of the inertial oscillations; at other frequencies the wind motion was rectilinear. The wind and 20 m depth inertial-period fluctuations were statistically coherent at 95% confidence with approximately zero phase difference (Fig. 14), but the 20 m and 50 m depth current records were not coherent at the inertial frequency (Fig. 14), suggesting that the 50 m VACM was usually below the mixed layer (see, e.g., Webster, 1972). Although we can speculate that the large amplitude inertial-period motion at 50 m was produced by nonlinear dynamic instabilities occurring at the bottom of the mixed layer, the temperature distribution (Fig. 8) does not substantiate this speculation and the generation of the large inertial oscillations at 50 m is unknown.

The amplitudes of the 20 m inertial currents were twice as large at A than at C suggesting that, perhaps, a resonant air-sea interaction process occurred at Site A which increased the

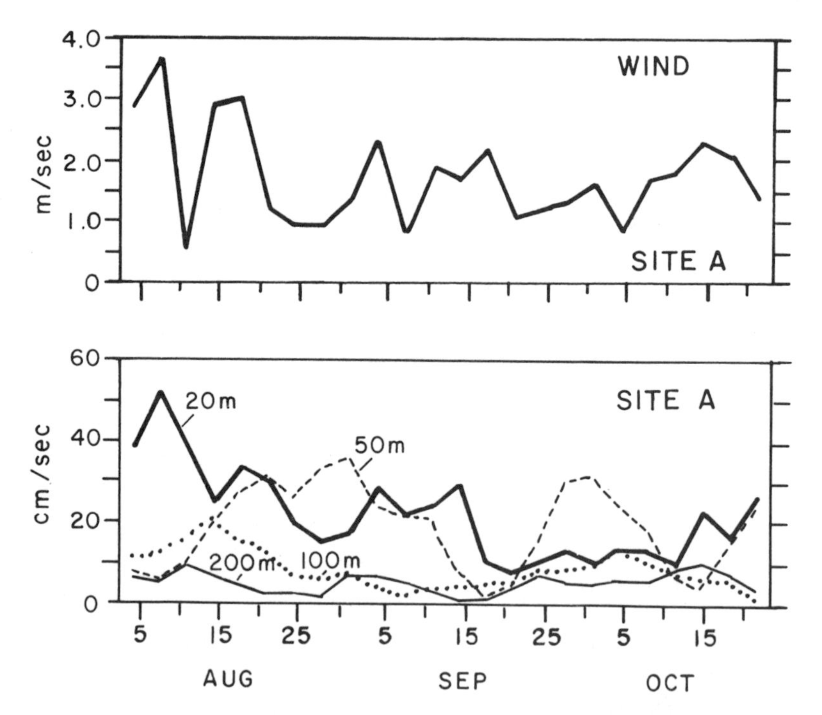

Fig. 13. Time variations of the r.m.s. amplitudes of the inertial-period motion of the wind and current measurements at Site A. The curve for the 300 m record was very similar to the 200 m curve and is not shown.

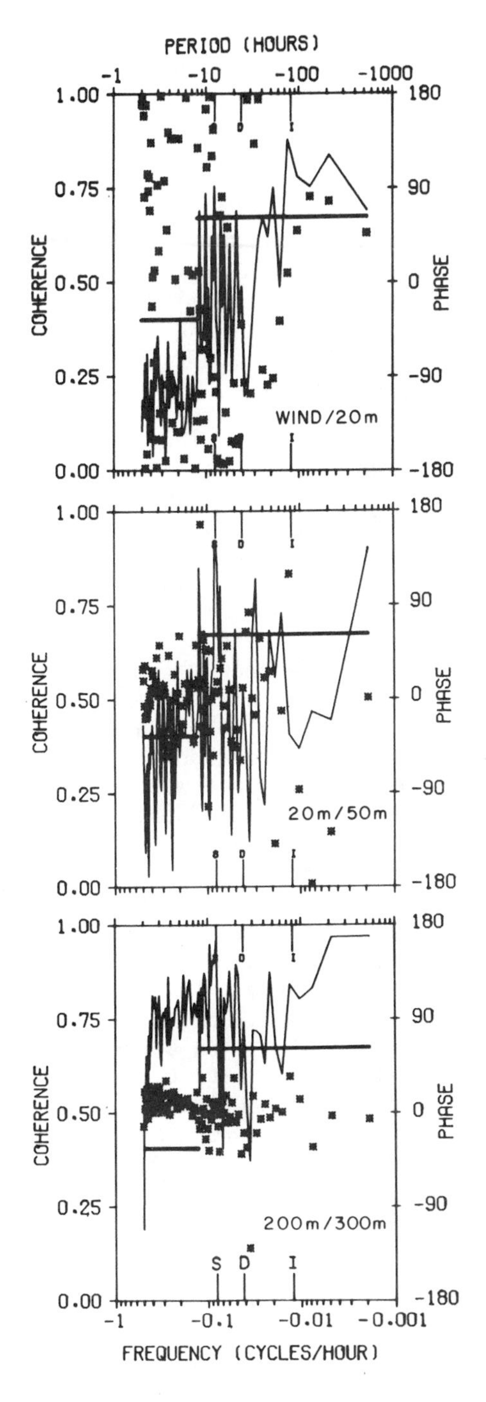

Fig. 14. Magnitude (solid curve) and phase difference (asterisks) of clockwise component (i.e., negative frequencies) of the rotary coherence between (upper) the wind and 20 m currents, (middle) 20 and 50 m currents, and (lower) 200 m and 300 m currents recorded at Site A. A positive phase difference between the wind and 20 m currents corresponds to the wind variation occurring earlier than the 20 m current variation. The 2 horizontal lines in each diagram represent the 95% confidence limits of the coherence estimate. The inertial, diurnal and semidiurnal frequencies are indicated by I, D and S.

efficiency of the transfer of wind momentum to the ocean. At Site A the local inertial-period and the period of the easterly waves propagating in the atmosphere between 5°N and 10°N (Wallace, 1971) were in near-equivalence. Additional evidence of increased efficiency of generation of inertial oscillations in the mixed layer by winds in the atmospheric equatorial trough zone was reported by Halpern (1979b).

At Site A the current fluctuations at the inertial-period were not coherent in the 50 m to 200 m layer. However, below the thermocline in the 200 m to 300 m layer the current fluctuations were coherent with nearly zero phase difference at the inertial-period and also throughout the frequency range (Fig. 14). The reduction of coherence in the thermocline was partly due to variations of the inertial wave phase velocity caused by the variable currents and changes in the density gradient. Were the inertial waves observed below the thermocline generated by the local winds, as presumed by Hayes (1979)? The decrease in the amplitude of the inertial oscillation from 200 m to 300 m does not necessarily imply a downward propagation of energy with ensuing dissipation because inertial wave energy propagates along rays oblique to the vertical, i.e., the distribution of energy along a vertical is dependent upon the slope of the rays and the location of the generating source. Although the statistical confidence of individual phase difference values in Fig. 14 is small and only a weak trend can be obtained from two adjacent phase difference values, inspection of the phase difference between the 200 m and 300 m current records near the inertial frequency shows that the phase difference was positive, i.e., the inertial oscillations measured at 200 m occurred earlier in time than at 300 m. This indication of downward phase propagation corresponds to upward energy propagation, which suggests that the inertial waves below the thermocline were not generated by the wind but, perhaps, by a nonlocal source. However, if the waves were due to randomly distributed sources at lower latitudes then, according to Philander (1978b), the amplitude of the waves would be greater at Site A than at Site C, which was not the case (Table 4). The roughness of the ocean bottom is greater at Site A than at C making the interpretation of the relative sizes of the amplitudes dubious. Further studies of the directivity of inertial wave energy propagation in the deep ocean are needed to determine the importance of nonlocal sources of inertial wave production.

Semidiurnal-Period Fluctuations

For a vertical displacement η the potential energy (PE) per unit volume is $\frac{1}{2}\rho_0 N^2 \eta^2$ where ρ_0 and N are the mean density and the Brunt-Väisälä frequency [$N^2 = \frac{-g}{\rho_o} \frac{\partial \rho}{\partial z}$, where g and ρ represent gravity and density]. For a horizontal velocity $(u^2 + v^2)^{\frac{1}{2}}$ the kinetic energy (KE) per unit volume is $\frac{1}{2} \rho_o (u^2 + v^2)$. The vertical and horizontal components of the semidiurnal-period fluctuations were coherent with zero phase difference throughout the 20 m - 300 m interval (Fig. 8 and 14). If these motions were internal gravity waves the predominant vertical mode number was presumably mode 1 indicating that the vertical wavelength of the internal waves was equivalent to the water depth. Then, to a first approximation, the depth-varying N is given by the virtually constant value of N occurring beneath the shallow thermocline and throughout 97% of the water column. For $N = 4.36 \times 10^{-3}$ rad sec^{-1}, $\eta = 500$ cm and $(u^2 + v^2)^{\frac{1}{2}} = 1.6$ cm sec^{-1}, the PE/KE ratio is ~ 1.9 which was only 2 times larger than the value computed from the theoretical PE/KE ratio which Fofonoff (1969, equation 40) derived for linear internal gravity wave motion in an incompressible, constant N fluid of zero mean motion. The relatively good agreement between the observations and Fofonoff's model indicated that the semidiurnal-period fluctuations were, to a first approximation, internal gravity wave motion. Further analysis of the structure of the internal wave motions will be presented elsewhere.

Aspects of Upper Ocean Current Variability and Deep-Ocean Mining of Manganese Nodules

One method of mining manganese nodules from the deep-sea floor involves a ship towing a 10^7 g collector with about a 6×10^5 cm long cable and pipe assembly wide enough for water, air, nodules and bottom sediment to flow through. Other mining systems differ in the manner of bringing the nodules to the surface. Because the ship will move at slow speeds [e.g., between 25 cm sec^{-1} and 150 cm sec^{-1} according to Herbich and Flipse (1978)] parallel to bottom contours, variations of wind, sea state and currents throughout the water column present considerable difficulties in maintaining a uniform configuration of the cable and pipe assembly between the ship and collector even with assistance of bow and stern thrusters. In contrast to the classical concept of the vertical distribution of tropical currents, the large current speeds and reversals in current direction encountered beneath the thermocline present interesting criteria for optimum design of a cable and pipe assembly which must minimize current drag to reduce fuel consumption of the ship but which must be sturdy enough to withstand large stresses produced by the current shears.

The effluent of a nodule-mining operation consists of stray

nodules, bottom and interstitial water, bottom sediment, benthic biota and other materials, and the behavior of the discharge upon entering the ocean is unknown, *e.g.*, will the fine material flocculate and sink more rapidly than as individual particles. Although the vertical velocity of the semidiurnal-period internal waves was at least 100 times larger than typical settling velocities computed by Ozturgurt *et al.* (1978) for bottom sediments falling through the thermocline, the residence time of the sediments passing through the thermocline will not be affected if the internal waves do not break. It is not clear whether internal waves will enhance dispersion of mining debris in the ocean but it is likely that stirring in the thermocline is intensified by internal wave breaking. The presence of internal waves will cause *in situ* measurements of settling velocity and of the corresponding identity of the effluent to be spurious if measurements were made over a short time interval compared to the semidiurnal-period.

Conclusions

The richness of time and space (vertical) variability of the current measurements was probably the most distinguishable aspect of the upper ocean current measurements recorded at DOMES Sites A, B and C. Prominent fluctuations occurred with long-periods greater than 20-days, with inertial-periods of about 2- to 4-days and with the semidiurnal tidal period. Large amplitudes were associated with the long-period variations, and *vice-versa*. The data clearly showed that the usual perception of the North Equatorial Current and North Equatorial Countercurrent in the central Pacific as slowly-varying zonal currents with speeds decreasing below the mixed layer to a negligible speed at the bottom of the thermocline was not appropriate. In regard to measurement techniques used in further studies, our results indicate that the trajectories of drifters drogued in the mixed layer will not be representative of the flow in the thermocline and that nearly all shipboard measurements (*e.g.*, hydrography, biological, chemical), if made for short-time intervals, will contain errors of omission (Gardenhire, 1964).

For time-scales greater than the inertial-period the wind and near-surface layer of the tropical ocean were coupled. In midlatitudes the ocean-wind interactions have been observed primarily at the inertial-period (*e.g.*, Pollard and Millard, 1970) and the correspondence between the monthly averaged Ekman transport and the transport measured in the mixed layer has yet to be documented in regions remote from coastal boundaries (*e.g.*, Halpern, 1976).

The upper ocean current time-series measurements made at DOMES Sites A, B and C were the first such measurements in the tropical current systems of the central Pacific. Some results of the data were similar to results obtained from other techniques; other results indicated the need for further studies. The role of standing waves in the vertical and the magnitude of the albedo of the bottom *vis-a-vis* wind-generated inertial waves needs to be determined. The representativeness of climatological-mean values, of the geostrophic balance as the first-order approximation of the dynamics, and of the neglect of time-dependence and nonlinear terms in the equations of motion require additional investigation. The absence of discussion of eddy motion occurred because adequate measurements of currents, temperature or salinity to describe horizontal variations were not made; it does not imply that eddies did not occur (see, *e.g.*, Hansen *et al.*, 1978).

Acknowledgements

I am grateful to a large number of people who participated in the field program, in particular, Eugene Duley, Hugh Milburn and Andrew Shepherd of PMEL, Douglas Fenton and Margi McCarty of the University of Washington, and the officers and crew of the NOAA Ship Oceanographer (Captain Kelly Taggart, Commanding Officer); and to Paul Freitag (PMEL), who processed the data, kept track of all the numbers and assisted in the analysis. The helpful comments of Drs. George Philander and Klaus Wyrtki were greatly appreciated. This research was funded by the Environmental Research Laboratories through PMEL and the Deep-Ocean Mining Environmental Studies division of the Marine Ecosystem Analysis Office. This is Contribution Number 397 from the NOAA/ERL Pacific Marine Environmental Laboratory.

References

Berteaux, H.O., (1975) Buoy Engineering, John Wiley, New York.

Bubnov, V.A., V.M. Vasilenko and V.D. Egorikhin, (1979) Some features of water dynamics in the tropical and equatorial Atlantic during GATE, submitted to Deep-Sea Research.

Bryden, H.L., (1977) Geostrophic comparison from moored measurements of current and temperature during the Mid-ocean Dynamics Experiment, Deep-Sea Research, 24, 667-682.

Davis, R.E. and D. Halpern, (1978) The MILE current meter intercomparison, Transactions of the American Geophysical Union, 59, 1093. (Abstract)

Davis, R.E. and R. Weller, (1977) SIO propeller current meters, Polymode News, 40. (Unpublished manuscript available from Woods Hole Oceanographic Institution, Woods Hole, MA.)

Day, C.G., (1970) Wind measurements from moored buoys, Technical Report 70-19, Woods Hole Oceanographic Institution, Woods Hole, MA, 9 pp. (Unpublished manuscript)

Düing, W., P. Hisard, E. Katz, J. Meincke, K. Moroshkin, G. Philander, A. Ribnikov, K. Voigt, and R. Weisberg, (1975) Meanders and long waves in the equatorial Atlantic, Nature, 257, 280-284.

Earle, M.D., (1975) Current measurements in the eastern central North Pacific Ocean, Deep-Sea Research, 22, 875-881.

Ekman, V.W., (1905) On the influence of the earth's rotation on ocean currents, Arkiv. Mat. Astron. Fysik, 12, 1-52.

Fofonoff, N.P., (1969) Spectral characteristics of internal waves in the Ocean, Deep-Sea Research, Supplement to Vol. 16, 59-71.

Gardenshire, L.W., (1964) Selecting sampling rates, Instrument Society of America Journal, 11, 59-64.

Gould, W.M., W.M. Schmitz and C. Wunsch, (1974) Preliminary field results for a Mid-Ocean Dynamics Experiment (MODE-0), Deep-Sea Research, 21, 911-932.

Halpern, D., (1974a) Summertime surface diurnal period winds measured over an upwelling region near the Oregon coast, Journal of Geophysical Research, 74, 2223-2230.

Halpern, D., (1974b) Observations of the deepening of the wind-mixed layer in the northeast Pacific Ocean, Journal of Physical Oceanography, 4, 454-466.

Halpern, D., (1976) Structure of a coastal upwelling event observed off Oregon during July 1973, Deep-Sea Research, 23, 495-508.

Halpern, D., (1978) Transport of the North Equatorial Counter-current in the central Pacific. In, NORPAX Quarterly Report, edited by D. Cutchin, Scripps Institution of Oceanography A-030, La Jolla, CA. (Unpublished manuscript)

Halpern, D, (1979a) Moored current measurements in the upper ocean. In, Instruments and Methods in Air-Sea Interaction, R.E. Davis, F.W. Dobson and L. Hasse, editors, Plenum Press, in press.

Halpern, D., (1979b) Variability of mixed layer currents during the GATE C-Scale experiment. Submitted to Deep-Sea Research.

Halpern, D. and R.D. Pillsbury, (1976) Near-surface moored current meter measurement, Marine Technology Society Journal, 10 (7), 32-38.

Halpern, D. and R.M. Reynolds, (1973) Comparison between FNWC-derived surface winds and near surface wind measurements in the Northeast Pacific, Transactions of the American Geophysical Union, 54, 142. (Abstract)

Halpern, D., R.L. Smith and R.K. Reed, (1978) On the California Undercurrent over the continental slope off Oregon, Journal of Geophysical Research, 83, 1366-1372.

Hansen, D., W.C. Patzert and G.J. McNally, (1978) The near-surface flow of the tropical Pacific Ocean observed using satellite-tracked drifting buoy trajectories, Transactions of the American Geophysical Union, 59, 1112. (Abstract)

Hayes, S.P., (1977) Study of the benthic boundary layer in the north equatorial Pacific. In, Proceedings of the Offshore Technology Conference, Dallas, 387-394.

Hayes, S.P., (1979) Benthic current observations in the tropical North Pacific Ocean, (this volume).

Herbich, J.B. and J.E. Flipse, (1978) Technological gaps in deep ocean mining. In, Oceans' 78, Marine Technology Society, Washington, D.C., 606-610.

Kendall, T.R., (1970) The Pacific Equatorial Countercurrent, International Center for Environmental Research, Laguna Beach, CA.

Knauss, J.A., (1961) The structure of the Pacific Equatorial Countercurrent, Journal of Geophysical Research, 66, 143-155.

McCullough, J.R., (1975) Vector-averaging current meter speed calibration and recording technique, Technical Report 75-44, Woods Hole Oceanographic Institution, Woods Hole, MA. (Unpublished manuscript)

Madden, R.A. and P.R. Julian, (1972) Description of global-scale circulation cells in the tropics with a 40-50 day period, Journal of the Atmospheric Sciences, 29, 1109-1123.

Meyers, G, (1975) Seasonal variation in transport of the Pacific North Equatorial Current relative to the wind field. Journal of Physical Oceanography, 5, 442-449.

Ozturgut, E., G.C. Anderson, R.E. Burns, J.W. Lavelle and S.A. Swift, (1978) Deep Ocean mining of manganese nodules in the North Pacific: Pre-mining environmental conditions and anticipated mining effects, NOAA Technical Report, U.S. Government Printing Office, in press.

Philander, S.G.H., (1978a) Instabilities of zonal equatorial currents, 2, Journal of Geophysical Research, 83, 3679-3682.

Philander, S.G.H., (1978b) Forced oceanic waves, Reviews of Geophysics and Space Physics, 16, 15-46.

Pollard, R.T. and R.C. Millard, (1970) Comparisons between observed and simulated wind-generated inertial oscillations, Deep-Sea Research, 17, 813-821.

Ramp, S.R., (1976) The measurement and analysis of low-frequency currents in the deep central North Pacific Ocean, M.Sc. thesis, Department of Oceanography, University of Washington. (Unpublished manuscript)

Reid, J.L., (1961) On the geostrophic flow at the surface of the Pacific Ocean with respect to the 1000-decibar surface. Tellus, 13, 489-502.

Reid, J.L., Jr., (1965) Intermediate Waters of the Pacific Ocean, Johns Hopkins Press, Baltimore.

Reid, J.L., Jr., (1973) The shallow salinity minima of the Pacific Ocean, Deep-Sea Research, 20, 51-68.

Smith, R.L., (1974) A description of current, wind, and sea-level variations during coastal upwelling off the Oregon coast, July-August 1972, Journal of Geophysical Research, 79, 435-443.

Sverdrup, H.U., (1947) Wind-driven currents in a baroclinic ocean; with application to the equatorial currents of the eastern Pacific, Proceedings of the National Academy of Sciences, 33, 318-326.

Sweers, H.E., (1971) A comparison of methods used to calculate sigma-t specific volume anomaly and dynamic height, Marine Technology Society Journal, 5, 7-26.

Tsuchiya, M., (1968) Upper Waters of the Intertropical Pacific Ocean, Johns Hopkins Press, Baltimore, 50 pp.

Tsuchiya, M., (1974) Variation of the surface geostrophic flow in the eastern intertropical Pacific Ocean, Fishery Bulletin, 72, 1075-1086.

Tsuchiya, M., (1975) Subsurface countercurrents in the eastern equatorial Pacific Ocean, Journal of Marine Research, Supplement, 33, 145-175.

Wallace, J.M., (1971) Spectral studies of tropospheric wave disturbances in the tropical western Pacific, Reviews of Geophysics and Space Physics, 9, 557-612.

Webster, F., (1972) Estimates of the coherence of ocean currents over vertical distances, Deep-Sea Research, 19, 35-44.

Wooster, W.S. and B.A. Taft, (1958) On the reliability of field measurements of temperature and salinity of the ocean. Journal of Marine Research, 17, 552-566.

Wunsch, C., (1978) Observations of equatorially trapped waves in the ocean: A review prepared for equatorial workshop, July 1977. In, Review Papers of Equatorial Oceanography FINE

Workshop Proceedings. (Unpublished manuscript available from Physical Oceanographic Laboratory, Nova University, Dania, Florida)

Wyrtki, K., (1965) Surface currents of the eastern tropical Pacific Ocean, Inter-American Tropical Tuna Comm. Bulletin, 9, 271-304.

Wyrtki, K., (1966) Oceanography of the eastern equatorial Pacific Ocean, Oceanography and Marine Biology Annual Reviews, 4, 33-68.

Wyrtki, K., (1967) Oceanographic observations during the Line Islands expedition, February-March, 1967. Technical Report 67-17, Hawaii Institute of Geophysics, University of Hawaii, Honolulu.

Wyrtki, K., (1974) Equatorial currents in the Pacific 1950 to 1970 and their relations to the Trade Winds, Journal of Physical Oceanography, 4, 372-380.

Wyrtki, K., (1975) Fluctuations of the dynamic topography in the Pacific Ocean, Journal of Physical Oceanography, 5, 450-459.

Wyrtki, K., (1978a) Monitoring the strength of equatorial currents from XBT sections and sea level, Journal of Geophysical Research, 83, 1935-1940.

Wyrtki, K, (1978b) Lateral oscillations of the Pacific Equatorial Countercurrent, Journal of Physical Oceanography, 8, 530-532.

Wyrtki, K. and R. Kendall, (1967) Transports of the Pacific Equatorial Countercurrent, Journal of Geophysical Research, 72, 2073-2076.

BENTHIC CURRENT OBSERVATIONS AT DOMES SITES A, B, AND C

IN THE TROPICAL NORTH PACIFIC OCEAN

S. P. Hayes

Pacific Marine Environmental Laboratory
Seattle, Washington 98105

Abstract

Benthic current measurements are reported from three locations (A: 8°27'N, 150°49'W; B: 11°42'N, 138°24'W; C: 14°38'N, 125°29'W) in the eastern tropical Pacific. Near-bottom stratification was weak at all sites. The measurements were of 4- to 6-month duration. Mean currents were small and to the northwest; however, low frequency fluctuations dominated the records. These fluctuations had dominant periods for the meridional component of 2 months at all sites; the zonal component had periods varying from 2 months at the easternmost site to about 5 months at the other sites. Low frequency kinetic energy increased from west to east. Vertically, the low frequency motions were coherent over the bottom 200 m. A small speed increase was observed from 200 m to 30 m off the bottom; below this the speed decreased. Interpretation of these data in terms of an Ekman layer showed counterclockwise (looking down) veering between 30- to 6-m levels, which was consistent with the expected layer thickness of 25 m.

High frequency inertial-internal wave oscillations were also investigated. The inertial oscillations were intermittent and showed evidence for downward energy propagation. Mean energy level and lack of correlation between low frequency currents and internal wave energy suggests that the bottom topography is not a strong source of high frequency internal wave energy in our data.

*Contribution No. 398 from the NOAA/ERL Pacific Marine Environmental Laboratory.

Introduction

To the physical oceanographer, the abyssal ocean represents a largely unexplored and potentially important region. Mean currents in deep water result from the thermohaline general circulation modified by topography. Superposed on this flow, motions with a spectrum of frequencies provide large variability that can yield individual current measurements far greater than the mean. Observed water mass properties are the net result of advection and diffusion. The exchange processes that occur within the boundary layers adjacent to the surface, the continental margins, and the bottom modify the circulation and water characteristics of the interior of the ocean. Thus, the currents and stratification of the benthic boundary layer are of interest, as are their possible effects on the general circulation.

In this paper we report near-bottom measurements made at three sites in the eastern tropical Pacific Ocean. These data were collected as part of the Deep Ocean Mining Environmental Study (DOMES) program of NOAA. The physical oceanographic characteristics of this region were largely unknown; therefore, the experiment was designed to provide a preliminary description of the stratification and current fields. The presented analysis focuses on temporal scales of the currents, differences among the sites, and some investigation of potentially important internal wave processes.

Review

The study locations (Fig. 1) are in a region of abyssal hills between the Clarion and Clipperton Fracture Zones. A gradual slope to the east decreases the mean depth from site A (151°W) to site C (125°W) by about 500 m. Bathymetric charts (Chase *et al.*, 1970) indicate a few nearby seamounts: 75 km east of A, a mount rises about 1000 m; 120 km east of C, a mount rises about 2000 m. However, in the immediate vicinity of the arrays, the relief is confined to vertical changes of about 200 m over a few kilometers horizontally.

Historical data of near-bottom currents in this area of the tropical Pacific are scarce. Most of the information on mean flow direction has been obtained from water mass analysis and theoretical models of abyssal circulation. Wong (1972) postulated an eastward bottom velocity south of Hawaii by tracing observations of anomalous values of oxygen, potential temperature, and salinity. This result agrees with near-bottom potential temperature distributions shown by Mantyla (1975) and with theoretical circulation models of Stommel and Arons (1960) and Kuo (1978). Superimposed on existing mean flow, one expects time-dependent fluctuations with frequencies ranging up to the local Brünt-Vaisälä frequency (about 0.2 cph). Low frequency (less than inertial frequency) mesoscale eddies presumably contribute a significant fraction of the total variance. Wyrtki *et al.* (1976) produced maps of near-surface

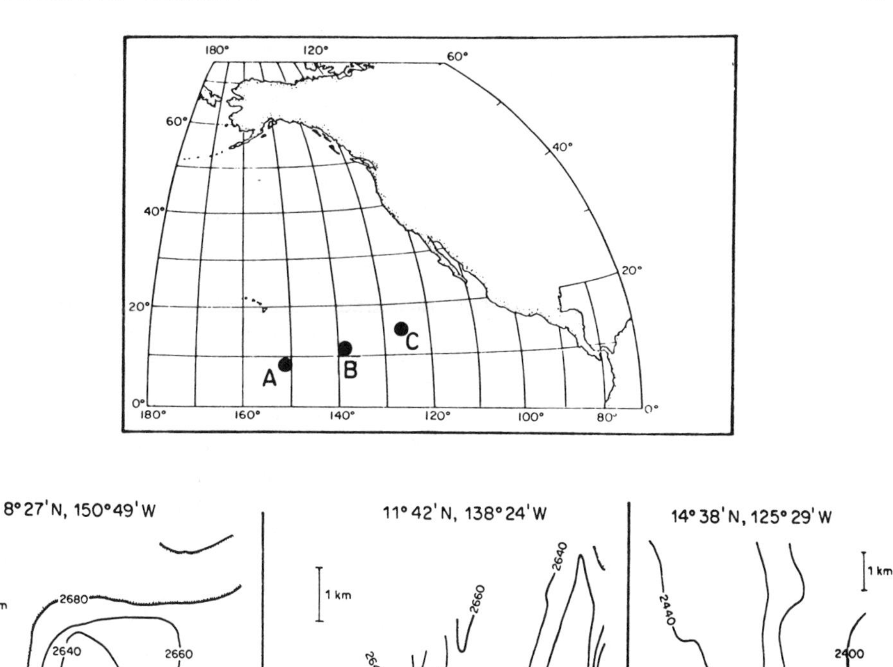

Fig. 1. Location of moored current meter measurements in tropical Pacific Ocean. Sites were located in a region of abyssal hills. Detailed bathymetry near each site indicates our best estimate of mooring location relative to topography. Contours are in fathoms.

eddy kinetic energy per unit mass for the world oceans. These maps indicated an increased energy in the north equatorial region. Within our study area, surface eddy kinetic energy increased by more than a factor of 2 proceeding from 15°N to 8°N. It is not known whether these surface features are associated with a deep eddy structure. Direct current measurements of sufficient length to resolve the mean or low frequency currents do not exist for this region. Near the eastern site (C) Amos et al. (1976) reported 1-month measurements of near-bottom currents. Two other long-term deep-current measurements have also been reported in the tropical Pacific; Taft et al. (1974) report a 4.5-month record from 1°S, 150°W, and Harvey and Patzert (1976) report 2-month records from about (0°, 95°W). We will compare these data to our results.

Information on the characteristics of the benthic boundary layer that have been obtained in other locations may be pertinent to the tropical Pacific. Continuous vertical profiles of potential temperature, salinity, and density that show an apparently well-mixed layer adjacent to the bottom have been presented for many oceanic locations. Amos et al. (1971) discussed such layers near the Blake-Bahama Outer Ridge in the Atlantic Ocean. Biscaye and Eittreim (1974) and Eittreim et al. (1975) studied suspended particulate matter and excess radon profiles taken over the Blake-Bahama Outer Ridge and over the Hatteras Abyssal Plain; the lack of near-bottom stratification in these parameters as well as in potential temperature indicates active mixing. In the Hatteras Abyssal Plain and to the east (55°W), Armi and Millard (1976) and Armi (1978) have found bottom layers thicker than 50 m. These studies in regions of weak mean flow, which are remote from large topographic features, might be expected to be similar to the tropical Pacific. The dynamics of the bottom mixed layers have been investigated by several authors (Wimbush and Munk, 1971; Weatherly, 1972, 1975; Thompson, 1973; Csanady, 1974; Armi and Millard, 1976). In general, some form of a modified Ekman boundary layer is assumed. The presence of topographic gradients and time-dependent velocities complicates the interpretation. Armi and Millard (1976) were able to correlate the mixed-layer thickness with low frequency near-bottom velocity. Such a relation would be expected if Ekman dynamics held; however, the thickness observed over a flat bottom was six times greater than the turbulent Ekman layer thickness.

Coupling between mixing and low frequency flow can effect the dynamics of mesoscale circulation. In addition, topography-current interaction can modify the flow found within a few kilometers of the bottom. Results from the Mid-Ocean Dynamics Experiment (MODE-I Dynamics Group, 1975) and subsequent current measurements in the North Atlantic (Schmitz, 1978) show effects that may be related to bottom topography. In MODE, the structure of mesoscale eddies with periods longer than the inertial period were studied. Near bottom these eddy motions had smaller time scales and larger horizontal kinetic energy than they did at mid-depths (1500 m). The spatial scales were also contracted near bottom, and localized currents were observed near some major topography. Rhines (1977) interpreted the general MODE results in terms of topographic influences on baroclinic Rossby waves. This picture of energetic, small-scale benthic currents, which data and theory suggest, contrasts with the intuitive notion of a sluggish, homogeneous abyssal flow.

In addition to modifying mesoscale currents, the bottom topography may have a dominant influence on energy dissipation. Rhines (1977) points out that a search for energy sinks that dissipate mesoscale eddies is critical. Conventional drag laws indicate that over a smooth bottom friction can contribute only a small amount. However, Bell (1975) investigated the generation of internal waves by the interaction of low frequency deep ocean currents with bottom topography. Using abyssal hills and a velocity amplitude of 5 cm/s for the low frequency currents as an example, he found an internal wave drag equivalent to about 0.5 dyn/cm^2. This production mechanism could contribute a significant fraction of the internal wave energy observed in the ocean. Few observational tests of Bell's conclusions have been made. In a study of several deep current meter records in the North Atlantic, Wunsch (1976) observed that the most significant internal wave inhomogeneity observed was associated with a seamount. Thus, bottom topography may be a source; however, its relative importance has not been established.

Description of the Experiment

Figure 1 gives the locations of our moored array measurements. At each site, a single mooring was deployed with current meters at 6 m, 30 m, 50 m, and (at C) 200 m above the bottom. Deployments were from May to November 1976 at A and B and from July to December 1977 at C. The detailed bathymetry at each location derived from local ship surveys represents our best estimate of the mooring location. Both A and B appear to be part-way up a small hill, while C is in a long valley. The local gradients (maximum ∿.05) shown in these figures are typical of the area around the arrays.

Current meters used were vector averaging current meters (VACM) manufactured by AMF Electrical Products Division, Alexandria, Virginia. These instruments sense speed with a Savonius rotor (16-cm diameter), direction with a vane (17 cm), and temperature with a bead thermistor. As the rotor turns, the instrument internally resolves east-west and north-south velocity components and stores the data on magnetic tape. The effective threshold is about 2 cm/s (McCullough, 1975). Individual rotor and vane calibrations which we made in a tow tank showed variation in this threshold; however, most instruments followed a standard calibration curve (given in Halpern et al., 1974) for speeds higher than 2 cm/s. In the data discussed here, instruments recorded every 15 minutes. During processing, any 15-min interval with an indicated speed less than 2 cm/s was assigned speed and direction of zero. This procedure probably underestimates mean currents. Threshold problems were most severe at A, where over 50% of the recorded values were below threshold and periods of several days contained no measurements above 2 cm/s. At the other two locations, the records do not appear to be seriously contaminated by the instrument thresholds; however, during periods of weak low frequency currents, estimates of internal wave energy may be erroneous.

The current-meter temperature sensors were calibrated as described in Halpern et al. (1974). Absolute temperature calibration had an expected error of ±.01°C; relative temperature changes were sensed with a resolution better than .001°C. In view of the weak near-bottom temperature stratification, the temperature measurements were not accurate enough to establish gradients; however, temperature changes at each level were resolved.

In addition to moored arrays, continuous profiles of conductivity, temperature, and depth (CTD) were taken with a Neil Brown Instrument System Mark III CTD. These profiles were processed as described in Fofonoff et al. (1974) in order to obtain temperature and salinity profiles. During the November 1976 deployment cruise, detailed studies were made in the vicinity of A and B. Unfortunately, subsequent processing revealed two problems with the CTD system: a) temperature-dependent noise level in the temperature measurements, which at cold temperatures yielded a noise level about ten times that expected from the instrument quantizing increment (.0005°C); and b) temperature instabilities of a few millidegrees. The first problem was traced to a component in the temperature-sensing electronics; the second problem may be related to a high pressure leak caused by a manufacturer's defect in the conductivity cell. In any case, these relatively small temperature errors made precise near-bottom profiles impossible. These problems were corrected prior to the site C CTD measurements that were made in July 1977. Also, site A was reoccupied in March 1978; these data are presented here on the assumption that the near-bottom water mass structures are reasonably constant.

Figure 2 shows representative profiles of potential temperature (θ) at each location. Polynomials developed by Bryden (1973) were used to calculate θ. Bottom potential temperature increased to the east (0.98°C) at A, 1.03°C at B, and 1.09°C at C) in agreement with the historical data (Wong, 1972; Mantyla, 1975). Potential temperature gradient decreased at the bottom. Over the lowest 200 m, this gradient was about 2×10^{-5}°C/m. This value is one order of magnitude less than the background stratification observed by Armi and Millard (1976) over the Hatteras Abyssal Plain. The weak mean gradients may account for the lack of an obvious bottom mixed layer in our data; the layer would have to exceed 50 m in thickness in order to be measureable.

Potential temperature-salinity diagrams for the abyssal water at sites A and C are shown in Fig. 3. These diagrams show the linear θ-S relation for the deep Pacific; note that due to water depth differences, the site-C curve is essentially a continuation of the site-A curve at higher temperatures. No distinct near-bottom water mass is seen.

Velocity and Temperature Time Series

Time series of velocity and temperature are shown in Fig. 4. The visual impression of these records is similar at all sites. Low frequency oscillations with periods of several weeks were superimposed on the high frequency inertial and tidal signal. The currents were vertically coherent. At site A, as mentioned above, extended periods (e.g., early September) of low current occurred, during which the speeds were below threshold. At site B, only the

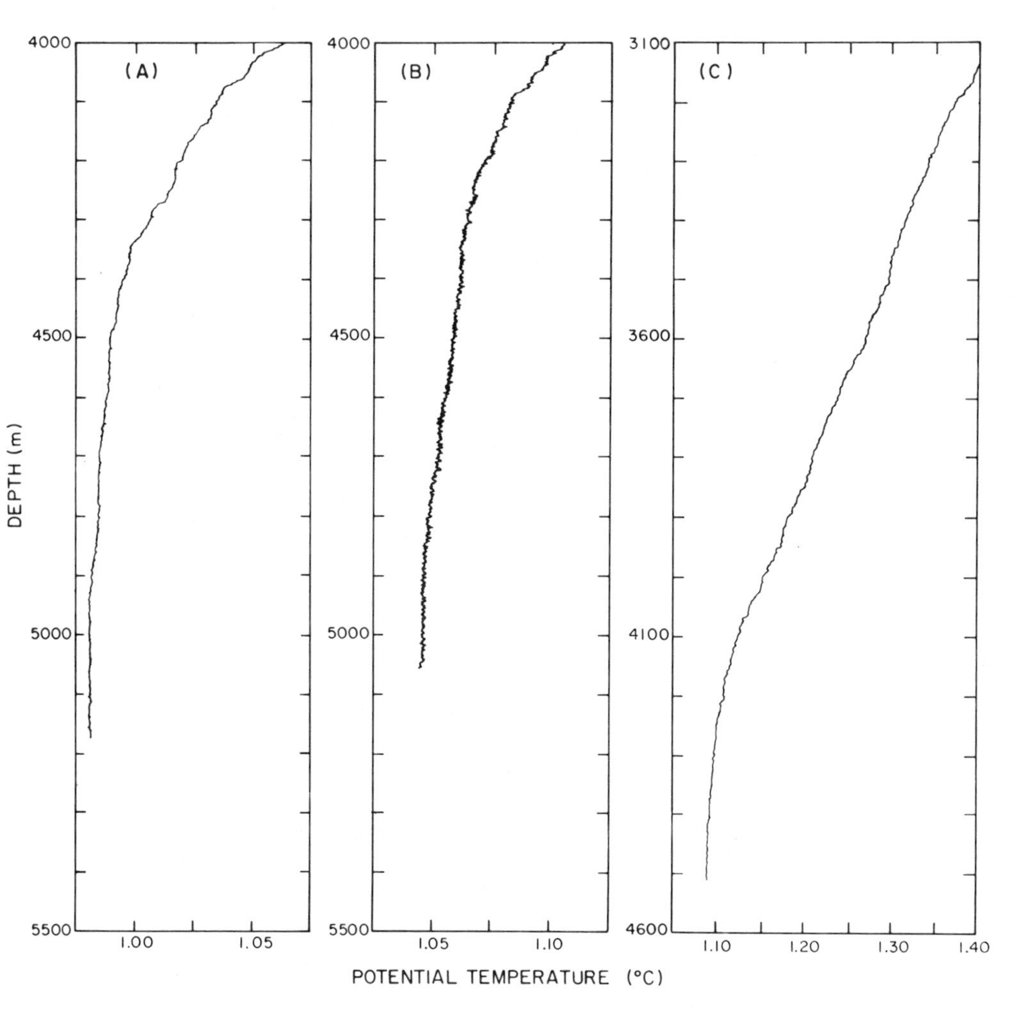

Fig. 2. Benthic potential temperature versus depth profiles near each mooring. Note scale change at each site.

30-m current meter worked throughout the deployment. Low frequency structure here was quite pronounced with a rapid direction change observed in mid-July. Site C had the most complete data set with velocity measurements up to 200 m above the bottom. The vertical coherence of much of the structure over this depth interval is obvious. Also, non-stationarity in the high frequency signals was pronounced. Near the beginning of October, large oscillations are evident. Later analysis showed that these signals have frequencies near the inertial.

Because of nearly uniform bottom-water temperature and relatively large errors in absolute calibration (±.01°C), temperature time series from the current meters are plotted in Fig. 4 relative

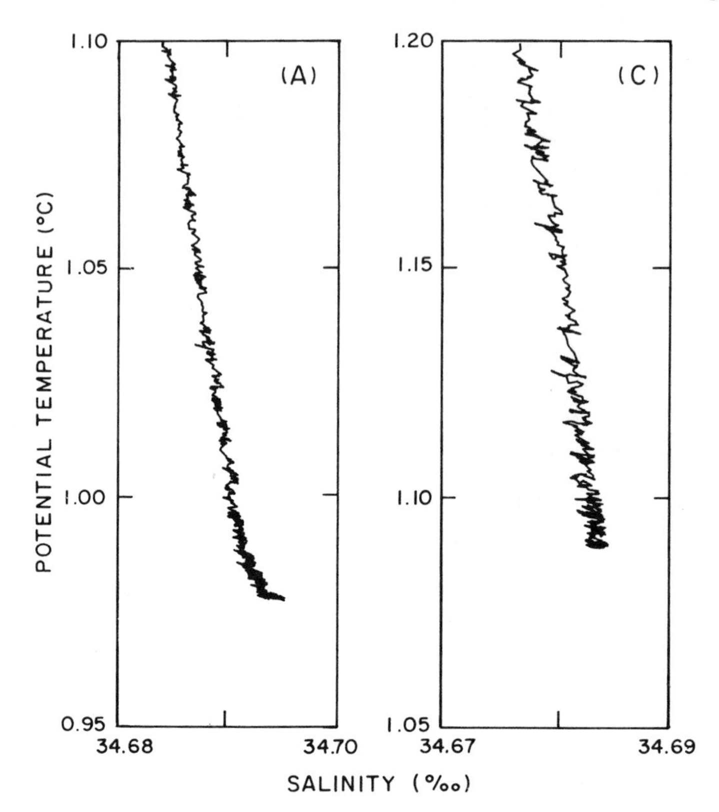

Fig. 3. Benthic potential temperature versus salinity diagrams for sites A and C. The structure in the bottom few hundred meters varied between casts so that the small salinity increase (0.002 ppt) seen at A is not considered significant.

to an arbitrary zero. Temperature changes at each site appeared vertically coherent. Site A had the smallest temperature variance over the record. This observation is consistent with the weak mean temperature gradients found in CTD data and the relatively weak currents at this location. At B, bottom temperature increased over the deployment period. Most of this increase occurred after the current change in mid-July; the warming probably represented lateral advection. The site B and C records show more visual correlation

---

Fig. 4a. Hourly time series of currents and temperature at locations A and B. The origin for the currents is indicated by the horizontal line at each depth; the bar scale gives the speed. Temperatures are plotted relative to an arbitrary origin and only temperature changes are significant. Values in meters are heights above bottom.

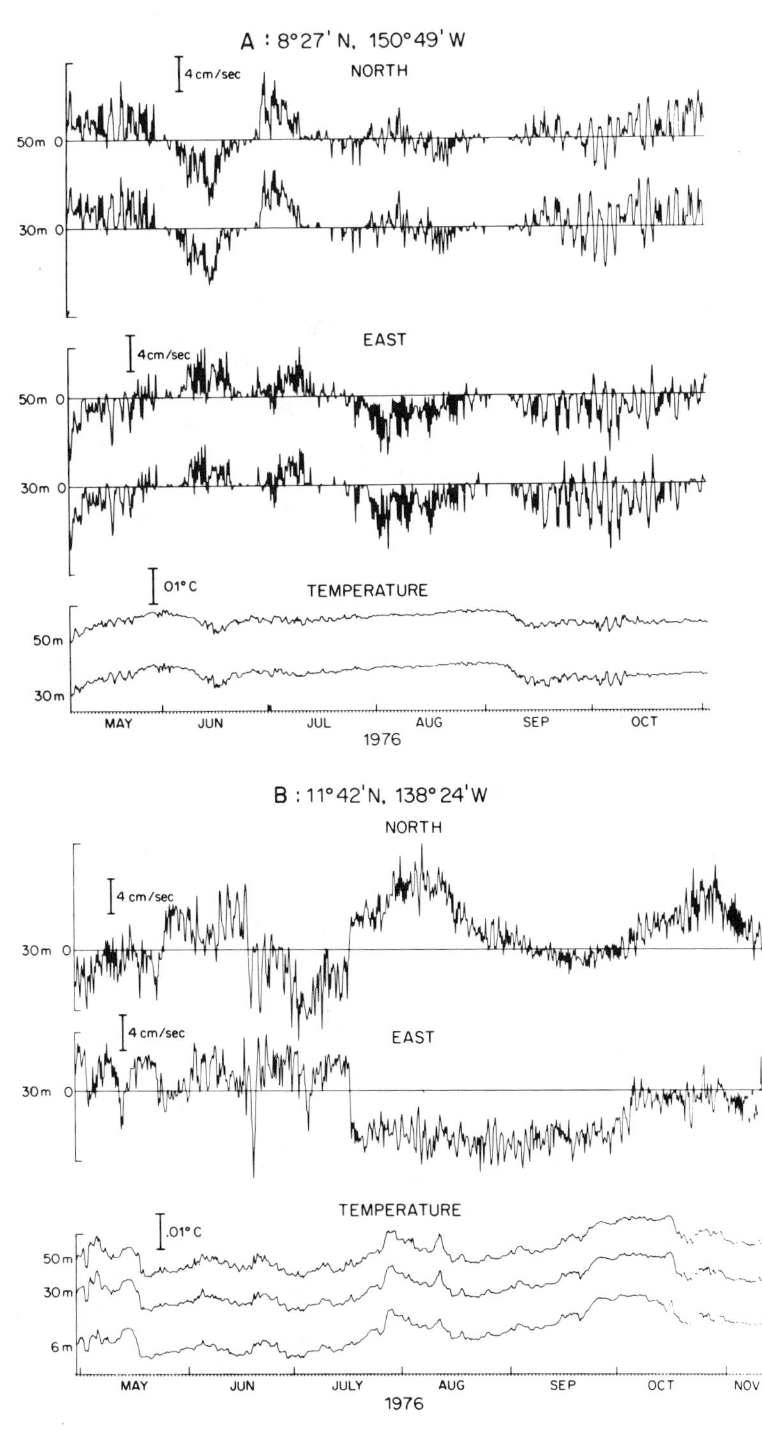
A : 8°27' N, 150°49' W
NORTH
4 cm/sec
50m 0
30m 0
EAST
4 cm/sec
50m 0
30m 0
TEMPERATURE
.01° C
50m
30m
MAY
JUN
JUL
AUG
SEP
OCT
1976
B : 11°42'N, 138°24'W
NORTH
4 cm/sec
30m 0
EAST
4 cm/sec
30m 0
TEMPERATURE
.01° C
50m
30m
6m
MAY
JUN
JULY
AUG
SEP
OCT
NOV
1976

Fig. 4a

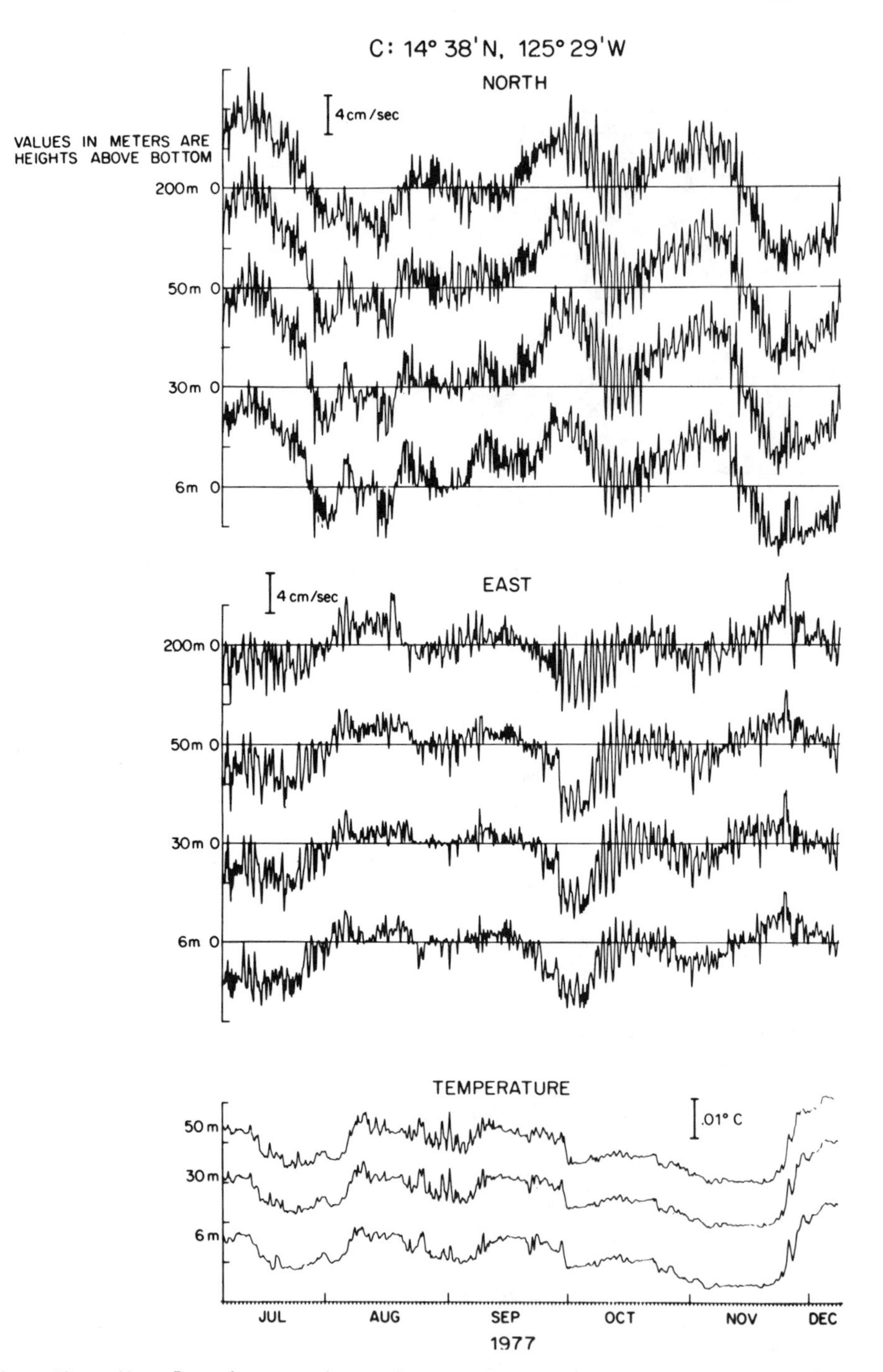

Fig. 4b. Hourly time series of currents and temperature at site C. Same plotting convention as Fig. 4a.

between temperature and velocity than the record at A. In particular, at C both low and high frequency changes appear related. For example, the initial low frequency velocity change from northeast to southwest currents is accompanied by a decrease in bottom temperature. In addition, a sharp drop in temperature on about 1 October appears to accompany the onset of high frequency oscillations mentioned above.

The separation between low frequency and inertial-internal wave energy regimes can be seen clearly in horizontal kinetic energy spectra. Figure 5 shows variance preserving (area under the curve is proportional to the variance) spectra for the 30-m record at each location. Inertial frequencies at A, B, and C are $f_A = 0.012$ cph, $f_B = 0.017$ cph, and $f_C = 0.021$ cph. Note that the energy scales are different for each site. These spectra show characteristic structures similar to most deep-ocean data sets. There is high energy in the tidal and near-inertial frequency bands, a low energy region (spectral gap) for frequencies between inertial and about 0.05 cph, and a low frequency rise in energy below the gap. Most of the total variance in a long current record comes from the low frequency motions. The existence of the spectral gap facilitates filtering the time series into a low frequency sub-inertial time series and a high frequency inertial-internal wave time series. The relative amplitudes of the two components are shown graphically in Fig. 6, where low- and high-pass filtered series of the north-south component of velocity at site C (200-m level) are plotted. Clearly, instantaneous current measurements will have significant contributions from both spectral regions, and any attempt to study the characteristics of a region must resolve the low frequency flow.

Low Frequency Motions

To describe motions with frequencies less than the inertial frequency, the time series were filtered with a symmetric Gaussian filter (Schmitz, 1976). The filter width was chosen so that at each site the half-power frequency was one-half the inertial frequency. Fig. 7 is a plot of the data at all locations in vector form. The length of each stick represents current speed and the angle indicates direction. Both A and C exhibit high vertical coherence.

The relative importance of low frequency oscillations compared to the mean flow (averaged over the record length) is seen in Fig. 7 or Table 1. In the latter, $\overline{KE} = \frac{1}{2}(\overline{u}^2 + \overline{v}^2)$ where a bar over the velocities indicates record length averages. $\overline{KE}$ is the kinetic energy (per unit mass) of the mean flow. Primed velocities are defined as the standard deviation of the low-pass filtered data, again computed over the record length. KE', kinetic energy (per unit mass) of the low frequency flow, is given by $KE' = \frac{1}{2}(u'^2 + v'^2)$.

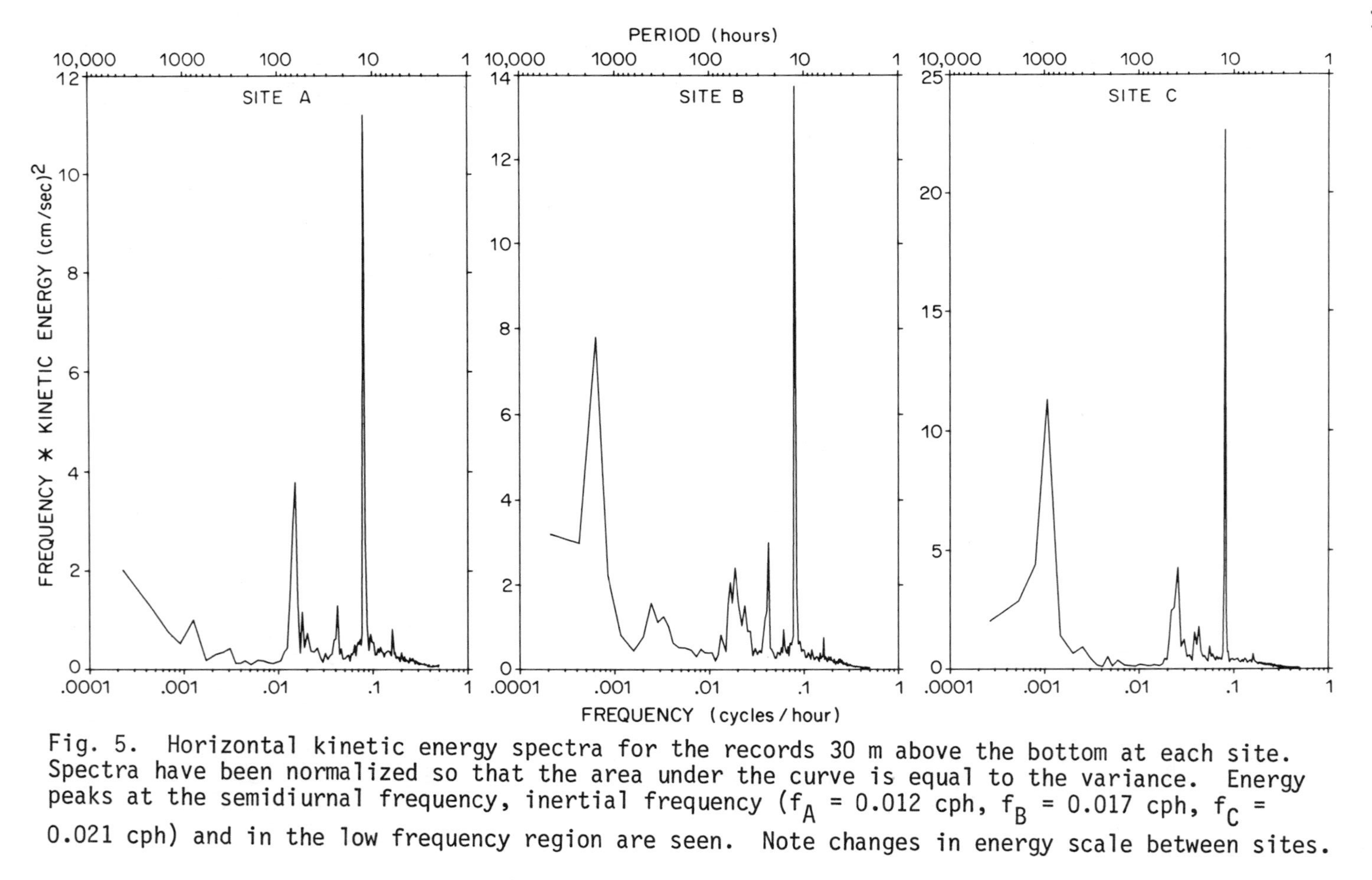

Fig. 5. Horizontal kinetic energy spectra for the records 30 m above the bottom at each site. Spectra have been normalized so that the area under the curve is equal to the variance. Energy peaks at the semidiurnal frequency, inertial frequency (f_A = 0.012 cph, f_B = 0.017 cph, f_C = 0.021 cph) and in the low frequency region are seen. Note changes in energy scale between sites.

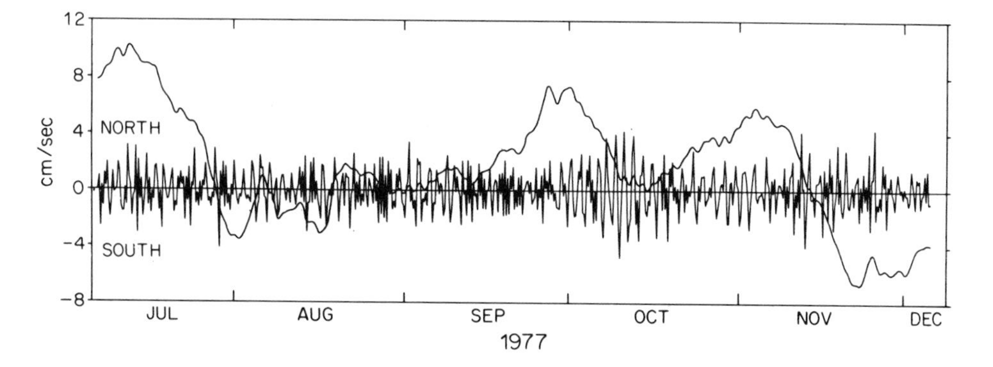

Fig. 6. Meridional velocity component at the 200-m level of site C, showing the contribution of low frequency (Gaussian low-pass filtered time series as described in text) and inertial-internal wave band oscillations.

The double-primed energy relates to internal wave contributions and is defined as one-half of the difference between the total variance and the low frequency variance. Table 1 shows that KE' is, on the average, about ten times larger than $\overline{KE}$. Mean velocity at all sites was to the northwest; however, the record lengths are insufficient for these mean velocities to be statistically significant. Interestingly, near-bottom currents appear somewhat larger than the currents above. At site C, mean speed increased by a factor of 3 between 200 m and 6 m. The low frequency speed increased between 200 m and 30 m, then decreased from 30 m to 6 m. KE' (6 m) roughly equals KE' (200 m) and is about 80% of KE' (30 m). North-south velocity differences [(v(200) - v(30)] are shown in Fig. 8 along with the velocity at 200 m. The southward mean velocity difference indicates that the 200 m northward velocity is smaller than the 30 m northward velocity. This difference reflects the baroclinic nature of the low frequency flow, rather than a constant bottom current. When the 200-m velocity is large, the shear is large; when the 200-m velocity is near zero, the shear is also small.

The time scale of the low frequency motion is obtained from autocorrelation functions of each velocity component (Fig. 9). These functions have an interesting pattern. At all locations the v component has a first zero crossing at about 15 days (corresponding to a dominant period of 60 days); the u component at sites A and B has a much longer time scale. For example, at site B, the zero crossing indicates a dominant period of order 180 days, which is essentially the record length. At site C, the time scales of u and v are almost equal. This same information is shown in spectral form in Fig. 10. Note that the energy scales differ at each location. All sites have a variance peak in the v component

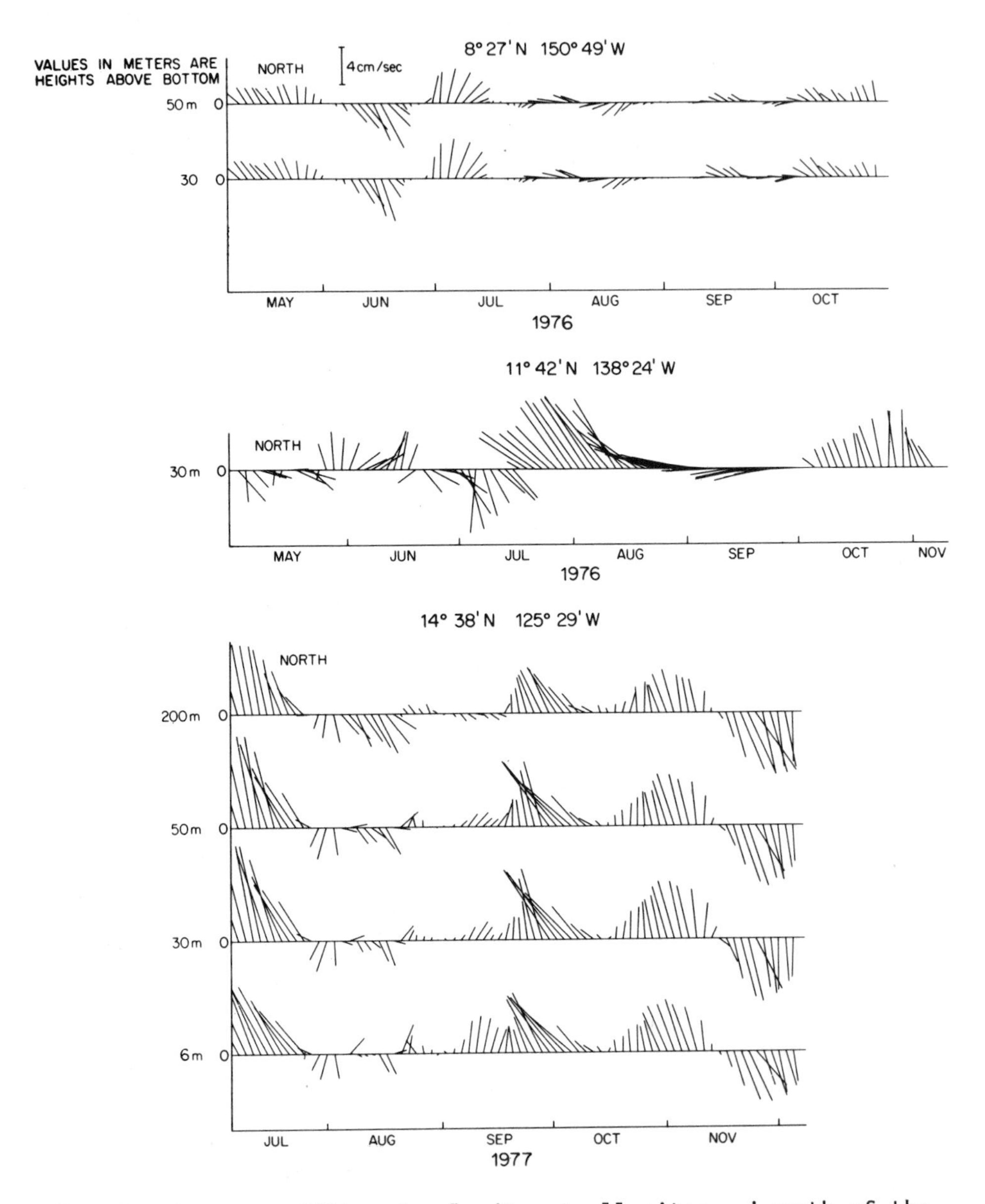

Fig. 7. Low-pass filtered velocity at all sites. Length of the sticks are proportional to speed and orientation of stick indicates direction.

TABLE 1

Summary of the current meter data at all locations

Location	Height Above Bottom (m)	Record Length (days)	$\overline{u}$ cm/s	$\overline{v}$ cm/s	u' cm/s	v' cm/s	$\overline{KE}$ cm^2/s	KE' cm^2/s	KE" cm^2/s
Site A 8°27'N	50	143	-.29	.19	1.38	1.54	.1	2.2	2.2
150°49.1'W	30	143	-.49	.29	1.29	1.56	.2	2.1	1.9
Site B 11°42'N 138°24'W	30	197	-1.68	1.37	3.38	3.29	2.4	11.1	3.1
Site C 14°38'N	200	156	-.11	.59	1.54	3.66	.2	7.9	3.3
125°29'W	50	156	-.46	1.61	1.93	3.98	1.4	9.8	3.1
	30	156	-.62	1.79	1.82	4.17	1.8	10.4	3.0
	6	156	-.81	1.83	1.97	3.43	2.0	7.8	2.5

Bars indicate averages over record length. Primes indicate low-pass filtered currents. The kinetic energy (KE) per unit mass is defined in the text.

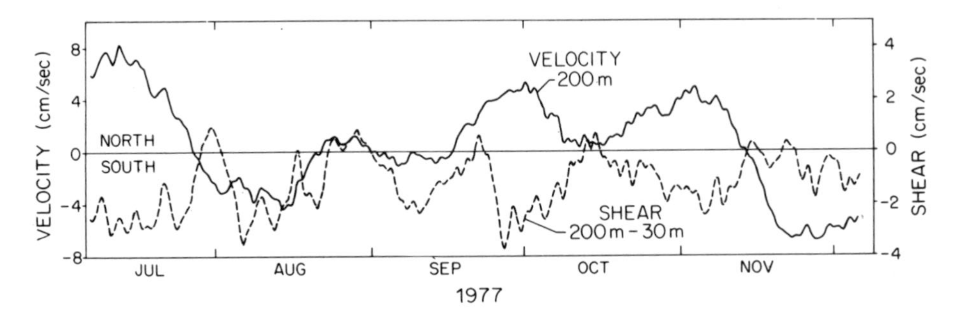

Fig. 8. Low-pass filtered meridional velocity at 200-m level at site C (solid) and velocity difference between 200-m and 30-m level (dashed). Note when difference is negative northward component at 30 m exceeds component at 200 m.

at about .0007-.001 cph (periods of 42 to 60 days). In this spectral region, north-south oscillations dominate east-west motions. At lower frequencies the north-south energy decreases, while at A and B, the east-west energy continually increases (a red spectrum). At site C the u component variance is flat throughout the low frequency region. It should be mentioned that prior to calculating the Fourier transform, these records were detrended by joining the end points (Frankignoul, 1974a). This detrending reduces the contamination of the spectra by oscillations with periods longer than the record length.

Table 1 shows that the eddy kinetic energy KE' increases from A to C. At the only common depth (30 m), B and C have essentially the same eddy kinetic energy, whereas A is lower by a factor of 5. The record lengths are too short, when compared to the dominant periods of the motion, to attach statistical significance to this energy increase. In the frequency band of the variance peak in meridional velocity (.0005-.001 cph), the site C north-south spectral level exceeds that at site A by a factor of 7. This difference is significant at the 95% level. A small increase in meridional energy level (30%) from B to C is not statistically significant.

The observed current statistics can be compared with other measurements in this area, in the central Pacific, and in the North Atlantic. Amos et al. (1976) reported preliminary data from a 33-day record of near-bottom currents near C. At 20 m (200 m) they found mean velocities u = -3 cm/s (-2.3 cm/s), v = .5 cm/s (0 cm/s). Inspection of Fig. 4 shows that 1-month periods with similar mean speeds could be found. The dominance of western flow is in agreement with our records; however, the small north-south velocity is somewhat anomalous. It is encouraging to note that the near-bottom speeds increased in agreement with our observations. Amos' measurements were about 80 km away from our C array, so the bottom inten-

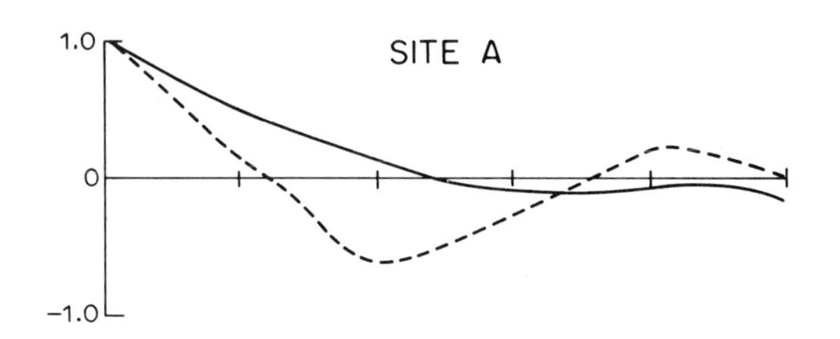

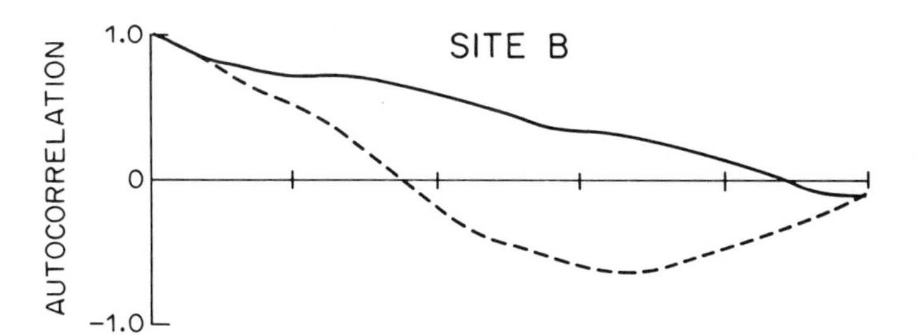

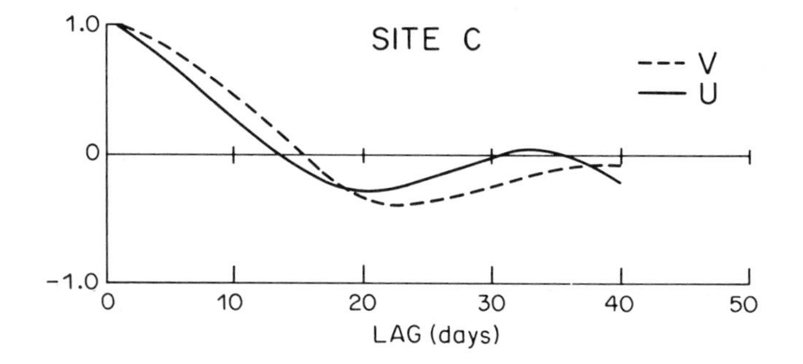

Fig. 9. Autocorrelation functions of low-pass filtered velocity components at 30-m level at each site.

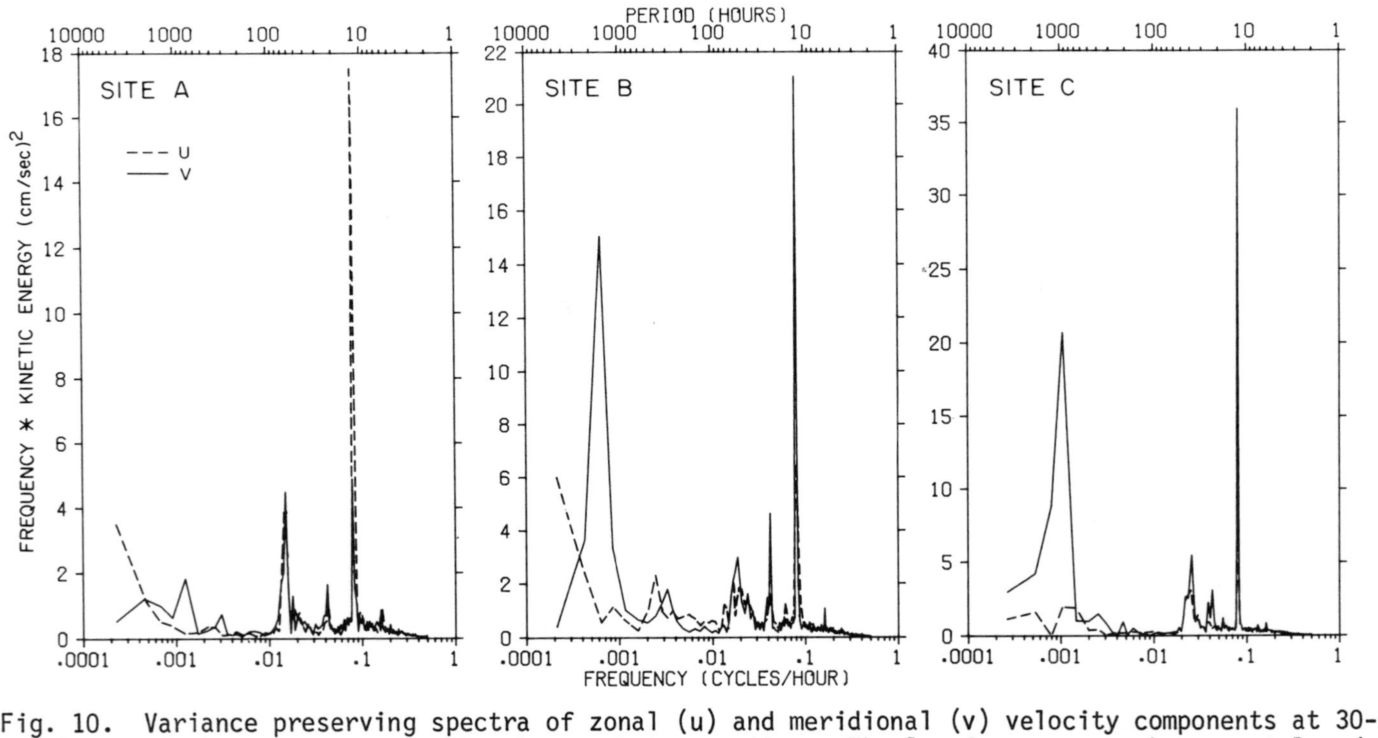

Fig. 10. Variance preserving spectra of zonal (u) and meridional (v) velocity components at 30-m level. Note that the energy scale changes between sites. The low frequency peak seen at B and C is almost exclusively in the meridional component.

sification is probably not a local orographic effect.

In the equatorial Pacific at 0°30'S, 95°W, Harvey and Patzert (1976) reported measurements 10 m above bottom at two sites separated by 110 km. The low frequency structures were coherent over this separation and showed an oscillation with a period of about 25 days that was propagating westward. These data were interpreted as a first-mode baroclinic Rossby wave trapped at the equator. Further west, Taft et al. (1974) reported measurements at 1°02'S, 149°51'W (approximately 1000 km south of our site A) at 1500 m above the bottom. The mean flow was predominantly eastward throughout the record; however, the north-south velocity component had more low frequency variance. A prominant feature of the record was a meridional oscillation with a period greater than 2 months.

In the central Pacific, Ramp (1976) reports measurements made near 30°N, 158°W for a 9-month period. His data were from current meters 100 m off bottom at three sites where the water depth was about 5900 m. Eddy kinetic energy averaged over the three sites was 2.5 erg/cm^3. This energy was divided equally between u and v components. These observations are similar to our site A measurement. In Ramp's data, zonal and meridional time scales determined from autocorrelation functions were approximately equal. They indicate dominant periods of 135-145 days. The shorter meriodional time scale found for the tropical Pacific does not appear to occur further north.

Measurements during the Mid-Ocean Dynamics Experiment (MODE) and subsequent experiments (POLYMODE) in the North Atlantic represent a much more detailed study of low frequency motions than the results presented here. However, our preliminary survey can be compared with some results of MODE. For example, both bottom intensification and meridional dominance in the deep water were part of the MODE results (MODE Group, 1978). Currents were not measured close to the bottom during MODE; the intensification was observed from 1500-m to 4000-m depth and may not be related to the feature we see. The ratio of meridional to zonal variance was about 1.6 at 4000 m at the MODE central site (MODE Dynamics Group, 1975). In our data this ratio varied from 1.5 at site A to 4.2 at site C. Considering the uncertainties in the low frequency kinetic energy determination, Atlantic and Pacific measurements are similar. The mean low frequency kinetic energy (KE') determined at our three sites can also be compared with the North Atlantic observations. Again, record length precludes a quantitative comparison; but the Pacific level is of the same order of magnitude as the values measured near the MODE site. These values are one order of magnitude smaller than the eddy kinetic energy seen near the Gulf Stream (Schmitz, 1976).

Inertial-Internal Wave Variability

In Table 1, KE" represents the horizontal kinetic energy contribution of frequencies higher than the local inertial frequency. In general, this high frequency kinetic energy was less than KE'. From the spectra in Fig. 5, most variance in the high frequency band comes from near-inertial and semidiurnal tidal oscillations. Small peaks at the diurnal period and near 6 hr are also seen. The latter is presumably associated with a tidal harmonic. Component spectra in Fig. 10 indicate that at site A the semidiurnal tide is predominantly east-west, whereas at sites B and C tidal motions are aligned meridionally. This result is substantiated by principal axes calculations (Gonella, 1972; Mooers, 1973) which show statistically significant orientations of 83°, 5°, and 14° at sites A, B, and C respectively. Although local topographic effects may be responsible for this anisotropy, the results at site C agree with measurements by Amos _et al_. (1976) which were 80 km away.

Our description of the inertial-internal wave band will focus on the nonstationarity of these signals and the possibility of near-bottom internal wave production. Deep-ocean measurements often show nonstationary inertial energy (e.g. Webster, 1968; Halpern, 1974). Near the surface these pulses of inertial energy can often be correlated with wind events (Pollard and Millard, 1970; Halpern, 1974); at deeper levels several measurements (Frankignoul, 1974b; Leaman and Sanford, 1975) indicate downward propagation. Near-bottom inertial currents could be caused by this downward propagation coupled with possible bottom reflection. On the other hand, inertial energy could be generated by the time-varying low frequency flow interacting with bottom topography. If near-bottom Ekman layers are important, then bottom generation of inertial oscillations might be similar to surface generation.

To study nonstationarity, the unfiltered data record was broken into pieces that were approximately two inertial periods long. These pieces were Fourier-transformed after detrending by joining the end points (Frankignoul, 1974a). To improve time resolution and to uncover spurious peaks, the Fourier analysis was repeated with the pieces shifted by one inertial period (i.e., half of the piece length). Time series of the energy in each of several bands were then constructed from the piecewise spectra. The inertial and semidiurnal bands consisted of one periodogram point each. Internal wave bands centered at periods of 10 hr and 5 hr were formed by frequency-averaging over 5 and 6 periodogram estimates respectively. Figure 11 is a plot of these horizontal kinetic energy time series as well as the horizontal kinetic energy associated with the low frequency flow for the 50-m level at site C. Note the relatively distinct structure seen in the inertial band about mid-October. This peak rises by almost a factor of 5 above the average level. This particular event was mentioned earlier since it is quite distinct in the time series plots (Fig.4). The vertical

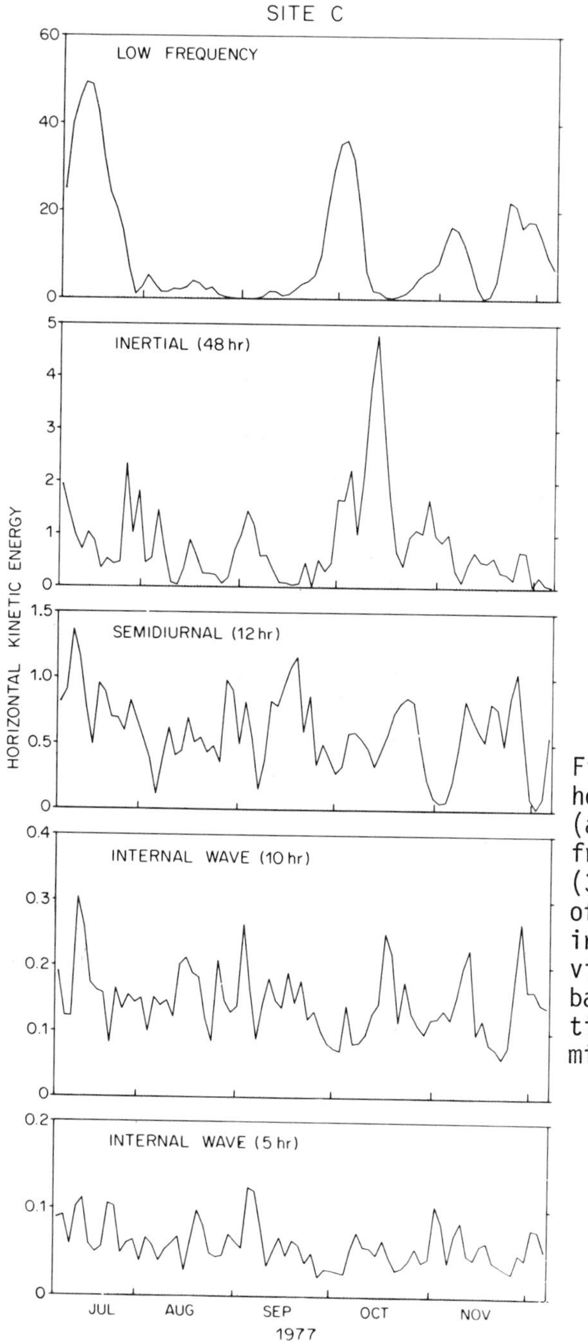

Fig. 11. Time series of horizontal kinetic energy (arbitrary units) in five frequency bands at site C (30-m level). The method of calculation is discussed in the text. Note the low visual correlation between bands and the large inertial energy burst seen in mid-October.

energy distribution is seen in Fig. 12. The event appears first at 200 m above bottom and then at 50-, 30-, and 6-m levels. No time delay can be discerned between the lower three levels (time resolution is about 2 days). The energy within the peak decreases toward the bottom. Based on these time series, it appears that the energy flux is downward from the upper waters rather than upward from the bottom. The observed delay of about 4 days between 200 m and 50 m corresponds to an energy propagation speed of about .04 cm/s.

The large time step (1 day) used in these calculations contributes to the uncertainty of this group velocity estimate. Mid-ocean observations by Frankignoul (1974b) in the central North Atlantic suggested a downward propagation speed of 0.5 cm/s — considerably faster than that observed here. Kroll (1975) pointed out that the propagation velocity depends on the frequency of the wave packet, the wavelengths of the oscillations, and the local stratification. In addition, he showed that the horizontal propagation velocity was generally much greater than the vertical propagation velocity; so that, for waves of a constant frequency, the rays travel about 100 km horizontally for each 1 km downward. Thus, the source of near-bottom inertial waves may be quite distant from the site where they are observed.

The second problem which we addressed in our study of the high frequency currents is the possibility of near-bottom internal wave generation. Such generation is related to low frequency currents interacting with near-bottom topography; thus, one expects a correlation between internal wave energy and low frequency energy. Figure 11 shows time series of low frequency, inertial, semidiurnal, and internal wave energy. No significant cross-correlation was found between internal wave energy and any lower frequency band. In addition, the mean energy level in the 10-hr internal wave band was within a factor of 2 of the energy level predicted by the Garrett and Munk (1975) internal wave model. This model describes mid-ocean internal waves in regions that are remote from energy sources or sinks. Our measurements, therefore, indicate no evidence for near-bottom production of internal waves.

The Bottom Boundary Layer

The velocity time series were examined to determine whether these data offered evidence of a near-bottom Ekman-like layer. In particular, most theories with turbulent Ekman-like dynamics (e.g. Weatherly, 1972, and references therein) predict a decrease in speed approaching the bottom and an angular veering counterclockwise looking down. Several techniques for estimating the veering are possible (Kundu, 1976); the most important consideration is to avoid instrumental effects in determining angular differences. The VACM has an angular resolution of 2.8°. However, at low speeds the vane direction is erratic. To eliminate this source of error,

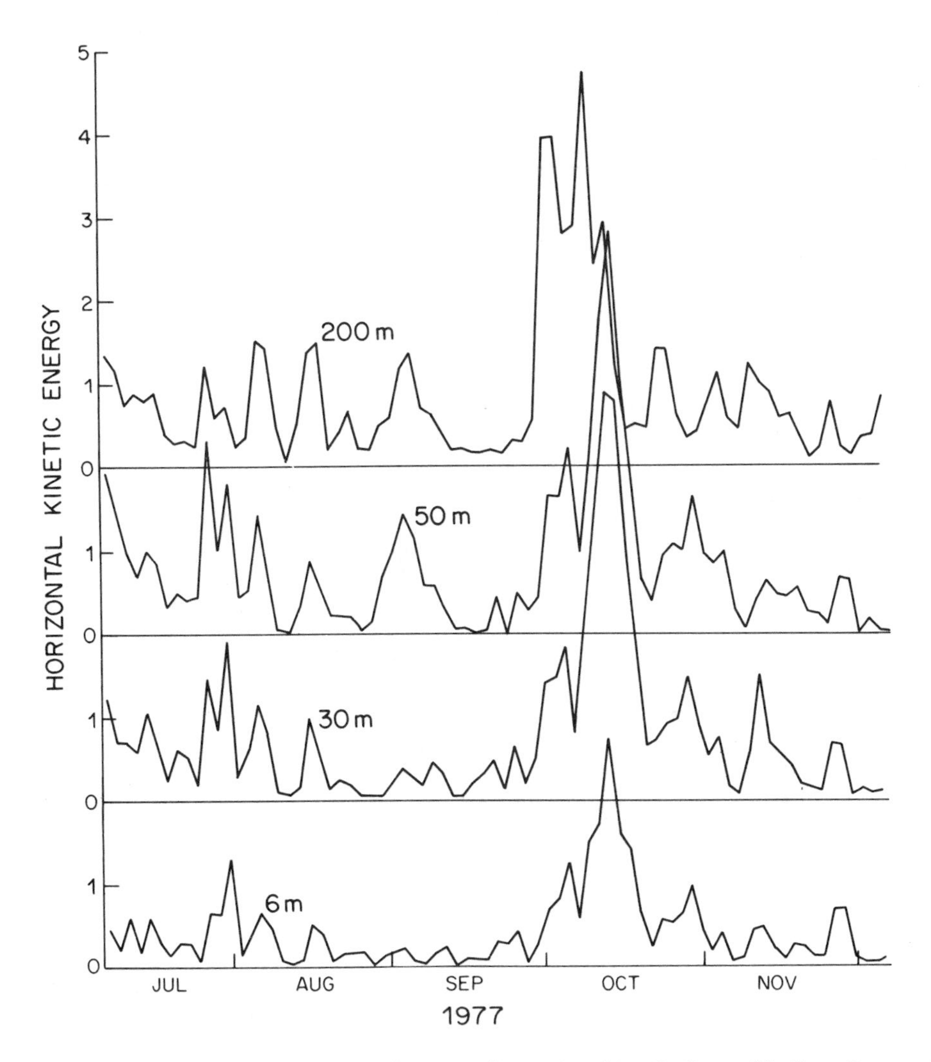

Fig. 12. Inertial energy time series at site C for all levels. Energy scale (arbitrary units) is the same for all plots; origin for successive levels is offset by 2 units. Time lags indicate that the energy burst in October appeared first at the 200-m level.

we selected periods when the low-pass filtered velocity exceeded 3 cm/s. The analysis was only applied to the site C data, since this record was the most complete. Four periods with speeds above our criterion occurred: 7-25 July; 23 September-8 October; 31 October-10 November; 18 November-5 December. Combined, these periods total 64 days. Daily values of speed and direction were found by subsampling the low-pass filtered data at each depth. Table 2 summarizes the results; angles are given relative to true north. In the first three intervals, the mean speed increased from 200 m to 30 m and then decreased to 6 m. In these cases, the direction change from 200 m to 30 m was negligible; but from 30 m to 6 m a change of 7° to 10° counterclockwise was recorded. In the last interval, speed consistently decreased from 200 m to 6 m; however, above 30 m the direction changed in a clockwise sense (looking down), whereas from 30 m to 6 m the change was counterclockwise. Average veering from 30 m to 6 m was 9° ±2°; average sheer was 0.8 ±0.2 cm/s. These results indicate that the current meters at 50 m and 200 m above bottom were above the influence of an Ekman layer; the 6-m level appears to be in the layer. An accurate estimate of the boundary layer thickness is not possible because of inadequate vertical resolution.

Theoretical calculations of boundary-layer thickness are probably not applicable in regions with topographic variations as large as those observed in this region. However, for lack of a better indicator, the expected thickness of a turbulent Ekman layer over a flat bottom in unstratified water can be calculated. In accordance with Armi and Millard (1976), the height of a turbulent Ekman layer h_e is

$$h_e = 0.4\ u_*/f \qquad (1)$$

This relation is derived from the experiments of Caldwell *et al.* (1972) and Howroyd and Slawson (1975). The local inertial frequency is f; u_* is the friction velocity at the bottom. For a smooth bottom, this velocity is given by Csanady (1967) as

$$u_* = 1/30\ u \qquad (2)$$

where u is the velocity outside the boundary layer. Then, combining (1) and (2) and considering u to be the mean speed at 30-m level (7 cm/s) during the four intervals considered gives

$$h_e = 25\ \text{m}. \qquad (3)$$

The measurements are consistent with an Ekman-like behavior confined to a 25-m thick boundary layer. The turning observed is similar to that measured on continental shelves; Weatherly (1972) found a mean veering 10° under the Florida current, and Kundu (1976) found a veering of 6° off the Oregon coast.

TABLE 2

Average speed and direction at site C for time periods indicated

Time Period	200 m		50 m		30 m		6 m	
	S	Θ	S	Θ	S	Θ	S	Θ
7 July-25 July	6.1 cm/s	343°	8.3 cm/s	340°	8.8 cm/s	339°	7.9 cm/s	328°
23 September-8 October	4.5 cm/s	331°	6.8 cm/s	329°	7.0 cm/s	331°	6.2 cm/s	324°
31 October-10 November	4.1 cm/s	347°	5.1 cm/s	347°	5.7 cm/s	348°	5.2 cm/s	341°
18 November-5 December	6.3 cm/s	163°	5.5 cm/s	165°	5.5 cm/s	170°	4.5 cm/s	160°

Periods selected had low-pass filtered velocity greater than 3 cm/s at all depths. Directions given with respect to true north.

Summary and Conclusions

Benthic current measurements have been described at three sites in the eastern tropical Pacific Ocean. Vertical stratification in this region was weak and the potential temperature-salinity relation was fairly uniform. Mean currents over the record lengths (4-6 months) were small and to the northwest at all sites. These mean velocities were not statistically significant because of large, low frequency (less than inertial) variance. These low frequency fluctuations contributed 50-80% of the total variance. Time scales and horizontal kinetic energy (per unit mass) varied among the sites. At all locations the meridional velocity component had an integral time scale of about 15 days (60-day period). At the eastern site (C), the zonal velocity time scale was similar; however, at A and B east-west oscillations had a dominant period nearly equal to the record length. Low frequency horizontal kinetic energy increased from west to east by a factor of 5. Vertical coherence on each mooring was large. However, between 200 m and 30 m both mean speed and amplitude of the low frequency oscillations increased. Below 30 m the speed decreased. Measurements of longer duration and greater vertical extent are required to relate the low frequency oscillations to the large-scale circulation of the tropical Pacific; however, our measurements point out the importance and some major characteristics of these motions.

High frequency oscillations contributed 20-50% of the total variance. Inertial and tidal motions were most important. Inertial oscillations were intermittent with brief periods where the energy in this frequency band exceeded the background level by one order of magnitude. These bursts of inertial energy were observed to propagate downward, and some evidence for near-surface generation was found. Energy in the high frequency internal wave bands was fairly uniform in time and had a gradual decrease toward the bottom. No correlation between internal wave energy and low frequency currents was found. Also, the mean energy level was within a factor of 2 of the "universal" spectrum for mid-ocean internal waves described by Garrett and Munk (1975). These observations suggest that over abyssal hills such as those that occur in the tropical Pacific, bottom topography is not a strong source of internal wave energy.

In the final section of this study, we compared bottom current measurements with theoretical ideas on Ekman boundary layers. Current veering between 30 m and 6 m was about 9° in the correct sense for an Ekman layer. Rough calculations indicated an expected layer thickness of 25 m, in agreement with observation. Our measurements did not show the very thick (about 6 h_e) bottom layers reported by Armi and Millard (1976) in the Atlantic.

This description of abyssal currents in the eastern tropical Pacific is preliminary and emphasizes that many aspects of the circulation are unknown. Present studies are underway to refine the description of the boundary layer and the internal wave generation mechanisms. Further experiments will rely on long-term deployments to characterize the low frequency fluctuations.

Acknowledgements

I wish to thank D. Halpern for helpful discussions in the early stages of this research. This study was supported in part by the Deep Ocean Mining Environmental Studies (DOMES) Program of NOAA.

References

Amos, A. F., A. L. Gordon, and E. D. Schneider, (1971) Water masses and circulation patterns in the region of the Blake-Bahama Outer Ridge, Deep Sea Res. 18, 145-165.

Amos, A. F., O. A. Roels, and A. Z. Paul, (1976) Environmental baseline conditions in a manganese nodule province in April-May 1975, Offshore Technology Conference, Houston, Texas, OTC-2456.

Armi, L., (1978) Some evidence for boundary mixing in the deep ocean, J. Geophys. Res. 83, 1971-1979.

Armi, L., and R. C. Millard, Jr., (1976) The bottom boundary layer of the deep ocean, J. Geophys. Res. 81, 4983-4990.

Bell, T. H., Jr., (1975) Topographically generated internal waves in the deep ocean, J. Geophys. Res. 80, 320-327.

Biscaye, P.E., and S.L. Eittreim, (1974) Variations in benthic boundary layer phenomena: Nepheloid layer in the North Atlantic Basin, In: Suspended Solids in Water, R.J. Gibbs, ed., Plenum Publ. Corp., New York 227-260.

Bryden, H. L.,(1973) New polynomials for thermal expansion, adiabatic temperature gradient, and potential temperature of sea water, Deep Sea Res. 20, 401-408.

Caldwell, D. R., C. W. Van Atta, and K. N. Helland, (1972) A laboratory study of the turbulent Ekman layer, Geophys. Fluid Dyn. 3, 125-160.

Chase, T. E., H. W. Menard, and J. Mammerickx, (1970) Bathymetry of the North Pacific. Scripps Institution of Oceanography, IMR Technical Report, Tr-13.

Csanady, G.T. (1967) On the 'resistance law' of a turbulent Ekman layer, J. Atmos. Sci. 24, 467-471.

Csanady, G.T., (1974) Equilibrium theory of the planetary boundary layer with an inversion lid, Boundary Layer Meteorol. 6, 63-79.

Eittreim, S., P.E. Biscaye, and A. F. Amos, (1975) Benthic nepheloid layers and the Ekman thermal pump, J. Geophys. Res. 80, 5061-5067.

Fofonoff, N.P., S.P. Hayes, and R.C. Millard, Jr., (1974) WHOI/Brown CTD microprofiler: methods of calibration and data handling, Woods Hole Oceanogr. Inst. Tech. Rept. 74-89, Woods Hole Mass.

Frankignoul, C.J., (1974a) A cautionary note on the spectral analysis of short internal wave records, J. Geophys. Res. 79, 3459-3463.

Frankignoul, C.J., (1974b) Preliminary observations of internal wave energy flux in frequency, depth-space, Deep Sea Res. 21, 895-910.

Garrett, C., and W. Munk, (1975) Space-Time scales of internal waves: a progress report, J. Geophys. Res. 80, 291-297.

Gonella, J., (1972) A rotary-component method for analyzing meteorological and oceanographic vector time series, Deep Sea Res. 19, 883-846.

Halpern, D., (1974) Observations of the deepening of the wind-mixed layer in the northeast Pacific Ocean, J. Phys. Oceanogr. 4, 454-466.

Halpern, D., J.R. Holbrook, and R.M. Reynolds, (1974) A compilation of wind, current, and temperature measurements: Oregon, July and August 1973, CUEA/IDOE Tech. Rept. 6, Dept. of Oceanogr., Univ. of Washington, Seattle.

Harvey, R.R., and W.C. Patzert, (1976) Deep current measurements suggest long waves in the eastern equatorial Pacific, Science, 193, 883-885.

Howroyd, G.C., and P.R. Slawson (1975) The characteristics of a laboratory produced turbulent Ekman layer, Boundary Layer Meteorol. 8, 210-219.

Kroll, J. (1975) The propagation of wind-generated inertial oscillations from the surface into the deep ocean, J. Marine Res. 33, 15-51.

Kundu, P. K. (1976) Ekman veering observed near the ocean bottom, J. Phys. Oceanogr. 6, 238-242.

Kuo, H. H., (1978) Topographic effects on the deep circulation and the abyssal oxygen distribution, J. Phys. Oceanogr. 8, 428-436.

Leaman, K. D., and T. B. Sanford, (1975) Vertical energy propagation of internal waves: a vector spectral analysis of velocity profiles, J. Geophys. Res. 80, 1975-1978.

Mantyla, A. W., (1975) On the potential temperature in the abyssal Pacific Ocean, J. Marine Res. 33, 341-354.

McCullough, J. R., (1975) Vector averaging current meter speed calibrations and recording techniques, Woods Hole Oceanogr. Inst. Tech. Rept. 75-44, Woods Hole, Mass.

MODE Group, (1978) The mid-ocean dynamics experiment, Deep Sea Res. 25, 859-910.

MODE-I Dynamics Group, (1975) Dynamics and the analysis of MODE-I, The Mid-Ocean Dynamics Experiment, MODE-I, Mass. Inst. of Technol., Cambridge.

Mooers, C.N.R., (1973) A technique for the cross spectrum analysis of complex valued time series with emphasis on properties of polarized components and rotational invariants, Deep Sea Res. 20, 1129-1141.

Pollard, R. T., and R. C. Millard, Jr., (1970) Comparisons between observed and simulated wind-generated inertial oscillations, Deep Sea Res. 17, 813-821.

Ramp, S. R., (1976) The measurement and analysis of low frequency currents in the deep central North Pacific Ocean, Unpublished thesis, Univ. of Washington, Seattle.

Rhines, P. B., (1977) The dynamics of unsteady currents, In: The Sea, Vol. 6: Marine Modeling, E. D. Goldberg, I.N. McCave, J. J. O'Brien, and J. H. Steele, eds., Wiley, New York.

Schmitz, W. J., Jr., (1976) Eddy kinetic energy in the deep western North Atlantic, J. Geophys. Res. 81, 4981-4982.

Schmitz, W.J., Jr., (1978) Observations of the vertical distribution of low frequency kinetic energy in the western North Atlantic, J. Marine Res. 36, 295-310.

Stommel, H., and A.B. Arons, (1960) On the abyssal circulation of the world ocean, Deep Sea Res. 6, 140-154.

Taft, B.A., B.M. Hickey, C. Wunsch, and D.J. Baker, Jr., (1974) Equatorial undercurrent and deeper flows in the central Pacific, Deep Sea Res. 21, 403-430.

Thompson, R.O.R.Y., (1973) Stratified Ekman boundary layer models, Geophys. Fluid Dyn 5, 201-210.

Weatherly, G.L., (1972) A study of the bottom boundary layer of the Florida Current, J. Phys. Oceanogr. 2, 54-72.

Weatherly, G.L., (1975) A numerical study of time-dependent turbulent Ekman layers over horizontal and sloping bottoms, J. Phys. Oceanogr. 5, 288-299.

Webster, F., (1968) Observations of inertial-period motions in the sea, Rev. Geophys. 6, 472-490.

Wimbush, M. and W. Munk, (1971) The benthic boundary layer, In: The Sea, v. 4, part I, A.E. Maxwell, ed., Wiley, N.Y., 731-758.

Wong, C.S., (1972) Deep zonal water masses in the equatorial Pacific Ocean inferred from anomalous oceanographic properties, J. Geophys. Res. 77, 7196-7202.

Wunsch, C., (1976) Geographical variability of the internal wavefield: a search for sources and sinks, J. Phys. Oceanogr 6. 471-485.

Wyrtki, K., L. Magaard, and J. Hager, (1976) Eddy energy in the oceans, J. Geophys. Res. 81, 2641-2646.

NUTRIENT CHEMISTRY IN THE TROPICAL NORTH PACIFIC

DOMES SITES A, B, AND C

James J. Anderson

Department of Oceanography
University of Washington
Seattle, Washington 98195

Abstract

The distributions of nutrient chemistry in the DOMES area are controlled by a balance between advective and diffusive mixing and biological reactions. The intensity of mixing changes with depth, defining 6 unique layers in terms of a T-S diagram. These include: the mixed layer, thermocline, oxygen minimum layer (OML), upper deep water (UDW), North Pacific deep water (NPDW), and Pacific bottom water (PBW).

Within each layer chemical profiles deviate slightly from linearity and at the layer interfaces the profiles have distinct changes in slope. Horizontal gradients are weak, except for oxygen in the OML where a tongue-like distribution of low oxygen extends westward along $15^{o}N$. No clear temporal patterns are evident when concentrations are referenced to density surfaces.

The interface concentrations of oxygen in the upper layers are principally controlled by the surface saturation value, thermocline diffusion and respiration in the OML. The interface concentrations of nitrate and phosphate in the upper layers are principally controlled by uptake in the thermocline and mixed layer and horizontal flux near the OML-UDW interface.

Mixed layer productivity can be depressed by a reduced nitrate flux into the layer caused by productivity associated nitrate uptake in the thermocline. With productivity occurring within the thermocline, the thermocline acts as a nutrient sink and with little or no productivity in the thermocline it acts as a nutrient filter.

The oxygen minimum tongue in the OML can be established by horizontal diffusion and respiration with the boundaries set by subsurface zonal flows north and south of the tongue.

The phosphate and nitrate maximums at the OML-UDW interface, between 800 and 1000 m, are produced by a depth decreasing oxidative nutrient profile and a depth increasing preformed nutrient profile. Horizontal advection and diffusion along density surfaces are important in maintaining the vertical nutrient maximum.

A silicate maximum in the middle of the NPDW (3000 m) is produced by a combination of deep-water circulation and a high silica dissolution rate throughout the water column. A high silicate content is acquired as the PBW flows north into the high latitudes of the North Pacific and upwells into the level of the NPDW. The NPDW flows south out of the North Pacific creating a silicate maximum at 3000 m in the DOMES area.

A high oxygen and lower nutrient content in the bottom layer are maintained by the northward flow of PBW.

Ammonia near the top of the thermocline is produced by excretion from zooplankton and is consumed by photosynthesis or by nitrifying bacteria.

The nitrite peak in the thermocline is probably produced by leakage of nitrite from phytoplankton, and consumed by phytoplankton. Ammonia and nitrite peaks are not normally observed in the mixed layer, probably because of larger vertical mixing in the layer.

In extremely low oxygen waters, nitrite is produced as an intermediate in the reduction of nitrate to molecular nitrogen in denitrification. In the process the nitrogen compounds are used in place of oxygen as terminal electron acceptors in bacterial respiration.

Introduction

The tropical North Pacific (TNP) is a uniquely interesting region of the oceans in part because of the intense and stable features in the chemical distributions through the water column. Because of the geographical isolation of the region a limited number of expeditions have been made in the region and consequently the definition and understanding of the chemical features have been slowly developed.

The first major chemical observations in the TNP were made on the *Carnegie* (Moberg, 1930; Graham and Moberg, 1944) and indicated the region contained some of the world's most intense low oxygen water. The extent and importance of this oxygen minimum zone was emphasized by Richards in his chapter on oxygen in the oceans in the "Treatise on Marine Ecology and Paleoecology", edited by Hedgpeth (1957). In the same volume Barnes reviewed the available work on nutrients and emphasized that the TNP contains some of the ocean's highest concentrations of nutrients in the deep water and some of the lowest concentration in the mixed layer above the strong thermocline. Barnes also suggested that the nutrient profiles can be discussed in terms of layers of minimum and maximum values and gradients.

The next major contributions to the chemical knowledge of the area appeared 10 years later. Reid (1965), analyzing temperature, salinity, oxygen and phosphate over what he describes as the intermediate water of the Pacific Ocean, identified the origins of oceanic features between 400 and 800 m. Tsuchiya (1968), using temperature, salinity and oxygen, investigated the oceanic setting of the region between the bottom of the thermocline and about 200 m. Barkley (1968) discussed the temperature-salinity characteristics of the Pacific Ocean down to 1000 m.

The relative seasonal stability of the chemical distributions was established by the EASTROPAC expeditions, which covered the eastern TNP every two months between January 1967 and February 1968 (Love, 1972). In the past, several expeditions have collected chemical observations within the eastern TNP but little information has been collected in the central TNP. This situation was remedied in 1975 and 1976 when the area of the central TNP shown in Figure 1 was intensely studied as part of the Deep Ocean Mining Environmental Study (DOMES). The project included studies of the chemical, physical, biological, and geological

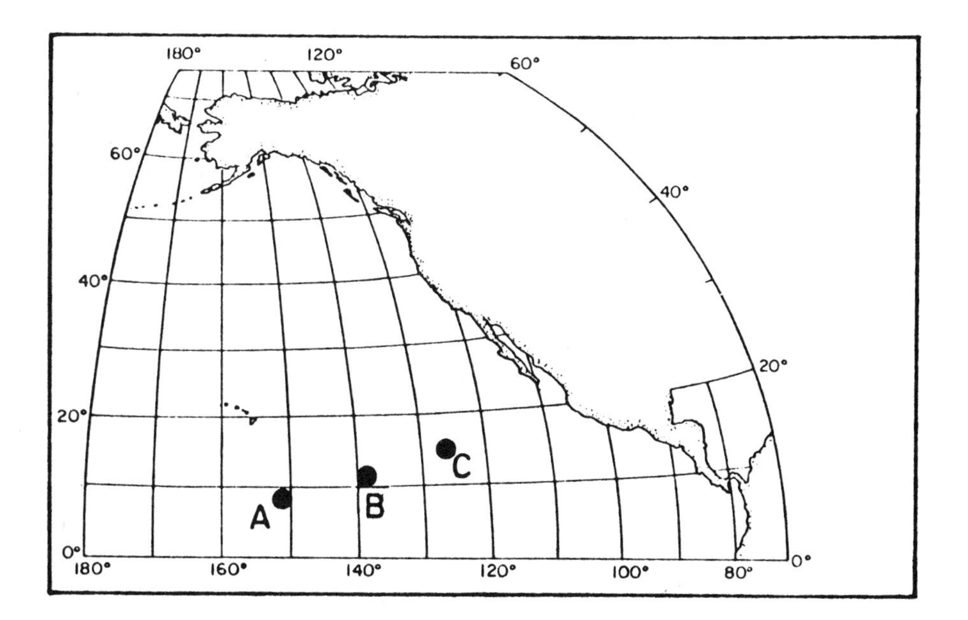

Fig. 1. Location of DOMES site stations A, B, and C.

characteristics of the water column and bottom.

The water column chemical data collected on DOMES is the topic of this chapter and the focus is on describing the detailed features of oxygen and nutrients and identifying the processes controlling the features. Emphasis is placed on the upper layers, including the mixed layer, thermocline, and oxygen minimum layer.

Following the notion of Barnes the chemical distributions can be envisioned in terms of layers that reflect the density and current structures. The water column of the DOMES area can be

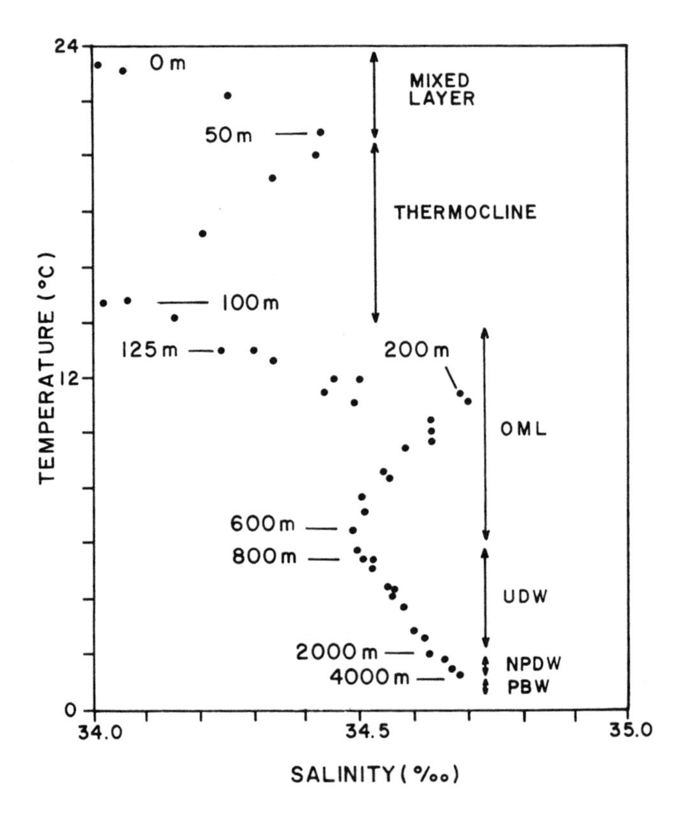

Fig. 2. T-S diagram for site B, April 1976. Including depths of interfaces between T-S segments and intervals of layers including the mixed layer, thermocline, oxygen minimum layer (OML), upper deep water (UDW), North Pacific deep water (NPDW) and Pacific bottom water (PBW).

divided into 6 vertical layers that are defined by temperature and salinity relationships (Fig. 2). The physical characteristics of the layers are as follows:

The mixed surface layer is well developed with temperatures of about 26° ± 2°C, and salinities of about 34.5°/₀₀. Generally temperature is uniform while salinity increases slightly with depth in the mixed layer (Fig. 3). The thickness of the mixed layer has both seasonal and spatial patterns (Robinson, 1976). In the autumn-winter period the mixed layer is thinnest and a thermal ridge, defined by a shallow thermocline, extends east-west along about 10°N. In the spring-summer period the mixed layer is thicker and the thermal ridge is not well developed (Fig. 4).

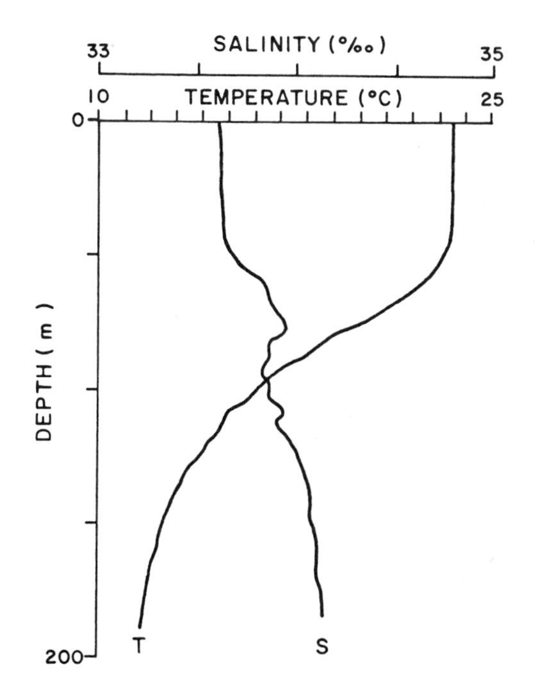

Fig. 3. Averages of 6 temperature and salinity profiles taken at site C between Feb. 20-21, 1976.

On the north side of the thermal ridge the North Equatorial Current (NEC) flows to the west between about 10° and 20°N, and south of the thermal ridge the North Equatorial Countercurrent (NECC) flows east between 10° and 5°N. Both currents are weak

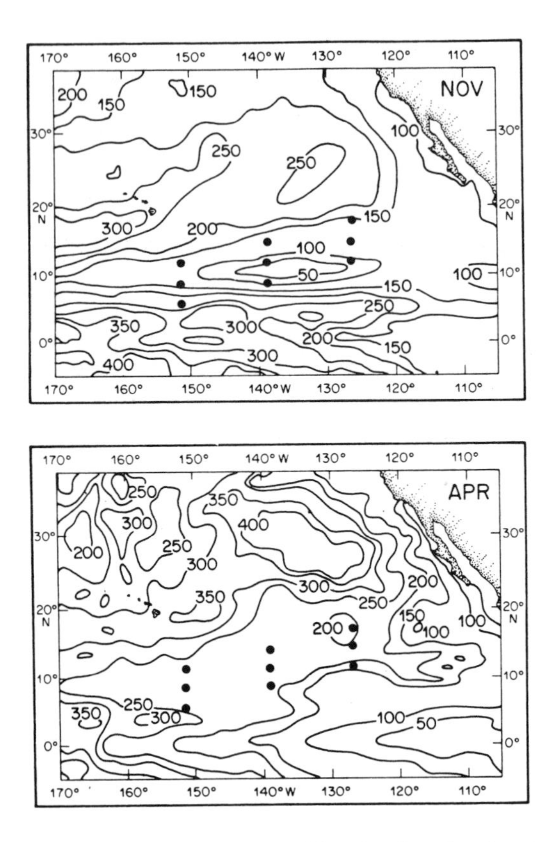

Fig. 4. Average depth (in ft.) of mixed layer for November and April. Dots indicate station locations from DOMES. (From Robinson, 1976).

in the summer when the ridge is weakly developed and strong in the winter when the ridge is well developed (Wyrtki and Kendall, 1967).

The thermocline in the DOMES area is well developed throughout the year with temperatures typically decreasing from 25 to 13°C between the top and bottom of the thermocline, a distance of about 70 m. The bottom of the thermocline is distinguished by a change in the temperature gradient, which corresponds closely with the depth of the 13° isotherm (Fig. 3). Tsuchiya (1968) showed the salinity maximum at the top of the thermocline is formed at the surface in the subtropical South Pacific while the minimum near the bottom of the thermocline presumably is formed in the eastern North Pacific and enters the DOMES area with the California Current along 17 to 20°N. Changes in the depth of the thermocline correspond with variations in the thickness of the mixed layer.

The oxygen minimum layer (OML) is below the thermocline and is characterized by low oxygen and a linear TS relationship extending from the salinity maximum, of South Pacific origin, to the salinity minimum between 600 and 800 m, which is of Antarctic origin (Fig. 2). The layer is bordered at the north and south by high oxygen, low salinity, and zonal flows extending to at least 500 m. At the southern boundary, 5°N, the North Equatorial Subsurface Countercurrent flows east with average maximum geostrophic velocities of 27 cm/s (Tsuchiya, 1975). At the northern boundary (17 to 20°N) the westward extension of the California Current flows west (Roden, 1974). At the top of the OML (corresponding with the 160 cl/t isanosteric surface) a westward geostrophic current maintains the salinity maximum (Tsuchiya, 1968). At the center of the OML (corresponding with the 125 cl/t surface) the geostrophic current is essentially zero (Reid, 1965). Horizontal advection and diffusion, or both, at the bottom of the layer (corresponding with the 80 cl/t surface) maintain the Antarctic salinity minimum. Reid (1965) in his dynamic topography map indicates a broad eastward flow in the DOMES area at about 800 m. Reid and Mantyla (1978) suggest that at 1000 db in the DOMES area the flow is eastward at 10°N and westward at 15°N. Burkov (1972) suggests that the 800 m level is dominated by an eddy-like current structure which produces a series of weak north and south flows through the area. The explanations are not mutually exclusive and suggest that a weak and complicated flow regime produces significant large scale horizontal diffusion or weak currents at 800 to 1000 m. Thus the picture emerges that the core of the OML is a stagnant zone bounded north and south by opposing zonal flows, and above by a westward flow and below by large scale horizontal diffusion produced by large scale eddies, or a weak adjacent eastward and westward flow.

Below the OML an approximate linear T-S relationship extends between 800 and 2000 m (Fig. 2). This layer is identified by Tusnogai et al. (1973) as the upper deep water (UPW) and is considered to be involved with the wind derived circulation. In the DOMES area the UDW flow is probably weak and variable because the region is between the large anticyclonic gyres of the North and South Pacific.

The North Pacific Deep Water (NPDW) lies between about 2000 and 4000 m and below this is the Pacific Bottom Water (PBW). Evidence suggests that the NPDW is involved with a southward flow of water from the North to the South Pacific. This flow is apparently driven by the northward flowing PBW, which originates in the Antarctic and flows north along the western side of the South Pacific and into the North Pacific. In the North Pacific the PBW apparently upwells, is modified into NPDW and returns south across the equator at a level just above the PBW layer (Mantyla, 1975).

The distributions and rates of growth and consumption of

bacteria, phytoplankton and zooplankton affect the distributions of chemical constituents and the constituents in turn affect the growth and distributions of the organisms. Phytoplankton are most important in regulating the distributions of nutrients and oxygen in the mixed layer and thermocline. In the DOMES area the rate of primary productivity is on the order of 100 mg C/m^2d, and seasonal variations are generally within a factor of 2 (Owen and Zeitzschel, 1970; El-Sayed et al., 1977). The average depth of the 1% light level is about 90 m while the depth of the mixed layer is about 50 m, so a significant amount of light extends into the thermocline. Consequently, up to 50% of the primary productivity has been observed in the thermocline. Less than 5% of productivity occurs below the 1% light level (El-Sayed et al., 1977).

Below the euphotic zone bacteria and zooplankton are the dominant organisms altering the chemical distributions. In aerobic waters the rate of oxygen consumption is about 10 $\mu\ell/\ell$ and is about equally divided between zooplankton and bacteria (Menzel, 1975). In low oxygen waters, the bacterial fraction can be larger and the total respiration rate can be elevated to about 30 $\mu\ell/\ell$ yr (King et al., 1978). When the oxygen concentration is on the order of 5 mg-at/m^3 or less, heterotrophic bacteria switch from using oxygen as the terminal electron acceptor in respiration to using nitrate (Devol, 1978). This process is known as denitrification and through a series of bacterial enzymatic reactions nitrate is reduced to nitrite, to nitric oxide, to nitrous oxide and finally to molecular nitrogen (Thauer, Jungerman and Decker, 1977). In the balance of reactions a small quantity of nitrite is maintained in the oxygen minimum zone, but nitrous oxide (Cohen and Gordon, 1978), and apparently nitric oxide, are not present in significant quantities. In the eastern tropical North Pacific east of the DOMES area over 10 mg-at/m^3 of excess nitrogen are evident in the oxygen minimum zone (Codispoti and Richards, 1976).

Methods

The purpose of the DOMES chemical study was to document the mean values and standard deviations of routinely measured chemical constituents in the deep ocean mining area. To accomplish this task three central sites were occupied by the R/V *Oceanographer* during the summer of 1975 and the winter of 1976 (Fig. 1). At each site, vertical profiles of nutrients were made about every 3 to 6 hours for two days. Additionally, water samples were collected for special chemical analysis. A lesser sampling routine was carried out at stations up to plus and minus 3 degrees north and south of the site stations. At all stations, pump casts for continuous sampling of nutrients were made to 200 m and Niskin bottle samples were collected between the surface and

1000 m. At the site stations in the winter of 1976 deep water sampling was carried out with sample depths every 500 m between 1500 m and the bottom (approximately 4000 and 4500 m deep). In the bottom layer, ten additional samples were taken at each site station with sampling depths of 10, 20, 60, 100, 130, 200, 250, 300, 350, and 500 m off the bottom.

The vertical pumping system was lowered at a rate of 2 to 5 m/min to 200 m depth and water was continuously pumped to the shipboard laboratory. A small flow was diverted from the effluent and fed directly into a Technicon AutoAnalyzer for continuous analysis of nitrate, nitrite, ammonia, phosphate and silicate. On most of the pump cast, the pump effluent was fed into a flow-through fluorometer and an in-line polarographic oxygen sensor to measure chlorophyll fluorescence and the percent oxygen saturation.

At sites A and B on the winter 1976 cruise, temperature and salinity were measured with the pump cast by attaching the ship's CTD to the pump head and lowering the pump and CTD together on separate cables. The oxygen sensor was calibrated with oxygen measurements from the pump effluent at the surface and 200 m. No temperature compensation was applied to the oxygen sensor, which introduced some error in the measurements where the temperature change between the mixed layer and thermocline was large. It is probable that the measured oxygen profiles are smoothed in this region and that the change in slope of the oxygen profile across the mixed layer thermocline interface is larger than shown in the pump profiles.

The vertical resolution of profiles is a function of the mixing in the pump hose, instrument response time, and the rate of vertical profiling. At 4 m/min, vertical features, with scales of 10 m, are attenuated between 20 and 40%. Thus the maximum observed concentration in a peak with a thickness of 10 m will be 20 to 40% less than its real value. Larger features are attenuated less than 10% (Anderson and Richards, 1977a).

The following chemical methods were used on the DOMES cruises: Oxygen samples taken from water bottles and from the pump effluent for sensor calibration were measured by the method of Carpenter (1965). Nutrient analyses from the pump cast and water bottles were measured with a Technicon AutoAnalyzer II. Phosphate was measured by the method of Hager et al. (1968); nitrate, nitrite, and silicate were measured by the methods in Armstrong et al. (1967) and ammonia was measured by the method of Slawyk and MacIsaac (1972). Electron transport activity (ETS) is a measure of the potential rate of microbial respiration and was measured on particulate material collected on Gelman type-A glass fiber filters by the method of Packard (1971). Oxygen respiration, R, was calculated from ETS assuming an R/ETS ratio of 0.43 (King et al., 1978). Adenosine triphosphate (ATP) is the ubiquitous cellular energy currently found in all living organisms and is typically found in a constant ratio with total cellular carbon (Holm-Hansen, 1970). Material for ATP analysis was collected on 0.45-μ membrane filters and measured by the method of Holm-Hansen and Booth (1966). Particulate organic carbon (POC) and particulate nitrogen (PN) were collected on 0.45-μ Selas Flotronics membrane filters and analyzed according to the method of Sharp (1974) on a Hewlett-Packard Model 185B CHN Analyzer. Dissolved organic carbon (DOC) was measured on water filtered through the silver filters used for the POC analysis. The wet oxidation method of Menzel and Vaccaro (1964) was used, employing an Oceanography International Corporation carbon analyzer.

The chemical and biochemical observations made on the DOMES cruises in 1975 and 1976 are presented in reports by Anderson and Richards (1977a and b).

Chemical Distributions

In the DOMES area, the dominant temporal and spatial patterns in the chemical distributions are correlated with patterns in density and current fields. Specifically, distinct chemical regimes are evident in the vertically defined layers which include: the mixed layer, thermocline, oxygen minimum layer (OML), upper deep water (UDW), North Pacific deep water (NPDW), and the Pacific bottom water (PBW). Additionally, in the oxygen minimum layer the chemical distributions are strongly affected by horizontal flows at the north and south boundaries of the DOMES area. In each layer the chemical distributions are generally smooth and close to linear. At the interfaces between layers the chemical profiles typically have distinct and often sharp changes in the vertical gradients. The average values of constituents at layer interfaces are similar between the site stations and seasons and in many cases the day-to-day variations are of the same order as the seasonal and area-wide spatial variations. Thus the major patterns in the chemical constituents are in the vertical direction and horizontal and temporal variations are small.

Oxygen

Typical vertical profiles of oxygen have well-defined changes in the gradients at the layer interfaces (Figs. 5 and 6). The general features of the vertical profiles are summarized by the concentrations at the layer interfaces given in Table 1. Differences between seasons are evident but no significant pattern emerges.

At the top of the mixed layer, oxygen values are within 5% of saturation and at the bottom of the mixed layer values are about 95% of the surface values. As with all chemical constituents in the mixed layer, the oxygen profile is smooth and nearly vertical. Through the thermocline, oxygen decreases in nearly linear manner although several profiles show inversions and step structures which are persistent for several days (Fig. 6). These inversions correspond with inversions and steps in the profiles of phosphate, nitrate and silicate and with steps in the density structure (Fig. 7). At the bottom of the thermocline a distinct change in the oxygen gradient is associated with a change in the gradients of temperature, salinity and density. The change in slope is also evident in the profiles of nitrate, phosphate and silicate and suggests that the bottom of the thermocline is a well-defined interface between the thermocline and the OML (Fig. 7). At this interface the lowest oxygens are found at site C and the highest are found at site A (Table 1). This trend corresponds with the distance of the site stations from the intense core of low oxygen water in the OML.

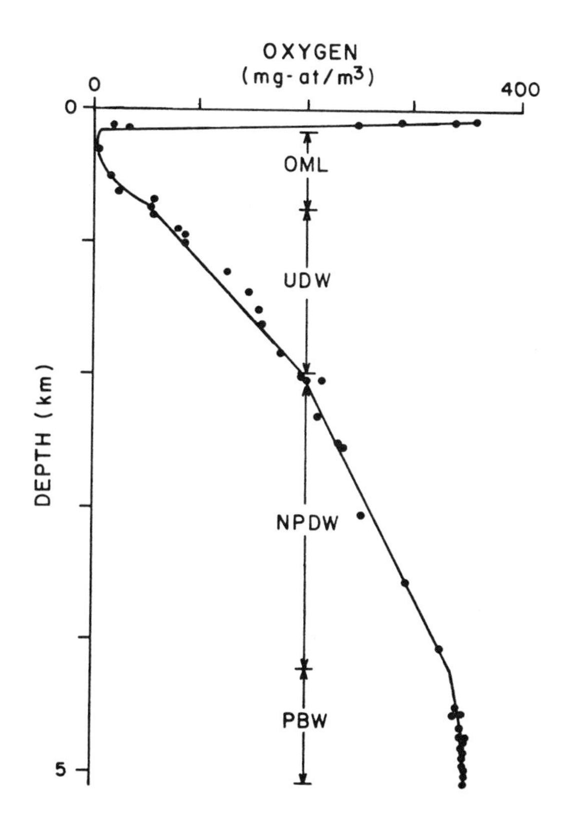

Fig. 5. Oxygen profile from site B, April, 1976. Layers are indicated by line segments.

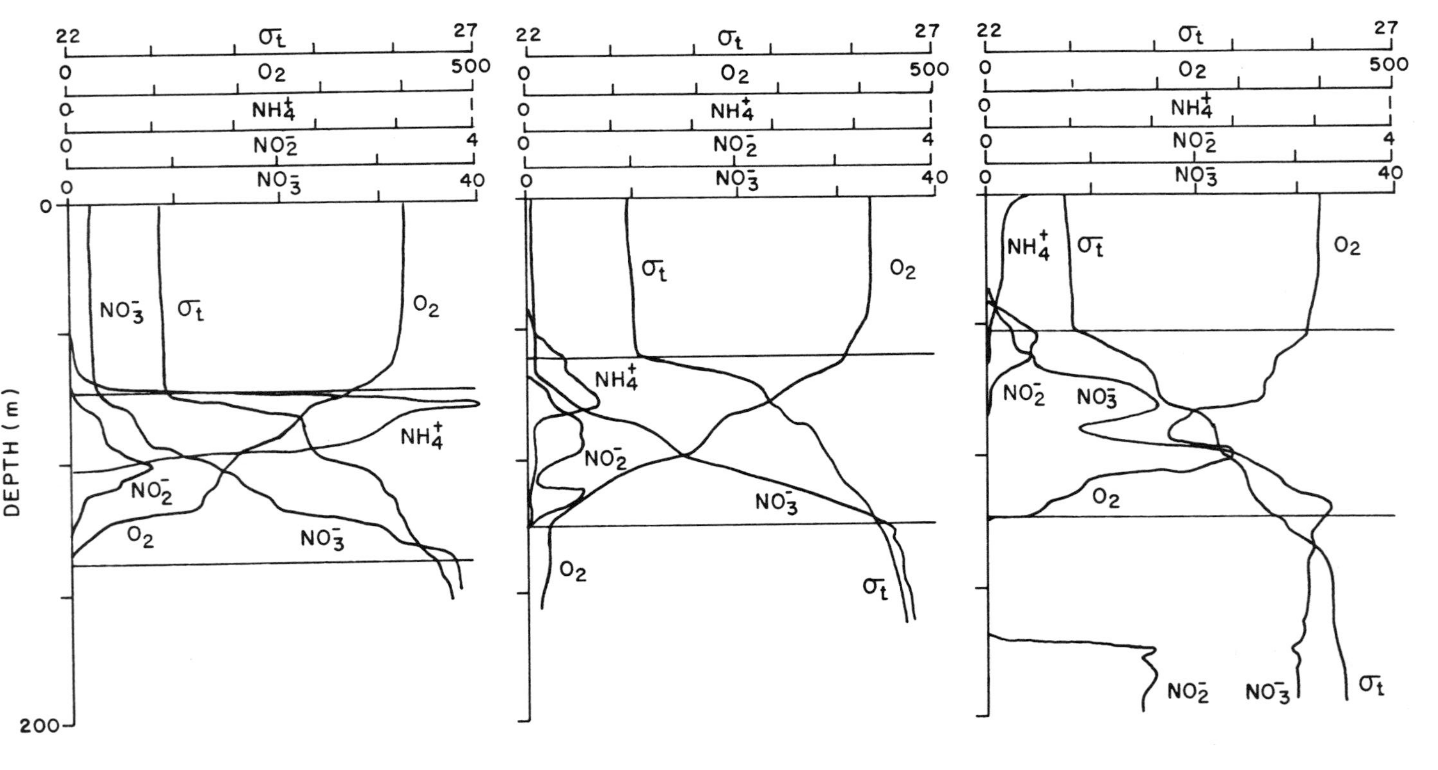

Fig. 6. Pump profiles of nutrients and oxygen with units of mg-at/m^3 and density in sigma-t. Fig. 6a (left) from site A, March 28, 1976. Fig. 6b (middle) from site B, April 6, 1976. Fig. 6c (right) from site C, February 20, 1976. Horizontal lines at about 60 and 120 m demark the mixed layer-thermocline and thermocline-OML interfaces.

TABLE 1

Oxygen at layer interfaces for site stations in winter and summer and for all stations. Numbers are mean values and standard deviations in mg-at/m^3 and number of data points. N column is values from all stations normalized to the maximum value

Interface and approximate depth in m	Summer 1975			Winter 1976			All	N
	A	B	C	A	B	C		
Surface (0 m)	393	410	-	412	470	405	418	1.0
	43	106		20	41	12	64	
	2	6		5	6	4	23	
Mixed layer/ thermocline (50 m)	377	376	-	397	460	383	399	0.95
	42	56		77	25	18	53	
	2	6		5	6	4	23	
Thermocline/OML (120 m)	42	62	-	1	14	0	24	0.06
	16	30		1	7	0	27	
	3	6		4	6	4	23	
OML/UDW (800 m)	32	66	25	74	54	17	45	0.12
	-	-	-	-	-	-	23	
	1	1	1	1	1	1	6	
UDW/NPDW (2000 m)	-	-	-	209	199	191	200	0.45
				4	9	5	9	
				5	4	3	12	
NPDW/PDW (4000 m)	-	-	-	358	342	329	343	0.81
				5	3	7	18	
				35	35	35	105	

The low oxygen core in the OML extends along about 15°N as a tongue from the coast of Central America to well past the Hawaiian Islands (Fig. 8). The zone is most intense near the coast where oxygen concentrations are essentially zero and denitrification is occurring as revealed by a build-up of nitrite. The denitrification zone is shown as the hatched area in Fig. 8. The lowest oxygen values in the DOMES area are found near site C, which has evidence of denitrification. The highest oxygen values in the core of the OML are found near site A where oxygen values are above 20 mg-at/m^3 (Fig. 9). At the bottom of the OML, about 800 m, oxygen values are about 20 mg-at/m^3 at the east end of the DOMES area and about 50 mg-at/m^3 at the west end.

The oxygen profile is essentially linear in each layer below the OML (Fig. 5). The UDW and NPDW have no consistent horizontal oxygen gradients but at the interface between the NPDW and PBW oxygen decreases to the northeast (Table 1).

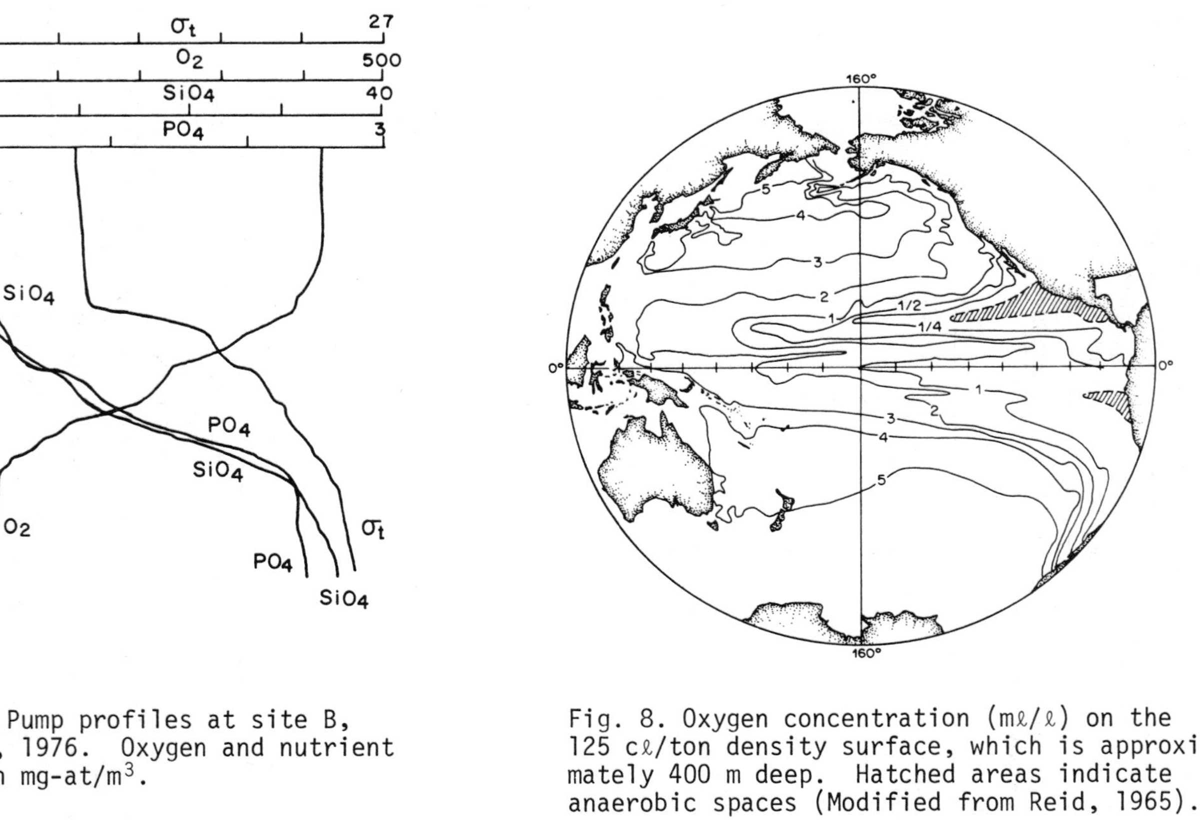

Fig. 7. Pump profiles at site B, April 6, 1976. Oxygen and nutrient units in mg-at/m^3.

Fig. 8. Oxygen concentration (mℓ/ℓ) on the 125 cℓ/ton density surface, which is approximately 400 m deep. Hatched areas indicate anaerobic spaces (Modified from Reid, 1965).

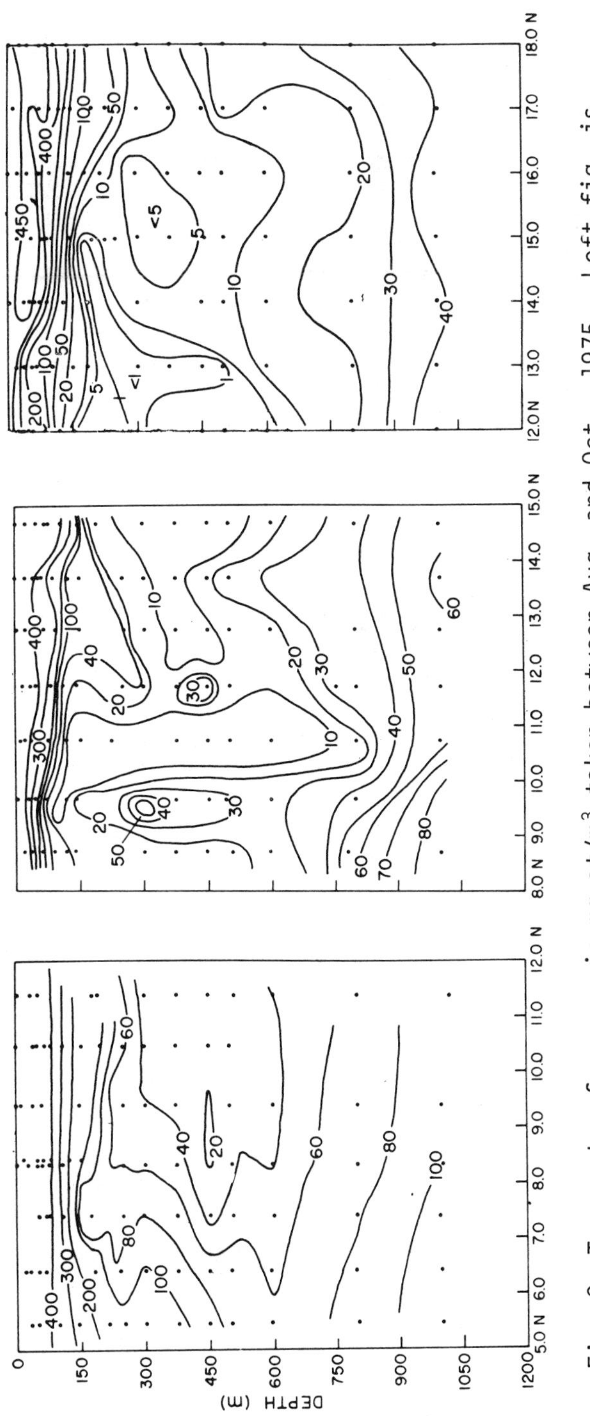

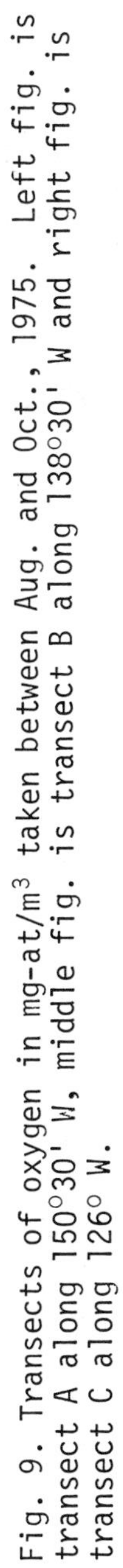
Fig. 9. Transects of oxygen in mg-at/m³ taken between Aug. and Oct., 1975. Left fig. is transect A along 150°30' W, middle fig. is transect B along 138°30' W and right fig. is transect C along 126° W.

Nitrate

The distribution of nitrate is strongly coupled to the physical regime and, as with oxygen, distinct breaks in the profiles are evident at layer interfaces (Fig. 10), but no obvious temporal patterns are evident.

The mixed layer nitrate concentration is typically less than 1 mg-at/m^3 (Fig. 6) and generally increases to the south (Fig. 11). This pattern is common to phosphate and silicate also.

In the thermocline nitrate increases with depth and typically the profile is linear, as at station A, or concave, as at station C (Fig. 6). Rarely are thermocline profiles convex upwards in the DOMES area. The concave profiles are generally associated with low nitrate concentrations in the mixed layer while linear or convex profiles are associated with higher nitrate concentrations in the mixed layer. At the bottom of the thermocline nitrate values are highest at site A and lowest at site C (Table 2).

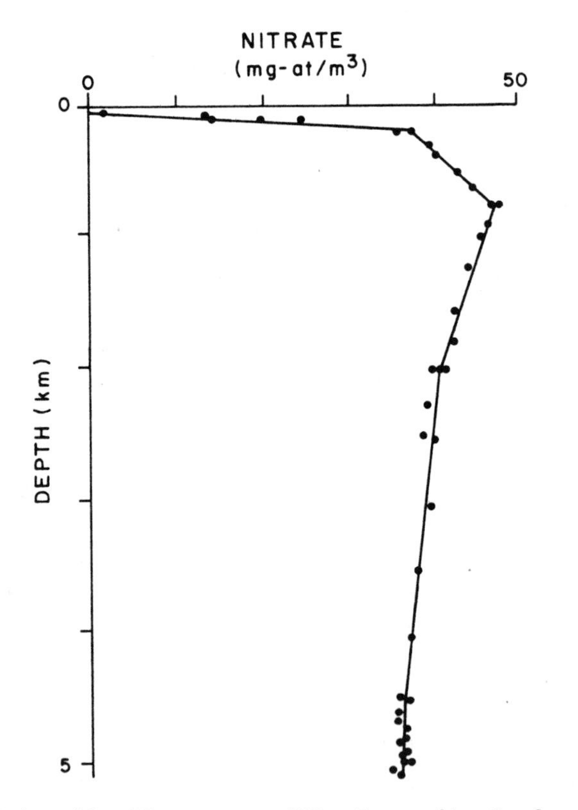

Fig. 10. Nitrate profile from site B, April, 1976.

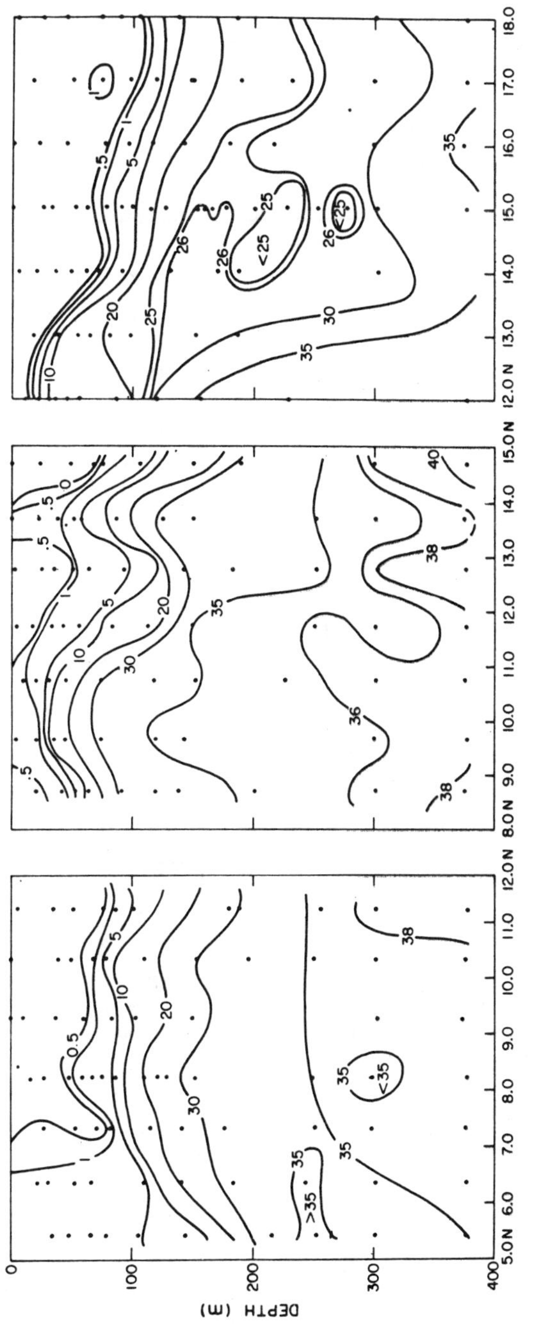

Fig. 11. Transects of nitrate in mg-at/m^3 taken between Aug. and Oct., 1975. Left fig. is transect A along 150°30' W, middle fig. is transect B along 138°30' W, and right fig. is transect C along 126° W.

TABLE 2

Nitrate at layer interfaces for site stations in winter and summer and for all stations. Numbers are mean values and standard deviations in mg-at/m^3 and number of data points. N column is values from all stations normalized to maximum value

Interface and approximate depth in m	Summer 1975 A	Summer 1975 B	Summer 1975 C	Winter 1976 A	Winter 1976 B	Winter 1976 C	All	N
Surface	0	0.7	0.2	2.0	0.6	0.2	0.8	.02
(0 m)	0	0.5	0.2	0.4	0.1	0.2	1.1	
	4	8	6	10	7	6	57	
Mixed layer/	.13	2.4	0.5	2.5	0.9	0.5	1.4	.03
thermocline	.15	1.6	0.4	0.5	0.2	0.4	1.4	
(50 m)	4	7	6	10	7	6	58	
Thermocline/	37.8	35.6	33.6	37.5	36.9	33.6	34.9	.78
OML	2.7	1.2	0.5	0.7	1.7	0.5	3.0	
(120 m)	4	7	6	10	7	6	59	
OML/UDW	47.4	44.4	41.4	44.3	46.9	45.4	44.5	1.0
(800 m)	-	-	-				2.8	
	1	1	1	1	1	1	27	
UDW/NPDW				40.9	40.3	41.9	41.0	.92
(2000 m)				0.3	0.9	0.5	0.8	
				4	3	2	9	
NPDW/PBW				35.8	36.5	36.7	36.3	.82
(4000 m)				0.4	1.4	0.7	0.5	
				35	35	35	105	

Nitrate increases gradually through the OML except near site C where a local minimum is evident between 150 and about 400 m (Figs. 6 and 11). Outside of the extremely low oxygen region high nitrate values are associated with low oxygen values. This correspondence results in a horizontal nitrate maximum associated with the core of the oxygen minimum for the waters west of site C.

The local nitrate minimum at site C (25 mg-at/m^3) is about 10 mg-at/m^3 less than waters of the same density but with slightly higher oxygen content. This nitrate deficit is formed by denitrification.

The vertical nitrate maximum is between 800 to 1000 m, which is near the interface between the OML and the UDW (Fig. 10). No significant horizontal or seasonal variations occur at this maximum or below it (Table 2). Below the maximum, nitrate decreases in a linear manner to the PBW where no significant vertical or horizontal gradients are evident.

Nitrite and Ammonia

Nitrite and ammonia are essentially zero in the mixed layer although traces of ammonia are occasionally observed in one profile and absent from another made several hours later at the same location.

In 19 out of 28 ammonia profiles, made during the DOMES project, a well-defined ammonia peak was observed near the top of the thermocline (Fig. 6). On the average, the depth of maximum concentration was 4 ± 12 m below the mixed layer-thermocline interface, but in some profiles the peak was in the mixed layer. No persistent diurnal pattern was evident in the peaks (Fig. 12), and the average maximum concentration was 0.31 ± 0.34 mg-at/m^3 as determined from 28 profiles.

A distinct nitrite peak was evident near the middle of the thermocline in all profiles taken in the DOMES area (Fig. 6). The peak has no apparent diurnal pattern (Fig. 12), and the average maximum concentration from 51 profiles is 0.4 ± 0.28 mg-at/m^3.

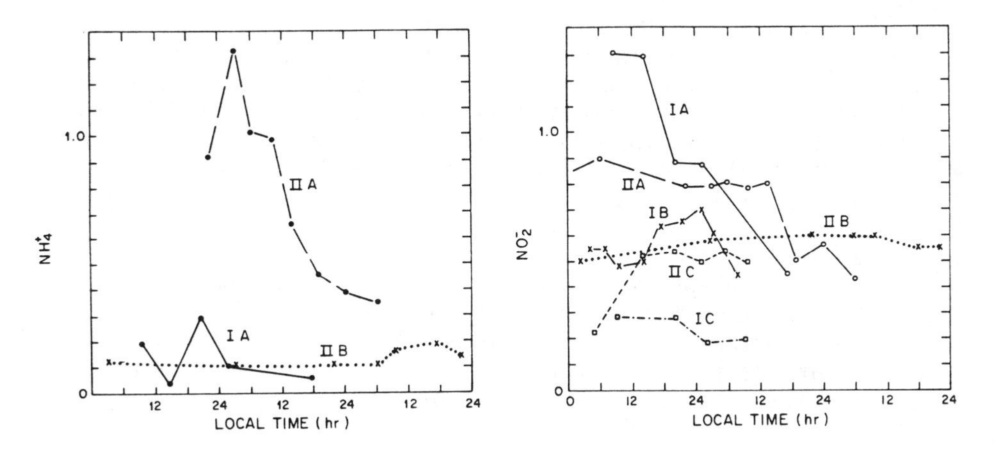

Fig. 12. The maximum ammonia and nitrite concentrations in mg-at/m^3 in the thermocline peaks as a function of local sun time, for sites A, B and C. Observations taken in Aug., Sept., and Oct., 1975 are designated I, and observations taken in Feb., March, and April, 1976 are designated II.

In the thermocline nitrate, nitrite and ammonia profiles are related. The most striking relationship is between nitrite and ammonia, where the maximum nitrite concentration is located near the bottom of the ammonia peak, where the ammonia concentration approaches zero (Fig. 6). This relationship is persistent and for an average of 24 profiles, covering the DOMES area in summer and winter seasons, the depth of the nitrite maximum is 1.4 ± 2.8 m above the lower boundary of the ammonia peak. The top of the nitrite peak is coupled to the mixed layer-thermocline interface and the kink in the nitrate profile near the top of the thermocline. A majority of the time, the three features almost coincide but when the kink in the nitrate profile is below the mixed layer-thermocline interface so is the top of the nitrite peak (see Fig. 6a and b).

Below the thermocline ammonia values were below the limit of detection ($\sim$0.1 mg-at/m^3). According to sensitive measurements by Wada and Hattori (1972), the deep water nitrite concentration is on the order of 0.03 mg-at/m^3. The exception is near site C where nitrite concentrations of up to 2.6 mg-at/m^3 were observed in the core of the OML (Fig. 6c). These high nitrite values coincide with a local nitrate minimum (25 mg-at/m^3) and oxygen values less than 5 mg-at/m^3 (Figs. 9 and 11).

Phosphate

The phosphate distribution is similar to the nitrate distribution except that no minimum occurs in the extreme low oxygen water near site C. In the mixed layer highest concentrations are in the south. Maximum concentrations are between 800 and 1000 m and below this level phosphate decreases linearly to the PBW in which no vertical or horizontal phosphate gradients are evident (see Figs. 7, 13 and Table 3).

Silicate

The distribution of silicate is similar to the phosphate distribution except that the maximum concentration is at about 3000 m, which is in the middle of the NPDW. In the PBW vertical gradients are essentially zero but silicate increases to the northeast (see Figs. 7, 14 and Table 4).

ETS

ETS activities are highest in the mixed layer and decrease through the thermocline. On the average, activities in the mixed layer were 50% higher in the summer of 1975 than in the winter of 1976. In deeper layers no seasonal variations were evident. The activity decreases exponentially in the upper 100 m of the OML. Below the OML ETS activity is essentially uniform (Table 5).

ATP

The distribution of ATP is similar to the ETS distribution,

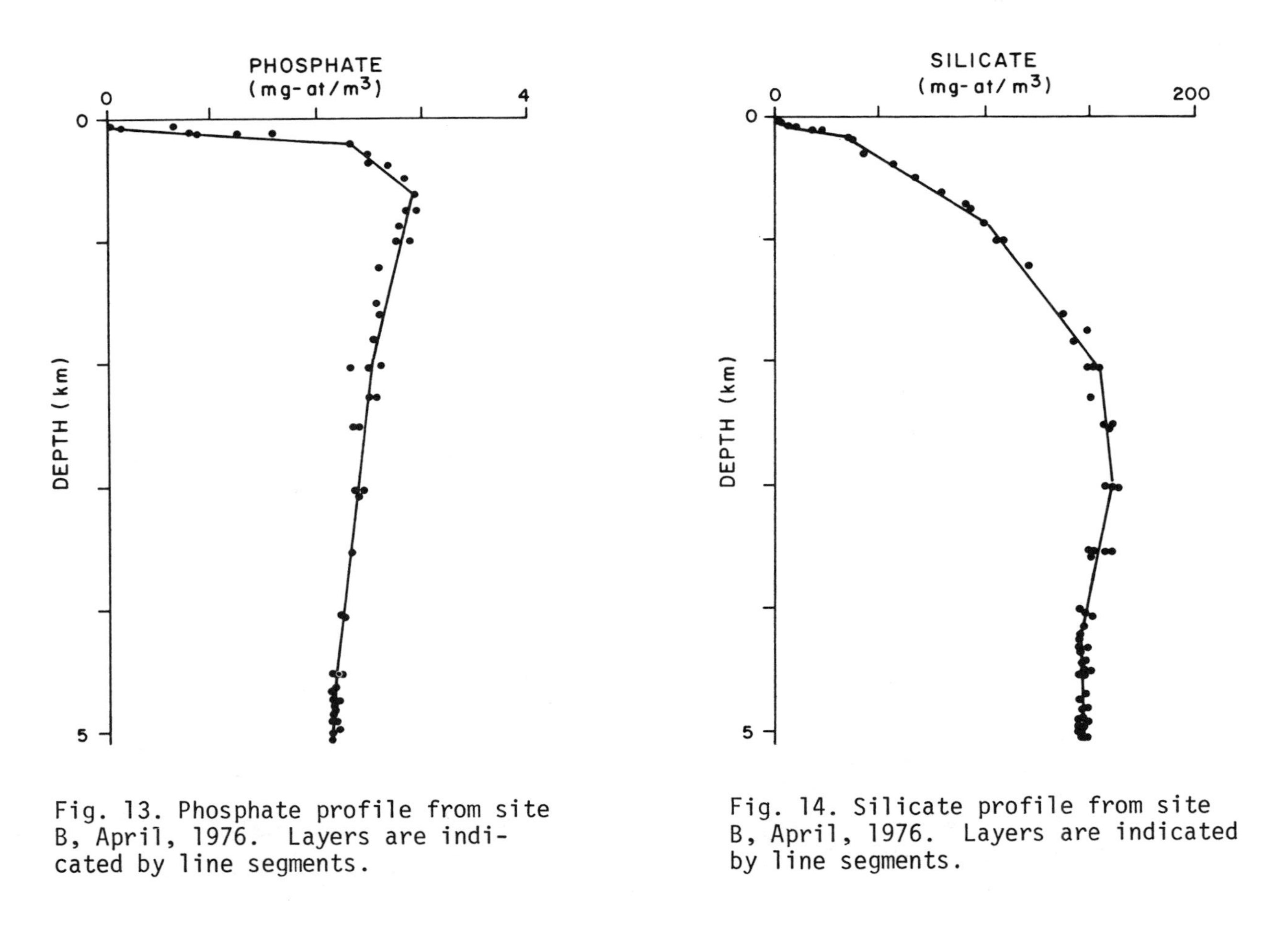

Fig. 13. Phosphate profile from site B, April, 1976. Layers are indicated by line segments.

Fig. 14. Silicate profile from site B, April, 1976. Layers are indicated by line segments.

TABLE 3

Phosphate at layer interfaces for site stations in winter and summer and for all stations. Numbers are mean values and standard deviations in mg-at/m^3 and number of data points. N column is values from all stations normalized to maximum value

Interface and approximate depth in m	Summer 1975 A	B	C	Winter 1976 A	B	C	All	N
Surface	0.11	0.17	0.10	0.13	0.07	0.02	0.12	.04
(0 m)	0.06	0.08	0.03	0.03	0.01	0.01	0.08	
	5	8	4	10	7	6	60	
Mixed layer/	0.15	0.22	0.15	0.17	0.11	0.04	0.14	.05
thermocline	0.07	0.08	0.04	0.05	0.02	0.01	0.09	
(50 m)	5	8	4	10	7	6	60	
Thermocline/OML	2.58	2.56	2.69	2.45	2.42	2.43	2.51	0.81
(120 m)	0.03	0.08	0.08	0.03	0.04	0.08	0.15	
	5	8	3	10	7	6	61	
OML/UDW	3.11	3.13	2.97	2.98	2.93	2.96	3.09	1.0
(800 m)	-	-	-	-	-	-	0.13	
	1	1	1	1	1	1	28	
UDW/NPDW				2.46	2.45	2.56	2.49	0.81
(2000 m)				0.09	0.13	0.03	0.06	
				5	4	4	13	
NPDW/PBW				2.11	2.11	2.18	2.13	.69
(4000 m)				0.05	0.03	0.04	0.04	
				35	35	35	105	

with the major change occurring in the upper few hundred meters of the water column (Table 5). Mixed layer values are lower than in Southern California coastal waters east of the DOMES area. Typically, ATP concentrations in the coastal waters are between 100 and 300 ng/ℓ (Holm-Hansen, 1969), while mixed layer ATP values in the DOMES area average 62 ng/ℓ.

POC, PN, DOC

The distributions of particulate organic carbon, particulate nitrogen and dissolved organic carbon have maximum concentrations in the mixed layer and decrease through the thermocline and upper 100 m of the OML. Below the OML the concentrations are essentially uniform throughout the water column (Table 5).

In general, the POC, PN, and DOC values in the DOMES area are similar to values observed in other parts of the oceans. For DOC

TABLE 4

Silicate at layer interfaces and in middle of NPDW layer for site stations in winter and summer and for all stations. Numbers are mean values and standard deviations in mg-at/m^3 and number data points. N column is values from all stations normalized to maximum value

Interface and approximate depth in m	Summer 1975 A	Summer 1975 B	Summer 1975 C	Winter 1976 A	Winter 1976 B	Winter 1976 C	All	N
Surface (0 m)	0.7 1.0 5	2.4 0.7 6	0.7 0.6 4	3.1 1.1 10	2.0 0.3 7	2.6 0.3 6	2.2 1.3 56	0.01
Mixed layer/ thermocline (50 m)	1.4 1.1 5	4.5 1.1 7	1.4 1.1 4	3.3 1.0 10	2.6 0.3 7	2.6 0.3 6	2.8 1.4 57	0.02
Thermocline/OML (120 m)	27.3 1.6 5	28.9 1.4 8	32.9 2.5 3	32.0 1.5 10	31.4 2.2 7	29.5 3.1 6	30.0 2.6 58	0.19
OML/UDW (800 m)	76.8 - 1	80.2 - 1	84.3 - 1	83.8 - 1	92.7 - 1	81.4 - 1	83.4 6.5 28	0.53
UDW/NPDW (2000 m)				146.6 8.7 5	147.9 3.2 4	148.2 2.7 4	147.6 6.8 13	0.94
Middle of NPDW (3000 m)				157.0 11.8 3	157.5 3.2 3	157.0 2.5 2	157.2 8.8 8	1.00
NPDW/PBW (4000 m)				137.5 4.4 35	144.0 2.0 35	147.2 2.1 35	142.9 4.9 105	0.91

the observed range of 0.5 to 1.5 mg/ℓ is within the range reported for other regions of the Pacific and Atlantic (Holm-Hansen et al., 1966; Sharp, 1974; Menzel and Ryther, 1968 and 1970). The range of POC is similar to the ranges observed by Wangersky (1976) along 150°W, which is in the DOMES area, and Gordon (1971) near Hawaii. Wangersky reported a mixed layer range of 20 to 34 μg/ℓ and a thermocline range of 12 to 19 μg/ℓ. Gordon reported surface water values between 5 and 35 μg/ℓ. For PN, Gordon reported an average surface water value of 2 μg/ℓ and a range of 0.1 to 3.5 μg/ℓ over the water column.

TABLE 5

Biochemical properties through the water column. Mean values, $\bar{x}$, are averaged over both summer and winter cruises, s is standard deviation and n is number of samples. The bottom of the thermocline is taken as the 26 to 26.5 sigma-t interval

		$\frac{mg\text{-}at}{m^3 d}$ ETS	$\frac{ng}{\ell}$ ATP	$\frac{mg}{\ell}$ DOC	$\frac{\mu g}{\ell}$ POC	$\frac{\mu g}{\ell}$ PN
Mixed layer	$\bar{x}$	0.47	62	1.03	23	3.3
	s	0.33	42	0.23	16	1.7
	n	100	63	117	95	106
Bottom of thermocline	$\bar{x}$	0.08	39	0.83	14.2	1.8
	s	0.07	44	0.22	7.6	1.2
	n	40	37	43	53	51
Middle of OML (400±50 m)	$\bar{x}$	0.05*	14	0.73	10.0	1.2
	s	0.03	8	0.24	3.5	0.7
	n	90	40	29	24	24
OML/UDW (800 m)	$\bar{x}$	0.02*	6	0.68	7.6	1.1
	s	0.02	3	0.22	3.4	0.7
	n	30	30	24	20	20
UDW/NPDW (2000 m)	$\bar{x}$	0.004	3	0.55	6.7	0.56
	s	0.004	2	0.12	3.6	0.27
	n	8	8	8	12	12
PBW (bottom 500 m)	$\bar{x}$	0.002	3	0.70	8	0.72
	s	0.001	2	0.23	4	0.62
	n	35	35	35	35	35

* Values are uncertain and possibly high

Discussion

In general the vertical distributions of chemical constituents have simple shapes within the layers defined by T-S characteristics. At the interfaces between layers the profiles have distinct changes in slope. Horizontal variations in the distributions are less distinct and are generally of the same order as the seasonal variations, so in most cases both seasonal and horizontal patterns are small compared to the well-defined vertical patterns.

To investigate the controlling processes we will define the distribution of a chemical constituent S in terms of a balance of mixing and net in situ reactions as

$$\frac{\partial S}{\partial t} = K_h \frac{\partial^2 S}{\partial x^2} + K_h \frac{\partial^2 S}{\partial y^2} + K \frac{\partial^2 S}{\partial z^2} - u \frac{\partial S}{\partial x} - v \frac{\partial S}{\partial y} - (w - \frac{\partial K}{\partial z}) \frac{\partial S}{\partial z} + R \qquad (1)$$

where u, v and w are velocities in the x, y and z directions. K_h and K are horizontal and vertical diffusion coefficients and R is the rate of in situ reaction and is positive for net production of S and negative for net consumption.

To investigate the processes controlling the shapes of the profiles the dominant terms in (1) will be considered for a particular layer. To investigate the flux and concentration at the interfaces between layers equations for the layers will be coupled by concentration and flux requirements. The analysis will proceed first with the oxygen distribution. Nitrate and phosphate will be considered together and the silicate distribution will be considered separately. Processes controlling the peaks of nitrite and ammonia in the thermocline will be discussed briefly.

Oxygen

We begin by considering oxygen in the mixed layer. Typical distributions for the DOMES area (Fig. 6) show nearly linear profiles with slight increases in slope near the bottom of the mixed layer. To explain this shape we will assume the distribution is at steady state and mixing and reactions are uniform throughout the mixed layer. The main features are in the vertical direction and thus the significant processes are vertical diffusion and reaction, so equation (1) reduces to

$$K \frac{\partial^2 S}{\partial z^2} + R = 0 \qquad (2)$$

which has the solution

$$S(z) = S(0) + Cz - Rz^2/2K \qquad 0 \leq z \leq Z_1 \qquad (3)$$

where S(0) and C are the oxygen concentration and gradient at the sea surface and Z_1 is the depth at the bottom of the mixed layer. The shape of S(z) is nearly linear near z = 0 and as z approaches Z_1 the profile changes as a function of z. These characteristics are qualitatively the same as are observed in the oxygen profiles in the mixed layer and if the concentrations at the top and bottom of the layer are given, then the mixed layer profile in Fig. 15 is fit with R/K = 0.028. Supplying the boundary conditions S(0) and $S(Z_1)$ and the center point concentration $S(Z_1/2)$ equation (3) can be rearranged to give the reaction-diffusion ratio as

$$R/K = -4[S(0) - 2 S(Z_1/2) + S(Z_1)]/Z_1^2 \qquad (4)$$

At the surface the oxygen concentration S(0) is controlled

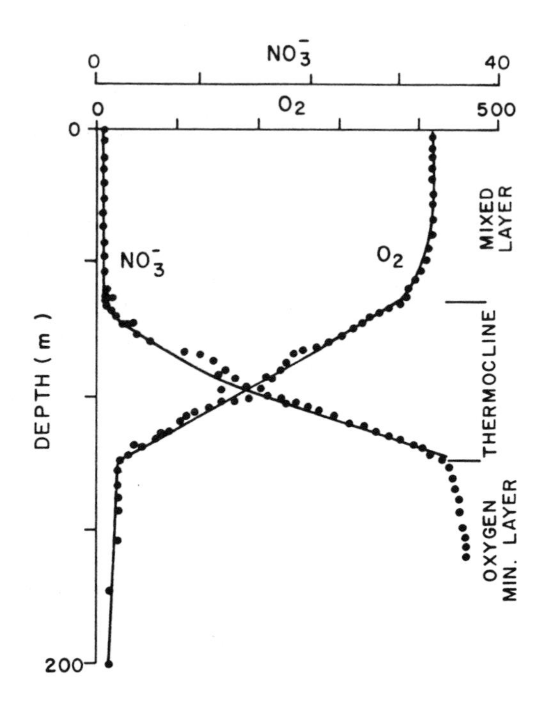

Fig. 15. Fit of model to pump profile of oxygen and nitrate from site B, April, 1976. Profile in (•) model fit in (—). Units are mg-at/m^3.

by the exchange across the air-sea interface and in the DOMES area S(0) is within $\pm$ 5% of the saturation value. From equation (3) the concentration at the bottom of the mixed layer is

$$S(Z_1) = S(0) + S'(Z_1)Z_1 + RZ_1^2/2K_1 \tag{5}$$

where $S'(Z_1)$ is the gradient at the bottom of the mixed layer. The reaction term R is positive with photosynthesis, so increasing R increases the concentration at the bottom of the mixed layer. To determine the factors that control $S'(Z_1)$ we can define an equation for the thermocline and couple this to the mixed layer equation.

In the thermocline a scaling of equation (1) suggests that vertical advection, diffusion and reaction are all potentially of the same order of magnitude while horizontal terms are less important. Observations indicate the oxygen distribution in the thermocline is relatively stable, so assuming a steady state, equation (1) for the thermocline reduces to

$$K \frac{d^2S}{dz^2} - w^* \frac{dS}{dz} + R(z) = 0 \tag{6}$$

where $w^* = w - \partial K/\partial z$ is the combined vertical transport resulting from vertical velocity and a depth variable diffusion coefficient. The reaction term is the difference between photosynthesis and respiration and if the distribution of all respiring organisms follows the distribution of phytoplankton or if the effects of phytoplankton are dominant, then R is expected to decrease with depth as light intensity decreases. Upwelling w is expected to decrease towards the top of the thermocline while $\partial K/\partial Z$ is expected to increase, as K decreases in response to an increasing density gradient towards the top of the thermocline. The net flux from upwelling and the changing diffusion coefficient are both upwards and as one increases with depth the other decreases. Thus, qualitatively, there are grounds for assuming w^* is approximately constant through the thermocline and the solution of (7) with $R(z) = R_0 e^{-bz}$ is

$$S(z) = C_1 + C_2 e^{az} - C_3 e^{-bz} \qquad 0 \leq z \leq Z_2 \tag{7}$$

where

$$a = w^*/K, \qquad \text{and } C_3 = \frac{R_0}{bK(a+b)}$$

The depth coordinate is relative, with $z = 0$ being the top of the thermocline, and Z_2 the bottom, R_0 is the net reaction rate at the top of the thermocline and b describes the exponential decrease in R. The constants C_1 and C_2 control the concentrations at the top and bottom of the thermocline.

The shape of the profile from (7) can assume concave or convex forms. Convex upward profiles result when upwelling ($w < 0$) or respiration ($R < 0$) dominate. Concave downward profiles result when photosynthesis ($R > 0$) or downwelling ($w > 0$) dominate the equation. A linear profile suggests that vertical diffusion is the dominant process controlling the distribution of S. If photosynthesis is large at the top of the profile and upwelling is large at the bottom, then the profile can assume a concave shape at the top and a convex shape at the bottom of the thermocline.

Observations suggest that linear or concave downward profiles are most typical in the DOMES area (Fig. 6). These types of profiles can be fit with the parameters $K = 1\ m^2/d$, $w = -1$ m/y, $R_0 = 0.1$ mg-at/m^3d corresponding to a net primary productivity of about 30 mg C/m^2d and $b = 0.03\ m^{-1}$ corresponding to a 90% change in R over the thermocline.

We can further reduce the thermocline equation by

eliminating the features of equation (7) that the profile is least sensitive to as measured by the concentration at the center of the thermocline, $S(Z_2/2)$. If w* is made vanishing small and the rest of the parameters are retained, then $S(Z_2/2)$ changes by only 5%. Thus, to a good first order approximation upwelling can be eliminated from the thermocline equation. The effect of the exponential rate term also has a small effect on the profile and if the exponential rate term is replaced by a depth constant term, which is the depth integrated average of the exponential rate expression such that $R = 0.39\ R_0$, then $S(Z_2/2)$ remains unchanged. Thus a good first order approximation of the essential thermocline balance can be expressed with a reaction-diffusion equation of the type given by (2) with the solution given by equation (3) where the reaction term R is the depth averaged net rate.

The thermocline oxygen profile in Fig. 15 is fit using $R/K = 0.014$ obtained from equation (4).

The mixed layer and thermocline can be coupled using equation (3) for each layer requiring concentration and flux continuity at the interface, so

$$S_1(Z_1) = S_2(0) \quad \text{and} \quad K_1 \frac{dS_1(Z_1)}{dz} = K_2 \frac{dS_2(0)}{dz} \tag{8}$$

where the subscripts 1 and 2 refer to the mixed layer and thermocline parameters.

The oxygen gradient at the top of the thermocline can be expressed

$$C_2 = [S_3(0) - S_1(0)]/H - \Theta_1/H + \Theta_2/H \tag{9}$$

where $S_1(0)$ is the concentration at the air-sea interface, $S_3(0)$ is the concentration at the thermocline-OML interface,

$$\Theta_i = R_i Z_i^2/2K_i \quad \text{with } i = 1,2 \quad \text{and } H = Z_2 + Z_1K_2/K_1$$

From the flux continuity given by equation (8) we can write $S'(Z_1) = C_2K_2/K_1$ and equation (5) for the oxygen concentration at the mixed layer-thermocline interface becomes

$$S_1(Z_1) = S_1(0) + A_1[S_3(0) - S_1(0)] + A_1\Theta_2 + \Theta_1(1-A_1) \tag{10}$$

where $A_1 = (K_2Z_1/K_1Z_2)/(1 + K_2Z_1/K_1Z_2)$. The Θ_i terms in equation (10) can be obtained from oxygen profiles using equation (4) and Z_i. By rearranging equation (10), A_1 is given as

$$A_1 = \frac{S_1(Z_1) - S_1(0) - \Theta_1}{S_3(0) - S_1(0) - \Theta_1 + \Theta_2} \tag{11}$$

For the oxygen profile given in Fig. 15 $\Theta_1 = 50$ and $\Theta_2 = 29$ and $A_1 = 0.20$, so the numerical expression of equation (10) showing the contributions to the oxygen concentration at the mixed layer-thermocline interface is

$$S_1(Z_1) = 384 = 418 - 79 + 6 + 40$$

The oxygen concentration at the air-sea interface (+ 418) sets the general concentration of $S_1(Z_1)$ and the flux of oxygen into the thermocline (- 79) decreases its concentration in proportion to the ratio of the diffusion coefficients in the layers. Photosynthesis in the mixed layer (+ 40) has about half the effect of the flux and reactions within the thermocline (+ 6) have a small effect on the oxygen concentration at the interface.

The oxygen concentration at the thermocline-OML will be investigated by coupling the thermocline form of equation (3) to a reaction-diffusion equation describing the vertical distribution of oxygen in the OML.

Wyrtki (1962) suggested that advection is weak or absent within an oxygen minimum layer and in the DOMES area this situation appears to occur with the core of the low oxygen water lying within a shear zone of zero or weak currents. Currents can cause deformations of the shape of the core, but the intensity and basic shape is dependent on the distribution of respiration in the core and the shape of the boundary conditions, which are presumably maintained by circulation at the edges of the core.

In the DOMES area the thermocline limits the diffusion of atmospheric oxygen into the OML and is a significant boundary to the layer. The lower boundary concentration at the OML-UDW interface is presumably fixed by the circulation, which also maintains the Antarctic salinity minimum. The zonal flows near 10 and 20°N establish the horizontal boundaries with oxygen concentrations of about 90 mg-at/m^3(1 mℓ/ℓ).

The distribution of respiration in the OML decreases with depth in an approximate exponential manner (King et al., 1978). The rate is likely to decrease to the east in response to a decrease in primary productivity away from the coast of Central America as observed by Owens and Zeitzschel (1970).

To investigate how the boundary conditions and respiration distribution might interact to produce the distribution of oxygen in the OML, we will separate the vertical and horizontal regimes and investigate separately the balance of vertical advection and diffusion with respiration, and the balance between horizontal diffusion and respiration.

The justification for the separation of the equation was discussed by Wyrtki (1962). He suggested that a location in an oxygen minimum layer can be selected where horizontal effects are minimum and thus the vertical distribution can be described in terms of a balance between vertical fluxes and the vertical distribution of respiration. In a similar manner the condition for the application is $d^2S(x,y)/dx^2 + d^2S(x,y)/dy^2 \to 0$, which is approached at the center of the tongue-like core of the OML at 15° N. The horizontal distribution can be described in terms of a balance of horizontal diffusion and respiration. The condition for the application is $d^2S(z)/dz^2 \to 0$, which occurs at the vertical minimum of the OML, at about 400 m.

The vertical distribution of oxygen will be defined by equation (7) with the solution given by equation (8) where the top of the OML is z = 0 and the bottom is $z = Z_3$. To investigate the dominant parameter we begin with K = 6 m^2/d (0.7 cm^2/s) and w = 0.02 m/d (6 m/y), in accordance with studies by Wyrtki (1962), Munk (1966) and King et al. (1978), b = $3.5 \times 10^{-3} m^{-1}$, as obtained from ETS data of King et al. (1978); and $R_0 = -0.013$ mg-at/m^3d. With observed boundary conditions $S_3(0) = 25$ and $S_3(Z_3) = 40$ mg-at/m^3, the resulting profile has a minimum value of 6 mg-at/m^3 at a depth of 210 m. This is within the range of the observed profiles which have minimums between 0 and 20 mg-at/m^3 at depths of between 200 and 400 m. If the advection and diffusion parameters are of the right order, then because we expect the vertical model to account for only a portion of the flux into the OML, with horizontal flux accounting for the rest, the respiration rate needed to balance the vertical flux should be less than the observed rate that balances both vertical and horizontal fluxes. The rate given by R_0 is less than 30% of that observed by King et al. (1978), thus on a first order basis we expect that the vertical flux should be less than 30% of the total flux into the OML.

If the vertical velocity is made vanishing small and respiration rate is set constant with depth, while still retaining the same depth integrated rate, then a minimum concentration of 3.5 mg-at/m^3 occurs at a relative depth of 280 m (Fig. 16). Thus the principal features of the vertical profile are insensitive to vertical advection or the manner in which the respiration term changes with depth and the basic oxygen profile can be described in terms of a reaction-diffusion balance, given by equation (3), where the rate term is the average rate over the layer. With this equation the profile has a mid-depth minimum and the gradient of the profile increases with depth. With $S(Z_3/2) = 3.5$ equation (4) gives $R/K = -5 \times 10^{-4}$mg-at/m^5d for the profile in Fig. 16.

To determine what processes are important in controlling oxygen at the thermocline-OML interface we can couple the two layers in the same manner that the mixed layer and thermocline layers were coupled. The thermocline-OML interface concentration, $S_3(0)$, can be expressed in terms of the air-sea interface concentration $S_1(0)$ and the OML-UDW interface concentration, $S_4(0)$, by combining mixed layer-thermocline and thermocline OML forms of equation (10) to give

$$S_3(0) = \frac{(1-A_1)(1-A_2)[S_1(0)+\Theta_1]}{B} + \frac{(1+A_1)(1-A_2)\Theta_2}{B} + \frac{A_2\Theta_3}{B} + \frac{A_2S_4(0)}{B} \tag{12}$$

where $B = 1 - A_1(1-A_2)$. For the oxygen profiles given by Figs. 15 and 16 $\Theta_1 = 50$, $\Theta_2 = 29$, $\Theta_3 = -122$ $A_1 = 0.20$ and $A_2 = 0.78$. Taking boundary conditions $S_1(0) = 418$ and $S_4(0) = 40$, then the numerical expression of the contributions to $S_3(0)$ is

$$S_3(0) = 25 = 84 + 8 - 100 + 33$$

This calculation suggests that $S_3(0)(+25)$ is principally established by a balance between respiration in the OML (-100) and the flux of oxygen from the mixed layer via the thermocline (+84). The oxygen flux at the bottom of the OML (+33) is of secondary importance. Additionally, photosynthesis in the thermocline (+8) is insignificant in controlling the concentration at the thermocline-OML interface.

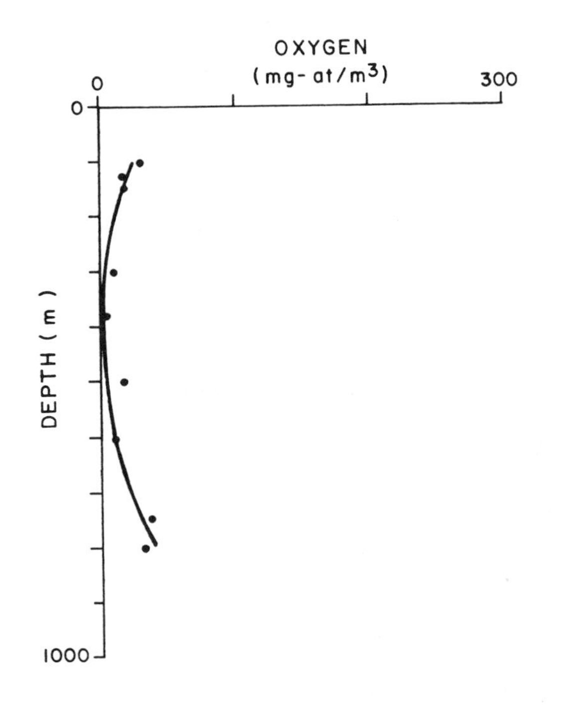

Fig. 16. Oxygen profile from site B, April, 1976 shown in (•). Line is model fit using equation (3).

To investigate the horizontal distribution of oxygen in the OML we define a balance between horizontal diffusion and respiration so

$$K_h\left[\frac{\partial^2 S}{\partial x^2} + \frac{\partial^2 S}{\partial y^2}\right] = R_0 e^{-ax} \tag{13}$$

The horizontal plane is taken along a density surface within the core of the OML, with x positive north and y positive west. The term, R_0, is the respiration rate at the east end of the OML, which is the coast of Central America and, a, describes the respiration decrease to the west. Boundary conditions are set at 10 and 20°N in accordance with the locations of the subsurface zonal flows. We assume the sediments on the Central American shelf have negligible respiration and that the western side of the ocean does not affect the distribution in the central and eastern Pacific. These boundary conditions are expressed

$$S(x,0) = S(x,Y) = S(\infty,y) = S_0 = 90 \text{ mg-at/m}^3$$

and

$$dS(0,y)/dx = 0$$

where x = 0, y = 0 is the southeast corner of the layer (approximately 10°N, 90°W) and x = 0, y = Y is the northeast corner of the layer (approximately 20°N and 90°W)(see Fig. 17). Equation (13) is nondimensionalized with $S' = (S_0 - S)/S_0$, $x' = x/Y$, and $y' = y/Y$, where Y = 10° lat is the distance between zonal flows. The solution of the nondimensional form of equation (13) is

$$S'(x',y') = \frac{\phi e^{-\alpha x'}}{\alpha^2}\left(1 - \frac{\cos\alpha y' + \cos\alpha(y'-1)}{1 + \cos\alpha}\right) + 2\alpha\phi \sum_{n=1}^{\infty} \frac{(1 - \cos n\pi)\sin n\pi y'}{n^2\pi^2(n^2\pi^2 - \alpha^2)} e^{-n\pi x'} \tag{14}$$

where $\alpha = aY$ and $\phi = -Y^2R_0/S_0K_h$ is the ratio of horizontal diffusion and reaction time scales.

A representative distribution of S is shown in Fig. 17 along with the oxygen distribution on the 125 cl/t isanosteric surface, which is approximately the core of the OML. The hatched area in the model distribution has negative oxygen values averaging 0.1 mℓ/ℓ (9 mg-at/m^3). The negative concentration reaches 0.2 mℓ/ℓ (18 mg-at/m^3) at the coast. The hatched area in the observed distribution map is a region of denitrification, with near zero oxygen concentration and about 1 mg-at/m^3 of nitrite. A representative nitrate deficit in the hatched area is 10 mg-at/m^3 (Codispoti and Richards, 1976) and is equivalent to a negative oxygen concentration of 25 mg-at/m^3.

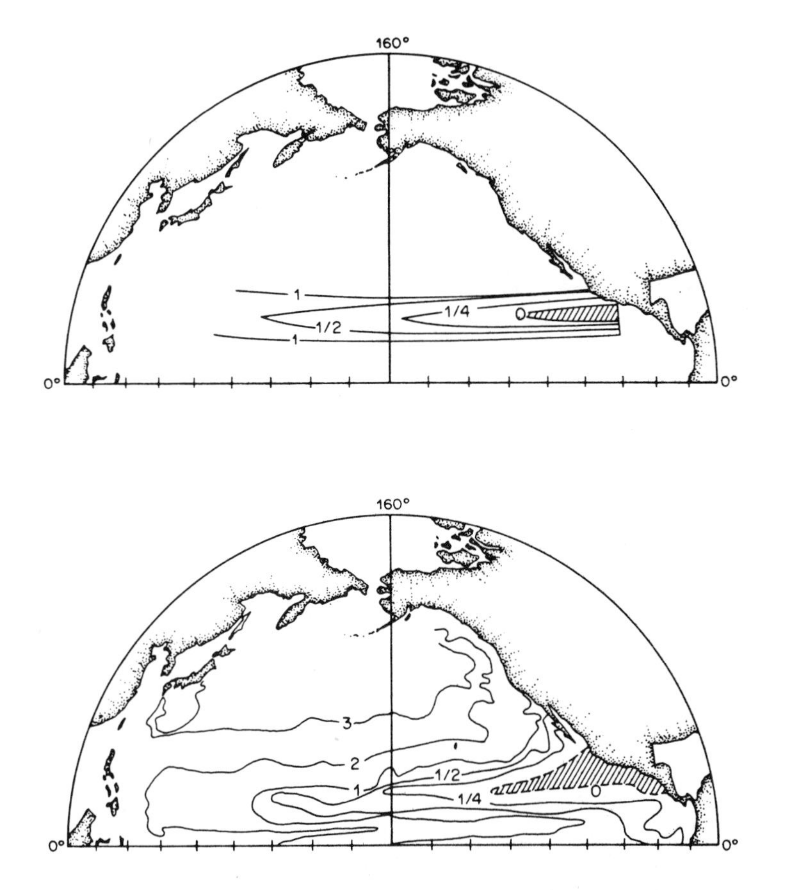

Fig. 17. Oxygen distribution (mℓ/ℓ) along the 125 cℓ/ton density surface, which is the core of the oxygen minimum zone. Upper figure is predicted from equation (14). Lower figure is observed distribution (From Reid, 1965). Hatched area indicates anaerobic space.

The distribution of S or S' according to equation (14) is dependent on α and ϕ . For Fig. 17 $\alpha = 0.1$ and $\phi = 10$ and the corresponding dimensional parameters are $Y = 10^6$m, $K_h = 8.6 \times 10^7$ m^2/d ($10^7 cm^2/s$), $R_0 = -0.007$ mg-at/m^3d (30 μℓ/ℓ y), $S_0 = 90$ mg-t/m^3 (1 mℓ/ℓ) and $a = 10^{-9} m^{-1}$, which gives a 1% decrease in R for every degree longitude away from the coast. The diffusion coefficient is within the range used in other studies of oceanic distributions while the respiration rate R_0 is slightly lower than the respiration rate at 400 m as inferred from ETS by King et al. (1978).

Equation (14) suggests that the distribution of S' down the axis of the core (15°N) follows the zonal distribution of respiration as described by α. Thus the zonal gradient of respiration as controlled by surface layer primary productivity might control the zonal oxygen gradient. Equation (14) also indicates that the size and intensity of the region of negative oxygen has a one to one correspondence with ϕ. Thus, an increase in ϕ by a factor of 2 will increase the maximum negative concentration by 2 and will cause the westward extent of the region to increase by 2.

These models on the vertical and horizontal distribution of oxygen show that with typical values for diffusion and respiration the OML will contain a substantial region of denitrification. Important in maintaining the region are, first, the sharp thermocline, which limits the diffusion of oxygen to possibly 30% of the total flux into the layer, and, second, the absence of significant currents. In the DOMES area the geostrophic currents suggest the core of the OML has zero or weak currents associated with a shear zone in the horizontal and possibly vertical directions.

Below the OML the oxygen profile has three nearly linear segments corresponding with the UDW layer (800 to 2000 m), the NPDW layer (2000 to 4000 m) and the PBW layer (4000 m to bottom at about 4500 m) (Fig. 5). It is difficult to establish a simple balance for the profile because the terms in equation (1) may be of equal importance in the DOMES area. The problem is not untenable though and in a qualitative manner several features of the deep water balance can be postulated. The most important and well-defined feature of the deep water is the northeastward current in the PBW, which has its origins in the Antarctic surface water. This flow maintains a well-mixed high oxygen regime within about 500 m of the bottom. Because the North Pacific has no deep water outflow passage, the return flow, driven by the PBW, must rise vertically in the high latitudes of the North Pacific and return to the South Pacific probably through the NPDW. In this scenario then, in the DOMES area, the NPDW layer has a southward flow and no vertical velocity. The distribution of temperature and salinity in the NPDW layer have convex upward curvature suggesting that upwelling is important in establishing the profile. Oxygen in the same layer is essentially linear suggesting a balance between upwelling, vertical diffusion and respiration (Wyrtki, 1962; Munk, 1966; Craig, 1971). These profiles can be reconciled with the assumed circulation pattern if the form of the vertical profiles is produced in the high latitudes of the North Pacific and is transported south intact in the NPDW layer. This assumes that a profile can acquire its vertical shape in one region and this shape can be transported into another region with horizontal currents.

In the UDW the temperature and salinity profiles are convex upward and the oxygen profile is linear or concave down. King et al. (1978) were able to show these forms can be produced in a balance between vertical advection, diffusion and respiration in the UDW of the eastern tropical North Pacific. As assumed for the NPDW layer, the features may have been produced in higher latitudes and through horizontal diffusion or currents essentially transported to the equatorial region. It is also possible that a significant vertical velocity component exists in the UDW in the equatorial Pacific.

Nitrate and Phosphate

In general, the nitrate and phosphate distributions in the DOMES area are mostly inverse to the oxygen distribution because nutrient uptake in photosynthesis is accompanied by oxygen production and nutrient regeneration in respiration and nitrification is accompanied by oxygen consumption.

In the mixed layer the distribution of nutrients can be described with the reaction-diffusion balance given by equation (2). Because nutrients are in an ionic form no significant flux occurs across the air-sea interface, so $C = 0$, and the vertical profile of nutrient S in the mixed layer is given by

$$S(z) = S(0) - Rz^2/2K \qquad 0 \leq z \leq Z_1 \qquad (15)$$

The shape of the profile is dependent on $RZ^2/2K$ and with uptake exceeding regeneration, $R < 0$ and the concentration increases with the square of the depth. A fit of equation (15) to nitrate in the mixed layer is shown in Fig. 15.

In the thermocline the nutrient profiles respond to advection and diffusion in the same manner as does oxygen, so the essential features of the nutrient profile can be described in terms of a reaction-diffusion balance described by equation (3) with the thermocline depth interval defined $z = 0$ to $z = Z_2$ and R being the average net reaction rate of the nutrient over the thermocline. A fit of equation (3) to nitrate in the thermocline is shown in Fig. 15.

The concentration of a nutrient at the mixed layer-thermocline interface can be expressed with equations (10) and (15) to give

$$S_1(Z_1) = S_3(0) + 2\Theta_1(1/A_1-1) + \Theta_2 \qquad (16)$$

where the first term is the concentration at the interface without reactions in the thermocline and mixed layer, the second term is the contribution of mixed layer reactions to the interface concentration and the third term is the contribution of thermocline reactions.

As was done for oxygen, using equation (11), A_1 can be estimated by arranging (16) to give

$$A_1 = 2\Theta_1/[S_1(Z_1) - S_3(0) + 2\Theta_1 - \Theta_2] \quad (17)$$

For nitrate at station B (Fig. 15) $\Theta_1 = -0.61$, $\Theta_2 = -23.4$, $S_1(0) = 0.9$, and $S_3(0) = 35.5$ giving $A_1 = 0.98$. This estimate of A_1 is a factor of 2 smaller than A_1 estimated from oxygen using equation (11). With the smaller value the conclusions about processes controlling oxygen at the interfaces are unchanged.

The nitrate concentration at the mixed layer-thermocline interface given by equation (16) can now be expressed numerically as

$$S_1(Z_1) = 0.9 = 35.5 - 11.2 - 23.4$$

This suggests that $S_1(Z_1)$ is controlled by diffusion across the thermocline (+35.5) and reactions within the mixed layer and thermocline, where the thermocline reaction (-23.4) dominates over the mixed layer reaction (-11.2) in Fig. 15.

The concentration of a nutrient at the thermocline-OML interface can be expressed by using equations (15) and (16) in equation (12) giving

$$S_3(0) = 2(1-A_1)(1-A_2)\Theta_1/A_1A_2 + 2(1-A_2)\Theta_2/A_1 + \Theta_3 + S_4(0) \quad (18)$$

The first term is the contribution from uptake in the mixed layer, the second term is the contribution from uptake in the thermocline, the third term is the contribution from regeneration in the OML and the fourth term is the contribution from the concentration at the OML-UDW interface.

For nitrate at station B, $\Theta_3 = 3$, $S_4(0) = 46.9$ and the numerical expression of equation (18) is

$$S_3(0) = 35 = -3 - 12 + 3 + 47$$

This suggests that at thermocline uptake (-12) is the most important negative contribution and regeneration in the OML (+3) accounts for about 10% of the input. The concentration at the OML-UDW interface is the dominant term (+47) and is the maximum concentration in the water column. Thus vertical diffusion away from the OML-UDW interface must be balanced by horizontal advection and diffusion along the interface. This conclusion is supported with the observation that the interface is also associated with a salinity minimum that can be traced along density surfaces to the surface waters of the Antarctic.

A similar analysis for phosphate indicates it is controlled by the same processes controlling nitrate in the upper layers.

Below the thermocline the phosphate and nitrate profiles are essentially linear in each layer and at the layer interfaces the profiles change slope (Figs. 7, 13, 14). A local nitrate minimum occurs at site C as a result of nitrate uptake by denitrification. The maximum nitrate and phosphate concentrations lie between 800 and 1000 m and are typically about 100 m below the salinity minimum demarking the interface between the OML and UDW.

The position of the nutrient maximum is due to the superposition of two chemically indistinguishable but distinct nutrient fields--the oxidative nutrient field and the preformed nutrient field. The oxidative nutrient is produced in respiration and is related to the apparent oxygen utilization (AOU) by a fixed ratio, known as the oxidative ratio $\Delta S/\Delta AOU$ where ΔAOU is a unit amount of oxygen consumed in the water since it left the surface and ΔS is a unit release of nutrient for the unit consumption of oxygen. The surface oxygen value is calculated from the saturation value as determined by the temperature and salinity of the water sample. The preformed nutrient is the amount in oxygen saturated water at the surface and is independent of respiration. In this manner it is a conservative property like salinity. The observed nutrient is related to the preformed nutrient and AOU according to the relation

$$S = S_{pre} + \frac{\Delta S}{\Delta AOU} AOU \qquad (19)$$

The oxidative ratio $\Delta S/\Delta AOU$ is essentially constant throughout the oceans and AOU increases when oxygen decreases. Thus the oxidative nutrient has a minimum at the surface and a maximum in the oxygen minimum, about 400 m. Between 400 m and the bottom it decreases with depth following the oxygen distribution in a reverse manner.

The vertical profiles of the preformed nutrients in the DOMES area increase with depth and are connected to a poleward increase in the surface concentration of preformed nutrients. The mechanism required to connect the two preformed distributions is mixing along density surfaces which are continuous between the surface in high latitudes and depth in low latitudes (Redfield et al., 1963). The superposition of the depth increasing preformed nutrient and depth decreasing oxidative nutrient fields, below 400 m, produce a maximum in the observed nutrient field that is deeper than the oxygen minimum.

Silicate

The silicate distribution in the waters above 800 m is qualitatively similar to the distributions of phosphate and nitrate. In the deep water the silicate distribution is different in that the maximum concentration is at about 3000 m (Fig. 14) instead of 800 m (Fig. 13). The deeper maximum appears to be produced by a combination of a more uniform regeneration rate with depth and the southward flow in the NPDW layer. Fiadeiro (1975) used a three-dimensional numerical model of the deep Pacific to investigate the distributions of silicate, nitrate, phosphate and oxygen. The model balances advection and diffusion in three-dimensions with fluxes from the sediments and an exponentially decreasing regeneration rate with depth. To fit the silicate distribution the rate has to attenuate about 80% between 1 km and 4 km depth. By contrast, to fit the distributions of oxygen, nitrate and phosphate the rate attenuation is on the order of 97%.

With this scenario PBW flows north into the high latitudes of the North Pacific and upwells into the NPDW acquiring additional silicate. The NPDW flows south over the PBW creating a silicate maximum at 3000 m.

Ammonia

The concentration of ammonia is near zero throughout the water column except near the top of the thermocline where a peak is frequently observed (Fig. 6). This distinct peak requires that ammonia production exceed uptake within some depth interval near the top of the thermocline. Observations support this hypothesis and indicate that in the tropical ocean zooplankton congregate near the mixed layer-thermocline interface (Longhurst, 1976; Hirota, 1977; Anderson, 1978) and excrete ammonia while grazing (Eppley et al., 1973; Biggs, 1977).

The sharp vertical gradients in the peak suggest that vertical diffusion is the major flux mechanism. The peak has sharper boundaries than is expected from diffusion alone suggesting that uptake occurs outside the interval of net production. This uptake can be driven by bacterial oxidation of ammonia to nitrite (Carlucci and Strickland, 1968; Carlucci et al., 1970) and assimilation in phytoplankton photosynthesis (Dugdale and Goering, 1967). Both of these mechanisms have been documented in numerous other studies. Thus the sharp peak suggests a depth interval where net ammonia production is imbedded in a larger interval with net uptake. The top of the ammonia peak often coincides with the mixed layer-thermocline interface and suggests that rapid diffusion in the mixed layer is also responsible for maintaining the sharp gradient in the upper part of the ammonia peak.

To obtain a first order estimate of the magnitudes of reaction and diffusion needed to produce the peak we can estimate the balance at the depth of the ammonia maximum by defining the second derivative with a three point approximation

$$\frac{d^2S}{dz^2} = \frac{4}{H^2} [S(0) - 2S(H/2) + S(H)] \tag{20}$$

where H is the thickness of the peak and the concentrations S are taken at the top, middle and bottom of the peak. The peak is nearly symmetrical, so the maximum concentration occurs at about H/2, and S(0) = S(H) = 0. Thus, the reaction-diffusion balance at the maximum concentration is essentially

$$R = \frac{8KS(H/2)}{H^2} \tag{21}$$

where R is the net rate of production at the ammonia maximum.

From observations S(H/2) = 0.3 mg-at/m^3 and H = 20 m. Assuming K = 1 m^2/d, then at steady state the net production is R = 6 x 10^{-3} mg-at/m^{3}d. Eppley et al. (1973) estimated a gross production rate of 5 to 9 x 10^{-3} mg-at/m^{3}d in the North Pacific mixed layer in November. Eppley et al. (1973 and 1977) also found the uptake of ammonia to be of the same order of magnitude as the production. Thus we expect the ammonia peak is formed in a precarious and uncoupled balance between zooplankton production and phytoplankton uptake.

Nitrite

The concentration of nitrite is essentially zero throughout the water column except for a stable peak in the thermocline and a secondary peak in extremely low oxygen waters (Fig. 6).

The peak in the thermocline can be produced by bacterial oxidation of ammonia and consumed by bacterial oxidation of nitrite to nitrate (Wada and Hattori, 1971, 1972; Miyazaki et al., 1973). Phytoplankton can also produce and consume nitrite. In phytoplankton production, nitrite is an intracellular intermediate in the photosynthetic reduction of nitrate to ammonia. Part of the nitrite is leaked to the environment when the reduction of nitrite to ammonia is reduced or blocked by environmental factors (Kiefer et al., 1976). In consumption, nitrite is apparently assimilated competitively with nitrate to meet the nitrogen requirements of photosynthesis (Eppley and Coatsworth, 1968). Phytoplankton biomass in the thermocline is expected to exceed that of bacteria, so phytoplankton should be responsible for producing the nitrite peak.

A first order estimate of the net production rate required to maintain the nitrite peak against diffusion can be estimated from equation (21). From observations the average maximum nitrite concentration in the DOMES area is 0.4 mg-at/m^3 and the thickness of the peak is about 20 m. Taking K = 1 m^{2}d, then the net production rate at the nitrite maximum is 8 x 10^{-3} mg-at/m^{3}d. Wada and Hattori (1971) measured a net rate of 6 x 10^{-3} mg-at/m^{3}d within the nitrite peak in the North Pacific.

In the OML at site C a secondary peak of nitrite is produced by denitrification. Because the distribution has a peak-like form, we can use equation (21) to estimate the order of the reaction rate at the peak center. For the peak at site C, $S(H/2) = 0(1)$ mg-at/m^3, $H = 0(300)$ m, and with $K = 0(10)$ m^2/d, then $R = 0(10^{-3})$ mg-at/m^{3}d. By comparison the oxygen respiration or the equivalent nitrate reduction rate in denitrification is on the order of 10^{-3} mg-at/m^{3}d as estimated from ETS measurements by King et al. (1978) and by Devol et al. (1976). Thus, with the assumed parameters the net production rate of nitrate is on the order of the rate of reduction of nitrate in denitrification.

Nutrient Flux into the Mixed Layer

The flux of nitrate and phosphate into the mixed layer can be defined by coupling equation (3) and (15) using equation (9). Noting that $Z_1 \simeq Z_2$, $S_3(0) >> S_1(0)$ and $K_1 >> K_2$, then the flux of S from the thermocline into the mixed layer is approximately

$$F = \frac{K_2}{Z_2}\,[S_3(0) + \Theta_2] \tag{22}$$

where $\Theta_2 = R_2 Z_2^2/2K_2$ is the relative contribution to F from reactions in the thermocline and $S_3(0)$ is a relative contribution of F without reactions. For station B (Fig. 15) the numerical expression of (22) for nitrate is

$$F = \frac{K_2}{Z_2}\,(35 - 24)$$

which indicates that the thermocline nitrate uptake (-24) substantially reduces F.

If the nutrient limits photosynthesis in the mixed layer, then the net rate of primary productivity integrated over the mixed layer can be defined

$$P_1 = a_1 S_3(0) K_2/Z_2 - P_2 a_1/2a_2 \tag{23}$$

where P_2 is the integrated net rate of primary productivity in the thermocline and a_1 and a_2 are nutrient recycling coefficients for the mixed layer and thermocline. Defining P_1 and P_2 in terms of carbon, then the recycling factor is defined

$$a_i = \frac{\delta C}{\delta N}\,\frac{R_i^*}{R_i} \tag{24}$$

where $\delta C/\delta N$ is the carbon to nitrogen ratio in phytoplankton and R_i^*/R_i is the ratio of net rate of total nitrogen uptake in photosynthesis (including nitrate, nitrite and ammonia) to the net rate of nitrate uptake alone.

The first term of equation (23) gives the potential P_1 and is controlled by the physical characteristics of the thermocline. The second term gives the decrease in P_1 from productivity in the thermocline, P_2.

If $P_2 \rightarrow 0$, then P_1 approaches a maximum value of

$$P_1(max) = a_1 S_3(0)\ K_2/Z_2$$

and the thermocline nutrient profile is nearly linear. In this condition the thermocline acts as a nutrient filter and P_1 is controlled by the physical properties of the thermocline.

As P_2 increases, P_1 decreases and the thermocline nutrient profile is concave downward as in Fig. 6a and b. In this condition the thermocline acts as a nutrient sink. The maximum thermocline productivity occurs when the nutrient uptake in the thermocline equals the flux into the thermocline. This implies $P_1 = 0$ and the maximum thermocline productivity is then approximately

$$P_2(max) = 2a_2 S_3(0) K_2/Z_2$$

Goering et al. (1970) measured nitrate and ammonia uptake in the mixed layer and thermocline of the eastern tropical North Pacific and determined in the mixed layer 95% of the nitrogen used in photosynthesis comes from ammonia while in the thermocline less than 50% comes from ammonia. Thus assuming a carbon to nitrogen ratio in phytoplankton of 7 to 1 by atoms, then for station B (Fig. 15) with Z_2 = 64 m equation (23) becomes

$$P_1 = 919K_2 - 5P_2 \qquad (25)$$

where K_2 is in m^2/d and P_1 and P_2 are in mg C/m^2d. This equation suggests that a relatively small amount of thermocline photosynthesis can significantly depress photosynthesis in the mixed layer.

Reaction Rates and Diffusion Coefficients

From the chemical distribution model estimates of the ratio of reaction to diffusion, R/K, can be estimated for the mixed layer, thermocline and oxygen minimum layer. Essentially the model couples the three layers and establishing a reaction rate or diffusion coefficient in any layer establishes the rates and coefficients of all the layers. We have individual estimates of rates and coefficients so a test of the model can be made by comparing predicted and observed rates and coefficients.

The respiration rate in the OML, R_3, is the least complicated term in the model in that it is essentially measured by ETS. Thus, R_3 will be used to establish temporal scales of the layers. For an average respiration in the OML R_3 = -0.01 mg-at/m^3d. This value is obtained from King et al. (1978) and represents an order of magnitude estimate of the rate that is balanced by vertical fluxes into the layer. We assume the vertical flux accounts for about 30% of the total flux of oxygen into the layer and that the rate is probably within a factor of 2. From the oxygen profile and equation (4) $R_3/K_3 = -5 \times 10^{-4}$, for station B, $K_3 = 20\ m^2/d$. From oxygen and nitrate we obtain the ratios of diffusion coefficients between layers and for station B $K_1/K_2 = 6$, $K_2/K_3 = 0.022$ and the vertical diffusion coefficients in m^2/d are $K_1 = 2.7$, $K_2 = 0.45$ and $K_3 = 20$, or in cm^2/s, $K_1 = 0.31$, $K_2 = 0.05$ and $K_3 = 2.31$.

With these coefficients net reaction rates can be inferred from the reaction:diffusion ratios in the layers. For oxygen at station B,$R_1/K_1 = 0.028$ and $R_2/K_2 = 0.014$, so the predicted reaction rates for oxygen are $R_1 = 0.07$ mg-at/m^3d for the mixed layer, and $R_2 = 0.006$ mg-at/m^3d for the thermocline. These rates are the difference between phytoplankton photosynthesis and respiration by bacteria, phytoplankton and nekton.

In the DOMES area the average primary productivity in the mixed layer is 66 ± 52 mg C/m^2d and for the thermocline the value is 73 ± 42 mg C/m^2d, as computed from C-14 uptake given in El-Sayed et al. (1977). These values represent the difference between phytoplankton photosynthesis and respiration, and assuming a carbon consumed to oxygen produced ratio of 1/2, then an average oxygen production rate for either layer is R(phyto) = 0.19 ± 0.13 mg-at/m^3d. It follows that R (phyto) > R_1 or R_2 and because the predicted values of R_1 and R_2 are smaller than R(phyto) by factors of 3 and 32 if the estimates are correct, a close balance may exist between photosynthesis and respiration, especially within the thermocline.

The diffusion parameters and primary productivity can be used with the distribution of nitrate to evaluate the model. For station B the reaction diffusion ratios for nitrate are $R_1/K_1 = -0.0003$ and $R_2/K_2 = -0.013$, so the nitrate uptake rates are $R_1 = -0.0008$ and $R_2 = -0.006$ mg-at/m^3d.

To estimate the nitrate uptake in photosynthesis we will assume a carbon to nitrogen uptake ratio of 7/1 by atoms. The nitrogen uptake rate in either layer is then on the order of R(phyto) = -0.013 ± 0.009 mg-at/m^3d. Using the ammonia nitrate uptake ratios of Goering et al. (1970) the uptake computed from C-14 might be $R_1 = -0.001$ for the mixed layer and $R_2 = -0.007$ mg-at/m^3d for the thermocline, and the chemical model and C-14 predicted rates are within a factor of 2 for both the thermocline and mixed layer.

Equation (25) gives the primary productivity in the mixed layer from K_2 and primary productivity in the thermocline. Using $K_2 = 0.45\ m^2/d$ and $P_2 = 73$ mgC/m^2d, then $P_1 = 48$ mgC/m^2d. The average observed rate is 66 + 52 mgC/m^2d, so the estimate of P_1 is of the right order and thus it appears the mixed layer productivity is significantly depressed by thermocline productivity.

For another comparison between the thermocline diffusion coefficient and a rate we note that using $K_2 = 0.45$, $S(H/2) = 0.4$ mg-at/m^3 and $H = 20$ m, then equation (23) gives a maximum net nitrite production in the thermocline peak of 3.6×10^{-3} mg-at/m^{3}d. An observed net rate determined by Wada and Hattori (1971) for the North Pacific is 6×10^{-3} mg-at/m^{3}d. Further, the rate for ammonia production is of the proper order if $K = O(1)$ m^2/d.

Thus, generally within a factor of 2 the chemical distribution models are in agreement with the vertical distributions of oxygen, nitrate, nitrite, ammonia, and observed rates in the upper layers of the DOMES area.

The diffusion coefficients predicted from the model can be compared to estimates of diffusion established from Tritium and radon distributions. The vertical diffusion coefficient can be estimated from a relationship with the density gradient developed by Sarmiento et al. (1976) using radon measurements near the ocean bottom. The relation is

$$K = (4 \times 10^{-6})/[(g/p)(dp/dz)] \qquad (26)$$

where K is given in cm^2/s with 1 cm^2/s = 8.64 m^2/d. From the density gradient at station B equation (26) gives the approximate diffusion coefficients in m^2/d; $K_1 = 2.16$, $K_2 = 0.07$, $K_3 = 2.33$, or in cm^2/s $K_1 = 0.25$, $K_2 = 0.008$ and $K_3 = 0.27$. These coefficients are about a factor of 10 lower than the coefficients that fit the chemical distributions with the rates.

Rooth and Östlund (1972) estimated a thermocline vertical diffusion coefficient of 1.73 m^2/d (0.2 cm^2/s) using Tritium distributions in the North Atlantic. Their estimate is 3.8 times larger than that determined in the chemical model and estimates of rates with this larger number are within the range of rates observed in the DOMES area. The lower value of K_2 predicted from equation (26) gives unreasonably low rates for the DOMES area.

Peng et al. (1974) estimated a winter mixed layer diffusion coefficient from Radon distributions in the North Pacific. The value, $K_1 = 1728$ m^2/d (200 cm^2/s), is two orders of magnitude larger than K_1 from chemical model or equation (26). This difference could be real and reflect two different mixing environments or the diffusion coefficients estimated from the chemical distributions in the mixed layer are in error. An error is probable, because the estimate of K_1 is obtained from concentrations at the top and bottom of the mixed layer, and these are near the resolution of the analytical methods. Thus the vertical gradients might be less and the vertical diffusion coefficient more. In this manner the mixed layer rates required to balance diffusion would remain unchanged and the conclusions of the chemical model remain essentially unchanged.

Summary

In summary the distributions of nutrients and oxygen in the water column of the DOMES area are strongly controlled by the density structure and currents. The most pronounced features are in the vertical direction and are associated with six layers in the water column which include; the mixed layer, thermocline, oxygen minimum layer, upper deep water, North Pacific deep water, and Pacific bottom water. These layers are defined by segments of the T-S diagram and represent distinct physical regimes.

Within each layer the chemical profiles deviate slightly from linearity with either convex or concave curvature while at the layer interfaces the profiles have pronounced changes in slope. Horizontal gradients are weak with a slight increase in nutrients to the south in the mixed layer and a possible increase to the north in the Pacific bottom water. In the oxygen minimum layer oxygen and nutrients have a tongue like distribution extending west along about 15°N. A temporal-spatial pattern in the depth of the upper layer interfaces is associated with the development of a thermal ridge along about 10°N in the winter, but no clear pattern is evident in the concentrations when oxygen and nutrients are referenced to density surfaces.

In the mixed layer the vertical profiles of nutrients and oxygen can be modelled with a balance of depth-independent reactions and vertical diffusion. In the model and observations the profiles show little curvature at the air-sea interface and an increase of curvature near the bottom of the mixed layer. In the model this is a function of the ratio of reaction to diffusion in the layer.

The vertical profiles in the thermocline have increased curvature with depth and the same model fits the profiles suggesting that upwelling and depth variations in the reaction rates are of secondary importance in controlling the shape of the thermocline profiles.

In the oxygen minimum layer both vertical and horizontal flux are significant in balancing consumption of oxygen in bacterial respiration. The vertical profiles of oxygen and nutrients in the oxygen minimum layer can be fit with a reaction-diffusion model with depth independent coefficients while the horizontal distribution of oxygen can be modelled with horizontal diffusion balancing a respiration rate that decreases westward in an exponential like manner. These models are applied to the center of the oxygen minimum tongue where symmetry conditions suggest that diffusion in the unaccounted for dimension should be minimum. The oxygen modelling suggests that the position of the tongue is controlled by the position of the zonal subsurface flows of the North Equatorial current and counter current while the intensity of the tongue is controlled by diffusion through the strong tropical thermocline and from the subsurface zonal flows and respiration in the tongue.

By coupling the reaction-diffusion models in the individual layers according to flux and concentration requirements, the dominant processes controlling the chemical concentrations at the interfaces between the mixed layer, thermocline and oxygen minimum layer can be inferred. The analysis suggests that the interface concentrations of oxygen are principally controlled by the surface saturation value, the thermocline diffusion rate and respiration in the oxygen minimum layer. The interface concentrations of nitrate and phosphate are primarily controlled by thermocline and mixed layer uptake and the horizontal flux of nutrients near the interface between the oxygen minimum layer and the upper deep water.

The phosphate and nitrate maximums between about 800 and 1000 m are produced by a depth decreasing oxidative nutrient profile and a depth increasing preformed nutrient profile. Horizontal advection and diffusion are important in maintaining the maximum.

The silicate maximum at about 3000 m is produced by a combination of deep water circulation and a high silicate dissolution rate at depth. A high silicate content is acquired as the Pacific bottom water flows north into the high latitudes of the North Pacific and upwells into the level of the North Pacific deep water. This flows south out of the North Pacific directly over the bottom water and produces the maximum.

If the level of primary productivity in the mixed layer is limited by nitrate then nitrate uptake in thermocline photosynthesis can affect mixed layer photosynthesis via the nitrate flux through the thermocline. If thermocline photosynthesis and nitrate uptake are small then the thermocline acts as a nitrate filter and mixed layer photosynthesis is dependent on the thermocline diffusion coefficient and thickness. If thermocline photosynthesis and nitrate uptake are appreciable then the thermocline acts as a nitrate sink and mixed layer productivity is limited in direct proportion to thermocline productivity. Because a majority of productivity in the mixed layer uses nitrate-nitrogen recycled as ammonia, for a given unit of thermocline productivity the mixed layer productivity is decreased by about 5 units. In the DOMES area this amounts to a significant depression of the mixed layer primary productivity.

An ammonia peak near the top of the thermocline is produced by zooplankton grazing. The intensity of the peak is limited by diffusion away from the peak and uptake of ammonia primarily by phytoplankton. A nitrite peak in the thermocline is probably produced by leakage of nitrite from phytoplankton. In the extremely low oxygen waters, nitrite is produced as an intermediate in the reduction of nitrate to molecular nitrogen in denitrification.

Acknowledgments

I wish to thank Drs. L.A. Codispoti and E.T. Baker for their useful comments, and D. Doyle and S. Patterson for their valuable technical assistance. This work was in part supported under contracts from the National Oceanographic and Atmospheric Agency (NOAA-03-06-22-35117) and the Office of Naval Research (N-00014-75-C-0502). Contribution No. 1056 from the Department of Oceanography, University of Washington, Seattle, Washington 98195.

References

Anderson, J.J., (1978) Deep ocean mining and the ecology of the tropical North Pacific, Dept. of Oceanog. Special Rept. No. 83, University of Washington, Seattle, WA.

Anderson, J.J. and Richards, F.A., (1977a) Continuous profiles of chemical properties in the euphotic zone of the DOMES study area in the Equatorial North Pacific, Dept. of Oceanog. Special Rept. No. 78, Univ. of Washington, Seattle.

Anderson, J.J. and Richards, F.A., (1977b) Chemical and biochemical observations from the DOMES study area in the Equatorial North Pacific, Dept. of Oceanog. Special Rept. 79, Univ. of Washington, Seattle.

Armstrong, F.A., Jr., Stears, C.R., and Strickland, J.D.H., (1967) The measurement of upwelling and subsequent biological processes by means of the Technicon AutoAnalyzer and associated equipment, Deep-sea Res. 14, 381-389.

Barkley, R.A., (1968) Oceanographic Atlas of the Pacific Ocean, Univ. of Hawaii Press.

Biggs, D.C., (1977) Respiration and ammonia excretion by open ocean gelatinous zooplankton, Limnol. Oceanog. 22, 108-117.

Burkov, V.A., (1972) General Circulation of the Pacific Ocean Waters, The Academy of Sciences U.S.S.R., Nauka Moskow (in Russian).

Carlucci, A.F., Hartwig, E.O. and Bowes, P.M., (1970) Biological production of nitrite in seawater, Mar. Biol. 7, 161.

Carlucci, A.F. and Strickland, J.D.H., (1968) The isolation, purification and some kinetic studies of marine nitrifying bacteria, J. Exp. Mar. Biol. Ecol. 2, 156-166.

Carpenter, J.H., (1965) The Chesapeake Bay Institute technique for the Winkler dissolved oxygen method, Limnol. Oceanog. 10, 141-143.

Codispoti, L.A., and Richards, F.A., (1976) An analysis of the horizontal regime of denitrification in the eastern tropical North Pacific, Limnol. Oceanog. 21, 379-388.

Cohen, Y., and Gordon, L.I., (1978) Nitrous oxide in the oxygen minimum of the eastern tropical North Pacific: evidences for its consumption during denitrification and possible mechanisms for its production, Deep-sea Res. 25, 509-524.

Craig, H., (1971) The deep metabolism: oxygen consumption in abyssal ocean water, J. Geophys. Res. 76, 5078-5086.

Devol, A.H., (1978) Bacterial oxygen uptake kinetics as related to biological processes in oxygen deficient zones of oceans, Deep-sea Res. 25, 137-146.

Devol, A.H., Packard, T.T. and Holm-Hansen, O., (1976) Respiratory electron transport activity and adenosine triphosphate in the oxygen minimum of the eastern tropical North Pacific, Deep-sea Res. 23, 963-973.

Dugdale, R.C. and Goering, J.J., (1967) Uptake of new and regenerated forms of nitrogen in primary productivity, Limnol. Oceanog. 12, 196-206.

El-Sayed, S.Z., Taguchi, S., Franceschini, G.A., Fryxell, G.A., and Gates, C.E., (1977) Phytoplankton and primary productivity studies (in connection with the DOMES study), Dept. of Oceanog. Texas A and M Univ., College Station, Texas.

Eppley, R.W. and Coatsworth, J.L., (1968) Uptake of nitrate and nitrite by Ditylum brightwellii--kinetics and mechanisms, J. Phycol. 4, 151-156.

Eppley, R.W., Renger, E.H., Venrick, E.L., and Mullin, M.M., (1973) A study of plankton dynamics and nutrient cycling in the central gyre of the North Pacific Ocean., Limnol. Oceanog. 18, 534-551.

Eppley, R.W., Sharp, J.H., Renger, E.H., Perry, M.J., and Harrison, W.S., (1977) Nitrogen assimilation by phytoplankton and other microorganisms in the surface waters of the Central North Pacific Ocean., Mar. Biol. 39, 111-120.

Fiadeiro, M.E., (1975) Numerical modeling of tracer distributions in the deep Pacific Ocean, PhD Thesis, Oceanography, Univ. Calif., San Diego.

Goering, J.J., Wallen, D.D., and Nauman, R.M., (1970) Nitrogen uptake by phytoplankton in the discontinuous layer of the eastern tropical Pacific Ocean, Limnol. Oceanog. 15, 789-796.

Gordon, D.C.,Jr., (1971) Distribution of particulate organic carbon and nitrogen at an oceanic station in the central Pacific, Deep-sea Res. 18, 1127-1134.

Graham, J.W., and Moberg, E.G., (1944) Chemical results of the last cruise of the Carnegie, Chemistry I, Carnegie Inst. Washington Publ.

Hager, S.W., Gordon. L.I., and Park, P.K., (1968, Oct.) (unpub. manuscript), A practical manual for use of Technicon AutoAnalyzer in seawater nutrient analysis, A final report to B.C.F., Contract 14-17-0001-1759, Ref. 68-33.

Hedgpeth, J.W., (ed.), (1957) Treatise on Marine Ecology and Paleoecology, v. 1, Memoir 67, Geological Soc. of New York.

Hirota, J., (1977) Hawaii Instit. of Marine Biol. and Dept. of Oceanog., Univ. Hawaii, Manoa.

Holm-Hansen, O., (1969) Determination of microbial biomass in ocean profiles, Limnol. Oceanog. 14, 740-747.

Holm-Hansen, O., (1970) ATP levels in algal cells as influenced

by environmental conditions, Plant Cell Physiol. 11, 689-700.
Holm-Hansen, O. and Booth, C.R., (1966) The measurements of adenosine triphosphate in the ocean and its ecological significance, Limnol. Oceanog. 11, 510-519.
Holm-Hansen, O., Strickland, J.D.H., and Williams, P.M., (1966) A detailed analysis of biologically important substances in a profile off southern California, Limnol. Oceanog. 11, 548-561.
Kiefer, D.A., Olson, R.J., and Holm-Hansen, O., (1976) Another look at the nitrite and chlorophyll maxima in the central North Pacific, Deep-sea Res. 23, 1199.
King, F.D., Devol, A.H., and Packard, T.T., (1978) Plankton metabolic activity in the eastern tropical North Pacific, Deep-sea Res. 25, 689-704.
Longhurst, A.E., (1976) Interactions between zooplankton and phytoplankton profiles in the eastern tropical Pacific Ocean., Deep-sea Res. 23, 729-754.
Love, C.M., (ed.) (1972) EASTROPAC Atlas 1, CNC 330, Natl. Mar. Fish. Serv., Washington, D.C.
Mantyla, A.W., (1975) On the potential temperature in the abyssal Pacific Ocean, J. Mar. Res. 33, 341-354.
Menzel, D.W., (1975) Primary productivity, dissolved and particulate organic matter, and the sites of oxidation of organic matter, In The Sea, v. 5, (E.D. Goldberg, ed.), Wiley, New York, 659-678.
Menzel, D.W., and Ryther, J.H., (1968) Organic carbon and the oxygen minimum in the South Atlantic Ocean, Deep-sea Res. 15, 327-337.
Menzel, D.W., and Ryther, J.H., (1970) Distribution and cycling of organic matter in the oceans, In Organic Matter in Natural Waters (D.W. Hood, ed.), Inst. Mar. Sci. Publ. No. 1, Univ. Alaska, 31-54.
Menzel, D.W., and Vaccaro, R.F., (1964) The measurement of dissolved organic and particulate carbon in seawater, Limnol. Oceanogr. 9, 138-142.
Miyazaki, T., Wada, E., and Hattori, A., (1973) Capacities of shallow waters of Sagami Bay for oxidation and reduction of inorganic nitrogen, Deep-sea Res. 20, 571-577.
Moberg, E.G., (1930) Distribution of oxygen in the Pacific, In Contr. to Marine Biol. IV, Stanford Univ. Press, Stanford Univ., Calif., 69-78.
Munk, W.H., (1966) Abyssal recipes, Deep-sea Res. 13, 707-730.
Owen, R.W., and Zeitzschel, B., (1970) Phytoplankton production: seasonal change in the oceanic eastern tropical Pacific, Mar. Biol. 7, 32-36.
Packard, T.T., (1971) The measurement of respiratory electron transport activity in marine phytoplankton, J. Mar. Res. 29, 235-244.
Peng, T.H., Takahashi, T., and Broecker, W.S., (1974) Surface radon measurements in the North Pacific Ocean Station Papa, J. Geophys. Res. 79, 1772-1730.

Redfield, A.C., Ketchum, B.H., and Richards, F.A., (1963) The influence of organisms on the composition of seawater, In The Sea , v. 2, (M.N. Hill, ed.) 26-77.
Reid, J.L., (1965) Intermediate Waters of the Pacific Ocean, John Hopkins Press, Baltimore.
Reid, J.L. and Mantyla, A.W., (1978) On the mid-depth circulation of the North Pacific, J. Phys. Oceanog. 8, 946-951.
Robinson, M.K., (1976) Atlas of North Pacific Ocean monthly mean temperatures and mean salinities of the surface layer, Naval Oceanographic Office Reference Publ. 2, Washington, D.C.
Roden, G.I., (1974) Thermocline structure, fronts and sea-air energy exchange of the trade wind region east of Hawaii, J. Phys. Oceanogr. 4, 168-182.
Rooth, C.G., and Ostlund, H.G., (1972) Penetration of tritium into the Atlantic thermocline, Deep-sea Res.19, 481-492.
Sarmiento, J.L., Feely, H.W., Moore, W.S., Bainbridge, A.E. and Broecker, W.S., (1976) The relationship between vertical eddy diffusion and buoyancy gradient in the deep sea, Earth Planet. Sci. Lett. 32, 357-370.
Sharp, J.H., (1974) Total organic carbon in seawater - comparison of measurements using persulfate oxidation and high temperature combustion, Mar. Chem. 1, 211-229.
Slawyk, F., and MacIsaac, J.J., (1972) Comparison of two automated ammonium methods in a region of coastal upwelling, Deep-sea Res. 19, 521-524.
Thauer, R.K., Jungerman, K. and Decker, K., (1977) Energy conservation in chemostrophic anaerobic bacteria, Bacterial Rev. 41, 100-180.
Tsuchiya, M., (1968) Upper waters of the intertropical Pacific Ocean, Johns Hopkins Oceanogr. Stud. 4, 50 pp.
Tsuchiya, M., (1975) Subsurface countercurrents in the eastern equatorial Pacific Ocean, J. Mar. Res. supplement 33, 145-175.
Tsunogai, S., Matsumoto, E., Kido, K., Nozoki, V., and Hattori, A., (1973) Two discontinuities in the deep water of the western North Pacific Ocean, Deep-sea Res. 20, 527-536.
Wada, E., and Hattori, A., (1971) Nitrite metabolism in the euphotic layer of the central North Pacific Ocean, Limnol. Oceanogr. 16, 772-776.
Wada, E., and Hattori, A., (1972) Nitrite distribution and nitrate reduction in deep-sea waters, Deep-sea Res. 19, 123-132.
Wangersky, P.J., (1976) Particulate organic carbon in the Atlantic and Pacific Oceans, Deep-sea Res. 23, 457-466.
Wyrtki, K., (1962) The oxygen minima in relation to ocean circulation, Deep-sea Res. 9, 11-23.
Wyrtki, K. and Kendall, R., (1967) Transports of the Pacific equatorial countercurrents, J. Geophys. Res. 72, 2073-2076.

CHEMICAL COMPOSITION, SIZE DISTRIBUTION AND PARTICLE MORPHOLOGY OF SUSPENDED PARTICULATE MATTER AT DOMES SITES A, B, AND C: RELATIONSHIPS WITH LOCAL SEDIMENT COMPOSITION[1,2]

Edward T. Baker and Richard A. Feely

Pacific Marine Environmental Laboratory
National Oceanic and Atmospheric Administration
Seattle, Washington 98115

Kozo Takahashi[3]

Department of Oceanography
University of Washington
Seattle, Washington 98195

Abstract

Detailed measurements of the concentration, composition, size, and morphology of suspended particulate matter (SPM) in the eastern equatorial Pacific Ocean were made during separate cruises in 1975 and 1976. SPM concentrations in the surface waters (0-300 m) were highly variable ($\sim 30 \pm 18$ μg/l) and primarily dependent on the concentration of biogenic material which accounted for > 90% of the total SPM. Mean Si/Al, K/Al, Ti/Al ratios were significantly higher than the corresponding values for the underlying sediments, indicating probable concentration of these elements in biogenic matter and other nonaluminosilicate phases. Mn/Al ratios, however, were close to or even less than the sediment values.

---

[1]Contribution No. 391 from the NOAA/ERL Pacific Marine Environmental Laboratory.
[2]Contribution No. 1052 from the Department of Oceanography, University of Washington.
[3]Now at: Department of Geology and Geophysics, Woods Hole Oceanographic Institution, Woods Hole, Mass. 02543.

Particle concentrations, mean size, and Si/Al, K/Al, Ti/Al, and Fe/Al ratios all decreased from the surface waters to the top of the bottom nepheloid layer (BNL), indicating progressive dissolution and a concomitant remineralization of associated elements. However, nonaluminosilicate phases still accounted for more than 40-70% of the Si, K, Ti, and Fe in the deep water zone. After deposition, most of this nonaluminosilicate phase is either transformed into authigenic smectites or dissolved and released back into the water column.

Within the BNL (approximately the bottom 400 m of the water column), concentration increases of Al, Si, K, Ti, Mn, and Fe averaged 201%, 74%, 46%, 78%, 200%, and 119%, respectively, greater than the overlying waters, whereas total SPM increased only 20-40%. Chemical mass-balance calculations suggest that this BNL concentration increase is due to resuspension of aluminosilicates and hydrogenous Mn from the ≤ 1 μm fraction of the local bottom sediments.

Introduction

Present-day sedimentation processes on the ocean floor cannot be fully understood without a thorough knowledge of the dynamics and character of the suspended particulate matter (SPM) in the overlying water column. Compositional and textural factors of SPM can be sensitive indicators of benthic processes and thus constitute sedimentological tools complementary to more established techniques such as current velocity measurement, bottom photography, acoustic stratigraphy, and analyses of bottom sediment texture, fabric, and composition. In addition, studies of SPM unlike companion sea floor studies, have the advantage of being unequivocally related to modern processes.

In the course of making an extensive baseline survey of SPM in the eastern equatorial Pacific manganese nodule province for NOAA's Deep Ocean Mining Environmental Study, we have had the opportunity to make detailed measurements of the concentration, composition, size, and morphology of particulate matter in the waters overlying this region. Our objective in these studies has been to relate these properties of the SPM to the dynamics of the ocean, particularly in the relatively active regions adjacent to the upper and lower boundaries of the water column. Our ultimate goal is to quantify the vertical transport of SPM in the ocean, from input and generation in the surface layers, along its downward course through various transport stages, to final removal from the system by dissolution or deposition.

Study Area

This paper covers samples and data collected during DOMES cruises RP-8-OC-75 (August 24-October 16, 1975) and RP-8-OC-76

(February 11-April 17, 1976) aboard the NOAA ship OCEANOGRAPHER. Stations were occupied every 1° of latitude (RP-8-OC-75) or 1.5° of latitude (RP-8-OC-76) along three north-south transects extending for 6° of latitude and centered at stations A, B, and C located in three representative manganese nodule mining areas (Table 1). Transects A, B, and C are all underlain by siliceous clay, although transects A and B have a higher proportion of biogenic silica than transect C (Piper _et al._, this vol.; Bischoff _et al._, this vol.). Water depths increase from 4500 m at C, to 4900 m at B, to 5000 at A. Data were collected from the upper 1000 m at all stations during both cruises. In addition, the entire water column (to within 10 m of the sea floor) was sampled at stations A, B, and C during cruise RP-8-OC-76.

TABLE 1

Station Locations

Sta.	Location	Sta.	Location	Sta.	Location
Cruise RP-8-OC-75, Aug. 24-Oct. 16, 1975					
A+3	11°27'N, 150°47'W	B+3	14°42'N, 138°28'W	C+3	18°02'N, 126°00'W
A+2	10°28'N, 150°48'W	B+2	13°42'N, 138°27'W	C+2	17°00'N, 126°20'W
A+1	9°27'N, 150°47'W	B+1	12°44'N, 138°24'W	C+1	15°58'N, 126°02'W
A	8°27'N, 150°47'W	B	11°47'N, 138°26'W	C	15°02'N, 126°02'W
A-1	7°22'N, 150°38'W	B-1	10°43'N, 138°24'W	C-1	14°10'N, 125°59'W
A-2	6°26'N, 150°41'W	B-2	9°40'N, 138°25'W	C-2	12°58'N, 126°21'W
A-3	5°29'N, 150°47'W	B-3	8°46'N, 138°24'W	C-3	12°01'N, 125°55'W
Cruise RP-8-OC-76, Feb. 11-Apr. 17, 1976					
A+3	11°26'N, 150°47'W	B+3	14°45'N, 138°24'W	C+3	17°59'N, 125°54'W
A+1.5	9°48'N, 150°51'W	B+1.5	13°16'N, 138°24'W	C+1.5	16°41'N, 125°57'W
A	8°22'N, 150°43'W	B	11°42'N, 138°22'W	C	14°59'N, 126°21'W
A-1.5	6°36'N, 150°35'W	B-1.5	10°12'N, 138°23'W	C-1.5	14°09'N, 126°14'W
A-3	5°30'N, 150°24'W	B-3	8°46'N, 138°21'W	C-3	12°21'N, 126°06'W

Methods

The light scattering measurements recorded during this study were made with an integrating nephelometer designed after a principle introduced by Beutell and Brewer (1949). The nephelometer consists of a flashing light source, a scattered light detector system, and the necessary battery power mounted in a self-contained, easily portable, deep-sea housing. As the instrument is lowered through the water column, it continually measures the light scattered from the SPM. The nephelometer information is fed into the mixer of a CTD unit and recorded in real time on deck. A complete description of a slightly earlier version of this instrument is given by Sternberg *et al*. (1974).

The nephelometer output was calibrated in terms of the mass concentration of particles. The most appropriate and simplest method of calibration appeared to be an empirical technique of comparing the relative light scattering values to the total mass concentration of simultaneously collected SPM samples. Using the relationship described by the data in Fig. 1, the light scattering transects discussed below can be interpreted in terms of the absolute concentration of SPM. Absolute quantification of the light scattering data is a complicated and necessarily inexact procedure, however, because the total scattering value of a parcel of water is a complex integral of all the properties of all the particles, of which particle mass concentration is only one, albeit major, factor (Jerlov, 1976).

Discrete sample collection occurred concurrently with CTD-nephelometer rosette casts so that the SPM distributions could be related to the hydrography. Water samples were collected in 30-liter PVC sampling bottles and filtered through preweighed 0.4 μm pore size Nuclepore filters. The filters were washed with three 10-ml aliquots of deionized, buffered (to pH 8), filtered water, dried in a desiccator, stored in plastic petri dishes, and returned to the laboratory. At the laboratory, the filters were reweighed on a seven-place electrobalance.

The elemental chemistry (Al, Si, K, Ti, Mn, Fe) of the particulate matter was determined by x-ray secondary emission (fluorescence) spectrometry utilizing the thin-film technique (Baker and Piper, 1976). The inherent broad band of radiation from a Ag x-ray tube was used to obtain a series of characteristic emission lines from a single element secondary target which then more efficiently excited the thin-film sample. A Se secondary target was used to analyze the samples for both major and trace elements. Standards were prepared by passing suspensions of finely ground USGS standard rocks (W-1, G-2, GSP-1, AGV-1, BCR-1, PCC-1) and NBS trace element standards through a 37 μm mesh polyethylene screen followed by collection of the < 37 μm fraction on filters identical to those used for sample acquisition.

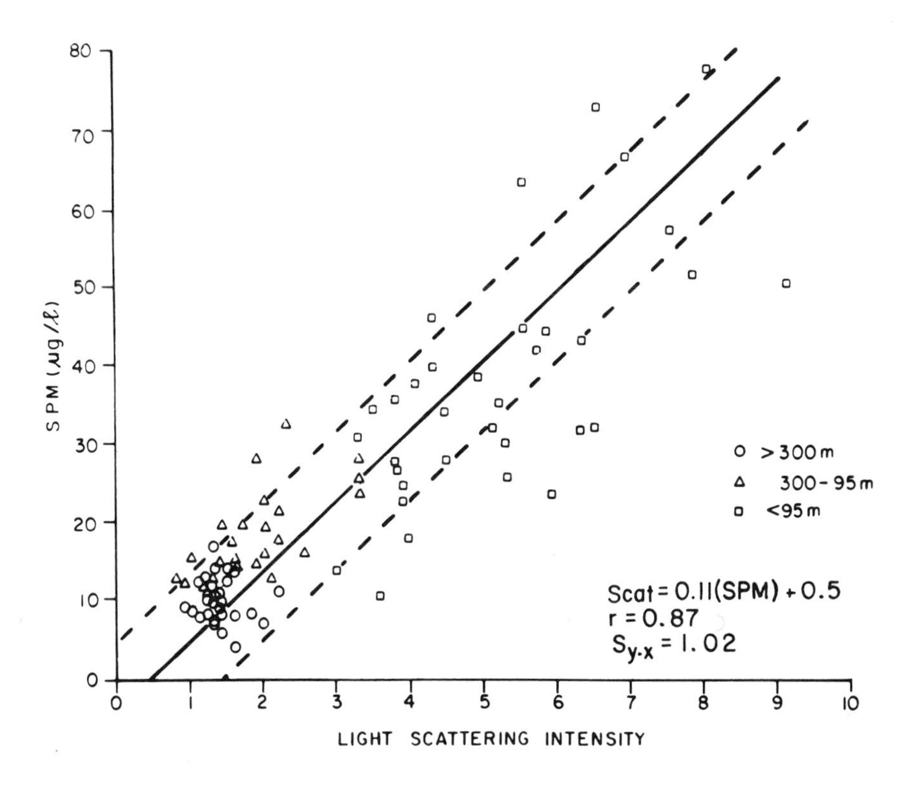

Fig. 1. Plot of nephelometer light scattering intensity (relative units) vs. the concentration of suspended particulate matter for the DOMES samples. Dashed lines show envelope of ± 1 standard error of estimate. Symbols show the depth intervals of different calibration samples. Note that the surface samples, high in particulate organic matter, show a relatively poorer correlation between particle mass and light scattering intensity than the deeper, more refractory samples.

Particle size distributions of selected sample filters were calculated using photomicrographs (1000x magnification) taken on a scanning electron microscope (SEM). Particles in each field were then sized and morphologically classified (coccoliths, diatoms, aggregates, and unidentifiable fragments) on an operator-controlled particle size analyzer. This device measured the area of each particle as it appeared on the photomicrograph and expressed it in terms of an equivalent circle radius (r). At 1000x magnification, particles of at least 0.5 µm (i.e., 0.5 mm on the photomicrographs) were routinely discernible, provided their contrast with the filter background was sufficiently distinct. The particles counted from each of the 30 samples analyzed from stations A, B, and C represent the total population (for $r \geq$ 0.65 μm) from a volume of sea water ranging from 0.60 to 2.34 ml ($\overline{X}$ = 1.38 ± 0.54 ml).

Results and Interpretations

Suspended Particulate Matter Concentrations

Surface waters (0-300 m). Figure 2 shows the light scattering data for the upper 300 m of transects A, B, and C for cruise RP-8-OC-75 (herein designated as summer). Superimposed upon the scattering contours are the 26.0 σ_t contour and the bottom of the surface mixed layer (Oztergut *et al.*, in press) as representative of the hydrography, and the location of the (fluorometric) chlorophyll maximum as measured by El-Sayed and Taguchi (this vol.). Figure 3 illustrates the same observations from cruise RP-8-OC-76 (herein designated as winter).

In the surface waters of the open ocean, far removed from significant terrestrial input sources, the concentration of particulate matter is a function of the standing crop of phytoplankton. Vertical and horizontal gradients, on the other hand, are controlled primarily by the physical processes of advection, diffusion, and gravitational settling (e.g., Ichiye *et al.*, 1972; Lal and Lerman, 1973). Observations of the SPM throughout the DOMES region illustrate these distribution functions, since temporal and spatial variations in the regional pattern are apparently linked to seasonal and geographical changes in the local hydrography.

Throughout the study area, the depth of the maximum light scattering value agreed closely with the depth of the (fluorometric) chlorophyll maximum (Figs. 2 and 3). The light scattering maximum occurred below the surface mixed layer, generally at the point within the thermocline of the stability maximum (where vertical diffusivity is minimal) (Fig. 4). The highest light scattering gradients were consistently found within the thermocline.

Measured values of the SPM concentration closely followed the light scattering pattern. The regional average concentration within the surface mixed layer was 47 ± 20 μg/l. Within the particle maximum just below the mixed layer, values increased to 50-70 μg/l, with occasional values up to 110 μg/l. Within the thermocline concentrations rapidly decreased to 10-20 μg/l at 200 m. These concentration trends are similar to those measured during the Atlantic GEOSECS program (Brewer *et al.*, 1976) and by Betzer *et al.* (1974) in a transect across the Atlantic at 11°N.

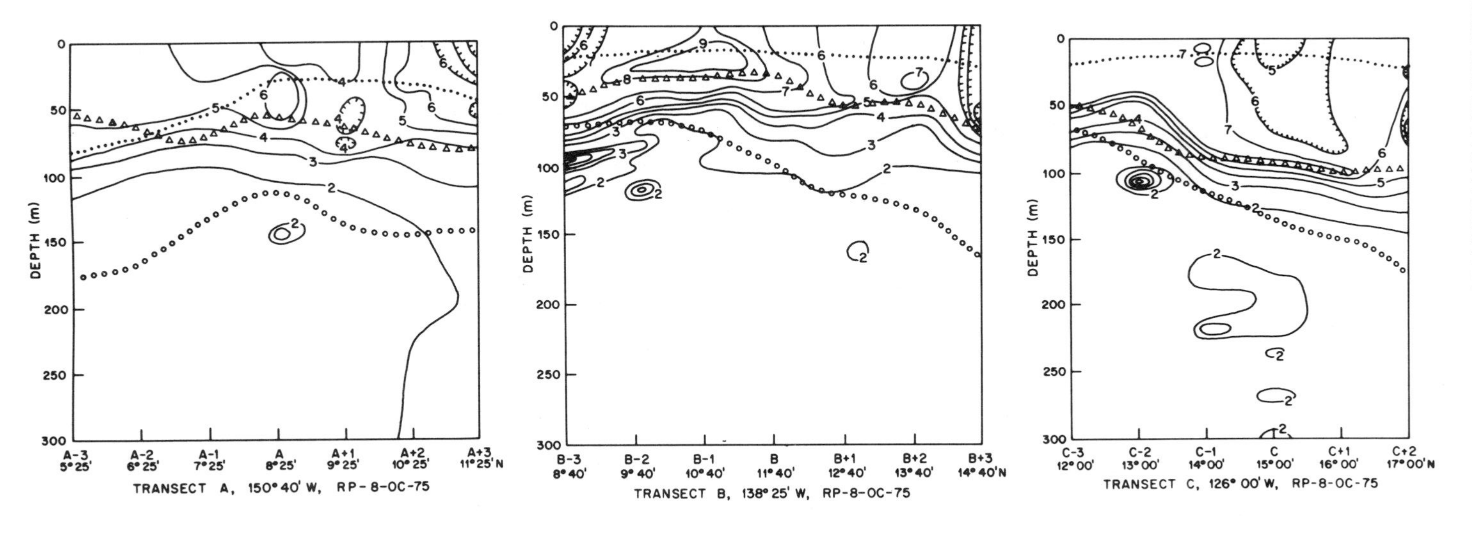

Fig. 2. Cross sections of light scattering intensities (relative units) along transects A, B, and C during the summer cruise (RP-8-OC-75). Dots (●) indicate the bottom of the surface mixed layer (Oztergut _et al._, in press), open circles (○) follow the 26.0 σ_t contour (Oztergut _et al._, in press), and triangles (△) follow the (fluorometric) chlorophyll maximum (El-Sayed and Taguchi, this volume).

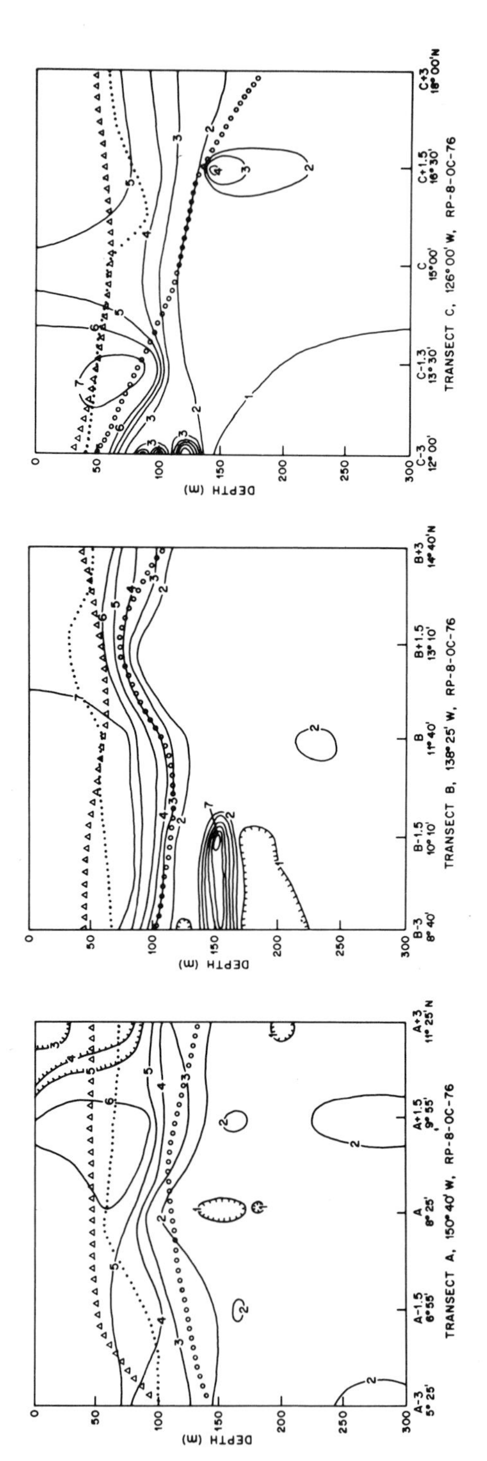

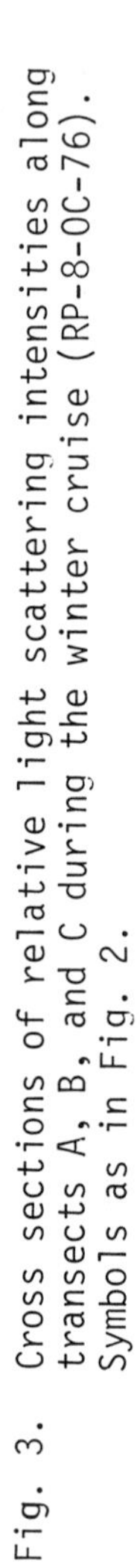
Fig. 3. Cross sections of relative light scattering intensities along transects A, B, and C during the winter cruise (RP-8-OC-76). Symbols as in Fig. 2.

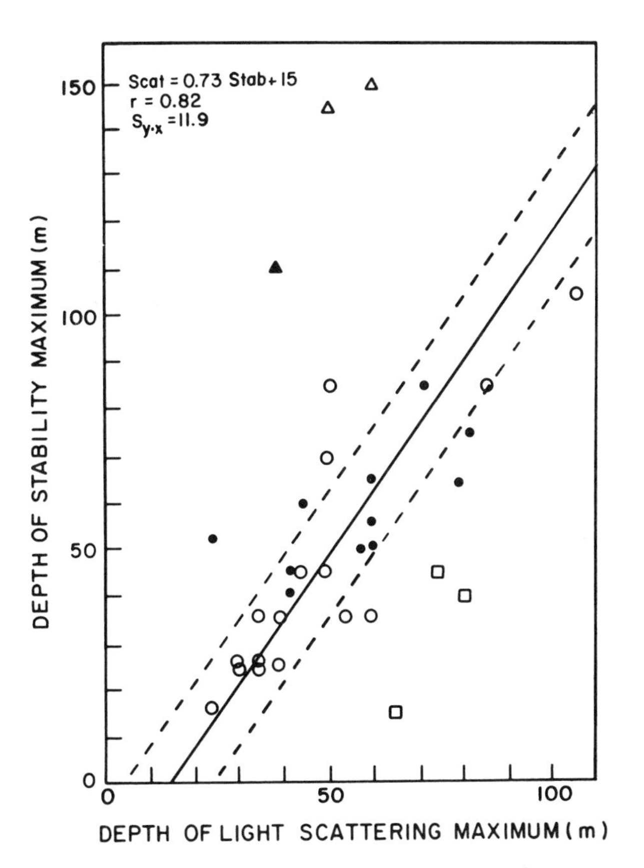

Fig. 4. Relationship between stability $\left(\sim 10^{-3} \frac{d\sigma_t}{dz}\right)$ and light scattering maximums for summer (open symbols) and winter (closed symbols) cruises. Triangles denote stations where the stability maximum was below the euphotic zone; squares denote stations with more than one prominent scattering maximum. Most stations (circles) show a close correlation between stability and scattering since the scattering particles (mainly biogenic) tend to concentrate at the top of the pycnocline. The regression line shown is for the circle stations only.

The actual depth and intensity of these scattering features varied geographically and seaonally. The density field (as described by the 26.0 σ_t contour and depth of the mixed layer) during the summer cruise (Fig. 2) suggests an upwelling in each transect, centered at ~8°25'N in transect A, ~10°N in B, and ~13°N in C. This feature was probably related to the divergence between the westward flowing North Equatorial Current and the easterly Equatorial Counter Current (Sverdrup et al., 1942).

Two distinctive features of the SPM distribution are associated with the inferred upwellings:

(1) High light scattering intensities with steep horizontal gradients (particularly along transect B). The elevated SPM concentrations may be the result of both increased particle production due to increased (i.e., upwelled) nutrient levels at these locations (Anderson, this vol.) and to longer retention of particles in the upper water layers due to an increase in upward water velocity. Comparison of the light scattering data with transects of chlorophyll *a* and primary productivity (El-Sayed and Taguchi, this vol.) suggests that the inferred upwelling regions, although not necessarily correlated with high chlorophyll *a* concentration values, do show a close association with high levels of primary productivity. Yoshida et al. (1953) constructed a mathematical model of upwelling in divergent zones of the equatorial current systems and deduced upward velocities of 10^{-3}-10^{-4} cm/sec, similar to the Stokes settling velocities of particles in the 2-6 μm diameter size range.

(2) Subsurface (120 m-150 m) light scattering maxima. These maxima were invariably found below the inferred upwelling region during the summer cruise, and generally corresponded to high levels of phaeo-pigments observed by El-Sayed and Taguchi (this vol.). The significance of these maxima is unclear, since the phaeo-pigment maxima are not correlated with any other hydrographic or chemical property (El-Sayed and Taguchi, this vol.).

Transect C was unique during the summer cruise in that significant scattering maxima were present below 150 m in the center of the transect (Fig. 2). This region also showed the most intense development of the O_2 minimum zone encountered in any of the transects (Anderson, this vol.).

The basic light scattering features along each transect during the winter (Fig. 3) were largely similar to those observed in the summer. North-south gradients in the density field, except for transect C, were subdued and the mixed layer was 20-30 m deeper along each transect. Transect A maintained the discernible uplift in the scattering contours centered at 8°25'N. The shoaling of the light scattering contours observed at ~10°N along transect B during the summer migrated north to ~13° during the winter cruise, although the high subsurface maxima found at the south end of the transect remained there during the winter observations. Transect C continued to show a shoaling of the light scattering contours and intense subsurface maxima at its southern end, but the complex scattering structure within the O_2 minimum zone (150-300 m) at the center of the summer transect was not observed during winter. This change may be related to a slight increase in the dissolved O_2 values during winter (Anderson, this vol.).

Waters below the surface zone (> 300 m). Compared to the surface layer, the distribution of SPM in the water column below 300 m was very uniform. SPM concentrations averaged 9.0 ± 2.3 μg/l between 450 m and ~400 m above the sea floor, where a slight increase in particulate concentrations (to a regional average of 12 ± 6 μg/l) occurs at all three stations (Fig. 5). The concentration increase within this bottom nepheloid layer (BNL) was not uniform among stations A, B, and C. Station B showed the most pronounced increase, from 9.3 ± 2.0 μg/l above to 14.1 ± 4.5 μg/l within the BNL (significant at $p < .01$ level), whereas at C the increase was only from 8.7 ± 2.2 μg/l to 11.9 ± 8.0 μg/l (significant at $p < 0.20$ level). Station A showed almost no variation: 9.0 ± 2.5 μg/l above to 10.5 ± 4.8 μg/l below. The nephelometer profiles indicated a correspondingly slight increase in light scattering intensity at each station in the bottom 300-500 m. These indications of a weak BNL agree with earlier measurements taken in the east central Pacific by Ewing and Connary (1970) and with measurements previously recorded at station C by Amos *et al.* (1976).

Elemental Composition of the Particulate Matter

There is very little published data on the elemental composition of suspended particulate matter in open-ocean waters of the North Pacific Ocean. A large body of chemical data from the North and South Pacific exists as a result of the GEOSECS (Geochemical Ocean Section Study) project. However, most of these data have not yet been published. Therefore, the following summarizes available published literature on the chemical composition of suspended matter in the oceans.

Little information is presently available on the concentration of particulate aluminum (*P*Al) in open waters of the North Pacific Ocean. Sackett and Arrhenius (1962) found *P*Al concentrations which ranged from 500-120,000 ng Al/l on two transects normal to the Southern California coast and concluded that the aluminum in the SPM consisted of aluminosilicate material derived from coastal runoff. Toyota and Okabe (1967) reported that concentrations of *P*Al in the western North Pacific were high and variable in surface waters (2,000-6,000 ng Al/l between 0 and 100 m) and decreased to low and constant values averaging about 1,300 ng Al/l at depths below 150 m. Chester *et al.* (1976) measured *P*Al concentrations in open-ocean and coastal surface waters from the North and South

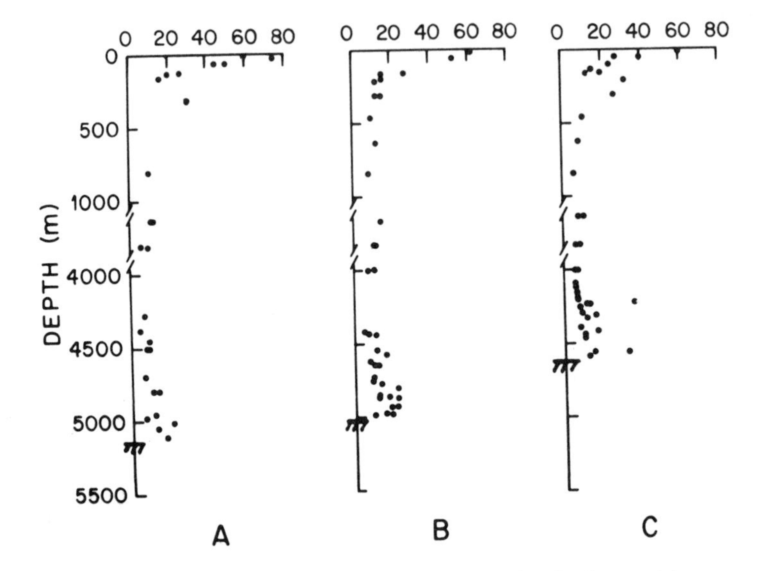

Fig. 5. Concentration of suspended particulate matter throughout the water column at stations A, B, and C during winter. Note the change in the depth scale between 1000 and 4000 m.

Atlantic, Indian Ocean, and China Sea. Concentrations of *P*Al ranged from < 150 ng Al/l in the South Atlantic to > 50,000 ng Al/l in the China Sea. The authors concluded that these widely varying *P*Al concentrations were directly related to the supply of aluminosilicate material, primarily from aeolian and riverine sources.

Available data for particulate silicon (*P*Si) in the North Pacific Ocean are even more scarce. Toyota and Okabe (1967) reported vertical distributions of *P*Si ranging from 3,000-37,000 ng Si/l in samples from the western North Pacific. Kido and Nishimura (1975) determined *P*Si concentrations in samples collected from the Oyashio and Kuroshio areas of the western North Pacific and found concentrations ranging from 3,000-50,000 ng Si/l in surface waters and 3,000-7,000 ng Si/l at depths below 3,000 m. The authors stated that the high concentrations of *P*Si in the surface waters were due to a relatively high production of organic matter in this region.

The concentrations of particulate iron (*P*Fe) and manganese (*P*Mn) are also poorly known. Betzer and Pilson (1970) reported values averaging about 180 ng Fe/l in shallow open-ocean water and 255 ng Fe/l in deep water in samples from the western Atlantic and Caribbean Sea. Concentrations of *P*Fe in large volume surface samples from the western North and South Atlantic ranged from 58-581 ng Fe/l with a mean of 230 ng Fe/l, whereas *P*Mn concentrations ranged from 0.6-23 ng Mn/l with a mean of 7 ng Mn/l (Krishnaswami and Sarin, 1976). Krishnaswami and Sarin estimated that about 40% of the *P*Fe in the surface waters was associated with aluminosilicates, 30% of the *P*Fe and 50% of the *P*Mn was associated with organic matter, and the remaining 30% of the *P*Fe was associated with other undetermined particulate phases, possibly including goethite and hematite. Along a transect from Rhode Island to Bermuda, concentrations ranged from 52-350 ng/l for *P*Fe and 1.5-29 ng/l for *P*Mn in the surface waters (Wallace et al., 1977). The authors stated that from an analysis of Fe/Al and Mn/Al ratios nearly all of the iron and half the manganese was associated with aluminosilicate material from aeolian sources, and that associations of *P*Fe and *P*Mn with organic matter accounted for most of the remaining iron and manganese in the suspended matter.

The data on the elemental composition of the suspended matter from the three transects discussed in this paper are given in Appendix I, summarized in Table 2, and graphically represented as depth profiles in Fig. 6. For convenience, the data in Table 2 have been grouped into four depth ranges: 0-300 m (surface), 450-1000 m (midwater), 2,000 m -< BNL (deep water), and the BNL, which encompasses approximately the bottom 400 m of the water column. Uncertainties are expressed as the standard deviation (± 1 S.D.) of the mean. For some elements, especially aluminum in near-surface samples, these uncertainties are large because filter loadings are low and the minimum detection limit of the procedure is approached (Table 3).

The *P*Al and *P*Si distribution patterns showed distinct differences between transects which are related to regional variations of the production of biogenic matter and to regional differences of the atmospheric input of terrestrial aluminosilicates. In the surface waters of the DOMES area, *P*Al and *P*Si concentrations were high and variable, with concentrations ranging from 50-1,020 ng/l for *P*Al and 420-7,590 ng/l for *P*Si. Concentrations of *P*Al and *P*Si were somewhat higher along transect C than either transect B or transect A (Table 2, Fig. 6).

In contrast to the surface samples, mid- and deep-water concentrations of *P*Al and *P*Si were low and constant at about 200

TABLE 2

Mean, standard deviation and sample size of elemental concentrations of particulate samples from selected depth ranges along DOMES transects A, B, and C

Sample Location and Geochemical Species	Mean Concentration 0-300 m (ng/l)	Mean Concentration 450-1000 m (ng/l)	Mean Concentration 2,000 m-Top of BNL (ng/l)	Mean Concentration BNL (ng/l)
Transect A				
TSM	33,230 ± 18,220 (85)	10,960 ± 2,700 (31)	8,450 ± 1,960 (6)	10,470 ± 7,930 (12)
POC[1]	15,950 ± 13,160 (126)	7,150 ± 3,840 (52)	7,410 ± 1,540 (15)	8,450 ± 3,910 (12)
PAl	137 ± 93 (32)	101 ± 35 (9)	81 (2)	252 ± 111 (4)
PSi	2,231 ± 1,499 (34)	1,309 ± 592 (9)	1,002 (2)	1,325 ± 538 (4)
PK	446 ± 233 (34)	205 ± 46 (9)	156 (2)	123 ± 51 (4)
PTi	38 ± 22 (34)	19 ± 6 (9)	18 (2)	33 ± 8 (4)
PMn	8 ± 4 (34)	4 ± 1 (9)	5 (2)	10 ± 5 (4)
PFe	495 ± 283 (34)	234 ± 86 (9)	135 (1)	356 ± 102 (4)
Transect B				
TSM	32,920 ± 19,790 (79)	11,980 ± 4,330 (33)	9,170 ± 2,810 (6)	14,060 ± 4,460 (22)
POC[1]	28,490 ± 38,070 (43)	8,460 ± 2,020 (12)	6,550 ± 3,870 (15)	7,370 ± 4,690 (11)
PAl	137 ± 44 (31)	85 ± 22 (9)	203 ± 133 (3)	575 ± 85 (9)
PSi	2,664 ± 337 (31)	1,222 ± 297 (9)	1,334 ± 630 (3)	2,504 ± 484 (9)
PK	397 ± 219 (31)	151 ± 32 (9)	147 ± 54 (3)	257 ± 45 (9)
PTi	36 ± 18 (31)	20 ± 9 (9)	28 ± 20 (3)	42 ± 5 (9)
PMn	8 ± 4 (31)	3 ± 1 (9)	3 (2)	16 ± 12 (7)
PFe	389 ± 189 (31)	175 ± 52 (9)	226 (2)	492 ± 95 (8)
Transect C				
TSM	29,610 ± 17,460 (58)	11,480 ± 3,120 (38)	7,190 ± 1,290 (6)	11,890 ± 8,040 (21)
POC[1]	15,550 ± 7,530 (142)	7,170 ± 3,960 (50)	6,560 ± 1,990 (5)	7,950 (2)
PAl	268 ± 240 (14)	95 ± 48 (10)	159 ± 76 (3)	426 ± 165 (10)
PSi	3,053 1,457 (15)	1,138 ± 526 (11)	803 ± 248 (3)	1,624 ± 602 (10)
PK	478 ± 232 (15)	169 ± 42 (11)	156 ± 28 (3)	222 ± 85 (10)
PTi	36 ± 16 (15)	21 ± 11 (11)	15 ± 6 (3)	30 ± 10 (10)
PMn	9 ± 4 (15)	4 ± 2 (11)	6 ± 2 (3)	10 ± 3 (10)
PFe	725 ± 816 (15)	213 ± 87 (11)	210 ± 52 (3)	373 ± 134 (10)

[1]Particulate organic carbon data from Anderson and Richards (1977).

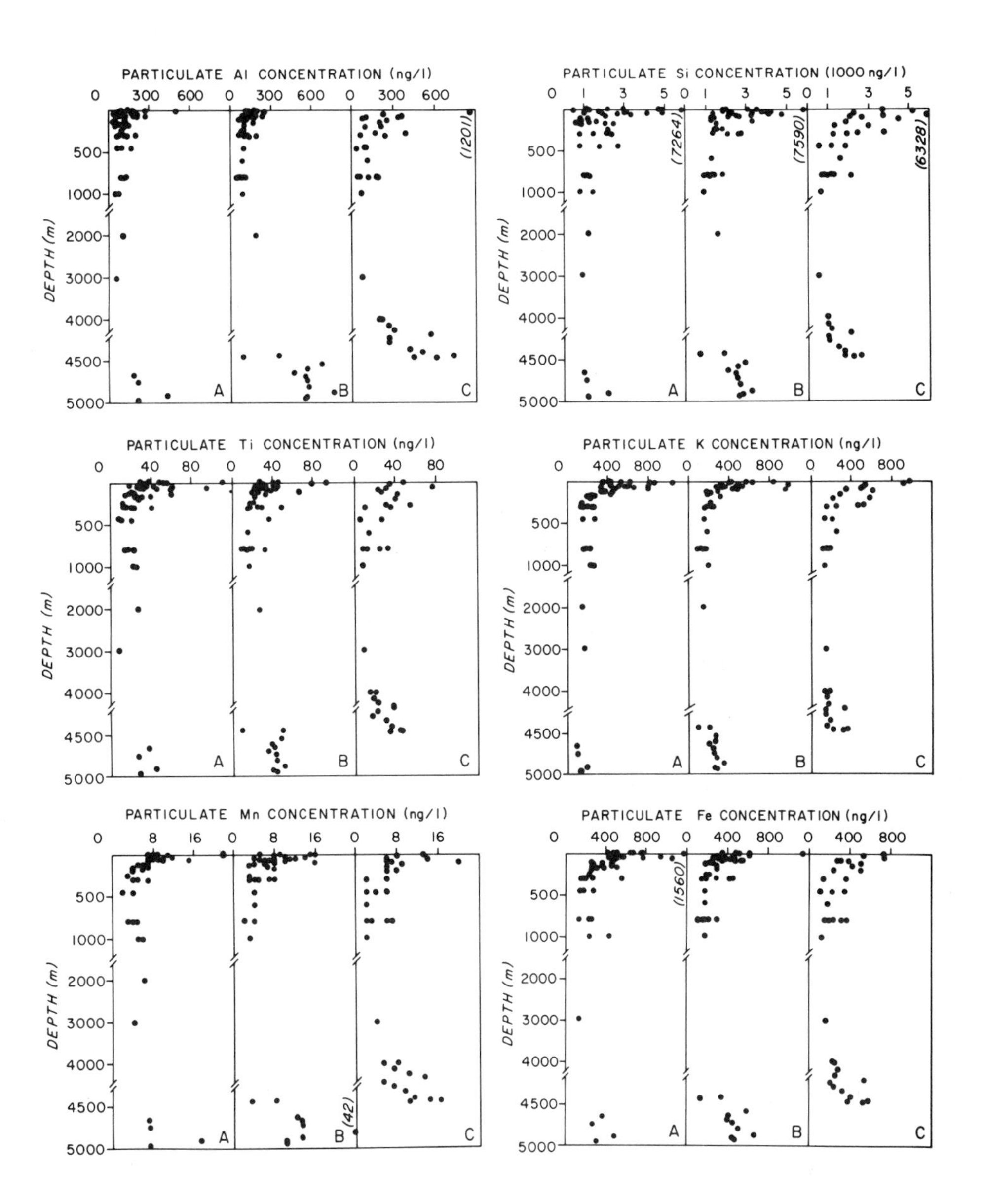

Fig. 6. Vertical profiles of particulate aluminum, silicon, titanium, potassium, manganese, and iron along transects A, B, and C. Note the change in depth scale between 1000 and 4000 m.

and 1180 ng/l, respectively. *P*Si concentrations were higher at these depths along transects A and B than along transect C.

TABLE 3

Minimum detection limits for elements analyzed by secondary emission x-ray fluorescence spectroscopy using a selenium secondary target

Al (ng)	Si (ng)	K (ng)	Ti (ng)	Mn (ng)	Fe (ng)
398	257	20	12	5	31

In the BNL, all samples were significantly enriched in *P*Al and *P*Si. Enrichments of *P*Al over mid- and deep-water samples ranged from increases by a factor of 2.6 at site A to a factor of 5.0 at site B, whereas *P*Si enrichments ranged from 1.1-2.0. These enrichments are significantly higher than corresponding enrichments for the SPM (1.1-1.6), which suggests that the solid phase composition of suspended matter within the BNL must be somewhat different from the suspended matter in the overlying water. A plot of the relationship between *P*Al and *P*Si (Fig. 7A) indicates that these elements are highly correlated (r = 0.89) in the BNL. The average Si/Al ratio (4.29) of the suspended matter in the BNL is only slightly higher than the Si/Al ratio (3.49) of the underlying sediments (Table 4), indicating that these elements are primarily associated with aluminosilicate minerals. Mean values of surface, midwater, and deep samples nearly all fall above the standard error of the regression of the data for the BNL and illustrate the predominance of biogenic Si in the water column above the BNL.

From similar suspended matter studies, Price and Calvert (1973) and Feely (1975) suggested that if the Si/Al ratio of pelagic clays is known, an estimate of biogenic Si content of the suspended matter could be obtained by assuming that any Si in excess of the amount expected from the Si/Al ratio of the suspended clays was biogenic. Bostrom et al. (1972) suggest that the Si/Al ratio in pelagic clays varies only slightly from a value of about 2.65. The mean Si/Al ratios from all depth intervals in the DOMES region were > 2.65 (Table 4), indicating that the SPM was composed of varying relative amounts of aluminosilicate minerals

TABLE 4

Comparison of mean element:aluminum ratios for DOMES particulate samples to ratios from the underlying sediments

Sample Description	0-300 m	450-1000 m	2000 m-<BNL	BNL	Underlying Sediments[1]
Transect A					
POC[2]/Al	116.42	74.26	91.48	33.53	.09
Si/Al	19.26	13.71	12.87	5.31	3.72
K/Al	4.42	2.28	2.14	0.49	0.40
Ti/Al	0.33	0.22	0.20	0.14	.06
Mn/Al	.09	.05	.06	.04	.06
Fe/Al	4.51	2.77	2.36	1.51	0.63
Transect B					
POC[2]/Al	207.96	99.53	32.27	12.81	.05
Si/Al	18.72	14.67	7.12	4.35	3.64
K/Al	2.87	1.83	0.84	0.45	0.41
Ti/Al	0.27	0.23	0.13	.07	.05
Mn/Al	.06	.04	.03	.03	.06
Fe/Al	2.82	2.08	1.20	0.87	0.58
Transect C					
POC[2]/Al	58.02	75.47	41.26	18.66	.03
Si/Al	14.61	13.47	5.53	3.84	3.11
K/Al	2.18	2.14	1.23	0.53	0.35
Ti/Al	0.18	0.24	0.10	.07	.06
Mn/Al	.05	.05	.04	.02	0.16
Fe/Al	3.13	2.53	1.50	0.89	0.72

[1]Data from Bischoff *et al.* (this volume).

[2]Data from Anderson and Richards (1977).

and biogenic Si. These ratios are combined with the data from Table 2 to estimate the concentrations of aluminosilicate matter and biogenic Si and silica in the suspended matter (Table 5). The relative proportion of biogenic Si was highest in the surface waters (75-86% of total Si) and gradually declined through the midwater levels (76-82% of total Si) and the deep waters (50-80% of total Si), probably due to slow dissolution of siliceous tests as suggested by the data of Kido and Nishimura (1975). The relative amount of biogenic Si dropped sharply within the BNL (28-45% of total Si) due to an increase in the concentration of aluminosilicate minerals.

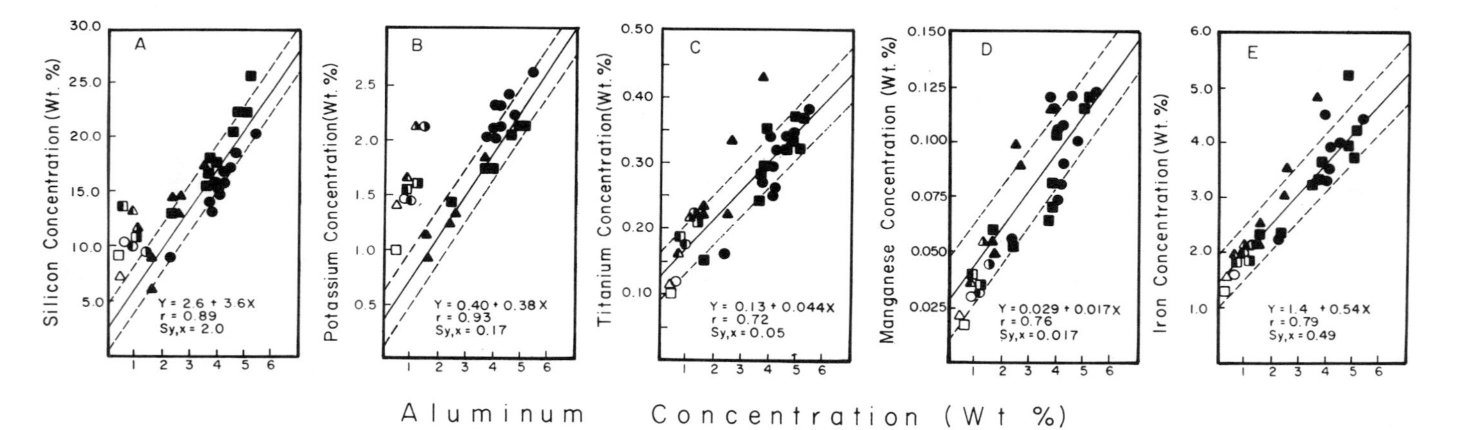

Fig. 7. Scatter diagrams of the (A) silicon, (B) potassium, (C) titanium, (D) manganese, and (E) iron particulate concentrations against the particulate aluminum concentration for selected depth intervals along transects A (triangles), B (squares), and C (circles). Mean values for each transect are given for samples from the surface layer (0-300 m; A:△, B:□, C:○), midwater (450-1,000 m; A:◭, B:◧, C:◐), and deep water (2,000 m-BNL; A:◮, B:◨, C:◑). Individual values in the bottom nepheloid layer are plotted for stations A (▲), B (■), and C (●). The solid line represents the least-squares fit of the bottom nepheloid layer data only, and the dashed lines represent the standard error of the regression.

TABLE 5

Comparison of mean estimated aluminosilicate matter and biogenic silica of particulate samples from selected depth ranges along DOMES transects A, B, and C

Sample Location and Geochemical Species	0-300 m	450-1,000 m	2,000 m-<BNL	BNL
Transect A				
Total suspended matter (μg/l)	33.2	11.0	8.4	10.5
Total aluminosilicate matter (μg/l)	1.2	0.9	0.7	2.1
Total biogenic Si (silica) (μg/l)	1.9 (4.0)	1.0 (2.2)	0.8 (1.8)	0.6 (1.4)
% aluminosilicate fraction	3.6	8.2	8.3	20.0
% biogenic Si (silica) fraction	5.7 (12.0)	9.1 (20.0)	9.5 (21.4)	5.7 (13.3)
Transect B				
Total suspended matter (μg/l)	32.9	12.0	9.2	14.1
Total aluminosilicate matter (μg/l)	1.2	0.7	1.7	4.9
Total biogenic Si (silica) (μg/l)	2.3 (5.0)	1.0 (2.1)	0.8 (1.7)	0.7 (1.5)
% aluminosilicate fraction	3.6	5.8	18.4	34.7
% biogenic Si (silica) fraction	7.0 (15.1)	8.3 (17.5)	8.6 (18.4)	5.0 (10.6)
Transect C				
Total suspended matter (μg/l)	29.6	11.5	7.2	11.9
Total aluminosilicate matter (μg/l)	2.3	0.8	1.4	3.6
Total biogenic Si (silica) (μg/l)	2.3 (5.0)	0.9 (1.9)	0.4 (0.8)	0.5 (1.1)
% aluminosilicate fraction	7.8	6.9	19.4	30.2
% biogenic Si (silica) fraction	7.8 (16.9)	7.8 (16.5)	5.5 (11.1)	4.2 (9.2)

The distributions of *P*K, *P*Ti, *P*Mn, and *P*Fe in the DOMES area are generally similar to the distributional patterns for *P*Al and *P*Si (Fig. 6). Concentration of these elements were high and variable in surface waters with concentrations ranging from 155-1,040 ng/l for *P*K, 18-112 ng/l for *P*Ti, 3-22 ng/l for *P*Mn and 194-1,560 ng/l for *P*Fe. There were no significant concentration differences between transects in this region of the water column. At mid- and deep-water depths, concentrations of these elements decreased to values averaging about 170, 37, 4, and 213 ng/l, respectively, for particulate K, Ti, Mn, and Fe. In the BNL all samples were significantly enriched in these elements (Table 4 and Appendix I). The highest enrichments were found at stations B and C, where the bottom nepheloid layers were most intense.

The relationships between particulate K, Ti, Mn, and Fe and *P*Al are represented as scatter diagrams in Figs. 7B-7E. Individual values from the BNL are plotted for stations A, B, and C. In addition, mean values for all stations from each transect are also plotted for surface, midwater, and deep water. The diagrams show that K, Ti, Mn, and Fe are highly correlated with Al in the BNL, indicating that these elements are primarily associated with aluminosilicate minerals in this region of the water column. Above the BNL, however, they appear to be associated with other solid phases in addition to the aluminosilicate minerals. The pattern of *P*K resembles that of *P*Si in the region above the BNL, indicating that it is closely associated with biogenic matter. The sharp decrease in the concentration of *P*K in the upper 500 m of the water column (Fig. 6) particularly supports such an association. The positive intercepts on the scatter diagrams of *P*Ti vs *P*Al (Fig. 7C), *P*Mn vs *P*Al (Fig. 7D), and *P*Fe vs *P*Al (Fig. 7E) indicate that at least one other nonaluminosilicate phase in the suspended matter contains these elements. For Mn and Fe this could be oxyhydroxide particles and coatings. Another possibility is that these elements are associated with biogenic tests. For example, Martin and Knauer (1973) found elevated concentrations of Ti, Mn, and Fe in the silica fraction of phytoplankton samples collected between California and Hawaii. Similar results were found by Riley and Roth (1971) from studies with cultured phytoplankton.

Particle Size and Morophology

Particulate samples for size and morphological studies were collected at all stations during the winter cruise, but discussion in this paper will be limited to stations A, B, and C where sampling extended down to the sea floor.

Cumulative percentage curves were plotted from the size distribution data and an average of 96.6 ± 1.5% of the total number of particles counted were $\leq$ 3 μm (expressed as the radius of the circular area equivalent to the actual particle area). The median radius of each sample is plotted in Fig. 8A and the mean median radius of the particle population at seven horizons in the water column, plus the (dispersed) bottom sediments, is given in Table 6. The mean median radius uniformly decreased from 1.22 ± 0.14 at 300 m to 0.72 ± .07 μm within the BNL; the bottom sediments were even finer at 0.43 ± 0.14 μm. Within the

TABLE 6

Variation of the mean median particle radius with depth

Depth (m)	Mean Median Radius ± 1 S.D. (μm)	Number of Samples
0-100	1.11 ± 0.15	5
100-200	1.00 ± 0.18	3
300	1.22 ± 0.14	3
800	1.06 ± 0.18	3
2000	1.01 ± 0.11	3
3000	0.88 ± .04	3
<400 above bottom	0.72 ± .07	5
bottom sediment	0.43 ± 0.14	3

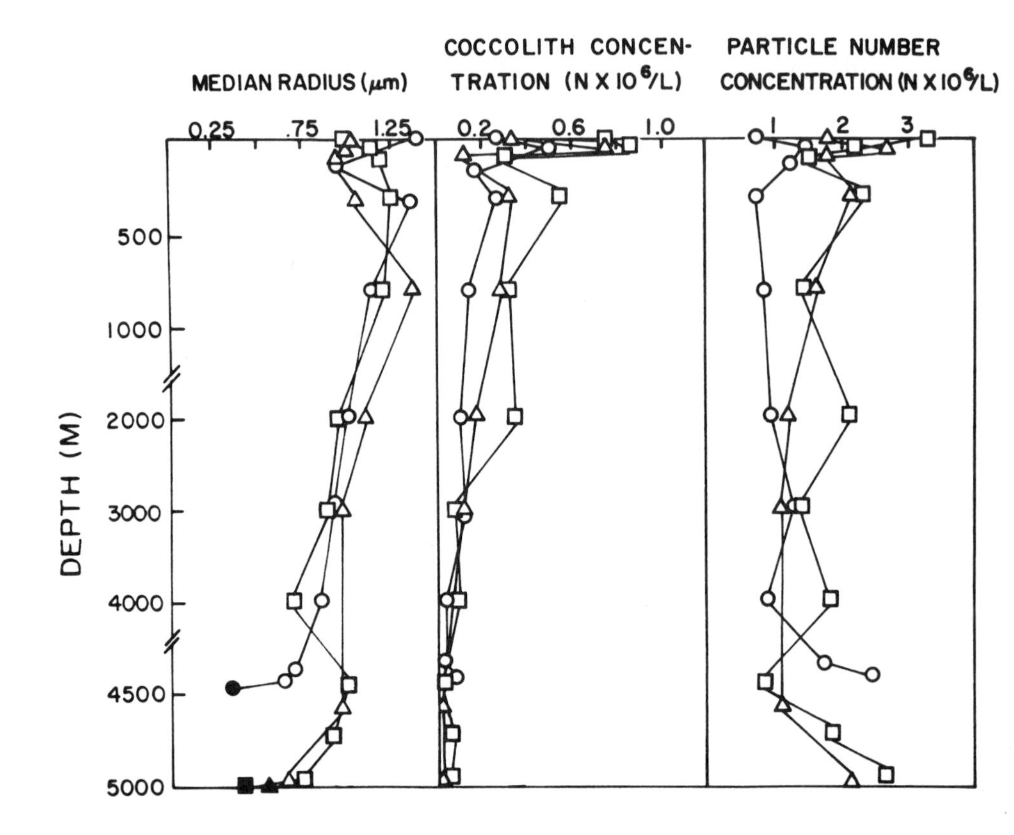

Fig. 8. Particle size and concentration characteristics for stations A (△), B (□), and C (○). A. Decrease of median size with depth. Closed symbols indicate the bottom sediment sample. B. Decrease in coccolith concentration with depth. Note that the subsurface maximum at 300 m is expressed as an increase in the median radius plot. C. Total particle number concentration against depth. Note that the bottom nepheloid layer increase is not reflected in the coccolith concentration but is reflected as a decrease in the median particle radius.

surface layers ($z \leq 300$ m) median particle size fluctuations were apparently influenced by variations in the concentration of coccoliths (Fig. 8B). The low value of the mean median radius between 100 and 200 m occurs at a regional coccolith minimum in the surface waters, and the high value at 300 m is associated with a prominent subsurface coccolith maximum.

Bader (1970), Carder and Schlemmer (1973), Carder et al. (1974), Lal and Lerman (1975a), Brun-Cottan (1976), and Lerman et al. (1977) have used a power law relationship of the following type to describe the distribution of fine-grained particles in the sea:

$$dN/dr = Ar^{-b}, \qquad (1)$$

where A and b are constants, N = particle number, and r = equivalent circle radius of the particle in μm. Starting with the smallest particle counted on the photomicrographs (r = 0.65 μm), the number of particles (ΔN) was determined for each of 11 size intervals (Δr) ranging in width from 0.17 to 0.48 μm and extending to particles as large as r = 4 μm. The ratio $\Delta N/\Delta r$ (particles cm^{-3} μm^{-1}) for each size interval was plotted at the midpoint (r, μm) of that interval on log-log graphs (Fig. 9). A least squares estimation of the best-fit line for all points with values of $r \geq 1.47$ μm was calculated for each sample on the basis of equation (1). The value of the constants A and b for each sample are given in Table 7.

The good fit of the data to the above relationship and the mean value of b for all samples (4.04 ± 0.49) agrees well with other data derived from Coulter Counter measurements of oceanic SPM (Brun-Cottan, 1976; Lerman et al., 1977) and microscopic measurements of oceanic SPM and bottom sediments (Lal and Lerman, 1975a). Lower values of b (≃3) have been identified with high productivity waters, such as regions of coastal upwelling (Carder and Schlemmer, 1973; Carder et al., 1974). The only discrepancy in the present data is the curvilinear trend for values of $r < 1.47$ μm, which will be discussed more fully below.

Although the total concentration of particles remains relatively constant throughout most of the water column (Fig. 8C), significant variations in both concentration and size distribution occur within the surface waters and the BNL. The particle concentrations in the surface waters at stations A and B are about 2x that of station C. Most of this deficiency is apparently in the finer grain sizes, as illustrated by the relatively low dN/dr values at station C, particularly for $r < 1.47$ μm (Fig. 9), as well as the lower b values (Table 7).

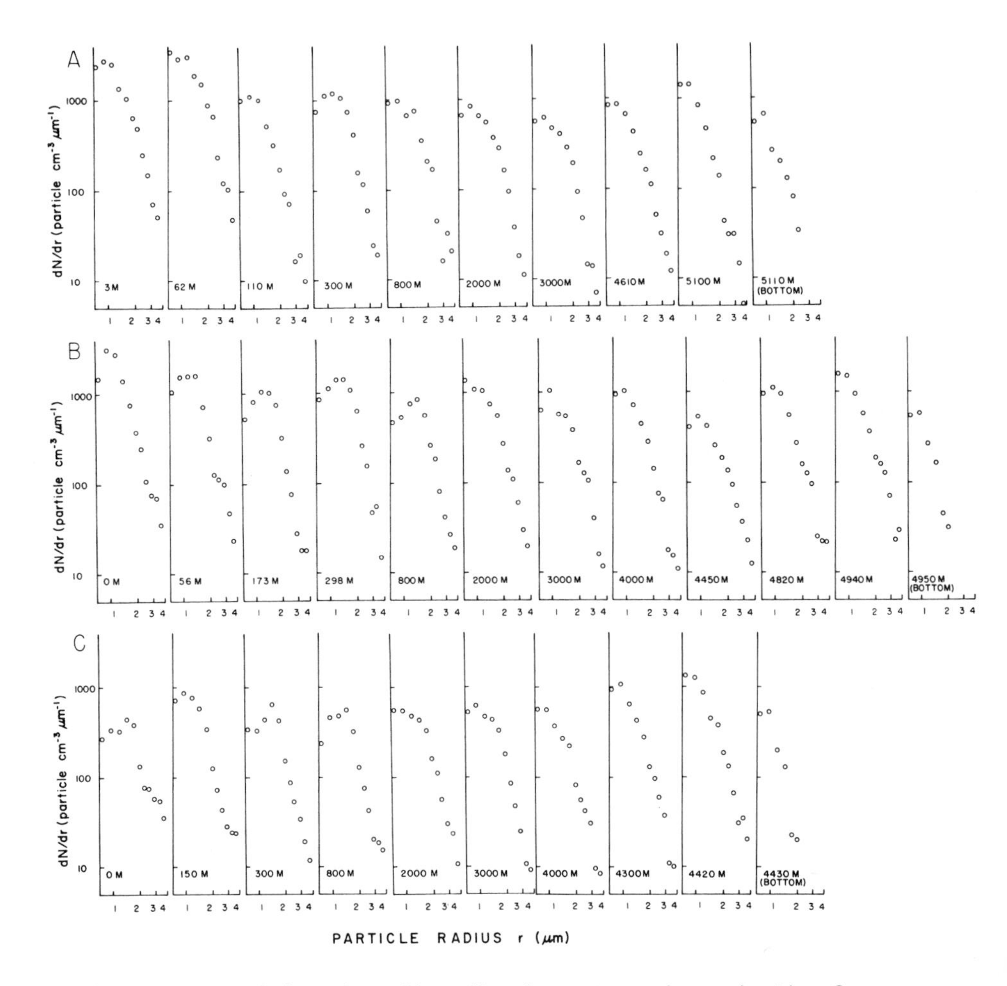

Fig. 9. Particle size distribution at various depths for stations A, B, and C. The dN/dr values for the bottom sediment samples are only relative; their slopes (b), however, are directly comparable to the suspended matter data. Mean particle radii (r in µm) of the plotted size intervals are 0.75, 0.94, 1.17, 1.47, 1.72, 1.97, 2.24, 2.55, 2.91, 2.31, and 3.77.

TABLE 7

Particle size distribution parameters A (particles cm^{-3} μm^{-1}) and b for values of $r \geq 1.47$ μm

Station A (5110 m) Depth (m)	A	b	Station B (4950 m) Depth (m)	A	b	Station C (4430 m) Depth (m)	A	b
3	3685	3.68	0	5766	3.89	0	1112	2.71
62	6303	4.13	56	6086	4.14	150	1825	3.65
110	3355	4.46	173	7881	4.91	300	3155	4.26
300	7807	4.59	298	14239	4.97	800	2244	4.05
800	3251	4.13	800	5061	4.31	2000	2278	3.93
2000	4388	4.39	2000	4178	4.02	3000	3161	4.50
3000	3347	4.65	3000	3588	4.26	4000	1408	3.89
4610	2002	3.84	4000	2427	4.17	4300	2481	4.15
5100	2486	4.47	4450	1115	3.22	4420	2080	3.57
Sediment	N.A.[1]	3.07	4820	2159	3.67	Sediment	N.A.	4.61
			4940	2181	3.36			
			Sediment	N.A.	4.03			
Mean[2]		4.26			4.06			3.81
±S.D.		0.34			0.53			0.51

[1]No applicable parameter.

[2]Mean of suspended samples only.

At all three stations, b increases systematically from the surface to the 300 m sample, then abruptly decreases at 800 m. The peak at 300 m may be related to the subsurface coccolith maximum observed at that depth (Fig. 8B), because an increase in the concentration of coccoliths (with a mean size of 1.60 ± 0.25 μm in these samples) and coccolith fragments would tend to increase b (i.e., steepen the slope).

The coccolith maximum between 200 and 800 m seems anomalous but might be explained on the basis of fecal pellet disintegration and reinjection of fecal material into the water column as hypothesized by Honjo (1976, 1977). According to Honjo (1976), degradation of the pellicle by bacteria begins shortly after pellet production and is highly dependent on the temperature of the ambient water. He found that fecal pellets produced by laboratory cultured zooplankton disintegrated in only a few minutes at 25°C, whereas at 15°C the pellets remained intact for several days. Using a rough estimate of 150 m/day for fecal pellet settling velocity (Smayda, 1969; Fowler and Small, 1972; Wiebe et al., 1976), a speculative but plausible model can be described where pellets are produced primarily in the euphotic zone during nighttime grazing, remain intact through the thermocline, and rapidly begin to break up after reaching a depth of 300 m in 2-3 days. Water temperature in the DOMES region decreases from 26°C at the top of the thermocline to 10°C at 300 m (Oztergut et al., in press); this range seems capable of preventing pellet disintegration for at least a few days based on Honjo's (1976) data.

As mentioned above, a conspicuous deviation in our dN/dr vs. r plots from similar studies (Carder and Schlemmer, 1973; Carder et al., 1974; Lal and Lerman, 1975a, Brun-Cottan, 1976; Lerman et al., 1977) is in the character of the size distribution for values of $r < 1.47$ μm. [Note that our microscopic data extend to finer-size classes than most of the Coulter Counter studies cited, which generally include only those particles with $r > 1.15$ μm. Brun-Cottan (1976), however, measured particles as fine as $r = 0.75$ μm.] All samples in this study show a curvilinear trend beginning at $r = 1.47$ or 1.17 μm, with maximum values of dN/dr usually at 0.94 or 1.17 μm rather than a linear increase in dN/dr through the finest grain sizes measured. In other words, our data consistently show a relative deficiency in the number of particles below a threshold size of 1.47 μm.

This deficiency remains relatively uniform down to the top of the BNL (~400 m above the sea floor). Two samples (z = 2,000 and 4,000 m) from station B possess an abnormally high number of particles in the finest grain sizes [and in the total particle concentration (Fig. 8C)]. Most of this increase is due to an increase in the microbial population within these samples (Takahashi and Baker, in prep.) and does not signify an addition of inorganic particles.

Within the BNL at all three stations the value of dN/dr for $r < 1.47$ μm increases substantially so that 10 m above the sea floor the plot of dN/dr vs. r becomes nearly linear for all r. This change in the curve indicates that the increase in the total particle concentration observed within the BNL is due principally to an addition of very fine particles, although the coarser size fractions generally show an increase as well. Counts of the microbial population at each station indicate that the BNL fine particle increase is not due to bacteria or other microorganisms (Takahashi and Baker, in prep.).

A relative deficiency of fine particles in the water column below the zone of calcium carbonate saturation has been postulated on a theoretical basis by Lal and Lerman (1975a, b). They show that an initial population of calcium carbonate particles with a size distribution described by equation (1) will develop a pronounced flattening of the dN/dr vs. r curve for small values of r as a result of the relatively faster dissolution of the fine particles (due to a larger surface-to-volume ratio) during settling. They also show that a relative depletion of fine particles could result from fragmentation of an originally uniform population, although in this case the value of b would be expected to increase by ~1 over values normally observed for suspended particle distributions.

Since Lal and Lerman (1975a) found no curvilinear trend in dN/dr values in the samples they examined, they concluded that dissolution of particles from the surface inward (i.e., dissolution which would change the size of a particle) did not affect the size distributions. A similar conclusion was reached by Lerman et al. (1977), who added that the constancy of b with depth argues for particle removal at a rate proportional to their number and cross-sectional area. They suggested that zooplankton filtration might provide the mechanism for such a particle removal.

Other experimental evidence exists that does agree with our observations of changes in the particle size distribution for small particles. Harris (1977), using samples from the deep waters of the Gulf of Mexico, has reported a comprehensive transmission electron microscope study of particle size distributions between .02 and 8 μm. Plotting the data in terms of equation (1), he found a distinct break in slope and decrease in the mean b value for particles of $r \lesssim 1$ μm. He attributed this relative deficiency of fine particles to control of their distribution by processes related to the surface area distribution (i.e., electrostatic binding, hydration, sorption, etc.), whereas the distribution of particles of $r \gtrsim 1$ μm is governed by the volume distribution (i.e., Stokes settling).

Another explanation of our results which must be considered is that the SEM counting techniques result in a systematic underestimation of the fine particle concentration, at least relative to Coulter Counter data. Although we have no evidence of such a problem, this kind of error could occur if many of the fine particles were not distinguishable from the filter background, or if the larger particles covered up many of the small particles. Nuclepore filters, however, are an excellent substrate for microscopic observation, and particles larger than $r = 1.25$ μm never covered more than 16% of the filter surface and normally covered much less than 10%. A detailed investigation at higher magnifications is required to more accurately resolve the distribution of particles with $r < 1.5$ μm. Nevertheless, it appears clear that there is at least a substantial relative increase in the number of fine particles in samples from the BNL, and that the increase is greatest for those samples closest to the sea floor.

Discussion

Sediment-Suspended Matter Relationships

As stated in the introduction to this paper, one of the main objectives of the present study was to compare the composition of pelagic SPM with the underlying sediments in order to ascertain some of the mechanisms controlling the supply and distribution of materials in marine sediments. The data from the DOMES region are particularly useful for this purpose because they encompass stations located near the Central American landmass and north of the usual position of the highly productive equatorial waters (Reid, 1962) (transect C), as well as those far from any landmass and close to the usual position of the equatorial upwelling of nutrient-rich subsurface water (transects A and B). This geographical difference is expressed in the surface sediments at each station by the increase in the concentration of biogenic silica debris from station C to stations A and B (Bischoff et al., this vol.).

The element/Al ratios from transects A, B, and C (Table 4) indicate the distinct compositional differences in the SPM which arise from the geographical position of each transect and are preserved in sedimentary record. The Si/Al and POC/Al ratios (above the BNL) are generally higher along transects A and B than along C. Similarly, the Si/Al and organic carbon/Al ratios of the sediments at stations A and B are higher than at station C (Bischoff et al., this vol.). These results suggest that higher proportions of (siliceous) biogenic matter are both produced and preserved at stations A and B than at station C. Part of this effect may be due to a greater dilution of the biogenic matter by terrestrial input at station C. According to Prospero and Bonatti (1969), aeolian dust, consisting primarily of plagioclase, quartz, and clay minerals, is transported towards the DOMES region from the arid lands of western and southern Mexico.

The composition of the sediments is also controlled by in-situ chemical changes in the SPM after it is introduced or produced in the surface waters. These changes can be examined by looking at variations in element ratios. By assuming that Al is exclusively associated with the aluminosilicate minerals in the

SPM and sediments, element/Al (Table 4) ratios can be used to determine the major phases which supply elements to the sediments.

With the exception of *P*Mn, all elements in the SPM samples have significantly higher element/Al ratios than the corresponding ratios for the sediments. Furthermore, the ratios decrease rapidly with depth to values approaching the ratios for the sediments. These results can be interpreted as evidence for concentration of Si, K, Ti, and Fe in nonaluminosilicate phases in the surface waters (0-300 m), followed by progressive dissolution of these phases and remineralization of the elements with depth. In the deep-water zone (2,000 m), the element/Al ratios are still higher than the corresponding bottom sediment ratios by factors ranging from 1.6-5.4. Since the underlying sediments are primarily composed of aluminosilicate minerals (Bischoff et al., this vol.), these data suggest that nonaluminosilicate matter (i.e., biogenic matter and chemical precipitates such as Fe and Mn oxyhydroxides) contribute most of the Si, K, Ti, and Fe to the deep-water SPM and probably the sediments as well. Furthermore, the relative importance of the biogenic component in the deep water at each station varies according to its distance from the equatorial productivity zone: nonaluminosilicate matter accounts for more than 70% of the Si, K, Ti, and Fe at station A, but only about 40% (except for K) at station C.

This rapid change from nonaluminosilicate dominated SPM to aluminosilicate dominated sediments suggests that much of the biogenic matter that falls to the sea floor dissolves and reforms within the sediment column. From an analysis of cores taken from the DOMES region, Hein et al. (in press) conclude that nearly 30% of the clay fraction of the sediments is composed of Fe-rich smectites which were authigenically formed from the combination of Fe hydroxides with Si released from dissolving biogenic matter. A similar authigenic formation of smectite was hypothesized by Heath and Dymond (1977) for samples from the eastern tropical Pacific between the East Pacific Rise and South America. If the composition of the SPM which reaches the sea floor is the same as the deep water SPM, then the suspended biogenic matter contributes nearly four times more Si than is required to form the authigenic smectites. Thus a large fraction of the remineralized Si must be released back to the overlying water column as well as recombining to form authigenic minerals. This conclusion follows Edmond's (1974) suggestion that dissolution of particulate Si at the sediment-seawater interface is the primary mechanism regulating the deep water distribution of dissolved Si.

As indicated previously, Mn/Al ratios for SPM are lower than Mn/Al ratios for the underlying sediments. The highest Mn/Al ratio in the sediments is from station C where ~75% of the Mn is supplied from sources other than the suspended matter. [The mean Mn/Al ratio at station C given by Bischoff et al. (this vol.) includes three (out of 43) samples with anomalously high Mn values apparently due to micronodules or nodule debris. Elimination of these samples would result in a mean Mn/Al ratio of 0.11 and imply that approximately 60%, rather than 75%, of the Mn is not accounted for by the SPM.] Calvert and Price (1977) suggested two sources for the enriched Mn in the surface sediments of this region: (1) precipitation of Mn from seawater, and (2) diagenetic remobilization of Mn from the sediment column. Although our data cannot be used to provide irrefutable evidence for the dominance of one mechanism over the other, the observed trend of decreasing importance of Mn suspended phases with increasing latitude (i.e., away from the equatorial productivity zone) is consistent with Krishnaswami's (1976) conclusion that the concentration of authigenically formed Mn increases as the sedimentation rate decreases.

Unfortunately, we have not as yet had an opportunity to examine the element relationships in the rare, large particles (e.g., fecal material) which may contribute a substantial percentage to the total sediment flux in oceanic waters (McCave, 1975; Bishop *et al.*, 1977). Rapid removal of fine particles from the water column by biogenic or physical aggregation into large particles may be an important factor in the depletion of certain elements with depth and their delivery to the sea floor.

Origin of the Bottom Nepheloid Layer

The chemical, size, and morphological data presented above suggest a sedimentation process in the deep sea whereby fine particles are introduced or formed in the surface waters, undergo some preliminary modification as they sink to the sea floor (often including some degree of biological or chemical recycling), and suffer more extensive compositional alteration after deposition at the sediment-seawater interface. Knowledge of the differences in the chemical and morphological composition of the SPM and the sediments should be useful in attempting to investigate resuspension processes in the near-bottom zone, where characteristics of the particulate material are transitional between the bulk of oceanic suspended matter and the recently deposited sediments. Bottom nepheloid layers, for instance, have been commonly observed throughout the world ocean (Ewing and Connary, 1970; Eittreim *et al.*, 1972; Biscaye and Eittreim, 1977), but the processes which lead to their formation are poorly understood. An "excess" accumulation of particles above the sea floor in the deep sea may be due to either or both of two effects: (1) accumulation of sinking particles from the overlying water column, and (2) resuspension of some fraction of the bottom sediment, either locally or at some distance up-current of the observed feature.

The accumulation hypothesis requires that the chemistry of the SPM undergoes a marked change within the BNL resulting in a relative increase in the more refractory (Al-rich) particles.

This increase could occur by a preferential loss of the Al-poor fraction, either by in-situ dissolution/oxidation or by differential settling. Neither of these processes are supported by the data. POC values show no decrease in the BNL (Table 2). Neither is there any substantial increase in the dissolution of the inorganic biogenic phases: the concentration of both biogenic silica (Table 5) and the number of coccoliths (Fig. 8B) remain relatively unchanged upon entering the BNL. Furthermore, the BNL is the site of an increase in both total particle mass and particle number concentration, so a chemical alteration by loss of a particular fraction cannot be the dominant process. Finally, density profiles of the bottom water (Hayes, this vol.) show no region where a change in settling velocity might lead to an increase in particle concentration or a fractionation of particle sizes.

Alternatively, resuspension can alter the BNL chemistry by selectively injecting a chemically distinct fraction of the bottom sediment. There is strong evidence from the size data (Fig. 9) that the BNL is characterized by an addition of fine-grained particles (chiefly with $r < 1.47$ μm), and the logical source of these particles is in the sediment. If the chemical composition of the appropriate size fraction of the sediment is known, it should be possible to predict the average chemistry of the BNL particles by combining the proper proportions of deep-water (2,000 m) particles from above the BNL with resuspended particles from the sea floor. The proportions of these components differ for each station and are based on the amount of resuspended material required to elevate the mean value of deep-water SPM concentration to that of the mean BNL concentration. The ratios, based on a 400 m thick BNL and expressed as the ratio of resuspended sediments:ambient deep-water particulates in the BNL at each station, are 0.24:1 at A, 0.53:1 at B, and 0.65:1 at C. Resuspension thus accounts for about 20-40% of the material in the BNL.

In order to test this model, samples of surficial box core sediments from each site were size-fractionated by settling in distilled water and then chemically analyzed by the methods used for the SPM samples. Since the greatest increase in particle number within the BNL is shown by those particles in the finest grain sizes (Fig. 9), we assume that the ≤ 1 μm fraction (47-76% of the total sediment by weight) is most likely to be resuspended. The possibility that resuspension is an important factor in generating the BNL in the DOMES region was tested by calculating the theoretical chemical composition of a BNL created by mixing deep-water SPM and the fine fraction of the bottom sediment in the proportions given above for each station (Table 8).

Agreement between the calculated values of each element and the observed BNL values is good in nearly all cases. A t-test performed on the hypothesis that there is no difference between the calculated and observed means of the elemental percentages in

TABLE 8

Chemical mass-balance calculations for the bottom nepheloid layer (BNL)

Element	Observed, above BNL (Wt. %)	Bottom sediment ($\leq 1\ \mu m$) (Wt. %)	Calculated, BNL (Wt. %)	Observed, BNL (Wt. %)
		Station A		
Al	1.1	7.6	2.3 ± 0.1	2.8 ± 0.6
Si	13.8	22.7	15.5 ± 3.3	14.8 ± 1.9
K	2.3	1.6	2.2 ± 1.1	1.4 ± 0.3
Ti	0.21	0.56	0.28 ± 0.08	0.39 ± 0.11
Mn	0.068	0.359	0.124 ± 0.012	0.102 ± 0.007
Fe	2.6	7.8	3.6	4.2 ± 1.1
		Station B		
Al	1.8	8.8	4.2 ± 0.5	4.4 ± 0.6
Si	12.2	24.8	16.5 ± 0.6	19.4 ± 4.5
K	1.4	2.2	1.7 ± 0.4	2.0 ± 0.2
Ti	0.23	0.68	0.39 ± 0.10	0.32 ± 0.04
Mn	0.057	0.288	0.137 ± 0.002	0.119 ± 0.070
Fe	2.3	8.5	4.4 ± 0.0	3.8 ± 0.7
		Station C		
Al	3.4	7.4	5.0 ± 0.5	4.0 ± 0.8
Si	13.3	19.6	15.8 ± 4.2	15.7 ± 2.8
K	2.6	2.2	2.4 ± 0.2	2.2 ± 0.3
Ti	2.4	0.66	0.41 ± 0.08	0.30 ± 0.06
Mn	0.092	0.403	0.215 ± 0.024	0.101 ± 0.021
Fe	3.5	8.3	5.4 ± 0.8	3.7 ± 0.6

the BNL resulted in acceptance at the $p < .05$ level of significance for all cases except Fe at station A and Ti, Mn, and Fe at station C. Our sampling to date in the deep-water zone has been sparse and the demarcation between BNL and non-BNL material is not precise. Nevertheless, these preliminary chemical mass-balance results clearly suggest that the concentration increases in the BNL in this area are due to resuspension of the fine-grained bottom sediments.

The capability of bottom currents in the DOMES area to actually erode the underlying sediments is presently only speculative. Near-bottom currents in the DOMES region were measured for 6 months at stations A and B (Hayes, this vol.) and for 1 month at station C (Amos et al., 1976). The mean and maximum (hourly average) current speeds at A were 2.1 and 10.1 cm/sec (30 m above the bottom); at B 5.2 and 13.4 cm/sec (30 m above the bottom); and at C 8.1 and 18.5 cm/sec (50 m above the bottom, 10 min. average). These currents are well correlated with the observed BNL parameters, station A having the weakest BNL and the lowest PAl values (i.e., the weakest inferred resuspension). Lonsdale and Southard (1974) have measured the threshold erosional velocities of a bed of Pacific red clay in a recirculating flume under various degrees of compaction and bed roughness. They found that rapid erosion could occur under current speeds as low as 12 cm/sec (1 m off the bed) when the sediment surface was roughened with manganese nodules and the water content increased to > 80%. Water content measured from the upper 2 cm of box cores collected in the DOMES area only averaged about 75% at stations A and B and about 65% at C (Richards, 1977). However, the in-situ water content of the surface microlayer where particle-by-particle erosion would actually occur has not been measured and may be considerably higher than that of the bulk sediment below, thereby enhancing its potential erodability. Thus, the resuspension inferred by our chemical mass-balance considerations and supported by the particle size and morphology analyses might be generated by long-term action of the relatively weak bottom currents, and/or the influence of brief and infrequent occurrences of abnormally high currents.

Summary and Conclusions

The geochemical and optical measurements of SPM described in this paper illustrate some of the processes controlling the supply and distribution of materials in marine sediments. Elements such as Si, K, Ti, Mn, and Fe occur primarily in nonaluminosilicate phases in the surface waters of the eastern equatorial Pacific Ocean, with element/Al ratios (excluding Mn) between 3.0 and 11.1 times greater than the corresponding values for the primarily aluminosilicate sediments. These enrichment factors uniformly decrease with depth as dissolution of the biogenic fraction occurs, but even in the deep water zone (2,000 m) the SPM is enriched in these elements (again excluding Mn) by factors of 1.6-5.4 over the sediments. Clearly, significant dissolution and some degree of mineral reformation must occur at or very near the sediment-seawater interface. Operating processes which result in a shift from nonaluminosilicate supply to aluminosilicate sediments probably include authigenically formed smectites and the release of large amounts of dissolved Si into the water column.

Mn was excluded in the preceding discussion because apparently it is not principally associated with the biogenic phase in the SPM. Mn/Al ratios in the SPM are generally lower in the SPM than the sediments and thus require authigenic precipitation of Mn in the surface sediments.

The character of the SPM is influenced by its geographical proximity to the equatorial upwelling zone as well as by its depth in the water column. For instance, nonaluminosilicate matter accounts for more than 70% of the Si, K, Ti, and Fe at station A (the farthest south), but only about 40% (except for K) at station C (well north of the upwelling zone). Similarly, the fraction of Mn supplied by the SPM to the bottom sediments falls off sharply moving from station A to station C.

One use of these data is to investigate resuspension processes in the deep sea, because even small additions of bottom sediment to the ambient SPM can be chemically resolved. We suggest that the BNL in this area, which has an elemental balance significantly different than either the overlying waters or the bottom sediments, is the result of resuspension of fine-grained ($\leq$ 1 μm) sediments which ultimately account for 20-40% of the mass of material in the BNL. Further investigations will be required to discover the frequency and intensity of the resuspension process.

Acknowledgments

We sincerely thank J. B. Nevins for field and laboratory assistance in all phases of this investigation. The assistance of G. J. Massoth in the x-ray fluorescence analyses is also appreciated. S. A. Swift and H. Curl, Jr. critically read the manuscript. This study was supported by NOAA's Deep Ocean Mining Environmental Study program and by NOAA's Environmental Research Laboratories.

References

Amos, A. F., Roels, O. A., and Paul, A. Z., (1976) Environmental baseline conditions in a manganese-nodule province in April-May 1975, Proc. Offshore Tech. Conf. p. 341-356.

Anderson, J. J., (1979) Distribution of chemical properties in the water column at Sites A, B, and C, (this volume).

Anderson, J. J., and Richards, F. R., (1977) Chemical and biochemical observations from the DOMES study area in the Equatorial North Pacific, Univ. of Wash., Dept. of Oceanography Special Report No. 79.

Bader, H., (1970) The hyperbolic distribution of particle sizes, J. Geophys. Res. 75, 2822-2830.

Baker, E. T., and Piper, D. Z., (1976) Suspended particulate matter: collection by pressure filtration and elemental analysis by thin-film x-ray fluorescence, Deep-Sea Res. 23, 181-186.

Betzer, P. R., Carder, K. L., and Eggimann, D. W., (1974) Light scattering and suspended particulate matter on a transect of the Atlantic Ocean at 11°N, In: Suspended Solids in Water (editor R. J. Gibbs) p. 295-314, Plenum Press, New York.

Betzer, P. R., and Pilson, M.E.Q., (1970) Concentrations of particulate iron in Atlantic open-ocean water, J. Mar. Res. 28, 251-267.

Beutell, R. G., and Brewer, A. W., (1949) Instruments for the measurement of the visual range, J. Sci. Instr. 26, 357-359.

Biscaye, P. E., and Eittreim, S. L., (1977) Suspended particulate loads and transport in the nepheloid layer of the abyssal Atlantic Ocean, Mar. Geol. 23, 155-172.

Bischoff, J. L., Heath, G. R., and Leinen, M., (1979) Geochemistry of deep-sea sediments from the Pacific manganese nodule province: DOMES Sites A, B, and C, (this volume).

Bishop, J.K.B., Edmond, J. M., Ketten, D. R., Bacon, M. P., and Silker, W. B., (1977) The chemistry, biology, and vertical flux of particulate matter from the upper 400 m of the equatorial Atlantic Ocean, Deep-Sea Res. 24, 511-548.

Bostrom, K., Joensun, O., Valdes, S., and Rieral, M., (1972) Geochemical history of South Atlantic Ocean sediments since late Cretaceous, Mar. Geol. 12, 85-122.

Brewer, P. G., Spencer, D. W., Biscaye, P. E., Hanley, A., Sachs, P. L., Smith, C. L., Kodar, S., and Fredericks, J., (1976) The distribution of particulate matter in the Atlantic Ocean, Earth Planet. Sci. Lett. 32, 393-402.

Brun-Cottan, J-C., (1976) Stokes settling and dissolution rate model for marine particles as a function of size distribution, J. Geophys. Res. 81, 1601-1606.

Calvert, S. E., and Price, N. B., (1977) Geochemical variation in ferromanganese nodules and associated sediments from the Pacific Ocean, Mar. Chem. 5, 43-74.

Carder, K. L., Betzer, P. R., and Eggimann, D. W., (1974) Physical, chemical and optical measures of suspended particle concentration: Their intercomparison and application to the West African Shelf, In: Suspended Solids in Water (editor R. J. Gibbs) p. 173-193, Plenum Press, New York.

Carder, K. L., and Schlemmer, F. C. II, (1973) Distribution of particles in the surface waters of the eastern Gulf of Mexico: An indicator of circulation, J. Geophys. Res. 78, 6286-6299.

Chester, R., Cross, D., Griffiths, A., and Stoner, J. H., (1976) The concentration of "aluminosilicates" in particulates from some surface waters of the ocean, Mar. Geol. 22, M59-M67.

Edmond, J. M., (1974) On the dissolution of carbonate and silicate in the deep ocean, Deep-Sea Res. 21, 455-480.

Eittreim, S., Gordon, A. L., Ewing, M., Thorndike, E. M., and Bruckhausen, P., (1972) The nepheloid layer and observed bottom currents in the Indian-Pacific Antarctic Sea, In: Studies in Physical Oceanography--A Tribute to George Wüst on His 80th Birthday (editor A. C. Gordon) p. 19-35, Gordon and Breach, New York.

El-Sayed, S. Z., and Taguchi, S., (1979) Spatial and seasonal variabilities in the phytoplankton standing crop and primary productivity in a manganese nodule province (tropical Pacific), (this volume).

Ewing, M., and Connary, S. D., (1970) Nepheloid layer in the North Pacific, In: Geological Investigations of the North Pacific (editor J. Hays), Geol. Soc. Amer. Mem. 126, 41-82.

Feely, R. A., (1975) Major element composition of the particulate matter in the near-bottom nepheloid layer of the Gulf of Mexico, Mar. Chem. 3, 121-156.

Fowler, S. W., and Small, L. F., (1972) Sinking rates of euphausiid fecal pellets, Limnol. Oceanogr. 17, 293-296.

Harris, J. E., (1977) Characterization of suspended matter in the Gulf of Mexico--II. Particle size analysis of suspended matter from deep water, Deep-Sea Res. 24, 1055-1061.

Hayes, S., (1979) Benthic current observations in the tropical North Pacific Ocean, (this volume).

Heath, G. R., and Dymond, J., (1977) Genesis and transformation of mettaliterous sediments from the East Pacific Rise, Bauer Deep, and Central Basin, northwest Nazca Plate, Geol. Soc. Am. Bull. 88, 723-733.

Hein, J. R., Yeh, H-W., and Alexander, E. R., (in press) Distribution, mineralogy, chemistry, and oxygen isotopes of clay minerals from the north equatorial Pacific manganese nodule belt, Clays and Clay Minerals.

Honjo, S., (1976) Coccoliths: production, transportation and sedimentation, Marine Micropaleont. 1, 65-79.

Honjo, S., (1977) Biogenic carbonate particles in the ocean: Do they dissolve in the water column? In: The Fate of Fossil Fuel CO_2 in the Oceans (editors N. R. Andersen and A. Malahoff), p. 269-294, Plenum Press, New York.

Ichiye, T., Bassin, N. J., and Harris, J. E., (1972) Diffusivity of suspended matter in the Caribbean Sea, J. Geophys. Res. 77, 6576-6588.

Jerlov, N. G., (1976) Marine Optics, Elsevier, Amsterdam.

Kido, K., and Nishimura, M., (1975) Silica in the sea--its forms and dissolution rate, Deep-Sea Res. 22, 323-338.

Krishnaswami, S., (1976) Authigenic transition elements in Pacific pelagic clays, Geochim. Cosmochim. Acta. 40, 425-434.

Krishnaswami, S., and Sarin, M. M., (1976) Atlantic surface particulates: composition, settling rates and dissolution in the deep sea, Earth Planet. Sci. Lett. 32, 430-440.

Lal, D., and Lerman, A., (1973) Dissolution and behavior of particulate biogenic matter in the ocean: Some theoretical considerations, J. Geophys. Res. 78, 7100-7111.

Lal, D., and Lerman, A., (1975a) Size spectra of biogenic particles in ocean water and sediments, J. Geophys. Res. 80, 423-430.

Lal, D., and Lerman, A., (1975b) Correction, J. Geophys. Res. 80, 4563.

Lerman, A., Carder, K. L., and Betzer, P. R., (1977) Elimination of fine suspensiods in the oceanic water column, Earth Planet. Sci. Lett. 37, 61-70.

Lonsdale, P., and Southard, J. B., (1974) Experimental erosion of North Pacific red clay, Mar. Geol. 17, M51-M60.

Martin, J. H., and Knauer, G. A., (1973) The elemental composition of plankton. Geochim. Cosmochim. Acta. 37, 1639-1653.

McCave, I. N., (1975) Vertical flux of particles in the ocean, Deep-Sea Res. 22, 297-310.

Oztergut, E., Anderson, G. C., Burns, R. E., Lavelle, J. W., and Swift, S. A., (in press) Deep Ocean Mining of Manganese Nodules in the North Pacific; Pre-Mining Environmental Conditions and Anticipated Mining Effects.

Piper, D. Z., Cook, H. E., and Gardner, J. V., (1979) Lithostratigraphy and acoustic stratigraphy of the equatorial North Pacific: DOMES Sites A, B, and C, (this volume).

Price, N. B., and Calvert, S. E., (1973) A study of the geochemistry of suspended particulate matter in coastal waters, Mar. Chem. 1, 169-189.

Prospero, J. M., and Bonatti, E., (1969) Continental dust in the atmosphere of the eastern Equatorial Pacific, J. Geophys. Res. 74, 3362-3371.

Reid, J. L., Jr., (1962) On circulation, phosphate-phosphorus content, and zooplankton volumes in the upper part of the Pacific Ocean, Limnol. Oceanogr. 7, 287-306.

Richards, A. F., (1977) Geotechnical testing of U.S. Geological Survey cores collected from the pelagic nodule area, central northwest Pacific Ocean, U.S.G.S. Open File Report 77-778, 267-350.

Riley, J. P., and Roth, I., (1971) The distribution of trace elements in some species of phytoplankton grown in culture, J. Mar. Biol. Ass. U. K. 51, 63-72.

Sackett, W. M., and Arrhenius, G., (1962) Distribution of aluminum species in the hydrosphere, I. Aluminum in the oceans, Geochim. Cosmochim. Acta. 26, 955-968.

Smayda, T. J., (1969) Some measurements of the sinking rate of fecal pellets, Limnol. Oceanogr. 14, 621-625.

Sternberg, R. W., Baker, E. T., McManus, D. A., Smith, S., and Morrison, D. R., (1974) An integrating nephelometer for measuring particle concentrations in the deep sea, Deep-Sea Res. 21, 887-892.

Sverdrup, H. U., Johnson, M. W., and Fleming, R. H., (1942) The Oceans, Prentice-Hall, New York.

Toyota, Y., and Okabe, S., (1967) Vertical distribution of iron, aluminum, silicon and phosphorus in particulate matter collected in the western North Pacific, Indian and Antarctic Oceans, J. Oceanogr. Soc. Japan 25, 1-9.

Wallace, G. T., Jr., Hoffman, G. L., and Duce, R. A., (1977) The influence of organic matter and atmospheric deposition on the particulate trace metal concentration of northwest Atlantic surface seawater, Mar. Chem. 5, 143-170.

Wiebe, P. H., Boyd, S. H., and Winget, S., (1976) Particulate matter sinking to the deep-sea floor at 2,000 m in the Tongue of Ocean, Bahamas, with a description of a new sedimentation trap, J. Marine Res. 34, 341-354.

Yoshida, K., Mao, H-L., and Horrer, D. L., (1953) Circulation in the upper mixed layer of the equatorial North Pacific, J. Marine Res. 12, 99-120.

VERTICAL DISTRIBUTION OF DIVERSE PHYTOPLANKTON COMMUNITIES IN THE CENTRAL PACIFIC

Greta A. Fryxell, Satoru Taguchi,* and Sayed Z. El-Sayed

Department of Oceanography
Texas A&M University
College Station, Texas 77843

*Hawaii Institute of Marine Biology
University of Hawaii at Manoa
P.O. Box 1346, Coconut Island
Kaneohe, Hawaii 96744

Abstract

Southeast of Hawaii in the central Pacific, phytoplankton samples were taken on three transects. An average of over 20,500 cells/liter, including flagellates, was found for all depths counted, with diatoms, dinoflagellates, and coccolithophorids together accounting (in about equal numbers) for half that number. Coccolithophorids were dominant at some depths. The profile of cell numbers with depth showed a subsurface maximum with decreasing numbers at greater depths. The empty cell counts followed the same profile, with counts that were almost an order of magnitude smaller, probably indicating much recycling within the euphotic zone. Diversity indices revealed the complexity of the phytoplankton communities on these transects, and the Shannon Weaver index (H') summed over all depths for each station appeared to approach an asymptote in most cases. Transect C (northeast of the other sites) showed a deeper mixed layer and greater diversity.

Introduction

The present study deals with quantitative analysis of marine phytoplankton from the water column above the central Pacific manganese nodules in the area of the Deep Ocean Mining Environmental Study (DOMES). These "blue water" stations can be classified as oligotrophic, but they are rich in the number of species present. Collections for the present study were taken in the fall of 1975 and early spring of 1976, during the seasons when the incident energy from the sun is approximately the same (Venrick, 1969). Some seasonal variation was noted, however, with winds from the Northeast stronger in the spring and weaker in the fall (Oztergut et al., 1978), with the atmospheric InterTropical Convergence Zone moving from 13°N in September, 1975, to 4°N in March, 1976.

The core of the North Equatorial Current (NEC) lies near the 16°N latitude (measured between 148-157°W, Seckel, 1975), with the northern boundary at 20-23°N and a more rapid current in the fall. However, only Site C was actually in the NEC (Oztergut et al., 1978), with both Sites A and B (the more southern sites) having eastward moving currents as part of the North Equatorial Counter Current (see fig. 1).

Mid-Pacific phytoplankton communities are supported by adequate illumination for photosynthesis at considerable depth in the water column. Sukhanova (1976) calculated the critical depth to be from 400-600 m (apparently from several cruises) in the equatorial waters, and Vernick (1969) reported a range of 175-209 m in the fall and 164-188 m in the spring at 25°N, a more northern location. Light is attenuated more rapidly near the coast. Reid et al.(1978), found the chlorophyll a (chl a) maximum layer for 100 km along the coast of California at a depth of 14-19 m. This maximum is in the thermocline and upper portion of the nutricline (the depth of sudden increase in amount of nutrients), usually at 3-30% of surface irradiance. Gundersen et al (1976) found the chlorophyll a (chl a) maximum layer much deeper, from 50-130 m, off the Hawaiian Islands with the 1% light level (often considered the base of the euphotic zone or the base of the layer where most of the photosynthesis can take place in the ocean) at 130-140 m.

Because the light supports photosynthesis to such depths in mid-Pacific waters, limiting factors could well be nutrients. Silicic acid rarely limits growth rate, at least of diatoms, in the eastern tropical Pacific (Thomas & Dodson, 1975), so nitrogen can be expected to control productivity in the DOMES area. Belyaeva (1971), however, related phosphate concentration to the groups of species present in the western tropical Pacific. Although individual species might have different nutrient requirements, it can be argued that a limiting factor can be taken up quickly by a population and thus show a negative, rather than positive, correlation to cell numbers.

Methods for studying phytoplankton in this area have varied widely, including the Hardy Plankton Indicator with a net mesh of 80 μm (Desrosieres, 1969), the classical inverted microscope sedimentation technique using water samples (e.g., Hasle, 1959; 1960a; 1960b; Marshall, 1970; 1972), and filtration techniques

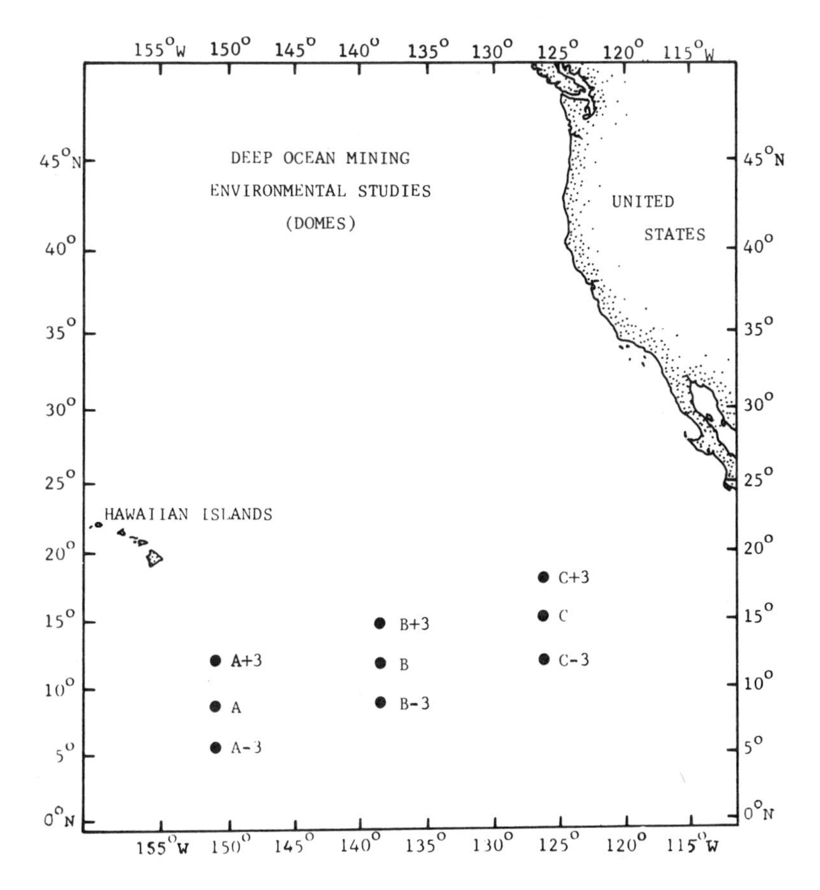

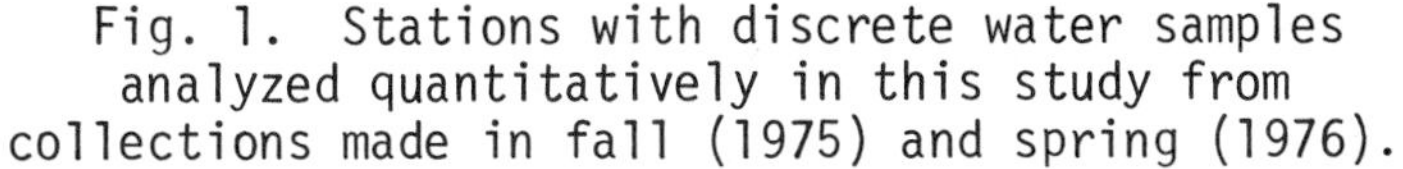
Fig. 1. Stations with discrete water samples analyzed quantitatively in this study from collections made in fall (1975) and spring (1976).

(Honjo & Okada, 1974). Vertical net hauls integrate the entire population within the water column sampled, and collect the larger rarer species in sufficient numbers for morphologic and taxonomic studies. Vertical net hauls, however, cannot be compared directly to settled water samples and are about an order of magnitude lower in phytoplankton quantitative estimates (Guillard and Kilham, 1977). Sukhanova (1976) sampled discrete depths with a closing net and noted that in the Indian and Pacific Oceans the species found in settled samples differed by as much as 90% from those in net samples (large and small organisms). Semina and Tarkhova (1972) found that the cell diameters of net-caught phytoplankton in the tropics were larger than those of bottle-caught cells, up to a ratio of 10:1. Steyaert (1973) studied a transect from Africa to Antarctica. She found a correlation coefficient of 0.47 for percentages of six species in the net and water bottle phytoplankton samples. Coccolithophorids and other small forms are not usually found in net phytoplankton (personal observations), and their calcareous scales can be destroyed by acidic preservations (SCOR Working Group 33, 1974).

Some studies from the DOMES area have dealt only with diatoms, e.g., Mann (1907) with mud samples and Venrick (1969) with water samples. Rampi (1952) found mostly dinoflagellates in his nets. Hasle (1960a) was the first to note coccolithophorids, often dominant in water samples from three stations in the equatorial Pacific. Marshall (1970, 1972) also noted the presence of coccolithophorids, and Okada and Honjo (1973) and Honjo and Okada (1974) studied only the coccolithophorids. Coccolithophorids slip through most nets and can be obtained best by use of preserved, neutralized water samples or filtered water samples. Gundersen et al. (1976) found no coccolithophorids in stations off the Hawaiian Islands, probably due to their sampling methods, but they noted the importance of silicoflagellates. In their studies pennate diatoms accounted for 88% of the cell numbers, but only 19% of the biomass, while silicoflagellates accounted for 3% of the cell count and 41% of the biomass. Thus different methods serve different purposes, with net hauls (used here only for reference and total numbers) smoothing out patchiness and selectively sampling the larger species present, and the settled water samples (to be reported here), providing quantitative information and diversity measurements, especially on the smaller forms amenable to preservation.

Oceanic species tend to be widely distributed in the world ocean. For instance, half of the diatoms found in this study were also found by Sournia (1970) in his list of species from the Mozambique Channel, although there were only two *Nitzschia* species in common in the two areas. The same was true of 57% of the dinoflagellates found in the DOMES region. Semina (1974) summarized the phytoplankton found in the Pacific Ocean, listing over 1,000 taxa. A comparison with other numbers found in recent studies is shown in Table 1. These studies varied in methods (see above), areas sampled, and objectives. However, they illustrate the large numbers of taxa, most of which are involved in primary production in the Pacific Ocean. Geographical studies such as these are dependent on accurate identifications at the specific level, and good identification often is based on sophisticated taxonomic research (e.g., Hasle, 1976).

Although species are widely distributed, small scale horizontal distribution patterns of oceanic diatoms were noted by Venrick (1972) from August to September, 1964, and significant non-random distributions were found, depending on species and depth; she noted that the diatom *Nitzschia turgiduloides* Hasle had a pattern of aggregation on a scale of about one mile, perhaps

TABLE 1

Number of phytoplankton taxa (mostly at the specific level) found in different studies of the Pacific Ocean

Reference	Location	Method	Diatoms	Dino-flagellates	Cocco-lithophorids	Other
Rampi, 1952	Equatorial, 165°-175°W	Surface net tows	40	169	-	6
Hasle, 1959	Equatorial, 145°W	Water samples	42	53	33	10
Smayda, 1966	Inshore tropical waters Gulf of Panama	Water samples	175	105	22	24
Venrick, 1969	North of Hawaii, oceanic & neritic 155°W, between 26° and 56°N	Net and water samples	110	-	-	-
Marshall, 1970	Peru Current, Ecuador to Gulf of Panama	Water samples	39	9	3	2
Marshall, 1972	Galapagos Islands to Ecuador	Water samples	57	31	16	7
Semina, 1974	Review of entire Pacific, oceanic and coastal	Net and water samples	272	683	48	21
Honjo & Okada, 1974	155°W, off Alaska to W of Tahiti	Filtered water samples	-	-	90	-
Present study	DOMES area	Net and water samples	163	122	48	15

reflecting internal wave action along the thermocline of the Subarctic Pacific.

Large scale geographic distribution patterns can be seen between coastal and oceanic waters in species present and biomass. The biomass is larger inshore, and the amount of phytoplankton offshore (and off-shelf) is related to the depth of the main pycnocline. Semina and Tarkhova, (1972) found a negative correlation (-0.69, $p < .01$) between the amount of phytoplankton from bottle samples and the depth of the pycnocline with the highest density gradient. They also found a positive correlation (+0.74, $p < .01$) between the amount of phytoplankton and size of density gradients in seasonal and main pycnoclines.

Phytoplankton also has an uneven vertical distribution at a given station. Hasle (1959) observed the equatorial area to be comparable to other tropical localities, with the abundance of phytoplankton greatest in the upper 50 m. Honjo and Okada (1974) found the most populated depths for coccolithophorids to be 50 m. Marshall (1972) stated that cells were most abundant to depths of 100 m, below which there was a rapid decline. Venrick (1969) found that the diatom maximum layer increased in depth from north to south (using many observations --- 409 samples from 50 stations). Hobson and Ketcham (1974) found decreasing phytoplankton carbon from 10 to 50 m, but the chl _a_ maxima were between 45-75 m. (The explanation seems to be that cells at greater depths contain more chl _a_). A sub-surface chl _a_ maxima is apparently the usual pattern in tropical and subtropical waters,and, at least nearshore, the flora of the surface and chl _a_ layers may be distinct (Reid _et al._, 1978).

The sub-surface maxima of cell numbers, chl _a_, and primary production is frequently noted in the literature with different causes cited (Saijo _et al._, 1969; Anderson _et al._, 1972; Hobson and Lorenzen, 1972; Takahashi _et al._, 1972; Venrick _et al._, 1973). Gundersen _et al._ (1976) found that the peak of the phytoplankton was at the upper boundary of the thermocline where the concentration of nutrient salts was close to limiting, and they also found that numbers of microzooplankton followed the phytoplankton curve. Longhurst (1976) found that vertical zooplankton distribution was irregular. Highest zooplankton concentrations were found to be closer to the maximum primary productivity layer than to the maximum layer of plant material as measured by chl _a_. Thus he concluded that grazing effects the numbers of primary producers in layers near maximum production, in addition to differential sinking (on cell number profiles) considered in earlier models. But there is not complete agreement that grazing is a major control of phytoplankton cell numbers at any one level. Venrick (1971) found that zooplankton biomass in the North Pacific never accounted for more than 4% of the variability of diatoms in her considerations of horizontal relationships.

The purpose of this paper is to summarize the vertical distribution of phytoplankton (full and empty cells) in the DOMES area and to consider the diversity within the water column in the three transects sampled in two different seasons.

Materials and Methods

Samples of water (500 ml size) were collected from discrete depths from three stations on each transect (Fig. 1) in the fall of 1975 and again in the spring of 1976. They were preserved in approximately 1% formalin (neutralized with hexamine)/seawater for later counting. Coccolithophorids and diatoms were well preserved, as were heavily thecate dinoflagellates; naked dinoflagellates and flagellates were fairly well preserved. Containers were shaken vigorously for at least 75 seconds to dislodge organisms adhering to the walls of the container before decanting into 50 or 100 ml settling chambers, and the aliquots settled for 24 or 30 hours, respectively. The quantity settled depended on the number of the cells/l expected; the goal was to cover approximately half of the chamber in counting at least 300 full cells, except in a few deep, sparse samples, because if a taxon is present at the 1% level in a population, there is a 95% probability that it will be found if 300 cells are counted (Shaw, 1964). Approximately half of the settling chamber was examined in parallel strips, in order to reduce the effect of non-random distribution near the margin of the settling chamber. Usually five depths from each station, including surface (=1m), 50% level of surface incident light, chl a maximum layer or near it, 1% and 0.1% level of surface indicend light, were counted on a Zeiss phase and bright-field inverted microscope at X200 (X500 as needed and rarely at X1250). Whenever possible, replicate subsamples were prepared in the same way, and major groups (diatoms, dinoflagellates, etc.) were counted to allow an average to be used for a total from any one depth. Cells per liter were calculated from the area covered, the amount of water settled, and the number of cells counted.

Empty cells were counted and assumed to be dead at the time of collection; it must be recognized that some cells without a sturdy theca (small flagellates, etc.) cannot be recognized after the cells start to deteriorate after death. Thus there were no "empty flagellates", but it was still considered valuable information to have the other empty cells recorded, especially since the recycling of thecal material can be an important resource. A program was written for entering the quantitative data on a computer in FORTRAN, and correlations with other variables were computed.

Vertical net hauls were made with a 35 μm mesh net, and preserved in 2% formalin (neutralized with hexamine)/seawater, and studied in a Zeiss phase-contrast light microscope (LM) in water mounts for identification of larger dinoflagellates and some diatoms including those that form distinctive chains. LM studies were used in assessing the total numbers of species presented in this paper, as well as aiding in the identification of rare species seen at difficult angles in the settled quantitative samples. To study diatoms with diagnostic fine structure, samples were cleaned of organic matter (Simonsen, 1974; Hasle and Fryxell, 1970) and mounted on cover slips glued to aluminum stubs and coated with gold-palladium for study in a JEOLCO JSM-U3 scanning electron microscope (SEM) for qualitative analysis and taxonomic studies not reported here (Fryxell, 1978).

The following formulae were used for diversity indices from full cells:

1) for the simple diversity index (Margalef, 1958), $d = \frac{S - 1}{\ln N}$

where: S = number of species in sample
N = total number of full cells

2) for the Shannon and Weaver (1949) index, $H' = -\sum_{1}^{S} P_i \log_2 P_i$

where $P_i = \frac{n_i}{N}$ or proportion of i'th species in total, N.

3) for the coefficient of variation, $C.V. = \frac{s}{X}$ where s is standard deviation and X is average abundance in cells/l.

A precipitate in samples collected from Transect A, 1976, prevented that transect from being analyzed.

Quantitative Results

Diatoms, coccolithophorids, dinoflagellates, and other taxa were found in moderately small numbers in our samples. The number of cells/l for each of the three well-defined groups was approximately comparable, with an overall average in the samples counted of about 3,080 diatom cell/l, 3,110 dinoflagellate cells/l, 4,800 coccolithophorid cells/l, and 9,600 cells/l of other organisms, including flagellates. Thus the overall average (not weighted for depth of sample) for the DOMES area was over 20,500 cells/l. This value does not agree well with the quantity of phytoplankton in "bathymetric" (Water sample?) assembalges mapped by Semina (1974, Fig. 10, p. 44) of 1,000-10,000 cells/l over much of the same area, although if the flagellates and other organisms are omitted, the agreement is much better.

The number of cells/l was not high in the DOMES stations, with numbers ranging from 83,500 cells/l recorded at the surface at Station B, 1976, to 1,720 cells/l from 117 m at the same station in 1975. The average number of cells/l of each station is shown in Table 2, with no weighting for the different depths counted. The highest numbers were recorded in the spring, with many flagellates. A complete print-out of the phytoplankton quantitative data found at the DOMES stations is presented in the Appendix.

TABLE 2

Average number of phytoplankton cells per liter collected during DOMES Cruises 1975-1976 (total number of samples and subsamples examined)

Station	1975 Cells/l.		1976 Cells/l.	
A+3	6,520	(10)	------	
A	13,900	(8)	------	
A-3	26,600	(10)	------	
B+3	12,700	(10)	43,600	(10)
B	23,100	(8)	38,200	(10)
B-3	15,200	(10)	40,000	(8)
C+3	4,980	(8)	24,600	(10)
C	6,490	(4)	11,500	(4)
C-3	11,900	(9)	28,700	(10)
Year mean	13,500		31,100	

On the middle graph of quantitative data for each station (Figs. 2-4), the density gradient is plotted (dotted line). It can be seen that Transects A and B generally had strong pycnoclines, deeper at the southern stations. But the two northern stations of Transect C had a fairly steady density gradient increase in 1975, and a poorly developed pycnocline was seen again in Station C+3 in 1976.

The nitrate curve is shown in the third graph for each station (Anderson & Richards, 1977), smoothed in the cases noted from few data points. Since nitrates are commonly believed to be limiting in oceanic environments, it is interesting to note that the chl _a_ maximum is at or above this nutricline, and the pycnocline and nutricline are closely related, except in Station A, 1975.

In general the profiles for the total cells/l show a subsurface maximum in or above the pycnocline and then a gradual diminution with depth (Figs. 2-4). In Transect A (1975), the maximum cell count was usually in or above the pycnocline

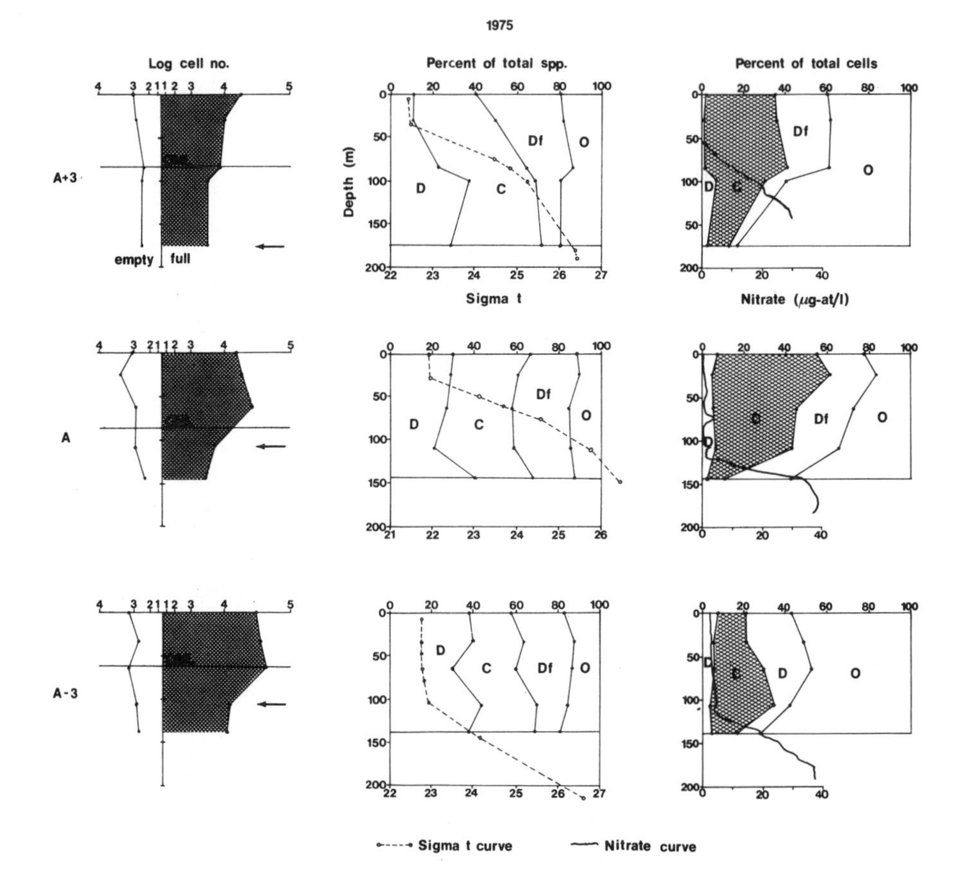

Fig. 2. Quantitative information for Transect A, 1975. Log cell numbers of full and empty cells (note modified log scale), percent of total species at each depth, and percent of total cells at each depth. Density (sigma t) and nitrate profiles are included. Arrow points to base of euphotic zone, or level of 1% of the incident light at the surface. D = diatoms, C = coccolithophorids, Df = dinoflagellates, and O = other taxa, CML = chlorophyll maximum layer. Depth the same on all graphs.

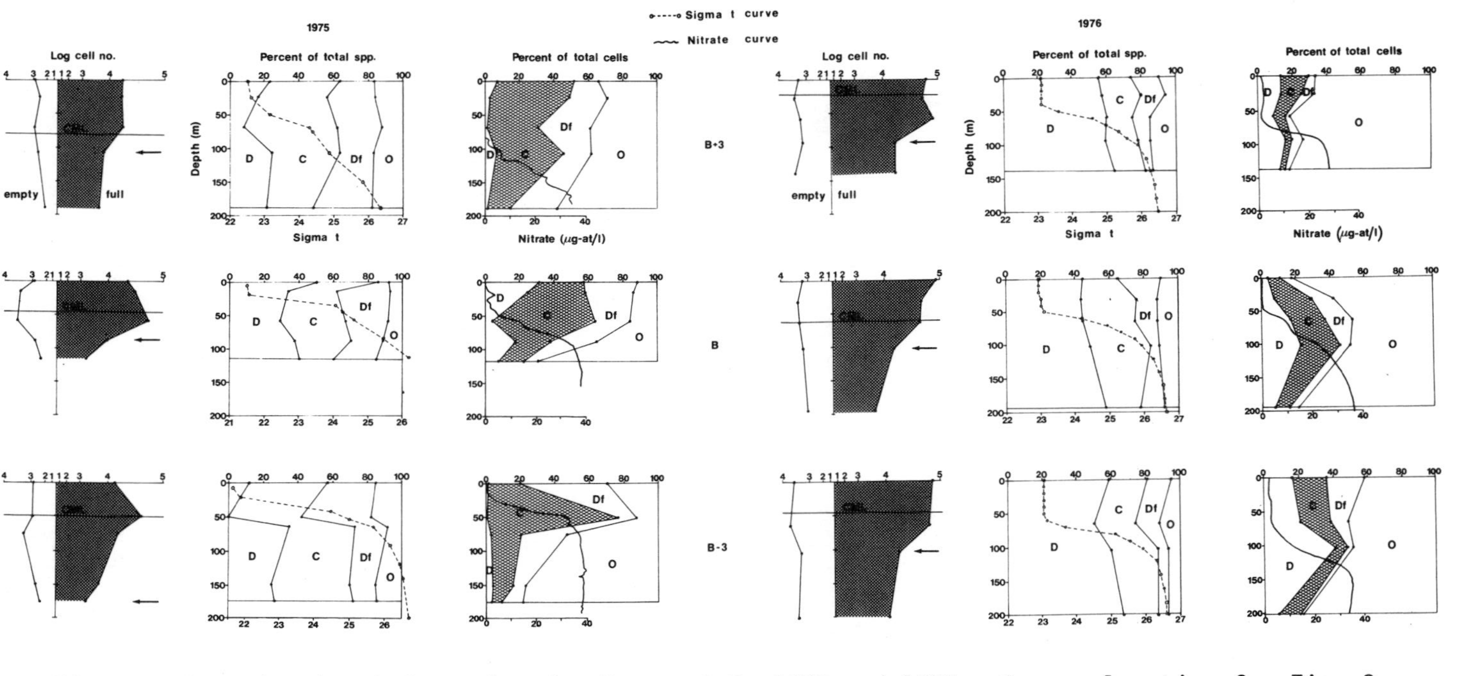

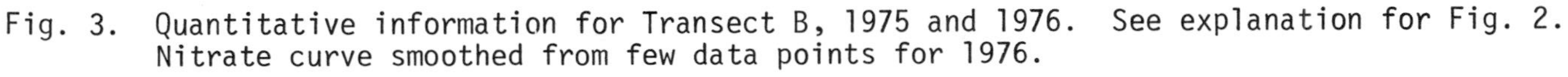
Fig. 3. Quantitative information for Transect B, 1975 and 1976. See explanation for Fig. 2. Nitrate curve smoothed from few data points for 1976.

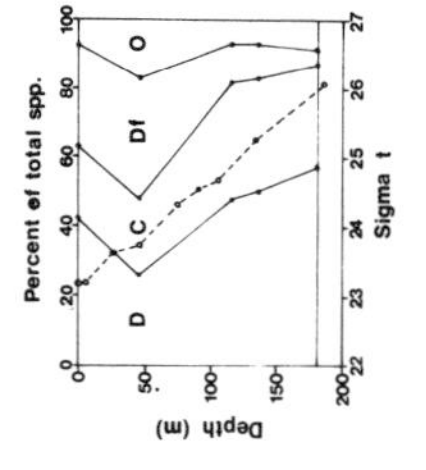
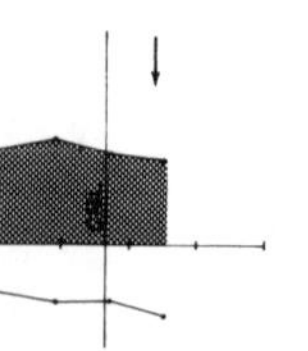
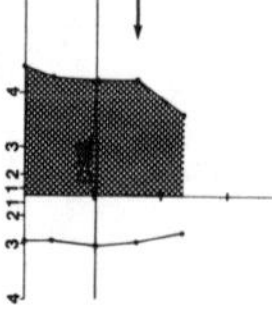
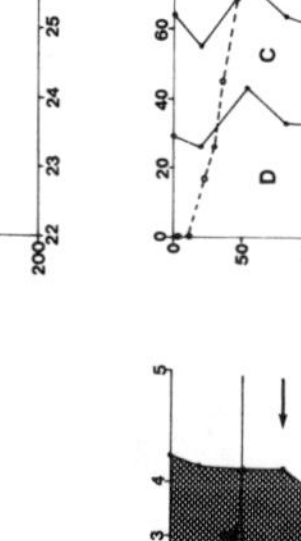
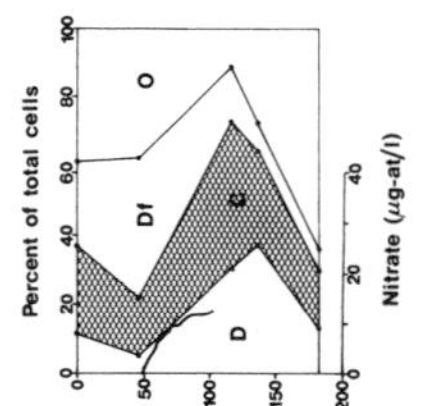
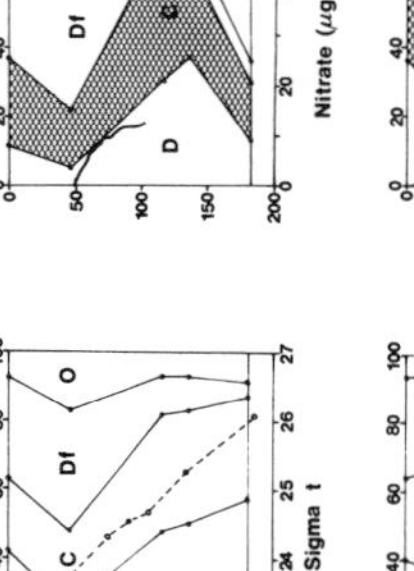
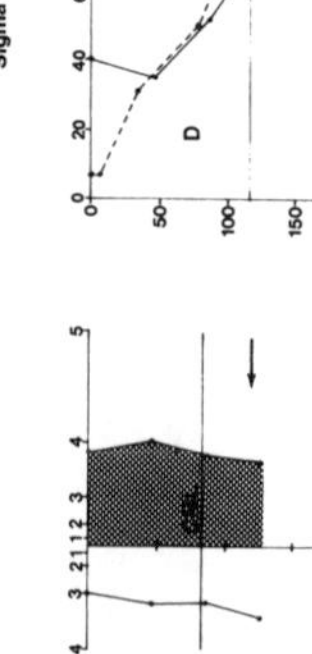
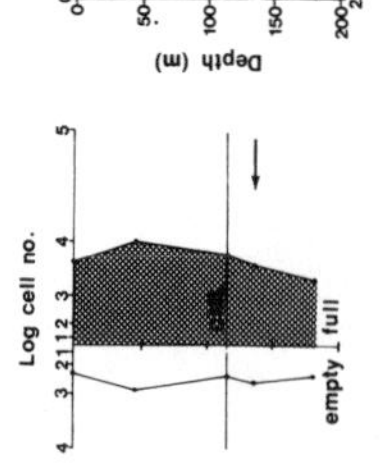

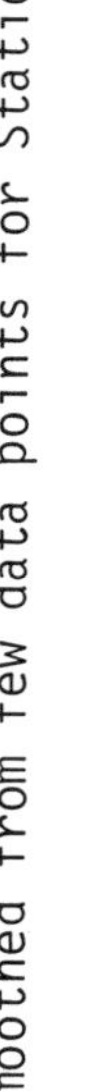
Fig. 4. Quantitative information for Transect C, 1975 and 1976. See explanation for Fig. 2. Nitrate curve smoothed from few data points for Stations C+3, C-3, 1976.

(Fig. 2), and the same is true for Transect B in both years (Fig. 3). However, the pycnocline in 1975 was not clear at Station C+3 or Station C, and the total cell number was quite evenly distributed. A very marked pycnocline was present at Station C-3 the same year, with the maximum population at the surface. In the spring of 1976, a very slight thermocline was in evidence at Station C+3, and the maximum concentration was just above it. The pycnocline was more marked at the southern stations, and Station C-3 once again showed a strong, shallow pycnocline, with maximum cell counts at the surface at this season. Transect C showed marked differences from the other transects, as well as north-south differences on the transect (Fig. 4).

The maximum cell concentration was usually above the chl a maximum layer (9 stations) (see El-Sayed & Taguchi, this volume), although occasionally at the same level (3 stations) or below the chl a maximum (3 stations). Samples were not collected from the chl a maximum depth in all cases for quantitative work; the exact depth was counted in more than half of the cases. In all cases, the bulk of the phytoplankton cells was above the base of the euphotic zone (arrows, Figs. 2-4), considered at the level of 1% incident light at the surface.

The percent of total taxa present from each group showed interesting changes. Few diatom species were recorded at some depths (middle graph for each station, Figs. 2-4), but generally they were well represented. Coccolithophorids were fairly even in the percent of total taxa, with the exception of Station C-3, 1976, which was dominated by diatom species at all depths. In general, there was a lower percentage of dinoflagellate taxa in the spring (1976). The percentages of other taxa remained fairly constant, mostly because flagellates and monads, etc., were counted as groups, and some were always present.

The percent of total numbers of full cells for each group is shown on the right-hand graph for each station (Figs. 2-4), with the coccolithophorids shaded. The coccolithophorids were consistently an important fraction of the total phytoplankton population, with the exception of some of the stations in the spring of 1976 with the large numbers of flagellates. The flagellates, monads, etc., in the "Other Taxa" group cannot be ignored when considering the cell numbers, increasing below the pycnocline in most cases. With the possible exception of Station C, 1975, this group was abundant and often dominant. Diatoms were in small numbers in Station A and two stations of Transect B, 1975, while the coccolithophorids were abundant at those stations. Dinoflagellates were never abundant at lower depths.

In general, the profiles of empty-cells/l showed a subsurface maximum and then a gradual diminution with depth (Figs. 2-4), following the full-cell curves with numbers smaller by almost an order of magnitude. (Note modified log scale.) The traditional view is that of empty cells slowly sinking. On this basis one would expect a "crossover" of profiles of full and empty cells, with empty cells being in greater numbers at greater depths. Such "crossovers" did occur below 100 m between chl _a_ and phaeopigments (El-Sayed & Taguchi, this volume). Fecal pellets as well as dead cells are involved with phaeopigments, and empty cells may have lost their pigments some time previously, so relationships may be complex. However, crossovers did not take place with total cell counts and empty cells, and were found only at four stations for coccolithophorids alone, and five stations for diatoms, at deep levels and in small numbers. The non-thecate cells, such as flagellates, probably burst after death and are recycled rapidly.

TABLE 3

Ratio of empty thecate cells to total full phytoplankton cells on DOMES 1975-1976 Cruises (with number of observations)

Year	Transect	Surface		Lowest Depth		Average	
1975	A	.090±.061	(6)	.073±.028	(6)	.071±.030	(30)
	B	.043±.008	(6)	.107±.059	(5)	.070±.023	(26)
	C	.094±.059	(5)	.295±.285	(4)	.156±.119	(21)
1976	B	.037±.018	(6)	.208±.032	(6)	.097±.024	(28)
	C	.113±.040	(6)	.186±.059	(5)	.158±.051	(25)
1975 Average		.076±.028	(17)	.158±.043	(15)	.099±.050	(77)
1976 Average		.075±.050	(12)	.197±.050	(11)	.127±.049	(53)

The ratios of empty thecate to total full cells were small, with an overall average of 0.110 ± 0.0666 (standard deviation of 130 observations), probably indicating few dead cells (Table 3). There might be some effect of sinking, though, as evidenced by the overall average ratio of 0.075 ± 0.048 (29) at the surface, increasing to 0.174 ± 0.139 (26) at the lowest depth sampled. If the ratio remained the same throughout the water column, including the surface, one could assume dead cells remained at the level where they had lived. However, increased mortality at low light levels is also a possible explanation of the increased ratio of empty/full cells with depth as well as sinking of dead cells, and so is a decreased rate of cycling in deeper, colder water. (It should be noted that the actual number of cells represented in the ratios is much smaller at the lowest depth than at the surface.) Station C-3, 1976, had the highest ratio of empty to full cells, with an average 0.211 ± 0.013, with empty diatom cells largely involved.

The integrated cell count (cells/m^2) throughout the euphotic zone ranged from 1.00 X 10^9 cells for Transect C, 1975, to 5.22 X 10^9 cells for Transect B, 1976; the overall average was 2.87 X 10^9 cells for both years (Table 4). These figures are averages for the transects; the lowest integrated cell count was 0.896 X 10^9 at Station C in 1975, and the highest was 5.89 X 10^9 at Station B-3, 1976. The higher counts came at stations with many flagellates recorded, although our method of preservation probably lost many of the athecate forms.

TABLE 4

Integrated phytoplankton quantitative data on each DOMES transect with standard deviations

Year	Transect	#Cells X10^{-9}/m^2 in euphotic zone*	#Cells below euphotic zone** / #Cells in euphotic zone
1975	A	2.81 ± 1.06	8.29% ± 2.40
	B	2.53 ± 0.75	15.5 % ± 14.8
	C	1.00 ± 0.13	19.1 % ± 8.30
1976	B	5.22 ± 0.63	18.9 % ± 4.26
	C	2.74 ± 0.18	15.5 % ± 10.3
Overall Average		2.87 ± 1.60	15.8 % ± 7.73

*1% - 100% incident light
**.1% - 1% incident light

Although the cell counts generally diminished with depth, full (and presumably living) cells were noted below the level of 1% of the incident light at the surface, the level generally considered to be the base of the euphotic zone (Table 4). The average depths of the euphotic zone in the stations used for quantitative data were 122 m ± 34 m in 1975 and 93 m ± 14 m in 1976. The overall average of the cells below the euphotic zone amounted to about 16% of those recorded for lighted layers (Table 4). The percent of chl a found between 1 and 0.1% of the surface incident radiation was 19% ± 17% in 1975 and 22% ± 18% in 1976 (El-Sayed & Taguchi, this volume). It appears that there is a sparse viable population, at least between the 1% and 0.1% light level, spread over a considerable depth (43 ± 20 m in 1975, 68 ± 27 m in 1976).

It is interesting to compare these findings with those from similar studies in the equatorial Pacific. Hasle (1959) recorded populations beneath 0.1 m^2, adapted in Table 5 to be comparable to our material (under 1 m^2). Hasle's counts were probably to depths much lower than 1% of the light level, and they ranged from 3.04 - 4.52 X 10^9 cells, similar to the upper range of the counts in Table 4. The lowest part of the water column studied by Hasle has 4-6% of the cells compared with the "upper" layers, although she worked with very deep samples (down to 434 m, in one case). It would seem that, in working with depths to 0.1% of the incident light level, the DOMES study did cover the depths with preponderant numbers of phytoplankton cells.

The ten most abundant taxa (that is, found in the greatest numbers overall) are listed in Table 6. Grouped together for the most abundant single category was the flagellate and monad group, which were not distinguished. The increased count of flagellates in 1976 might be attributed partly to experience at recognizing them. Another group was the Gymnodiniaceae, which includes all the dinoflagellates with non-rigid thecae that "round up" when preserved. Thus several species are no doubt included. "Yellow cells" (or "olive-green cells" in other works) are probably a similar grouping of resting or reproducing cells (Anderson, 1975), packed full of storage material, but with little morphological detail to distinguish them.

TABLE 5

Integrated phytoplankton cells/m^2, adapted from Hasle (1959)

Equatorial Pacific --- Cells X 10^9		
Station 88	Station 1	Station 2
To 159 m - 4.47	To 158 m - 4.52	To 175 m - 3.04
To 434 m - 4.74	To 318 m - 4.70	To 267 m - 3.17
Antarctic Pacific --- Cells X 10^9		
Station 25 (pre-maximum pop.)	Station 41 (maximum population)	Station 50 (post-max. pop.)
To 150 m - 4.01	To 100 m - 70.47	To 120 m - 0.2

Gephyrocapsa huxleyi [Reinhardt, 1972, = *Coccolithus huxleyi* (Lohmann) Kamptner] was the most abundant coccolithophorid, amounting to almost 60% of the total coccolithophorid cells (Table 7) and the most abundant species in the chl *a* maximum layers counted.

TABLE 6

Ten most abundant taxa in order of decreasing abundance determined by total number of cells counted in discrete water samples collected during DOMES Cruises 1975-1976

Flagellates and Monads
Gephyrocapsa huxleyi (Lohm.) Reinhardt
Gymnodiniaceae
Yellow Cells
Gephyrocapsa oceanica Kamptner
Nitzschia bicapitata Cleve
Oxytoxum variable Schiller
Nitzschia lineola Cleve
Nitzschia closterium (Ehrenberg) Wm. Smith
Gephyrocapsa ericsonii McIntyre & Bé

TABLE 7

Geophyrocapsa huxleyi
As Percent of Coccolithophorids
(with number of observations)

Station	1975	1976
A+3	50.3 (5)	-
A	31.2 (5)	-
A-3	52.8 (5)	-
B+3	58.8 (5)	74.1 (5)
B	62.6 (5)	80.7 (5)
B-3	86.4 (5)	32.4 (4)
C+3	26.1 (5)	68.2 (5)
C	45.6 (4)	43.7 (4)
C-3	62.9 (5)	88.9 (5)
Mean	52.9	64.7
Overall Mean	57.6	

Okada & Honjo (1973) showed it to be increasingly dominant among coccolithophorid species at all depths from about 30°N toward higher latitudes, and practically the only coccolithophorid N of 40°N into the Subarctic. They found it from 25-50% of coccolithophorids in their "Equatorial North" zone. G. oceanica, usually slightly larger than G. huxleyi, tended to be golden in color at least in these samples, possibly the result of preservation. The color, however, aided in identification in these samples. G. ericsonii is an extremely small species, usually less than 4 µm in diameter and often only 2 µm. Although it is more irregular in outline than G. huxleyi, the two species are enough alike that it is possible that one was counted as the other at magnification of X500 in the overlapping size range in this study.

The diatom Nitzschia bicapitata is omnipresent in oceanic samples and varies from a minute, stubby bicapitate frustule 5 µm long to larger forms about 35 µm long. Large cells of N. bicapitata appear similar to small cells of N. braarudii Hasle. Simonsen (1974) shows a range of sizes and shapes, but it can be noted that his micrographs show that the costae of this species are arcuate near the apices, while those of N. braarudii are straight. Since it was not always possible to distinguish N. bicapitata from N. braarudii between 30 and 40 µm in length, 35 µm was arbitrarily taken as the boundary between the two taxa, with N. braarudii being the larger, as defined in the literature (Hasle, 1960a).

Nitzschia lineola was found in permanent mounts prepared from net samples, but it is a very slender pennate diatom found in chains with over-lapping tips, and it could have been confused in water mounts with N. pseudodelicatissima. N. closterium was also commonly found, although there was much variation in its form, casting some doubt on the identification.

Oxytoxum variabile is also commonly found in the Gulf of Mexico (Fryxell, unpublished), and there is considerable doubt that it is photosynthetic. Attempts to grow others of its genus in culture have failed, and the swimming cells seemed to exhibit phagotrophic feeding behavior.

Species found at every station (Table 8) include two diatoms, three taxa of dinoflagellates, two coccolithophorids, and two other taxa. Not mentioned previously is another Nitzschia species, N. sicula, which has many of the appearances of the Fragilariopsis group of Nitzschia common in waters near Antarctica. Gymnodinium oceanicum also was often seen empty with a delicate theca. Ophiaster hydroideus, apparently not frequently observed by Okada & Honjo (1973), was present in every DOMES station. Perhaps it was not well preserved by their filtration methods, for the cell was often separated from long "jointed" coccoliths in our water samples.

Taxa that only were found at or below the chl a maximum included 31 diatoms, 19 dinoflagellates, 9 coccolithophorids, and 3

TABLE 8

Taxa present at every station in quantitative samples collected during DOMES Cruises 1975-1976

Diatoms
Nitzschia braarudii Hasle
Nitzschia sicula (Castr.) Hust.
Dinoflagellates
Gymnodiniaceae
Gymnodinium oceanicum Hasle
Oxytoxum variabile Schill.
Coccolithophorids
Gephyrocapsa huxleyi (Lohm.) Reinh.
Ophiaster hydroideus (Lohm.) Lohm.
Other
Flagellates and Monads
Yellow Cells

other organisms. However, it is difficult to assign any particular "shade flora" concept to these forms, since they were mostly found once or twice and could well have been rare forms instead of confined to deeper waters. One of us (GAF) ascribes some significance to the finding in deeper waters of *Asteromphalus sarcophagus* Wallich, a distinctive diatom previously noted only in deep samples or in vertical net tows in the Gulf of Mexico (Fryxell, unpublished), and from the Indian Ocean, e.g., in samples from depths of 60-100 m (Thorrington-Smith, 1970). It was originally described by Wallich (1860) from an Indian Ocean salp and may well be a deep-water form. Also significant is the presence of *Florisphaera profunda* (Okada and Honjo, 1973) considered a deep water species by its describers. They found it principally at or below 100 m (mostly at 150 and 200 m) and surmised that it preferred limited light intensity below the euphotic zone, and temperatures above 10°C. Thus, limitations of light intensity and temperature might control its distribution. *Anthosphaera oryza* (Schlauder) Gaarder, sometimes found near the surface, was dominant in the two deepest depths counted at each station almost as often as was *Gephyrocapsa huxleyi*.

Diversity

Diversity is a measure of the number of species present and the relative proportions of individuals in each species. It is greater as the number of species increases if the population is evenly distributed among the species. With most indices, the diversity approaches zero in an almost unialgal bloom, even though the number of individuals may be high.

In calculating a diversity index, it is generally assumed that species tend to be ecological equivalents, an assumption that has been examined and found generally to hold (Johnson and Raven, 1970). It must also be assumed that a small subsample of the entire community is a reliable predictor of its overall diversity, a real problem in phytoplankton because of irregular horizontal distribution and stratified vertical distribution. Because the total number of phytoplankton species cannot be known as exactly as the number of species of trees in a forest area, for instance, the number of species and diversity in samples of decreasing size are generally underestimated, and the evenness of distribution overestimated (Johnson and Raven, 1970).

However, the diversities of progressively pooled samples approach an asymptote (Wilhm, 1970; Pielou, 1966). Diversity indexes measured from the asymptote may thus be considered independent of size and comparable between different communities; our individual sample indices, which have been calculated on at least 300 cells from each depth, are not independent of size and can be compared best to similar counts. The pooled estimates of diversity can be used with greater confidence, because they have been pooled over all the depths analyzed for one column of water at a station. However, one must accept that the "communities" being measured is that of the whole water column, and not a particular depth (such as the chl a maximum layer).

The simplest measure of diversity is the number of species present, thus disregarding the evenness of distribution. In all the net hauls and discrete water samples in the entire area we observed: 163 diatom, 122 dinoflagellate, 48 coccolithophorid, and 15 other taxa, for a total of almost 350 taxa (most of them at the specific level) as noted in Table 1. Although Sukhanova (1976) found the total number of diatom and dinoflagellate species varied little with depth, the usual pattern in our study was a sub-surface maximum, followed by a slight decrease (Fig. 5). The average number of taxa at a discrete depth is 30 ± 9, and the average found in the water column of one station is 68 ± 12. The range for a discrete depth is from 12-58 species, and for a station is 45-93.

Both fall (1975) and early spring (1976) stations on Transect C (farthest east and north, Fig. 1) showed a higher number of species than stations on the other two transects (Fig. 5). The chl a

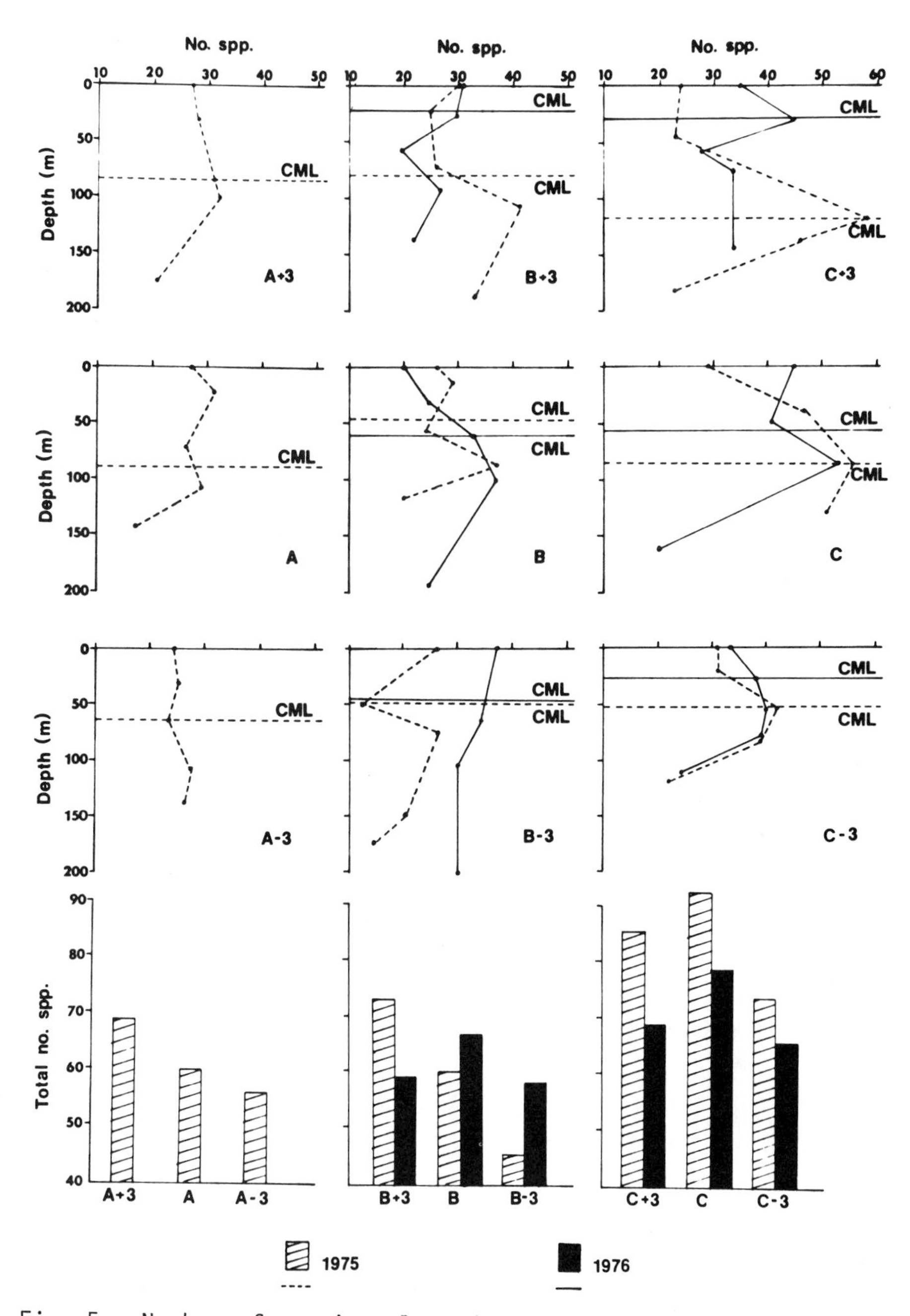

Fig. 5. Number of species plotted with depth and total number of species in DOMES stations from discrete water samples. Dashed and solid lines indicate chl a maximum layer (CML) in the fall of 1975 and spring of 1976, respectively.

maximum is either at or above the level of the highest number of species recorded. More diatom species were present on Transect C (Fig. 6) than on the transects farther west. While there were more coccolithophorid species found in the fall, this group was fairly consistent in the studied area.

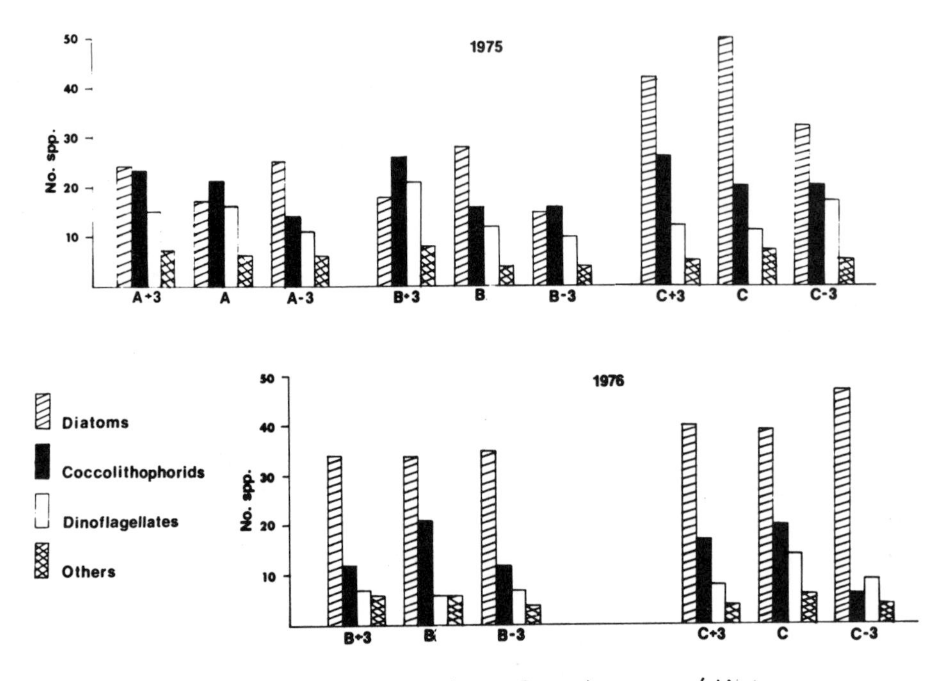

Fig. 6. Total numbers of species of each group (diatoms, coccolithophorids, dinoflagellates, and other taxa) for each station in 1975 and 1976 from discrete water samples.

The overall diversity, d (considering only the number of species and the number of individuals), for all depths was 2.89 for 1975 and 2.51 for the following spring. Once again, the overall pattern was one of a subsurface maximum and a gradual diminution with depth (Fig. 7), with the chl a maximum at or above the highest value for the index, d. The subsurface maximum of stations on Transect C, especially in 1975, stands out with high "d" values, in contrast to Station B-3 or A-3 in 1975.

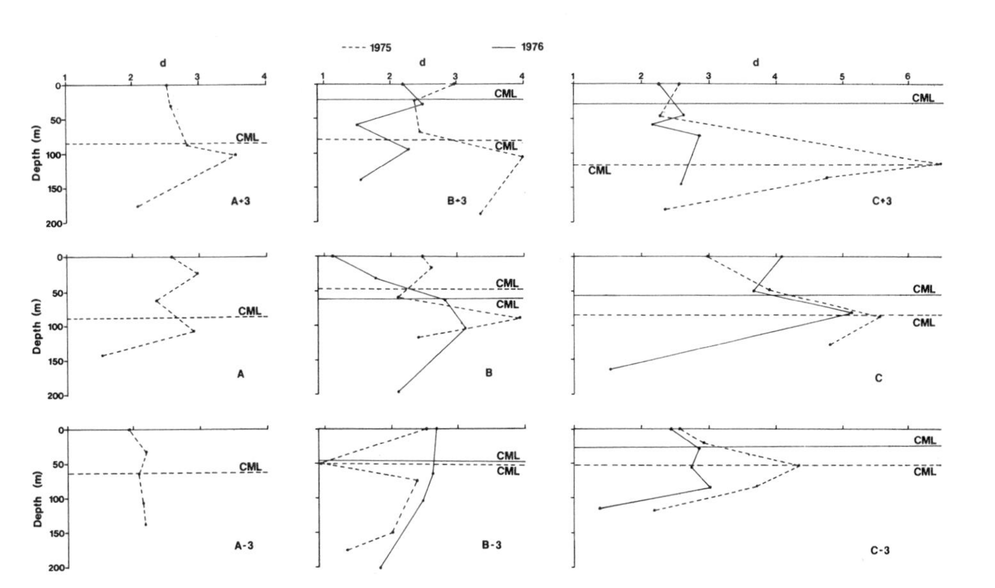

Fig. 7. Simple diversity index (see Methods section) plotted against depth for 1975 (dotted lines) and 1976 (solid lines). CML = chl a maximum layers for each year.

The Shannon-Weaver index, H', gives weighting to both the numbers of species and the probability that each species will be found. Once again, higher diversities were seen in the fall, with an overall average of H' = 3.24 as opposed to 2.77 for the spring when large numbers of flagellates were found. The individual transects varied from a low diversity index of 2.50 on Transect B in 1976 to a high of 3.70 on Transect C in 1975. There is a significant difference in H' at the 95% level between Transects B and C and at the 91% level in seasonal variation. The profiles of H' with depth are plotted on Fig. 8. The usual, but by no means uniform, pattern is one of a subsurface maximum, followed by lower values of the index with depth. The chl a maximum layer is at or above the maximum diversity, as was the case with the number of species and the simple diversity index, d.

The diversity values, H', of cumulatively pooled samples are shown in Fig. 9. In most cases, the diversity increased slightly

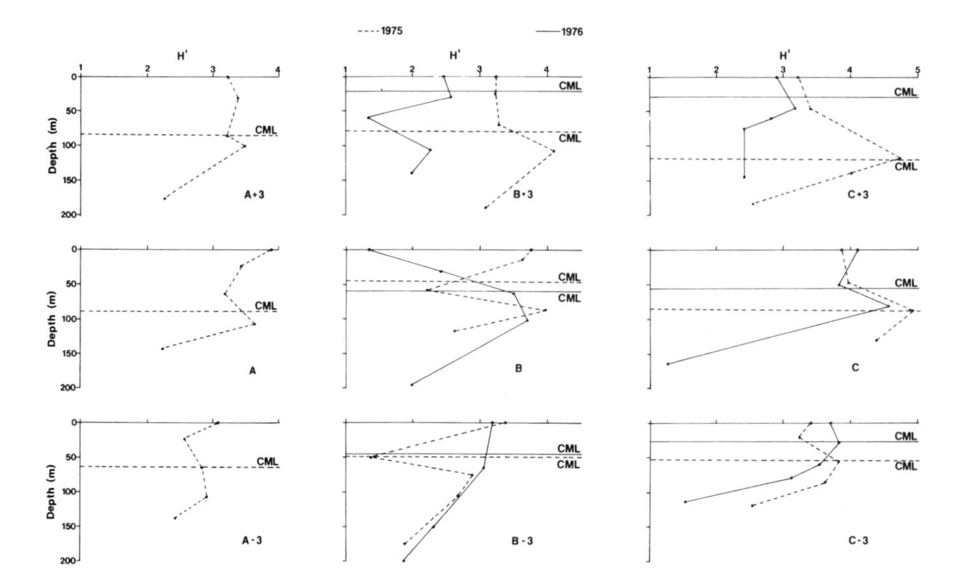

Fig. 8. Diversity index H' plotted against depth for DOMES stations in 1975 and 1976. CML = chl a maximum layer for each year.

with the cumulative data from additional depths. It may be that an asymptotic value can be estimated. The surface information from Transect A (1975) gave a good estimate of the asymptote to the H' curve of the whole station. Transect B showed variations in both years, and it is probable that more material was needed at Station B-3 (1975) and Station B+3 (1976) to approach the asymptote. Transect C had high diversity, with the asymptote approaching or exceeding H' = 4 both seasons. In 1975, sufficient material was not examined from Station C to indicate that the asymptote was reached. Hendey (1977) considers H' to vary typically between 1-4, and he worked out a rough scale on the assumption that lower numbers indicate pollution of (stress on?) the system. Few comparisons can be made with other oceanic work, but it is evident that this area has highly diverse populations of phytoplankton, with the highest diversity in the waters in the farthest north and east of the area studied in the North Equatorial Current.

The coefficient of variation (CV) takes into account the evenness of distribution --- that is, it varies directly with the standard deviation (as well as inversely with the average abundance). It gives a different kind of information to augment

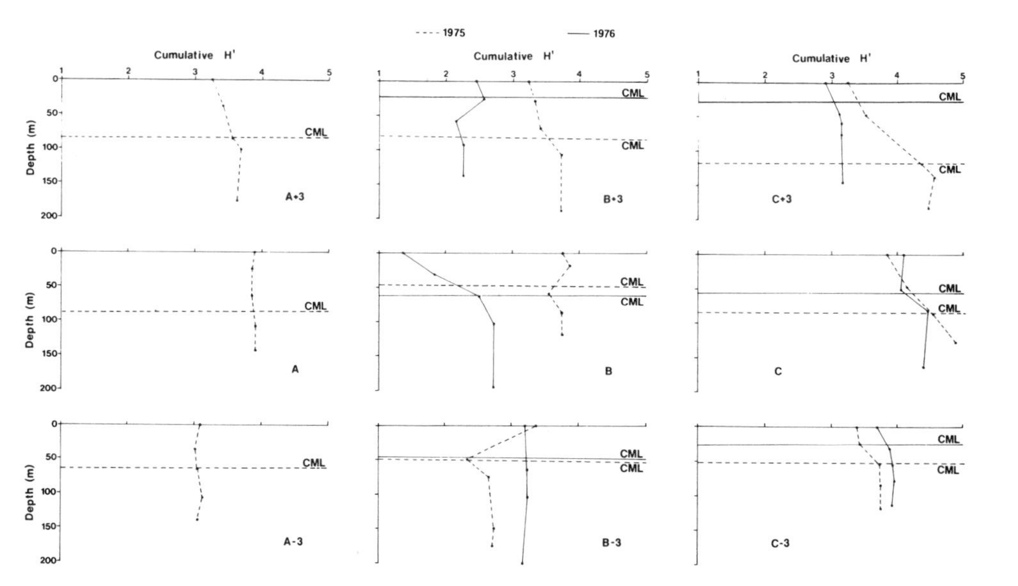

Fig. 9. Diversity index H' calculated cumulatively with depth. That is, H' was calculated fro the surface sample, as above. Then the top two samples were calculated as one population for the next point, etc., and the bottom point was calculated from the entire water column. In most cases, and symptote (H' value) may have been reached.

diversity indices. Although there was sometimes a sub-surface maximum, the deeper samples tended to have a higher coefficient of variation (Fig. 10). The average of the surface CV was 1.85 ± 0.57 (15 observations) and of the deepest layer counted was 2.28 ± 0.42 (15). The overall average for all stations is 2.05, and Transect C had a low of 1.72 in 1975, indicating an even distribution. Transect B had a high average of 2.67 in 1976, indicating uneven numbers in different taxa --- in this case, a high number of flagellates in relation to the rest of the community. A fairly frequent occurrence was a mid-depth peak, indicating many taxa at that depth with similar counts.

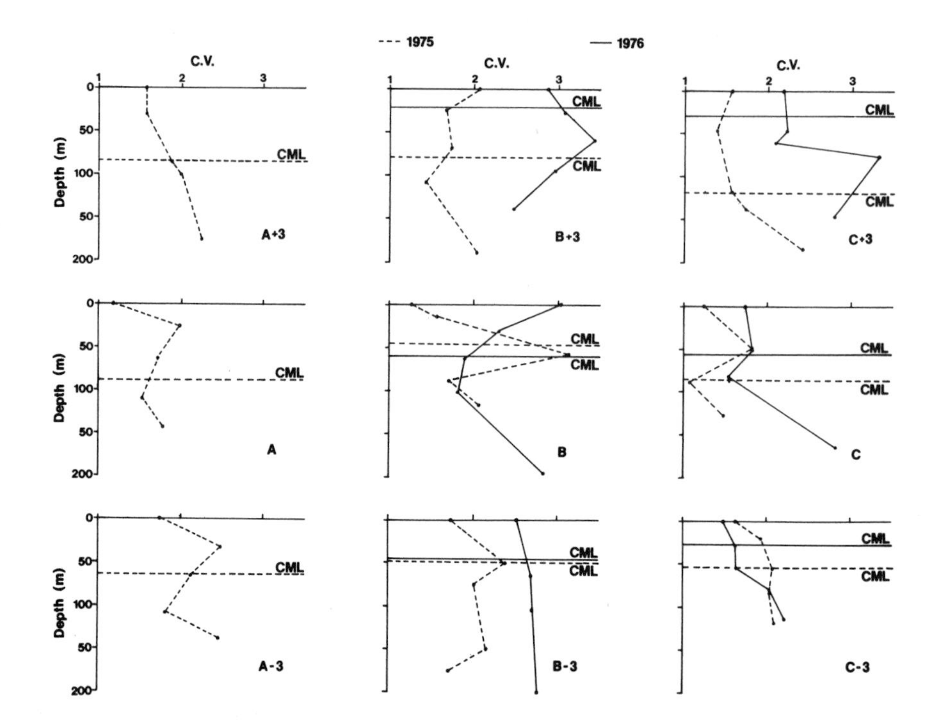

Fig. 10. Coefficient of variation C.V. (see Methods section) plotted against depth for 1975 and 1976. CML = chl a maximum layer.

Discussion

Transect C was different from the other two transects in phytoplankton community. The cell numbers were lower (Table 3), with higher diversity (Figs. 7 and 8), than in the other transects. The difference in the mean diversity, H', between Transects B and C is significant at the 95% level. Okada & Honjo (1973) stated that the North Equatorial Current (NEC) was broad (between 25° and 10°N), while the Equatorial Counter-current was swift and narrow toward the east. Transect C was the only one in the NEC and only the southern station of the transect, C-3, had a strong pycnocline. In the northern two stations of Transect C, the surface mixed layer appears to extend below the euphotic zone, while in most other DOMES stations, a fairly strong pycnocline (with a following nutricline --- nitrates, in this case), is evident. Shoaling of the pycnocline in Station C-3 in both years might have been a factor in the larger surface populations, as suggested by Longhurst (1976). However, a similar feature was seen in 1975 at Stations B and B-3 with a sub-surface maximum.

Since chl *a* is a measure of phytoplankton standing stock, one would expect the CML to be located at the level of the maximum cell concentration. However, the CML was at or above the maximum cell concentration, suggesting a sinking, and thus possibly senescent, population. However, the ratio of empty to full cells was quite even throughout the mixed layer and did not increase with depth, in contrast to the findings of Ohwada (1972) in the Sea of Japan, where there were more living cells at the surface and more dead cells, in general, below 50 m. Thus it would appear that there is considerably more recycling in the oceanic euphotic zone than in areas of higher surface populations.

Longhurst (1976) proposes that in regions of stable production, phytoplankton profiles are determined more by grazing pressures than by sinking rates. He cites the zooplankton subsurface maximum at the bottom of the mixed layer ("planktocline"), concentrated closer to the maximum primary production layer than to the CML. Longhurst (1976) does not discount differential sinking rates invoked in earlier models, but he assumes that herbivore grazing may be the primary determining factor of phytoplankton profiles within a water column. Mortality has been variously estimated in the literature, from 10-15% in coastal waters (Smayda, 1966, Gulf of Panama) to an estimate of 40-70% of phytoplankton in the water column consumed each day in tropical waters of the western Pacific (Sorokin & Tsvetkova, 1972, calculated from diurnal changes in relative photosynthesis in a water column). Gundersen *et al*. (1976) point out that below the chl *a* maximum layer there is a NO_2 maximum layer at about the level of the increase in NO_3, with the microzooplankton concentrated close to the chl *a* maximum. There can be little doubt that there is dynamic interaction between zooplankton and phytoplankton in the water column in the levels of the nutricline, pycnocline, and chl *a* maximum. Dramatic reduction in phytoplankton numbers below the chl *a* maximum (lacking in Transect C) could be explained better by Longhurst's grazing hypothesis than by sinking.

Semina & Tarkova (1972) found a negative correlation between the amount of phytoplankton in the water column and the depth of the main pycnocline and a positive correlation between the standing stock and the size of the density gradient. We found only weak agreement on our data, with a correlation of -0.228 between the average number of cells/l (including flagellated cells) in the euphotic zone and the depth of the maximum density gradient, and a correlation of +0.278 between the average number of cells/l in the euphotic zone and the maximum rate of change of the density gradient. Semina & Tarkova (1972) pointed out that a small density gradient favors greater loss of cells by their sinking below the lower limit of the biotope, and it probably also favors a more diffuse distribution of the zooplankton.

We also found a strong correlation (+0.928) of the number of cells in the surface sample with the average number of cells/l in the water column, figured on the unweighted samples from the four or five depths counted. Such a correlation is significant for remote sensing of oceanic areas. There was a negative correlation (-0.562) of the total number of cells integrated over the euphotic zone with the depth of the chl *a* maximum, indicating higher populations when the nutrients in the water column can support growth in layers with more light.

Other correlations were run on all our data with physical factors (Table 9), giving, in general, better correlation in the spring of 1976 than in the preceding fall. There was negative correlation with those factors directly related to depth --- salinity and the density. There was a positive correlation with those factors inversely related to depth --- light and temperature.

TABLE 9

Matrix of correlation coefficients between phytoplankton cell number and physical factors from 1975-1976 DOMES Cruises

1975	Depth	Light	Temperature	Salinity	Sigma T
Log cell #	-0.656	0.347	0.682	0.006	-0.606
(Prob.)	0.0001	0.0225	0.0001	0.9733	0.0001
(# Obs.)	43	43	40	38	38
1976					
Log cell #	-0.769	0.540	0.693	-0.403	-0.690
(Prob.)	0.0001	0.0030	0.0001	0.0371	0.0001
(# Obs.)	28	28	26	27	26

The phytoplankton cell numbers were negatively correlated with nutrients --- nitrates, phosphates, and silicates --- in both seasons (Table 10). However, the correlation was stronger, again, in the spring of 1976 than in the preceding fall. Thus, if the phytoplankton numbers were high, the nutrients had been removed from that part of the water column. A strong negative correlation would seem to illustrate that, depending on the order of magnitude involved and complicated by the need for light for photosynthesis, each of the nutrients recorded could be limiting to the phytoplankton growth, and recycling within the water column was essential to maintenance of the population.

TABLE 10

Matrix correlation coefficients between phytoplankton cell number and nutrients from 1975-1976 DOMES Cruises

1975	Nitrate	Phosphate	Silicate
Log cell #	-0.481	-0.322	-0.526
(Prob.)	0.0034	0.0591	0.0012
(# Obs.)	35	35	35
1976			
Log cell #	-0.711	-0.722	-0.692
(Prob.)	0.0001	0.0001	0.0001
(# Obs.)	28	28	28

There was a slight positive correlation with phytoplankton cell numbers and chl *a* in the fall of 1975 and a much stronger one in the following spring (Table 11). There was no correlation between phaeopigments and phytoplankton cell numbers in the fall and a weak negative relationship in the spring. Primary production was positively correlated with cell numbers, more strongly in the spring.

TABLE 11

Matrix of correlation coefficients between phytoplankton cell numbers related variables from 1975-1976 DOMES Cruises

1975	Chlorophyll *a*	Phaeopigments	Primary production
Log cell #	0.296	0.0327	0.398
(Prob.)	0.0573	0.8373	0.0082
(# Obs.)	42	42	43
1976			
Log cell #	0.711	-0.320	0.715
(Prob.)	0.0001	0.1038	0.0001
(# Obs.)	27	27	28

The species we found seem to fit better into the water masses farther north than equatorial tropical waters. In comparison with species found in other studies (Table 12), the "Subtropical Complex" of Jousé et al. (1970) were found at more of the DOMES stations than were the "Tropical Complex". Their samples were from

TABLE 12

Number of stations in DOMES area in which common species from other studies were observed

Species as listed by Jousé, Kozlova, & Muhina, 1970	Numbers of DOMES stations (both seasons included)
"Subtropical Complex" from sediments	
Coscinodiscus radiatus	0
Nitzschia bicapitata	15
N. interrupta	0
N. sicula	15
Pseudoeunotia doliolus	7
Thalassionema nitzschioides	12
Thalassiosira decipiens	0
T. lineata	10
Roperia tessellata	7
"Tropical Complex" from sediments	
Coscinodiscus crenulatus	1
C. nodulifer	2
*Hemidiscus cuneiformis	1
*Nitzschia marina	0
Rhizosolenia bergonii	0
Thalassiosira oestrupii	8

Species as listed by Okada & Honjo, 1973 Central North Pacific "Zone C" from water column	Number of DOMES Stations
Anthosphaera quadricornu	3
Cyclococcolithus fragilis	4
Discosphaera tubifera	7
Florisphaera profunda	2
Gephyrocapsa huxleyi (=Emiliania huxleyi)	15
G. oceanica	14
Rhabdosphaera stylifer (=R. clavigera)	3
Thorosphaera flabellata	7
Umbellosphaera irregularis	10
U. tenuis	2
Umbelicosphaera sibogae	10

*Dominant in complex

sediments, and this may only indicate that rarer species can be concentrated in the sediments but not be recorded from discrete water samples.

Okada & Honjo (1973), working on filtered samples from the water column, found many of the same species in their "Zone C" in the Central North Pacific (Table 12) that we found. However, they found large populations of Gephyrocapsa huxleyi farther north, and we did not find the large populations of Umbellosphaera irregularis that they did in latitudes similar to those of the DOMES collections (although it was present in 10 of 15 stations). It is possible that U. irregularis is better preserved on filters than in water samples or that we were really dealing with different water masses. They found the highest diversity, H', at 33°N for coccolithophorids, and diminishing in the DOMES latitudes.

The overall species diversity index, H', usually decreased with depth, although a high index was found at the chlorophyll maximum layer if diatoms were abundant. Individual transects varied, Transect C in the fall of 1975 having the highest diversity index; and Transect B in the spring having the lowest. The small flagellate numbers were higher in the spring and decreased the diversity. There is a significant difference at the 91% level of probability between the diversity in the fall and the spring.

The higher species diversity in the fall is associated with a lower standing crop (chl a = 9.25 ± 3.5 mg/m^2) as the lower species diversity in the spring is associated with a higher standing crop (chl a = 19.86 ± 7.92 mg/m^2). This is consistent with the results observed during the spring bloom of phytoplankton in coastal water (Platt & Subba Rao, 1970). However, the negative relationship between the standing crop and species diversity index seems to be at variance with the thesis of Margalef (1968) that increases in biomass can be expected to be associated with increases in diversity. Margalef (1977) states that thermodynamic considerations anticipate an inverse correlation between temperature and diversity; this concept may be easier to apply on land than in the stratified tropical sea. He feels that low diversity is associated with high primary producer/carnivore ratios and considers productivity usually inversely related to diversity in less productive waters. In the DOMES region, high diversity was related to a low standing crop. Copeland and Bechtel (1971) found a similar relationship inshore in the Gulf of Mexico.

The relationship between H' and number of species, simple diversity index d, and coefficient of variation, are shown on Fig. 11. As one would expect, there is a positive correlation of H' with number of species and with simple diversity index--apparently good enough to have some predictive value. The negative correlation with the coefficient of correlation is not quite as intuitively obvious, but since it is a measure of lack of evenness of

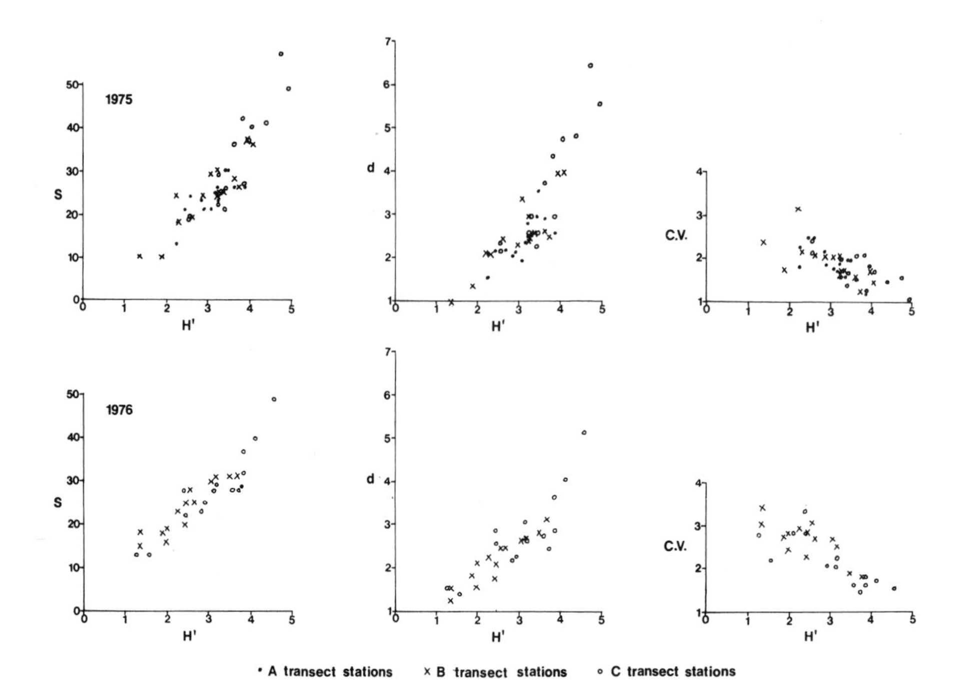

Fig. 11. Relationship between the diversity index, H', and number of species, S, simple diversity index, d, and coefficient of variation, C.V., for stations from Transects A, B, and C in 1975 and 1976.

distribution, it is reasonable that field samples that are high in diversity will be low in coefficient of variation.

Diversity of macroalgae and diatoms depend to some extent upon the grazing pressure from herbivores (Lubchenco, 1978; Patrick, 1970, 1978). Herbivores have been shown either to increase or decrease diversity, depending on the intensity of grazing. That is, intense grazing lowers the population, decreasing numbers and eliminating marginal species. But the near-absence of grazing can also lower diversity by allowing the dominant alga to out-compete other species for the nutrients and light needed for maintenance of a population. Thus moderate grazing can allow the less competitive algae to survive, increasing the evenness of the distribution of numbers among the plant species and thus increasing the diversity. Since unialgal "blooms" are rare in the open ocean, and certainly none was observed in the DOMES area, it appears that a balance is achieved between plants and herbivores with resultant high diversity.

Summary and Conclusions

Diatoms, coccolithophorids, and dinoflagellates were present in approximately the same numbers of cells/liter in these samples from the water column and together equalled the number of (preservable) cells of flagellates and others. *Gephyrocapsa huxleyi* accounted for more than half of the coccolithophorids and was the dominant organism found at the chl *a* maximum layer. At greater depths most of the species found below the euphotic zone were in small numbers and perhaps better termed "rare species" than truly "shade flora". However, *Asteromphalus sarcophagus* (diatom), *Florisphaera profunda*, and *Anthosphaera oryza* (both coccolithophorids) were considered deep water species.

The north equatorial phytoplankton represents a diverse community of low biomass. This study supports the growing awareness that coccolithophorids are an important group in the ecosystem of the tropical and subtropical waters. They are present in roughly the same numbers as diatoms and dinoflagellates, although most of the cells are small. Little is known about the reproduction rate of most of the species. It could be slower than the smallest flagellates but considerably faster than the larger dinoflagellates, and thus a renewable resource to be considered in the overall ecosystem of the DOMES area.

In general, there was a subsurface maximum in cell numbers and diversity of the phytoplankton population, and a decrease with greater depth. Transect C, farthest east, showed differences with higher diversity and lower numbers of cells. The bulk of the cells was above the base of the euphotic zone, with more than 10^9 cells under 1 m^2, and about 16% of that number was estimated to be lower, between the depths of 1% and 0.1% of the incident light (surface) measured at noon.

The depth of the chl *a* maximum was negatively correlated with the total number of cells (integrated) in the water column, and there was a stronger correlation between the number of cells in the surface layer and the average number found in samples from the water column. The cell number was negatively correlated with the amount of nutrients and positively correlated with primary production and chl *a*, especially in the spring.

The community studied is sparse, diverse, and spread over great depths in the DOMES area. Empty cells, found to occur at an order of magnitude less than full cells, followed the full cells in numbers and did not indicate much sinking effect in these small populations. Recycling of organic nutrients rather than sinking of dead cells seems to be the common pattern.

Acknowledgments

We wish to express our appreciation for the cooperation of

many in the collecting of the samples used in this study and the supporting data used, especially K. R. Buck, D. A. Stockwell, and R. A. Warner. M. Riggs and N. Fisher aided in the computer and statistical analysis with skill and good humor. A. S. Dreyfus helped to put the original data in order and prepare it for computer analysis. A. M. Ketterer drafted the figures and aided in the final preparation of the manuscript, as did N. Kimble and P. Turner. P. A. Fryxell critically read the manuscript.

This work was supported by NOAA Contract #03-6-022-35115.

References

Anderson, G. C., B. W. Frost and W. K. Peterson, (1972) On the vertical distribution of zooplankton in relation to chlorophyll concentration, Biological Oceanography of the Northern North Pacific, A. Y. Takenouti (ed.), Idemitsu Shoten, Tokyo, 341-345.

Anderson, J. J. and F. A. Richards, (1977) Continuous profiles of chemical properties in the euphotic zone of the DOMES study area in the Equatorial North Pacific, Special Report No. 78, NOAA Contract NOAA-03-6-022-35117, University of Washington, Seattle.

Anderson, O. R., (1975) A possible origin of some olive-green cells in aphotic marine environments, Limnol. Oceanogr. 20, 1000-1004.

Belyaeva, T. V., (1971) Quantitative distribution of planktonic diatoms in the Western Tropical Pacific (in Russian, English summary), Okeanologiya 11, 687-694.

Copeland, B. J. and T. J. Bechtel, (1971) Species diversity and water quality in Galveston Bay, Texas, Water, Air, and Soil Pollution 1, 89-105.

Desrosieres, R., (1969) Surface macrophytoplankton of the Pacific Ocean along the equator, Limnol. Oceanogr. 14(4), 626-632.

El-Sayed, S. Z. and S. Taguchi, Phytoplankton standing crop and primary productivity in the tropical Pacific, this volume.

Fryxell, G.A., (1978) Chain-forming diatoms: three species of Chaetoceraceae, J. Phycol. 14, 62-71.

Guillard, R. R. L. and P. Kilham, (1977) The ecology of marine planktonic diatoms, in The Biology of Diatoms, D. Werner, ed., University of California Press, Berkeley, 372-469.

Gundersen, K. R., J. S. Corbin, C. L. Hanson, M. L. Hanson, R. B. Hanson, D. J. Russell, A. Stollar, and O. Yamada, (1976) Structure and biological dynamics of the oligotrophic ocean photic zone off the Hawaiian Islands, Pacific Science 30(1), 45-68.

Hasle, G. R., (1959) A quantitative study of phytoplankton from the equatorial Pacific, Deep-Sea Res. 6(1), 38-59.

Hasle, G. R., (1960a) Plankton coccolithophorids from the subantarctic and equatorial Pacific, Nytt Mag. Bot. 8, 77-92.

Hasle, G. R., (1960b) Phytoplankton and ciliate species from the tropical Pacific, Skr. Norske Vidensk. Akad. I. Mat.-Nat. Kl. No. 2, 1-58.

Hasle, G. R., (1976) The biogeography of some marine planktonic diatoms, Deep-Sea Res. 23, 319-338.

Hasle, G. R. and G. A. Fryxell, (1970) Diatoms: cleaning and mounting for light and electron microscopy, Trans. Amer. Microsc. Soc. 89(4), 469-474.

Hendey, N. I., (1977) The species diversity index of some in-shore diatom communities and its use in assessing the degree of pollution insult on parts of the north coast of Cornwall, Nova Hedwigia, Beih. 54, 355-378.

Hobson, L. A. and D. E. Ketcham, (1974) Observations on subsurface distributions of chlorophyll a and phytoplankton carbon in the Northeast Pacific Ocean, J. Fish. Res. Bd. Can. 31, 1919-1925.

Hobson, L. and C. J. Lorenzen, (1972) Relationship of chlorophyll maxima to density structure in the Atlantic Ocean and Gulf of Mexico, Deep-Sea Res., 19, 297-306.

Honjo, S. and H. Okada, (1974) Community structure of coccolithophores in the photic layer of the mid-Pacific, Micropaleontology 20(2), 209-230.

Johnson, M. P. and P. H. Raven, (1970) Natural regulation of plant species diversity, in Evolutionary Biology Vol. 4, T. Dobzhansky, M. K. Hecht, and W. C. Steere, eds., Meredith Corp., New York, 127-162.

Jouse, A. P., O. G. Kozlova, and V. V. Muhina, (1970) Distribution of diatoms in the surface layer of sediment from the Pacific Ocean, in Micropaleontology of the Oceans, B. M. Funnell and W. R. Riedel, eds., Cambridge, 263-269.

Koblentz-Mishke, O. I. and V. I. Vedernikov, (1973) Tentative comparison of primary production and quantity of phytoplankton on the ocean surface, Oceanology 13(1), 55-62.

Lloyd, M., J. H. Zar, and J. R. Karr, (1968) On the calculation of information-theoretical measures of diversity, American Midland Naturalist 79(2), 257-272.

Longhurst, A. R., (1976) Interactions between zooplankton and phytoplankton profiles in the eastern tropical Pacific Ocean, Deep-Sea Res., 23, 729-754.

Lubchenco, J., (1978) Plant species diversity in a marine intertidal community: importance of herbivore food preference and algal competitive abilities, American Naturalist 112(983), 23-39.

Mann, A., (1907) Report on the diatoms of the Albatross voyages in the Pacific Ocean, 1888-1904, Contributions from the U. S. National Herbarium 10(5), 221-443.

Margalef, R., (1958) Information theory in ecology, Genet. Systems 3, 36-71.

Margalef, R., (1968) Perspectives in Ecological Theory, University of Chicago Press, Chicago.

Margalef, R., (1977) Ecosystem diversity differences: poles and tropics, in Polar Oceans, M. J. Dunbar, ed., Arctic Institute of North America, Calgary, Alberta, 367-375.
Marshall, H. G., (1970) Phytoplankton in tropical surface waters between the coast of Ecuador and the Gulf of Panama, J. Wash. Acad. Sci., 60(1), 18-21.
Marshall, H. G., (1972) Phytoplankton composition in the southeastern Pacific between Ecuador and the Galapagos Islands (Archipielago de Colon), Proc. Biol. Soc. Wash. 85(1), 1-38.
Ohwada, M., (1972) Vertical distribution of diatoms in the Sea of Japan, in Biological Oceanography of the Northern Pacific Ocean, A. Y. Takenouti, ed., Idemitsu Shoten, Tokyo, 145-163.
Okada, H. and S. Honjo, (1973) The distribution of oceanic coccolithophorids in the Pacific, Deep-Sea Res., 20, 355-374.
Ozturgut, E., G. C. Anderson, R. S. Burns, J. W. Lavelle, and S. A. Swift, (1978) Deep ocean mining of manganese nodules in the North Pacific: pre-mining environmental conditions and anticipated mining effects, Preliminary Draft, DOMES Final Report, NOAA DOMES Program, Seattle.
Patrick, R., (1970) Benthic stream communities, Amer. Sci., 58, 546-549.
Patrick, R., (1978) Effects of trace metals in the aquatic ecosystem, Amer. Sci., 66, 186-191.
Patten, B. C., (1962) Species diversity in net phytoplankton of Raritan Bay, J. Mar. Res., 20(1), 57-75.
Pielou, E. C., (1966) The measurement of diversity in different types of biological collections, J. Theoret. Biol., 13, 131-144.
Platt, T. and D. V. Subba Rao, (1970) Energy flow and species diversity in a marine phytoplankton bloom, Nature, 227(5262), 1059-1060.
Rampi, L., (1952) Ricerche sul Microplancton di superficie del Pacifico tropicale, Bull. de L'Institut Oceanographique, No. 1.014, 1-16.
Reid, F. M. H., E. Stewart, R. W. Eppley, and D. Goodman, (1978) Spatial distribution of phytoplankton species in chlorophyll maximum layers off southern California, Limnol. Oceanogr., 23(2), 219-226.
Reinhart, P., (1972) Coccolithen, Die Neue Brehm-Bücherei, A. Ziemsen Verlag, Wittenberg Lutherstadt.
Saijo, Y., S. Izuka, and O. Asaoka, (1969) Chlorophyll maxima in Kuroshio and adjacent area, Mar. Biol., 4, 190-196.
SCOR Working Group 33, (1974) A review of methods used for quantitative phytoplankton studies, UNESCO Technical Papers in Marine Science 18, 1-27.
Seckel, G. R., (1975) Seasonal variability and parameterization of the Pacific north equatorial current, Deep-Sea Res., 22, 379-401.
Semina, H. J., (1974) Pacific Phytoplankton, Publishing House Nauka, Moscow.

Semina, H. J. and I. A. Tarkhova, (1972) Ecology of phytoplankton in the north Pacific Ocean, in Biological Oceanography of the Northern North Pacific Ocean, A. Y. Takenouti, ed., Idemitsu Shoten, Tokyo.

Shannon, C., and W. Weaver, (1949) The mathematical theory of communication, University of Illinois Press, Urbana.

Shaw, A. B., (1964) Time in Stratigraphy, McGraw-Hill, New York.

Simonsen, R., (1974) The diatom plankton of the Indian Ocean Expedition of RV "Meteor" 1964-1965, "Meteor" Forsch.-Ergebnisse, Reihe D., No. 19, 1-66, 41 pl.

Smayda, T. J., (1966) A quantitative analysis of the phytoplankton of the Gulf of Panama 111. General ecological conditions, and the phytoplankton dynamics at 8°45'N, 79°23'W from November 1954 to May 1957, Bull. Inter-American Tropical Tuna Commission, 11(5), 355-612.

Sorokin, Yu. I. and A. M. Tsvetkova, (1972) Marine biology: vertical structure and productivity of a phytoplankton community in the western Pacific, Oceanology, 12(5), 870-878.

Sournia, A., (1970) A checklist of planktonic diatoms and dinoflagellates from the Mozambique Channel, Bull. Mar. Sci., 20, 678-696.

Steyaert, J., (1973) Distribution of plankton diatoms along the African-Antarctic transect, Inv. Pesq., 37(2), 295-328.

Sukhanova, I. N., (1976) Vertical structure of phytocoenosis of some regions of the Indian and Pacific Oceans, Mar. Sci. Communications, 2(6), 375-386.

Takahashi, M., K. Satake and N. Nakamoto, (1972) Chlorophyll profile and photosynthetic activity in the north and equatorial Pacific Ocean, J. Oceanogr. Soc. Japan, 28, 27-36.

Thomas, W. H. and A. N. Dodson, (1975) On silicic acid limitation of diatoms in near-surface waters of the eastern tropical Pacific Ocean, Deep-Sea Res., 22, 671-677.

Thorrington-Smith, M., Some new and little-known planktonic diatoms from the west Indian Ocean, Nova Hedwigia, Beih, 31, 815-835.

Venrick, E., (1969) The distribution and ecology of oceanic diatoms in the North Pacific, Ph.D. Dissertation, Univ. of California, San Diego, University Microfilms, Ann Arbor.

Venrick, E. L., (1971) Recurrent groups of diatoms species in the North Pacific, Ecology, 52(4), 614-625.

Venrick, E. L., (1972) Small-scale distribution of oceanic diatoms, Fish. Bull., 70(2), 363-372.

Venrick, E. L., J. A. McGowan and A. M. Mantyla, (1973) Deep maxima of photosynthetic chlorophyll in the Pacific Ocean, Fish. Bull., 71, 41-51.

Wallich, G. C., (1860) On siliceous organisms found in the digestive cavities of the Salpae, Trans. Micr. Soc. Lond., n.s. 8, 36-55, 2 pl.

Wilhm, J. L., (1970) Effect of sample size on Shannon's formula, Southwestern Naturalist, 14(4), 441-445.

PHYTOPLANKTON STANDING CROP AND PRIMARY PRODUCTIVITY IN THE TROPICAL PACIFIC

Sayed Z. El-Sayed and Satoru Taguchi*

Department of Oceanography
Texas A&M University
College Station, Texas 77843 U.S.A.

*Hawaii Institute of Marine Biology
University of Hawaii at Manoa
P.O. Box 1346, Coconut Island
Kaneohe, Hawaii 96744 U.S.A.

Abstract

Phytoplankton standing crop and primary production were studied during the summer of 1975 and the winter of 1976, in the Deep Ocean Mining Environmental Study (DOMES) area between 5° - 20°N and 128° - 155°W. Chlorophyll *a* values (used as a measure of phytoplankton standing crop) averaged 0.063 mg/m^3 in summer and 0.17 mg/m^3 in winter. In the euphotic zone average values were 9.8 mg/m^2 in the summer and 21 mg/m^2 in the winter. The vertical distribution of chlorophyll *a* at the majority of the stations occupied was characterized by the presence of a subsurface chlorophyll maximum layer (CML). CML was found at a deeper depth (about 60 ± 24 m) in summer than in winter (about 54 ± 30 m). This layer occurred either at the middle or the bottom of the maximum density gradient (pycnocline) in summer whereas in winter it was found either at the top or the middle of the pycnocline. CML was also found to be closely associated with the distribution of the nutricline especially that of NO_3-N.

Chlorophyll maximum was found to correspond to the depth of about 10% and 20% of surface light level in summer and winter, respectively. The concentration in the chlorophyll maximum exceeded the surface concentration by a factor of 3.6 (± 1.9) in summer and by a factor of 1.8(±0.8) in winter. At the chlorophyll maximum layer the proportion of chlorophyll *a* to the total pigments (i.e. chlorophyll *a* plus phaeopigments) was lower (0.44 ± 0.06) in summer than in winter (0.56 ± .13). Following the chlorophyll *a* maximum, there was a general decrease in this algal pigment. In summer, 6.61% of chlorophyll *a* content in the water column occurred below the depth of 1% light level (average: 19 ± 17%). In winter the chlorophyll concentration ranged between 8 and 78% with an average of 22 ± 18%. The coccolithophrid *Gephyrocapsa huxleyi* is the dominant species in the chlorophyll maximum layer. Our data suggest that the chlorophyll maximum layer is due either to zooplankton grazing or to slower growth of larger phytoplankton cells. Our data also permit us to draw the conclusion that the phytoplankton standing crop is likely to be controlled by the depth of the surface mixed layer.

The contribution of the nannoplankton (organisms <20 μm) to total chlorophyll in the water column averaged 84% for both summer and winter cruises.

Primary production in the surface water averaged 1.5 mgC/m^3/day in summer and 2.7 mgC/m^3/day in winter. The daily primary production in the euphotic zone was 120 mgC/m^2/day in summer and 144 mgC/m^2/day in winter. Nannoplankton contribution to primary production averaged about 77% for both summer and winter cruises. The phytoplankton below the euphotic zone contributed about 5% of the production in the water column. Unlike the phytoplankton standing crop, primary production was not controlled by the depth of the surface mixed layer. Analysis of variance showed spatial and seasonal variations in the phytoplankton standing crop and primary production. The degree of variability (which varied by a factor of two) was of the same magnitude found in adjacent regions by other investigators.

Introduction

The first study of primary production and phytoplankton standing crop in the eastern tropical Pacific was made during the EASTROPIC Expedition (King *et al*., 1957; Holmes, 1958). However, it was largely due to the extensive investigations carried out during the Eastern Tropical Pacific Survey (EASTROPAC) in 1967 and 1968 that our knowledge of the physical, chemical and biological oceanography of that region has been greatly enhanced (see Love, 1970-1975; Beers and Stewart, 1971; Blackburn *et al*., 1970; Longhurst, 1976). The phytoplankton in the central gyre of the North Pacific was studied by Venrick (1971). Later Venrick *et al*. (1973) compared the phytoplankton of the Central Pacific Gyres to that of the equatorial regime, as did Marumo (1970). The present study is based primarily on the data collected in an area approximately intermediate (between 5°-20°N and 128°-155°W) to these two regimes. The material was gathered in connection with the Deep Ocean Mining Environmental Study (DOMES) with a 13x10^6 km^2 area of the Pacific that is currently considered a primary target for commercial mining of manganese nodules.

The general physical structure in the study area indicates that a divergence zone running east-west at about 9° to 10°N between the westward flowing North Equatorial Current and the eastward flowing North Equatorial Counter Current controls the mixed layer depth which varies from 20 to 120 m in the area. A strong pycnocline (density gradient) at the base of the mixed layer is controlled by the location of the thermal ridge (an oceanographic feature running east-west between about 9° to 10°N) and has thickness between 50 and 100 m.

The nutrient concentrations in the DOMES area are low everywhere (Anderson and Richards, 1977). The concentrations of phosphate and silicate are about 0.3 and 1.6 μg.at./l, respectively. Representative concentrations of nitrogen compounds in the mixed layer are 0.1 μg.at./l for nitrate, 0.01 μg.at./l for nitrite and 0.05 μg.at./l for ammonia. The nutrient content increased sharply with depth within and through the pycocline. Between the thermocline and 400 m representative concentrations of the nutrients are 35 μg.at./l for nitrate, 2.8 μg.at./l for phosphate, and 35 μg.at./l for silicate. Detailed account of the nutrients chemistry of the DOMES area is given in Anderson and Richards (1977). In the present investigation we are primarily interested in studying the short-term, vertical, spatial and temporal variabilities of chlorophyll *a*, phaeopigments and primary production in the tropical Pacific. Data on solar radiation and light penetration are given by Franceschini (this volume) and the species composition and species diversity of the phytoplankton samples collected are discussed by Fryxell *et al.* (also this volume).

Materials and Methods

All field operations were carried out at seven stations in 1975 and five stations in 1976 along each of three longitudes, approximately 126°W (Transect C), 138°W (Transect B) and 151°W (Transect A). Each transect extended over six degrees of latitude. They consisted of one centrally located "site station" with the other stations to the north and south of each site station separated by one degree of latitude in the summer of 1975 and 1.5 degrees of latitude in the winter of 1976 (Fig. 1). Water samples were collected at various depths with Niskin bottles shortly after measurements of submarine light penetration were made. The depths correspond to 100, 50, 25, 12, 6, 3, 1, 0.1 and 0.01% of light level incident at sea surface (I_0). Dates, time of sampling and depths of the water sampled are given in Appendix I, II and III.

Time-series observations were made over a 30-hr period in order to study the short-term variability of chlorophyll *a* and phaeopigments in the water column. Water samples were collected

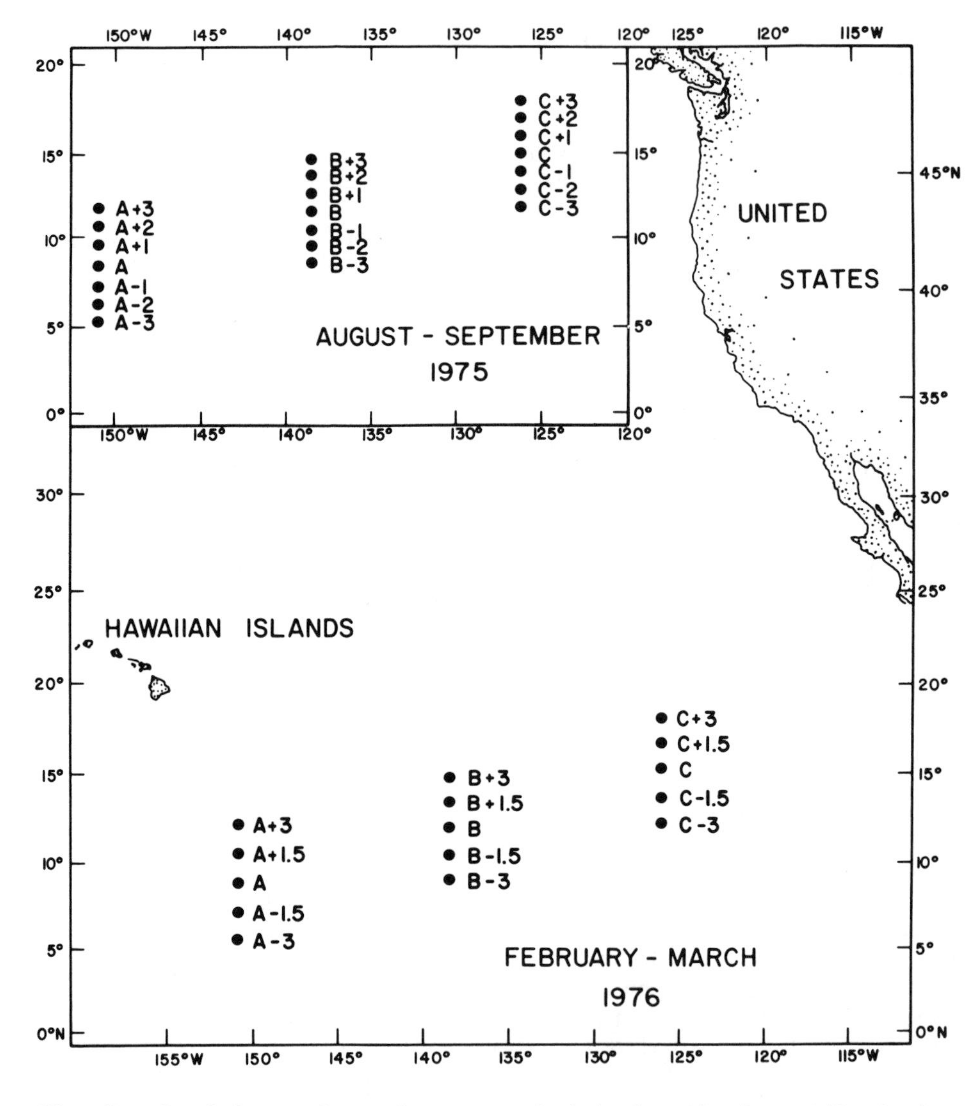

Fig. 1. Positions of stations occupied during the August/September 1975 and February/March, 1976 cruises.

every four hours at Sites A, B and C in the summer of 1975 and the winter of 1976 to investigate the possibility of diurnal variation in the pigment contents. Special series of experiments also were conducted at several locations around Sites A and B to study the spatial variability of chlorophyll _a_ and phaeopigments at these two sites.

Standing Crop of Phytoplankton

Chlorophyll _a_, used as a measure of the phytoplankton standing crop, together with phaeopigments (degradation product of chlorophyll) were determined fluorometrically (Yentsch and Menzel, 1963, with modification of Holm-Hansen _et al_., 1965) using a Turner Designs fluorometer (Model 10).

Water samples (2l) for nannoplankton chlorophyll measurements were fractionated by passing the water first through a 20-μm nytex screen and then through Whatmann GF/C glass-fiber filter. The same amount of water was filtered directly through Whatmann GF/C glass filter for total chlorophyll measurements. The coefficient of variation between 6 replicate measurements was 10%.

Primary Production

Primary production was determined by _in situ_ experiments using the ^{14}C isotope uptake technique of Steemann Nielsen (1952) as modified by Strickland and Parsons (1970). _In situ_ buoys were deployed for about 7 hours (between local apparent noon to sunset).

During the summer cruise, three light bottles and one dark bottle (125 ml each) were used in the _in situ_ experiments. Two light bottles and one dark bottle were filled with unfractionated water (i.e. total plankton) and one light bottle was filled with fractionated water (nannoplankton) using a 20-μm nytex-net screen.For the winter cruise, four light bottles and two dark bottles (125 ml each) were filled with unfractionated water during the _in situ_ experiments. In the winter experiments, two light bottle samples from each depth were fractionated by passing the water first through a 20-μm nytex-net screen and then through an HA Millipore filter. The samples in the remaining bottles were filtered through an HA Millipore filter (total plankton). The radio-activity of the filters was assayed twice for 10 minutes by a liquid-scintillation counter (Packard Model 3550). The coefficient of variation between replicate measurements was 10%.

Primary productivity values have been corrected to full day values on the basis of the amount of photosynthetically active radiation (PAR) available to the water mass (total incident light, I_0, minus reflected light) during the incubation period compared to that received during the full day. This calculation was made on the assumption that the uptake of ^{14}C is in direct proportion to the amount of light available during the day. Values for production per unit light energy (mgC/E) were also calculated in order to evaluate the photosynthetic potential of the stations on a more equitable basis.

Results

Phytoplankton Standing Crop

Table 1 summarizes the minimum, maximum, average, standard deviation and number of observations of surface and integrated values (to the depth penetrated by 1% surface radiation, considered here as the bottom of the euphotic zone; to 0.1 and to 0.01% I_0) of chlorophyll a, phaeopigments and primary production taken during the 1975 and 1976 cruises. Chlorophyll a concentrations in the study area averaged 0.063 mg/m^3 in the summer and 0.17 mg/m^3 in the winter. In the euphotic zone, average values were 9.8 mg/m^2 in the summer and 21 mg/m^2 in the winter.

The phaeopigments concentrations collected in the water column in 1975 and 1976 showed very similar trends to those of chlorophyll a; namely, higher average values in winter than in summer (Table 1).

The percent contribution of nannoplankton-chlorophyll a to total phytoplankton values for each of the light levels studied (i.e. 100, 50, 25%, etc.) varied between 65% and 100% for the three transects in summer and between 33% and 98% for the winter transects, with no significant differences between transects or seasons. Nannoplankton contributed an average of 84 and 89% to the total chlorophyll a and phaeopigments for the combined summer and winter cruises, respectively (Table 2).

Short-term variability of chlorophyll a concentration. Time-series experiments were conducted over a 30-hr period at sites A, B and C in 1975 and 1976 in order to study the short-term variability of chlorophyll a. The chlorophyll maximum layer (CML) ranged between 30 and 100 m in summer and between 20 and 80 m in winter (Fig. 2). Average depth of CML in summer was deepest (86 $\pm$ 16 m) at Site C and shallowest (49 $\pm$ 14 m) at Site B. It was also deepest at Site C (73 $\pm$ 9.5 m) in winter, and shallower at Site A (65 $\pm$ 16 m) and shallowest at Site B (54 $\pm$ 22 m). The short-term variability in the depth of CML was largest at Site B in both summer (coefficient of variation, standard deviation divided by mean: 0.29) and winter coefficient of variation: 0.41).

The chlorophyll a concentration at CML ranged from 0.076 to 0.676 mg/m^3 in summer and from 0.084 to 0.20 mg/m^3 in winter (Fig. 2). Maximum concentration of chlorophyll a at CML was observed in early morning (0400-0800 hr local time) in both seasons. As to the concentration of chlorophyll a in the water column (to 160 m), the highest values were recorded in early morning (0400-0800 hr) in summer and in the afternoon (1200-1600 hr) in winter. The

TABLE 1

Minimum, maximum, average, number of observations and standard deviation (S.D.) for surface and integrated values to 1% light level (1), 0.1% light level (2), and 0.01% light level (3) of chlorophyll a, phaeopigments and primary production taken during the summer and winter cruises. Units are mg/m³ for surface and mg/m² for integrated values.

		Summer 1975					Winter 1976				
		MIN.	MAX.	AVERAGE	No. of Obs.	S.D.	MIN.	MAX.	AVERAGE	No. of Obs.	S.D.
Chlorophyll a											
Surface	($mg \cdot m^{-3}$)	0.0201	0.127	0.0628	22	0.0403	0.0855	0.277	0.172	16	0.0598
Integrated to:											
euphotic depth	($mg \cdot m^{-2}$)	5.72	18.3	9.80	22	3.54	8.64	34.6	21.0	16	6.29
0.1% level	($mg \cdot m^{-2}$)	6.00	21.6	11.5	21	4.01	9.33	51.1	26.3	16	9.21
0.01% level	($mg \cdot m{-}2$)	6.20	22.4	12.0	19	4.25	-	-	27.6	1	-
Phaeopigments											
Surface	($mg \cdot m^{-3}$)	0.0186	0.122	0.0541	22	0.0359	0.0545	0.223	0.116	16	0.0539
Integrated to:											
euphotic depth	($mg \cdot m^{-2}$)	4.88	31.5	12.5	22	6.03	8.29	34.5	19.3	17	6.24
0.1% level	($mg \cdot m^{-2}$)	9.99	33.2	16.3	22	6.02	12.8	90.4	38.9	17	20.4
0.01% level	($mg \cdot m^{-2}$)	10.5	34.8	18.7	20	6.68	-	-	40.1	1	-
Primary Production											
Surface	($mg \cdot m^{-3}$)	0.158	5.64	1.52	20	1.33	0.755	5.67	2.66	16	1.7
Integrated to:											
euphotic depth	($mg \cdot m^{-2}$)	42.8	265	120	20	65.9	59.2	283	144	15	64.3
0.1% level	($mg \cdot m^{-2}$)	46.9	268	132	19	59.5	63.5	303	153	15	69.5
0.01% level	($mg \cdot m{-}2$)	48.9	268	135	18	60.3	119	153	133	3	18.4

TABLE 2

Percent contribution of nannoplankton to total phytoplankton throughout the euphotic zone in the summer of 1975 and the winter of 1976.

Parameter	Summer			Winter			Total		
	Mean	S.D.	No. of Obs.	Mean	S.D.	No. of Obs.	Mean	S.D.	No. of Obs.
Chlorophyll *a*	80.5	10.8	22	88.1	8.2	17	83.8	10.4	39
Phaeopigments	83.8	14.8	22	96.5	10.1	17	89.3	14.3	39
Primary Production	86.3	33.7	20	63.0	9.9	14	76.7	28.8	34

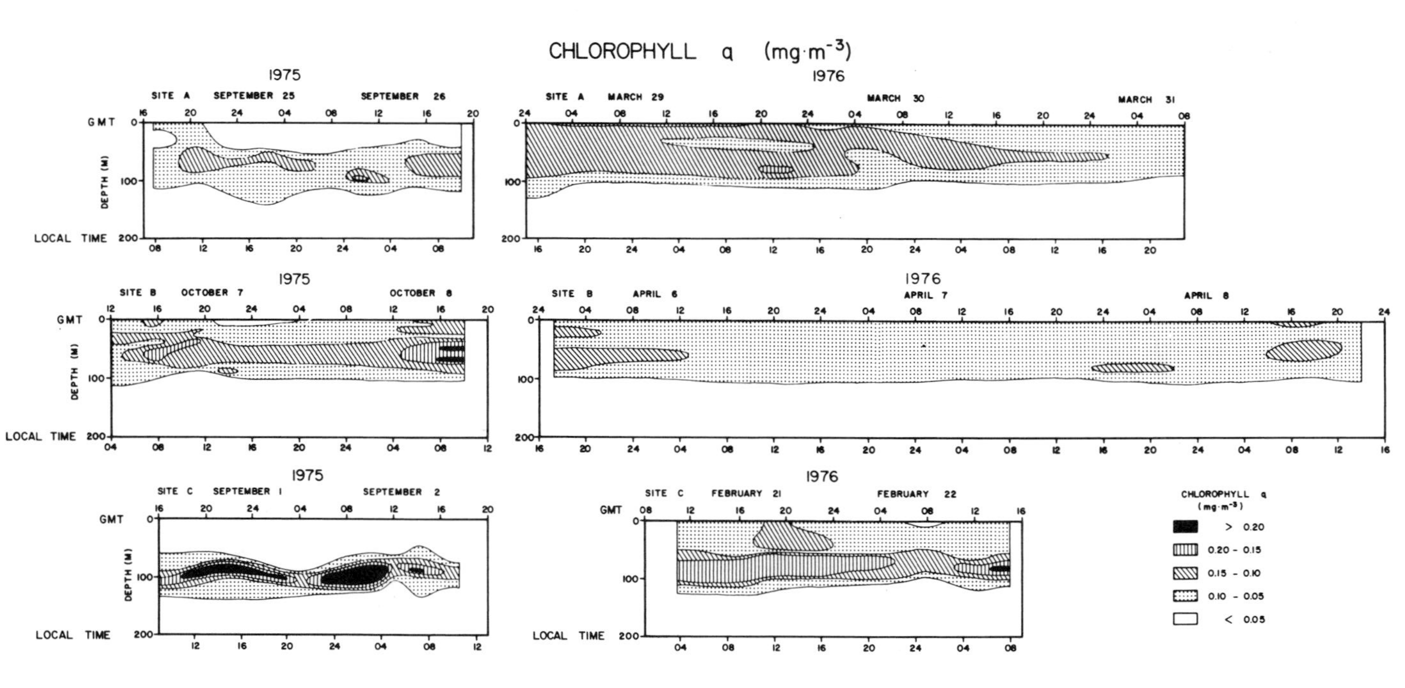

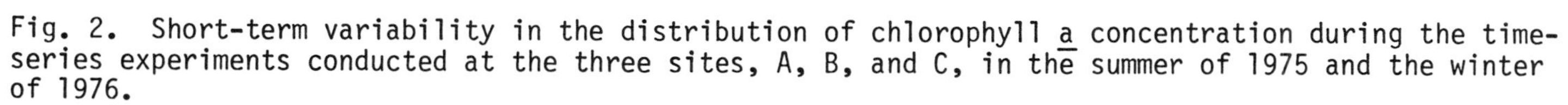

Fig. 2. Short-term variability in the distribution of chlorophyll a concentration during the time-series experiments conducted at the three sites, A, B, and C, in the summer of 1975 and the winter of 1976.

highest average concentrations were observed at Site C for both summer and winter (0.278 ± 0.202 mg/m^3, and 0.174 ± 0.022 mg/m^3), respectively.

Spatial variability of chlorophyll a concentration. Special series of experiments were carried out at seven locations around site Station A in October 1975 and at five locations around site Station B in November 1975. Average concentrations of chlorophyll a and phaeopigments at these locations are given in Table 3. Subsurface chlorophyll maximum layer was found at 12% light level at Site A and at 6% light level at Site B. Average integrated values (i.e., between surface and 0.1% light penetration) of chlorophyll a were 14.4 ± 8.33 and 23.0 ± 8.57 mg/m^2 for Sites A and B, respectively. Spatial variability of chlorophyll a at the former site (coefficient of variation: 0.578) was greater than at the latter (coefficient of variation: 0.373) as shown in Table 3. Mean values of chlorophyll a are similar to those found in the spring of 1976 (Table 1).

Vertical distribution of chlorophyll a and phaeopigments along Transects A, B, and C. The vertical distribution of chlorophyll a was characterized by the presence of subsurface maxima (Figs. 3 and 4). These maxima occurred at a deeper depth (about 69 ± 24 m) in the summer than in winter (about 54 ± 30 m). They occurred above, or in the middle of the maximum density gradient (usually at about the 24.5 pycnocline) as shown in the above figures. These maxima correspond to the depth of about 10% and 20% of surface light level in summer and winter, respectively. The concentration in the chlorophyll maximum exceeded the surface concentration by a factor of 3.6 (± 1.9) in summer and by a factor of 1.8 (± 0.8) in winter.

At the chlorophyll maximum layer the proportion of chlorophyll a to the total pigments (i.e. chlorophyll a plus phaeopigments) was lower (0.44 ± 0.06) in summer than in winter (0.56 ± 0.13). Following these chlorophyll a maxima, there was a general decrease in the concentrations of the photosynthetic pigment to the depth of about 150 m; below this depth the quantity of chlorophyll a was very much reduced. In summer, 6.61% of chlorophyll a content in the water column occurred below the depths of 1% light level (average: 19 ± 17%). In winter the chlorophyll concentration ranged between 8 and 78% with an average of 22 ± 18%.

The vertical distribution of phaeopigments showed a more or less similar distribution to that of chlorophyll a (Figs. 3 and 4). Maximum concentrations of phaeopigments either coincided with that of chlorophyll a or were located at some deeper depth.

TABLE 3

Concentrations of chlorophyll *a* and phaeopigments at various light levels observed around site Stations A and B in October and November, 1975.

Site Station	Light Level	Depth (m)	Chlorophyll *a* (mg/m^3)	Phaeopigments (mg/m^3)
A	100	0	0.0623±0.0355	0.0510±0.0278
	50	26±6	0.0629±0.0307	0.0605±0.0282
	25	46±9	0.152 ±0.100	0.219 ±0.612
	12	67±20	0.194 ±0.128	0.270 ±0.180
	6	82±18	0.145 ±0.129	0.271 ±0.219
	1	121±21	0.0441±0.0239	0.171 ±0.179
	0.1	170±35	0.0277±0.0297	0.142 ±0.115
Σ(integrated)			14.4±8.33	29.8±17.1
B	100	0	0.0678±0.0160	0.0493±0.0104
	50	20±7	0.0969±0.0444	0.0788±0.0492
	25	41±5	0.148 ±0.0833	0.252 ±0.217
	12	55±9	0.175 ±0.0974	0.321 ±0.155
	6	69±7	0.250 ±0.114	0.776 ±0.493
	1	110±8	0.164 ±0.142	0.897 ±0.830
	0.1	154±7	0.0473±0.0277	0.288 ±0.154
Σ(integrated)			23.0±8.57	75.1 ±43.6

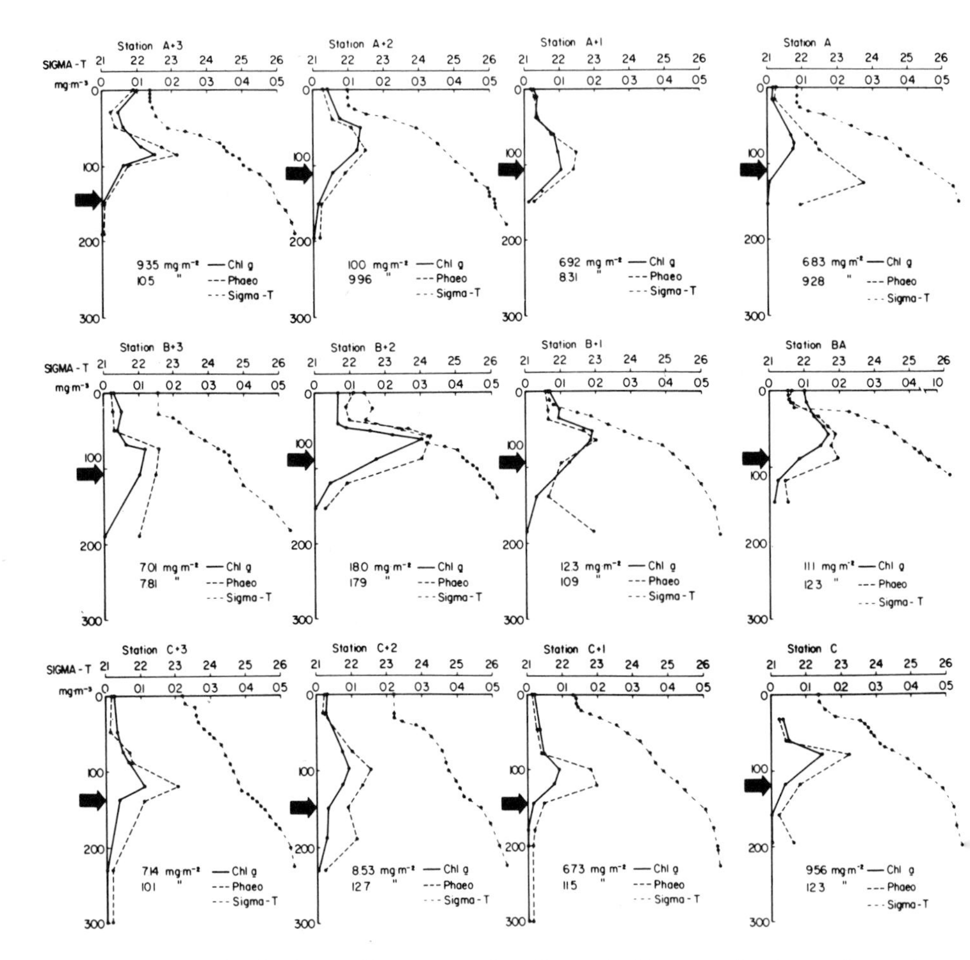

Fig. 3. Vertical distribution of chlorophyll a and phaeopigments concentrations with sigma-t at the stations occupied in the three Transects A, B, and C in the summer of 1975. Depth of the euphotic zone is shown by arrows.

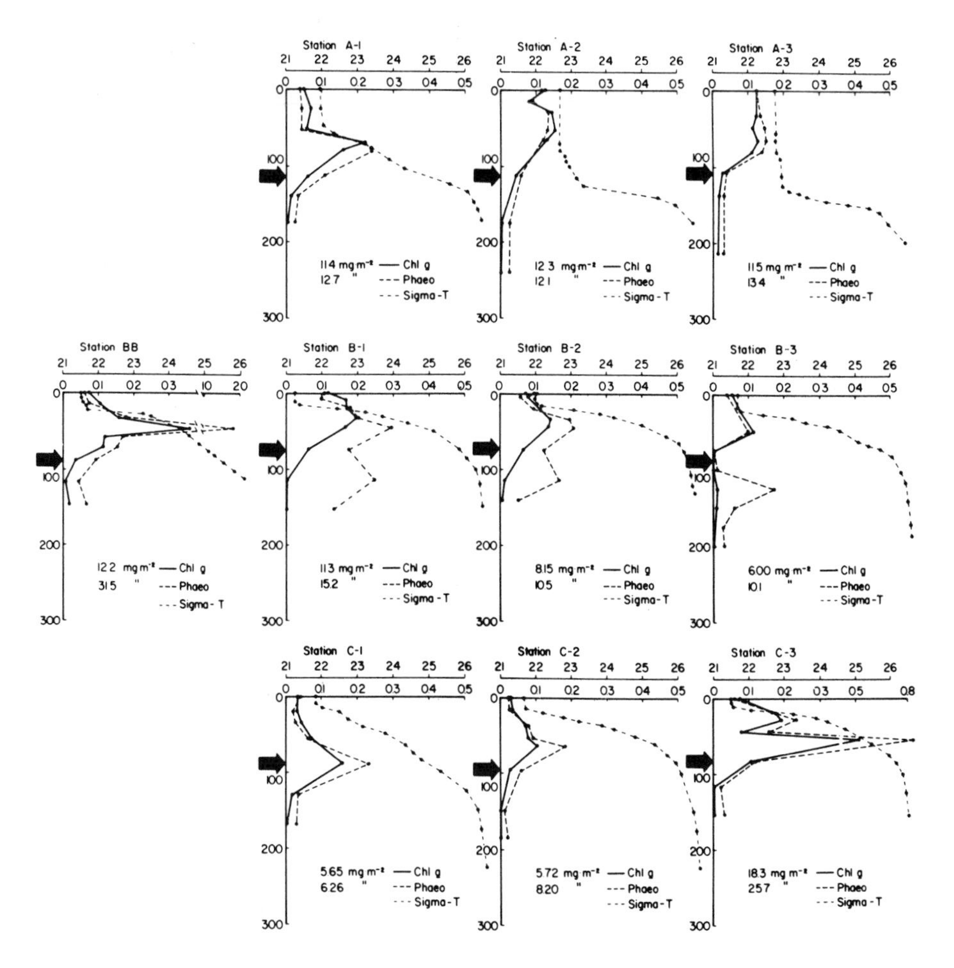
Station A-1
114 mg m⁻² — Chl g
12.7 " --- Phaeo
--- Sigma-T
Station A-2
12.3 mg m⁻² — Chl g
12.1 " --- Phaeo
--- Sigma-T
Station A-3
11.5 mg m⁻² — Chl g
13.4 " --- Phaeo
--- Sigma-T
Station BB
12.2 mg m⁻² — Chl g
31.5 " --- Phaeo
--- Sigma-T
Station B-1
11.3 mg m⁻² — Chl g
15.2 " --- Phaeo
--- Sigma-T
Station B-2
8.15 mg m⁻² — Chl g
10.5 " --- Phaeo
--- Sigma-T
Station B-3
6.00 mg m⁻² — Chl g
10.1 " --- Phaeo
--- Sigma-T
Station C-1
5.65 mg m⁻² — Chl g
6.26 " --- Phaeo
--- Sigma-T
Station C-2
5.72 mg m⁻² — Chl g
8.20 " --- Phaeo
--- Sigma-T
Station C-3
18.3 mg m⁻² — Chl g
25.7 " --- Phaeo
--- Sigma-T

Fig. 3. (cont'd).

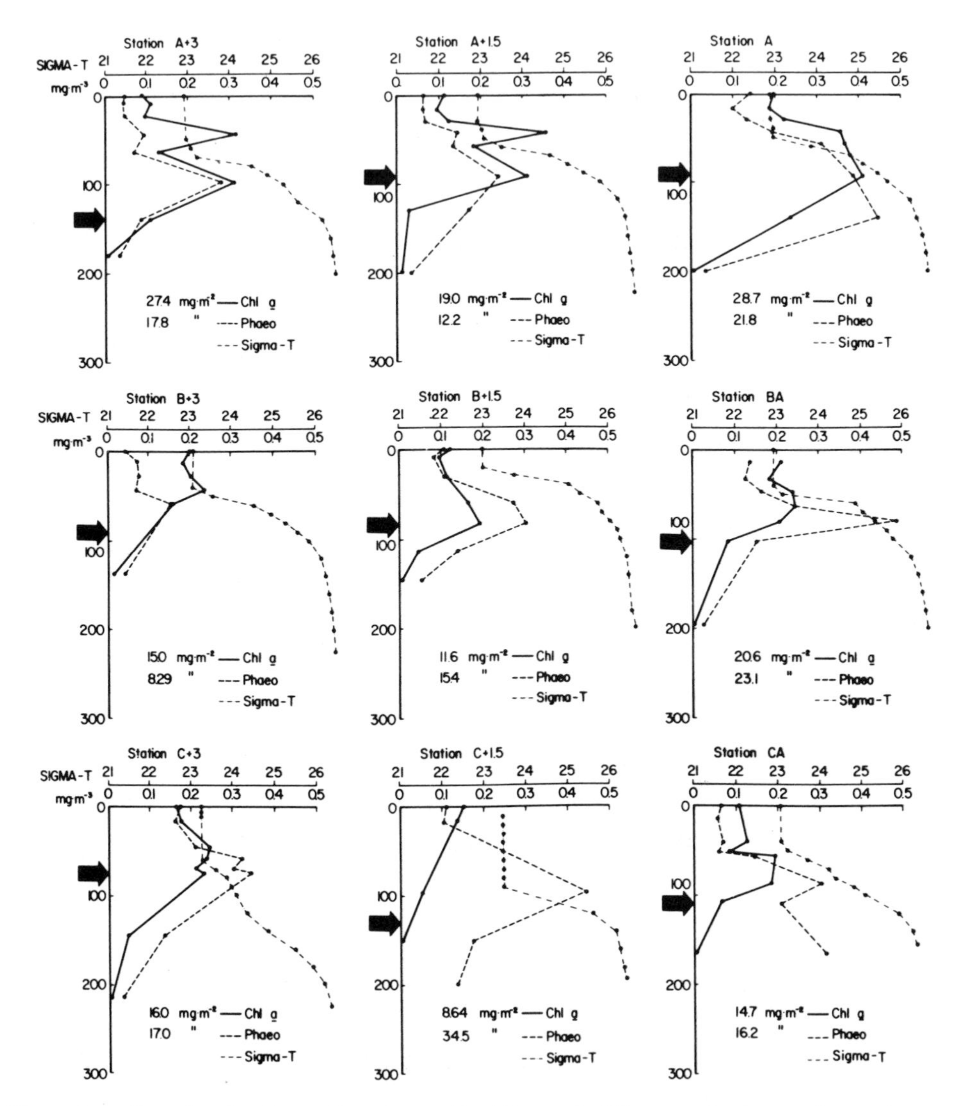

Fig. 4. Vertical distribution of chlorophyll a and phaeopigments concentrations with sigma-T at the stations occupied in the three Transects A, B, and C in the winter of 1976. Depth of the euphotic zone is shown by arrows.

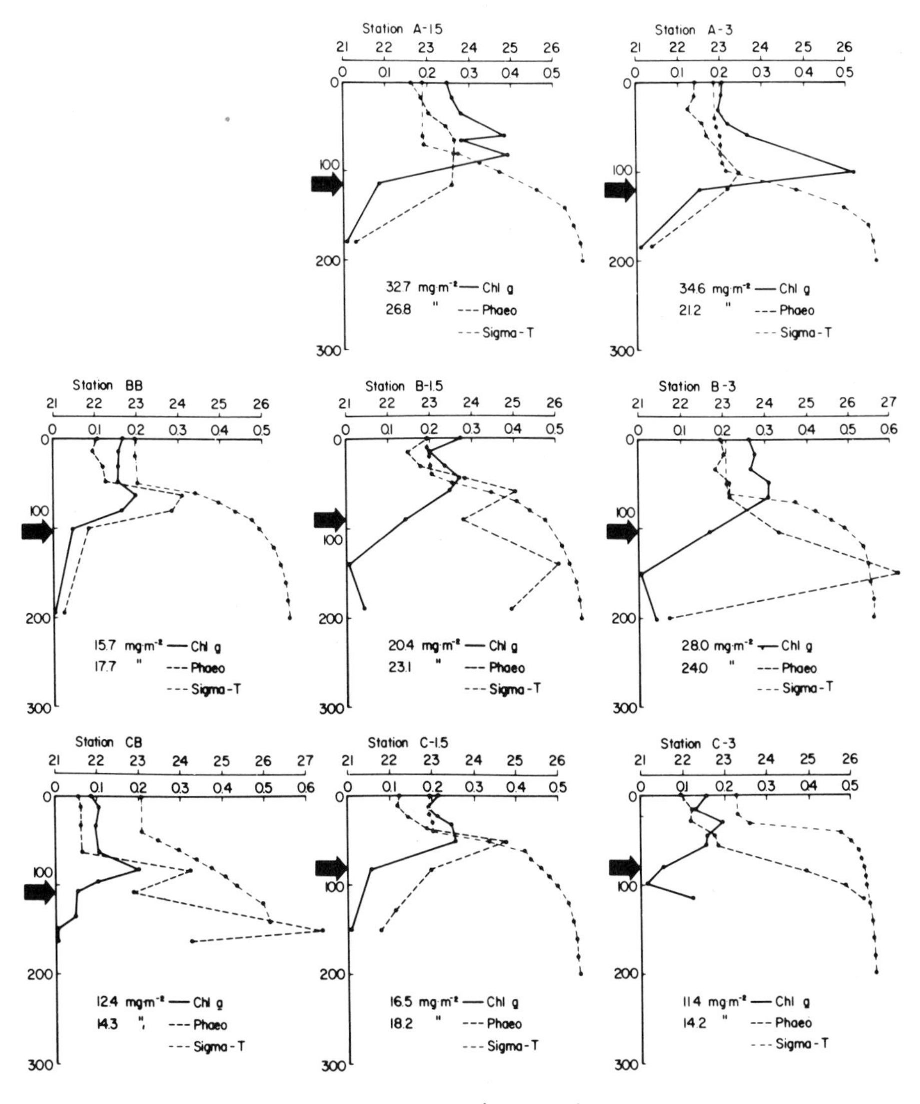
Station A-1.5
32.7 mg·m⁻² — Chl a
26.8 " --- Phaeo
--- Sigma-T
Station A-3
34.6 mg·m⁻² — Chl a
21.2 " --- Phaeo
--- Sigma-T
Station BB
15.7 mg·m⁻² — Chl a
17.7 " --- Phaeo
--- Sigma-T
Station B-1.5
20.4 mg·m⁻² — Chl a
23.1 " --- Phaeo
--- Sigma-T
Station B-3
28.0 mg·m⁻² — Chl a
24.0 " --- Phaeo
--- Sigma-T
Station CB
12.4 mg·m⁻² — Chl a
14.3 " --- Phaeo
--- Sigma-T
Station C-1.5
16.5 mg·m⁻² — Chl a
18.2 " --- Phaeo
--- Sigma-T
Station C-3
11.4 mg·m⁻² — Chl a
14.2 " --- Phaeo
--- Sigma-T

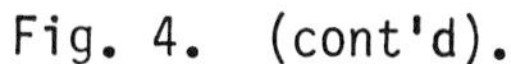
Fig. 4. (cont'd).

Spatial and temporal distributions of chlorophyll a and phaeopigments concentrations.

Chlorophyll a . The depth of CML in Transect C tended to be shallower with decreasing latitude, from about 100 m up to 60 m in summer (Fig. 5). Along Transect B, the depth of CML also shoaled gradually from 90 m at St. B+3 to 30 m at St. B-1; this was followed by a deepening of this layer to 70 m at St. B-3. In Transect A the depth of CML also shoaled from about 90 m at St. A+3 to about 60 m at St. A+2 and then sank deeper to 110 m at St. A-1. At St. A-2 the depth of CML shallowed to 50 m, then finally disappeared at St. A-3.

The vertical distribution of chlorophyll a concentrations in summer along 12°N displayed a maximum layer at a much shallower depth (60 m) in Transects C and B than at Transect A (90 m).

The winter distribution of chlorophyll a concentrations was markedly different from those of the summer (Fig. 5). At the northernmost station (C+3), CML was observed at about 50 m. No CML was discernible at the next station (C+1.5). The depth of the maximum layer shoaled from 60 m at Site C to 30 m at the southernmost station (C-3). Along Transect B the depth of that layer deepened from 40 m at St. B+3 to 90 m at St. B+1.5; it then shoaled to 70 m at Site B. The depth of CML was at the surface at St. B-1.5 and at 70m at St. B-3. However, the maximum extent of this layer at the three latter stations was very small. Along Transect A, the depth of CML at the two northern stations (A+3 and A+1.5) was about 50 m; it then deepened southward to 100 m at Site A. At St. A-1.5 the depth of CML shoaled to 80 m and deepened again to 100 m at the southernmost station (A-3).

The vertical distribution of chlorophyll a along 12°N in winter was characterized by a double chlorophyll maximum layer centered at the depth of 50 and 100 m at St. A+3. However, the deep CML at 100 m was altogether absent at Sts. B and C-3.

Phaeopigments. In summer the depth of the phaeopigments maximum layer (PML) along Transect C followed the same trend as the chlorophyll maximum layer; i.e., it tended to be shallower with decreasing latitude (from 120 m to 60 m) (Fig. 6). A similar trend was observed in Transect B where PML shoaled from 80 m at St. B+3 to 50 m at St. B-3. The depth of PML was at about 130 m at St. B-3. The depth of PML deepened from 80 m to 90 m between Sts. A+3 and A-1. It then shallowed to 30 m at St. A-2 and sank deeper to 70 m at St. A-3. The latitudinal distribution (12°N) of phaeopigments concentrations was characterized by the presence of PML at a depth of about 60 m at Sts. C-3, B and A+3.

The PML in winter was more discernible than in the preceding summer (Fig. 6). The depth of PML shallowed from about 80-90 m at St. C+3 to 50 m at St. C-1.5. This was followed by a sharp

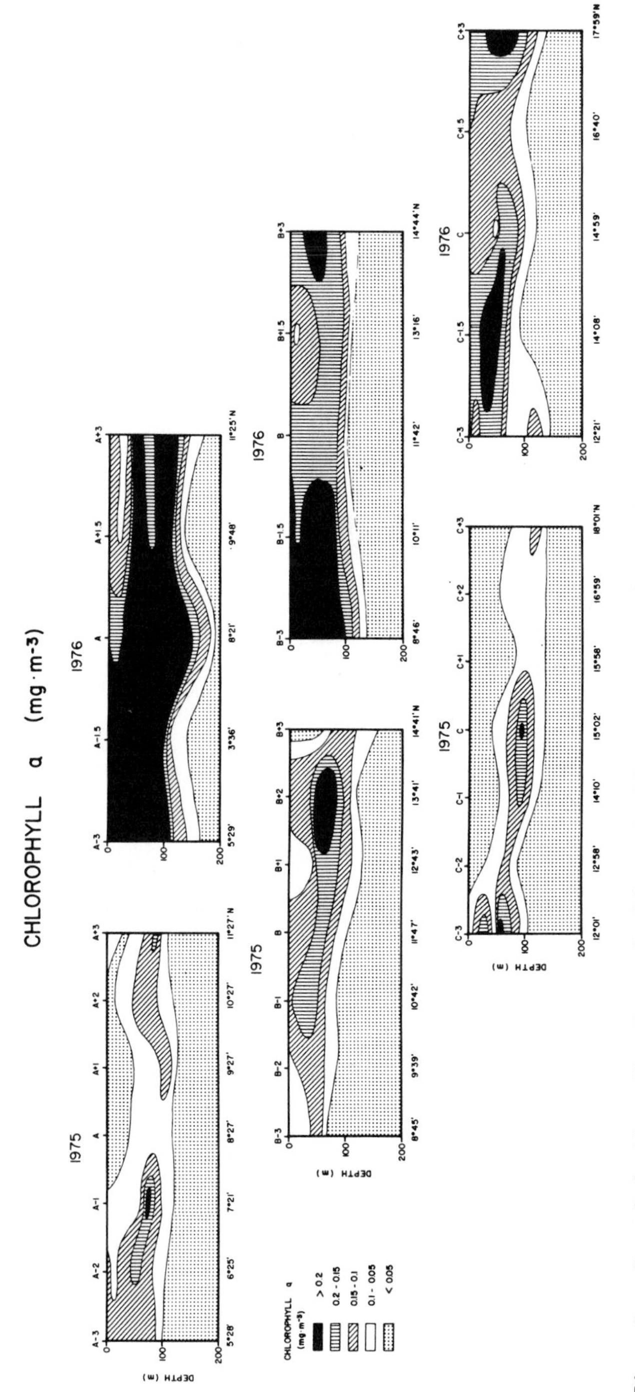

Fig. 5. Spatial distribution of chlorophyll a concentration along the three Transects A, B, and C in the summer of 1975 and the winter of 1976.

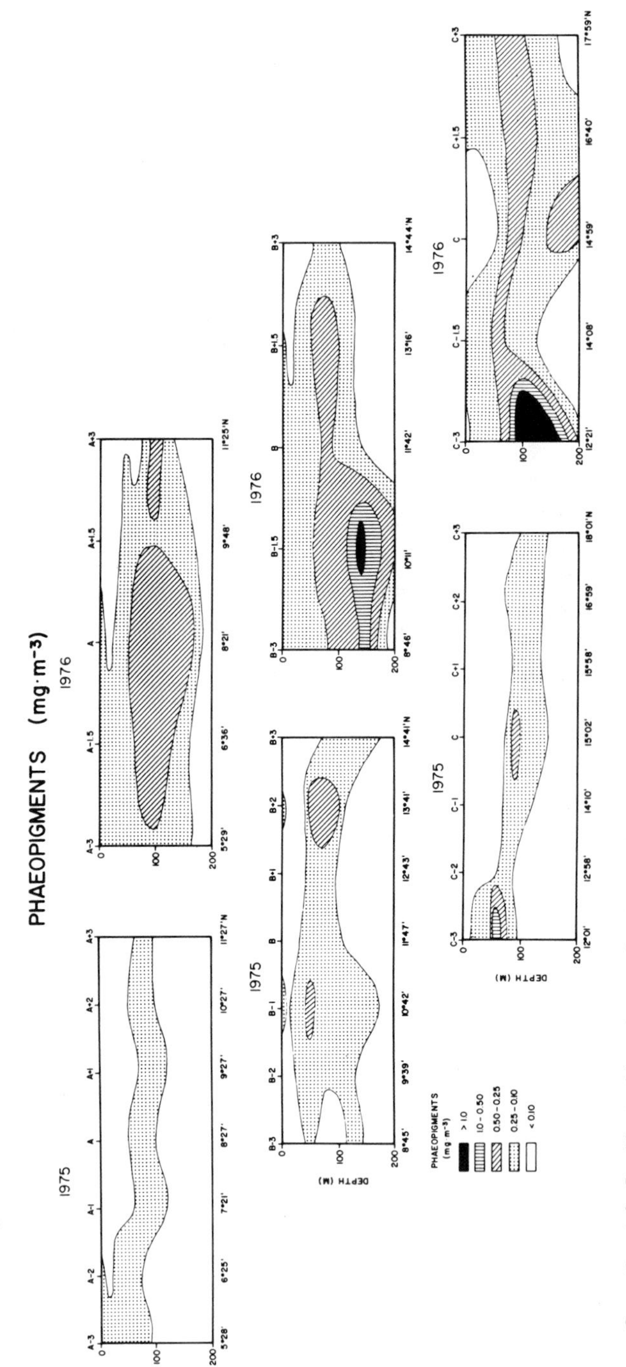

Fig. 6. Spatial distribution of phaeopigments concentration along the three Transects A, B, and C in the summer of 1975 and the winter of 1976.

deepening to 120 m at St. C-3. Along Transect B, the depth of PML was found at about 70-80 m between St. B+3 and Site B. It then deepened to 140 m at St. B-1.5 and stayed at about that depth at St. B-3. Along Transect A, the depth of PML was about 100 m at Sts. A+3 and A-1. It sank to 140 m at Site A and then shoaled to 70 m at St. A-1.5. At the southernmost station in this transect the depth of PML was about 100 m; it was not a conspicuous maximum, though.

The latitudinal distribution of phaeopigments concentrations in winter along about 12°N was characterized by the shoaling of PML from 120 m at St. C-3 to 80 m at Site B and a further deepening to 100 m at St. A+3.

Primary Production

Primary production values (either surface or integrated) of the study area are, in general, low (Table 1). The average values for surface water samples and for the euphotic zone were 9.09 $mgC/m^3/day$ and 132 $mgC/m^2/day$, respectively. The average of primary production in winter (144 $mgC/m^2/day$) was slightly higher than in summer (120 $mgC/m^2/day$).

The percent contribution of the nannoplankton to total primary production for each light level varied between 22% and 100% for the three transects studied in summer (Fig. 7) and between 22% and 80% for the winter transects (Fig. 8). No significant differences in the percent contributions to total primary production were found between the transects studied in winter; their average value was about 62%. In summer, however, the stations occupied along Transect C gave much lower values (77%) than at the stations occupied along Transects A and B (91% and 92%, respectively). The percent contribution of the nannoplankton to primary production was lower in winter (63%) than in summer (86.3%). On the whole nannoplankton contributed an average of 77% of primary production in the DOMES area.

Vertical distribution of primary production. The vertical distribution of primary production in the study area (Figs. 7 and 8), in general, shows a similar pattern to that of chlorophyll *a* (Figs. 3 and 4). Maximum primary production occurred either at the surface or at some subsurface depth. In summer, maximum production was recorded at 50% of surface light level, whereas in winter, maximum production was found at or near the surface. Following these maxima, there was a general decrease in the photosynthetic activities of the phytoplankton till the depth of the euphotic zone was reached (shown by arrows in Figs. 7 and 8).

On several occasions, however, primary production in the water column extended below the euphotic zone (average 104 m for all the DOMES stations). At St. C-1 in summer, for instance, about 18% of the production occurred below the euphotic depth. At St. A+1.5 in winter, the primary production below the euphotic

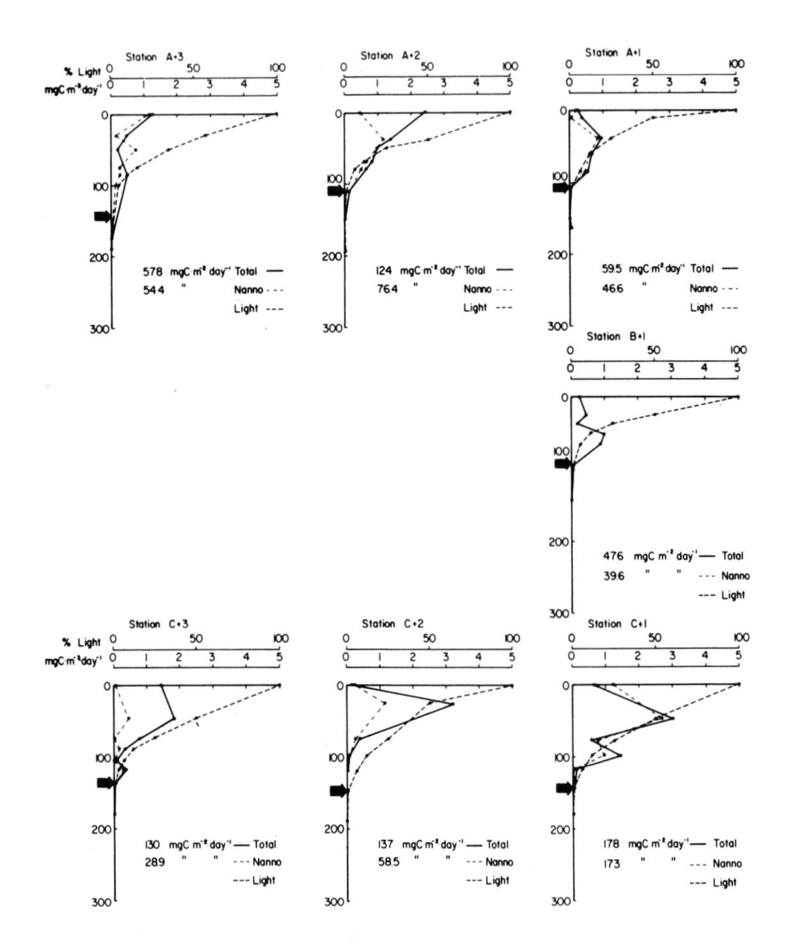

Fig. 7. Vertical distribution of total primary production, nanno-plankton production and percent light penetration at the stations occupied in the three Transects A, B, and C in the summer of 1975. Depth of the euphotic zone is shown by arrows.

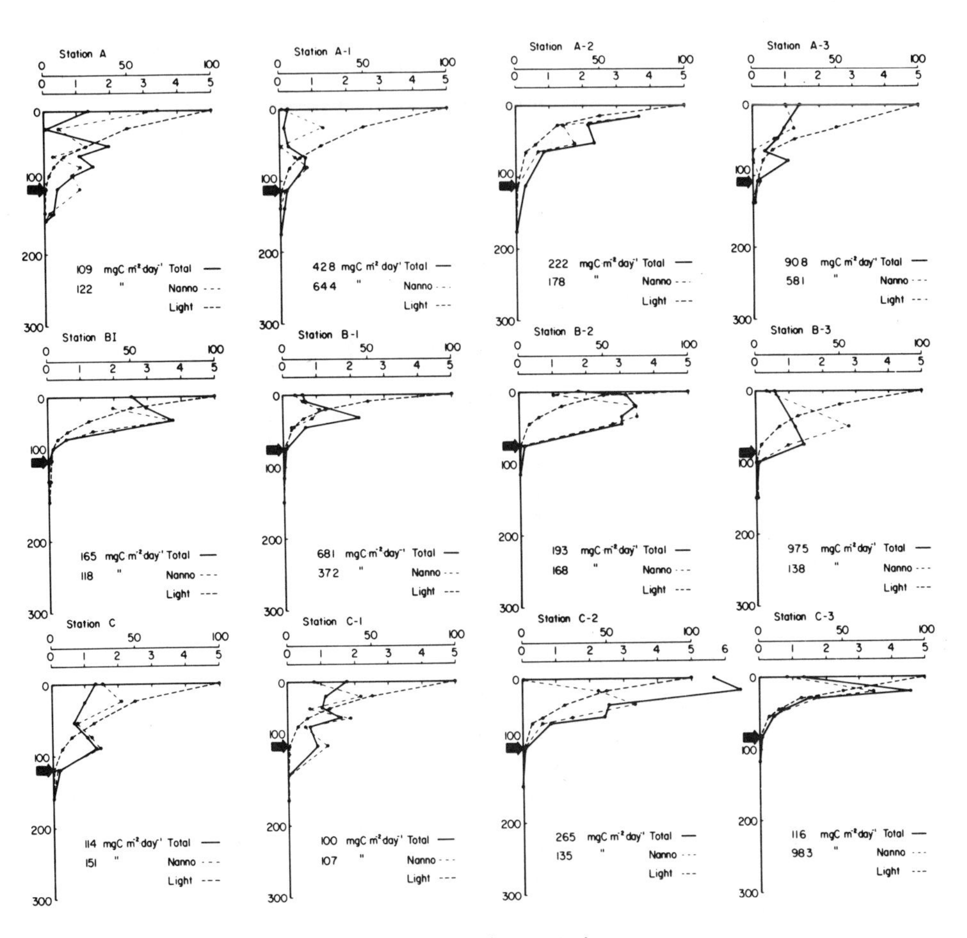

Station A
109 mgC m⁻² day⁻¹ Total
122 " Nanno
Light
Station A-1
428 mgC m⁻² day⁻¹ Total
644 " Nanno
Light
Station A-2
222 mgC m⁻² day⁻¹ Total
178 " Nanno
Light
Station A-3
908 mgC m⁻² day⁻¹ Total
581 " Nanno
Light
Station B1
165 mgC m⁻² day⁻¹ Total
118 " Nanno
Light
Station B-1
681 mgC m⁻² day⁻¹ Total
372 " Nanno
Light
Station B-2
193 mgC m⁻² day⁻¹ Total
168 " Nanno
Light
Station B-3
975 mgC m⁻² day⁻¹ Total
138 " Nanno
Light
Station C
114 mgC m⁻² day⁻¹ Total
151 " Nanno
Light
Station C-1
100 mgC m⁻² day⁻¹ Total
107 " Nanno
Light
Station C-2
265 mgC m⁻² day⁻¹ Total
135 " Nanno
Light
Station C-3
116 mgC m⁻² day⁻¹ Total
983 " Nanno
Light

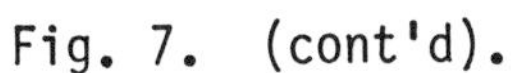
Fig. 7. (cont'd).

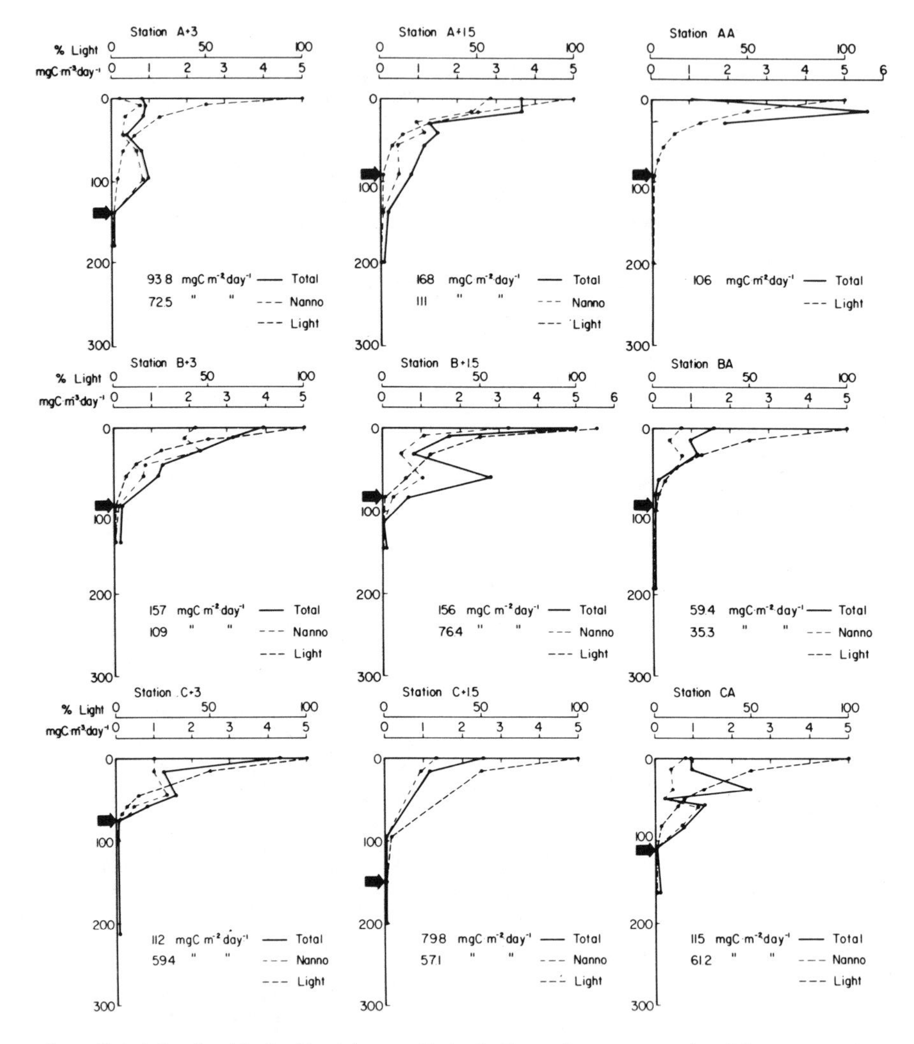

Fig. 8. Vertical distribution of total primary production, nanno-plankton production and percent light penetration at the stations occupied in the three Transects A, B, and C in the winter of 1976. Depth of the euphotic zone is shown by arrows.

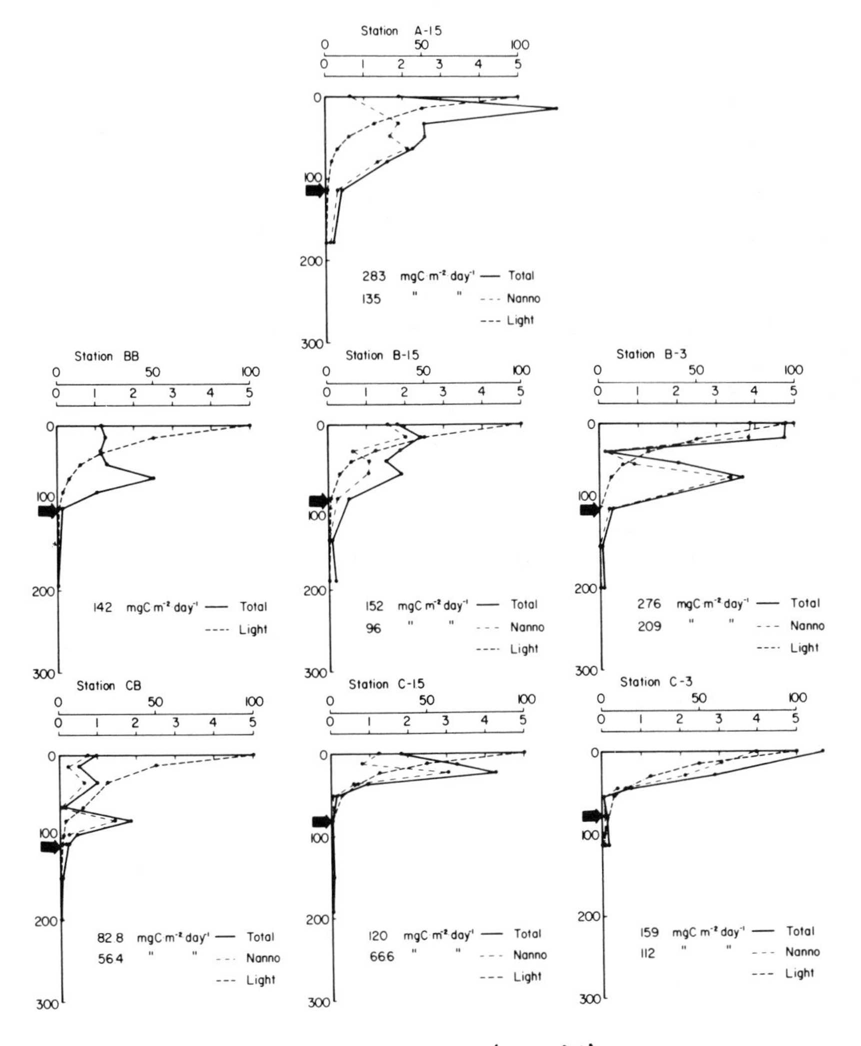
Station A-15
0 50 100
0 1 2 3 4 5
0
100
200
300
283 mgC m⁻² day⁻¹ — Total
135 " " Nanno
Light
Station BB
142 mgC m⁻² day⁻¹ — Total
Light
Station B-15
152 mgC m⁻² day⁻¹ — Total
96 " " Nanno
Light
Station B-3
276 mgC m⁻² day⁻¹ — Total
209 " " Nanno
Light
Station CB
82.8 mgC m⁻² day⁻¹ — Total
56.4 " " Nanno
Light
Station C-15
120 mgC m⁻² day⁻¹ — Total
66.6 " " Nanno
Light
Station C-3
159 mgC m⁻² day⁻¹ — Total
112 " " Nanno
Light

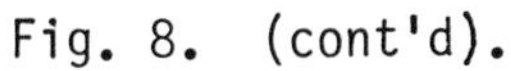
Fig. 8. (cont'd).

zone contributed about 17% of the production of the water column. The average contribution of phytoplankton below the euphotic zone to the production in the water column was 4.0 $\pm$ 0.6% in summer of 1975 and 6.0 $\pm$ 4.3% in the winter of 1976. There was no significant difference between these two values.

Spatial distribution of primary production. At Transect A in the summer there seems to be a symmetric distribution of primary production values centered around Site A (Fig. 9). Values less than 0.5 mgC/m^3/day were observed at the surface of Sts. A+1 and A-l. However, values higher than 1 mgC/m^3/day were found at the surface at Sts. A-3, A-2, A+2 and A+3. The least productive station was B+1. The depth of 1 mgC/m^3/day isopleth was about 50 m just south of St. B+1 to St. B-1; its depth then increased gradually to 90 m at St. B-3, corresponding to the bottom of the euphotic zone. Along Transect C the upper 50 m exhibited values higher than 1 mgC/m^3/day. The depth of the isopleth was nearly half that of the euphotic zone. At Sts. C-2 and C-3, production values of about 5 mgC/m^3/day were observed at 10 m or even at shallower depths.

The latitudinal distribution of primary production along 12°N in summer was characterized by: (1) the depth of isopleth (1 mgC/m^3/day) was at 40 m with a subsurface maximum (about 5 mgC/m^3/day) occurring at 10 m at St. C-3, (2) the depth of the isopleth (1 mgC/m^3/day) was deepest (60 m) with a subsurface maximum (about 3 mgC/m^3/day) at 30 m at Site B, and (3) the depth of the isopleth (1 mgC/m^3/day) was shallowest (10 m) with no subsurface maximum at St. A+3.

In winter along Transect A, the least productive water column the observed at St. A+3 (Fig. 9). Just south of this station was 1 mgC/m^3/day isopleth increased from 0 m to about 80 m at St. A+1.5; it then remained unchanged to about 90 m at Sts. A and A-1.5. The subsurface primary production maximum (about 6 mgC/m^3/day) was observed at 20 m south of St. A+1.5 to St. A-1.5. Along Transect B, the 1 mgC/m^3/day isopleth was almost identical to the depth of the euphotic zone, except at Site B. Isopleths >4 mgC/m^3/day were observed at the surface at Sts. B+3, B+1.5 and B-3. The upper 40 m of Transect C displayed values higher than 1 mgC/m^3/day. The depth of this isopleth was more than half that of the euphotic zone. At the southernmost stations (C-1.5 and C-3), primary production values of about 4 mgC/m^3/day were observed at 20 m or at even shallower depth. Similar observations were also found in the same region in the summer of 1975.

The latitudinal distribution of primary production along 12°N in winter was characterized by: (1) the depth of the 1 mgC/m^3/day isopleth was about 40 m at St. C-3 with a surface maximum (about 6 mgC/m^3/day), (2) the depth of the 1 mgC/m^3/day isopleth was about 40 m at Site B, and (3) the distinct

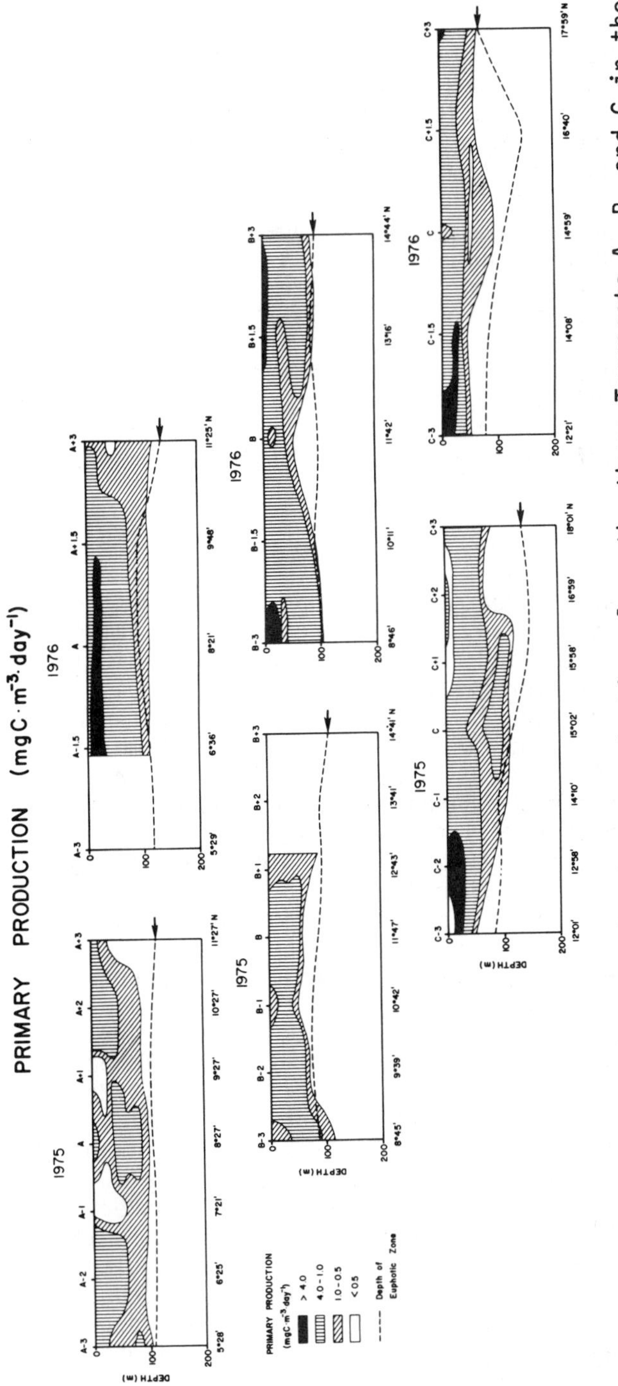

Fig. 9. Spatial distribution of primary production along the three Transects A, B, and C in the summer of 1975 and the winter of 1976.

stratification disappeared altogether at St. A+3.

Photosynthetic Index (PI). Values of the photosynthetic index, (calculated as the ratio of primary production to chlorophyll a concentrations mgC/mgChl.a/day), for each light level for Transects A, B and C for summer and winter cruises are plotted in Figure 10. Maximum productivity indexes (PI) were found at depths corresponding to either 100% or 50% of light level. Average values of PI were more than threefold higher in summer (78 ± 60 mgC/mgChl.a/day) than in winter (25 ± 13 mgC/mgChl.a/day). Following these maximum values, the average PI decreased with depth. Below the 25% light level, there were no significant differences in the mean of PI between the three transects. In summer Transect C showed the highest average PI (20 ± 13 mgC/mgChl.a/day) in the water column of the three transects studied. In winter, on the other hand, there was no significant difference in PI among the three transects.

Discussion

In the study area, the chlorophyll a values collected in 1975 and 1976 displayed, by and large, the low values one would normally expect to find in the oligotrophic tropical/subtropical oceanic waters. Average surface chlorophyll a was 0.117 mg/m^3, and the average value integrated through the water column (to the bottom of the euphotic zone) was only 19.6 mg/m^2. The surface value is less than the 0.18 mg/m^3 found by Ryther and Menzel (1959) in June for the Sargasso Sea, and about half the value found by El-Sayed (1972) for the Gulf of Mexico.

The vertical distribution of chlorophyll a at 37 of the 39 stations occupied was characterized by the presence of a subsurface maximum (Figs. 2 and 3). Light scattering data also showed subsurface maximum at about the depths of either chlorophyll or phaeopigments maximum layers (Baker and Feely, this volume). Our data corroborate the findings of other investigators that the chlorophyll maximum layer occurs over most of the tropical and subtropical waters of the world ocean (Steele and Yentsch, 1960; Yentsch, 1965; Anderson, 1969, 1972; Kawarada and Sano, 1969; Saijo et al., 1969; Vinogradov et al., 1970; Beers and Stewart, 1971; Hobson and Lorenzen, 1972; Takahashi et al., 1972; Taniguchi and Kawamura, 1972; Eppley et al., 1973; Venrick et al., 1973; Berman and Holm-Hansen, 1974; Gundersen et al., 1976; Kiefer et al., 1976; El-Sayed and Turner, 1978). It would seem then that the deep chlorophyll maximum is a regular feature in most tropical and subtropical waters of the world ocean.

In Table 4 we summarized the data on the depth of chlorophyll maximum layer, the corresponding light level and the

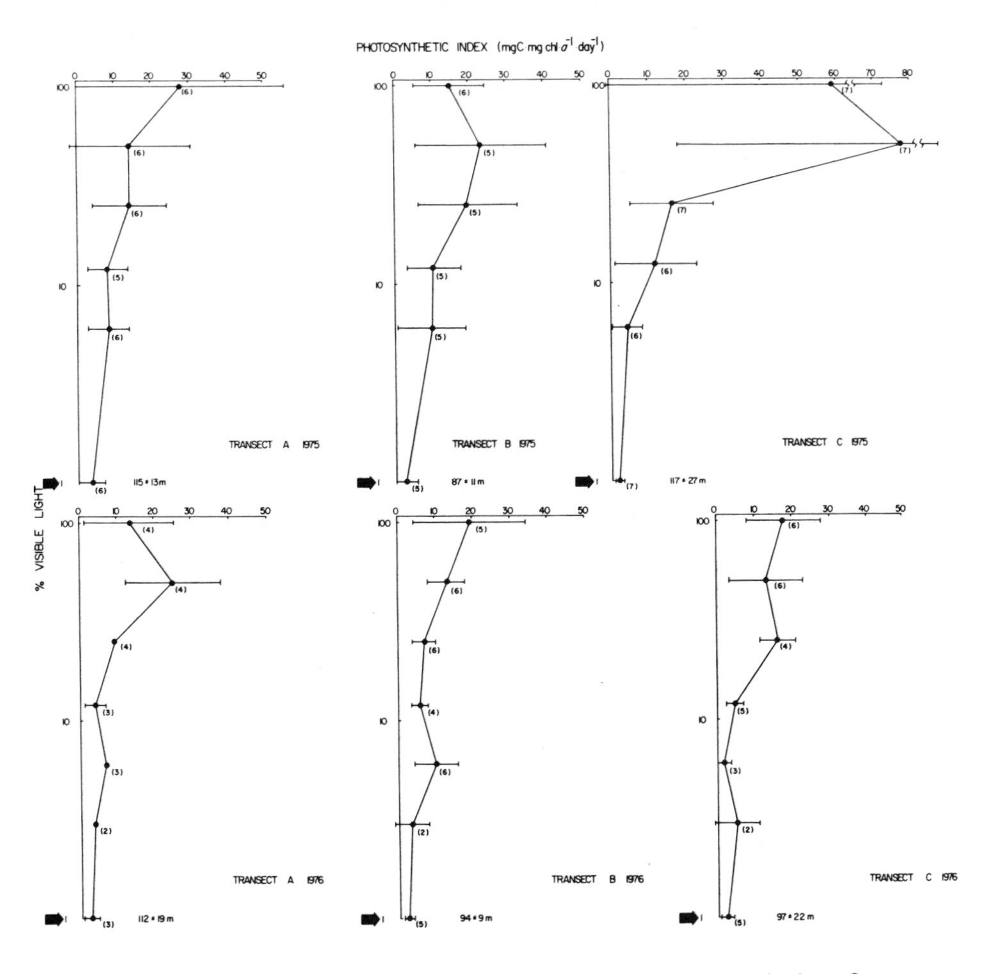

Fig. 10. Vertical profiles of average productivity index for Transects A, B, and C in the summer of 1975 and the winter of 1976. Depth of the euphotic zone is shown by arrows.

TABLE 4

Summary of the depth of chlorophyll maximum layer (CML), the relative light level at CML, the pigment ratio at CML, the depth difference between CML and phaeopigments maximum layer (PML) and the depth of euphotic zone at the three Transects A, B, and C in the summer 1975, and the winter, 1976.

Transect	Year	Depth of CML (m)	Light Level %	Chl. *a* / Chl. *a* + Phaeo.	Difference Between CML and PML (m)	Depth of Euphotic Zone (m)
A	1975	73.6±20.4	10.7±7.8	0.471±0.0583	7.0±27.9	115±13.3
B		48.9±15.2	11.6±6.3	0.481±0.0422	16.5±24.3	87±11.3
C		86.4±22.0	6.6±4.2	0.373±0.0261	2.7±7.18	117.±27.3
All stations in 1975		68.7±24.4	9.7±6.4	0.443±0.0648	8.7±21.3	103±21.8
A	1976	72.4±27.5	6.2±5.4	0.655±0.0988	27.0±32.0	112±19.5
B		54.7±15.7	9.2±4.7	0.549±0.143	48.3±58.8	94±8.7
C		45.8±31.8	28.3±36.4	0.501±0.124	64.5±41.7	97±41.7
All stations in 1976		54.2±29.8	20.2±30.9	0.563±0.133	47.8±46.1	101±18.6
All Stations in 1975 and 1976		62.4±27.5	14.3±21.3	0.546±0.282	25.9±39.1	104±20.7

difference in the depth between the chlorophyll maximum layer and the phaeopigments maximum layer at the three Transects A, B and C in the summer of 1975 and the winter of 1976. It is clear from the above table that there are discernible variabilities in the depth of the chlorophyll maximum layer both spatially (i.e. between transects) and seasonally (i.e. between 1975 and 1976).

The chlorophyll maximum layer occurred at either the middle or at the bottom of the pycnocline in summer (Fig.3) and at either the top or in the middle of the pycnocline in winter. The critical feature in the chlorophyll distribution is the occurrence of the chlorophyll maximum at about 24.5 σ_t. Interestingly, the profiles for total cells/l show a subsurface maximum in or above the pyncnocline (Fryxell _et al_., this volume).

Although the depth of the surface mixed layer (SML) was deeper (possibly due to stronger northeast winds) in winter (54 ± 19 m) than in summer (29 ± 21 m), the chlorophyll maximum occurred at deeper depths (69 ± 24 m) in summer than in winter (54 ± 30 m) as shown in Table 4. These chlorophyll maxima correspond to about 10% and 20% of the surface light level for the summer and the winter, respectively. These light values are much higher than those (0.1 - 0.26%) given by Venrick _et al_. (1973) but similar to those of Gundersen _et al_. (1976). It is likely that the development of the chlorophyll maximum layer at deeper depths than the surface mixed layer was due to unfavorable light regime at the bottom of the mixed layer in summer. In winter, on the other hand, the chlorophyll maximum developed at the bottom of the mixed layer owing, perhaps, to more favorable light conditions.

The depth of the chlorophyll maximum layer was also found to be closely associated with the distribution of the nutricline (nutrient gradient), especially that of NO_3-N. Mechanisms contributing to the formation and maintenance of the deep chlorophyll maximum layer included concentrations of detrital chlorophyll in the pycnocline (Lorenzen, 1965); increase in the chlorophyll/carbon ratio in plant cells (Steele, 1964); sinking of active or senescent cells from shallower depths (Steele and Yentsch, 1960; also see Venrick _et al_., 1973, for literature review). Recently Jamart _et al_., 1977, have shown that _in situ_ photosynthesis is a major factor in the maintenance of the chlorophyll maximum layer in higher latitudes of the northeast Pacific Ocean.

The coccolithophorid _Gephyrocapsa huxleyi_ is the dominant species in the chlorophyll maximum layer in the phytoplankton samples studied. This species also accounted for more than half of the coccolithophorid cells in the study area (Fryxell _et al_., this volume).

Recent study by Reid _et al_. (1978) showed that the phytoplankton species assemblages of CML and the surface layer were similar. Fryxell _et al_. (this volume), failed to show difference in the phytoplankton species assemblages between the chlorophyll maximum layer and the surface layer. These observations seem to support the theory of Steele and Yentsch (1960) that depletion of nutrients above the thermocline leads to a reduction in the buoyancy of phytoplankton, and that a subsurface chlorophyll maximum layer results from the accumulation of impoverished cells at the top of the nutricline. The availability of higher nutrient concentrations at the nutricline decreases the sinking velocity of these algal cells (Eppley *et al.* 1967).

Smayda (1970) showed a higher sinking velocity for larger diatoms. However, it is interesting to note that the net plankton (i.e. cells larger than 20 μm) contribution to the chlorophyll *a* of total plankton was similar between the chlorophyll maximum layer (12-30%) and the surface layer (5-22%). The net plankton contribution to the primary production of total plankton was also similar between the CML (0-37%) and the surface layer (8-45%) except at Transect A in the summer where net plankton contribution to primary production was 46% at CML and 8% in the surface layer.

In seeking an explanation for the similarity in the cell size of the phytoplankton assemblages between CML and the surface, it should be pointed out that phytoplankton profiles may reflect grazing pressures more than sinking rates of cells (Longhurst, 1976). Longhurst showed that the subsurface zooplankton maximum at the base of the surface mixed layer was closer to the maximum primary production layer than to the chlorophyll maximum layer. It is likely, therefore, that the CML is due to either zooplankton grazing (Vinogradov *et al.*, 1970; Anderson *et al.*, 1972; Hobson and Lorenzen 1972; Mullin and Brooks 1972) or to slower growth of larger cells (Eppley and Sloan, 1966; Taguchi, 1976). One may conclude, therefore, that there would be no difference in cell size distribution between the chlorophyll maximum and the surface layer even though large cells sink faster than small cells (Smayda, 1970).

The phaeopigments showed distribution more or less similar to that of chlorophyll *a*, just as empty cells counts followed the full cell counts (see Fryxell *et al.*, this volume). The phaeopigments maximum layer either coincided with the chlorophyll maximum or was located at deeper depths. The depth differences between the chlorophyll and phaeopigments maximum were much less (9 ± 21 m) in summer than in the samples studied in winter (48 ± 46 m) (Table 4). This is more likely due to the more stable depth of the phaeopigments maximum relative to that of chlorophyll.

Cell counts were positively correlated with chlorophyll *a* values although much more strongly in winter (r = 0.71) than in summer (r = 0.30), (Fryxell *et al.*, this volume). Below the euphotic zone the cell counts ranged between 6 and 16% and averaged 14% for the summer cruise and 18% for the winter cruise.

As to whether chlorophyll *a* found below the euphotic zone was photosynthesizing or not, our data showed significant positive correlation between chlorophyll *a* and primary production values. The correlation coefficients were 0.63 for summer and 0.76 for winter, respectively. The amount of primary production below the euphotic zone was 5% of the total primary production, a figure which is lower than the 7 - 20% found by Venrick *et al*. (1973).

As to the spatial distribution of chlorophyll *a*, phaeopigments and primary production along the three transects studied in summer and winter cruises, one notes that the winter distributions are markedly different from those of the preceding summer (Figs. 5 and 9). In order to test the significance in the variabilities of the parameters studied, the fact that seven stations were occupied per transect during the summer and only five stations were occupied per transect during the winter, made it imperative to resort to some rigorous statistical treatment of the data. This necessitated the development of a statistical model in an effort to determine the significance of the variations of the parameters studied. Further, the analysis of the time-series data with sampling at irregular intervals also required the development of a "diurnal model". These models are discussed at some length in El-Sayed *et al*. (1977); however the results of the analysis of variance tests performed (Tables 5-8) on the data obtained along the three transects in summer and winter are summarized below.

Between Stations

Using individual depths (Table 5). Chlorophyll *a* concentrations and primary production were statistically different ($P > 0.05$). Pigment ratios and productivity indexes were also significantly different ($P > 0.01$).

Using integrated values (Table 6). None of the parameters were significantly different.

Between Transects

Using individual depths (Table 5). Chlorophyll *a* concentrations and primary production values were significantly different ($P > 0.05$). Pigment ratios and productivity indexes were also significantly different ($P > 0.05$).

Using integrated values (Table 6). Chlorophyll *a* concentrations were significantly different ($P > 0.01$).

Between Seasons

Using individual depths (Table 7). All parameters except PI were significantly different ($P > 0.05$).

TABLE 5

Results* of analysis of variance of chlorophyll a, phaeopigments, pigment ratios, primary production and productivity index taken at the individual depth at the stations occupied in Transects A, B and C in the summer of 1975 and the winter of 1976.

Parameter	Analysis of variance between	
	Stations	Transects
Chlorophyll a	$F\binom{30}{234} = 3.063$*	$F\binom{2}{30} = 3.630$*
Phaeopigments	$F\binom{30}{239} = 1.285$	$F\binom{2}{30} = 1.437$
Pigment ratio	$F\binom{30}{285} = 2.019$**	$F\binom{2}{30} = 4.218$*
Primary production	$F\binom{27}{220} = 3.278$*	$F\binom{2}{27} = 0.483$
Productivity index	$F\binom{30}{252} = 1.642$**	$F\binom{2}{30} = 0.185$

*Null hypothesis of homogeneous distribution between stations or transects is rejected at the 95% significance level (*) and at the 99% level (**).

Using integrated depths (Table 8). Chlorophyll a and phaeopigments concentrations were significantly different ($P > 0.05$).

Thus from the results of the individual depths, our data suggest a heterogeneous spatial distribution of chlorophyll a and primary production. The results also strongly suggest that the standing crop of phytoplankton, phaeopigments and primary production display significant seasonal variabilities. The degree

TABLE 6

Results* of analysis of variance of integrated values for chlorophyll *a*, phaeopigments and primary production at the stations occupied in Transects A, B and C in the summer of 1975 and the winter of 1976.

Parameter	Analysis of variance between	
	Stations	Transects
Chlorophyll *a*	$F\binom{30}{3} = 3.820$	$F\binom{2}{30} = 9.299$**
Phaeopigments	$F\binom{30}{3} = 0.556$	$F\binom{2}{30} = 0.039$
Primary production	$F\binom{26}{3} = 2.529$	$F\binom{2}{26} = 0.237$

*Null hypothesis of homogeneous distribution between stations or transects is rejected at the 95% significant level (*) and at the 99% level (**).

of variability was by a factor of two which correspond to the findings of Blackburn *et al.* (1970); Owen and Zeitschel (1970; and Blackburn (1973). However, it was smaller than the seasonal variation (2-3 fold) in the Sargasso Sea (Hulbert *et al.*, 1960).

In the present investigation the study of the physiological state of the phytoplankton populations in the pre-mining operations is of special ecological significance. Since the chlorophyll: phaeopigments ratio is sometimes considered as an indicator of the physiological state of the phytoplankton, we have plotted the percent of these ratios against the percent of visible

TABLE 7

Results* of analysis of variance of chlorophyll *a*, phaeopigments, pigment ratio, primary production and productivity index for individual depths at the stations occupied in the summer of 1975 and the winter of 1976.

Parameter	Analysis of variance between 1975 and 1976
Chlorophyll *a*	$F\binom{1}{239}$ = 59.11*
Phaeopigments	$F\binom{1}{239}$ = 13.741**
Pigment ratio	$F\binom{1}{285}$ = 14.516**
Primary production	$F\binom{1}{220}$ = 4.923*
Productivity index	$F\binom{1}{252}$ = 0.422

*Null hypothesis of no seasonality between the summer of 1975 and the winter of 1976 is rejected at the 95% significance level (*) and at the 99% level (**).

light, and the corresponding depths in summer and winter cruises (Fig. 11). The relationship between these ratios and light corroborates Yentsch's (1965) suggestion that the relationship is not linear. Throughout the upper 50 m, phaeopigments rarely exceeded 60% of total pigments (i.e. chlorophyll *a* plus phaeopigments). Below 50 m, chlorophyll *a* and phaeopigments were about the same. However, at deeper depths (i.e. below 100 m) the percent concentration of phaeopigments was greater than chlorophyll *a*.

Since the pigment ratio is a function of light (Yentsch, 1965), the photosynthetic index (calculated as the ratio of primary production to chlorophyll *a*) is also a function of the

TABLE 8

Results* of analysis of variance of integrated (to the 1% light level) chlorophyll *a*, phaeopigments and primary production of the stations occupied in the summer of 1975 and the winter of 1976.

Parameter	Total plankton
Chlorophyll *a*	$F\binom{1}{30} = 61.787$*
Phaeopigments	$F\binom{1}{30} = 316.26$**
Primary production	$F\binom{1}{26} = 1.025$

*Null hypothesis of no seasonality between the summer of 1975 and the winter of 1976 is rejected at the 95% significance level (*) and at the 99% level (**).

light regime. The physiological state of phytoplankton from different areas or seasons can be compared directly by examining the light utilization efficiency. The latter is calculated as a ratio of daily primary production to net downward flux of photosynthetically active radiation (PAR). By using this ratio one eliminates the effect of differential radiation. The spatial distribution of light utilization efficiency between stations was more variable in the summer of 1975 than in the winter of 1976 (Tables 9 and 10). The average light utilization efficiency of phytoplankton for the summer and winter was 3.6 ± 1.7 mgC/E and 3.4 ± 1.2 mgC/E, respectively. These values are not significantly different. When the efficiency is considered as a function of chlorophyll *a*, Transect C consistently displayed the highest efficiency in summer 1975 (0.57 ± 0.33 mgC/E/mg Chl.*a*/m^2) and in winter 1976 (0.24 ± 0.08 mgC/E/mg Chl.*a*/m^2). The efficiency ratios per unit of chlorophyll were found to be significantly higher in summer (0.42 ± 0.27 mgC/E/mg Chl.*a*/m^2) than in winter (0.21 ± 0.07 mgC/E/mg Chl *a*/m^2).

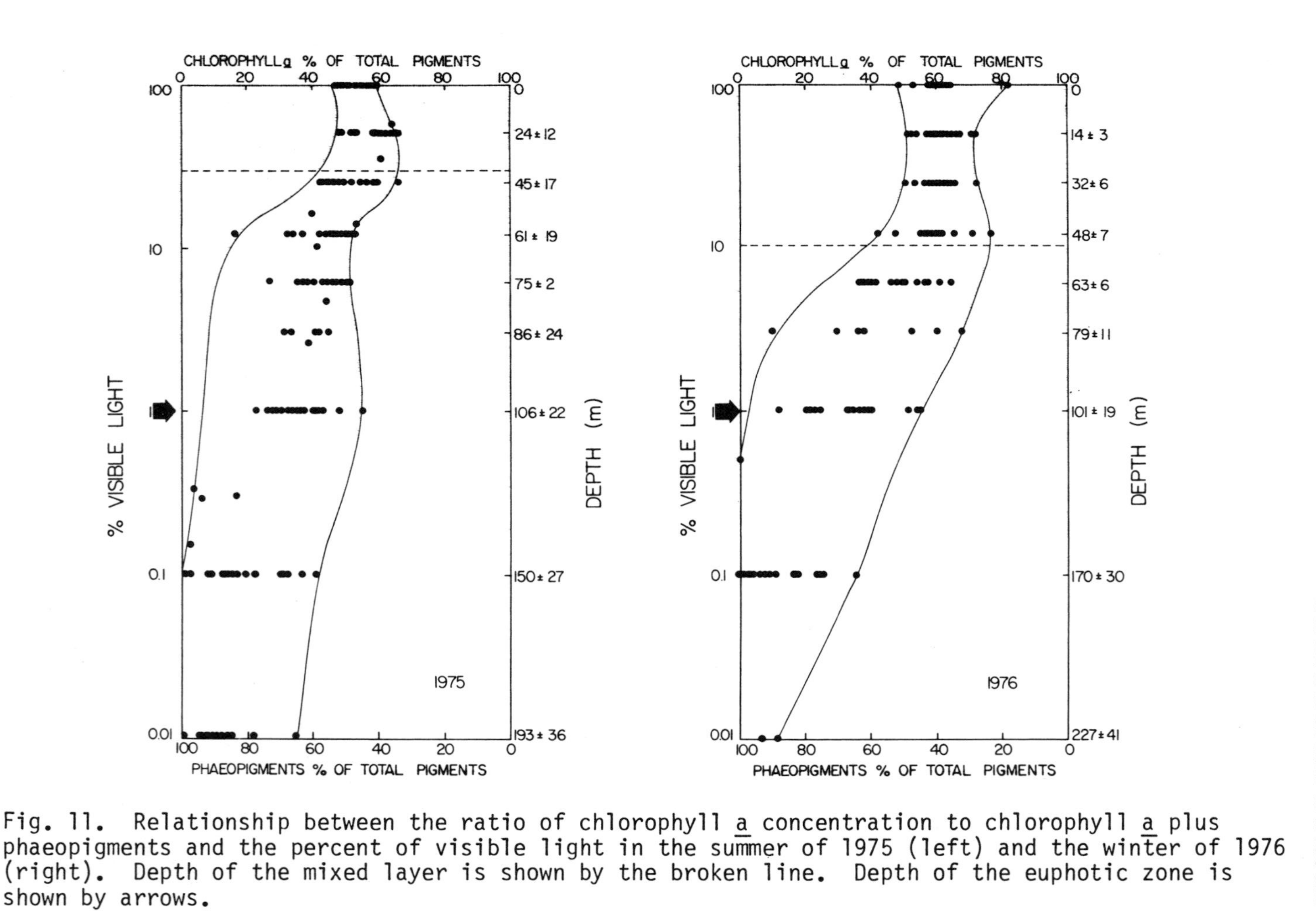

Fig. 11. Relationship between the ratio of chlorophyll a concentration to chlorophyll a plus phaeopigments and the percent of visible light in the summer of 1975 (left) and the winter of 1976 (right). Depth of the mixed layer is shown by the broken line. Depth of the euphotic zone is shown by arrows.

TABLE 9

Net downward flux of photosynthetically active solar radiation (PAR), daily primary production, chlorophyll *a*, ratio of primary production to PAR and ratio of latter to chlorophyll *a* concentration at the stations occupied in the summer of 1975.

Station	PAR ($E \cdot m^{-2} \cdot day^{-1}$)	Primary Production ($mgC \cdot m^{-2} \cdot day^{-1}$)	Chl.a ($mg \cdot m^{-2}$)	Primary Production / PAR	Primary Production / PAR/Chl. *a*
C+3	22.9	130.2	7.14	5.69	0.796
C+2	34.5	137.3	8.53	5.60	0.657
C+1	49.7	178.3	6.73	3.59	0.533
C	52.9	114.0	9.55	2.16	0.226
C-1	46.8	100.1	5.87	2.14	0.364
C-2	40.3	264.8	5.72	6.57	1.150
C-3	26.4	115.8	18.31	4.39	0.240
mean	39.1±11.6	148.6±57.0	8.83± 4.40	4.31± 1.76	0.567± 0.333
B+3	41.9	-	7.001	-	-
B+2	41.9	-	18.05	-	-
B+1	50.1	47.6	11.97	0.95	0.079
BA	-	164.7	11.07	-	-
BB	45.3	208.9	12.24	4.61	0.416
B-1	41.6	68.1	11.30	1.64	0.145
B-2	40.4	192.8	8.15	4.77	0.585
B-3	-	91.5	5.58	-	-
mean	43.5±3.61	128.9±68.5	10.67± 3.87	2.98± 1.60	0.326± 0.187
A+3	21.8	57.8	9.35	2.65	0.284
A+2	30.8	123.6	10.00	4.01	0.401
A+1	42.9	59.5	6.29	1.39	0.200
A	41.6	108.7	6.82	2.61	0.383
A-1	-	42.8	11.37	-	-
A-2	32.5	192.4	12.31	5.92	0.481
A-3	38.7	93.0	11.53	2.40	0.208
mean	34.7±7.97	96.8±51.4	9.76± 2.20	2.86± 1.66	0.293± 0.137
Total	39.1±8.89	124.6±60.0	9.25± 3.05	3.59± 1.75	0.420± 0.268

TABLE 10

Net downward flux of photosynthetically active solar radiation (PAR), daily primary production, chlorophyll a, ratio of primary production to PAR and ratio of latter to chlorophyll a concentration at the stations occupied in the winter of 1976.

Station	PAR ($E \cdot m^{-2} \cdot day^{-1}$)	Production ($mgC \cdot m^{-2} \cdot day^{-1}$)	Chl.a ($mg \cdot m^{-2}$)	Primary Production / PAR	Primary Production / PAR/Chl. a
C+3	23.4	111.9	16.96	4.72	0.296
C+2	32.0	79.8	8.64	2.47	0.286
CA	41.5	115.5	15.01	2.77	0.185
CB	44.3	82.8	12.37	1.86	0.151
C-1.5	42.1	120.2	16.46	2.48	0.173
C-3	40.4	159.1	11.38	3.92	0.344
mean	37.5±7.94	111.6±28.9	13.30± 3.04	3.10± 1.04	0.239± 0.079
B+3	50.5	157.0	14.97	3.11	0.208
B+1.5	44.7	155.6	11.65	3.48	0.299
BA	41.5	59.2	23.41	1.43	0.061
BB	52.5	147.0	15.87	3.54	0.223
B-1.5	55.6	151.9	20.35	2.73	0.134
B-3	47.5	275.6	27.97	5.80	0.207
mean	48.7±5.19	157.7±68.9	19.04± 6.03	3.35± 1.43	0.189± 0.082
A+3	22.0	93.8	27.40	4.26	0.156
A+1.5	46.2	168.1	19.05	3.64	0.191
A	53.2	-	28.74	-	-
A-1.5	52.3	283.4	32.75	5.42	0.165
A-3	-	-	34.00	-	-
mean	43.4±14.6	181.8±95.5	28.51± 6.04	4.44± 0.904	0.171± 0.018
Total	43.2±9.90	144.1±64.3	19.86± 7.92	3.47± 1.22	0.205± 0.075

Semina and Tarkhova (1972) found a negative correlation between the amount of phytoplankton and the depth of the pycnocline with the highest density gradient. Fryxell _et al_. (this volume) also found a similar relationship, although the correlation was not significant (Table 11). The same table shows that daily primary production is also not significantly related to the surface mixed layer. This may be due to large day to day variations in primary production at the same station rather than due to regional variation in productivity as shown by Platt and Filion (1973). On the other hand, Table 11 shows a significant positive correlation between chlorophyll _a_ and the surface mixed layer. It is likely that the replenishment of nutrient supply (from the deep mixed layer) increases the chlorophyll _a_ content in the algal cells and contributes to the high pigment ratios (Fig. 11). One may conclude that the phytoplankton biomass and its physiological state is likely to controlled by the depth of the surface maximum layer.

Hirota (1976) showed a significant negative relationship between the night zooplankton standing crop and the depth of the bottom of the thermocline (Table 11). A close coupling between primary production in the mixed layer, grazing and excretion rates of zooplankton was shown in the North Pacific Central gyre by Eppley _et al_. (1973). If we assume that nutrient supply from below is dependent on the depth of the mixed layer, and the excretory product of zooplankton is proportional to the abundance of zooplankton, then the total amount of nutrient available to primary production would be more or less the same regardless of the depth of the mixed layer. We may conclude, therefore, that unlike the phytoplankton biomass, primary production is independent of the depth of the mixed layer.

Although only two seasons (August-October and February-March) were studied, they correspond to seasonal maximum and minimum of phytoplankton and zooplankton in the eastern tropical Pacific (Blackburn _et al_., 1970). Ratios of winter to summer values are 2.1 for the phytoplankton chlorophyll _a_ and 1.6 for the night zooplankton standing crop, respectively (Hirota, 1976). This relationship is consistent with the theory advanced by Cushing (1959), Heinrich (1962), and confirmed by Blackburn (1973) that the delay period is short, that the phytoplankton is fully utilized by the zooplankton and that the seasonal cycle of the plankton community is balanced.

Summary and Conclusions

In the present study we were primarily interested in assessing the magnitude of the spatial and seasonal variabilities of the standing crop of phytoplankton and primary production at stations occupied along three transects in the tropical Pacific

TABLE 11

Correlation coefficients between the depth of SML and total cell number, chlorophyll a concentrations, daily primary production, light utilization efficiency (LUE) and the night zooplankton standing crop. (1) was calculated with the depth of pycnocline (Fryxell *et al*., this volume). (2) was calculated with the depth of bottom of thermocline (Hirota, 1976). ** indicates a 99% significance level.

	Cell Counts (1)	Chlorophyll a	Primary Production	LUE	Zooplankton (2)
SML	- 0.23	0.44**	- 0.17	-0.08	- 0 .61**

during the period of August/October 1975 and February/March 1976. The study area (located between 50°- 20°N and 128° - 155°W) is characterized by the low chlorophyll a values one would normally expect to find in the oligotrophic tropical/subtropical oceanic waters. Average surface chlorophyll a was 0.117 mg/m^3, and the average value integrated through the water column (to the bottom of the euphotic zone) was only 19.6 mg/m^2. A subsurface chlorophyll a maximum layer (CML) is a characteristic feature of the area investigated. This maximum layer was found at a somewhat deeper depth (about 69 $\pm$ 24 m) in summer than in winter (about 54 $\pm$ 30 m). The dominant phytoplankter at the chlorophyll maximum layer was the coccolithophorid *Gephryocapsa huxleyi* which also accounted for more than half of the coccolithophorid cells in the study area.

The chlorophyll maximum layer occurred either in the middle or at the bottom of the pycnocline in summer and at either the top or in the middle of the pycnocline in winter. These chlorophyll maxima were found to correspond to about 10% and 20% of the surface light level for the summer, and the winter, respectively. The depth of the chlorophyll maximum layer was also found to be closely associated with the distribution of the nutricline, especially that of nitrate. Following the chlorophyll maximum, there was a general decline in the concentration of chlorophyll. At many stations substantial quantities of chlorophyll a were found below the euphotic zone, contributing an average of 20% of the chlorophyll in the water column for both summer and winter cruises. Our data suggest that the chlorophyll maximum layer is due either to zooplankton grazing or to slower growth of larger phytoplankton cells. Our observations corroborate the findings of other investigators that the chlorophyll maximum layer is a regular feature in most tropical and subtropical waters of the world ocean.

Our primary productivity data area agree quite well with those found by other investigators in the same or adjacent regions. Primary production values (either surface or integrated) were, in general, low. The average value in winter (144 mgC/m^2/day) was slightly higher than in summer (120 mgC/m^2/day). Maximum primary production was found at 50% of surface light level in summer, but was also found at or near the surface in winter. Phytoplankton production below the euphotic zone (average depth about 104 m) contributed an average of 5% of the total production in the water column. The contribution of the nannoplankton (organisms <20 μm) to total primary production was lower in winter (63%) than in summer (83.3%). On the whole nannoplankton contributed an average of 77% of primary production in the study area.

Spatial, diel and seasonal variabilities in the chlorophyll a, phaeopigments and primary production are well documented in the present study. The degree of seasonal variability was about a factor of two and was of the same magnitude as that found by other investigators in the adjacent regions.

Our data allowed us to draw the conclusion that whereas the phytoplankton standing crop is likely to be controlled by the depth of the surface mixed layer, primary production seems to be independent of the depth of the mixed layer.

Phytoplankton and zooplankton data collected suggest that the area investigated is at a near steady state for an interval of, at best, a few days. This means that the feeding of the zooplankton or phytoplankton apparently balances production by phytoplankton.

Acknowledgement

We wish to express our sincere thanks to Kurt Buck, Dean Stockwell, Robert Warner and Michael Meyer, all graduate students in the Oceanography Department, and Norman Fisher and Mark Riggs, graduate students, Institute of Statistics, Texas A&M University, for their efforts in data collecting/analysis. We are indeed grateful to Professor Charles Gates, Professor of Statistics, Texas A&M University for his advice and assistance in carrying out the statistical analysis and statistical model used in this study. Grateful acknowledgement is expressed to Dr. G. C. Anderson, Dr. Robert Burns, Mr. George Poore and Mr. Barrett Ericson for their efforts in organizing and coordinating the field activities of the DOMES project, and to Captain K. E. Taggut, Commanding Officer of the NOAA ship *Oceanographer* for his excellent cooperation. Grateful acknowledgement is expressed to the National Oceanic and Atmospheric Administration for support of this work through research contract NOAA #03-6-022-35155.

References

Anderson, G. C., (1969) Subsurface chlorophyll maximum in the northeast Pacific Ocean, *Limnol. Oceanogr*. 14, 386-391.

Anderson, G. C., (1972) Aspects of marine phytoplankton studies near the Columbia River, with special reference to a subsurface chlorophyll maximum. In *The Columbia River Estuary and Adjacent Ocean Waters*, A. T. Pruter and D. L. Alverson, (editors), University of Washington, 219-240.

Anderson, J. J. and F. A. Richards, (1977) Chemical and biochemical observations from the DOMES study area in the Equatorial North Pacific, *Special Report* No. 79, University of Washington, Seattle, 1-95.

Baker, E. T. and R. A. Feely, Distribution and composition of the suspended particulate matter in the waters of the DOMES region, *Revised Final Report*, Pacific Marine Environmental Laboratory, NOAA, 1-62, this volume.

Beers, J. R. and G. L. Stewart, (1971) Microplankters in the plankton communities of the upper waters of the eastern tropical Pacific, *Deep-Sea Res*. 18, 861-883.

Berman, T. and O. Holm-Hansen, (1974) Release of photoassimilated carbon as dissolved organic matter by marine phytoplankton, *Mar. Biol*. 28, 305-310.

Blackburn, M., (1973) Regressions between biological oceanographic measurement in the eastern tropical Pacific and their significance to ecological efficiency, *Limnol. Oceanogr*. 18, 552-563.

Blackburn, M., R. M. Laws, R. W. Owen and B. Zeitschel, (1970) Seasonal and areal changes in standing stocks of phytoplankton, zooplankton and micronekton in the eastern tropical Pacific, Mar. Biol. 7, 14-31.

Cushing, D. H., (1959) The seasonal variation in oceanic production as a problem in population dynamics, J. Cons. perm. int. Explor. Mer 24, 298-338.

El-Sayed, S. Z., (1972) Primary productivity and standing crop of phytoplankton. In Chemistry, Primary Productivity and Benthic Algae of the Gulf of Mexico, Folio 22, American Geographical Society, V. Bushnell, (editor), 8-13.

El-Sayed, S. Z., S. Taguchi, G. A. Franceschini, G. A. Fryxell and C. E. Gates, (1977) Phytoplankton and primary productivity studies (in connection with Deep Ocean Mining Environmental study), Final Report, Texas A&M Research Foundation, 1-443.

El-Sayed, S. Z. and J. T. Turner, (1978) Productivity of the Antarctic and tropical subtropical regions: a comparative study. In Proceedings of SCAR/SCOR Polar Oceans Conference, Montreal, Canada, May, 1974, Arctic Inst. of North Amer., J. J. Dunbar, (editor), 463-504.

Eppley, R.W. and P. R. Sloan, (1966) Growth rate of marine phytoplankton: correlation with light absorption by cell chlorophyll a, Physiol. Plant. 19, 47-59.

Eppley, R. W., R. W. Holmes and J.D.H. Strickland, (1967) Sinking rates of marine phytoplankton measured with a fluorometer, J. of Exp. Mar. Biol. Ecol. 1, 191-208.

Eppley, R., W. E. H. Renger, E. L. Venrick and M. M. Mullin, (1973) A study of plankton dynamics and nutrient cycling in the central gyre of the North Pacific Ocean, Limnol. Oceanogr. 18, 534-551.

Franceschini, G. A., The solar radiation environment in the eastern tropical N. Pacific Ocean. In Marine Geology and Oceanography of the Central Pacific Manganese Nodule Province, J. L. Bischoff and D. Z. Piper, (editors), Pergamon Press, New York, this volume.

Fryxell, G. A., S. Taguchi and S. Z. El-Sayed, Vertical distribution of diverse phytoplankton populations in the central Pacific. In Marine Geology and Oceanography of the Central Pacific Manganese Nodule Province, J. L. Bischoff and D. Z. Piper, (editors), Pergamon Press, New York, this volume.

Gunderson, K. R., J. S. Corbin, C. L. Hansen, M. L. Hanson, R. B. Hanson, D. J. Russell, A. Stollar and O. Yanada, (1976) Structure and biological dynamics of the oligotrophic ocean photic zone off the Hawaiian Island, Pacific Science, 30, 45-68.

Heinrich, A. K., (1962) The life histories of plankton animals and seasonal cycles of plankton communities in the oceans, J. Cons. perm. int. Explor. Mer 27, 15-24.

Hirota, J., 1976 DOMES preliminary report on zooplankton, (University of Hawaii, Honolulu; Hawaii), (unpublished manuscript) 1-170.

Hobson, L. A. and C. J. Lorenzen, (1972) Relationships of chlorophyll maxima to density structure in the Atlantic Ocean and Gulf of Mexico, Deep-Sea Res. 19, 297-306.

Holmes, R. W., (1958) Size fractionation of photosynthesizing phytoplankton. In Physical, Chemical and Biological Oceanographic Observations Obtained on Expedition SCOPE in the Eastern Tropical Pacific, Nov.-Dec. 1956, Part I, Methods Station Data, U.S. Fishery and Wildlife Service, Special Scientific Report, Fishery, No. 279, 1-71.

Holm-Hansen, O., C. J. Lorenzen, R. W. Holmes and J.D.H. Strickland, (1965) Fluorometric determination of chlorophyll, J. Cons. perm. int. Explor. Mer 30, 3-15.

Hulbert, E. M., J. H. Ryther and R.R.L. Guillard, (1960) The phytoplankton of the Sargasso Sea off Bermuda, J. Cons. perm. int. Explor. Mer 27, 15-24.

Jamart, B. M., D. F. Winter, K. Banse, G. C. Anderson and R. K. Lam, (1977) A theoretical study of phytoplankton growth and nutrient distribution in the Pacific Ocean off the northwestern U.S. coast, Deep-Sea Res. 24(8), 753-773.

Kawarada, Y. and A. Sano, (1969) Distribution of chlorophyll and phaeopytin in the western north Pacific, Oceanographic Magazine 21, 137-146.

Kiefer, D. A., R. J. Olson and O. Holm-Hansen, (1976) Another look at the nitrate and chlorophyll maxima in the central North Pacific, Deep-Sea Res. 23, 1199-1208.

King, J. E., T. S. Austin and M. S. Doty, (1957) Preliminary report on expedition EASTROPAC, Special Scientific Report, U.S. Fish and Wildlife Service, Fishery, No. 201, 1-155.

Longhurst, A. R. (1976) Interactions between zooplankton and phytoplankton profiles in the eastern tropical Pacific Ocean, Deep-Sea Res. 23, 729-754.

Lorenzen, C. (1965) A note on the chlorophyll and phaeophytin content of the chlorophyll maximum, Limnol. Oceanogr. 10(3), 482-483.

Love, C. M., (editor), (1970) EASTROPAC Atlas 2: Biological and nutrient chemistry data from principal participating ships, NOAA, National Marine Fishery Service, Circular 330.

Love, C. M., (editor), (1971) EASTROPAC Atlas 4: Biological and nutrient chemistry data from principal participating ships, NOAA, National Marine Fishery Service, Circular 330.

Love, C. M., (editor), (1972) EASTROPAC Atlas 6: Biological and nutrient chemistry data from principal participating ships, NOAA, National Marine Fishery Service, Circular 330.

Love, C. M., (editor), (1974) EASTROPAC Atlas 8: Biological and nutrient chemistry data from principal participating ships, NOAA, National Marine Fishery Service, Circular 330.

Love, C. M., (editor), (1975) EASTROPAC Atlas 10: Biological and nutrient chemistry data from principal participating ships, NOAA, National Marine Fishery Service, Circular 330.

Marumo, R., (editor), (1970) Preliminary report of the Haku Maru cruise KH-69-4, Ocean Research Institute, University of Tokyo, 1-68.

Mullin, M. M. and E. R. Brooks, (1972) The vertical distribution of juvenile Calanus (Copepods) and phytoplankton within the upper 50 m of water off La Jolla, California. In Biological Oceanography of the Northern North Pacific Ocean, A. Y. Takenouti (editor), Idemitsu Shoten, Tokyo, 347-354.

Owen, R. W. and B. Zeitschel, (1970) Phytoplankton production: seasonal change in the oceanic eastern tropical Pacific, Mar. Biol. 7, 32-36.

Platt, T. and C. Filion, (1973) Spatial variability of the productivity biomass ratio of phytoplankton in small marine basin, Limnol. Oceanogr. 18, 743-749.

Reid, F.M.H., E. Stewart, R.W. Eppley and D. Goodman, (1978) Spatial distribution of phytoplankton species in chlorophyll maximum layers off southern California, Limnol. Oceanogr. 23(2), 219-226.

Ryther, J. H. and D. W. Menzel, (1959) Light adaptation by marine phytoplankton Limnol. Oceanogr. 4, 492-502.

Saijo, Y. , S. Izuka and O. Asaoka, (1969) Chlorophyll maxima in Kuroshio and adjacent area, Mar. Biol. 4(3), 190-196.

Semina, H. J. and I. A. Tarkhova, (1972) Ecology of phytoplankton in the northern Pacific Ocean. In Biological Oceanography of the Northern Pacific Ocean, A. Y. Takenouti, (editor), Idemitsu Shoten, Tokyo.

Smayda, T. J., (1970) The suspension and sinking of phytoplankton in the sea, Oceanogr. Mar. Biol. Annual review, 8, 353-414.

Steele, J. H. (1964) Some problems in the study of marine resources, ICNAF Environ Symp. Rome, Contrib. No. C-4, 1-11.

Steele, J. H. and C. S. Yentsch, (1960) The vertical distribution of the chlorophyll, J. of Mar. Biol. Ass. of U.K. 39, 217-226.

Steemann Nielsen, E. (1952) The use of radio-active carbon for measuring organic production in the sea, J. Cons. perm. int. Explor. Mer 18, 117-140.

Strickland, J.D.H. and T. R. Parsons, (1970) A practical handbook of seawater analysis, Bull. Fish. Res. Bd. of Can. 167, 1-311.

Taguchi, S., (1976) Relationship between photosynthesis and cell size of marine diatoms, J. Phycol. 12, 185-189.

Takahashi, M., K. Satake and N. Nakamoto, (1972) Chlorophyll profile and photosynthetic activity in the north and equatorial Pacific Ocean, J. Oceanogr. Soc. of Japan, 28, 27-36.

Taniguchi, A. and T. Kawamura, (1972) Primary production in the Oyashio region with special reference to the subsurface chlorophyll maximum layer and the phytoplankton relationships. In Biological Oceanography of the Northern North Pacific Ocean, A. Y. Takenouti, (editor), Idemitsu Shoten, Tokyo, 245-252.

Venrick, E. L., (1971) Recurrent groups of diatom species in the North Pacific, Ecology, 52, 614-625.

Venrick, E. L., J. A. McGowan and A. W. Mantyla, (1973) Deep Maxima of photosynthetic chlorophyll in the Pacific Ocean, Fishery Bulletin, U.S. Department of Commerce, 71, 41-52.

Vinogradov, M. E., I. I. Gitelzon and Yu. I. Sorokin, (1970) The vertical structure of a pelagic community in the tropical ocean, Mer. Biol. 6, 187-194.

Yentsch, C. S., (1965) Distribution of chlorophyll and phaeophytin in the open ocean, Deep-Sea Res. 12, 653-666.

Yentsch, C. S. and D. W. Menzel, (1963) A method for the determination of phytoplankton chlorophyll and phaeophytin by fluorescence, Deep-Sea Res. 10, 221-231.

ABYSSAL COMMUNITY STRUCTURE OF THE BENTHIC INFAUNA OF THE EASTERN EQUATORIAL PACIFIC: DOMES SITES A, B, AND C

Barbara Hecker and Allen Z. Paul

Lamont-Doherty Geological Observatory
Columbia University
Palisades, New York 10964

Abstract

Eighty 0.25m^2 box cores were collected and analyzed to determine the faunal composition, species diversity and sample similarity of the benthic communities of three sites in the eastern Equatorial Pacific. (The results are part of a baseline survey of the fauna of the manganese nodule province of the Pacific.) The bottom topography of this region is characterized by abyssal hills, with a relief of 50 to 300 meters, at a depth of 4350 to 5150 meters.

Macrofaunal density ranged from 36-268 individuals per m^2. A significantly higher density was found at site C, when compared to sites A or B. Numerically, the fauna was dominated by polychaetes (40.1%), tanaids (19.5%), isopods (11.7%) and bivalves (8.4%). Meiofauna was present in high abundances (62% of the total metazoan fauna), even though the screen size used (300 μm) was too large to retain the majority of them. Biomass of the macrofauna ranged from 0.0156-1.5708 grams wet weight per m^2.

Deposit feeders comprised 81% of the total macrofauna, while suspension feeders accounted for 19%. This predominance of deposit feeders was also reflected in the trophic structure of the polychaetes, accounting for approximately 89% of the total number of individuals. Site C had the highest percent of suspension feeders, while site A had the lowest.

Species diversity was found to be extremely high and comparable to the results obtained from other deep sea investigations. A very obvious feature of the fauna was the rarity with which most of the species were encountered. As a result reliable estimates of the total number of species in these communities could not be made.

Species similarity analysis suggests there is a higher degree of faunal affinity within a site than between sites. The only obvious environmental difference between the sites is a possible east-west gradient in organic flux.

Introduction

Research in recent years has resulted in substantial advances in our understanding of deep-sea ecosystems. These advances have been facilitated by the development of sophisticated sampling and observational techniques. Results obtained using the epibenthic sled, anchor dredge, and box corer demonstrate that deep-sea benthic communities have a low standing crop and are composed of an exceptionally diverse assemblage of species (Sanders et al., 1965; Sanders, 1968; Hessler and Sanders, 1967; Hessler and Jumars, 1974; Jumars, 1976; and Jumars and Hessler, 1976). The finding of a high species diversity in the deep sea was particularly surprising, because such high diversity had not been predicted by the prevalent ecological theory. The assumption was that a uniform environment with few food resources could support only a limited number of species by precluding a large number of possible niches (Hutchinson, 1953). The deep sea has environmental constancy and few food resources, and yet supports a very large number of species. The discrepancy between theory and observation led to speculation concerning the mechanism causing and maintaining the high diversity. These speculations are primarily concerned with which biological interaction, interspecific competition or predation, plays the dominant role in maintaining community structure.

Slobodkin and Sanders (1969) proposed that the long-term environmental stability of the deep sea enables many species to coexist by allowing them to specialize to the extent that they are not competitively excluded. Sanders (1968, 1969) argues that interspecific competition plays a major role in structuring deep-sea benthic communities, thus making them biologically controlled. Dayton and Hessler (1972), however, suggest that the biological controlling agent in maintaining the high diversity is disturbance, such as predation. They argue that predation prevents species from reaching densities high enough to necessitate a great degree of niche diversification. Grassle and Sanders (1973) favor a compromise theory. They point out that the two views are not mutually exclusive, and that neither one taken by itself can fully account for the observed high diversity. They suggest that biological disturbance creates small-scale heterogeneity and thereby allows highly localized and long-term succession. The temporal constancy of the environment permits the species to finely adapt to various stages of the successional sequence, thus producing a highly diverse community composed of small-scale mosaics of successional stages. Jumars (1975 a, b; 1976) using a spade or box core subdivided into 25 smaller compartments, has shown that the factors controlling diversity in the deep-sea are probably of a very small scale. He found that very small-scale environmental heterogeneity does exist in the deep sea and that the death of an individual or the excretion of a single fecal pellet provides the resources for colonization by other organisms.

The present study reports on the findings of the benthic baseline study of the DOMES (Deep Ocean Mining Enivornmental Survey) Project. Eighty box cores were taken to determine the faunal composition of the benthic community of the abyssal eastern Equatorial Pacific. Interest in this geographic locality was generated by concern over the possible consequences of mining the manganese nodules found in this region.

Methods

Sampling Equipment

A modified United States Naval Electronics Laboratory (USNEL) 0.25m^2 box corer, as described in Hessler and Jumars (1974) was used. The box corer allows for replication by sampling a precise area of the ocean floor. Obtaining replicate samples is particularly important when studying deep-sea benthic communities, because of the low standing crop and the high diversity of the organisms. The box core has fewer biases associated with it than other sampling devices by allowing for accurate replication. The chances of sampling across intercommunity boundaries is greatly reduced, because replicates can be obtained close to a given location. For this reason the box core method of sampling greatly increases the probability of detecting differences in community structure on a much finer scale than can be obtained with epibenthic sleds or dredge trawls. Such precise sampling is particularly valuable in the deep-sea, because the factors maintaining community structure are probably on a very small scale (Jumars, 1975 a & b).

Sampling Procedure

The samples were taken with a 0.25m^2 box corer. The corer was provided with a pinger on the trawlwire to monitor its position in the water column. (Details for this procedure can be found in Hessler and Jumars (1974).) At approximately 100 meters off the bottom, the descent rate was reduced from 60 m/min. to 30 m/min. or less, in order to minimize the bow wave effect, thus reducing the probability of washing surface fauna away from the area to be sampled (Menzies and Rowe, 1968). Bottom contact and closure was monitored by the pinger. After an initially slow pullout, the speed of the winch was increased to 60 m/min., and was again decreased as the corer approached the surface.

Only relatively undisturbed box cores were retained for biological examination. As soon as the corer was back on deck, the overlying water was drained off and passed through a 300 μm mesh screen. Some cores were subsampled for sediment analysis and future meiofaunal study. The area sampled by the core was reduced accordingly. The epifauna on the sediment and the nodules were removed and preserved in 10% buffered formalin and sea water. The top 20 cm. of sediment was transferred to a sediment washer with 300 μm mesh sides. The sediment was then washed with sea water, while compressed air was bubbled through it, for approximately two hours. This process aided in suspending and removing the fine particles. The residuum was preserved in 10% buffered formalin and sea water.

Laboratory Processing

Rose bengal, a protoplasmic stain (Walton, 1952), was used as an aid in separating the living fauna from the sediment. The sediment was then sorted, a small aliquot at a time, under a dissecting microscope at a minimum power of 18x. All organisms were removed and transferred to 70% isopropyl alcohol with the exception of the nematodes, which were kept in buffered formalin. We later noticed that some of the specimens were decalcifying and therefore we subsequently transferred all the samples into 90% isopropyl alcohol. Wet weight of the macrofauna was determined using a Mettler H-15 Balance. Organisms were blotted and air dried in groups for approximately one minute prior to weighing.

The organisms were initially sorted to major taxon. Whereever possible, all individuals were identified to species (with the exception of the nematodes). Some organisms were damaged and could only be identified to higher taxon. These individuals were omitted from the species diversity and community analysis, but not from the standing crop estimates. Meiofaunal organisms retained by the 300 μm mesh screen were classified in a similar manner, but were ignored in the community analysis, because we have no estimate of how representative of the total meiofauna they are. Thus far, only the polychaetes and the harpacticoid copepods have been examined by taxonomic specialists. Consequently, the species level sorting leaves room for error. It is hoped that any undersorting will have been balanced by oversorting and that the estimates of diversity are not seriously in error.

One problem was encountered with the polychaetes in these samples. Some of the soft bodied polychaetes were damaged during the sample handling and could not be identified. However, it was noted that even if these polychaetes could have been identified they would probably not have significantly increased the number of species in the samples.

Statistical Analyses

Because abyssal benthic sample sizes are small and diversity is high, we chose a method for estimating diversity that retains the good small sample characteristics of Simpson's (1949) diversity measure while giving weight to the rarer species in the community. This method was first proposed by Hurlbert (1971) as an improvement over Sanders' (1968) "rarefaction" technique, which tends to overestimate diversity. The Hurlbert diversity measure consists of a group of diversity measures for species richness that can be plotted as a curve. These measures are based on the expected number of species encountered in hypothetical subsamples, drawn at random without replacement, from the original sample. As is noted by Smith and Grassle (1977) the Hurlbert measure for a subsample size of 2 is equivalent to Simpson's diversity measure, while for larger subsamples it is increasingly sensitive to the contribution of rarer species. Hessler and Jumars (1974) point out that the Hurlbert curves can also be used to estimate species evenness and dispersion pattern of the species.

For estimates of faunal similarity between the samples we used an index that retains good small-sample characteristics and is sensitive to the contribution of rarer species. The index is a normalized-expected-species-similarity measure (NESS) developed by Grassle and Smith (1976). The calculations are a logical extension of Hurlbert's (1971) and Smith and Grassle's (1977) methods for estimating the species richness of a population. This similarity index is based on the expected number of species shared between hypothetical subsamples of a fixed size (m) drawn from the two populations. The number of species shared is determined by repeating this subsampling many times. At large values of m the rarer species contribute to the index because of their greater probability of appearing in the subsample. For the formulae used to compute NESS and a further explanation of the method the reader should refer to Grassle and Smith (1976).

Sample Locations and Environment

Three sampling sites in the eastern Equatorial Pacific were studied, each representing a different sedimentary regime. The sites were defined as 2° squares. Site A was centered at 8°27'N, 150°47'W under the North Equatorial Countercurrent and the North Equatorial Current in a zone of siliceous clay. Site B was centered at 11°42'N, 138°24'W under the North Equatorial Current

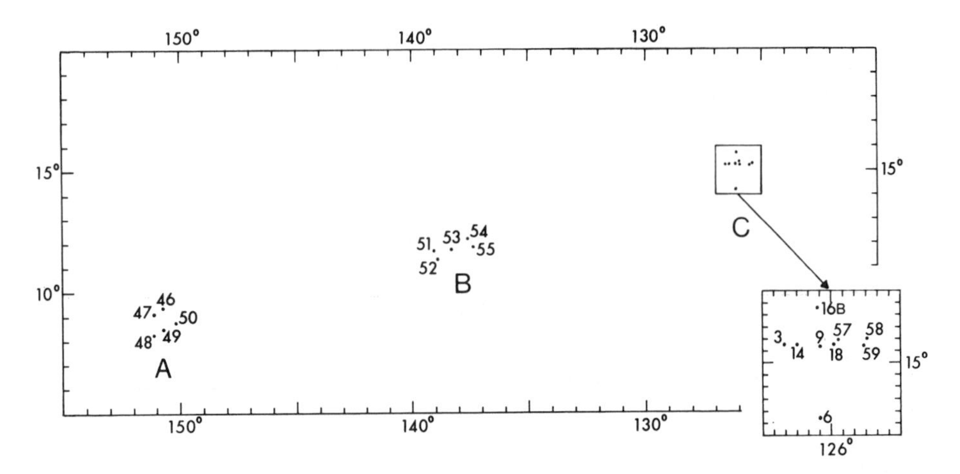

Fig. 1. Map of locations of sampling sites in the DOMES region.

in a zone of mixed siliceous clay and red clay. Site C was centered at 16°N, 126°W under the North Equatorial Current in a zone of red clay. Within each site five or seven stations were chosen at random (Fig. 1). Four or five replicate box cores were taken as close as possible at each station. The location and depth of each box core is shown in Table 1. Most of the samples at site C were taken in May and June of 1975 during RP-6-OC-75 of OSS-01 NOAA Ship OCEANOGRAPHER. Three additional cores at site C (stations 57 and 58) and the cores at sites A and B were taken in October and November of 1975 during RP-8-OC-75 of the same vessel.

The bottom water characteristics were as follows: Site A - temperature 0.982°C, salinity 34.68-34.69%, oxygen 3.9-4.1 ml/l., bottom current 2 cm/sec. oscillating between east and west; Site B - temperature 1.034°C, salinity 34.68-34.69%, oxygen 3.75-3.9 ml/l., bottom current 5 cm/sec. again oscillating between east and west; Site C - temperature 1.46-1.50°C, salinity 34.69%, oxygen 3.65-3.78 ml/l., bottom current 2 cm/sec. or less to the southwest (Hayes and Anderson, personal communication). There appears to be a warming trend towards the east, which is in agreement with other reports of Pacific bottom water (Mantyla, 1975). The salinity and oxygen content does not vary appreciably between the sites. The biochemical properties of the bottom waters, basically, are equivalent at the three sites (Anderson, personal communication).

Table 1

Box core locations and depths

RP-8-OC-75 Site A

Sta.	BC	Lat. (N)	Long. (W)	Depth (m)
46	1	9°20.7'	150°50.7'	5160
46	4	9°23.8'	150°52.8'	5155
46	6	9°24.0'	150°49.9'	5229
46	8	9°19.1'	150°50.4'	5108
46	9	9°20.0'	150°48.4'	4839
47	10	9°00.3'	151°10.9'	5108
47	11	9°02.5'	151°10.6'	5053
47	14	9°04.0'	151°14.2'	5110
47	15	9°04.8'	151°11.1'	5115
47	18	9°04.8'	151°09.4'	4978
48	19	8°16.5'	151°07.3'	5043
48	20	8°14.3'	151°14.3'	5043
48	22	8°16.0'	151°11.3'	5117
48	23	8°18.1'	151°09.5'	5144
49	24	8°27.5'	150°46.7'	5027
49	25	8°28.5'	150°44.5'	5005
49	26	8°27.4'	150°50.2'	4971
49	27	8°30.2'	150°47.8'	5039
50	28	8°41.7'	150°18.7'	5009
50	29	8°43.8'	150°18.7'	5033
50	31	8°41.4'	150°15.0'	4969
50	32	8°43.1'	150°14.1'	4948

RP-6-OC-75 Site C

Sta.	BC	Lat. (N)	Long. (W)	Depth (m)
3	3	15°15.1'	126°40.4'	4183
3	5	15°15.1'	126°39.5'	4666
3	6	15°16.8'	126°37.2'	4429
3	8	15°15.4'	126°38.6'	4690
3	9	15°18.04'	126°36.5'	4400
5	1	15°15.7'	125°31.5'	4509
5	2	15°15.0'	125°33.2'	4493
5	5	15°14.2'	125°33.5'	4487
5	7	15°15.0'	125°33.3'	4462
5	9	15°17.2'	125°29.4'	4370
6	1	14°16.2'	126°11.0'	4430
6	2	14°17.1'	126°08.2'	4462
6	3	14°14.2'	126°12.4'	4430
6	4	14°14.7'	126°11.6'	4410
6	5	14°13.2'	126°13.8'	4372
9	1	15°14.8'	126°11.8'	4630
9	3	15°15.3'	126°13.5'	4582
9	4	15°16.3'	126°10.4'	4520
9	5	15°16.5'	126°11.0'	4490
9	6	15°14.9'	126°10.5'	4480

TABLE 1 (cont'd)

Sta.	BC	Lat. (N)	Long. (W)	Depth (m)
RP-8-OC-75 Site B				
51	33	11°41.6'	139°11.0'	4881
51	34	11°42.4'	139°10.8'	4925
51	35	11°43.1'	139°08.9'	4914
51	36	11°43.7'	139°08.2'	4914
52	37	11°13.7'	139°09.9'	4872
52	40	11°16.3'	139°04.2'	4859
52	41	11°16.5'	139°07.7'	4850
52	42	11°15.5'	139°03.3'	4871
53	43	11°42.2'	138°23.4'	4997
53	44	11°43.3'	138°22.4'	5001
53	45	11°43.9'	138°22.2'	4985
53	46	11°44.3'	138°21.2'	4930
54	47	12°10.4'	137°44.1'	4930
54	48	12°11.0'	137°41.0'	4930
54	51	12°08.9'	137°44.6'	3934
54	52	12°09.6'	137°42.4'	4934
55	53	11°48.6'	137°28.3'	4859
55	54	11°50.8'	137°26.7'	4830
55	55	11°48.2'	137°26.3'	4853
55	56	11°48.6'	137°24.3'	4896

Sta.	BC	Lat. (N)	Long. (W)	Depth (m)
RP-6-OC-75 Site C				
14	1	15°12.5'	126°31.0'	4521
14	2	15°14.3'	126°30.1'	4521
14	3	15°15.1'	126°29.6'	4530
14	4	15°14.7'	126°29.8'	4467
14	5	15°15.1'	126°29.7'	4529
16B	1	15°45.7'	126°01.0'	4554
16B	2	15°46.7'	126°09.7'	4628
16B	3	15°43.7'	126°10.7'	4546
16B	4	15°45.1'	126°12.1'	4569
16B	5	15°44.5'	126°10.6'	4660
18	1	15°14.5'	125°59.3'	4200
18	2	15°14.5'	125°58.73'	4370
18	3	15°14.5'	125°50.1'	4383
18	4	15°14.01'	125°50.24'	4375
18	5	15°14.4'	125°50.4'	4365
RP-8-OC-75 Site C				
57	57	15°17.8'	125°55.3'	4520
57	59	15°20.4'	125°54.1'	4625
58	60	15°17.8'	125°28.4'	4375

Results and Discussion

Standing Crop and Faunal Composition

Eighty box cores were taken, 22 at site A, 20 at site B, and 38 at site C. Two cores were excluded from the standing crop estimates, the surface of core 49-24 was disturbed when it was collected and was judged only marginally acceptable and part of the sample of core 53-44 dried out in the laboratory. The remaining 78 cores yielded 9-67 macrofaunal individuals per $0.25m^2$. The average standing crop was 32 individuals per $0.25m^2$. This number may be an underestimate of the actual standing crop, as a result of some of the polychaetes having been destroyed during the sample processing. Approximately 45% of the polychaetes were lost (Fauchald, personal communication), thus a more realistic estimate of standing crop would be 42 macrofaunal individuals per $0.25m^2$ or 168 per m^2. Our results are within the range of densities reported for comparable depths. Hessler and Jumars (1974) found densities of 21-40 individuals per $0.25m^2$, in the oligotrophic gyre of the central North Pacific. Sanders, Hessler and Hampson (1965) reported densities of 33-264 individuals per m^2, with an average of 109.4, for the Sargasso Sea. Comparable densities have also been reported for other locations at depths in excess of 4000 meters (Rowe, 1971).

There were, however, differences among the sites. At site A the average number of individuals per $0.25m^2$ was 24.76 (S.E.= 1.83). The average number of individuals per core at site B was 28.47 (S.E.=2.62). The highest density was found at site C with an average of 38.11 (S.E.=2.13) individuals per $0.25m^2$. When tested with a two-level, nested analysis of variance (Sokal and Rohlf, 1969) the results indicate that there is significant difference ($P<0.01$) among the sites. The density of macrofauna at site C was higher than at sites A or B, but there was not a significant difference in the standing crop between sites A and B.

An east-west gradient in nutrient flux could explain this observed difference in standing crop among the sites. The primary productivity of the surface waters in the region does not vary enough to explain the higher density of macrofauna at site C. This is in contradiction to the reported coupling of benthic standing crop and surface primary productivity (Rowe, 1971; Sokolova, 1971). However, differences in the proximity of the three sites to the Eastern Boundary Current System of the North Pacific could account for the higher density at site C. Site C is directly downstream of the productive California Current, whereas sites A and B are further removed from this source water. (Malone, personal communication). We propose that the productivity of the surface water at the three sites might not reflect the amount of organic material reaching the benthos. We feel that the proximity of site C to the Eastern Boundary Current System indicates that more nutrients are reaching the bottom at site C allowing it to support a higher standing crop.

TABLE 2

Faunal composition by number of individuals and their percentage

Macrofaunal taxa	Site A #	Site A %	Site B #	Site B %	Site C #	Site C %	Total #	Total %
Polychaeta	189	38.6	239	46.4	542	38.2	970	40.1
Tanaidacea	121	24.7	77	15.0	274	19.3	472	19.5
Isopoda	57	11.6	30	5.8	197	13.9	284	11.7
Bivalvia	40	8.2	73	14.2	90	6.4	203	8.4
Gastropoda	13	2.7	25	4.9	23	1.6	61	2.5
Ectoprocta	25	5.1	8	1.6	97	6.8	130	5.4
Porifera	4	0.8	16	3.1	55	3.9	74	3.1
Hydrozoa	3	0.6	2	0.4	3	0.2	8	0.3
Stephanoscyphus	1	0.2	10	1.9	2	0.1	13	0.5
Actiniaria	3	0.6	-	-	15	1.1	18	0.7
Brachiopoda	10	2.0	9	1.7	31	2.2	50	2.1
Hemichordata	-	-	1	0.2	1	0.1	2	0.1
Sipunculoidea	3	0.6	4	0.8	14	1.0	22	0.9
Echiuroidea	-	-	-	-	3	0.2	3	0.1
Ophiuroidea	9	1.8	-	-	10	0.7	19	0.8
Echinoidea	-	-	3	0.6	1	0.1	4	0.2
Crinoidea	1	0.2	-	-	7	0.5	8	0.3
Holothuroidea	1	0.2	-	-	2	0.1	3	0.1
Aplacophora	2	0.4	2	0.4	2	0.1	6	0.2
Polyplacophora	1	0.2	-	-	5	0.4	6	0.2
Monoplacophora	1	0.2	-	-	-	-	1	-
Scaphopoda	1	0.2	-	-	1	0.1	2	0.1
Oligochaeta	-	-	-	-	8	0.6	8	0.3
Pycnogonida	-	-	-	-	3	0.2	3	0.1
Cumacea	-	-	4	0.8	3	0.2	7	0.3
Amphipoda	2	0.4	5	1.0	14	1.0	21	0.9
Cirripedia	-	-	-	-	3	0.2	3	0.1
Ascidacea	3	0.6	7	1.4	7	0.5	17	0.7
Unknown	-	-	-	-	4	0.3	4	0.2
Total	490	99.9	515	100.2	1417	100.0	2422	99.9
Total per core	22		25		37			
Meiofaunal taxa								
Nematoda	1116	87.3	1486	87.0	709	69.1	3311	82.5
Ostracoda	77	6.0	82	4.8	226	22.0	385	9.6
Copepoda	84	6.6	138	8.1	81	7.9	303	7.5
Acarina	-	-	2	0.1	8	0.8	10	0.2
Turbellaria	2	0.2	-	-	1	0.1	3	0.1
Kinorhyncha	-	-	1	0.1	1	0.1	2	-
Total	1279	100.1	1709	100.1	1026	10010	4014	99.9
Total per core	58		85		27			

The total number of individuals and the percent faunal composition found at the three sites are listed in Table 2. Polychaetes were by far the most abundant constituents of the benthic macrofauna. They were followed by tanaids, isopods and bivalves. These four taxonomic groups accounted for approximately 80% of the total fauna. All the other taxa accounted for 20%, with most of them contributing less than 2%. The percentage composition of the fauna is similar to that reported from other studies at abyssal depths, with the exception of the polychaetes. Sanders, Hessler and Hampson (1965) studied the benthic fauna of the Sargasso Sea and obtained similar percentages. Hessler and Jumars' (1974) study of the central North Pacific fauna also yielded similar results. The polychaetes in our samples accounted for 40.1% of the total macrofauna. This value is significantly lower than that found by Hessler and Jumars (1974) or that found by Sanders, Hessler and Hampson (1965), which were 55.1% and 55% respectively. This result may not reflect a real difference in the fauna of the DOMES region, because some of the polychaetes were destroyed during the sample processing.

Although the 300 μm mesh screen used in this study was too large to retain most of the meiofauna, their abundance still exceeded that of the macrofauna. Nematodes, harpacticoid copepods and ostracods were by far the most abundant meiofaunal constituents (Table 2). The nematodes alone accounted for approximately half of the total metazoal fauna.

Table 4 lists the weights of the taxa and their percent contribution to the total biomass. The biomass was disproportionately influenced by the occassional larger epifauna retrieved by the corer. At sites A and C anemones (found attached to nodules) accounted for 42.4% and 49.9% of the biomass respectively, even though numerically they only accounted for 0.6% and 1.1% of the fauna. At site A this result was attributed to one large anemone weighing 0.3358 grams and at site C to 15 smaller anemones. At site B approximately half of the biomass was accounted for by two epifaunal individuals, one echinoid (0.1887 gm) and one ascidacean (0.2350 gm). Including the weight of the larger epifauna, site A had an average biomass of 0.144 gm/m^2, site B 0.1911 gm/m^2, and site C 0.6435 gm/m^2. Without including the large epifauna the averages were 0.083 gm/m^2, 0.091 gm/m^2 and 0.323 gm/m^2 respectively. Biomass varied with numerical abundance and site C had the highest biomass. By weight the meiofauna accounted for less than 1% of the total biomass, whereas by number they were approximately 60%. Our biomass measurements are somewhat higher than those reported by Hessler (1975) for central ocean gyres (0.01-0.05 gm/m^2), but are within the values reported for open oceans (Thiel, 1975). This is probably the result of more nutrients reaching the bottom in the DOMES area when compared with the central North Pacific gyre.

One very obvious feature of the macrofauna was the rarity with which most of the species were encountered. Of a total of

TABLE 3
Faunal composition by weights and their percentage

Macrofaunal taxa	Site A wt.(gm)	Site A %	Site B wt.(gm)	Site B %	Site C wt.(gm)	Site C %
Polychaeta	0.1696	21.4	0.2920	30.6	1.5724	25.7
Tanaidacea	0.0154	1.9	0.0175	1.8	0.0886	1.5
Isopoda	0.0303	3.9	0.0300	3.1	0.2030	3.3
Bivalvia	0.0117	1.5	0.0258	2.7	0.1836	3.0
Gastropoda	0.0175	2.2	0.0153	1.6	0.1198	2.0
Ectoprocta	0.0202	2.5	0.0090	0.9	0.3068	5.0
Porifera	0.0015	0.2	0.0200	2.1	0.2434	4.0
Hydrozoa	0.0003	-	-	-	0.0039	0.1
Stephanoscyphus	0.0001	-	0.0039	0.4	0.0002	-
Actiniaria	0.3362	42.4	-	-	3.0498	49.9
Brachiopoda	0.0028	0.4	0.0066	0.7	0.0521	0.9
Hemichordata	-	-	-	-	0.0001	-
Sipunculoidea	0.0015	0.2	0.0219	2.3	0.0241	0.4
Echiuroidea	-	-	-	-	0.0003	-
Ophiuroidea	0.1097	13.8	-	-	0.0531	0.9
Echinoidea	-	-	0.2510	26.3	0.0014	-
Crinoidea	0.0006	0.1	-	-	0.0275	0.4
Holothuroidea	0.0101	1.3	-	-	0.0014	-
Aplacophora	0.0004	0.1	0.0001	-	0.0109	0.2
Polyplacophora	0.0005	0.1	-	-	0.0055	0.1
Monoplacophora	0.0088	1.1	-	-	-	-
Scaphopoda	0.0005	0.1	-	-	0.0002	-
Oligochaeta	-	-	-	-	0.0013	-
Pycnogonida	-	-	-	-	0.0014	-
Cumacea	-	-	0.0102	1.1	0.0008	-
Amphipoda	0.0097	1.2	0.0046	0.5	0.0199	0.3
Cirripedia	-	-	-	-	0.0541	0.9
Ascidacea	0.0449	5.7	0.2477	25.9	0.0809	1.3
Unknown	-	-	-	-	0.0070	0.1
Total	0.7926	100.0	0.9556	100.0	6.1135	100.0
Total per core	0.0360		0.0448		0.1609	
Meiofaunal taxa						
Nematoda	0.0088	39.1	0.0164	49.7	0.0168	20.7
Ostracoda	0.0100	44.4	0.0112	33.9	0.0611	75.2
Copepoda	0.0035	15.6	0.0051	15.5	0.0029	3.6
Acarina	-	-	0.0002	0.6	0.0003	0.4
Turbellaria	0.0002	0.9	-	-	0.0001	0.1
Kinorhyncha	-	-	0.0001	0.3	-	-
Total	0.0225	100.0	0.0330	100.0	0.0812	100.0
Total per core	0.0010		0.0016		0.0021	

381 macrofaunal species encountered, 131 were represented by only one individual, 58 species by two individuals, 51 species by three individuals and 36 species by four individuals. Thus at least 34% of the total number of species probably have an average abundance of less than one individual per $20m^2$, and only 28% have an average abundance of more than one individual per $4m^2$.

Trophic Composition

Most of the individuals encountered in our samples were deposit feeders. The major taxa that are potential suspension feeders are sponges, cnidarians, ectoprocts, brachiopods, ascidaceans, crinoids, eumellibranch bivalves and some of the polychaetes (Jumars and Fauchald, 1977). Of the total macrofauna approximately 81% were deposit feeders and 19% were suspension feeders. The lowest percent of suspension feeders was found at site A (14%), an intermediate percent at site B (17.7%), and the highest percent at site C (21.9%). These values follow the same trend found for numerical abundance and biomass, with the highest values at site C and the lowest at site A. This further substantiates that more nutrients are probably reaching the ocean floor at site C, because suspension feeders are dependent on particles suspended in the water column. Hessler and Jumars (1974) found that suspension feeders accounted for approximately 7% of the total macrofauna of the North Pacific oligotrophic gyre, where the nutrients reaching the bottom as particles are quite sparse due to the low primary productivity.

Similar results are reflected in a closer examination of the trophic structure of the polychaetes. This group was chosen because of its numerical dominance. The polychaetes were assigned to three feeding type categories; filter feeders, surface deposit feeders, and burrowing deposit feeders; according to the classification of Jumars and Fauchald (1977). The numerically most abundant families were surface deposit feeders, the cirratulids with 221 individuals (23.5%) and the spionids with 122 individuals (13%). Next in abundance were the paraonid polychaetes with 65 individuals (6.9%), which are burrowing deposit feeders. In total 74% of the polychaetes were surface deposit feeders, 15% were burrowing deposit feeders and 11% were filter feeders. Thus, the majority of the polychaetes (89%) were deposit feeders. These findings are also in agreement with the results of Hessler and Jumars (1974) for the North Pacific.

Faunal Diversity

The box cores for each station were pooled for estimates of species diversity. The justifications for this procedure are essentially those of Hessler and Jumars (1974), that the standing crop is very low and the individual to species ratio is also very

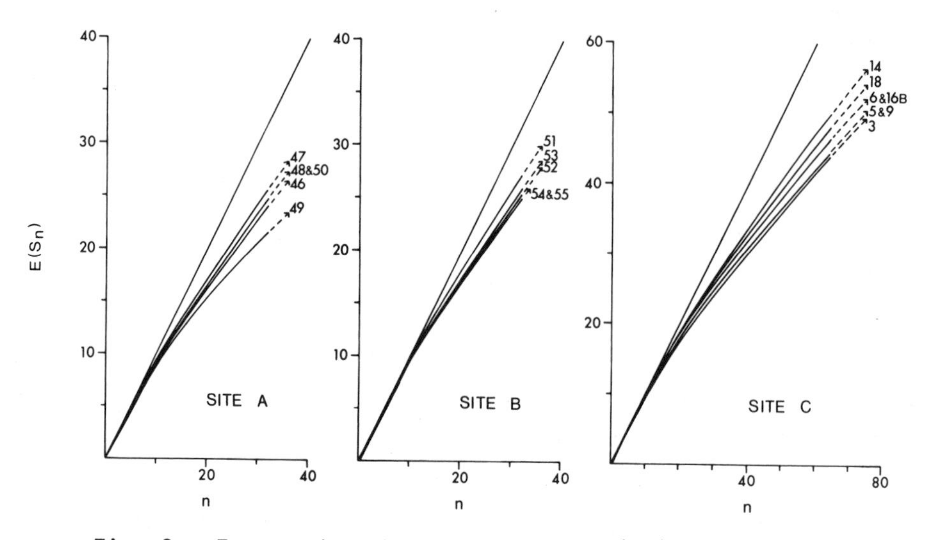

Fig. 2. Expected number of species, $E(S_n)$, for total macrofauna, calculated by the Hurlbert method (1971).

low. Additionally, since the distances between the box cores within a station are much smaller than between the stations it is highly unlikely that we are pooling over major environmental gradients (see also results of cluster analysis).

When the diversity of the total macrofauna is considered (Fig. 2) it is obvious that the sampling is far from the saturation point. The slopes of the curves have not leveled off, indicating that the addition of more individuals would cause the number of species to increase. The species diversity is uniformly high at all three sites. The 45° line represents maximal diversity where each individual encountered is a different species and at all three sites the curves lie very close to this line.

Considering the diversity of the polychaetes (Fig. 3) it can be seen that the curves are beginning to level off, but that the saturation point has not yet been reached. The polychaete diversity at site B is uniformly high, at site A it is low and at site C it is more variable. The diversity of the polychaetes is similar to that reported for the San Diego Trough (Jumars, 1976) and for the central North Pacific (Hessler and Jumars, 1974), with the exception of station 49 which is quite low and similar to the Aleutian Trench (Jumars and Hessler, 1976). How these diversity estimates were affected by the destruction of some of the poly-

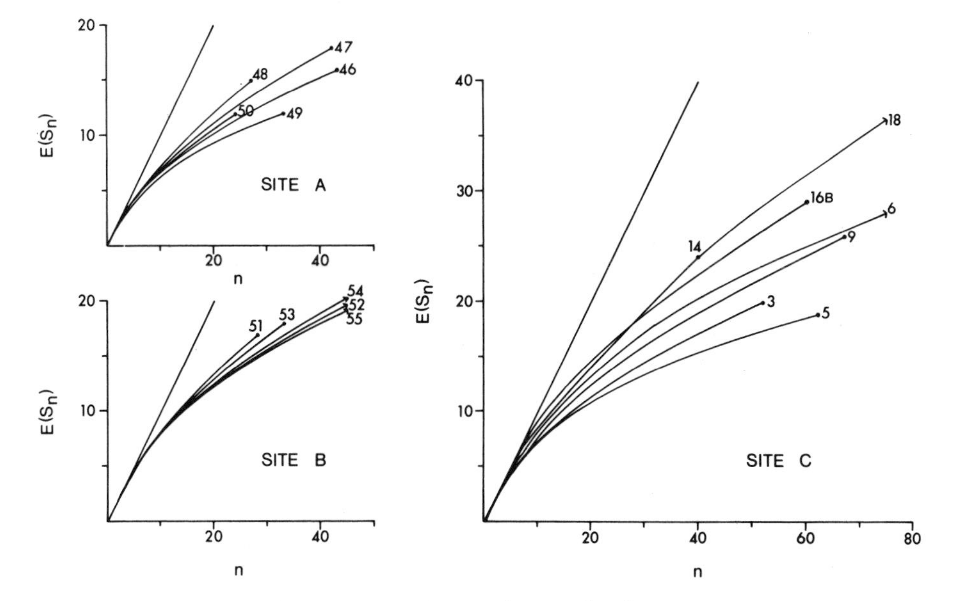

Fig. 3. Expected number of species, $E(S_n)$ for polychaetes, calculated by the Hurlbert method (1971).

chaetes during sample processing is not known. If the destruction was uniform or random for all species the diversity measures would not be affected, because they are based on the relative proportions within the sample.

We then pooled all the polychaetes within a site. The resulting expected-number-of-species curves are shown in Fig. 4(a). The differences between sites B and C are probably not significant, but the intersection of their curves tends to indicate that site B has a slightly greater species evenness and that site C has a slightly greater species richness. Again, site A shows a lower diversity. The polychaetes were also divided into two categories based on their motility (Jumars, 1975 a & b). Small-ambit polychaetes are relatively sessile and large-ambit ones are relatively motile. Jumars reasons that if the factors maintaining diversity are on a small scale, then species with a small ambit would be more diverse than ones with a large ambit, because they would encounter fewer organisms and would thus be experiencing a coarser grained environment. The diversity curves generated by grouping the polychaetes according to ambit are shown in Fig. 4(b). The curves for sites A and B substantiate Jumars'

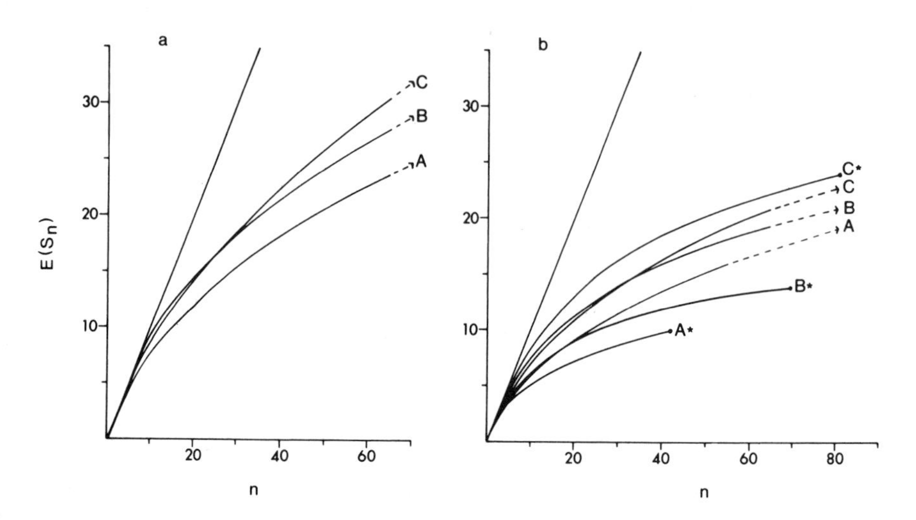

Fig. 4(a). Hurlbert species diversity curves for polychaetes pooled by sites. 4(b). Hurlbert species diversity curves for polychaetes grouped according to large and small ambits. Large ambit polychaetes are denoted by an asterisk(*).

hypothesis and agree with his results from the San Diego Trough. However, site C does not follow the same trend. The diversity of the large-ambit polychaetes at site C is slightly greater than the diversity of the small-ambit ones. At this time we can think of no obvious reason for this difference. This difference does, however, suggest that the factors maintaining diversity at site C might be different from those maintaining diversity at the other two sites.

Figure 5 shows the diversity curves generated by the peracarid crustaceans (isopods, tanaids, amphipods and cumaceans). At sites A and C the diversity of the crustaceans covers a wide range, whereas at site B it is consistently high. One station at site B (station 51) had a 1-to-1 ratio of species to individuals. On closer examination the high diversity of the crustaceans can be attributed to three of the taxa. Cumaceans had a 1-to-1 ratio of species to individuals, amphipods had 9 species and 17 individuals, and isopods had 79 species and 221 individuals. However, these taxa are so sparsely distributed that it is impossible to determine their comparative diversities from the available data.

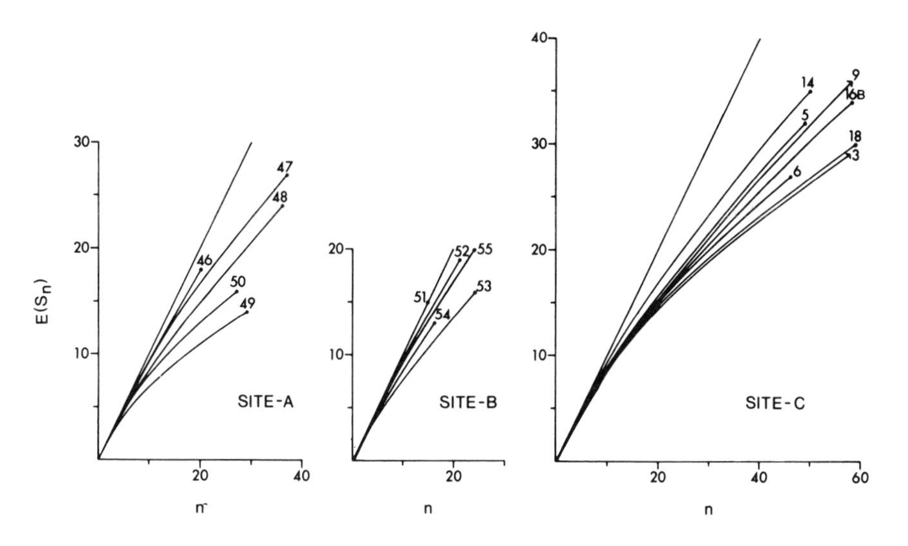

Fig. 5. Expected number of species, E(S_n), for peracarid, crustaceans, calculated by the Hurlbert method (1971).

The saturation level of the diversity curves, for any of the taxa, has not been reached. Thus, there is no way of estimating the number of species in these communities. This problem is due to the extremely sparse distribution of most of the species. Overall, the samples from sites B and C indicate a slightly greater diversity than those from site A. The largest difference appears to be between sites A and C.

Faunal Affinity

To test if discernible groupings of species exist in the sampling area NESS was calculated for every possible pair of stations. The resulting similarity matrix was then clustered using the arithmetic unweighted pair-group method (UPGMA) as described in Sokal and Sneath (1963). The results of clustering NESS at a value of m=25, using all the macrofaunal species, are shown in Fig. 6(a). NESS was also calculated without the most common species, a cirratulid polychaete, and the results are shown in Fig. 6(b).

Both with and without the most common species the stations within a site tended to cluster together. This result indicates

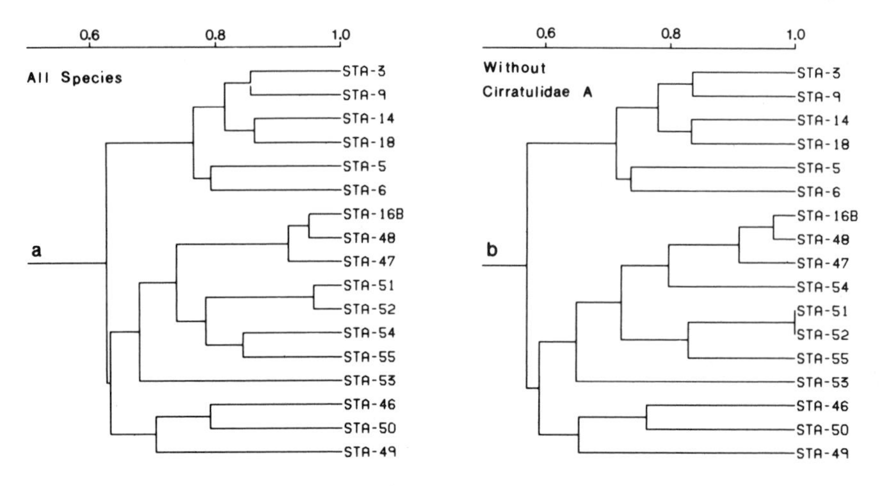

Fig. 6(a). Cluster diagram of NESS similarity at m=25 for all species, using arithmetic unweighted pair-group (group-average) sorting. The top axis is NESS similarity. 6(b). NESS at m=25 without the most common species, Cirratulidae A.

that they had a relatively high degree of species similarity. Additionally, stations within a site that were in close geographical proximity tended to share more species. With the exception of station 16B, which clustered with two stations from site A, the other stations from site C (stations 3-18) tended to cluster together. At site B four out of the five stations clustered rather closely. The stations at site A did not form a distinct cluster. Station 16B from site C appears to have a low faunal affinity to its geographically nearby neighbors, and more closely resembles two stations from site A (47 and 48). At site A stations 47 and 48 were quite similar to each other, but were not similar to the other three stations at that site.

Removal of the most common species, Cirratulidae A, did not change the general pattern of faunal affinity. It did, however, effect some minor changes. Generally, it lowered the similarity between clusters. Conversely, removal of Cirratulidae A raised the value of NESS between stations 51 and 52 to one, meaning that without counting the most common species their species composition was virtually identical. This difference was due to the fact that Cirratulidae A occurred at station 52 but not at station 51. The similarity between stations 54 and 55, also, changed with the removal of the most common species. Apparently, its presence was responsible for their clustering together.

When NESS was calculated at a value of m=1, which is identical to Morisita's index of similarity, no distinct clusters were distinguishable. The Morisita index is heavily biased towards the more dominant species in the populations. Thus, the species composition groupings by site are due, to a large extent, to the rarer species. This result is not surprising in view of the fact that only 6% of the species were present in high enough abundances to be represented by one individual per station; however, they accounted for half of the number of individuals.

The results obtained from the species similarity analysis are interesting for several reasons. They show that discernible community structure, in terms of species composition, does exist at the three sites. Hessler and Jumars (1974) did not find a discernible difference in the polychaete fauna of locations 90 miles apart, that had similar sediment characteristics and bottom water. However, our sites were located approximately 600 miles apart, with the largest distance between sites A and C being 1200 miles. Considering the differences in the fauna among the sites, the finding of distinguishable species groupings is not surprising. The standing crop estimates and the percent of suspension feeders indicate that there is a higher flux of organic material at site C. It is reasonable to assume that different species would be adapted to different levels of food availability (Dickinson and Carey, 1978). Further elucidation of which species account for the observed clustering might hold clues to how these organisms interact with their environment.

Summary

Standing crop, in terms of both abundance and wet weight, was lower at sites A and B than at site C, but the values all fall within the range reported for other abyssal communities. Faunal composition was also similar to that reported for other deep-sea areas, with polychaetes, tanaids, isopods and bivalves accounting for the majority of the macrofauna.

Deposit feeders dominated the fauna, accounting for 81% of the total macrofauna, suspension feeders comprised 19% of the total number of individuals. The dominance of deposit feeders was also reflected in the trophic structure of the polychaetes.

Species diversity was uniformly high at all three sites. The most obvious feature of the macrofauna was the rarity with which most of the species were encountered. Species composition groupings were discernible. Stations within a site tended to have a similar fauna, while the faunal affinity between the sites was not as pronounced.

Acknowledgments

We are indebted to the following individuals for their aid and expertise: R.R. Hessler and P.A. Jumars for advice on sampling procedure and analysis; F.J. Rohlf for help with programming and the cluster analysis; J.F. Grassle and W. Smith for their species diversity program; K. Fauchald for identifying the polychaetes; and F. urRehman for identifying the copepods. We are grateful to J.T. Durazzi, R.Y. George, R.R. Hessler, M.A. Rex, and H.L. Sanders for advice on taxonomy. We wish to thank the crew and officers of the R/V OCEANOGRAPHER for their help in the field sampling.

Special thanks go to P.R. Gibson, J.G. Makar and M.K. Evensen for their patient sorting of the samples and all their technical assistance in relation to this project.

This paper is Lamont-Doherty Geological Observatory Contribution No. 2737. The work on this project was supported by National Oceanic and Atmospheric Administration contract 03-06-022-35141.

References

Dayton, P.K. and R.R. Hessler. (1972) Role of biological disturbance in maintaining diversity in the deep sea. Deep-Sea Res. 19, 199-208.

Dickinson, J.J. and A.G. Carey, Jr. (1978) Distribution of gammarid Amphipoda (Crustacea) on Cascadia Abyssal Plain (Oregon). Deep-Sea Res. 25, 97-106.

Grassle, J.F. and H.L. Sanders. (1973) Life histories and the role of disturbance. Deep-Sea Res. 22, 643-659.

Grassle, J.F. and W. Smith. (1976) A similarity measure sensitive to the contribution of rare species and its use in investigation of variation in marine benthic communities. Oecologia 25, 13-22.

Hessler, R.R., (1975) The structure of deep benthic communities from central oceanic waters. In: The Biology of the Oceanic Pacific. C. Miller, ed. Proc. 33rd Ann. Biol. Colloq., Oregon State University.

Hessler, R.R. and P.A. Jumars. (1974) Abyssal community analysis from replicate box cores in the central North Pacific. Deep-Sea Res. 21, 185-209.

Hessler, R.R. and H.L. Sanders. (1967) Faunal diversity in the deep sea. Deep-Sea Res. 14, 65-78.

Hurlbert, S.H., (1971) The nonconcept of species diversity: A critique and alternative parameters. Ecology 52, 577-586.

Hutchinson, G.E., (1953) The concept of pattern in ecology. Proc. Acad. Nat. Sci. Phila. 105, 1-12.

Jumars, P.A., (1975a) Methods for measurement of community structure in deep-sea macrobenthos. Mar. Biol. 30, 245-252.

——— (1975b) Environmental grain and polychaete species diversity in a bathyal benthic community. Mar. Biol. 30, 253-266.

——— (1976) Deep-Sea species diversity: does it have a characteristic scale? J. Mar. Res. 34, 217-246.

Jumars, P.A. and K. Fauchald. (1977) Between-community contrasts in successful polychaete feeding strategies. pp. 1-20 In: Ecology of Marine Benthos, B.C. Coull, ed. Belle W. Baruch Lib. Mar. Sci. #6, Univ. South Carolina Press.

Jumars, P.A. and R.R. Hessler. (1976) Hadal community structure: implications from the Aleutian Trench. J. Mar. Res. 34, 547-559.

Mantyla, A.W., (1975) On the potential temperature in the abyssal Pacific Ocean. J. Mar. Res. 33, 341-354.

Menzies, R.J. and G.T. Rowe. (1968) The LUBS, a large undisturbed bottom sampler. Limnol. Oceanogr. 13, 708-714.

Morisita, M., (1959) Measuring of interspecific association and similarity between communities. Mem. Fac. Sci. Kyushu Univ. Ser. E 2, 215-235.

Rowe, G.T., (1971) Benthic biomass and surface productivity. In: Fertility of the Sea, J.D. Costlow, ed., Gordon and Breach 2, pp.441-454.

Sanders, H.L., (1968) Marine benthic diversity: a comparative study. Amer. Natur. 102, 243-282.

——— (1969) Benthic marine diversity and the stability-time hypothesis. Brookhaven Symp. Biol. 22, 71-81.

Sanders, H.L., Hessler, R.R. and G.R. Hampson. (1965) An introduction to the study of deep-sea benthic faunal assemblages along the Gay Head-Bermuda transect. Deep-Sea Res. 12, 845-867.

Simpson, E.H., (1949) Measurement of diversity. Nature 163, 688.

Slobodkin, L.B. and H.L. Sanders. (1969) On the contribution of environmental predictability to species diversity. Brookhaven Symp. Biol. 22, 82-93.

Smith, W. and J.F. Grassle. (1977) Sampling properties of a family of diversity measures. Biometrics 33, 283-292.

Sokal, R.R. and F.J. Rohlf. (1969) Biometry, Freeman Press, San Francisco.

Sokal, R.R. and P.H.A. Sneath. (1963) Principles of Numerical Taxonomy, Freeman Press, San Francisco.

Sokolova, M.N., (1971) Trophic structure of deep-sea macrobenthos. Mar. Biol. 16, 1-12.

Thiel, H., (1975) The size structure of the deep-sea benthos. Int. Revue ges. Hydrobiol. 60, 575-606.

Walton, W.R., (1952) Techniques for recognition of living Foraminifera. Contributions of the Cushman Foundation for Foraminiferal Research 3, 56-60.

LITHIC AND ACOUSTIC STRATIGRAPHY OF THE EQUATORIAL NORTH PACIFIC: DOMES SITES A, B, AND C

D.Z. Piper, H.E. Cook, and J.V. Gardner

U.S. Geological Survey
345 Middlefield Road
Menlo Park, California

Abstract

In surface sediment at the three DOMES sites in the equatorial North Pacific, the percentage of reworked fauna and the thickness of the acoustically transparent uppermost layer increases from east to west (Sites C to B to A) and the percentage of volcanogenic products decreases. On the basis of lithology and acoustic signature (3.5 kHz) this sediment can be correlated with surface sediment recovered in Deep Sea Drilling Project cores from this same area. It has been assigned to the radiolarian ooze unit of the Clipperton Ocean Formation. The underlying Marquesas Oceanic Formation, well documented in DSDP cores, was not observed in DOMES cores but was identified in the acoustic records as an acoustically transparent to weakly reflective unit. In the study area, the Clipperton Formation is commonly deeply eroded and unconformable with the sea-floor surface, but the Marquesas Formation is present everywhere and has a fairly uniform thickness.

Introduction

The primary objectives of this study were: (1) to establish an acoustic stratigraphy for the DOMES area, (2) to determine the mapability of this acoustic stratigraphy, and (3) to determine the relation between the acoustic stratigraphy and oceanic formations as they have been identified in DSDP cores.

The sedimentary sequence in the equatorial North Pacific, where exploitation of ferromanganese nodules is most likely to

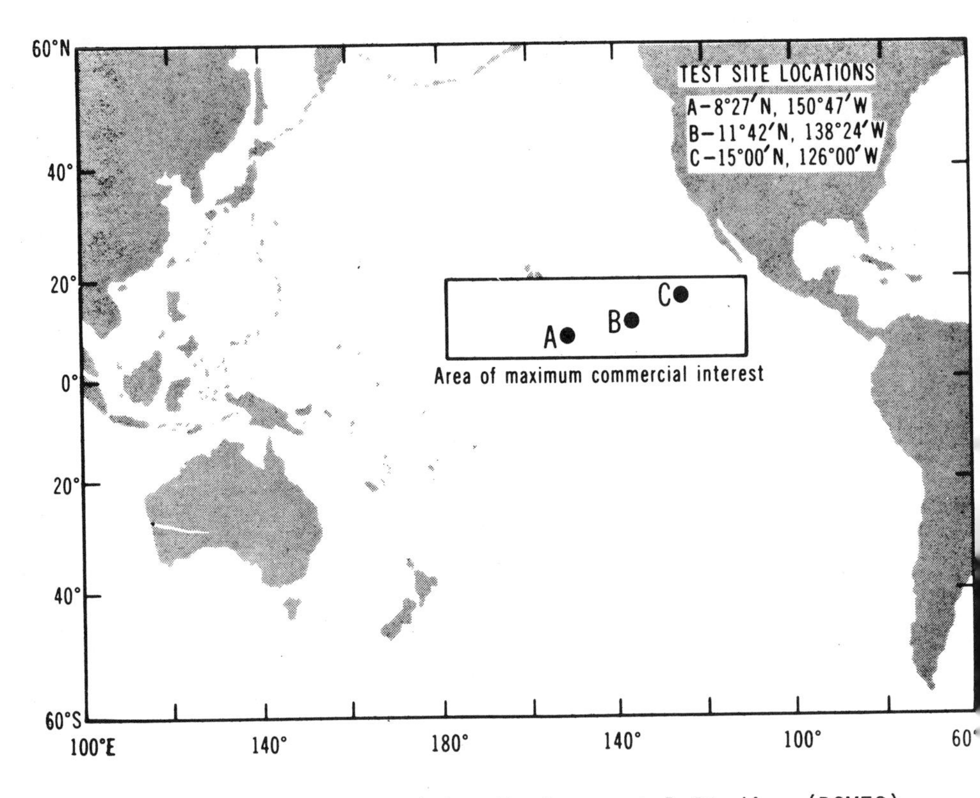

Fig. 1. Deep Ocean Mining Environmental Studies (DOMES) project area, Sites A, B, and C

occur (Fig. 1), is represented by an east-west-trending prism of sediment (Fig. 2) that has a maximum thickness of about 500 m at approximately 2 to 3°N (Ewing et al., 1968; Tracey et al., 1971; Hayes et al., 1972; Winterer, 1973, and Cook, 1975). The sediment thins to less than 100 m at higher latitude, both north and south of the equator. Currently, the zone of highest sediment accumulation, however, lies along the equator (Fig. 3). This zone immediately underlies the area of high primary productivity (Fig. 4) that results from upwelling of water rich in the biologically essential nutrients, combined nitrogen, PO_4^{3-}, and SiO_4^{4-}, into the photic surface water (Reid, 1962). Calcareous and, to a lesser extent, siliceous plankton fourish within these nutrient-rich surface waters.

The areal distribution of biogenic ooze (Fig. 5), as delineated by Horn et al. (1973), approximately coincides with areas of highest primary productivity and high rates of sediment accumulation (Fig. 3), and this coincidence demonstrates the link between biological processes within surface waters and accumulation of

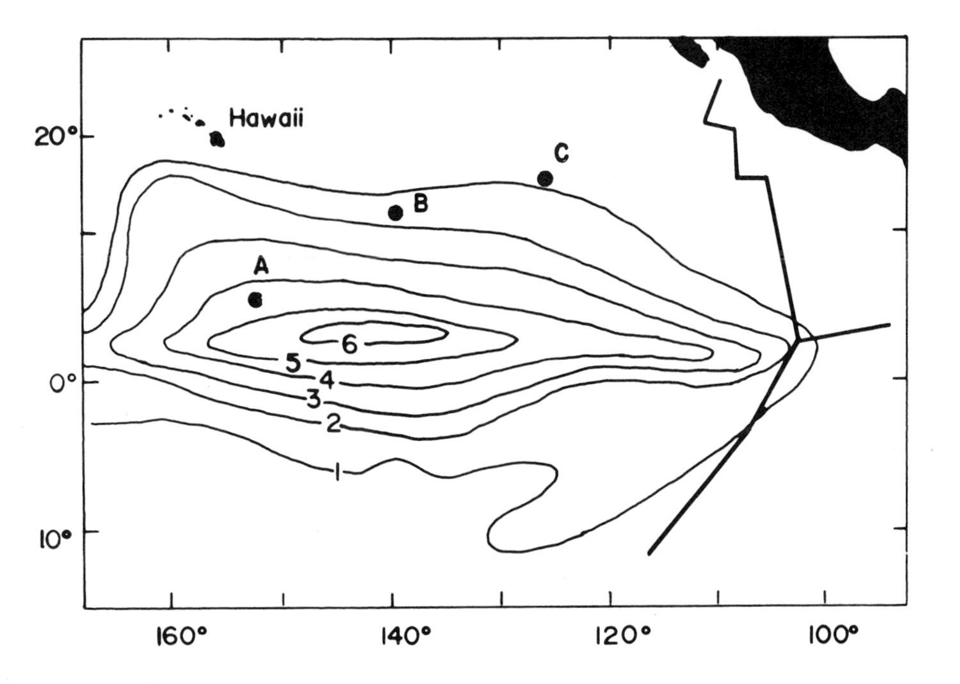

Fig. 2. Isopach map of sediment in meters above basement, east of 130°W, or above acoustically opaque Eocene chert, west of 130°W. Modified from van Andel, Heath et al. (1973). DOMES sites are shown by dots.

sediment on the sea floor. Also, comparison of the distribution of $CaCO_3$ ooze with productivity and sediment accumulation rates demonstrates the dominant role of $CaCO_3$ in pelagic sediment in this area. Carbonate ooze grades into siliceous ooze at latitudes higher than approximately 5°N, and at latitudes of 10° to 15°N, biogenic ooze becomes subordinate to pelagic clay, which, at 15° to 20°N, is virtually barren of all biogenic debris (Fig. 5).

High biological productivity along the equator exerts perhaps the dominant control on the distribution of sediment types, but other factors are also important.

1. Water depth - $CaCO_3$ deposited below the carbonate compensation depth (CCD) is attacked by dissolution. The sea floor within 3° of the equator is above the CCD owing to the high flux of $CaCO_3$ to the bottom. The sea floor away from the equator is generally below the CCD (the CCD north of 5°N is at approximately 4400 m). Although tests composed of $CaCO_3$ may be present in the surface veneer of sediment at higher latitudes, they are actively being dissolved. Silica apparently does not exhibit a compensation depth in the mid-ocean (Heath, 1974). Its accumulation rate

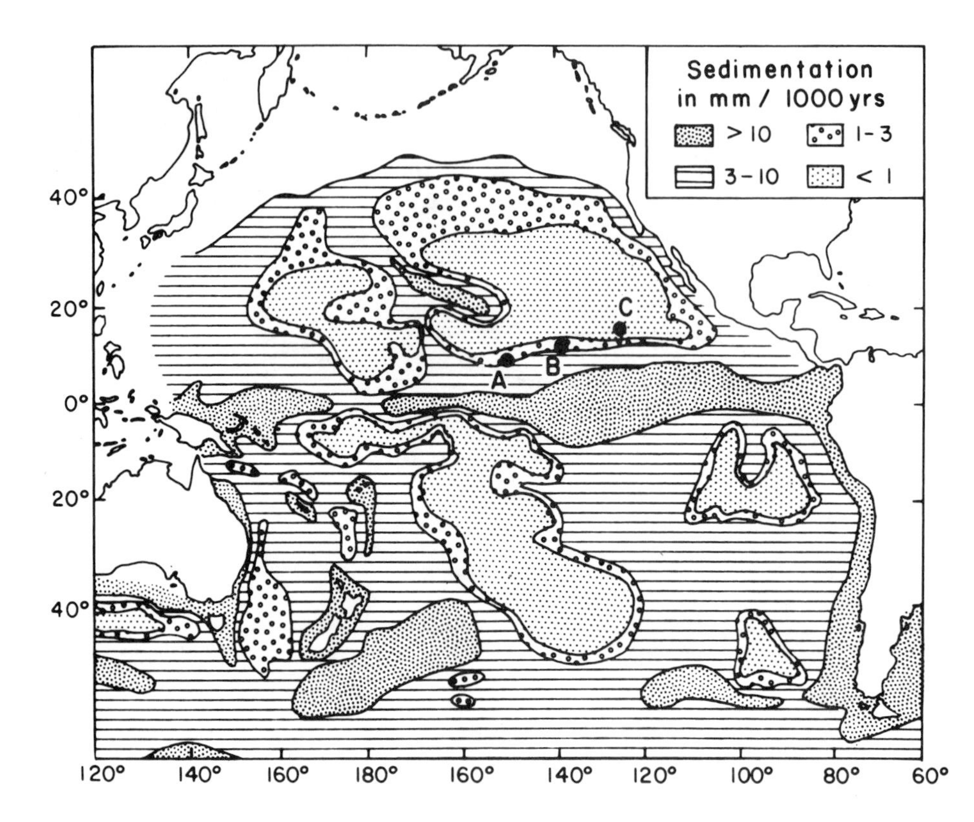

Fig. 3. Sedimentation rates for the Pacific Ocean (Lisitzin, 1972) and DOMES Sites A, B, and C.

closely parallels surface productivity (Fig. 4) except in the eastern equatorial Pacific where silica dissolution appears to be related to a high accumulation rate of terrigenous material (T. Johnson, 1976).

2. Bottom currents and sediment redistribution - Bottom currents may effectively remove and redistribute sediment on the sea floor so that some areas have little or no net accumulation (D. Johnson, 1972a, 1972b). Hiatuses are evident in Deep Sea Drilling Project (DSDP) cores within the Tertiary (van Andel et al., 1975). Also, the presence of Miocene and older sediment at or near the surface in piston cores and DSDP cores (Hayes et al., 1969; Ryan and Heezen, 1976; Fig. 6) suggests a hiatus during Pliocene-Pleistocene time. The areal extent of these hiatuses seems to correlate with the pattern of circulation of Antarctic Bottom Water. Potential-temperature isotherms of bottom water (Fig. 7) strongly suggest that this water mass intrudes into the equatorial Pacific through passages in the Line Island Ridge (Reid, 1969;

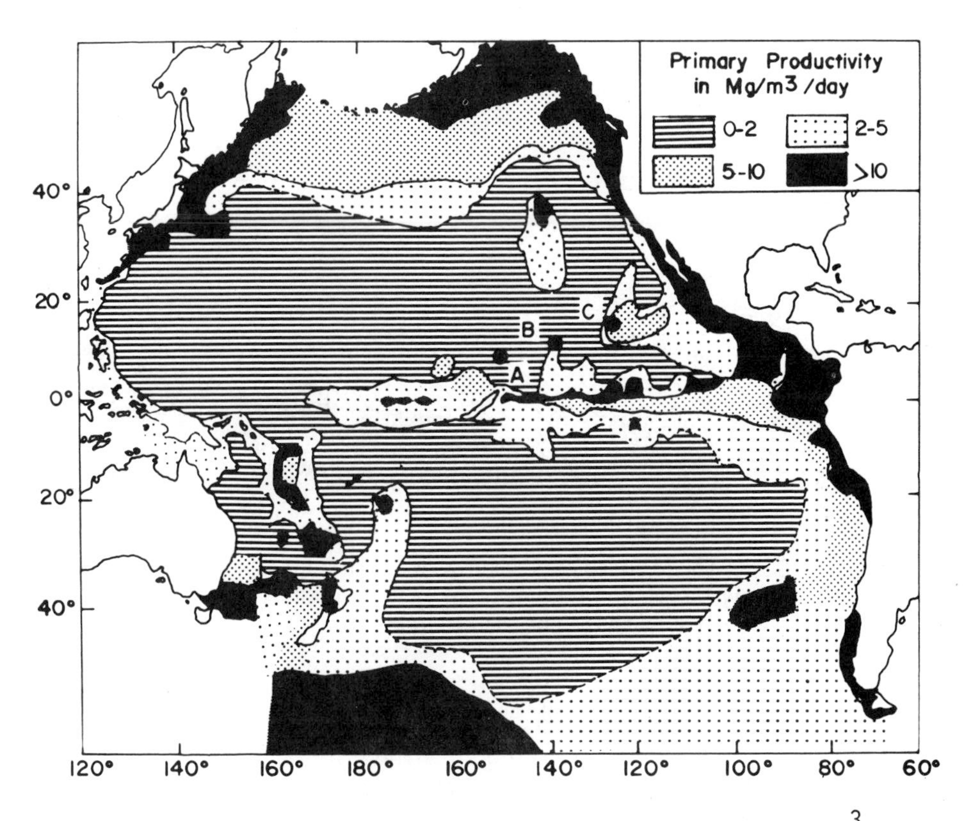

Fig. 4. Primary productivity in the Pacific Ocean, in mg/m^3/day. Area 1, 0-2; area 2, 2-5; area 3, 5-10; area 4, 10-20; area 5, greater than 20. Figure is from Bogorov (1969). A, B, C location of DOMES sites.

Gordon and Gerard, 1970; Edmond et al., 1971). Measured bottom currents, however, are seldom greater than 10 cm/sec and usually less than 5 cm/sec (Hayes, 1977), so currents necessary to erode sediment are probably episodic. Perhaps the most direct evidence of sediment redistribution is the occurrence of mixed fauna, particularly within Quaternary sediments.

3. Dilution effects - An influx of inorganic sediment, either of submarine or eolian origin, can dilute and significantly modify the composition of pelagic sediment, even in areas of high primary productivity. Biogenic sediment along the equator grades eastward into sediment composed largely of eolian volcanogenic material (Cook, 1972, 1975). Biogenic sediment grades latitudinally into material of largely terrigenous origin, as suggested by the distribution of quartz and clay minerals (Rex and Goldberg, 1958;

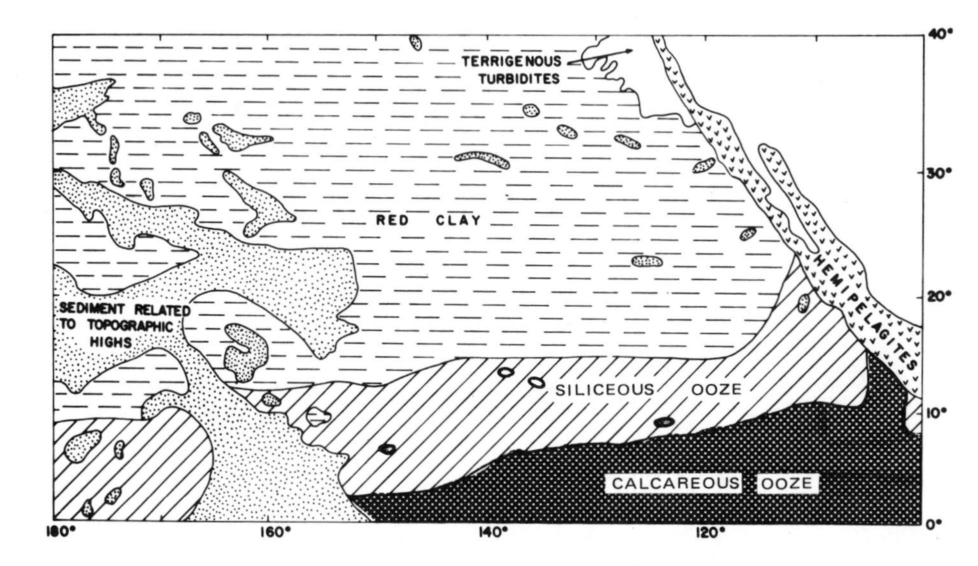

Fig. 5. Distribution of sediments within the equatorial North Pacific Ocean, from Horn et al. (1973). DOMES Sites A, B, and C are within the siliceous ooze band.

Griffin and Goldberg, 1963, Windom, 1969).

4. Diagenesis - Most prominent are the modifications resulting from selective remobilization of biogenic phases, organic matter(Romankevich, 1976), $CaCO_3$ (Bramlette, 1961; Berger, 1970), and SiO_2 (Schlanger and Douglas, 1974), and formation of authigenic phases such as phillipsite (Bonatti, 1963), barite (Church, 1970), smectite (Lyle et al., 1977), and ferromanganese nodules (Price and Calvert, 1970; Piper and Williamson, 1977).

5. Sea-floor spreading - This process moves the sea floor into different depths of water, different regions of biological productivity, different regimes of bottom currents, and areas of varying sediment dilution. One major effect in the equatorial Pacific since the Oligocene has been to shift the axis of the prism of maximum sediment thickness approximately 2 to 3° north of the equator (Ewing et al., 1968; Winterer, 1973; Bukry, 1973; van Andel, Heath et al., 1973; Cook, 1975). The axis of maximum sediment accumulation has been shifted as much as possibly 10° to the north during the past 45 m.y. (van Andel et al., 1975).

Two major sedimentary regimes prevailed in the central equatorial Pacific during the Tertiary - to the west a dominantly pelagic regime, and to the east a sediment regime that received abundant volcanogenic debris in addition to a pelagic component. Heezen et al. (1973) constructed the following model of sedimentation for the western Pacific based largely on DSDP cores:

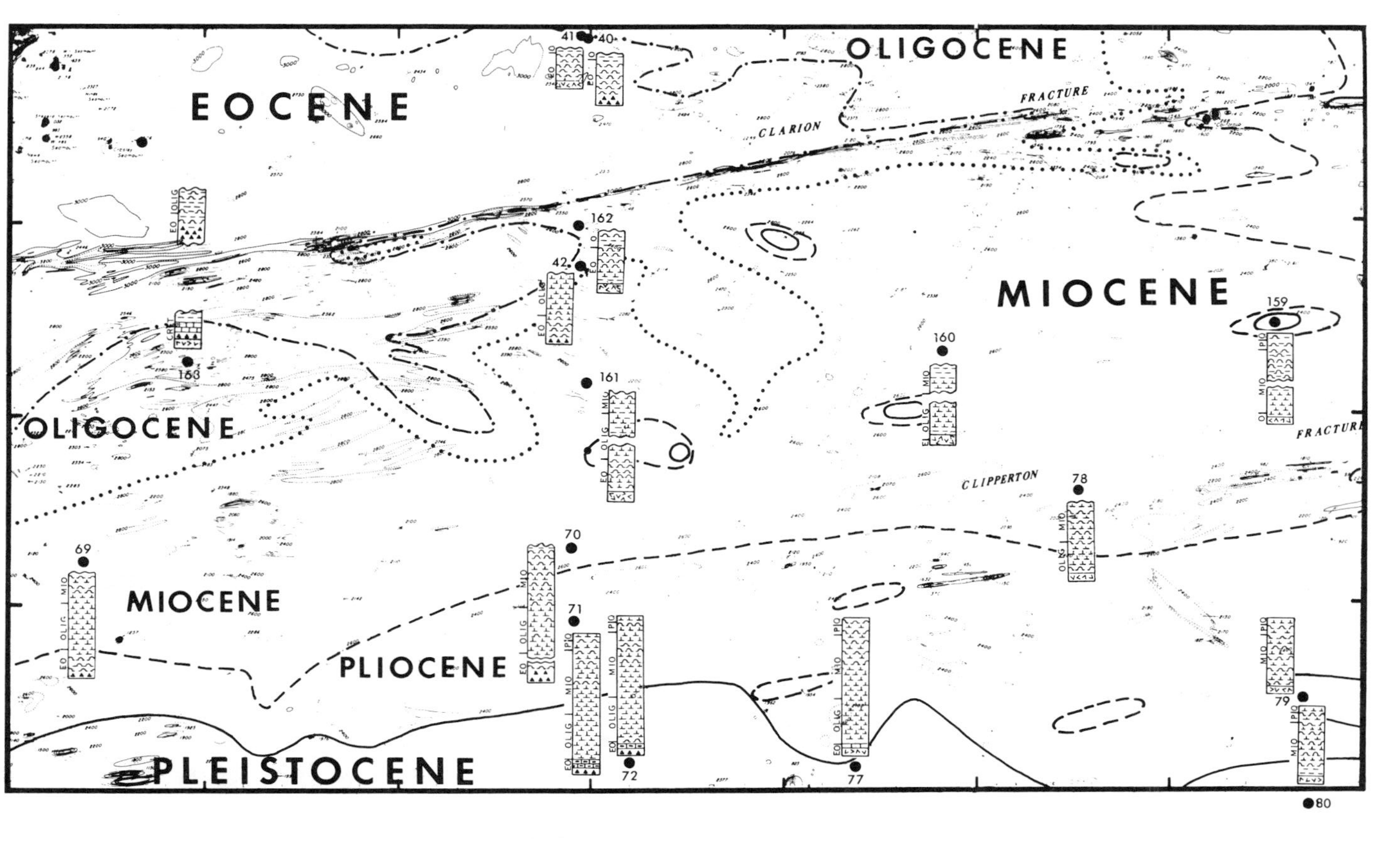

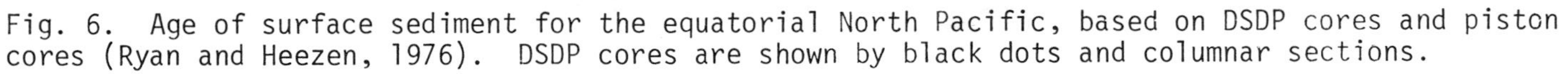

Fig. 6. Age of surface sediment for the equatorial North Pacific, based on DSDP cores and piston cores (Ryan and Heezen, 1976). DSDP cores are shown by black dots and columnar sections.

assuming initial sedimentation on a spreading center above the CCD, the basal sediment would be carbonate ooze succeeded by siliceous ooze and finally by abyssal clay. This progression reflects the movement of the lithospheric plate away from the spreading center, into water below the CCD (Heezen et al., 1973), and out of the zone of high primary productivity. Heezen et al. did not recognize the basal metalliferous sediment that commonly underlies calcareous ooze and immediately overlies basaltic basement (von der Borch and Rex, 1970; Tracey et al., 1971; Cook, 1972; and Cronan, 1973).

Tracey et al. (1971) and Cook (1975) synthesized this sedimentary sequence of the central equatorial Pacific into formations and mapped their areal and stratigraphic distribution (Figs. 8, 9, 10).

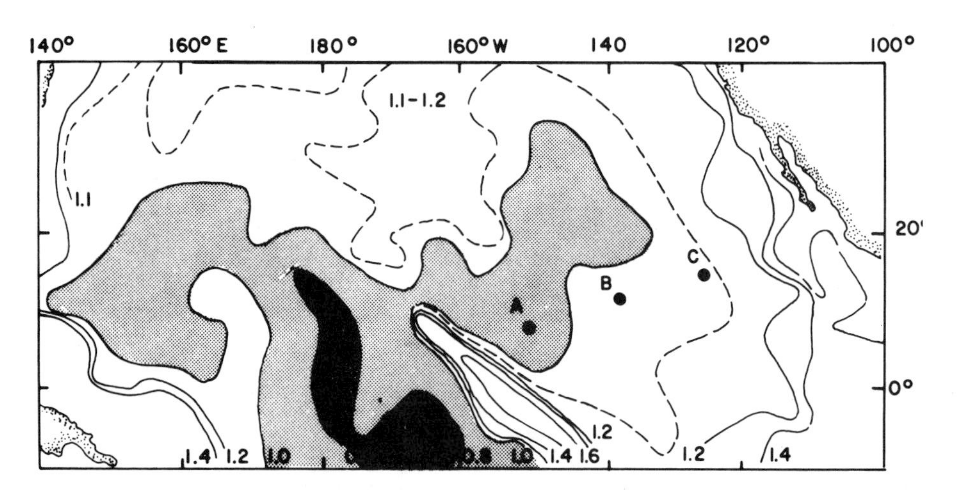

Fig. 7. Isotherms of bottom water potential temperature, degrees celsius, from Gordon and Gerard (1970).

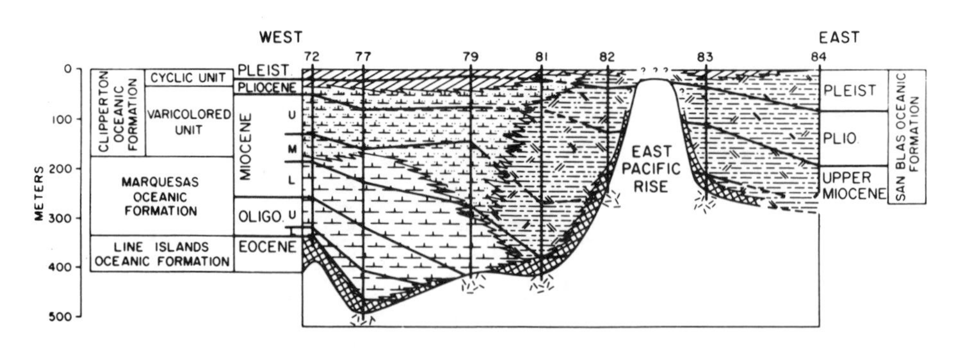

Fig. 8. Combined chronostratigraphic and rock-stratigraphic cross-section across the central equatorial Pacific from JOIDES Deep Sea Drilling Project Sites 72 to 84 (Cook, 1975).

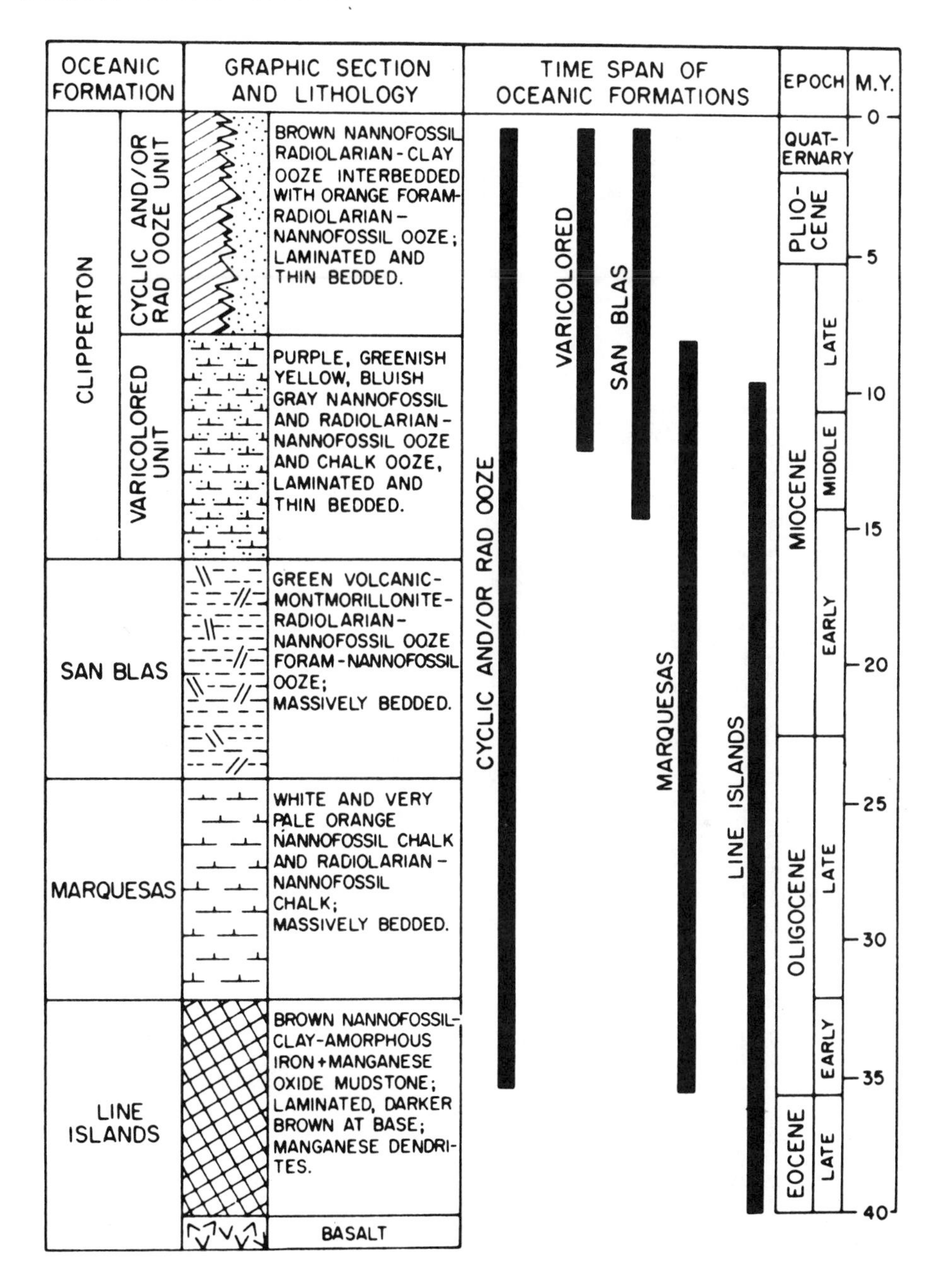

Fig. 9. Composite stratigraphic section of equatorial Pacific oceanic formations and their time span (Cook, 1975).

OCEANIC FORMATION		DOMINANT COLOR(S)	BEDDING	DOWNHOLE TREND OF $CaCO_3$%	DOWN-HOLE TREND OF PYRO-CLASTIC %	DOWNHOLE TREND OF CLAY AND/OR Fe-Mn OXIDE	REMARKS
CLIPPERTON	CYCLIC UNIT	Dusky yellowish-brown, grayish orange, very pale orange	Laminated, 5-25 cm thick beds, interbedded colors				Cyclic interbedding of thin brown and orange beds characteristic; common higher natural gamma count than underlying units
CLIPPERTON	VARICOLORED UNIT	Pale purple, pale greenish yellow, white, bluish gray, very dusky purple	5-100 cm thick beds, laminated purple layers				Pastel colors characteristic; usually has lower $CaCO_3$ %, higher radiolarian %, and thinner beds than Marquesas
SAN BLAS		Greenish black, dark greenish gray, grayish olive, yellowish gray, very dusky purple	5-100 cm thick beds, intense burrowing in Sites 83, 84				Abundant clay minerals; darker green, more montmorillonite, more pyroclastics, more burrowing to east; higher natural gamma count than Clipperton or Marquesas
MARQUESAS		White, bluish white, very pale orange, light gray	25-200 cm thick beds, rare laminations				Usually has higher $CaCO_3$ %, lower radiolarian %, and thicker beds than Clipperton
LINE ISLANDS		Dusky brown, pale yellowish brown, pale orange	Laminations, darker brown towards base, flattened burrows, manganese dendrites				Commonly darker brown towards base; stratigraphically on basalt

Fig. 10. Stratigraphic and petrographic characteristics used to recognize equatorial Pacific oceanic formations (Cook, 1975).

The stratigraphic sequence in the equatorial Pacific from basaltic basement upward begins with the Line Islands Formation. This formation is a clay-ooze-chert-limestone sequence with a dark brown metalliferous ooze at the base. In many areas only the metalliferous ooze is present. The Line Islands is overlain unconformably by the white Marquesas Formation, which contains variable but high percentages of calcium carbonate. Overlying the Marquesas is the Clipperton Formation, which is divided into three informal units -- a varicolored unit, a dark-brown and white cyclic unit, and a dark-brown radiolarian ooze unit. At latitudes higher than about 5°N and S, the varicolored unit of the Clipperton Formation is not present (Fig. 11). East of about 110°W longitude, the Marquesas and most of the Clipperton Formations are replaced by the greenish San Blas Formation, which is rich in volcanogenic material (Fig. 8). The lithostratigraphy at DOMES Sites A, B, and C includes the Line Islands Formation, the Marquesas Formation, and the cyclic and radiolarian ooze units of the Clipperton Formation (Fig. 12).

Moberly and Klein (1976) observed that certain colors in deep-sea sediment may change with time, once this sediment is

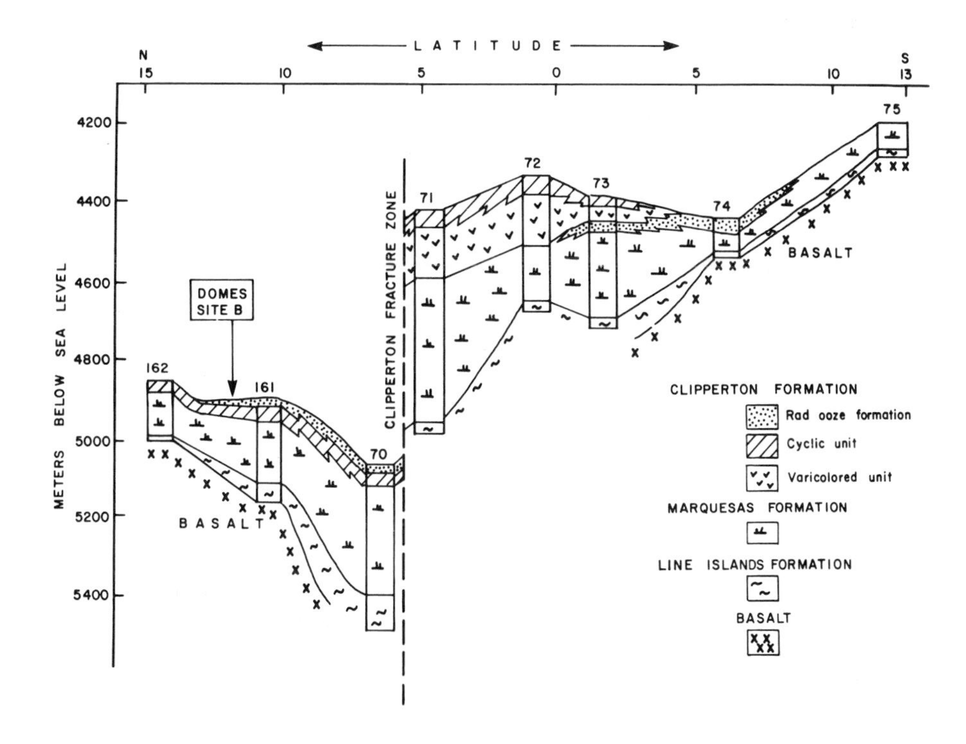

Fig. 11. Stratigraphic cross section along 135°W to 140°W showing oceanic formations at DSDP sites and location of DOMES Site B (modified from Tracey _et al_., 1971, Fig. 4, p. 18).

exposed to subaerial conditions. Such changes become important because, in addition to stratigraphic position and sediment type, sediment color is used to distinguish the above-mentioned formations (Cook, 1975). Moberly and Klein found that brown sediment, characteristic of the Line Islands Formation and the cyclic and radiolarian ooze units of the Clipperton Formation, is very stable and exhibits no color change even after 3 or 4 years of exposure. This also is true of gray and white sediment, characteristic of the Marquesas Formation. On the other hand, cores in the varicolored unit of the Clipperton Formation, which originally were pastel tints of green, blue, and purple, changed to white and yellowish-gray. Thus, in fresh cores there is a clear color distinction between the varicolored unit of the Clipperton Formation and the Marquesas Formation but in exposed cores this color distinction breaks down. There is no stratigraphic problem when examining cores from the area of DOMES Sites A, B, and C which have been exposed to the air because the varicolored unit of the Clipperton Formation does not extend this far northward (Fig. 11).

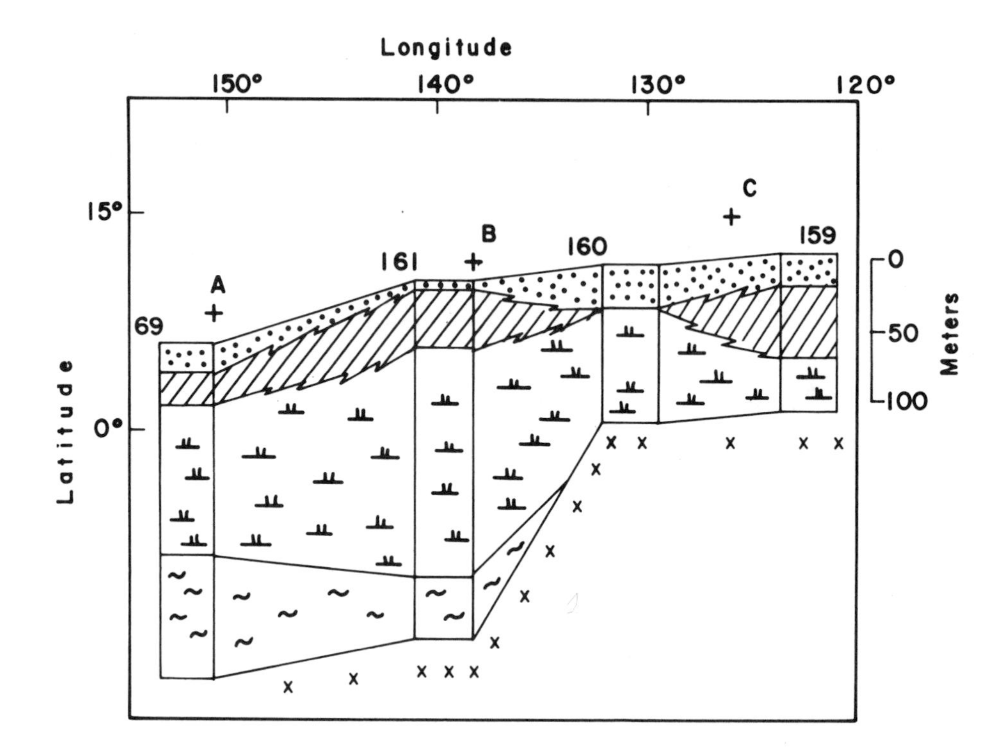

Fig. 12. Stratigraphic cross section along $5^{O}N$ to $12^{O}N$ showing oceanic formations at DSDP sites and location of DOMES sites. Legend same as Fig. 11.

The brown cyclic and radiolarian ooze units of the Clipperton Formation at these latitudes are directly underlain by the white Marquesas Formation. These colors are stable, and these units are clearly distinguishable from one another.

The sedimentary sequence in the study area (Fig. 8) broadly supports the model of Heezen et al. (1973). In detail, however, facies maps for successive periods of time (Cook, 1975, Figs. 10 through 17, p. 826-829) show that the gradual evolution of the oceanic formations in the equatorial Pacific was extremely complex. Simple eastward movement of the sea floor into deeper water and below the CCD and a northward component of movement out of the equatorial zone of high productivity alone do not adequately explain details of the facies patterns. In general, however, facies patterns such as that for the late Miocene (Fig. 13), seem to be explained by accumulation of sediment on a sea floor that is either above the CCD or north or south of the zone of high productivity, possibly coupled with effects of cyclic changes in global climate.

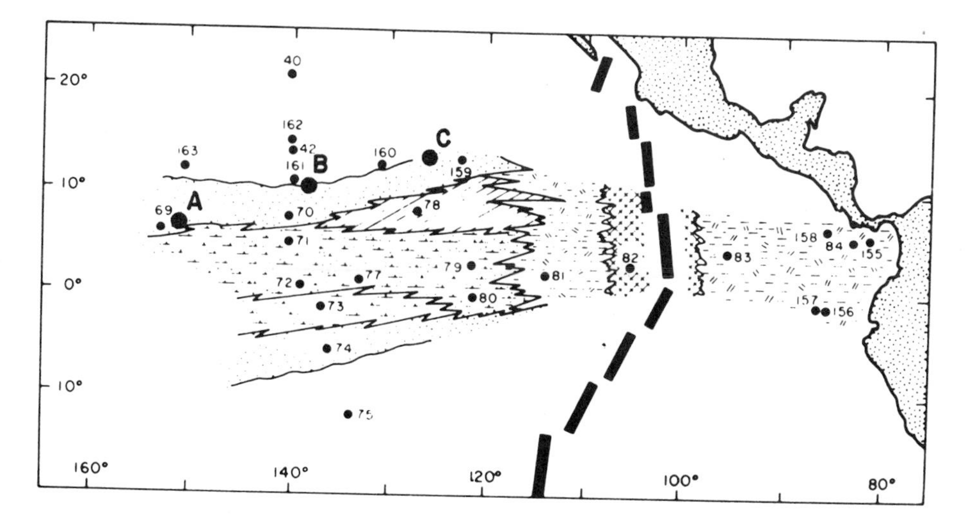

Fig. 13. Late Miocene paleogeologic facies map with location of DSDP sites and DOMES sites A, B, and C. Wavy lines, as between Sites 70 and 161, are traces of unconformities and represent wedge-edge boundary of that unit. Sharp irregular lines, as between Sites 70 and 71, represent interpreted interfingering between units. Stippled patterns for oceanic formations same as in Fig. 9, (modified from Cook, 1975).

DOMES sites A, B, and C, examined in this study, lie between 8° and 15°N latitude and 125° and 151°W longitude (Fig. 1). Water depths range from about 4350 to 5150 meters with average depths at Site A>B>C. Thus, all three sites are situated below the CCD and in the area of very low sediment accumulation, but marginal to the region of high primary productivity.

These sites also are located in the general area where sediment of late Miocene to Oligocene age is observed at or near the surface (Fig. 7). Many piston cores and gravity cores from this area, however, may have a few centimeters to a meter of Plio-Pleistocene sediment at the surface (Riedel and Funnell, 1964).

Data collected at DOMES sites A, B, and C consist of 3.5 kHz seismic records, box cores, and gravity cores. Box cores collected relatively undisturbed sediment to a depth of approximately 45 cm. Gravity cores, all from Site C, penetrated 1.5 m of sediment. The 3.5 kHz records extend to a depth of as much as 250 m. All of these data are compared to nearby DSDP cores and gravity and piston cores.

The dark-brown radiolarian ooze unit of the Clipperton Formation is the stratigraphically highest unit at all of the DSDP sites near the DOMES sites (Fig. 12) and all of the surface cores

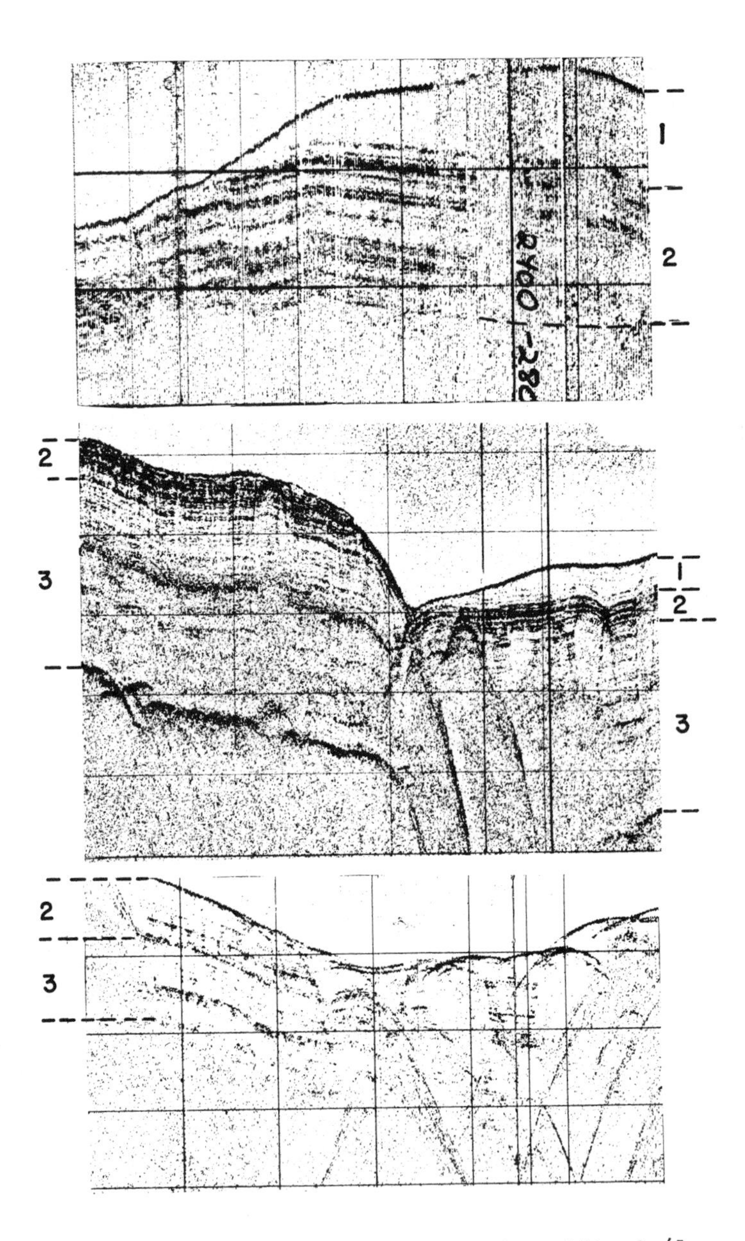

Fig. 14. Comparison of 3.5 kHz profiles from Site C (lower), B (middle), and A (upper). Acoustic units 1 and 2 are possibly the Clipperton Formation; acoustic unit 3 may be the Marquesas Formation. The distance between horizontal lines equals 36.6 m (20 f) in seawater.

at the DOMES sites are lithologically correlative to the radiolarian ooze unit. Stratigraphic units that underlie our cores at nearby DSDP sites are shown in Figure 12. The 3.5 kHz records at DOMES Site C coupled with the cored stratigraphic sequence at the nearby DSDP Site 159 were used as a datum for setting up the acoustic stratigraphy. Profiles of the DOMES 3.5 kHz records (Fig. 14) can be divided into three homotaxial units. These three acoustic units are regional in extent and can be mapped from DOMES Sites A to B to C. We interpret our acoustic unit 1 to represent the radiolarian ooze unit of the Clipperton Formation, acoustic unit 2 to be the upper part of the cyclic unit of the Clipperton Formation, and acoustic unit 3 to include the lower part of the cyclic unit and the Marquesas Formation (Fig. 15). Acoustic unit 3 at DOMES Sites A and B may include the Line Islands Formation.

Lithology and Petrography Procedures

Three-inch diameter cores, taken from 0.25 m^2 box cores, and three inch diameter gravity cores were split longitudinally and visually described, noting such characteristics as color, general appearance, nature of contacts, mottling, and distribution of manganese nodules. Five box cores were described from Site A, four from Site B, and ten from Site C (Table 1). These represent only a few of the total taken from the DOMES sites (Figs. 16, 17 and 18; App. A).

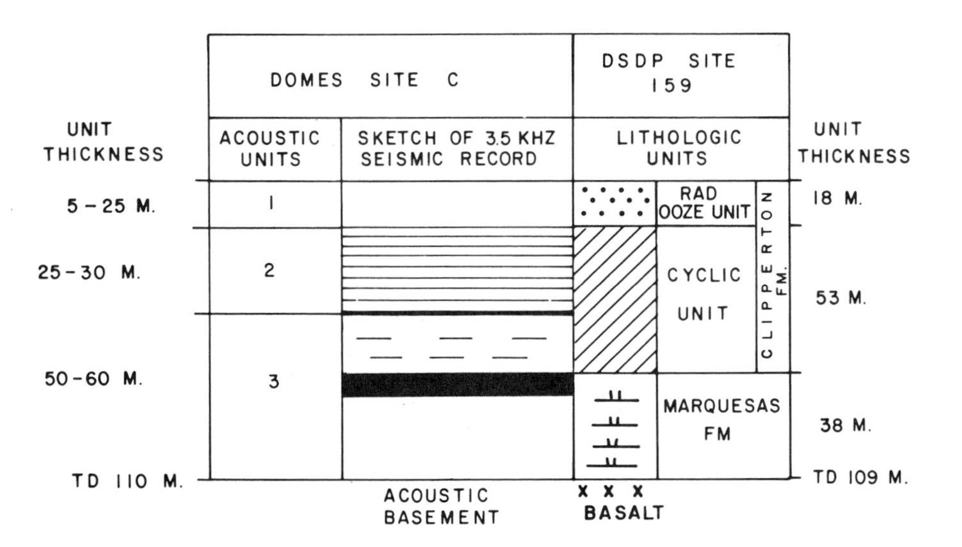

Fig. 15. Schematic seismic record at DOMES Site C showing the interpreted acoustic stratigraphy in relation to the lithostratigraphy at the nearby DSDP Site 159.

Table 1

Box Core and Camera Station Locations, Within Areas A, B, and C

Ship's Station Box Core Number	Latitude North	Longitude West	Depth, m	Ship's Bottom Photo Station Number	Latitude North	Longitude West
			CRUISE RP-6-OC-75			
3 - 5	15°15.9'	126°38.4'	4675	1	16°01.5'	126°40.3'
5 - 10	15°14.7'	126°32.1'	4493	2	15°45.5'	126°10.0'
6 - 18	14°15.0'	126°11.5'	4430	4	15°15.1'	125°59.2'
9 - 21	15°16.4'	126° 9.6'	4630	5	15°16.5'	125°31.4'
11 - 53	16°00.7'	126°46.3'	4603	6	14°17.5'	126°15.4'
15B - 50	15°45.6'	126°00.5'	4552	8	15°16.4'	126°52.0'
16B - 49	15°46.7'	126°11.2'	4552	9	15°16.6'	126°09.2'
18A - 36	15°15.7'	126°00.0'	4339	10	15°15.3'	125°01.5'
18B - 37	15°12.2'	125°58.6'	4406			
20 - 27	15°59.3'	126°11.6'	4672			
23B - 30	14°45.9'	125°58.6'	4667			
24B - 29	14°15.4'	126°01.4'	4468			
25 - 51	14°14.5'	124°58.5'	4561			
27 - 52	16°00.5'	124°59.6'	4406			
			CRUISE RP-8-OC-75			
46 - 1	9°20.7'	150°50.7'	5160	1	9°19.0'	150°51.8'
46 - 3	9°19.0'	150°52.5'	4900	2	9°01.6'	151°12.2'
46 - 5	9°21.3'	150°51.4'	5102	3	8°16.0'	151°10.8'
46 - 7	9°19.6'	150°50.7'	5116	4	11°15.4'	139°00.9'
47 - 10	9°00.3'	151°10.9'	5108	5	11°40.5'	138°25.9'
47 - 11	9°02.5'	151°10.6'	5053	6	12°08.7'	137°45.4'
47 - 12	9°03.5'	151°11.1'	5005			
47 - 13	9°02.3'	151°11.2'	5039			
47 - 14	9°04.0'	151°14.2'	5110			
48 - 19	8°16.5'	151°07.3'	5043			
48 - 21	8°17.6'	151°13.2'	5097			
49 - 25	8°28.5'	150°44.5'	5005			
49 - 27	8°30.2'	150°47.8'	5039			
50 - 28	8°41.7'	150°18.7'	5009			
50 - 29	8°43.8'	150°18.7'	5033			
50 - 30	8°41.1'	150°15.1'	4928			
50 - 32	8°43.1'	150°14.1'	4948			
51 - 33	8°43.1'	150°14.1'	4881			
51 - 34	11°42.4'	139°10.8'	4925			
52 - 37	11°13.7'	139°09.9'	4872			
52 - 39	11°14.8'	139°04.1'	4831			
52 - 42	11°15.5'	139°03.3'	4871			
53 - 45	11°43.9'	138°22.2'	4985			
54 - 47	12°11.4'	137°44.1'	4930			
54 - 49	11°15.5'	139°03.3'	4942			
54 - 50	12°08.5'	137°46.0'	4978			
54 - 51	12°08.9'	137°44.6'	4934			
55 - 53	11°48.6'	137°28.3'	4859			

Table 1. (cont'd)

			CRUISE RP-8-OC-76			
1 - 6	15°13.9'	125°58.4'	4380	1	15°14.5'	126°03.6'
2 - 7	15°14.5'	126°02.6'	4539	2	14°13.8'	125°56.6'
3 - 8	15°15.4'	125°59.3'	4363	3	15°08.4'	126°06.3'
4 - 9	15°14.6'	126°01.6'	4490	4	15°06.5'	125°57.6'
5 - 10	15°13.0'	125°55.9'	4465			
6 - 11	15°10.9'	125°53.6'	4442			
7 - 12	15°12.3'	125°52.1'	4601			
8 - 13	15°12.4'	125°57.2'	4367			
9 - 14	15°11.3'	126°01.5'	4444			
10 - 15	15°11.4'	126°02.2'	4516			
11 - 16	15°11.7'	126°03.9'	4524			
12 - 18	15°11.6'	126°05.9'	4451			
13 - 19	15°08.2'	126°06.4'	4481			
14 - 20	15°08.2'	126°03.7'	4586			
15 - 21	15°08.4'	126°02.7'	4501			
16 - 22	15°07.9'	125°59.4'	4442			
17 - 23	15°08.3'	125°57.3'	4376			
18 - 24	15°06.3'	125°57.4'	4426			
19 - 25	15°08.4'	126°00.9'	4465			
21 - 27	15°05.9'	126°02.1'	4522			
22 - 28	15°06.4'	126°02.8'	4641			
			CRUISE RP-7-OC-77			
7 - 1	15°13.8'	125°57.7'	4394			
5 - 2	15°12.2'	126°02.0'	4527			
6 - 3	15°13.8'	126°05.5'	4445			
4 - 4	15°12.2'	126°01.7'	4437			
2 - 5	15°12.6'	126°00.1'	4292			
10 - 6	15°11.9'	125°55.5'	4546			
9 - 7	15°11.5'	125°50.8'	4586			
8 - 8	15°06.2'	125°54.3'	4419			
3 - 9	15°08.3'	125°57.6'	4396			
1 - 10	15°06.8'	126°03.4'	4636			

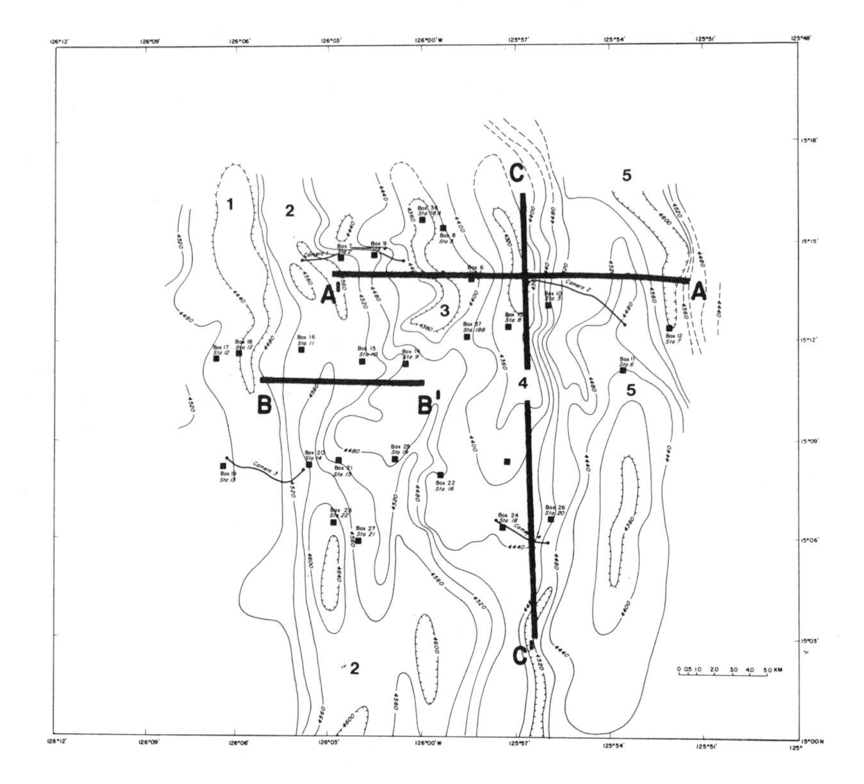

Fig. 16. Bathymetry, core locations, camera flights, and location of 3.5 kHz profiles for a region of Site C. The three labeled segments of cruise track give the locations of the three 3.5 kHz profiles in Fig. 21.

Ten gravity cores, which average 150 cm in length, were taken at Site C (App. A). Unfortunately, the uppermost layer of sediment in the gravity cores was lost, or badly disturbed, during recovery of the core. This sediment may be the very "soupy" surface sediment observed in all Site C box cores which usually had a thickness of approximately 10 cm, on the basis of visual examinations and vane shear measurements (Simpson et al., 1977).

Petrographic descriptions of the cores are based on smear slides of all representative lithologies. Estimates of the relative abundances of the constituents by this method do not yield absolute percentages. This technique is useful, however, for recognizing relative differences and establishing abundance trends of the inorganic and organic constituents. Smear-slide and core descriptions are in Appendix B.

The sediment names discussed and tabulated below are based upon the following rules:

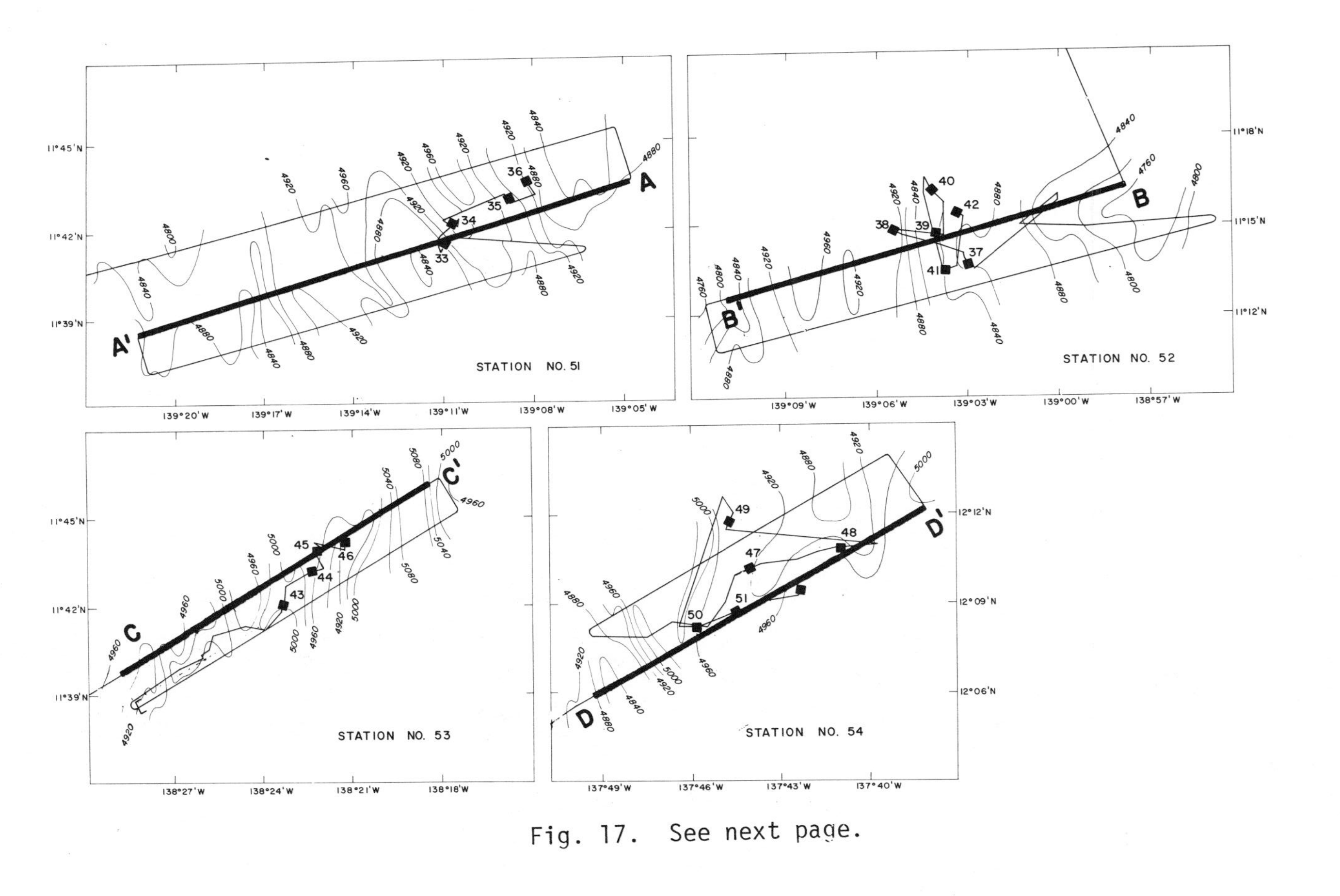

Fig. 17. See next page.

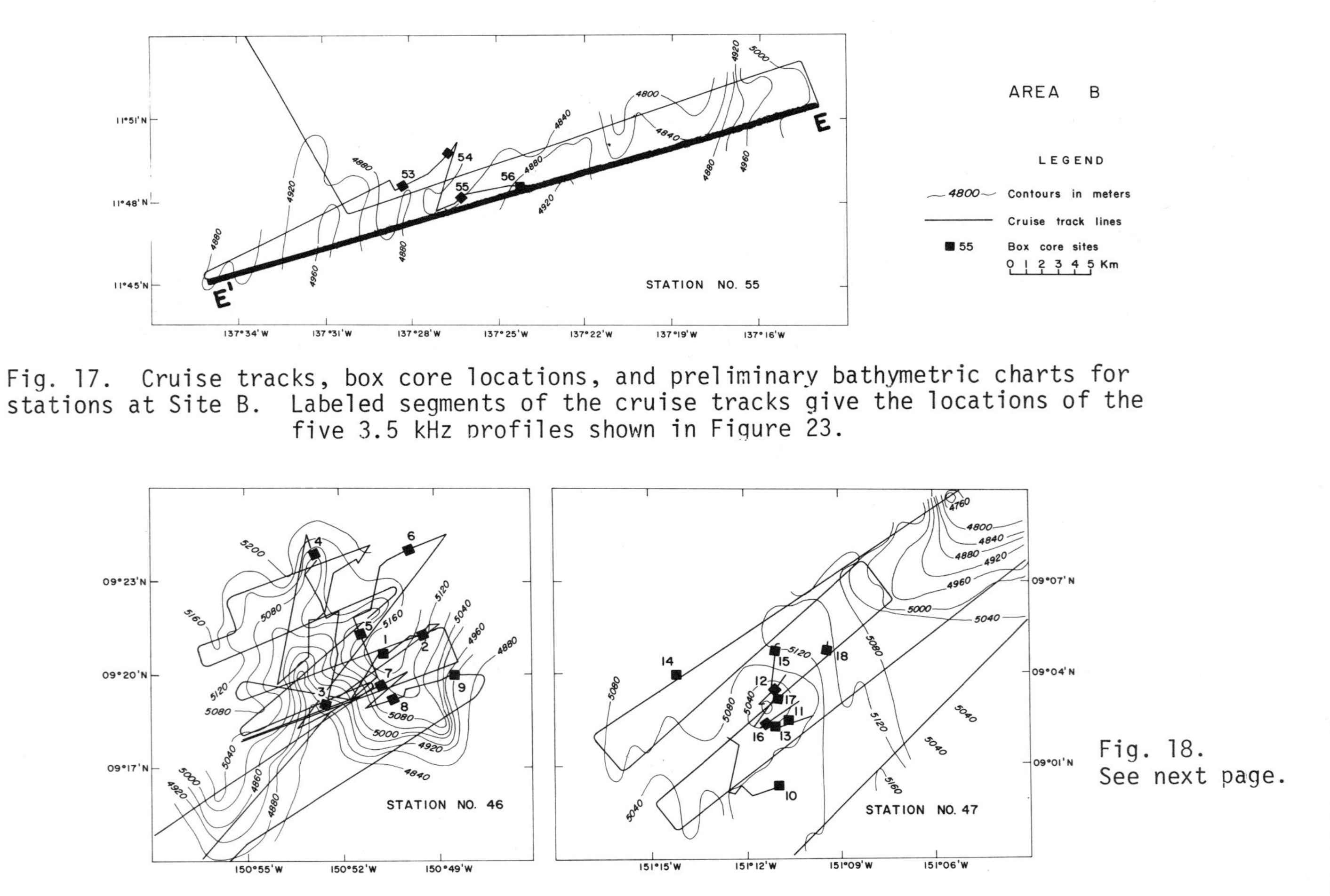

Fig. 17. Cruise tracks, box core locations, and preliminary bathymetric charts for stations at Site B. Labeled segments of the cruise tracks give the locations of the five 3.5 kHz profiles shown in Figure 23.

Fig. 18.
See next page.

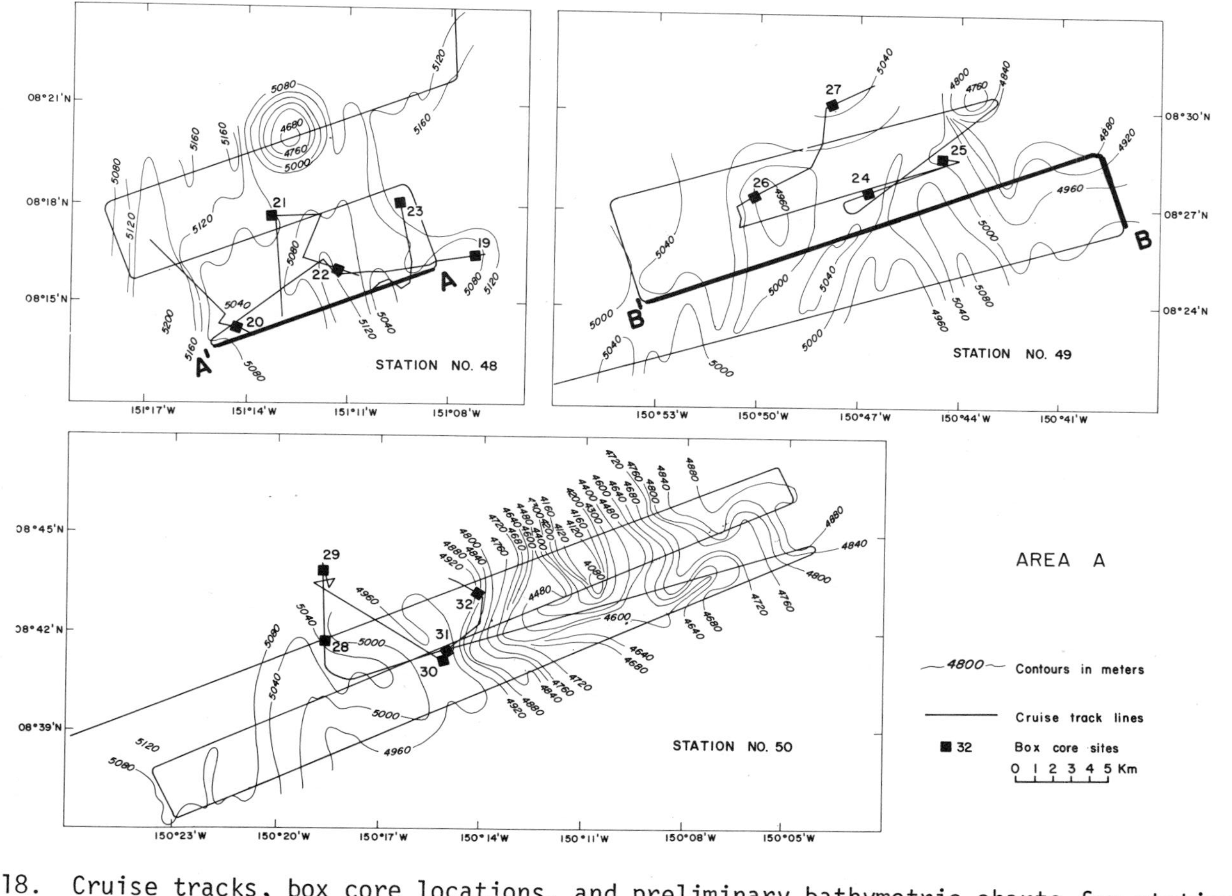

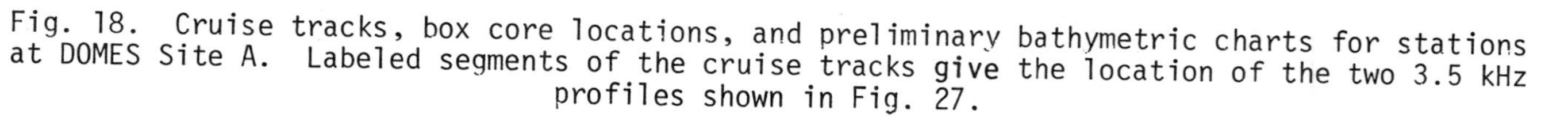
Fig. 18. Cruise tracks, box core locations, and preliminary bathymetric charts for stations at DOMES Site A. Labeled segments of the cruise tracks give the location of the two 3.5 kHz profiles shown in Fig. 27.

Rule 1: The word order for names is: color-constituent (s) - induration modifier (s).

Rule 2: Color is determined using GSA color chart names followed by color code number in (parentheses).

Rule 3: A constituent should be >5% to be included in a name.

Rule 4: Ranges used for smear-slide descriptions are:

<5% rare (R)
5-25% common (C)
25-75% abundant (A)
>75% dominant (D)

Rule 5: If a constituent is common (C), the modifier "bearing" is used after the constituent name. If a constituent is abundant (A) or dominant (D), the constituent name alone is used. Constituents are listed in the order of their abundance in the sediment name.

Rule 6: If component C in Figure 19 is greater than 75%, the sediment is classified a clay with preceding constituent modifier (s). If component C is less than 75%, the sediment is called an ooze with appropriate modifier.

Rule 7: Induration Rule: In as much as all box cores examined are not indurated, all sediment is either an "ooze" or a "clay". Hypothetical sediment notations without color designation are as follows:

>75% Clay	1-5% Clay
1-5% Micronodules	5-25% Radiolarians
1% Radiolarians	25-75% Calc. Nannofossils
-- Pelagic Clay	-- Radiolarian-bearing Nannofossil Ooze

The box cores consist mainly of pelagic clay and clayey siliceous ooze. Only a few of the cores contain calcareous ooze. Most of the cores show a distinct interbedding of orange and dark brown layers. A subordinate number of box cores, all from Site C, are devoid of bedding and consist of massive dark-brown sediment. All box cores we examined contain manganese nodules at the sediment-water interface and several from A and B have nodules at depth in the cores. A number of cores collected at Sites A and B, but not examined by us, did not contain nodules. Several of the gravity cores from Site C also did not contain nodules. A summary of the petrographic characteristics of the cores is shown in Figures 19 and 20.

DOMES Site C

The bathymetry at Site C shows a strong north-south "grain" (Fig. 16). Cruise tracks that trend east-west change slope more frequently than do north-south tracks (Piper, 1976). This bathymetry is quite similar to an area 2^{0} to the south which has been surveyed by deep-tow (Mudie and Grow, 1972).

D. Johnson (1972b) and Luyendyk (1970) have found that slopes of abyssal hills commonly exhibit scarps, a few meters high, that have slopes as great as 60°. Such escarpments were observed in bottom photographs on two camera flights. They were associated with angular rock fragments up to 1m across that may be basalt or ferromanganese slabs. The large beam-angle generated by the ship-mounted 3.5 kHz transducer precludes detection of such topographic features on acoustic records. On one camera flight, however, during which the ship's course was up the slope of an abyssal hill and the ship's speed was approximately 2 knots (see Fig. 8 in Piper et al., this volume), the profile of the bottom clearly shows disconformities in slope.

Box cores from Site C commonly are massively bedded, but several cores show cyclic bedding (see Cook et al., 1977; Fig. 14C). These cycles consist of pale-yellowish-brown (10YR 5/4) sediment interbedded with dusky-brown (5YR 2/2) sediment. Bedding ranges from 5 to 20 cm in thickness. Box cores from Site C are all classified as pelagic clay (Figs. 19 and 20). The estimated clay

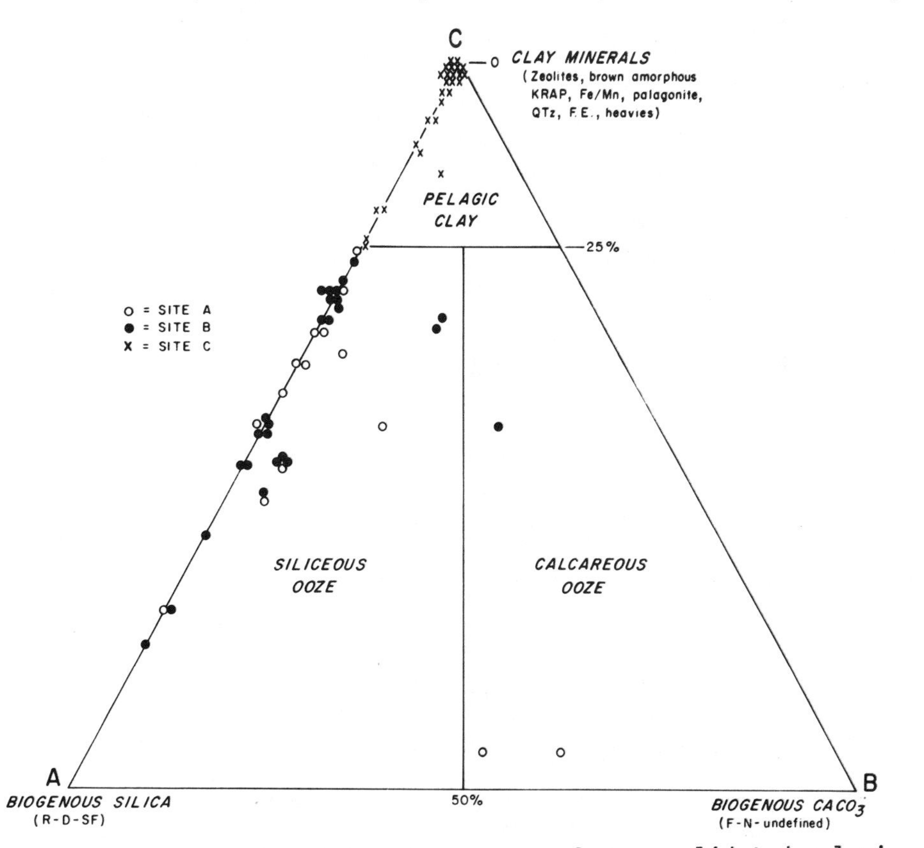

Fig. 19. General classification scheme of unconsolidated pelagic sediment and of sediment types from DOMES Sites A, B, and C cores.

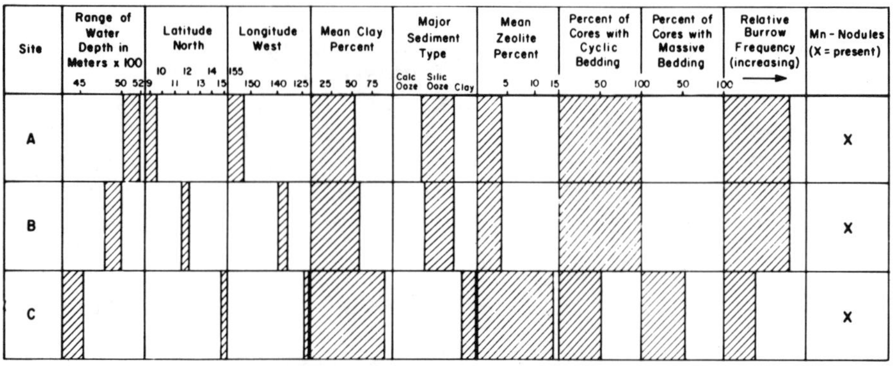

Fig. 20. Summary of major geologic parameters of cores from DOMES Sites A, B, and C.

content ranges from about 75 to virtually 100 percent (Figs. 19 and 20). Site C cores also appear to have a relatively high percentage of zeolite. Lithologies typical of cyclically bedded cores from Site C include:

1. Dark-yellowish-brown (10YR 4/2) zeolite-bearing pelagic clay.
2. Moderate-yellowish-brown (10YR 5/4) radiolarian-diatom-zeolite-bearing pelagic clay.
3. Dark-yellowish-brown (10YR 4/2) and dusky-brown (5YR 2/2) palagonite-bearing pelagic clay.
4. Dark-yellowish-brown (10YR 4/2) pelagic clay.

The massively bedded cores are typically moderately yellowish-brown (10YR 5/5) to dark-yellowish-brown (10YR 4/2). Usually only a few burrow traces are evident. It is possible, however, that burrowing is so intense that all remnants of bedding have been completely destroyed, but this is doubtful because burrows are usually very distinct. Alternatively, the rather uniform color of the beds may result in underestimating the degree of burrowing.

Lithologies of the massively bedded Site C cores are similar to the cyclically bedded sediment:

1. Grayish-orange (10YR 7/4) zeolite-bearing pelagic clay.
2. Dark-yellowish-brown (10YR 4/2) pelagic clay.
3. Moderate-yellowish-brown (10YR 5/4) radiolarian-bearing pelagic clay.
4. Moderate-yellowish-brown (10YR 5/4) diatom-bearing pelagic clay.

All box cores from Site C contained surface nodules, and none had buried nodules. Several of the gravity cores did not have nodules. However, in all 10 gravity cores, the surface layer of sediment, approximately the uppermost 10 cm, was badly disturbed, or totally lost, during recovery.

Except for box cores 18-36 and 22-28, and gravity cores 4G-4 and 2G-5, the pelagic clay contains a poorly preserved unmixed assemblage of Quaternary radiolarians (Quinterno and Theyer, this volume). Within box core 18-24 a well-preserved mid-Quaternary assemblage of coccolithophora is present at 12-14 cm. In box core 18-36, at 16-18cm, the sediment contains an unmixed nannofossil assemblage of Early Miocene age from the Triquetrorhabdulus carinatus Zone, and the Discoaster drugii subzone. The assemblage is similar in species composition and state of preservation to those of the same subzone from nearby DSDP cores 160 and 161.

Gravity cores 4G-4 and 2G-5 penetrate through the surface layer of zeolitic pelagic clay and into nannofossil ooze. The nannofossil ooze is typically moderate yellowish brown (10YR 5/4) to grayish orange (10YR 7/4), homogeneous, burrowed, and stiff. Nannofossils from this facies also yield lower Miocene ages within the Triquetrorhabdulus carinatus Zone (D. Bukry, personal communication, 1978), similar to that cored at DSDP Sites 159 to 161. Core 2G-5 also contained a mixed assemblage, of mid-Miocene to mid-Eocene age, above the early Miocene unit (Quinterno and Theyer, this volume).

Profiles of 3.5 kHz records taken at Site C show a series of acoustically opaque layers alternating with transparent layers. The section of sediment observed can be divided into three homotaxial acoustic units (Fig. 14). The uppermost sediment layer (acoustic unit 1) is acoustically transparent, but with occasional weak reflectors. It has a highly variable thickness (Fig. 21), with a maximum thickness of approximately 25 m on the southern part of the centrally located abyssal hill (Figs. 16 and 22). Unit one appears to be present as a continuous surface layer only in this area although its thickness even there is highly variable. Box core 18-24 recovered a Quaternary nannofossil assemblage from this area (Fig. 22). Unit 1 is locally observed as a wedge-shaped deposit up to 15 m thick (Fig. 21) in the channels to the east and west of this hill. It occurs elsewhere in the area as a thin layer, usually less than 5m thick. It may, however, cover most of the area as a thin sediment veneer that cannot be resolved from the underlying unit. This interpretation is based on the occurrence of a purely Quaternary assemblage of radiolarians in box cores that were taken where older acoustic units 2 and 3 appeared to be at the surface.

Acoustic unit 2 is a series of closely spaced (approximately 5m) opaque reflectors with a strong reflector at the base (Figs. 14 and 21). The unit has a rather uniform 25- to 30-m thickness. It is present at, or very near, the surface on the top of the central abyssal hill, where it has an east-southeast-trending slope. The small channel that dissects the central abyssal hill appears to follow the strike of this unit. This unit also occurs throughout much of the eastern and western channels (Fig. 22).

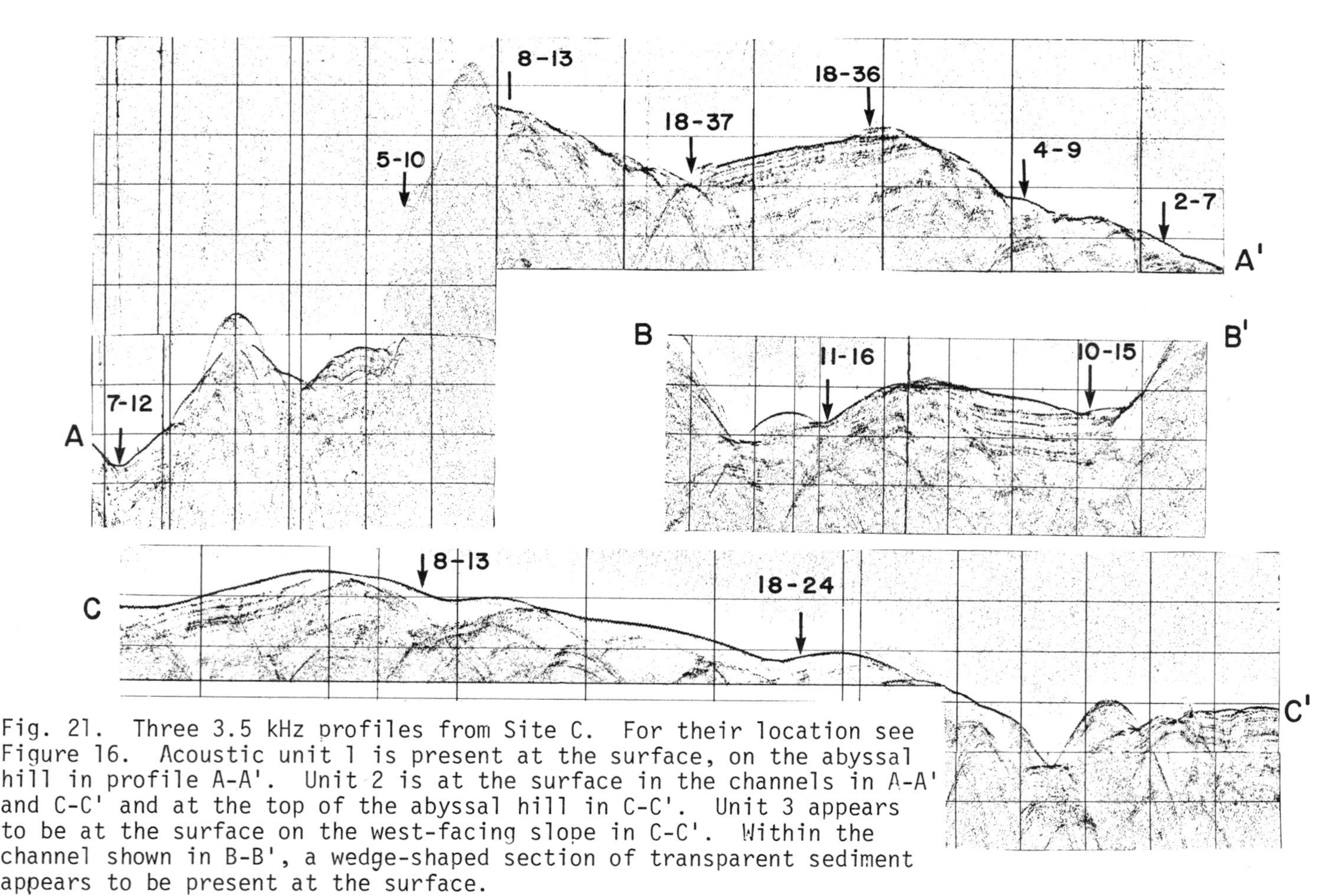

Fig. 21. Three 3.5 kHz profiles from Site C. For their location see Figure 16. Acoustic unit 1 is present at the surface, on the abyssal hill in profile A-A'. Unit 2 is at the surface in the channels in A-A' and C-C' and at the top of the abyssal hill in C-C'. Unit 3 appears to be at the surface on the west-facing slope in C-C'. Within the channel shown in B-B', a wedge-shaped section of transparent sediment appears to be present at the surface.

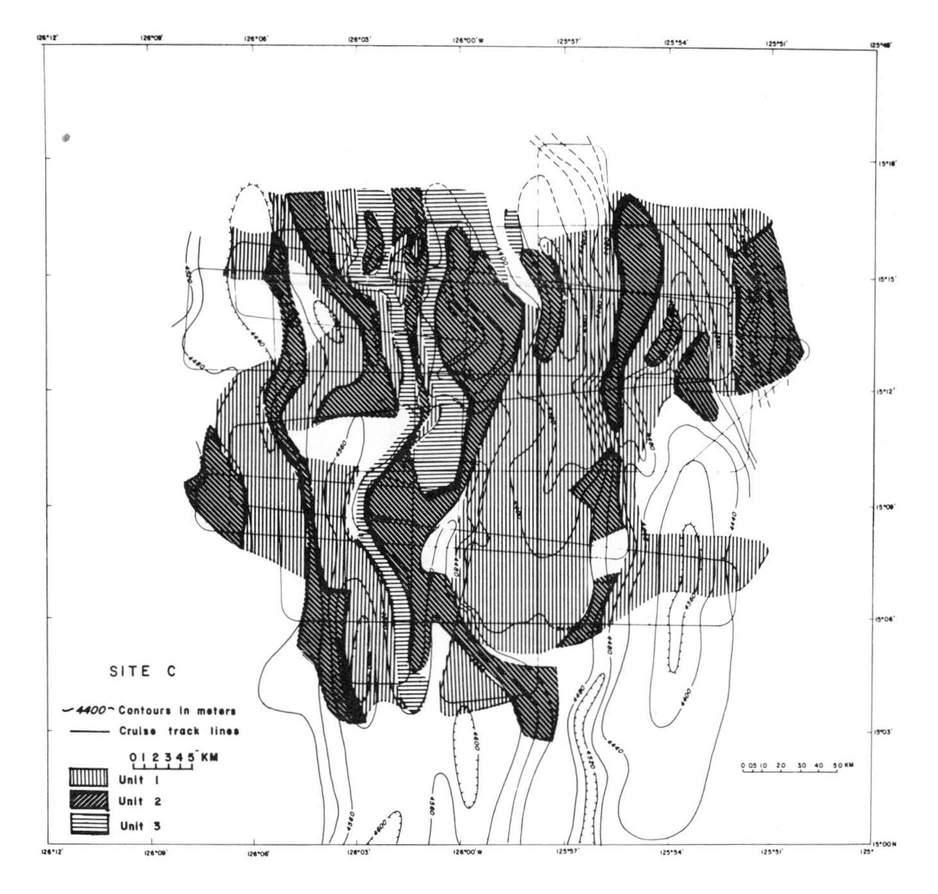

Fig. 22. Surface distribution of acoustic units at Site C. For acoustic character of units, see Figure 14.

Gravity core 2G-5, located on the western slope of the central abyssal hill, contains a mixed nannofossil assemblage of mid-Miocene to mid-Eocene age. Box core 18-36, located near the top of the central abyssal hill and in an area where units 2 and 3 appear to be in contact at the surface (Fig. 22), contains unmixed early Miocene nannofossils.

Unit 3 is acoustically much less reflective than unit 2, but with a 10-m-thick diffuse reflector about 25m from the top of the unit. This unit appears to be exposed along the slope of the western channel and central abyssal hill (Fig. 22).

Beneath unit 3 is a very poorly defined reflector, observed only locally, which is considered to be acoustic basement. The combined thickness of sediment, assuming that the three units represent the entire sedimentary sequence at Site C, is 110 m. The poor definition of acoustic basement certainly makes our assumption and estimate of sediment thickness somewhat speculative. DSDP Sites 159 and 160, located near Site C (van Andel, Heath *et al.*, 1973, Fig. 2), however, had sediment thicknesses of 108 and 109 m respectively, approximately the same as that inferred from the 3.5 kHz records (Fig. 21).

Our division of the sedimentary column into three acoustic units is similar to that made by Richter and Schlüter (1973). They subdivided our unit 3 into two units, however, possibly at the depth where we observed the 10-m-thick diffuse reflector.

The sediment column at DSDP Site 159 was divided into the following three lithologic units by van Andel, Heath *et al.* (1973; p. 238, 942):

1. 0-18 m. A grayish-orange to moderate-yellow-brown siliceous clay (the radiolarian ooze unit of the Clipperton Formation). The upper 9 m contain a mixed assemblage of Quaternary and older radiolarians. The bottom of the unit is middle Miocene.

2. 18-70 m. Alternations of clay, calcareous clay, and nannofossil marl ooze (cyclic unit of the Clipperton Formation). Above approximately 38 m the unit is dominantly clay and calcareous clay. Below this depth it is largely marl ooze. A brief hiatus may occur at 47 m. The age is middle to early Miocene.

3. 70-108 m. Nannofossil marl ooze (Marquesas Formation). Its age is early Miocene to late Oligocene.

Although correlation of DSDP sediment cores with 3.5 kHz records from Site C is somewhat speculative, we interpret acoustic unit 1 at Site C to correlate with the upper 18 m of sediment at DSDP Site 159. Thus, this uppermost acoustically transparent layer may represent the radiolarian ooze unit of the Clipperton Formation and have an age of middle Miocene to Holocene. The unit at Site C is a siliceous clay rather than radiolarian ooze. Acoustic unit 2 may represent the upper part of DSDP 159 - unit 2 (i.e., from 18 m to 38-47 m) and represent approximately the upper half of the cyclic unit of the Clipperton Formation. It would have an age of early to middle Miocene. Acoustic unit 3 may correlate with the lower 61 to 70-m of sediment at DSDP 159, the lower half of the Clipperton Formation and all of the Marquesas Formation. Its age would then be late Oligocene to early Miocene. Solely on the basis of sediment thicknesses, the boundary between these two formations in DSDP core 159 may correlate with the 10-m-thick diffuse reflector within acoustic unit 3 (Fig. 14).

Our interpretation of these acoustic records and of the sediment recovered in box core 18-36 and gravity cores 2G-5 and 4G-4 correlates the fossil assemblage in these cores with the *Triquetrorhabdulus carinatus* Zone of DSDP 159.

DOMES Site B

The acoustic records at Site B strongly suggest that the abyssal hills are elongate in a north-south direction, similar to

Site C. The quantity of records at any one station, however, is insufficient to permit detailed mapping of part of this site.

Box cores from Site B consist of weakly to moderately bedded layers of light- and dark-brown siliceous ooze (see Cook et al., 1977, Fig. 14b). The light-colored layers are usually grayish orange (10YR 7/4), and the dark beds dark yellowish brown (10YR 4/2). Beds range from 5 to 20 cm in thickness, with irregular bedding contacts that have as much as 5 cm of relief. This irregularity of bedding contacts is probably due to burrowing. Cores commonly have a mottled appearance; the lighter colored beds are speckled with cylindrical blebs of darker colored sediment from superjacent and subjacent beds.

About half of all box cores contained surface nodules, and five box cores had nodules at depth in the cores.

Sediment from Site B cores are virtually all clayey siliceous ooze (Figs. 19 and 20) with an estimated clay content that ranges from about 25 to 75% (Fig. 20). Biogenic components are mainly radiolarians and diatoms. Dominant lithologies are:

1. Dark-yellowish-brown (10YR 4/2) clay-diatom-bearing radiolarian ooze.
2. Grayish-orange (10YR 7/4) radiolarian-clay-bearing diatom ooze.
3. Grayish-orange (10YR 7/4) clay-bearing diatom ooze.

The radiolarian assemblage of Site B box cores is strongly mixed and usually contains less than 50% early Miocene to Eocene radiolarian. The main exception is core 51-33, which has greater than 99% early Miocene to middle Eocene fauna at the base of the core.

No area within Site B has been examined in as much detail as the 700 km^2 area within Site C. Acoustic records collected at Site B (Figs. 14 and 23), however, are much better and appear to be strikingly similar to the Site C records.

The youngest acoustic unit is transparent. It ranges in thickness from 0 to 40 m. It is quite widespread, much more so than at Site C, and has an average thickness of approximately 25 m. However, its distribution at each station is somewhat different. At station 51 (Fig. 23) it is thickest on the tops of hills and on west-facing slopes. At Stations 52, 53, and 54 it appears to be present throughout these areas, except on the steeper slopes and on a prominent hill at Station 54 (Fig. 24). It is very thin, or absent, at Station 55.

Acoustic unit 2 is essentially identical to unit 2 at Site C, with regard to its acoustic signature and thickness. It is observed throughout the area, except on a few prominent hill tops and steep slopes.

Acoustic unit 3 is easily recognized in most records. It can be divided possibly into the same two units recognized by Richter and Schlüter (1973). The upper one consists of a series of weak reflectors, with a broad diffuse reflector at the base. The lower part is transparent. The total thickness of unit 3 is 110 m and

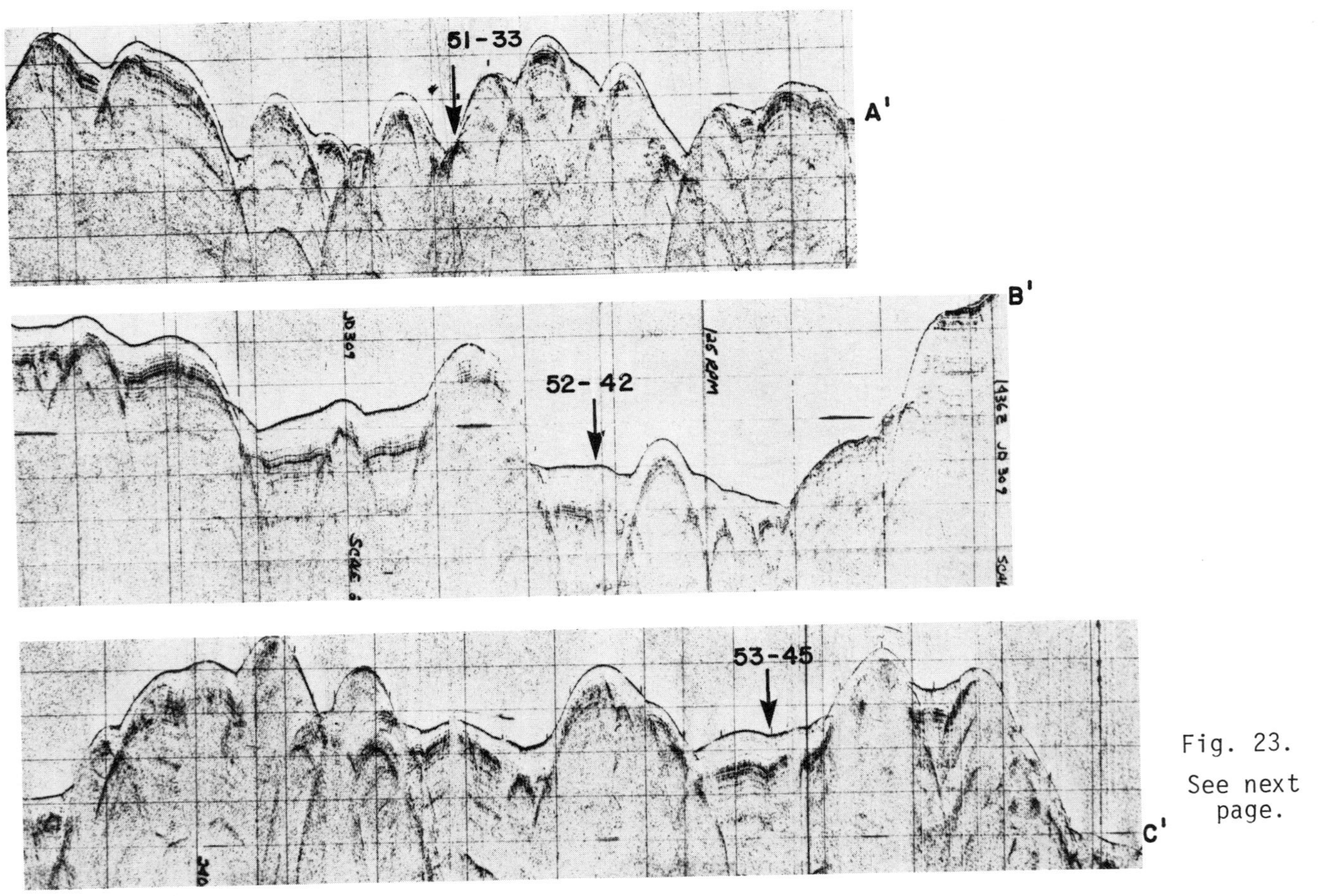

Fig. 23. See next page.

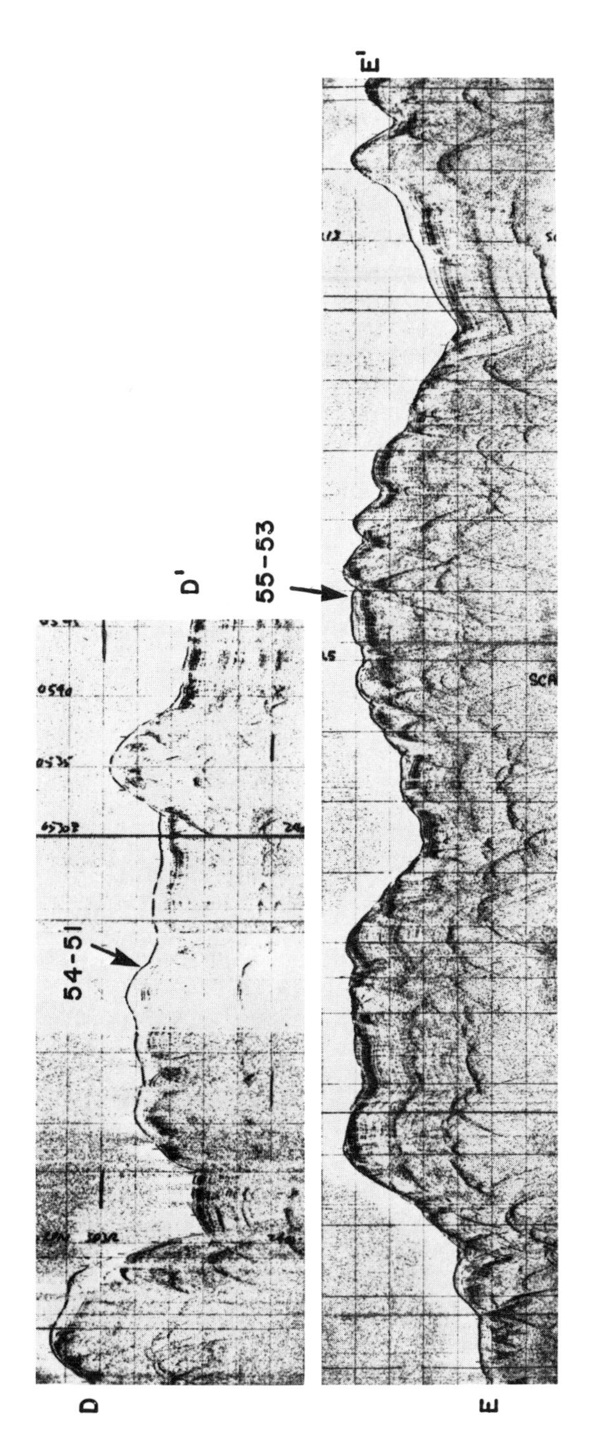

Fig. 23. Five 3.5 kHz profiles from Site B. For their location, see Fig. 17.

quite uniform.

Total sediment thickness at Site B above a well-defined acoustic basement, which may be basaltic basement, (Fig. 23) is approximately 160 m. The thickness of sediment at DSDP Site 161 (10°14.25'N and 139°59.21W) is 244 m (van Andel, Heath et al., 1973). The upper 2 m of sediment in this core (radiolarian ooze unit of the Clipperton Formation) is Quaternary in age and may correspond to the transparent layer of sediment at Site B. Gravity and piston cores collected at Site B and examined by Riedel and Funnell (1964) contained possibly as much as 45 cm of Quaternary sediment at the surface. The sequence of acoustic reflectors immediately below this transparent layer (Fig. 14) may correlate with unit 2 at DSDP Site 161 (cyclic unit of the Clipperton Formation). Its age would then be late Oligocene to early Miocene. Apparently no box cores sampled this unit because all box cores contain an abundant assemblage of Quaternary radiolarians (Quinterno and Theyer, this volume). The occurrence of mixed fauna in these cores of Eocene to early Miocene age, however, strongly suggests that unit 2, as well as unit 3, is exposed within the immediate area, as is suggested by the acoustic records (Figs. 23 and 24). The lowermost acoustic unit at Site B, unit 3, may correlate with the white nannofossil chalk of the Marquesas Formation, unit 3 at DSDP Site 161, of early to late Oligocene age. It was not possible to identify a basal unit in the acoustic records that might correlate with the Line Islands Formation, 44 m of which were recovered in DSDP Core 161.

DOMES Site A

The bathymetry at Site A is poorly known, as it is at Site B. The few records we do have, however, do not indicate any trend to the bottom relief.

Cores from Site A consist of light- and dark-brown siliceous ooze interbedded in 5- to 40-cm-thick beds (see Cook et al., 1977, Fig. 14a), similar to Site B box cores. The lighter colored beds are typically grayish orange (10YR 7/4) or pale yellowish orange (10YR 8/6), whereas the darker colored beds are commonly dark yellowish brown (10YR 4/2) to dusky yellowish brown (10YR 2/2). Contacts between the beds are slightly to highly irregular, probably owing to burrowing. There are some suggestions that these contacts were once horizontal and planar. Planar contacts are the exception, however, and most contacts exhibit varying degrees of disruption. Abundant evidence of burrowing within individual beds gives a mottled appearance to the cores. Horizontal burrows in several cores exposed at the surface have been partially removed (Fig. 25). Manganese nodules occur at the sediment surface of approximately 80% of the cores. According to ship's logs, ten box cores had nodules at depth in the cores.

Most of the sediment from Site A is siliceous ooze (Figs. 19 and 20) with an estimated clay content from 25 to 75%. Only two

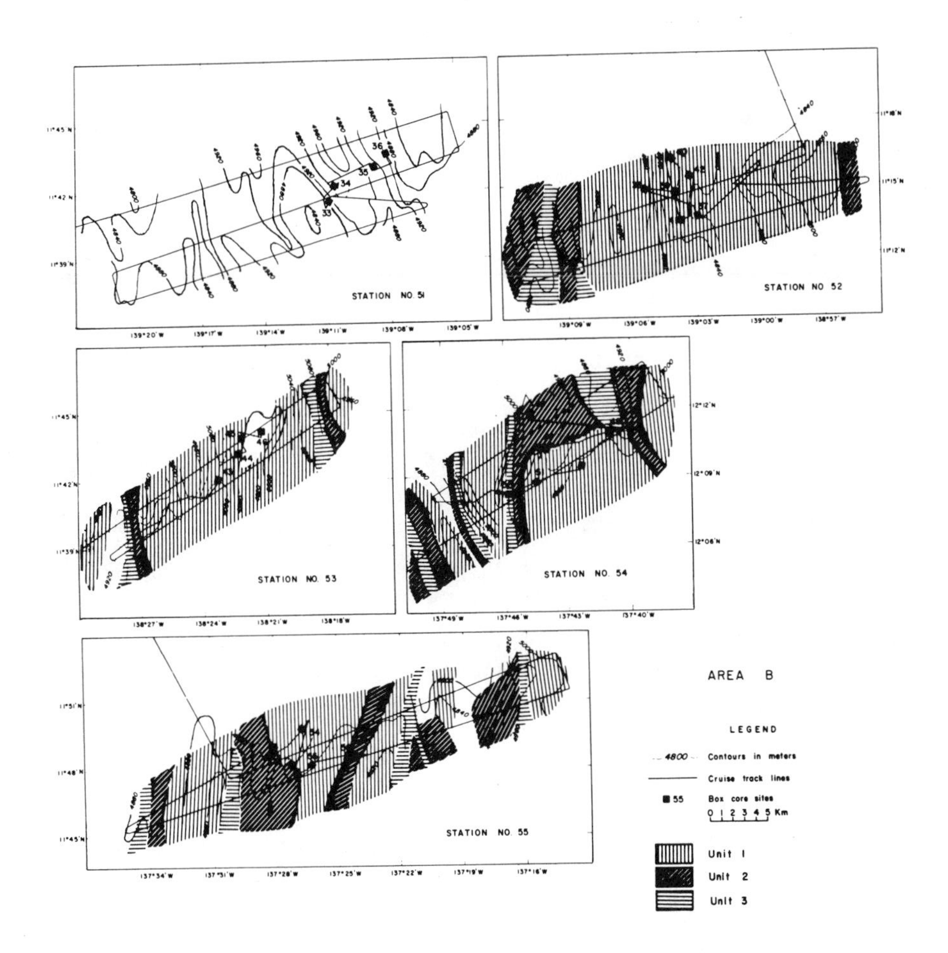

Fig. 24. Surface distribution of acoustic units at Site B. For acoustic character of units, see Figure 14.

samples are calcareous ooze. Typical lithologies at Site A are:

1. Moderate-yellowish-brown (10YR 5/4) clay-bearing radiolarina ooze.
2. Grayish-orange (10YR 7/4) clay-foraminifer-diatom-radiolarian-bearing coccolith ooze.
3. Grayish-orange (10YR 7/4) clay-radiolarian-bearing diatom ooze.

The radiolarian faunal assemblage of Site A cores is characterized by a high percentage of species of early Miocene to Eocene age (Quinterno and Theyer, this volume). The abundance of Quaternary radiolarians decreases sharply and continuously with depth, from approximately 80% at the surface to 1-50% in the lower part of the cores. Box core 49-25 is an exception; it contains a dominantly Quaternary population throughout the core.

Fig. 25. Photography of DOMES box core from Site A. A sinuous horizontal burrow is in the lower center part of the figure. Black sheroids are manganse nodules.

The acoustic records for Site A, although of poorer quality than those for Site B and less numerous than for Site C, suggest a stratigraphy that is similar to that at the other two sites. The transparent layer has an extensive areal distribution (Fig. 26). Locally, it is significantly thicker, up to 55 m thick, than at Sites B and C (Fig. 27). The underlying series of reflectors, unit 2 at Sites B and C, also appears to be thicker than at these other two sites. The contact between these two units commonly is strongly unconformable, more so than appears to be the case at Sites B and C. Acoustic basement could not be identified at Site A.

Acoustic stratigraphy at Site A may correlate with the lithostratigraphy at DSDP Site 69 (Tracey et al., 1971). The radiolarian ooze unit of the Clipperton Formation in this core overlies the cyclic unit. Its age is early to middle Miocene. However, almost 3 m of surface sediment which may represent Quaternary sediment, was washed out during coring at DSDP Site 69.

Discussion

Cores from DOMES Sites A, B, and C are relatively close to DSDP Sites 69, 70, 78, 159, 160, and 161 in the central equatorial Pacific (Figs. 11, 12, and 13). All of these DSDP sites penetrated

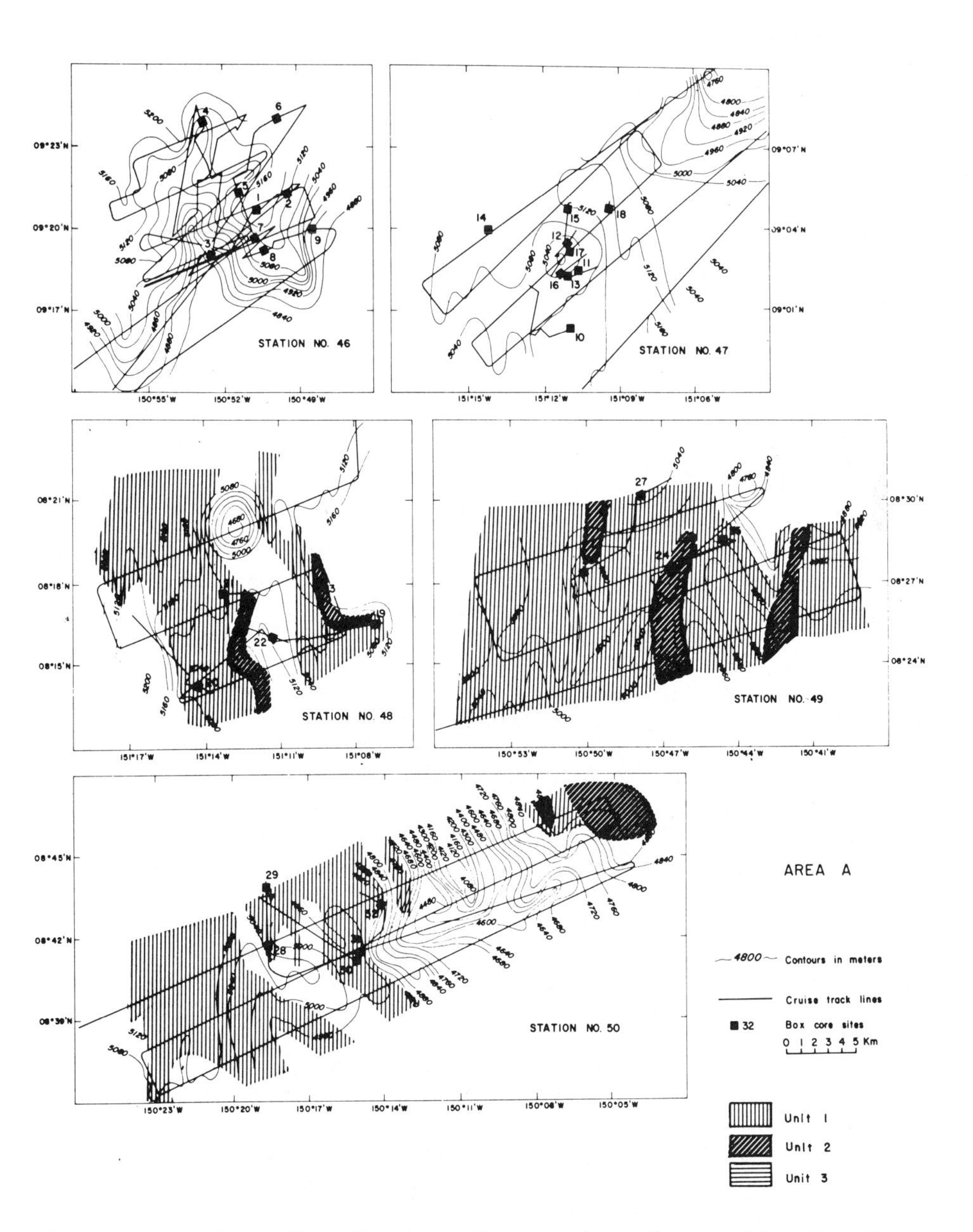

Fig. 26. Surface distribution of acoustic units at Site A. For acoustic character of units, see Figure 14.

cyclically bedded or massively bedded brown and orange pelagic clay and siliceous ooze at the sediment-water interface. This lithologically distinct type of sediment has been correlated for thousands of kilometers, from 90°W longitude to at least 150°W longitude. These lithologic units have been informally designated the cyclic and radiolarian ooze units of the Clipperton Oceanic Formation (Tracey *et al.*, 1971; Cook, 1975).

Cores from DOMES Sites A, B, and C are similar in lithology, bedding characteristics, and stratigraphic position to the radiolarian ooze unit of the Clipperton Oceanic Formation. These cores strengthen the concept that this sediment type forms a widespread, lithologically and stratigraphically distinct unit in the central equatorial Pacific.

Comparison of the acoustic data from the three DOMES sites strongly suggests that the acoustic units can be correlated with the stratigraphic units of DSDP cores and also can be correlated between DOMES sites. At each of the sites, however, the rather uneven distribution of the uppermost acoustically transparent layer (acoustic unit 1), which corresponds to the radiolarian ooze unit of the Clipperton Formation identified in DSDP cores (Tracey *et al.*, 1971; Cook, 1975), suggests that sediment accumulation is currently very uneven. The variability in thickness of this unit may be attributed to erosion and redeposition of early Tertiary sediment (D. Johnson, 1972a), i.e., of acoustic units 2 and 3.

The intensity of these processes apparently increases from east to west. This interpretation is based on the greater thickness of acoustic unit 1 at Site A and the greater percentage of reworked early Tertiary radiolarians in the surface sediment at Site A (Quinterno and Theyer, this volume). At Site C acoustic unit 1 is very thin, usually less than 5 m thick, and the box cores contain an unmixed assemblage of Quaternary radiolarians.

Although no direct measurements of sediment accumulation rates have been made, the greater thickness of acoustic unit 1 at Site A implies a higher sedimentation rate than at Site C. The relatively high percentage of reworked Tertiary radiolarians in Site A sediment suggests that the bulk of the nonbiogenic fraction may also be reworked. The rate of accumulation of primary material, therefore, is much less at Site A than the rate of sedimentation would suggest, and possibly even less than at Site C.

The mechanism whereby sediment is redistributed is unclear. No relation exists between the distribution of acoustic unit 1 and the bathymetry. The decrease of Quaternary radiolarians with depth in box cores and the frequent occurrence of 5- to 10-cm-high mounds of sediment that cover nodules suggest that bioturbation contributes to sediment resuspension. However, sediment resuspension during bioturbation, coupled with its redistribution by the rather weak bottom currents observed near the sea floor (Hayes, this volume), seem to be inadequate to account for the distribution of acoustic unit 1.

If our lithologic interpretation of these acoustic units is

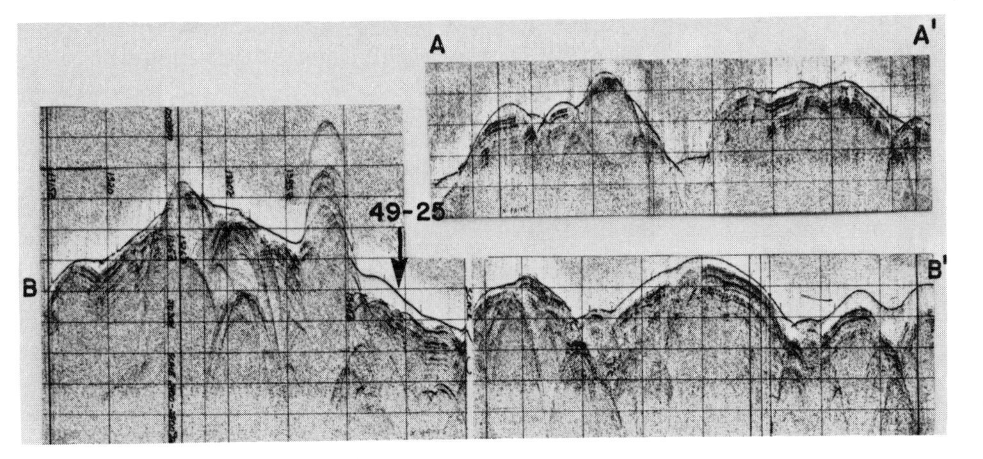

Fig. 27. Two 3.5 kHz profiles from Site A. For their location see Figure 18.

correct, our acoustic stratigraphy provides a powerful tool for mapping and interpreting lithostratigraphic variations in the equatorial Pacific and for contributing to our understanding of sedimentation in the pelagic environment.

Acknowledgments

We are grateful to J. Tracey, Jr., for many helpful and constructive comments and to J. Henning for typing and editing the final manuscript.

References

Berger, W.H., (1970) Planktonic foraminifera: selective solution and the lysocline, Mar. Geol. 8, 111-138.

Bogorov, V.G., (1969) The Pacific Ocean, Biology of the Pacific Ocean, Part I - Plankton, (English translation): U.S. Oceano. Office, Washington, D.C., 435 p.

Bonatti, E., (1963) Zeolites in Pacific pelagic sediment: Trans. N.Y. Acad. Sci. 25, 938-948.

Bramlette, M.N., (1961) Pelagic sediments, in Oceanography: Washington, D.C., Am. Assoc. Adv. Sci. 345-366.

Bukry, D., (1973) Coccolith stratigraphy, eastern equatorial Pacific, in van Andel, Tj. J., Heath, G.R. et al., Initial Reports of the Deep Sea Drilling Project, Washington, D.C.,

U.S. Govt. Printing Office 16, 915-936.
Church, T.M., (1970) Marine Barite. PhD thesis, Scripps Inst. Oceano., La Jolla, Calif., 100 p.
Cook, H.E., (1972) Stratigraphy and sedimentation, In Hayes, J.D., (ed)Initial Reports of the Deep Sea Drilling Project, vol. 9, Washington, D.C. (U.S. Govt. Printing Office), p. 933-943.
Cook, H.E., (1975) North American stratigraphic principles as applied to deep sea sediments: Am. Assoc. Petrol. Geol. Bull. v. 59, 817-837.
Cook, H.E., Piper, D.Z., and Gardner, J.V., (1977) Geologic and oceanographic framework of DOMES Sites A, B, and C: Central equatorial Pacific: In: Piper, D.Z. et al., (eds.), Deep Ocean Environmental Study: Geology and Geochemistry of DOMES Sites A, B, and C, equatorial North Pacific, U.S. Geol. Survey Open-file Report 77-778, 15-81.
Cronan, D.S., (1973) Basal metalliferous sediment cored during Leg 16, Deep Sea Drilling Project, In van Andel, Tj, H., Heath, G.R. et al., 1973, Initial Reports of the Deep Sea Drilling Project 16, Washington, D.C.(U.S. Govt. Printing Office) 610-604.
Edmond, J.M., Chung, Y., and Sclater, J.G., (1971), Pacific bottom water penetration east of Hawaii, Jour. Geophys. Res., v. 76, 8089-8097.
Ewing, J.I., Ewing, M., Aitken, T., and Ludwig, W.J., (1968) North Pacific sediment layers measured by seismic profiling, Am. Geophys. Union Geophys. Mon. 12, 147-173.
Gordon, A.L., and Gerard, R.D., (1970) North Pacific bottom potential temperature, In Hayes, J.D., (ed.) Geological investigations of the North Pacific, Geol. Soc. Amer. Mon. 126, 23-39.
Griffin, J.J., and Goldberg, E.D., (1963) Clay-mineral distributions in the Pacific Ocean, In Hill, M.N., (ed.) The Sea 3, New York, Interscience, 728-741.
Hayes, S., (1977) Study of benthic boundary layer, N.E. Pacific, Proceed. 9th Ann. Offshore Tech. Conf., 387-394.
Hayes, J.D., Saito, T., Opdyke, N.D., and Burckle, L.H., (1969) Pliocene-Pleistocene sediment of the eastern equatorial Pacific: their paleomagnetic, biostratigraphic, and climatic record, Geol. Soc. Amer. Bull. 80, 1481-1514.
Hayes, J.D. et al., (1972) Initial Reports of the Deep Sea Drilling Project 9, Washington (U.S. Govt. Printing Office), 1205 p.
Heath, G.R., (1974) Dissolved silica and deep sea sediments: In Hay, W.W., (ed.), Studies in Paleooceanography: Soc. Econ. Paleontologists and Mineralogists Spec. Pub. 20, 77-93.
Heezen, B.C. et al., (1973) Diachronous deposits, a kinetic interpretation of the post-Jurassic sedimentary sequence on the Pacific Plate: Nature, 241, 25-32.
Horn, D.R., Delach, M.N., and Horn, B.M., (1973) Metal content of ferromanganese deposits of the ocean: IDOE Tech. Report, no. 4, Natl. Sci. Found. Washington, D.C., 57 p. (unpublished).

Johnson, D.A., (1972a) Eastward flowing bottom currents along the Clipperton Fracture Zone: Deep Sea Research 19, 253-257.
Johnson, D.A., (1972b) Ocean-floor erosion in the Equatorial Pacific: Geol. Soc. Amer. Bull. 83, 3121-3144.
Johnson, T.C., (1976) Controls on the preservation of biogenic opal in sediments of the eastern Tropical Pacific: Science, 192, 887-890.
Lisitzin, A.P., (1972) Sedimentation in the World Ocean (English translation), Soc. Econ. Paleon. Min., Sp. Publ. no. 17, 218 p.
Luyendyk, B.P., (1970) Origin and history of abyssal hills in the northeast Pacific Ocean: Geol. Soc. Amer. Bull. 81, 2237-2260.
Lyle, M., Dymond, J. and Heath, G.R. (1977) Copper-nickel-enrichment in ferromanganese nodules and associated crusts from the Bauer Basin, northwest Nazca Plate, Earth Planet. Sci. Lett. 35, 55-64.
Moberly, R. and Klein, G.D., (1976) Jour. Sed. Pet. 46, Ephemeral color in deep sea cores, 261-225.
Mudie, J.D., and Grow, J.A., 1972, Near bottom observations on the detailed topography of an area of abyssal hills in a manganese nodule province (Abs.), In Horn, D.A. (ed.) Ferromanganese deposits on the ocean floor, IDOE Tech. Rept., Natl. Sci. Found. Washington, D.C., 293 p.
Piper, D.Z., 1976, Physiographic characteristics of the equatorial North Pacific and of DOMES Site C: In Bischoff, J.L. et al., Deep Ocean Mining Environmental Study, N.E. Pacific Nodule Province, Site C, Geology and Geochemistry, U.S. Geol. Survey Open-file Report 75-548, 16-62.
Piper, D.Z., and Williamson, M.E., (1977) Composition of Pacific Ocean ferromanganese nodules, Mar. Geol. v. 23, 285-303.
Piper, D.Z., Leong, K., and Cannon, W.L., (1979) Manganese nodule and surface sediment compositions (this volume).
Price, N.B., and Calvert, S.E., (1970) Composition variation in Pacific Ocean ferromanganese nodules and its Relationship to Sediment Accumulation Rates: Mar. Geol. 9, 145-171.
Quinterno, P. and Theyer, F., (1979) Biostratigraphy of the equatorial North Pacific DOMES Sites A, B, and C (this volume).
Reid, J.L., (1962) On the circulation, phosphate-phosphorus content and zooplankton volumes in the upper part of the Pacific Ocean: Limnol. Oceano. 7, 287-306.
Reid, J.L., Jr., (1969) Preliminary results of measurements of deep currents in the Pacific Ocean, Nature 221, 848.
Rex, R.W. and Goldberg, E.D., (1958) Quartz contents of pelagic sediment of the Pacific Ocean: Tellus, 10, 153-159.
Richter, H. and Schlüter, H.U., (1973) VALDIVIA exploration for manganese nodules, 1973, preliminary results of seismic reflection survey, In Morgenstein, M. (ed) The Origin and Distribution of Manganese Nodules in the Pacific and Prospects for Exploration, Honolulu, Hawaii, Inst. Geophys., 139-144.
Riedel, W.R. and Funnell, B.M., (1964) Tertiary sediment cores and microfossils from the Pacific Ocean floor, Geol. Soc. London Quart. Jour. 120, 305-368.

Romankevich, E.A., (1976) Organic matter of sediments to the east of Japan and its effect on oxidation-reduction processes in Volkov, I.I. ed. Biogeochemical Diagenesis of Marine Sediment, Moscow, Nayka, 5-19.
Ryan, W.B.F. and Heezen, B.C., (1976) Smothering of Deep Sea Benthic Communities from Natural Disasters: National Oceanic and Atmospheric Administration, Tech. Rept. Cont. 03-6-022-35120, 132 p.
Schlanger, S.O. and Douglas, R.G., (1974) The ooze-chalk-limestone transition and its implications for marine stratigraphy, In Hsu, K.J., and Jenkyns, H.C., (eds.) Pelagic sediments on land and under the sea, Internatl. Assoc. Sed. Spec. Publ. 1, London, Blackwell.
Simpson, F., Wallin, C.S., Crawford, H.M., and Springer, P.C., (1977) Geotechnical properties of NE Pacific sediments from DOMES Sites A, B, and C, in Piper, D.Z. et al., (eds), Deep Ocean Environmental Study: Geology and Geochemistry of DOMES Sites A, B, and C, Equatorial North Pacific, U.S. Geol. Survey Open-file Rept. 77-778, 351-492.
Tracey, J.I., Sutton, G.D. et al., (1971) Initial Reports of the Deep Sea Drilling Project 8, Washington (US Govt. Printing Office), 1037 p.
van Andel, Tj. H., Heath, G.R. et al., (1973) Initial Reports of the Deep Sea Drilling Project 16, Washington (US Govt Printing Office), 949 p.
van Andel, Tj. H., Heath, G.R., Moore, T.C., Jr., (1975) Cenozoic History and Paleooceanography of the Central Equatorial Pacific Ocean, Geol. Soc. Amer. Mem. 143, 134 p.
von der Borch, C.C., and Rex, R.W., (1970) Amorphous iron oxide precipitates in sediments cored during Leg 5, Deep Sea Drilling Project, in McManus, D.A., Burns, R.E. et al., Initial Reports of the Deep Sea Drilling Project 5, Washington, D.C., (US Govt Printing Office) 541-544.
Windom, H.L., (1969) Atmospheric dust records in permanent snowfields, Implications to marine sedimentation, Geol. Soc. Amer. Bull. 80, 761-782.
Winterer, E.L., (1973) Sedimentary facies and plate tectonics of the equatorial Pacific, Am. Assoc. Petrol. Geol. Bull. 57, 265-282.

BIOSTRATIGRAPHY OF THE EQUATORIAL NORTH PACIFIC DOMES SITES A, B, AND C

Paula Quinterno
U. S. Geological Survey
Menlo Park, California

and

Fritz Theyer
Hawaii Institute of Geophysics
University of Hawaii
Honolulu, Hawaii

Abstract

Radiolarians from box cores and gravity cores indicate that Quaternary sediment covers the sea floor at DOMES Sites A, B, and C; however, at Sites A and B dissolution-resistant, reworked Tertiary species (predominantly middle and late Eocene and early Miocene) generally outnumber the fragile Quaternary forms. In contrast, at Site C an early Miocene calcareous nannofossil flora underlies the Quaternary layer in three cores, and specimens of different ages are not mixed; barren sediment generally underlies the Quaternary in the remainder of cores from Site C. The abundant admixture of Tertiary radiolarians in Quaternary sediment at Sites A and B is most likely caused by an increase in bottom current activity during the Pleistocene, resulting in several cycles of erosion and redeposition of Tertiary sediment. This physical process of erosion, aided by selective dissolution of fragile radiolarians, explains the dominance of dissolution-resistant forms in the mixtures.

Introduction

Sediment analyzed in this study is from box cores and gravity cores collected for the DOMES project during 1975 and 1976 at test

Sites A, B, and C in the equatorial Pacific (Fig. 1 and Table 1). The objectives of this study are (1) to determine the age of the sediment based on microfossils, (2) to summarize briefly previous micropaleontological studies in the equatorial Pacific which are relevant to the DOMES project, and (3) to compare the results of our DOMES study with the previous studies.

Methods

The box cores represent the uppermost half meter of the ocean floor and are virtually undisturbed by sampling techniques. The original box cores were subsampled by pushing plastic pipes about 8 cm in diameter into the sediment column. These plastic pipes were then cut in half and biostratigraphic samples taken at 2-cm intervals. Calcareous fossils were not anticipated because of the greater-than-4000 m water depths from which the cores were collected. Most age determinations are based on radiolarians.

The sediment was first washed over a 62-micrometer sieve, the residues boiled in hydrogen peroxide, sieved again, and dried. Strewn microslides were prepared from a portion of the residues. Abundances of stratigraphically important radiolarians were determined by counting all specimens (approximately 300-500) found within a specific area of each slide; these areas were defined by settings of the microscope stage. Smear slides were prepared for calcareous nannoplankton to supplement the radiolarian datings. Foraminifers were studied using the 62-micrometer residues from the radiolarian samples. Radiolarian, coccolith, and foraminiferal species found are listed in Appendix A.

Biostratigraphic Results

Radiolarians

The radiolarian residues contained an abundant fauna in all samples from Sites A and B but meager assemblages in those from Site C. Consequently, the Quaternary radiolarian zones of Nigrini (1971) and the datum levels of Johnson and Knoll (1975) were used to determine the ages of samples from the former two sites, whereas only more generalized age determinations were possible at Site C.

Tables 2-4 summarize the radiolarian stratigraphy. Although the ages are based on samples taken at 2-cm intervals throughout the cores, intervals of similar composition and age are grouped together in the tables. A Quaternary age (Pleistocene and Holocene, 1.8 m.y. BP to present) is consistently indicated in all samples that contain radiolarians. In addition, most cores also contain some intervals of late Quaternary age (younger than 0.2 m.y. BP), indicated by the presence of *Buccinosphaera invaginata* Haeckel, a diagnostic species for this age in the Pacific (Johnson

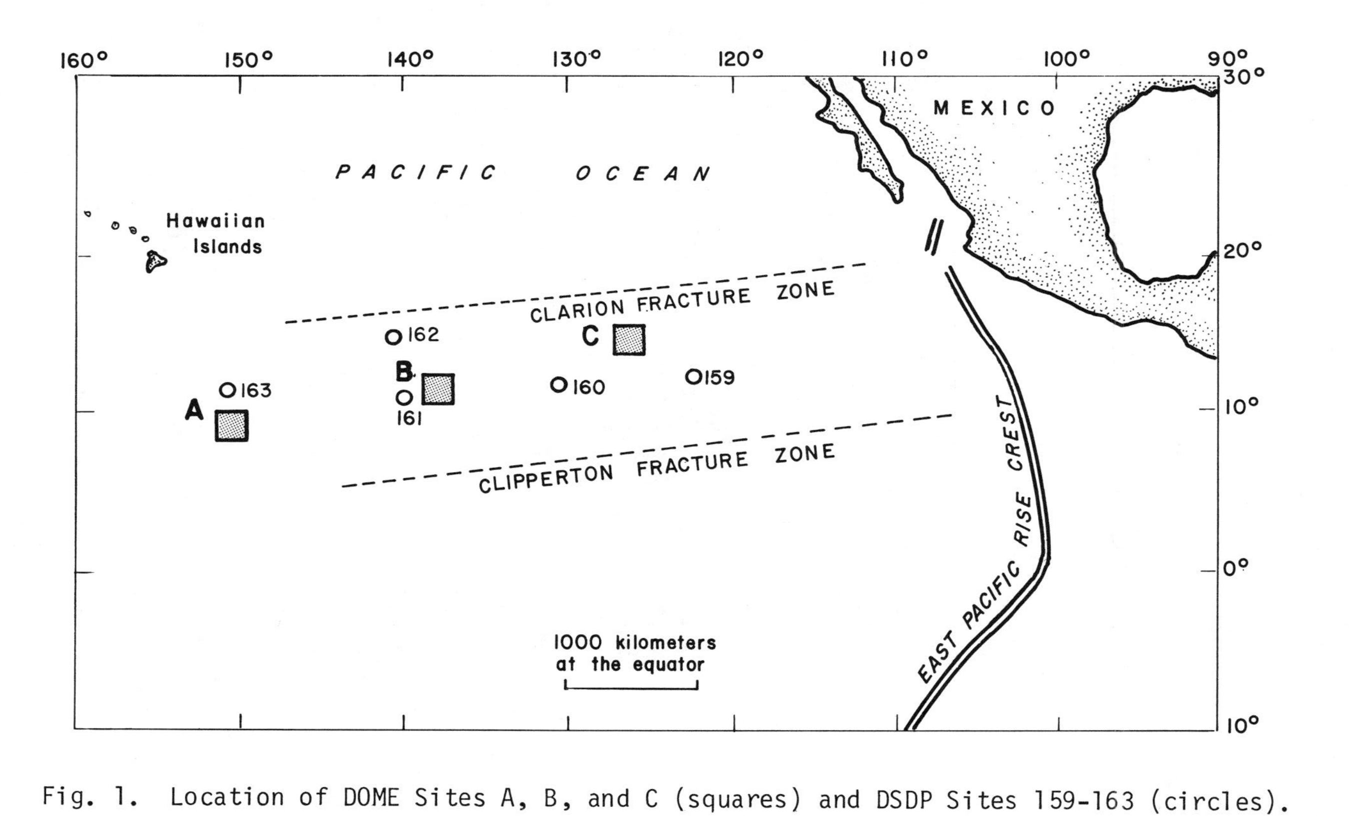

Fig. 1. Location of DOME Sites A, B, and C (squares) and DSDP Sites 159-163 (circles).

TABLE 1

Core locations, DOMES Sites A, B, and C

	Station, Box Core Number	Lat. N	Long. W	Depth, m
Site A	46 - 1	9°20.7'	150°50.7'	5160
	47 -10	9°00.3'	151°10.9'	5108
	48 -19	8°16.5'	151°07.3'	5043
	49 -25	8°28.5'	150°44.5'	5005
	50 -28	8°41.7'	150°18.7'	5009
Site B	51 -33	11°41.6'	139°11.0'	4881
	52 -42	11°15.5'	139°03.3'	4871
	53 -45	11°43.9'	138°22.2'	4985
	54 -51	12°08.9'	137°44.6'	4934
	55 -53	11°48.6'	137°28.3'	4859
Site C	11 -53	16°00.7'	126°46.3'	4603
	15B-50	15°45.6'	126°00.5'	4552
	16B-49	15°46.7'	126°11.2'	4552
	18 -24	15°06.3'	125°57.4'	4426
	18A-36	15°15.7'	126°00.0'	4339
	18B-37	15°12.2'	125°58.6'	4406
	20 -27	15°59.3'	126°11.6'	4672
	23B-30	14°45.9'	125°58.6'	4667
	24B-29	14°15.4'	126°01.4'	4468
	25 -51	14°14.5'	124°58.5'	4561
	27 -52	16°00.5'	124°59.6'	4406
	Gravity core			
	2G-5	15°12.6'	126°00.1'	4292
	4G-4	15°12.2'	126°01.7'	4437

and Knoll, 1975). The coherent early and late Quaternary radiolarian stratigraphy in a few cores suggests that a continuous Quaternary stratigraphic sequence is present in the layers penetrated by these cores. Box core 42 from Station 52, Site B, is particularly clear in this respect.

Abundant Tertiary radiolarians are reworked into virtually all intervals sampled at Sites A and B (Tables 2 and 3). The percentages of reworked specimens are quite high (up to 99%), in some samples nearly masking the Quaternary assemblages. Several studies, beginning with the pioneering work by Riedel and Funnell (1964), have described similar radiolarian reworking in sediment samples from the central equatorial Pacific. Admixtures in the DOMES samples predominantly are middle and late Eocene and early Miocene dissolution-resistant radiolarians. Similar admixtures were noted in previous studies in the equatorial Pacific by Johnson and Johnson (1970), Moore (1970), and Johnson and Parker (1972).

Radiolarians are more abundant in samples from Sites A and B (Tables 2 and 3) than in samples from the northeasterly Site C (Table 4). The percentage of reworked fossils generally increases down-core at Sites A and B. Reworking is less pronounced at Site B than at Site A, although Core 51-33 is an exception.

Calcareous Nannoplankton and Diatoms

Quaternary nannoplankton occur in samples from DOMES Sites A and C (Fig. 1), and in some of these samples the presence of *Gephyrocapsa oceanica* indicates an age of middle Quaternary or younger (Tables 2 and 4). Tertiary nannofossils are present only at Site C where coccolith ooze from Cores 18A-36 (14-16 cm), 4G-4 (58 and 112 cm) and 2G-5 (105 cm) is assigned to the earliest Miocene *Triquetrorhabdulus carinatus* Zone of Bukry (1973). Older species are neither reworked nor mixed with younger species in these assemblages, which are similar to those sampled at nearby DSDP Sites 159-161. Higher in DOMES Core 18A-36, at 10-12 cm, an early Miocene assemblage is dominated by *Discoaster deflandrei* and *D. druggii*. This assemblage is mixed with a Quaternary assemblage containing *Ceratolithus cristatus*, *C. simplex*, and *C. telesmus*. At 6-8 cm this core contains only a trace of the dissolution-resistant discoasters which are remnants of mixing from the Miocene deposits. Calcareous nannofossils are absent above 6 cm. A warm-water diatom assemblage of Quaternary age is present in the upper 8 cm of Core 18A-36; species include *Asteromphalus imbricatus*, *Coscinodiscus excentricus*, *Ethmodiscus rex*, *Nitzschia marina*, *Roperia tesselata*, *Thalassiosira oestrupii* and *Thalassiothrix* sp.

Foraminifers

Only a few DOMES cores contained calcareous foraminifers, because all samples were apparently collected from below the modern carbonate compensation depth (CCD). Calcareous faunas from Site

TABLE 2

Ages of cores from DOMES Site A. The sampling interval is 2 cm throughout. Intervals of similar composition and age are grouped together. Most ages are based on radiolarians; those based on calcareous nannofossils, and their respective intervals, are in parentheses. Percentages of reworked radiolarian specimens are calculated from countings of whole assemblages.

Core Station-number Water depth	Interval (cm)	Sediment age	Reworked radiolarians %	Age
46-1 5160 m	0-2	latest Quaternary	85	early Miocene, Oligocene, and late and middle Eocene
	4-6	Quaternary	85	early Miocene, Oligocene, and late and middle Eocene
	6-41	Quaternary	85-95	early Miocene, Oligocene, and late and middle Eocene
47-10 5108 m	0-2	late Quaternary	20	early Quaternary, Pliocene, and early Miocene
	2-22	late Quaternary	30-50	early Miocene and late and middle Eocene
	22-24	late Quaternary	80	early Miocene and late and middle Eocene
	24-26	late Quaternary	40	early Miocene and late and middle Eocene
	28-30	late Quaternary	80	early Miocene and late and middle Eocene

TABLE 2 (continued)

Core Station-number Water depth	Interval (cm)	Sediment age	Reworked radiolarians %	Age
	30-50	Quaternary	70-99	early Miocene and late and middle Eocene
48-19 5043 m	0-14	latest Quaternary	40-70	Miocene and late and middle Eocene
	16-41	Quaternary	50-99	early Miocene and late and middle Eocene
49-25 5005 m	0-44	Quaternary	1-5	middle and early Miocene and late and middle Eocene
	(0-14)	(middle Quaternary or younger)		
	(22-24)	(Pliocene or younger)		
50-28 5009 m	0-47	latest Quaternary	10-48	early Miocene and late and middle Eocene
	(20-22)	(middle Quaternary or younger)		
	(30-32)	(middle Quaternary or younger)		
	(38-40)	(middle Quaternary or younger)		

TABLE 3

Ages of cores from DOMES Site B. (See Table 2 for explanation.)

Core Station-number Water depth	Interval (cm)	Sediment Age	Reworked radiolarians %	Age
51-33 4881 m	0-4	latest Quaternary	25-40	early Miocene and late and middle Eocene
	4-22	Quaternary	90-99	early Miocene and late and middle Eocene
52-42 4871 m	0-6	latest Quaternary	2-4	early Miocene and late and middle Eocene
	6-45	late Quaternary	4-10	early Miocene, Oligocene, and late and middle Eocene
	45-49	early Quaternary	10	early Miocene, Oligocene, and late and middle Eocene
53-45 4985 m	0-4	Quaternary	1-4	early Miocene, late Oligocene, and late and middle Eocene
	4-30	latest Quaternary	1-4	early Miocene, late Oligocene, and late and middle Eocene
	30-45.2	Quaternary	1-4	early Miocene, late Oligocene, and late and middle Eocene
54-51 4934 m	0-2	Quaternary	5	early Miocene, Oligocene, and late and middle Eocene

TABLE 3 (continued)

Core Station-number Water depth	Interval (cm)	Sediment Age	Reworked radiolarians %	Age
	2-4	latest Quaternary	10	early Miocene and late and middle Eocene
	4-41	Quaternary	10-15	early Miocene, Oligocene, and late and middle Eocene
55-53 4859 m	0-2	late Quaternary	1	middle and early Miocene and late Eocene
	2-4	late Quaternary	2	Pliocene, late Miocene, early Miocene, late Oligocene, and late Eocene
	4-8	late Quaternary	1	early Miocene, Oligocene, and late Eocene
	8-34	Quaternary	8-40	early Miocene and late and middle Eocene

TABLE 4

Ages of cores from DOMES Site C. (See Table 2 for explanation.)

Core Station-number Water-depth	Interval (cm)	Sediment Age	Reworked radiolarians %	Age
11-53 4603 m	0-14	Quaternary	0	
	14-41	Barren		
15B-50 4552 m	0-40	Quaternary with barren layer at 26-30 cm	0	
16B-49 4552 m	0-16	Quaternary	0	
	16-20	Barren		
18-24 4426 m	(12-14)	(middle Quaternary or younger)		
	(20-22)	(Quaternary)		
18A-36 4339 m	0-10	Quaternary	0	
	(6-8)	(Trace of Miocene)		
	10-16	Barren		
	(10-12)	(Quaternary and Miocene)		
	(14-16)	(early Miocene)		
18B-37 4406 m	0-10	Quaternary	0	
	10-20	Barren		
20-27 4672 m	0-6	Quaternary	0	
	6-8	Barren		
	8-12	Quaternary	0	
	12-14	Barren		
	14-16	Quaternary	0	
	16-18	Barren		

Table 4 (continued)

Core Station-number Water-depth	Interval (cm)	Sediment Age	Reworked radiolarians %	Reworked radiolarians Age
23B-30 4667 m	0-18	late Quaternary	1-20	Quaternary; trace of Miocene
	18-26	Quaternary	0	
	26-30	late Quaternary	10	Quaternary
	30-44	Quaternary	0	
24B-29 4468 m	0-16	Quaternary		Trace of Quaternary
25-51 4561 m	0-14	late Quaternary		Trace of Quaternary
	14-27	Barren		
27-52 4406 m	0-16	late Quaternary		Trace of Quaternary
	16-20	Barren		
Gravity Cores				
2G-5 4292 m	(65 cm)	(middle Miocene to middle Eocene)		
	(87 cm)	(Barren)		
	(105 cm)	(early Miocene)		
4G-4 4437 m	(20 cm)	(Barren)		
	(58 cm)	(early Miocene)		
	(112 cm)	(early Miocene)		

A, Core 50-28 (26-28 cm and 40 cm), are similar to those from Site C, Cores 18-24 (12-14 cm) and 18A-36 (8-10 cm). Species in these samples are typical of deep sea deposits and include Laticarinina pauperata, Cassidulina subglobosa, Pullenia bulloides, Pullenia quinqueloba, Gyroidina zealandica, and species of Eponides and Epistominella. Although some of these occur as early as Eocene, all are known to extend to the present.

A similar fauna is present in DOMES Core 18A-36 at 10-16 cm, the deepest interval sampled, but additional forms are present there as well. These taxa and their ranges, as reported in the literature, include the following:

Baggatella sp. (of Loeblich and Tappan, 1964)--middle Eocene to late Oligocene
Globigerina senni (of Postuma, 1971)--early to middle Eocene
Planulina sp. (of Parker, 1964)--Miocene
Stilostomella rugosa (of Loeblich and Tappan, 1964)--Miocene

Considered together, these species help to bracket the age of Core 18A-36 (10-16 cm) between Eocene and Miocene. The Eocene and Oligocene species may be reworked, or their actual ranges may be greater than reported thus far and may extend into the Miocene. If so, the age of this interval would be Miocene, in agreement with the coccolith age determination.

Agglutinated foraminifers are present in very low numbers in most of the DOMES cores, probably because of the small sample size. They are typical of assemblages previously reported from surficial deep-sea sediment; taxa include Ammodiscus tenuis, Cyclammina sp., Eggerella propinqua, Karreriella bradyi, Recurvoides sp., Reophax bacillaris, R. cylindricus and Rhabdammina sp.

Discussion

Several studies report a thin Quaternary sediment cover unconformably overlying Tertiary sediment in the general area of the DOMES sites (Riedel and Funnell, 1964; Johnson and Johnson, 1970; Moore, 1970; Johnson, 1972; Johnson and Parker, 1972; van Andel and others, 1973). Quaternary sediment thickness at nearby DSDP Sites (Fig. 1) ranges from a very thin veneer at Sites 161, 162, and 163 to more than 3 meters at Sites 159 and 160 (Dinkelman, 1973). Samples from DOMES Sites A and B (Tables 2 and 3) and nearby DSDP Sites 160 and 163 are similar to samples from most previous studies in that they contain abundant reworked Tertiary radiolarians in the Quaternary layer, with middle and late Eocene and early Miocene radiolarians being the dominant admixtures. Within a specific sample, the reworked fossils generally are a concentration of dissolution-resistant forms from different Eocene and Miocene radiolarian zones. In the DOMES cores, middle and late Eocene forms typically predominate. In contrast, Quaternary sediment at DOMES Site C and DSDP Site 159 generally contains no reworked Tertiary radiolarians.

Johnson (1972) presents a comprehensive discussion of the mechanisms that might explain the hiatuses in the sedimentary record and the abundance of reworked radiolarian fossils in equatorial Pacific surface sediments. He notes that, unlike Quaternary sediment, Tertiary outcrops and near-outcrops in the area typically show little or no mixing of faunas of different ages, and infers that seafloor erosion and redeposition took place mainly during the Pleistocene. Johnson suggests that bottom currents were intensified during the Pleistocene glacial stages, resulting in significant erosion of Tertiary deposits and redeposition of the sediment and associated fossils. During reworking, these fossils would be subjected to a greater degree of dissolution than those deposited during quieter depositional periods; the more dissolution-resistant specimens would tend to be concentrated, as is clearly the case in the DOMES cores. Moore (1970) has indicated that, even if initially present in small amounts, the dissolution-resistant radiolarians will eventually dominate the assemblage, given enough time for selective dissolution.

An alternate explanation for the abundant reworking is vertical mixing by bioturbation. Although this mechanism may be important in some instances, it probably has less impact in the DOMES area of the equatorial Pacific than erosion and redeposition by Pleistocene bottom currents. Johnson and Johnson (1970, Table 1) list reworked radiolarians in 23 cores from the central Pacific. Their data demonstrate that Eocene radiolarians are particularly abundant in the Quaternary sediments but are commonly absent in the underlying Miocene intervals. If bioturbation were a dominant mechanism causing the mixing, then Eocene fossils should be present throughout the cores. Quaternary sediment at DSDP Sites 160 and 163 (Fig. 1) contains abundant reworked radiolarians as old as late Eocene and early Oligocene. At both sites, the Quaternary is underlain by tens of meters of late Oligocene and Miocene sediment. It is unlikely that the abundant reworked radiolarians now in the surficial sediments were bioturbated through this thick sequence of Oligocene and Miocene sediment.

The thinness of the Quaternary sediment layer present in the general area of the DOMES cores may be significant with regard to the origin and distribution of manganese nodules. Glasby (1978) reports that nearly 50% of manganese nodules from DSDP cores are from Pliocene and Pleistocene sediment and only 43% are from the older Tertiary. He suggests that manganese nodules in younger deep-sea sediment are more abundant where rates of sediment accumulation are low, as high accumulation rates cause nodules to be buried and they cease to grow. He further suggests that low accumulation rates are due to the increase in bottom-current velocities that began approximately 3.5 m.y. ago. An increase in bottom-current velocities would contribute to low net sediment accumulation rates, at least in some areas, and provide a favorable environment for the formation of manganese nodules.

The average concentration of manganese nodules at DOMES Site

C (Fig. 1) is greater than at Sites A and B (Piper, Cannon, and Leong, 1977) and may reflect a slightly lower sediment accumulation rate. Theyer (1977) determined the Quaternary rate of sediment accumulation at Site C to be approximately 0.1 $cm/10^3$ years and 0.1 to 0.3 $cm/10^3$ years at Sites A and B. These rates are in general agreement with those of Lisitzin (1972) for these areas. In addition to erosion and dissolution, biologic productivity also influences the accumulation rate of deep-sea sediment. A belt of high organic productivity lies along the equator and provides a high rate of biogenic input in this area. Biogenic silica increases relative to carbonate at about 5° N, and a decrease in productivity contributes to a decrease in the rate of sediment accumulation at latitudes higher than 10° to 15° N. A lower rate of biogenic sediment accumulation is therefore to be expected at Site C because it is located farther to the north (at 15° N), away from the equatorial belt of high productivity, than Sites A and B.

Acknowledgments

We thank David Bukry (USGS) for providing calcareous nannofossil age determinations; P. Lineberger (Univ. of Hawaii) for assisting with the radiolarian analyses; Harry Cook (USGS) and James Gardner (USGS) for reviewing the manuscript; and James Ingle (Stanford Univ.) and Helen Tappan (Univ. of Calif.) for reviewing an earlier version of the manuscript. Contribution number 952 of the Hawaii Institute of Geophysics.

References

Bukry, David, (1973) Coccolith stratigraphy, eastern equatorial Pacific Leg 16, Deep Sea Drilling Project, In: van Andel, T. H., Heath, G. R., _et al_., Initial Reports Deep Sea Drilling Project, 16, Washington D.C., U.S. Govt Printing Office, 653-711.

Dinkelman, M. G., (1973) Radiolarian stratigraphy: Leg 16, Deep Sea Drilling Project, In: van Andel, T. H., Heath, G. R., _et al_., Initial Reports Deep Sea Drilling Project, 16, Washington, D.C., U.S. Govt Printing Office, 747-813.

Glasby, G. P., (1978) Deep-sea manganese nodules in the stratigraphic record: evidence from DSDP cores, _Marine Geology_ 28, 51-64.

Johnson, D. A., (1972) Ocean-floor erosion in the equatorial Pacific, _Geol. Soc. Am. Bull._ 83, 3121-3144.

Johnson, D. A., and Johnson, T. C., (1970) Sediment redistribution by bottom currents in the Central Pacific, Deep-sea Res. 17, 157-169.

Johnson, D. A., and Knoll, A. H., (1975) Absolute ages of Quaternary radiolarian datum levels in the equatorial Pacific, Quatern. Res. 5, 99-110.

Johnson, D. A., and Parker, F. L., (1972), Tertiary radiolaria and foraminifera from the equatorial Pacific, Micropaleontology 18, 129-143.

Lisitzin, A. P., (1972) Sedimentation in the World Ocean, Soc. Econ. Paleon. and Mineral., Spec. Publ. No. 17, 218 p.

Loeblich, A. R., and Tappan, H., (1964) Treatise on Invertebrate Paleontology, R. C. Moore (ed.), Pt. C, Protista, Lawrence, Kansas, Geol. Soc. Am. and Univ. Kansas Press, 900 p.

Moore, T. C., (1970) Abyssal hills in the central equatorial Pacific: sedimentation and stratigraphy, Deep-sea Res. 17, 573-593.

Nigrini, C. A., (1971) Radiolarian zones in the Quaternary of the equatorial Pacific Ocean, In: The Micropaleontology of the Oceans (Funnell, B. M. and Riedel, W. R., eds.), Cambridge Univ. Press, 443-461.

Parker, F. L., (1964) Foraminifera from the experimental Mohole drilling near Guadalupe Island, Mexico, Jour. Paleo. 38, no. 4, 617-636.

Postuma, J. A., (1971) Manual of Planktonic Foraminifera, Amsterdam, Elsevier Publ. Co., 420 p.

Piper, D. Z., Cannon, W., and Leong, K., (1977) Composition and abundance of ferromanganese nodules at DOMES Sites A, B, and C: relationship with bathymetry and stratigraphy, In: Piper, D. Z. et al., (compilers) Deep Ocean Environmental Study: Geology and Geochemistry of DOMES Sites A, B, and C, equatorial North Pacific, U.S. Geol. Survey Open-file Rept. 77-778, 217-266.

Riedel, W. R., and Funnell, B., (1964) Tertiary sediment cores and microfossils from the Pacific Ocean floor, Quart. Geol. Soc., London, 120, 305-368.

Theyer, F., (1977) Micropaleontological dating of DOMES Project box cores from Test Areas A and B, Tropical Pacific, In: Piper, D. Z. et al., (compilers) Deep Ocean Environmental Study: Geology and Geochemistry of DOMES Sites A, B, and C, equatorial

North Pacific, U.S. Geol. Survey Open-file Rept. 77-778, 179-194.

van Andel, T. H., Heath, G. R. _et al_., (1973) Initial Reports of the Deep Sea Drilling Project 16, Washington, D.C., U.S. Govt. Printing Office, 949 p.

MINERALOGY AND DIAGENESIS OF SURFACE SEDIMENTS FROM DOMES AREAS A, B, AND C

James R. Hein, C. Robin Ross, and Elaine Alexander
Pacific-Arctic Branch of Marine Geology
U.S. Geological Survey
345 Middlefield Road
Menlo Park, California 94025

and

Hsueh-Wen Yeh
Hawaii Institute of Geophysics
University of Hawaii
Honolulu, Hawaii 96822

Abstract

Box cores were collected between 10 to 15°N latitude and 126 to 151°W longitude in the North Pacific. Sediments are primarily siliceous fossil rich mud, but span the range from siliceous ooze to red clay; there is also one bed of early Miocene nannofossil ooze. Terrigenous debris consist of quartz, feldspar, illite, and chlorite + kaolinite; andesitic volcanic glass shards, biotite, and other associated volcanic materials are also present in minor amounts and probably originated from Central and South American explosive volcanism. In general, terrigenous minerals decrease in abundance seaward and in pre-Quaternary deposits.

Siliceous microfossils -- radiolarians, diatoms, and silicoflagellates -- are the most abundant biogenic component, up to 50% by volume. These microfossils undergo extensive dissolution within the upper one-half meter of sediment. Spicules, the most robust siliceous microfossil, are the only forms remaining in the lower parts of most cores. Superimposed on the overall gradual decrease in siliceous fossils with depth, which is the result of dissolution, is a cyclic fluctuation in siliceous debris; the cyclic variation may result from changes in productivity in surface waters. Previous studies show that dissolution of biogenic opaline silica releases Si, Al, Cu, Ni, and perhaps Mn and Zn which in DOMES deposits are rapidly incorporated into authigenic smectite, zeolites, and ferromanganese nodules. Other biogenic components are nannofossils, fragments of planktic and benthic foraminifers, fish debris, fragments of worm tubes, and fecal pellets.

Authigenic minerals are important sediment builders in the equatorial North Pacific. Authigenic Fe-rich smectite, on the average, makes up 10% of the sediment. Minor authigenic barite, apatite, hematite, clinoptilolite, phillipsite, and perhaps atacamite, also occur in DOMES cores. The smectite most likely forms from the chemical combination of Fe oxyhydroxide and Si at low temperatures. The Fe oxyhydroxide is probably derived from volcanic activity on the East Pacific Rise and is dispersed in colloidal form throughout the Pacific. We suggest that the Si, Al, Cu, Ni, and Zn in the smectite are derived from the dissolution of biogenic silica and from displacement of minor elements adsorbed on Fe-Mn oxyhydroxides.

Formation of Fe-rich smectite may be the mechanism that fractionates Fe and Mn (combined in oxyhydroxides) in the sediment and nodule phases respectively. Silica may be the controlling constituent. In areas of siliceous ooze, Fe smectite forms relatively rapidly and, consequently, Mn nodules in the same deposits have high Mn/Fe ratios. In contrast, nodules formed in red clay areas have low Mn/Fe ratios possibly because of the relatively limited silica supply.

Locally abundant metalliferous sediment (deposits enriched in Fe, Mn, Cu, and Ni) probably formed by the extensive dissolution of biogenic debris. Dissolution would leave a metal-rich residue whose position may be useful for paleo-oceanographic studies.

Minor authigenic atacamite, barite, apatite, hematite, clinoptilolite, and phillipsite also occur in DOMES cores.

Introduction

This study is part of a multidisciplinary, multi-institutional investigation of the environment of formation of manganese nodules in the equatorial North Pacific (Fig. 1). Our part in the DOMES (Deep Ocean Mining and Environmental Study) project has been to determine the mineralogy of the detrital, authigenic, volcanic, and biogenic components of the sediment, and to describe changes in these components that are the result of diagenesis. In addition, we discuss the mineralogic data in terms of regional sedimentology as presented by previous investigators and discuss the role that sediment mineralogy may play in nodule chemistry.

Cores were collected from three areas centered at 9°N, 151°W (Area A); 12°N, 138.5°W (Area B); and 15°N, 126°W (Area C) (Fig. 1). Each area measures 2 degrees on a side. Areas A, B, and C

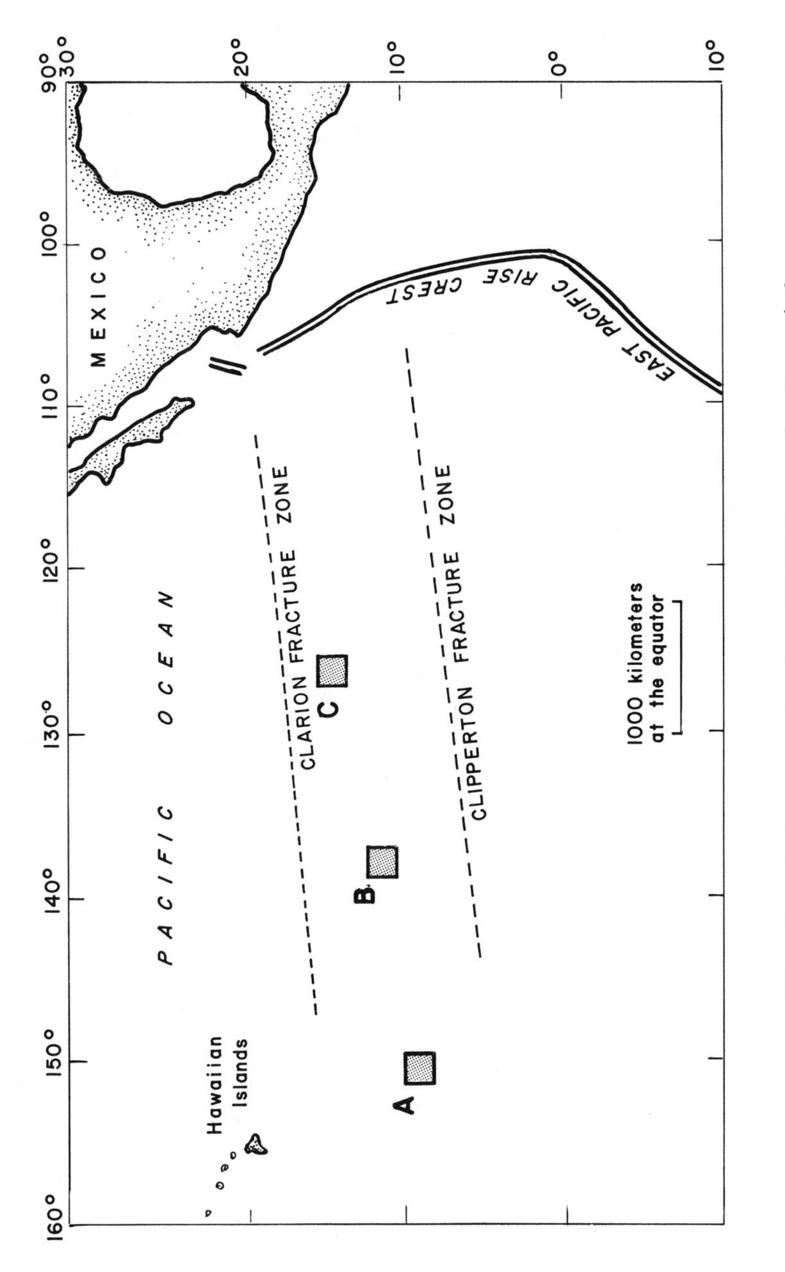

Fig. 1. Index map showing DOMES Areas A, B, and C.

are within a larger area (bounded by 5°N to 20°N latitude and 110°W to 180°W longitude) that is of commercial interest for mining of deep-sea nodules. Area A is farthest seaward and is separated from Area C by more than 2500 km.

Within each 2° area, a number of stations were occupied and several box cores were taken at each station. Box cores measure 50 cm x 50 cm, and 20 to 50 cm of sediment was penetrated and recovered at each station. Stations 46 through 50 are in Area A, 51 through 55 in Area B, and 11 through 27 in Area C. Every 2-cm interval of the box core was sampled at stations 47, 52, 16, 18B, 10, and 25, and other box cores were sampled at greater intervals.

Methods

Sediment components were studied using X-ray diffraction, grain mounts, smear slides, scanning electron microscopy (SEM), energy dispersive analysis by X-rays (EDAX), and refractive index oils (RI). Smear slides of bulk sediment and grain mounts of the (>44 μm) fraction were studied with the petrographic and binocular microscopes respectively. X-ray diffractograms of the >2 μm size fraction, the <2 μm size fraction (clay mineralogy), and selected <0.1 μm, <1 μm, and bulk samples were analyzed. Procedures used for clay mineralogy, SEM, EDAX, and RI analyses are discussed by Hein and Scholl (1978a), Hein et al. (1976), Hein and Jones (1977), and Hein et al. (in press). The EDAX technique is not a quantitative analytical procedure; however, the relative abundance of elements can be determined by EDAX and their element ratios used to characterize different samples.

Sediment Classification

DOMES sediments are classified according to DSDP (Deep Sea Drilling Project) recommendations (Burns, Andrews, et al., 1973). The sediment classification in Appendix 2 is based on X-ray diffractograms, smear slides, and clay/sand-silt ratios. The major constituent is listed last and represents more than 25 percent of the sediment, for example, "siliceous mud" means that both siliceous microfossils and mud (fine-grained quartz, feldspar, and clay minerals) are major sediment builders. Sediment particles of decreasing importance are listed progressively earlier in the name; other terms used are "major", greater than 25%, "rich", 10-25%, and "bearing", 2-10%.

DOMES deposits span the range from mostly siliceous ooze or siliceous fossil-rich clay to red clay; one sample is calcareous ooze. This classification is consistent with the location of the deposits within a narrow belt of siliceous biogenic deposits that occur between zones of thick calcareous ooze found at the equator and red clay in the North Pacific. Area C is located near the boundary between red clay and siliceous deposits.

Results

Clay Mineralogy

Smectite dominates the clay mineral suite in Areas A and B, and illite is of secondary importance (Table 1). In Area C, illite is dominant in the upper deposits, but in the lower part of most cores smectite dominates the suite. Smectite also increases with depth at stations 47, 48, and 55 in Areas A and B. Also, in general, smectite increases in abundance from Area C to Area A.

The chemistry and oxygen isotopic composition of clay minerals are discussed by Hein et al. (in press) and, therefore, will only be summarized here. As determined by atomic absorption analysis, the smectite is an Fe-rich montmorillonite containing significant amounts of Cu, Zn, Mn, Ti, and to a lesser extent Ni. The structural formula for this smectite, derived by the technique of Ross and Hendricks (1945), is
$Ca_{.14}K_{.22}Na_{.41})(Ni_{.004}Cu_{.01}Zn_{.01}Mn_{.03}Ti_{.04}Mg_{.59}Fe^{3+}_{.53}Al_{.75})$

$(Al_{.52}Si_{3.48})O_{10}(OH)_2$ (Hein et al., in press). The oxygen isotopic composition ($\delta^{18}O$) of smectite in the <1μm size fraction is about +29.6‰, a value that determines quite clearly that the smectite is authigenic and must have formed near the sediment-water interface in a deep ocean environment (Hein et al., in press).

The proportion of illite interlayers in the S/I mixed layer phase (10 to 35%) reported for the DOMES samples by Hein and Jones (1977) may be too large. The standard techniques used for measuring the amount of mixed layers in the S/I phase, those of Perry and Hower (1970) and Reynolds and Hower (1970), may not give true values if the smectites are very fine grained and formed authigenically (I. Srodoń, Polish Academy of Sciences, and Eric Eslinger, West Georgia College personal communications, 1978; see also Srodon and Hower, 1976), as are the smectites in DOMES cores. Thus, we infer that few if any illite interlayers occur on the basis of the preliminary work that has been completed on clay minerals similar to those found in DOMES deposits, that is, smectite formed authigenically and relatively rapidly at low temperatures.

Coarse Fraction (>44 μm)

Siliceous microfossils. Siliceous microfossils dominate the coarse fraction of all Area A and B samples (75% to greater than 99%) as well as the upper deposits in Area C (Appendix 1). Radiolarians, diatoms, spicules, and rare silicoflagellates make up the siliceous microfossil assemblage (Fig. 2a). There is a general and relatively gradual increase with depth (Appendix 1) in the proportion of siliceous sponge spicules and the rarer

TABLE 1

Semiquantitative clay mineralogy of samples from DOMES areas A, B, and C. Percentages represent the relative amount of each clay mineral in the less than 2 μm size fraction and for each sample are set to equal 100%

Sample No.*	K+C	K	C	I	S	Sample No.	K+C	K	C	I	S
					Area A						
46-1-1	16	8	9	33	51	47-10-18	10	5	6	16	74
46-1-7	17	7	9	33	50	47-10-19	10	5	6	15	75
46-1-12	17	7	10	32	51	47-10-20	9	4	5	14	77
46-1-15	17	7	10	32	51	47-10-21	11	5	6	17	73
46-1-20	16	7	9	30	54	47-10-22	10	5	6	15	75
47-10-1	17	7	10	30	53	47-10-23	9	4	5	16	75
47-10-2	16	6	9	32	53	47-10-24	10	4	6	16	74
47-10-3	16	6	10	28	56	48-19-1	12	5	7	24	64
47-10-4	15	7	8	28	57	48-19-7	11	5	6	18	71
47-10-5	17	7	10	29	54	48-19-12	7	3	4	7	86
47-10-6	17	7	10	29	54	48-19-15	5	2	3	8	87
47-10-7	14	6	8	27	59	48-19-19	5	2	3	6	89
47-10-8	16	6	9	27	57	49-25-1	17	7	10	30	53
47-10-9	16	7	9	31	54	49-25-7	15	6	9	29	56
47-10-10	15	6	8	26	60	49-25-12	16	6	10	34	50
47-10-11	14	6	8	23	64	49-25-16	18	7	11	35	48
47-10-12	15	6	8	22	64	49-25-21	16	7	9	37	47
47-10-13	14	6	8	23	62	50-28-1	19	8	11	39	42
47-10-14	10	5	6	17	73	50-28-11	17	8	9	32	52
47-10-15	11	5	6	18	72	50-28-16	17	7	10	32	51
47-10-16	11	5	6	17	72	50-28-20	17	7	10	32	51
47-10-17	11	5	6	15	74	50-28-24	17	8	9	34	49
					Area B						
51-33-1	18	?	?	37	45	52-42-13	21	9	12	43	37
51-33-6	12	12	0	8	79?	52-42-14	22	9	12	39	39
51-33-9	8	?	?	13	80	52-42-15	20	8	12	41	38
51-33-11	18?	18?	0?	14?	69?	52-42-16	20	9	12	40	39
52-42-1	20	9	11	43	37	52-42-17	22	9	13	41	38
52-42-2	19	9	10	42	39	52-42-18	18	7	11	37	45
52-42-3	19	8	10	43	39	52-42-19	21	9	12	40	39
52-42-4	20	8	12	40	39	52-42-20	18	8	10	40	42
52-42-5	21	8	12	43	37	52-42-21	19	8	11	41	41
52-42-6	21	9	12	44	35	52-42-22	19	8	11	38	43
52-42-7	19	8	11	42	39	52-42-23	19	8	12	41	40
52-42-8	17	7	10	44	39	52-42-24	17	7	10	41	42
52-42-9	19	8	11	45	36	52-42-25	19	8	11	38	44
52-42-10	21	9	13	45	34	52-42-26	19	8	11	38	44
52-42-11	20	9	11	41	39	53-45-1	21	9	12	39	40
52-42-12	19	9	11	41	40	53-45-10	20	9	11	38	42

TABLE 1 (continued)

Sample No.	K+C	K	C	I	S	Sample No.	K+C	K	C	I	S
					Area B (continued)						
53-45-14	20	9	12	36	44	54-51-21	19	8	11	41	40
53-45-17	19	7	12	37	44	55-53-1	17	5	12	30	53
53-45-22	20	9	11	38	42	55-53-7	10	?	?	14	76
54-51-1	20	9	11	37	44	55-53-11	8	4	4	10	82
54-51-9	20	8	11	44	36	55-53-14	9	5	4	8	83
54-51-14	19	8	11	42	39	55-53-17	7	4	3	11	82
54-51-17	18	8	10	37	46						
					Area C						
11-53-1	23	10	13	50	27	20-27-1	23	10	13	47	30
11-53-8	22	9	13	45	32	20-27-2	24	11	13	56	20
11-53-9	23	10	13	47	30	20-27-3	27	10	17	49	24
11-53-10	23	10	13	42	35	20-27-4	22	9	13	48	29
11-53-14	21	9	12	42	38	20-27-5	22	9	13	52	26
15B-50-1	26	11	15	44	30	20-27-6	22	10	13	40	38
15B-50-8	25	11	14	44	31	20-27-7	28	11	16	42	30
15B-50-9	23	10	13	45	32	20-27-8	26	11	15	41	33
15B-50-10	25	11	14	41	34	23B-30-1	24	10	14	44	32
15B-50-13	23	10	13	44	34	23B-30-8	35	14	21	35	29
16B-49-1	25	11	15	47	27	23B-30-9	25	11	15	45	30
16B-49-2	26	11	15	43	31	23B-30-10	24	10	13	44	32
16B-49-3	28	12	17	41	31	23B-30-15	20	9	10	34	46
16B-49-4	25	11	15	41	34	24B-29-1	28	12	15	44	28
16B-49-5	24	11	13	39	37	24B-29-3	25	10	15	41	34
16B-49-6	22	10	12	38	40	24B-29-5	24	10	14	46	30
16B-49-7	21	9	12	27	52	24B-29-6	25	11	14	47	28
16B-49-8	19	8	11	25	57	24B-29-7	23	10	13	45	32
16B-49-9	18	8	9	26	57	24B-29-8	22	10	12	38	40
16B-49-10	19	9	10	27	54	24B-29-9	26	11	16	39	35
18A-36-1	26	11	15	51	23	25-51-1	33	14	19	53	14
18A-36-3	24	11	14	45	31	25-51-2	33	15	18	48	19
18A-36-4	27	11	15	48	25	27-52-1	28	13	16	50	22
18A-36-6	25	10	15	39	36	27-52-8	27	13	14	49	24
18A-36-8	20	9	11	22	58	27-52-9	26	12	14	48	26
18B-37-1	26	11	16	48	26	27-52-10	26	12	13	48	26
18B-37-2	25	11	14	46	29	27-52-12	25	11	14	44	31
18B-37-3	25	11	14	43	32						
18B-37-4	25	10	15	46	28						
18B-37-5	22	10	12	38	40						
18B-37-6	22	9	13	41	37						
18B-37-7	18	8	11	30	51						
18B-37-8	7	3	4	16	77						
18B-37-9	13	6	7	26	61						

TABLE 1 (continued)

*The first number in the sample number is the station, the next is the box core number, and the last is the interval in the core. Each interval is 2 cm; 1 is 0-2 cm, 2 is 2-4 cm, and so on.

K, kaolinite; C, chlorite; I, illite; S, smectite.

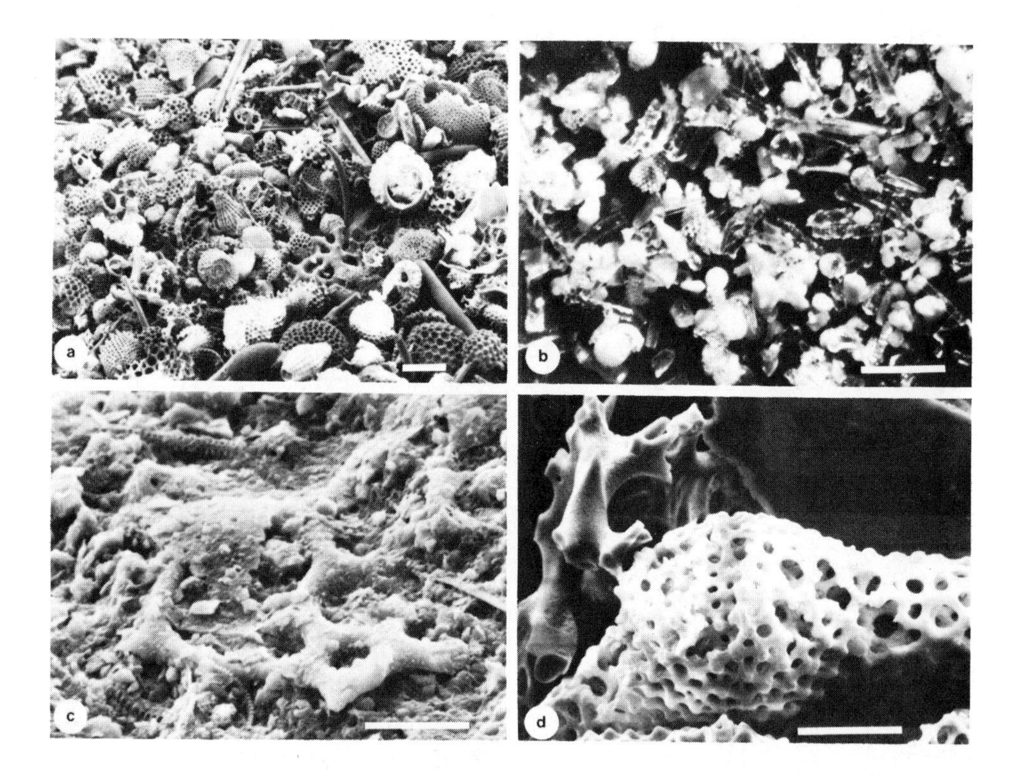

Fig. 2. DOMES siliceous biogenic debris. (a) SEM, >44 μm size fraction, 46-1-20, note many fragmented tests but only mild dissolution. (b) Photomicrograph grain mount, 47-10-24, extensive dissolution. Compare number of spicules relative to other microfossils with (a). (c) SEM, untreated sample 53-45-17, advanced dissolution of radiolaria and diatoms. (d) SEM, close-up of (a) showing mild dissolution; note enlargement of pore openings. Scale bars represent 100, 250, 10 and 25 μm respectively.

orosphaeride fragments, which resist solution (Johnson, 1974; Hein et al., 1978b), relative to other siliceous microfossils (Appendix; see also Fig. 2b). Because siliceous debris is the most abundant coarse-fraction material, the relative increase in spicules due to dissolution of the radiolarians and diatoms (Figs. 2c and 2d) is accompanied by a decrease in the coarse fraction (Appendix 1, Fig. 3). Superimposed on the general overall decrease in coarse-fraction material with depth is a fine-scale fluctuation in the amount of coarse debris (Appendix 1, Fig. 3). This fluctuation, on the scale of 2 to 8 cm, is well developed at Station 52 (Appendix 1, Fig. 3) and is also evident at Stations 47, 54, and 18A. The time needed to deposit the 2 to 8 cm layers is about 20,000 to 80,000 years (Lisitzin, 1972).

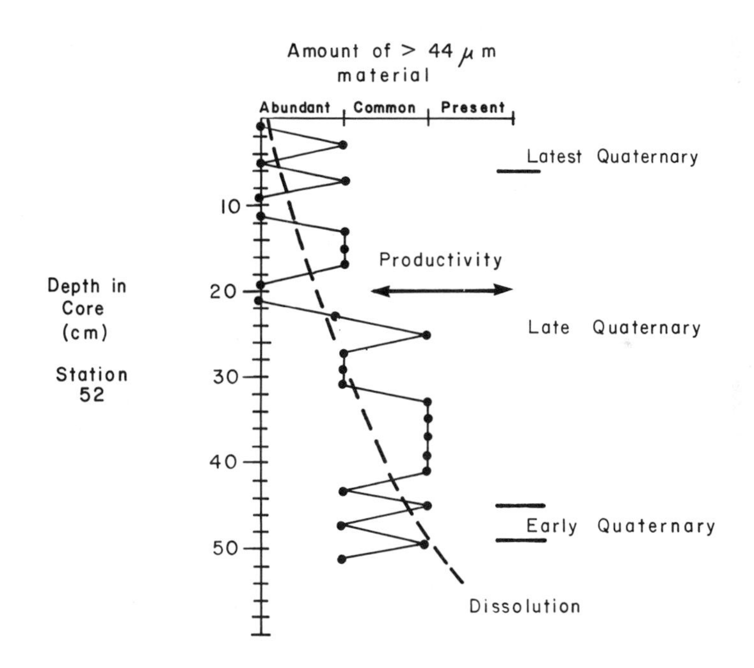

Fig. 3. Qualitative distribution of siliceous microfossils with depth at DOMES station 52. Although the graph is constructed with only three arbitrary abundance grades, the character of the deposits is an overall gradual decrease in the amount of >44 μm material (siliceous fossils). This overall decrease is the result of dissolution and is marked by the dashed line. The cyclic variation in the amount of >44 μm material may result from either changes in the productivity of siliceous organisms in surface waters or from dissolution (see text).

Flat plates. Flat plates are translucent to transparent, brown, mica-like plates, some of which are bent or curved and most of which commonly have a sugary to glassy texture (Fig. 4). They have no single origin, and sources of many plates are still speculative. Flat plates measure up to several hundred micrometers and can constitute a major part of the coarse fraction (Appendix 1, Figs. 4 and 5a). At least three varieties of plates can be distinguished by SEM and EDAX. These three varieties are very difficult to tell apart with the binocular microscope and, therefore, are listed together in Appendix 1. The first kind of flat plate (Fig. 4a, 4b) is made of bits (<3 μm) of siliceous fossil debris and clay minerals. An EDAX spectrum reveals the

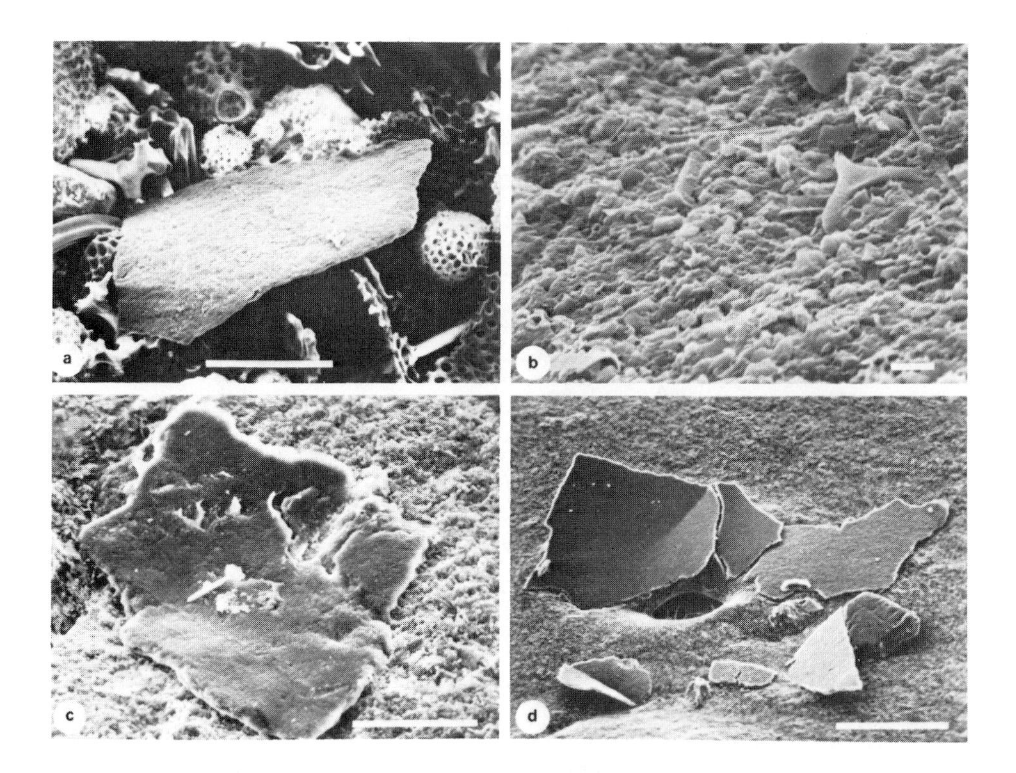

Fig. 4. Scanning electron micrographs of flat plates from the >44 μm size fraction. (a) 46-1-20, fragment of worm tube. (b) detail of (a); note that the worm tube fragment is made up of bits of siliceous microfossil debris and clay minerals. (c) 18A-36-1, flat plate of unknown origin. (d) 18A-36-4, very large, thin, translucent flat plate, maybe volcanic glass (Fig. 5a). Scale bars represent 250 and 100 μm respectively.

presence of Si, Al, Mg, Fe, K, Ca, and Ti (Si/Al = 3.27), in decreasing order of peak heights. This variety of plate is commonly large and bowed or curved, possibly from drying the samples. These pieces may be worm tube linings that were fragmented during sample preparation. Worm tube linings are fine-grained sediment particles cemented or bound perhaps by an organic glue. The second variety of flat plate appears thicker and is not

Fig. 5. (a) Photomicrograph, grain mount, 49-25-12, siliceous microfossils, flat plates with sugary texture, and two translucent flat plates (bottom center and center of northeast quadrant); microfossils are visible through these two plates. (b) SEM, typical fish debris in DOMES deposits, 18A-36-1. Scale bars represent 250 and 100 μm respectively.

composed of smaller particles (Fig. 4c). EDAX shows the presence, in decreasing order of peak height, of Si, Al, Fe, K, Mg, Na, and Ti (Si/Al=2.66). This variety is rare, and we do not know the origin. The third variety of flat plate is very thin but quite large and is commonly translucent (Fig. 4d and 5a). An EDAX spectrum shows that the grain in Fig. 4d contains Si, Al, Fe, K, Mg, Na, and Ti (Si/Al=3.13). These plates are also amorphous to X-rays. We suggest that they are hydrated, andesitic volcanic glass. The Si/Al ratio of the average andesite is 3.05, as compared to 4.84 and 2.77 for average rhyolite and basalt respectively (Turner and Verhoogen, 1960). Most samples apparently contain either predominantly the third variety or the first variety.

Fecal pellets. The amounts of fecal pellets listed in Appendix 1 are minimum values because the technique used to isolate the >44 μm size fraction tended to break them up. Selected samples prepared in a more time-consuming but gentler manner (for example, 18A, Appendix 1) showed fecal pellets to be the overwhelming coarse-fraction particle in contrast to the constituents listed in Appendix 1. Because abundant fecal pellets mask other grain types it is advantageous to disaggregate them so that minor sediment constituents valuable for determining source areas and for studying diagenesis can be identified.

Fish debris. Fish debris is ubiquitous and makes up a major part of the coarse fraction in some samples (Fig. 5b, Appendix 1). EDAX spectra show that Ca, P, and traces of Fe are present. Prismatic crystals of apatite, seen in smear slides, suggest that some fish debris may have undergone dissolution and precipitation.

Volanic glass. Traces of volcanic glass shards, of andesitic to rhyolitic compositions as determined by refractive index, are found in all DOMES cores. They are most evident in Area C, partly because Area C is closer to source areas in Central and South America. Shards are transparent to translucent and colorless to amber. Shards appear unaltered.

On the basis of RI, conchoidal fracture, X-ray amorphous character, and general appearance, Hein and Jones (1977) reported that abundant volcanic glass spheres occur in DOMES deposits (Fig. 6a). Because subsequent EDAX, microprobe, and atomic-absorption work gave element oxide sums as low as 30%, however, we searched for another explanation for these glassy spheres. We found that they are beads of cation-exchange resin from the deionizing water system used to prepare the samples and coincidentally have many of the physical properties of other amorphous substances, such as volcanic glass. The volcanic glass listed in Appendix 1 represents andesitic to rhyolitic glass shards only. Palagonite, which is rare in DOMES deposits, is listed in the "Others" column of Appendix 1.

Biotite. Traces of biotite occur in most Area C and many Area A and B samples. Biotite is noticeably altered only at Station 55 in Area B. Biotite maybe derived from the Central and

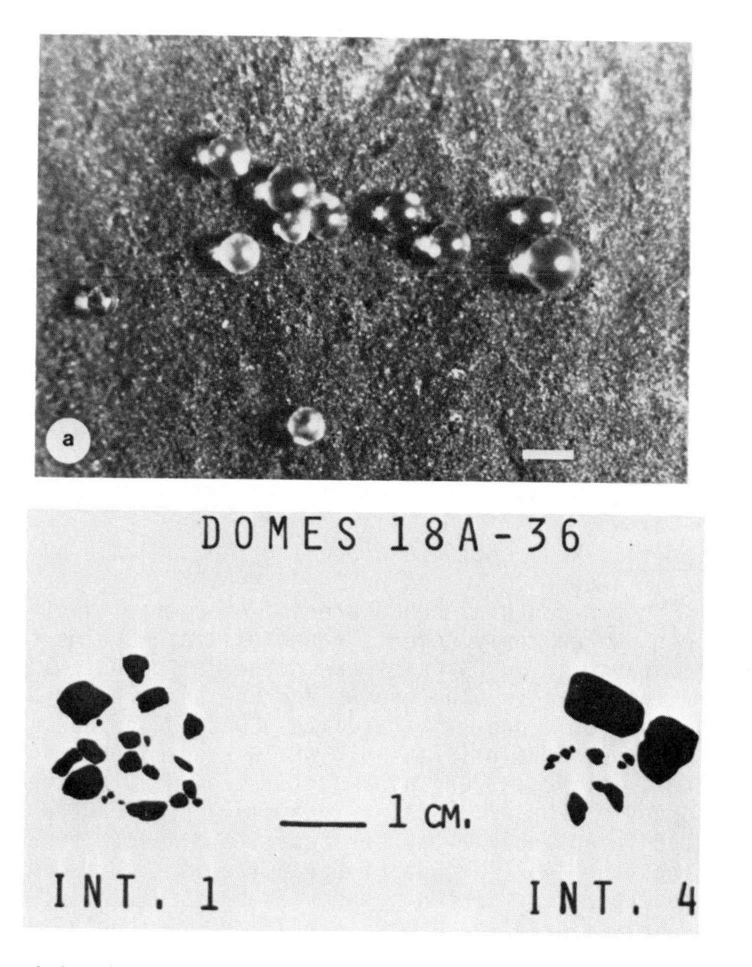

Fig. 6. (a) Photomicrograph grain mount, typical glassy spheres that occur in most samples, were found to be contamination by cation-exchange resin beads from the deionizing water system used to prepare the samples. Scale bar represents 250 μm. (b) Silhouettes of the larger ferromanganese micronodules isolated from 18A-36-1 and -4.

South American stratovolcanoes along with volcanic glass. Donnelly (1976) found that biotite is the dominant mafic mineral in DSDP Sites 320 and 321 (offshore Peru) ash beds.

Quartz and feldspar. Quartz is always more abundant than plagioclase. They increase in abundance (as do most other minerals) as siliceous debris decreases and co-vary in abundance. Heath (1969) showed that in the north equatorial Pacific both are eolian in origin. Rare, larger quartz (to 0.18 mm) and volcanic glass (to 0.1 mm) grains, however, may have been transported by other means.

Opaque Minerals. This category includes magnetite, hematite, and amorphous Fe hydroxides. The surface of most magnetite crystals is altered to hematite. These grains, however, are still magnetic and attract an unmagnetized steel probe. Most grains listed in Appendix 1 are hematite or at least hematite-stained ferromanganese grains (see below). X-rays show no reflections for this material, so we assume that these earthy, porous, and friable grains are mostly amorphous Fe hyroxides with only traces of hematite that give the red color.

Ferromanganese grains. These include micronodules (Fig. 6b) and fine irregular particles that are black or brown to yellow and porous. They make up the major part of deposits at Station 18A and occur in trace amounts at other stations. The particles may be derived from the abrasion of Mn nodules or from coalescence of RSO (discussed below).

Atacamite. We identified atacamite ($Cu_2(OH)_3Cl$) by X-ray diffraction in a core collected by Deep Sea Ventures, Inc. immediately to the south of DOMES areas. Atacamite is turquoise to blue-green and commonly cements nannofossils. Grains of similar appearance in DOMES cores are noted in Appendix 1. A preliminary look at five blue grains and one sea-green grain with EDAX showed that only one was atacamite (EDAX showed Cu, Cl; Fig. 7a). The sea-green grain is an unknown Zn-rich mineral (EDAX showed Si, Al, Zn, Fe, K, Ca, Na, Mg, Ti; Fig. 7b), and the other four were organic debris. Thus, only a small part of the blue grains listed in Appendix 1 may be atacamite. Also, it cannot be ruled out that the small amount of atacamite is a contaminant, formed by seawater corrosion of copper or brass equipment. On the other hand, this mineral has been identified previously in Red Sea metalliferous deposits (Mossman and Heffernan, 1978), and it may also occur naturally in DOMES deposits.

Others. Rutile, zeolites, chlorite, pyribole, and palagonite occur rarely in the coarse fraction. Fragments of agglutinated benthic foraminifers, and terrestrial plant fibers that are commonly coated with or replaced by Fe or Mn oxides occur in many places.

Smear Slides

Siliceous microfossils and mud (fine-grained quartz, feldspar, and clay minerals) make up the bulk of DOMES deposits. Locally RSO (defined below), volcanic glass, calcite, and zeolites are significant components of the sediment (Appendix 2). Fish debris and aggregates are consistently present in small amounts. The fine-grained fraction (<2 μm) makes up between 40% and 95% of the sediment, but is commonly about 65-75% (see also Cock et al., 1977). Of this fraction clay minerals make up 1/3 to 2/3 and

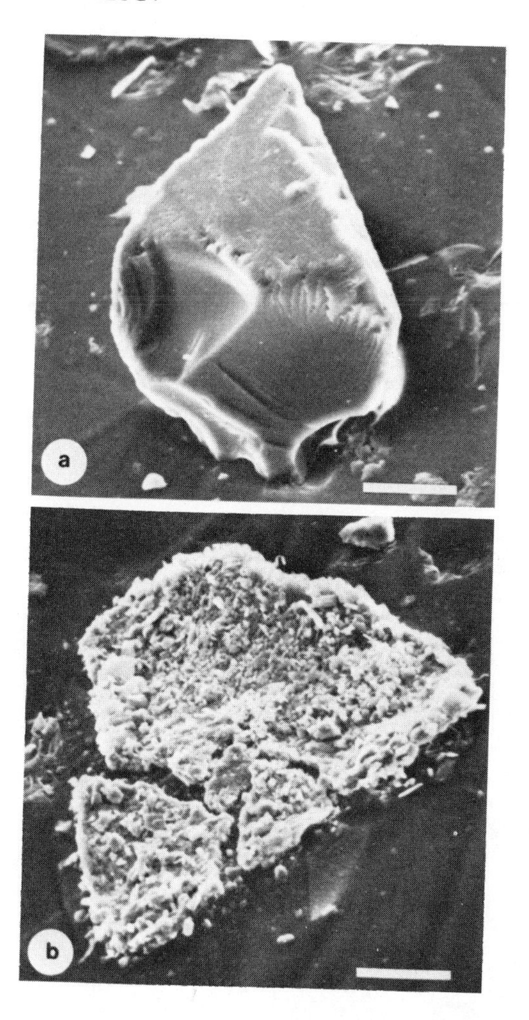

Fig. 7. (a) Scanning electron micrographs DOMES atacamite grain, EDAX registered Cu and Cl. (b) Unknown Zn-rich mineral, sea-green in color. Large grain is composed of minute clay-like flakes. Scale bars represent 30 μm.

quartz and feldspar the rest. As mentioned, quartz is more abundant than feldspar.

Siliceous microfossils. The loss of siliceous debris by dissolution is as evident in smear slides as it is in the grain mounts. Sponge spicules are the only siliceous fossil remains at the base of many cores. Spicules also decrease in size with depth of burial. Commonly, radiolarians and diatoms are severely fragmented, most likely by deposit feeders. Some frustules and tests are coated and filled with (and replaced by ?) smectite and

Fe-Mn oxyhydroxides.

Volcanic glass. Volcanic glass shards >10 µm in size are common in DOMES deposits. It is likely, however, that volcanic glass is over-estimated in Appendix 2 because fragments of siliceous microfossils are difficult to distinguish from fine-grained glass shards. Thus, the samples described as glass-bearing (Appendix 2) may actually contain less than 2% volcanic glass. Glass shards appear unaltered.

RSO. The term RSO is adopted from Quilty et al. (1976, p. 787) to describe "red-brown to yellow-brown, semi-opaque oxides". As with Quilty et al. (1976) and Cronan (1973, 1974), RSO from DOMES deposits occur in three general forms (Fig. 8): (1) tiny,

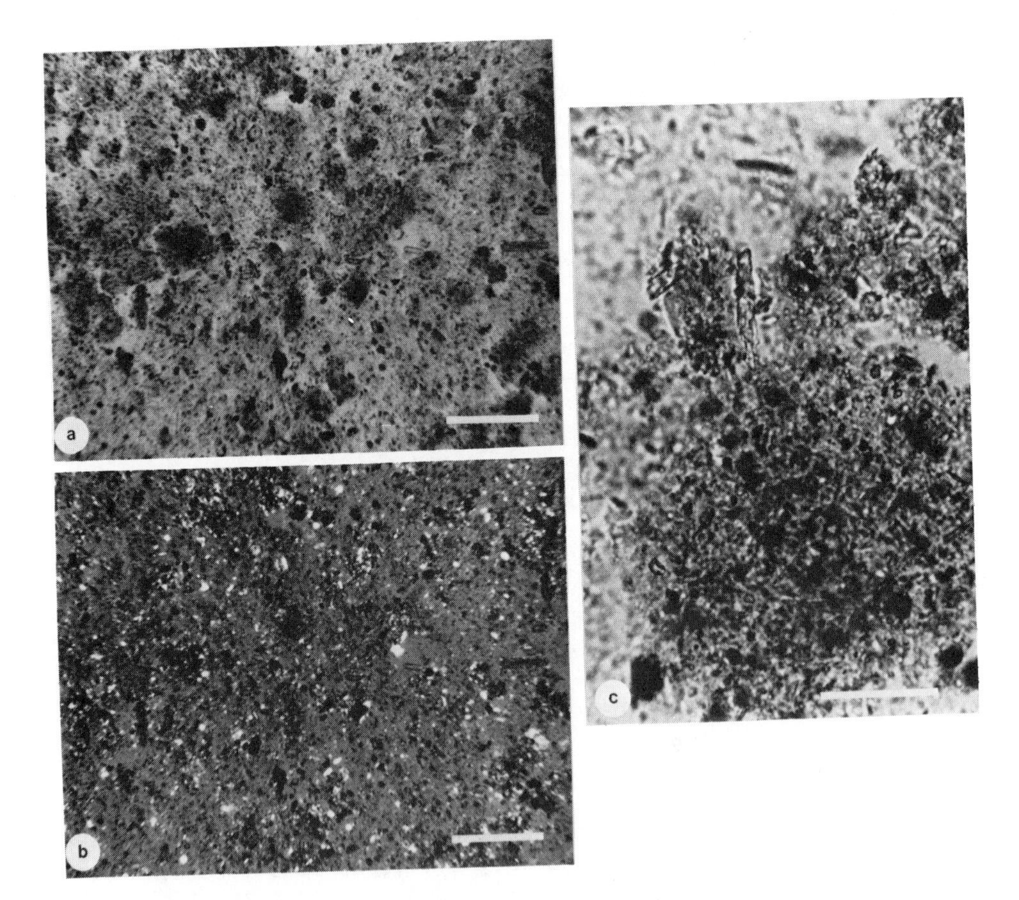

Fig. 8. Photomicrograph of smear slide, 18A-37-4, RSO-rich mud. (a) plane light, (b) crossed nicols, (c) detail, plane light. Scale bars are present 100, 100 and 31 µm respectively.

circular (spherical?), yellow to yellow-green, translucent, isotropic grains of presumanbly authigenic smectite that are never more than 2 μm in diameter, (2) irregular to circular, yellow-green, yellow-brown, red-brown, orange, semiopaque, 2 to 20 μm in diameter (most commonly to 10 μm) grains of Fe hydroxides or oxyhydroxides (Fig. 8), and (3) irregular to globular or botryoidal, orange, red, brown, black, mostly opaque aggregates that are 4 to 60 μm in diameter. These particles are amorphous to X-rays even when analyzed at very slow scans ($1/8^{0}$ 2θ/minute). We suggest that they are Fe-Mn oxyhydroxides.

In our samples, RSO particles have smaller upper size limits than do those described by Quilty et al. (1976) from the Nazca plate. The three groups of particles occur together and most commonly decrease in relative abundance from type 1 to type 3 in any given sample. RSO rarely fill diatom frustules and radiolarian tests. Iron or Mn hydroxides also coat the surface of (and replace?) siliceous microfossils.

Opaque Minerals. Opaques include trace amounts of euhedral magnetite that is fresh or partly altered to hematite and lesser amounts of irregular grains of hematite.

Aggregates. The most common types of aggregates are fecal pellets, casts of siliceous microfossils, and sediment lumps that were not disaggregated when smear slides were prepared. These types are often difficult to tell apart. Also included under this general heading are rare fragments of quartz-clay rock, chert, and altered volcanic rock.

Calcite. Most calcite occurs as jagged grains that are most likely foraminiferal test fragments, although these would be difficult to distinguish from other biogenic carbonate grains, for example echinoid spine fragments. Grains are as large as 80 μm, but most commonly are 30-40 μm. Coccoliths and discoasters are abundant but poorly preserved in some deposits (Fig. 9a); nannofossils dominate sample 18A-36-8 (Appendix 2). Because DOMES cores are from deposits below the calcium carbonate compensation depth (CCD), much of the calcite in Quaternary deposits was probably delivered to the seafloor in fecal pellets or was reworked from older deposits that accumulated when the seafloor was above the CCD. For example, the lower part (>10 cm) of core 18A-36 was dated as early Miocene by Quinterno and Theyer (this volume).

Authigenic Minerals. Clinoptilolite and phillipsite occur together as trace components of many deposits (Appendix 2). The zeolite in Figs. 9a (left of center) and 9b is phillipsite. EDAX spectra show Si, Al, K, Na, and Fe in both samples; Ca in the spectrum of Fig. 9a probably results from scatter from the carbonate host sediment. Si/Al ratios are 3.37 and 3.88 respectively, which is richer in Si (relative to Al) than phillipsite known to form from alteration of basaltic volcanic debris in the marine environment (see for example Kastner, 1976). In DOMES deposits, phillipsite, clinoptilolite, and authigenic

Fig. 9. Scanning electron micrographs. (a) 18A-36-8, nannofossil ooze, most grains are segments of coccolith plates. Coccolith in upper right is badly corroded. Zeolite just left of center is phillipsite. (b) 53-45-17, phillipsite crystal. Scale bars represent 3 μm

smectite (Hein et al., in press) are sinks for K and Na, as well as Si, Al, and Fe.

Traces of barite were detected by X-ray diffraction in a few samples but were not identified in smear slides (Appendix 2). The particles are no doubt very small. Apatite occurs as prismatic crystals and as fish debris and is a ubiquitous sediment component.

Discussion

Areas A, B, and C are located between the Clarion and Clipperton fracture zones, in abyssal-hill topography (Fig. 1). Sediment overlying basaltic basement is about 150 to 200 m thick (McManus, Burns, et al., 1970; Van Andel, Heath, et al., 1973), but may thin over basement highs. The upper 50 cm of sediment consists of mixtures of detrital and authigenic clay minerals, quartz, feldspar, siliceous microfossils, locally abundant volcanic glass, Fe-Mn oxyhydroxides (RSO), microscopic Mn nodules, agglutinated biogenic debris, minor apatite (fish debris), zeolites, barite, and detrital minerals. On the basis of radiolarian and nannofossil biostratigraphy, sediment is of Quaternary age but contains minor to abundant reworked Miocene and Eocene microfossils (Quinterno and Theyer, this volume). The one exception is Station 18A, where an upper Quaternary section (10 cm) unconformably overlies lower Miocene deposits.

The mineralogy of DOMES cores differs somewhat from that reported for equatorial North Pacific sites drilled during DSDP Legs 5, 8, 9, and 16. The upper meter or so of sediment is rarely recovered by the DSDP, which may account for the mineralogic differences. Specifically, barite, zeolites, and apatite are apparently far more abundant in DSDP cores (Rex and Murray, 1970; Cook and Zemmels, 1971, 1972; Zemmels, 1973). Deeper in the DSDP cores, palygorskite and pyrite occur; these are not found in DOMES deposits. Thus, the obvious mineralogic difference is a greater abundance and variety of authigenic phases at depth, which implies a continuing redistribution and mobilization of elements with increased burial and time; smectite also increases with depth in some DSDP cores, to a point where it is the sole clay mineral present. At greater depth in DSDP cores, silica released by dissolution of radiolarians, diatoms, and spicules forms porcelanite (opal-CT) and chert (quartz) beds rather than (or in addition to) smectite as occurs in surface deposits. Smectite formations is, no doubt, limited by the availability of Al and Fe. Abundant smectite does form at depth, however, from alteration of volcanic glass and associated minerals (Hein and Scholl, 1978a). As in DOMES cores, RSO are ubiquitous in DSDP deposits. Manganese nodules occur only in surface sediments.

Clay Mineralogy

The regional distribution of clay minerals in Pacific surface sediment has been delineated and discussed by Griffin and Goldberg (1963), Griffin et al. (1968), Aoki et al. (1975), and Gorbunova (1976). In addition, Gorbunova describes the distribution of clay minerals with depth in some Pacific DSDP cores. It is difficult, however, to compare the results of these earlier studies with ours because different techniques of sample preparation and data analysis were used. With these limitations in mind, however, our work is consistent with previous studies. The discrepancies, in part, can be attributed to the much more detailed -- and therefore more representative -- sampling in this work. In order to compare our work with the DSDP results we set the clay minerals equal to 100 percent and using Biscaye's (1965) weighting factors recalculated the <2 μm mineralogy from the Initial Reports of the DSDP for Legs 8, 9, and 16. The DSDP clay mineral values (recalculated) are much more variable than DOMES values. Minor chlorite and kaolinite often did not register on the DSDP diffractograms. Most variations between the DSDP and our work was in the percentages of illite, chlorite, and kaolinite; smectite values were often comparable to within ± 10%.

The primary difference between this study and previous ones concerns the interpretation of the origin of smectite. Previous workers considered that smectite was derived from the alteration of volcanic debris. This conclusion was reached not as much from hard evidence for a volcanic precursor as from the prevailing belief that smectite is the natural consequence of the alteration of volcanic debris. Rather, it is apparent, at least in DOMES deposits, that the smectite is probably formed at low temperatures from silica and Fe-oxyhydroxides (Hein et al., in press). Neither is the central Pacific smectite terrigenous; the oxygen isotopic composition of the smectite disallows this possibility (Hein et al., in press). Contrary to the statements of Griffin et al. (1968), smectite is an abundant constituent of only one major western North American river (the Columbia River; J.R. Hein, unpublished data), and smectite is not a major constituent of atmospheric dust (Windom, 1969). Consequently, terrigenous smectite would not be expected in the places where DOMES cores were collected.

Terrigenous and volcanic debris

It is well established that quartz, feldspar, illite, and chlorite + kaolinite in the equatorial North Pacific are terrigenous and that they are primarily eolian in origin (Hurley et al., 1959; Griffin and Goldberg, 1963; Heath, 1969; Windom, 1969; Savin and Epstein, 1970). We infer that these minerals in DOMES deposits are terrigenous and were transported by winds. The notion that some terrigenous debris is transported to the open Pacific in a benthic nepheloid layer (BNL) appears untenable because the primary deep circulation is from south to north (Knauss, 1962) and much of the suspended particulate matter in the BNL is the fine-grained fraction of bottom sediment that has been resuspended (Baker and Feely, 1978).

Volcanic glass shards are andesitic in composition as determined by their refractive index. The nearest sources of andesitic pyroclastic debris are Central and South American stratovolcanoes. Probably volcanic glass and associated minerals, such as biotite, were also transported by winds to DOMES Areas A, B, and C. Ninkovich et al. (1966) showed that major explosive volcanic eruptions can produce ash layers as far as 2000 km from the volcanic edifice. Thus, it is not unreasonable to find traces of volcanic glass dispersed in sediment 5000 km from the source volcano.

Palagonite is very uncommon in DOMES deposits. Presumably palagonite is derived from mass wasting of basal sediment or from erosion of basaltic basement that crops out locally.

Biogenic debris

DOMES areas lie to the north of the equatorial zone of high biological productivity (about 5°N to 5°S latitudes). High productivity and water depths less than the CCD allow for the accumulation of thick deposits of calcareous ooze. Accumulation rates there are 10-20 mm/1,000 yr (Van Andel et al., 1975), in contrast to less than 1 mm/1000 yr in DOMES areas (Lisitzin, 1972). Mostly siliceous microfossils are found in DOMES deposits because calcareous forms are removed by dissolution in the water column and at the seafloor. DOMES deposits are below the CCD, hence, the minor calcite found in some Quaternary cores must have been protected by being encased in fecal pellets. The abundant calcite in core 50-28 is explained by its occurrence near a topographic high where nannofossils may have moved by mass flow from the area of high relief. Nannofossil ooze at the base of core 18A was deposited during the early Miocene when this site was probably above the regional CCD. The minor calcareous microfossils in the upper part of core 18A occur either because it is located in an area slightly above the regional CCD or as the result of bioturbation.

In the central equatorial Pacific 92 to 99% of the siliceous biogenic debris dissolves before it reaches the seafloor, and much of that that is deposited dissolves before a meter of burial (Hurd, 1973; Johnson, 1976). DOMES samples support the observation of rapid dissolution in the upper few decimeters of sediment. Further, as reported in previous studies, spicules resist dissolution longer than other siliceous fossils (for example, Johnson, 1976; Hein et al., 1978b) and are in fact a good indicator of relatively advanced stages of dissolution.

As mentioned previously, superimposed on the general overall trend of decreasing biogenic silica with depth is a fine-scale fluctuation (20,000 to 80,000 years) in the amount of biogenic silica (Appendix 1; Fig. 3). Two explanations are possible. (1) The fluctuation may indicate that dissolution of siliceous debris at one stratigraphic level adds enough silica to the pore waters to prevent dissolution of an adjacent 2 to 8 cm layer. The amount of time needed to deposit the 2 to 8 cm intervals may reflect the time required for silica derived from dissolution of microfossils to be used in formation of authigenic minerals or to be released (by upward diffusion) into the overlying water column (20,000 to 80,000 years). This interpretation would mean that dissolution of siliceous debris continues throughout the burial history of the cores (an overall decrease in amount of coarse fraction) but that also dissolution on a finer scale alternates in intensity. The fine-scale fluctuations would be dissolution pulses similar to those described by Mikkelsen (1978; pulses at 7,000 and 18,000 years B.P.) and Johnson (1976; pulses at 150,000 to 200,000 years B.P.) for siliceous debris in the tropical Pacific, and by Thompson and Saito (1974) for foraminifers (Table 2). The latter authors attributed the pulses to intense dissolution during interglacial periods. Table 2 lists the time intervals where siliceous microfossils are relatively less abundant at DOMES station 52, assuming a sedimentation rate of 1mm/1000 years (Lisitzin, 1972; Quinterno and Theyer, this volume). Curiously, most of the DOMES pulses, based on siliceous organisms, correspond with those determined for calcareous microfossils (Table 2). Changes in the chemistry and temperature of Pacific bottom waters during interglacial stages may have influenced the rate of dissolution of siliceous debris in DOMES cores. (2) Alternatively, the pulses may represent changes in primary productivity of silica organisms in surface waters. Berger (1970) showed that the abundance of siliceous debris in Pacific sediment can co-vary with the primary productivity in surface waters. Although the present data show an overall dissolution of siliceous debris with depth in the cores, we cannot distinguish between the two explanations. Study of the distribution of robust versus delicate species of diatoms or radiolarians during the various pulses could solve this problem, but such an analysis is beyond the scope of this work. Finding the cause of the pulses could increase our understanding of Pacific Quaternary paleo-oceanographic events.

Dissolution of very large amounts of siliceous debris near the sediment-water interface (up to 50% by volume in the sediment column) releases structurally bound and adsorbed metals as well as Si into the environment. Leinen and Stakes (in press) state that siliceous microfossils alone could supply most of the Cu, Fe, and Zn found in equatorial Pacific sediment. F.T. MacKenzie and coworkers at Northwestern University (personal communication, 1978) found that Al co-varies with Si in pore waters of sediment containing siliceous microfossils. Their interpretation is that Al and Si were released by dissolution of diatoms and were then taken up by formation of an aluminosilicate. Apparently, the concentration of Al in seawater can be controlled by biological activity (Hydes, 1977; MacKenzie et al., 1978). Some Ni and Mn are also probably supplied by siliceous debris (Leinen and Stakes, In press). The process of release and uptake of elements is relatively rapid (tens of thousands of years) as is indicated by the extensive dissolution of biogenic silica and the subsequent formation of abundant authigenic smectite within the upper 50 cm

Table 2. Possible dissolution cycles at DOMES Station 52 compared with those determined by other selected studies.

This work ($x10^3$ years)	Other Studies ($x10^3$ years)
—	7[a]
—	10[b]
—	18[a]
30	—
70	—
	124[b]
130 to 170	175 to 200[c]
250	250[b]
320 to 410	342[b]
450	425[b]
490	485[b]

[a]Mikkelsen (1978), diatoms.
[b]Thompson and Saito (1974), foraminifera.
[c]Johnson (1976), Siliceous microfossils.

of sediment.

A locally important biogenic sediment component is fragments of the walls of worm tubes. The mineralogy and chemistry of the binding agent is not known. The presumably organic glue is amorphous to X-rays as are the tiny fragments of siliceous microfossils selectively used in wall construction. A detailed study of undisturbed box cores would shed light on the structure and composition of these features, their significance to the overall sediment chemistry, and the influence they might have on Mn nodule formation.

Metalliferous sediment

Metalliferous sediment was first described from the base of DOMES core 18B by Bischoff and Rosenbauer (1977). The 16-cm bed is enriched in Fe, Mn, Cu, and Ni, and depleted in Al relative to average pelagic clay. Bischoff and Rosenbauer suggested that the metalliferous sediment resulted from hydrothermal activity, perhaps on the Clarion fracture zone. Subsequent work, however, has shown that the metalliferous sediment is not a hydrothermal deposit (Hein et al., in press).

The most conspicuous mineralogic criteria for identifying metalliferous deposits are abundant RSO, and in some samples, abundant authigenic smectite. Authigenic smectite in metalliferous deposits may be more nontronitic or include two variations of smectite and contain much associated amorphous material in contrast to other DOMES smectite-rich deposits (J.R. Hein, unpublished data). High RSO contents indicate that the cores and intervals that contain metalliferous sediment are 18B-37-1 through 9 (especially interval 9); 18A-36-1 and 9; 11-53-9; 20-27-2, 3, 7, 8; 25-51-1, 2, 3; 27-52-12; and 48-19-12, 15, 19 (especially interval 19; Appendix 2). All these cores are from Area C except 48-19, which is from Area A.

We see three possible origins for these metal-rich deposits; (1) exposure of basal metalliferous sediment (see for example Cronan, 1976) due to the erosion of overlying deposits, (2) down-slope transport by density currents or slumping of basal metalliferous sediment exposed on topographic highs, and (3) extensive dissolution of biogenic debris, which would leave a metal-rich residue (Leinen and Stakes, in press). The first possibility can be eliminated because the thickness of the sediment where DOMES cores were taken is too great. If the metal-rich deposits were deposited by slumping or density currents, we would not expect the gradual and continuous change in clay mineralogy with depth in the cores evident from Table 1, but instead we would expect homogenization of the redeposited sediment and an abrupt change in clay mineralogy from redeposited to surrounding pelagic beds. Most metalliferous deposits in DOMES cores must have therefore formed during periods of extensive dissolution of deposits on a local scale. Similar metal deposits are attributed to the same mechanism of formation by Leinen and Stakes (in press) from DSDP site 159, which is located just southeast of DOMES Area C. The most obvious DOMES example is core 18A, where metalliferous sediment in the lower part of the core is separated from overlying deposits by hiatus of approximately 35 m.y. In other cores, less profound hiatuses, perhaps 1 or 2 m.y. long, could produce minor stratigraphic intervals with metal enrichment. DOMES areas have been subjected in varying degrees to regional and local hiatuses for the last 50 m.y. (Van Andel et al., 1975).

Smectite and Fe-Mn nodules

Two important very early diagenetic processes modify the composition of surface sediment, formation of smectite and formation of ferromanganese nodules. Both take place at or within centimeters of the sediment-water interface. The diagenetic processes profoundly influence the geochemical budgets of Si, Al, Fe, Mn, Ni, Cu, Co, Zn, and perhaps other elements.

Mn nodules are discussed elsewhere in this volume and we will only discuss their possible relation to authigenic smectite. By conservative estimate, there is about 20 x 10^9 cubic meters of metal-rich smectite in the upper 10 cm of sediment that connects the three DOMES areas (Hein et al., in press). More than 95% of the smectite in the <1 μm size fraction formed near the sediment-water interface at a maximum temperature of about 4^oC. The smectite probably formed from the chemical combination of Fe oxyhydroxide derived from volcanic activity at oceanic spreading centers, Si released by dissolution of opaline biogenic silica, and from Al presumably released by dissolution of biogenic silica and absorbed on Fe-Mn oxyhydroxides (Hein et al., in press). This mechanism of smectite formation (excluding the Al) was described in detail by Heath and Dymond (1977). Colloidal Fe-Mn oxyhydroxides that form at spreading centers are apparently dispersed throughout the ocean basins and form smectites wherever silica is available for chemisorption. The pan-Pacific distribution of Fe-Mn oxyhydroxides is inferred from the commonly reported occurence of RSO in Pacific DSDP cores. Fe-rich nontronite forms near the Fe source--the East Pacific Rise (EPR) for DOMES samples--whereas Fe montmorillonite forms at greater distances (Aoki et al., 1974; Heath and Dymond, 1977; Hein et al., in press.) Also, Cu, Zn, Mn, and Ni released from siliceous biogenic debris and absorbed on oxyhydroxides are incorporated into the authigenic smectite structure.

Iron and Mn are initially partly fractionated when they mix with oceanic bottom water near hydrothermal vents; this is when oxide and hydroxide phases first form. In open ocean deposits, formation of authigenic Fe-rich smectite may provide a second mechanism by which the Fe and Mn (bound as oxyhydroxides) fractionate, that is, smectite is the Fe sink and Mn nodules the Mn sink. The controlling factor may be the availability of silica. As outlined by Piper and Williamson (1977), Mn nodules from the DOMES areas have high Mn/Fe ratios compared to those from the North Pacific red clay areas. In deposits rich in siliceous microfossils, enough silica is available to use up much of the Fe by formation of smectite, and consequently the nodules are rich in Mn. In the North Pacific red clay areas, some authigenic smectite forms but probably at a slower rate that is determined by the availability of Si derived from dissolution of aluminosilicates, volcanic glass, and that contained in bottom waters. In the red clay environment the Fe-Mn fractionation is less, and nodules are Fe-rich. Lyle et al. (1977) also suggested that authigenic smectite formation fractionated Fe and Mn in the Bauer Basin; they showed this by comparing Mn/Fe ratios of Mn nodules that formed in conjunction with smectite with Mn crusts that formed outside this smectite influence. Similarly, Calvert et al. (1978) speculated that smectite acts as the Fe-Mn fractionator in Fe-poor ferromanganese nodules. Siliceous fossil debris apparently may not only control Mn/Fe ratios of nodules through smectite formation, it may also provide much of the Cu, Ni, and Zn found in nodules (Leinen and Stakes, in press). DOMES nodules contain more Ni and Cu than do those from the red clay areas (Piper and Williamson, 1977).

Thus, from what we now know, the formation of Fe-rich smectite is the best explanation for the fractionation of Mn and Fe between sediment and nodules. Also, it is probable that away from spreading centers, the availability of Si is the limiting factor in the fractionation process. Further, Cu, Zn, and Ni in the smectite and nodules may be derived primarily from siliceous biogenic debris. Testing this hypothesis may require a different direction in research than has heretofore been taken. Specifically, amorphous particles probably control early diagenetic processes, but virtually nothing is known about the structural or chemical nature of amorphous debris in marine deposits. Leaching experiments have revealed amorphous Fe-Mn oxyhydroxides in deep sea-sediments, but many other X-ray-amorphous compounds could exist in the deep-sea environment, for example, compounds that are chemically like sepioite, allophane, talc, silica, nontronite, and others.

Conclusions and Summary

DOMES deposits consist primarily of terrigenous minerals, biogenic debris, and authigenic minerals. Terrigenous minerals, quartz, feldspar, illite, chlorite, kaolinite, volcanic glass, biotite, and other materials are most likely transported by winds from the Americas and increase in abundance seaward of this source and in Quaternary deposits.

The three DOMES areas are at the northern boundary of the equatorial zone of high productivity in water mostly deeper than the CCD. Consequently, biogenic debris is siliceous, mostly radiolarians, diatoms, and spicules. Siliceous microfossils undergo extensive dissolution in the upper one-half meter of sediment. Dissolution leaves only the most solution-resistant spicules and fragments of large radiolarians in the lowest deposits in box cores. Fish debris is a ubiquitous but mostly minor component. Fragments of worm tubes and agglutinated benthic foraminifers also occur. Calcareous microfossils are minor parts of a few cores and the primary component of one core.

Smectite and ferromanganese nodules are the primary authigenic components. Smectite probably forms from the chemical combination of Fe oxyhydroxide and Si. Colloidal oxyhydroxides (Fe, Mn) that are created at oceanic spreading centers are dispersed throughout much of the ocean basins. Si and some Al are derived from dissolution of siliceous microfossils. By formation of Fe-rich smectite, Fe and Mn are fractionated into the sediment and nodule phases respectively. Thus, nodules formed in areas of active smectite generation have high Mn/Fe ratios. Further, the dissolution of biogenic silica not only provides Si and Al for smectite but also releases Cu, Ni, and Zn (Leinen and Stakes, in press) that are subsequently bound into both the smectite and nodule phases. Minor phillipsite and clinoptilolite occur in DOMES deposits, and because there is little if any altered volcanic debris present they probably also formed from oxyhydroxides, constituents released from dissolution of opaline silica, and elements in seawater. Hein et al. (1978b) showed that in the Bering Sea clinoptilolite formed from dissolution of large quantities of siliceous biogenic debris. DOMES zeolites are a sink for Si, Al, K, Na, and some Fe. Smectite is an important sink for Si, Al, Fe, K, Na, Mg, Cu, and Zn, and ferromanganese nodules are important sinks for Mn, Fe, Cu, Ni, Co, Zn, and perhaps other trace metals.

Authigenic smectite formation represents a major reorganization of elements in deep Pacific surface sediments. If the upper one-half meter of sediment is considered, authigenic smectite is a more important authigenic component by volume and weight than Fe-Mn nodules. Smectite contains as much Zn, but an order of magnitude less Cu and Ni (Hein et al., in press) than the nodules (Piper et al., 1977). However, the smectite is at least an order of magnitude more abundant, and hence harbors as much of these economically important metals as the nodules.

Further advances in our understanding of the origin and evolution of ferromanganese nodules may come from a detailed study of the amorphous materials in the nodules, the overlying bottom waters, and the surrounding sediment. The results will further the study of deep-sea authigenic deposits in general.

Acknowledgments

We acknowledge helpful discussions with Fred McKenzie, Northwestern University; James L. Bischoff, David Z. Piper, Paula Quinterno, U.S. Geological Survey; Debra Stakes, Massachusetts Institute of Technlogy; and Tom Johnson, University of Minnesota. Tom Johnson, David Piper, and Tracy Vallier, U.S. Geological Survey, reviewed the manuscript. Technical assistance was given by Christina Gutmacher, Jacqueline Miller, Marjorie G. Jones, and Jeanne Henning. Margaret Leinen, University of Rhode Island; Eric Eslinger, West Georgia College; J. Srodón, Polish Academy of Sciences; and Fred MacKenzie kindly furnished preprints or unpublished data.

REFERENCES

Aoki, S., Kohyama, N., and Sudo, T., (1974) An iron-rich montmorillonite in a sediment core from the northeastern Pacific, Deep Sea Res. 21, 865.

Aoki, S., Oinuma, K., and Kobayashi, K., (1975) Study of clay minerals in recent marine sediments, Publication of the faculty of Science, University of Tokyo, 161.

Baker, E.T., and Feely, R.A., (1978) Chemistry of oceanic particulate matter and sediments: Implications for bottom sediment resuspension, Science 200, 533.

Berger, W.H., (1970) Biogenous deep-sea sediments: Fractionation by deep-sea circulation, Geol. Soc. Am. Bull. 81, 1385.

Biscaye, P.E., (1965) Mineralogy and sedimentation of recent deep sea clay in the Atlantic Ocean and adjacent seas and oceans, Geol Soc. Am. Bull. 76, 803.

Bischoff, J.L., and Rosenbauer, R.J., (1977) Recent metalliferous sediment in the North Pacific manganese nodule province, Earth Planet. Sci. Lett. 33, 379.

Burns, R.E., and Andrews, J.E., et al., (1973) Explanatory notes, Initial Reports Deep Sea Drilling Project 21, 5.

Calvert, S.E., Price, N.B., Heath, G.R., and Moore, T.C., Jr., (1978) Relationship between ferromanganese nodule compositions and sedimentation in a small survey area of the equatorial Pacific, J. Mar. Res. 36, 161.

Cook, H.E., Piper, D.Z., and Gardner, J.V., (1977) Geologic and oceanographic framework of DOMES Sites A, B, and C: Central Equatorial Pacific, In: Deep ocean environmental study: Geology and geochemistry of DOMES A, B, and C, Equatorial North Pacific, Piper, D.Z. (Compiler), U.S.G.S. Open-File Report 77-778, 15.

Cook, H.E. and Zemmels, I., (1971) X-ray mineralogy studies-Leg 8, Initial Reports Deep Sea Drilling Project 8, 901.

Cook, H.E. and Zemmels, I., (1972) X-ray mineralogy studies-Leg 9, Initial Reports Deep Sea Drilling Project 9, 707.

Cronan, D.S., (1973) Basal ferruginous sediments cored during Leg 16, Initial Reports Deep Sea Drilling Project 16, 601.

Cronan, D.S., (1974) Authigenic minerals in deep-sea sediment , In: The Sea, Goldberg, E.D. (Ed.) 5, Wiley, New York.

Cronan, D.S., (1976) Basal metalliferous sediments from the eastern Pacific, Geol. Soc. Am. Bull. 87, 928.

Donnelly, T.W., (1976) Tertiary explosive volcanic activity in the eastern equatorial Pacific Ocean: Sites 320 and 321, DSDP Leg 34, Initial Reports of the Deep Sea Drilling Project 34, 605.

Gorbunova, Z.N., (1976) The history of accumulation of clay minerals in the Pacific Ocean during the Cenozoic according to deep-sea drilling data, Oceanology 15, 568.

Griffin, J.J. and Goldberg, E.D., (1963) Clay-mineral distributions in the Pacific Ocean, In: The Sea, Hill, M.N. (Ed.) 3, 728, Wiley, New York.

Griffin, J.J., Windom, H., and Goldberg, E.D., (1968) The distribution of clay minerals in the World Ocean, Deep Sea Res. 15, 433.

Heath, G.R., (1969) Mineralogy of Cenozoic deep-sea sediments from the equatorial Pacific Ocean, Geol. Soc. Am. Bull. 80, 1997.

Heath, G.R. and Dymond, J., (1977) Genesis and transformation of metalliferous sediments from the East Pacific Rise, Bauer Deep, and Central Basin, northwest Nazca plate, Geol. Soc. Am. Bull. 88, 723.

Hein, J.R., Scholl, D.W., and Gutmacher, C.E., (1976) Neogene clay minerals of the far NW Pacific and southern Bering Sea: Sedimentation and diagenesis. In: AIPEA Proceedings, 1975 International Clay Conference, Mexico City, Bailey, S.W. (Ed.), 71, Applied Publishing, Wilmette, Illinois.

Hein, J.R. and Jones, M.G., (1977) Mineralogy and diagenesis of DOMES Site A, B, and C, In: Deep ocean environmental study: Geology and geochemistry of DOMES Sites A, B, and C, Equatorial North Pacific, Piper, D.Z. (Compiler), U.S.G.S. Open-File Report 77-778, 146.

Hein, J.R. and Scholl, D.W., (1978a) Diagenesis and distribution of late Cenozoic volcanic sediment in the southern Bering Sea, Geol. Soc. Am. Bull. 89, 197.

Hein, J.R., Scholl, D.W., Barron, J.A., Jones, M.G., and Miller, J., (1978b) Diagenesis of late Cenozoic diatomaceous deposit and formation of the bottom simulating reflector in the southern Bering Sea, Sedimentology 25, 155.

Hein, J.R., Yeh, H-W, and Alexander, E.R., Origin of iron-rich montmorillonite from the manganese nodule belt of the north equatorial Pacific, Clays and Clay Minerals (In Press).

Hurd, D.C., (1973) Interactions of biogenic opal, sediment and seawater in the central equatorial Pacific, Geochim. Cosmochim. Acta 37, 2257.

Hurley, P.M., Hart, S.R., Pinson, W.H., and Fairbairn, H.W., (1959) Authigenic versus detrital illite in sediments, Geol. Soc. Am. Bull., 70, 1622.

Hydes, D.J., (1977) Dissolved aluminum concentration in seawater, Nature 268, 136.

Johnson, T.C., (1974) The dissolution of siliceous microfossils in surface sediments of the eastern tropical Pacific, Deep-Sea Res. 21, 851.

Johnson, T.C., (1976) Biogenic opal preservation in pelagic sediments of a small area in the eastern tropical Pacific, Geol. Soc. Am. Bull. 87, 1273.

Kastner, M., (1976) Diagenesis of basal sediments and basalts of Sites 322 and 323, Leg 35, Bellingshausen abyssal plain, Initial Reports of the Deep Sea Drilling Project 35, 513.

Knauss, J.A., (1962) On some aspects of the deep circulation of the Pacific, J. Geophys. Res. 67, 3943.

Leinen, M. and Stakes, D., Metal accumulation rates in the Central Equatorial Pacific during the Cenozoic, Geol. Soc. Am. Bull. 89 (In press).

Lisitzin, A.P., (1972) Sedimentation in the world ocean, Soc. Econ. Paleo. Min., Sp. Publ. No. 17.

Lyle, M., Dymond, J., and Heath, G.R., (1977) Copper-nickel-enriched ferromanganese nodules and associated crusts from the Bauer Basin, Northwest Nazca plate, Earth Planet. Sci. Lett. 35, 55.

MacKenzie, F.T., Stoffyn, M. and Wollast, R., (1978) Aluminum in seawater: Control by biological activity, Science 199, 680.

McManus, D.A., Burns, R.E., et al., (1970)

Initial Reports Deep Sea Drilling Project 5, 827.

Mikkelsen, Naja, (1978) Preservation of diatoms in glacial to Holocene deep-sea sediments of the equatorial Pacific, Geology 6, 553.

Mossman, D.J. and Heffernan, K.J., (1978) On the possible primary precipitation of atacamite and other metal chlorides in certain stratabound deposits, Chem. Geol. 21, 151

Ninkovich, D., Opdyke, N., Heezen, B.C., and Foster, J.H., (1966) Paleomagnetic stratigraphy, rates of deposition and tephrachronology in North Pacific deep sea sediments, Earth Planet. Sci. Lett. 1, 476.

Perry, E. and Hower, J., (1970) Burial diagenesis in Gulf Coast pelitic sediments, Clays and Clay Minerals 18, 165.

Piper, D.Z., Cannon, W. and Leong, K., (1977) Composition and abundance of manganese nodules from DOMES Sites A, B, and C: relationship with bathymetry and stratigraphy: In: Deep ocean environmental study: Geology and geochemistry of Domes Sites A, B, C, Equatorial North Pacific, Piper, D.Z. (Compiler), U.S.G.S. Open-File Report 77-778, 146.

Piper, D.Z. and Williamson, M.E., (1977) Composition of Pacific Ocean Ferromanganese nodules, Marine Geol. 23, 285.

Quilty, P.G., Sachs, H.M., Benson, W.E., Vallier, T.L., and Blechschmidt, G., (1976) Sedimentologic history, Leg 34, Deep Sea Drilling Project, Initial Reports of the Deep Sea Drilling Project 34, 779.

Quinterno, P. and Theyer, F., Biostratigraphy of the equatorial North Pacific DOMES sites A, B, and C (This volume).

Rex, R.W. and Murray, B., (1970) X-ray mineralogy studies, Initial Reports Deep Sea Drilling Project 5, 441,.

Reynolds, R.C., Jr., and Hower, J., (1970) The nature of interlayering in mixed-layer illite-montmorillonites, Clays and Clay Minerals 18, 25.

Ross, C.S. and Hendricks, S.B., (1945) Minerals of the montmorillonite group, U.S.G.S. Prof. Paper 205-B.

Savin, S.M. and Epstein, S., (1970) The oxygen and hydrogen isotope geochemistry of ocean sediments and shales, Geochim. Cosmochim. Acta 34, 43.

Srodón, J. and Hower, J., (1976) Mixed layer smectite/illites from carboniferous bentonites and tonsteins of Poland, Program and Abstracts, 25th Clay Mineral Conference, 36.

Thompson, P.R. and Saito, Tsunemasa, (1974) Pacific Pleistocene sediments: Planktonic foraminifera dissolution cycles and geochronology, Geology 2, 333.

Turner, F.J. and Verhoogen, J., (1960) Igneous and Metamorphic Petrology, McGraw-Hill, New York.

Van Andel, T.H., Heath, G.R., et al., (1973) Initial Reports Deep Sea Drilling Project 16, 949.

Van Andel, T.H., Heath, G.R., and Moore, T.C., Jr., (1975) Cenozoic history and paleooceanography of the Central Equatorial Pacific Ocean, Geol. Soc. Am. Memoir 143, Boulder, Colorado, 134.

Windom, H.L., (1969) Atmospheric dust records in permanent snowfields; Implications to marine sedimentation, Geol. Soc. Am. Bull. 80, 761.

Zemmels, I., (1973) X-ray mineralogy studies-Leg 16, Initial Reports Deep Sea Drilling Project 16, 529.

GEOCHEMISTRY OF DEEP-SEA SEDIMENTS FROM THE PACIFIC MANGANESE NODULE PROVINCE: DOMES SITES A, B, AND C

James L. Bischoff

U.S. Geological Survey
Menlo Park, California

G. Ross Heath and Margaret Leinen

Graduate School of Oceanography
University of Rhode Island
Kingston, Rhode Island

Abstract

Sediments at DOMES Sites A, B and C are mixtures of pelagic clay, biogenic ooze, hydrogenous metals, and metalliferous sediment. The relative proportions of these components and the importance of local volcanic contributions can be estimated from bulk analyses of Si, Al, Fe, Ca, Mn, Ti and Na. Analyses of these and 20 additional elements for 118 samples indicate that average sediment at Sites A and B contains about 11% biogenic SiO_2 and 70% pelagic clay, whereas that at Site C contains less than 2% biogenic SiO_2, 87% pelagic clay and 1.8% metalliferous sediment.

A partitioning model based on element fractionation and inter-element correlation indicates that most of the Mn, Mo, Ni, and Pb is in the hydrogenous fraction of the sediment. The bulk composition of the hydrogenous fraction shows little variation from sample to sample and is comparable to the composition of the DOMES manganese nodules. The pelagic clay fraction is also remarkably homogeneous and is almost identical to average Pacific pelagic clay with respect to all major oxides except K_2O.

Several cores at Site C have large components of metalliferous sediment. Fe_2O_3, SiO_2, MgO, and P_2O_5 in excess of the amounts in pelagic clay are present in remarkably uniform proportions in the metalliferous fraction.

At Sites A and B, MgO in excess of the pelagic clay contribution has been taken up by very slowly deposited biogenic SiO_2.

Factor analyses of the major oxides and partitioning by the Heath and Dymond (1977) model yield results in good agreement with the present partitioning model and provide additional insight into the distribution of minor elements not strongly confined to a single sediment fraction.

Introduction

Pacific deep-sea sediments are mechanical mixtures of several distinct fractions. These fractions are 1) biogenic ooze (calcareous and siliceous), 2) terrigenous or pelagic clay, 3) local volcanic debris (both altered and unaltered), 4) hydrogenous material (nodule-related heavy metals), and 5) metalliferous sediment (hydrothermal precipitates). The local predominance of one fraction over the others is due to local differences in relative rates of supply to the sea floor. Such variations should be reflected in the chemical composition of the sediment.

Quantitative apportioning of the sediment fractions on the basis of bulk chemical composition can be done only when relatively complete analyses are available. That is, the number of analyzed elements must be large enough that all fractions are well represented. When a statistically significant number of samples are so analyzed, inferences can also be drawn regarding the sediment fraction affiliation and distribution of certain of the minor elements.

We report here the results of chemical analyses of 118 samples taken from 29 box cores from the three DOMES (Deep Ocean Mining Environmental Study) Sites A, B and C. These sites represent potential mining sites for ferromanganese nodules. We have studied the chemistry of the surface sediment because of its potential importance to the genesis and distribution of the nodules.

Geologic Setting

DOMES Sites A, B and C are located in the equatorial North Pacific (Fig. 1) in the belt of manganese nodules of maximum commercial interest (see Cook *et al.*, 1977; Piper *et al.*, this volume). Each site covers an area approximately 1 degree square (10,000 km^2). All three sites are in water depths greater than the local carbonate compensation depth (CCD). Surface sediments at Sites A and B are clay-bearing siliceous oozes with clay contents ranging from 25 to 70 percent. Sediments at Site C contain a larger clay fraction (70 to 100 percent) and are classified as siliceous pelagic clay (Cook *et al.*, 1977). The clay mineral suites have smectite: illite:chlorite-kaolinite ratios of 52:31:17 at Site A, 38:42:20 at Site B and 40:50:10 at Site C (Hein *et al.*, 1976; Hein and Jones 1977).

Table 1

Core Locations, Depth, and Sample Intervals Selected for Chemical Analysis, DOMES Sites A, B, and C

Station-Core No.	Sample Intervals	Latitude N.	Longitude W.	Depth, m.
Site A (R/V Oceanographer cruise RP-8-OC-75)				
46-1	1,8,10,13	9°20.7'	150°50.7'	5160
47-10	1,8,10,13	9°00.3'	151°10.9'	5108
48-19	1,8,10,13	8°16.5'	151°07.3'	5043
49-25	1,8,10,13	8°28.5'	150°44.5'	5005
50-28	1,8,10,13	8°41.7'	150°18.7'	5009
Site B				
51-33	1,8,10,11	8°43.1'	150°14.1'	4881
52-42	1,8,10,13	11°15.5'	139°03.3'	4871
53-45	1,8,10,13	11°43.9'	138°22.2'	4985
54-51	1,8,10,13	12°08.9'	137°44.6'	4934
55-53	1,8,10,13	11°48.6'	137°28.3'	4859
Site C (R/V Oceanographer cruise RP-8-OC-76)				
2-7	1	15°14.5'	126°02.6'	4539
3-8	1	15°15.4'	125°59.3'	4363
4-9	1	15°14.6'	126°01.6'	4490
5-10	1	15°13.0'	125°55.9'	4465
10-15	1,3,5,7,9,80	15°11.4'	126°02.2'	4516
11-16	1	15°11.7'	126°03.9'	4524
12-18	1	15°11.6'	126°05.9'	4451
14-20	1	15°08.2'	126°03.7'	4586
15-21	1	15°08.4'	126°02.7'	4501
16-22	1	15°07.9'	125°59.4'	4442
18-24	1,3,5,9,15	15°06.3'	125°57.4'	4426
(R/V Oceanographer cruise RP-6-OC-75)				
11-53	1,8,10,14	16°00.7'	126°46.3'	4603
15B-50	1,8,10,13	15°45.6'	126°00.5'	4552
16B-49	1-10 inclusive	15°46.7'	126°11.2'	4552
18A-36	1,3,4,6,8	15°15.7'	126°00.0'	4339
18B-37	1-9 inclusive	15°12.2'	125°58.6'	4406
21-27	1-8 inclusive	15°59/3'	126°11.6'	4672
24B-29	1,3,5,6,7,8,9	14°15.4'	126°01.4'	4468
25-51	1,2,3	14°14.5'	124°58.5'	4561
27-52	1,8,10,12	16°00.5'	124°59.6'	4406

Metalliferous sediment was recovered at Site C. This sediment, which is primarily smectite, is enriched in Fe, Si, and Mg and depleted in Al compared to average pelagic clay. It is comparable to metalliferous sediment from the Bauer Depression (Bischoff and Rosenbauer, 1977, and Bischoff *et al.*, 1979).

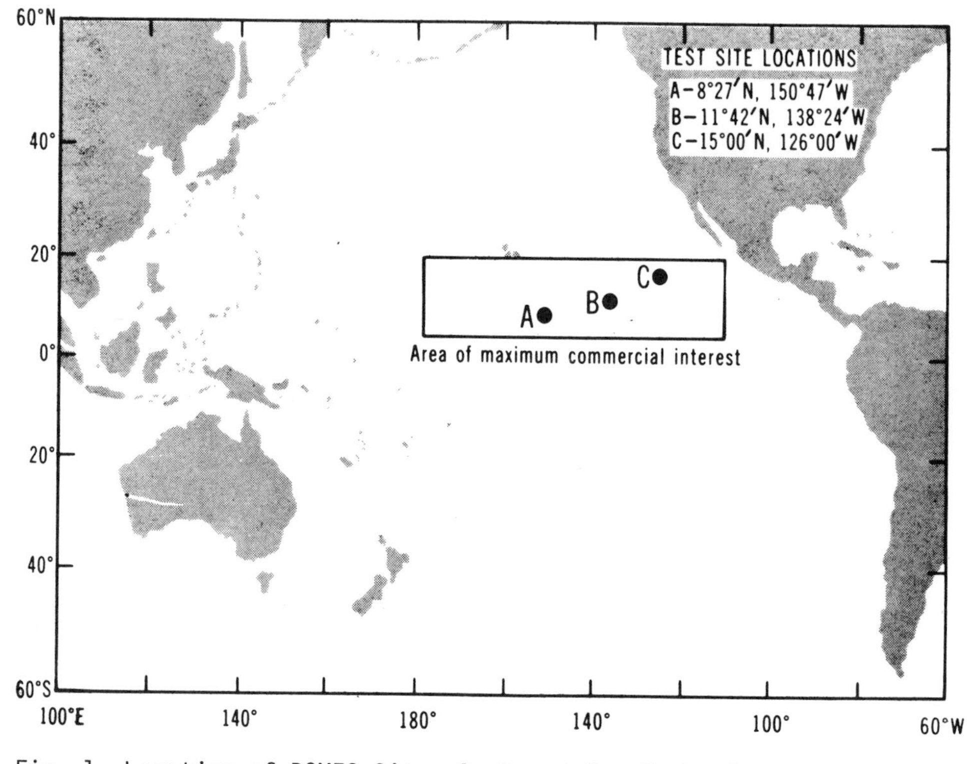

Fig. 1. Location of DOMES Sites A, B and C. Each site covers approximately one degree square

Sampling

Sediment was collected by box coring during three cruises of the R/V *Oceanographer* from 1975 to 1977 (Piper, 1977). Maximum sediment penetration was 50 cm. Sediment samples representing 2-cm vertical intervals were selected from 5 cores from Site A (20 samples), 5 cores from Site B (20 samples), and 19 cores from Site C (78 samples). The larger number from Site C reflects the greater degree of attention and cruise time devoted to this site (see Table 1). Sample number designation follows the DOMES convention

Table 2

Average chemical composition of sediments from DOMES Sites A, B, and C and average Pacific pelagic sediment (numbers in parenthesis are one standard deviation). For average Pacific pelagic sediment: Si, Al, Fe, Mg, Ca, Na, K, Ti, and B taken from Landergren (1964); P, Mn, Co, Cr, Cu, Mo, Ni, Pb, and V taken from Cronan (1969); Ba, Sc, Sr, and Y taken from Goldberg and Arrhenius (1958); Zn taken from Turekian and Wedepohl (1961); organic C taken from El Wakeel and Riley (1961). Zn and organic C refer to world average oceanic sediment.

Percent	Site A n=20	Site B n=20	Site C n=78	Pacific Pelagic Clay
SiO_2	48.3 (5)	51.5 (1.8)	50.7 (5)	54.9
Al_2O_3	11.3 (1)	12.5 (1.3)	14.5 (2)	16.6
Fe_2O_3	5.4 (.6)	5.4 (.8)	7.6 (1)	7.7
MgO	3.0 (.2)	3.0 (.1)	3.2 (.2)	3.4
CaO	4.7 (6)	1.5 (.6)	2.3 (3.6)	.7
Na_2O	5.5 (.7)	5.7 (.5)	2.9 (.8)	1.3
K_2O	2.9 (.3)	3.3 (.7)	3.2 (.3)	2.7
TiO_2	.55 (.1)	.59 (.1)	.72 (.1)	.78
P_2O_5	.54 (.2)	.51 (.3)	.42 (.1)	.25
MnO	.50 (.4)	.53 (.4)	1.2 (1.7)	.56
l.o.i.	14.5 (3.5)	11.2 (.8)	12.4 (4)	-
Sum	97.2 (5)	95.9 (3)	99.3 (2)	-
C_{org}	.11 (.2)	.15 (.3)	.2 (.15)	.27
CO_2	3.4 (4.5)	1.2 (.5)	1.05 (2.8)	-
parts per million				
B	178 (30)	167 (31)	145 (30)	100
Ba	2835 (650)	1505 (1373)	3926 (2015)	3900
Be	3 (.7)	3 (1)	3.5 (1)	-
Co	83 (31)	62 (27)	116 (90)	113
Cr	57 (12)	50 (20)	53 (16)	64
Cu	440 (160)	222 (70)	595 (1000)	230
Mo	12 (16)	8 (5)	24 (60)	10
Ni	183 (76)	112 (66)	341 (660)	210
Pb	34 (6)	26 (8)	61 (40)	34
Sc	33 (6)	30 (3)	21 (6)	25
Sr	175 (100)	343 (100)	317 (180)	710
V	89 (22)	99 (20)	102 (26)	117
Zn	243 (100)	95 (13)	160 (64)	165
Y	171 (90)	124 (60)	97 (32)	150

used in Bischoff (1976); Piper (1977), and other DOMES-related studies in this volume. In this convention, for example, sample 18B-37-2, refers to Station 18B, box core number 37, and interval 2, (i.e. 2-4 cm depth in core). Thus, data for individual samples from the various reports can be cross-referenced and compared.

Analytical procedures

Samples, including pore water salts, were dried at 110°C. The major elements Si, Al, Fe, Mg, Ca, Na, K, Ti, and P were analyzed by X-ray fluorescence spectroscopy after fusion with lithium tetraborate. Concentrations were determined by comparison with appropriate sediment and rock standards and were computer corrected for matrix effects. CO_2 released by acidification of solid carbonates was measured gasometrically on a LECO Corp. CO_2-analyzer. Organic carbon was estimated by difference after analyzing total CO_2 released by combustion on the same instrument. Precision (1 sigma), based on replicate analyses, is estimated to be 1 percent or less of the amount reported for the major elements. Loss on ignition (l.o.i.) was determined from weight loss at 1000°C and includes additional water loss above 110°C, CO_2, and loss of organic matter.

Ni and Cu were analyzed by atomic-absorption spectroscopy after complete sample dissolution, with estimated precisions of better than 3 percent (1 sigma). The remaining elements were analyzed by optical emission spectroscopy standardized against appropriate sediment and rock standards. Precision is estimated to be on the order of 20% (1 sigma).

Data

Major components for the 118 samples are reported as oxides in the form the sample would have at 1000°C in order to allow summation; minor and trace elements are reported in parts per million (Appendix 1). Averages for each area (Table 2) are presented to allow comparisons. Summations of major components are, in general, satisfactory; the totals average 97%, 96%, and 99% for Sites A and B and C, respectively. These results indicate not only that the major-component analyses are reliable, but also that the list of components is complete and that no component contributing a major fraction was overlooked.

Whether or not to include the pore-water salts in the chemical analyses presented a dilemna. If the sediment were rinsed in distilled water prior to drying, the composition of the exchangeable cations would be altered because of the dilution effect (Sayles and Mangelsdorf, 1977). In a typical pelagic clay, the fractions of exchangeable cations are approximately as follows: Na^+ .50, Mg^{+2} .30, K^+ .10, Ca^{+2} .10. Assuming a cation exchange capacity of

75meq/100g, Na_2O will account for about 1.2%, MgO 0.45%, K_2O 0.36%, and CaO 0.4% of the dry salt-free sediment. During rinsing, these values will be changed by substitution of Mg^{++} for Na^+, the extent of which depends on the original pore-water content and is unpredictable. If the sediment is not rinsed, the sea salts left in the dry sediment result in unpredictable amounts of Na_2O, depending on the original water content. A typical pelagic clay with 50 weight percent pore water will have, upon drying, a Na_2O content of approximately 1.35% owing to dried salts. The sea salts, however, will not add significant amounts of other components to the solids. Thus, either way, the Na_2O content becomes difficult to partition quantitatively. For example, the Na_2O contribution from unaltered volcanic material, the only other source of Na_2O, becomes impossible to assess. The only way to circumvent the problem is to determine either the original water content or the concentration of Cl in the dried sample so that the Na_2O from sea salt can be subtracted. Na_2O in excess of this amount can then be attributed to exchangeable plus volcanic Na_2O. In the absence of such data for the DOMES samples, we estimated that the sum of sea salts and exchangeable Na_2O should be approximately 2.5%. The average sediment from Sites A and B contains about 5.5% Na_2O (Table 2) and probably reflects the higher water content of siliceous oozes. In Site C sediment, the average of only 2.9% Na_2O suggests lower water content and negligible contributions of unaltered volcanic debris. The amount of sea salt in the sediment was approximated by the expression: Na_2O sea salt = 2.67 (Na_2O% - 1.2), where 2.67 is the reciprocal of the Na_2O fraction of sea salt and 1.2 represents the average expected cation exchange contribution of Na_2O.

Average bulk compositions

In order to compare the chemical composition of DOMES sediments to other Pacific pelagic sediments, we calculated a composition for average Pacific pelagic clay (Ppc) based on published analytical results (Table 2). The major oxides for Ppc (Si, Al, Fe, Mg, Ca, Na, K, Ti) are based on Landergren's (1964) optical spectrographic analyses of 450 samples from 8 cores from the equatorial Pacific. P, Mn, Co, Cr, Cu, Mo, Ni, Pb, and V are taken from Cronan's (1969) optical spectrographic analyses of 26 cores from the Central Pacific. Ba, Sc, Sr, and Y are taken from Goldberg and Arrhenius' (1958) optical spectrographic analyses of 22 pelagic sediment cores from the eastern Pacific. Zn and Organic C are taken from El Wakeel and Riley (1961) and Turekian and Wedepohl (1961) respectively, who give estimates for world average oceanic sediment composition.

Thus, our estimate of the composition of Ppc is based on a rather limited geographic coverage and primarily on optical spectrographic data, usually considered to be only semiquantitative. However, the resulting composition is the best estimate that can be made and serves as a basis for comparison, allowing us to determine how homogeneous Pacific pelagic sediments really are.

Another basis for comparison is the benthic nepheloid layer (BNL) for the DOMES sites, as determined by Baker and Feeley (this volume). BNL represents the bulk composition of the suspended matter of the lowermost 400 m of the water column, and analyses are given by Baker et al., this volume.

The average composition of sediments from Sites A, B and C is generally similar to Ppc (Table 2), but subtle differences are evident. SiO_2 concentration is approximately the same for all three areas but slightly lower than Ppc. This difference illustrates the difficulty often inherent in comparing absolute concentrations of single components from various studies. Site A is characterized as a siliceous ooze, yet its bulk SiO_2 is less than that of Site C and Ppc. Al_2O_3 contents at Site A are even lower than Ppc, so it is the SiO_2/Al_2O_3 ratio rather than absolute oxide contents that indicate Site A actually is richer in SiO_2 than Sites B or C, or Ppc. Thus, elemental or oxide ratios often reveal real differences between sediments having different proportions of sedimentary components and/or between groups of analyses that involved different sample-handling procedures and thus have differing absolute elemental abundances. Al is usually assigned to the clay-hydrolysate fraction (Landergren, 1964, but see Heath and Dymond, 1977) and so can be used for normalization of other elements associated with this fraction in the offridge DOMES samples. Mn can be used to normalize elements in the hydrogenous fraction (Krishnaswami, 1976). Elemental and oxide ratios are given in Table 3.

SiO_2/Al_2O_3 ratios show that Sites A and B are relatively richer in SiO_2 than Site C and Ppc (Table 3), a result consistent with the petrographic sediment classification. Moreover, relatively large variations in SiO_2 are indicated for Sites A and C by the somewhat large standard deviations (Table 2). If we make the assumption that the pelagic clay fraction of the DOMES sediment has the same absolute Al_2O_3 content as Ppc, we can infer that this fraction accounts for approximately 68%, 75%, and 87% of the sediments at Sites A, B, and C, respectively (Table 4). CaO is, for the most part, present as $CaCO_3$ and for Sites B and C represents about 3% of the sediment, as indicated by summing CaO and CO_2 (Table 2). The CaO/CO_2 ratio is close to, but slightly greater than, that of pure $CaCO_3$ for Sites A and B, significantly greater at Site C (Table 3). For Site C, CaO exceeds CO_2 by 0.9%. Apatite in Site C sediments as represented by P_2O_5 can account for 0.5 percent CaO, leaving 0.4% CaO still in excess. Exchangeable Ca^{++} was estimated above to contribute approximately 0.4% CaO, so all calcium appears to be accounted for.

Table 3

Selected Oxide and Elemental Ratios for Averaged DOMES Sediment from Sites A, B, and C and Average Pacific Pelagic Clay. Data calculated from Table 2.

	Site A	Site B	Site C	Average Pacific Pelagic Clay (Ppc)
SiO_2/Al_2O_3	4.3	4.1	3.5	3.3
Fe_2O_3/Al_2O_3	.48	.43	.52	.46
MgO/Al_2O_3	.27	.24	.22	.20
K_2O/Al_2O_3	.26	.26	.22	.16
TiO_2/Al_2O_3	.05	.05	.05	.05
P_2O_5/Al_2O_3	.05	.04	.03	.047
MnO/Al_2O_3	.04	.04	.08	.033
$B/Al_2O_3 (\times 10^4)$	16.	13.	10.	6.
$Ba/Al_2O_3 (\times 10^4)$	250.	120.	270.	234.
$Be/Al_2O_3 (\times 10^4)$	.26	.24	.24	-
$Cr/Al_2O_3 (\times 10^4)$	5.	4.	4.	4.
$Sc/Al_2O_3 (\times 10^4)$	3.	2.4	1.4	1.5
$Y/Al_2O_3 (\times 10^4)$	15.	10.	7.	9.
Co/Mn	.022	.015	.012	.026
Cu/Mn	.11	.054	.063	.053
Mo/Mn	.003	.002	.002	.002
Ni/Mn	.047	.027	.037	.048
Pb/Mn	.009	.006	.006	.008
Zn/Mn	.063	.023	.017	.038
$^*CaO/CO_2$	1.38	1.25	2.19	-
$Sr/CaO (\times 10^4)$	37.	228.	137.	1014.

$^*CaO/CO_2$ = 1.27 for pure $CaCO_3$.

Judging from the Sr/CaO ratio (Table 3) Sr is depleted relative to Ppc. The variability of the ratio suggests that much of the Sr is not associated with the $CaCO_3$ phase.

The ratio of TiO_2 to Al_2O_3 is essentially identical to the Ppc value. That the ratio of Fe_2O_3 to Al_2O_3 at Sites A and B is similar to Ppc, indicates a rather regular association with the clay

Table 4

Calculated Sediment Fractions for Average DOMES Sediments. Fractions are in weight percent and are calculated from average bulk compositions given in Table 2.

	A	B	C
Pelagic Clay ($6.06\ Al_2O_3\%$)	68.2	75.2	87.3
Biogenic $CaCO_3$ ($\%CaO + \%CO_2$)	8.1	2.7	3.3
Hydrogenous Metals (3.12 MnO)	1.6	1.6	3.7
Sea Salts ($2.67[Na_2O - 1.2]$)	11.5	12.0	4.5
Metalliferous Sediment 3.13 (Excess $Fe_2O_3\%$-.32 MnO%)	-	-	1.8
Biogenic SiO_2 (Excess $SiO_2\%$-1.74 [Excess $Fe_2O_3\%$ - .32 MnO%])	10.8	11.0	1.8
Sums	100.2	102.5	102.4

fraction (Table 3). The slight enrichment in Site C deposits reflects the contribution of the metalliferous fraction (see below). MgO/Al_2O_3 and K_2O/Al_2O_3 ratios are consistently higher for Sites A, B, and C than for Ppc.

Be/Al_2O_3, Cu/Al_2O_3, and Y/Al_2O_3 show little variation and are close to those for Ppc, whereas B/Al_2O_3 is significantly higher at all sites. Relative enrichment by a factor of two in the hydrogenous fraction for Site C compared to Sites A and B is indicated by both absolute MnO percentage and MnO/Al_2O_3 compared to A and B and Ppc. Krishnaswami (1976) found that the Mn in Ppc is in the hydrogenous fraction.

As a first approximation, we make the assumptions that all of the MnO of the DOMES sediment is in the hydrogenous fraction and that the hydrogenous fraction has the same absolute MnO content as the local manganese nodules (32%; Piper et al., this volume). We infer, therefore, that Sites A, B, and C contain approximately

Table 5

Inter-component correlation coefficients for chemical compositions of 118 sediment samples from DOMES Sites A, B, and C.

	SiO_2	Al_2O_3	Fe_2O_3	MgO	CaO	K_2O	TiO_2	P_2O_4	MnO	C	CO_2	B	Ba	Be	Co	Cr	Cu	Mo	Ni	Pb	Sc	Sr	V	Zn	Y
SiO_2		.76	.02	-.1	-.8	.66	.71	-.16	.18	.18	-.8	.37	.02	.39	-.16	.4	-.25	-.39	-.34	-.23	.13	-.7	.05	-.35	-.05
Al_2O_3			.24	.017	-.6	.70	.97	-.4	-.18	.23	-.62	.15	-.03	.51	.08	.46	-.11	-.23	-.13	.05	-.15	-.6	.26	-.3	-.23
Fe_2O_3				.65	-.2	.12	.3	-.04	.29	.16	-.3	-.35	.39	.15	.21	-.16	.18	.24	.18	.36	-.62	-.08	.2	.04	-.34
MgO					-.13	-.16	.03	.36	.38	.04	-.25	-.2	.23	-.09	.26	-.28	.27	.36	.28	.32	-.22	.25	.21	.19	.18
CaO						-.55	-.58	-.05	.01	-.01	.978	-.25	-.03	-.32	-.09	-.25	.03	.03	0	.004	-.19	.51	-.28	.23	-.12
K_2O							.67	-.49	-.3	.07	-.49	.33	.04	.3	-.08	.54	-.19	-.27	-.22	-.11	.067	-.54	.29	-.37	-.23
TiO_2								-.48	-.04	.3	-.6	.17	-.09	.47	.21	.52	-.02	-.1	.02	.18	-.21	-.57	.28	-.23	-.31
P_2O_4									-.16	-.21	-.14	-.16	.31	-.11	.04	-.58	.11	.14	.06	-.1	.28	.25	-.23	.2	.66
MnO										.08	-.05	-.12	.08	-.08	.89	-.14	.69	.93	.976	.81	-.24	.35	.15	.47	-.04
C											-.05	.115	.09	.165	.19	.227	.10	.001	.07	.23	-.4	-.19	-.26	.19	-.3
CO_2												-.19	-.14	-.38	-.16	-.18	.02	-.02	-.04	-.08	-.09	.47	-.23	.18	-.14
B													-.11	.24	.1	.62	-.008	-.08	-.03	.003	.44	-.21	.07	.12	.15
Ba														.38	.06	-.25	.01	.06	-.008	.08	-.28	-.006	-.19	.24	-.06
Be															.14	.33	-.09	-.09	-.06	.15	.04	-.34	.06	.002	.02
Co																.14	.60	.79	.91	.84	-.16	.2	.17	.48	-.01
Cr																	.04	-.12	-.01	.16	.2	-.26	.25	-.01	.15
Cu																		.56	.67	.61	-.14	.32	.078	.41	.006
Mo																			.94	.72	-.16	.32	.22	.47	.001
Ni																				.91	-.18	.30	.20	.49	.05
Pb																					-.4	.3	.13	.43	-.16
Sc																						.04	.3	.07	.7
Sr																							.07	.22	.21
V																								-.02	.21
Zn																									.2

1.6, 1.65, and 3.7 percent hydrogenous material, respectively (Table 4).

For the Mn-related metals, the ratios Mo/Mn, Pb/Mn, and Ni/Mn are remarkably constant and consistent with Ppc. The large variability in Co/Mn, Cu/Mn, and Zn/Mn ratios, however, suggests either

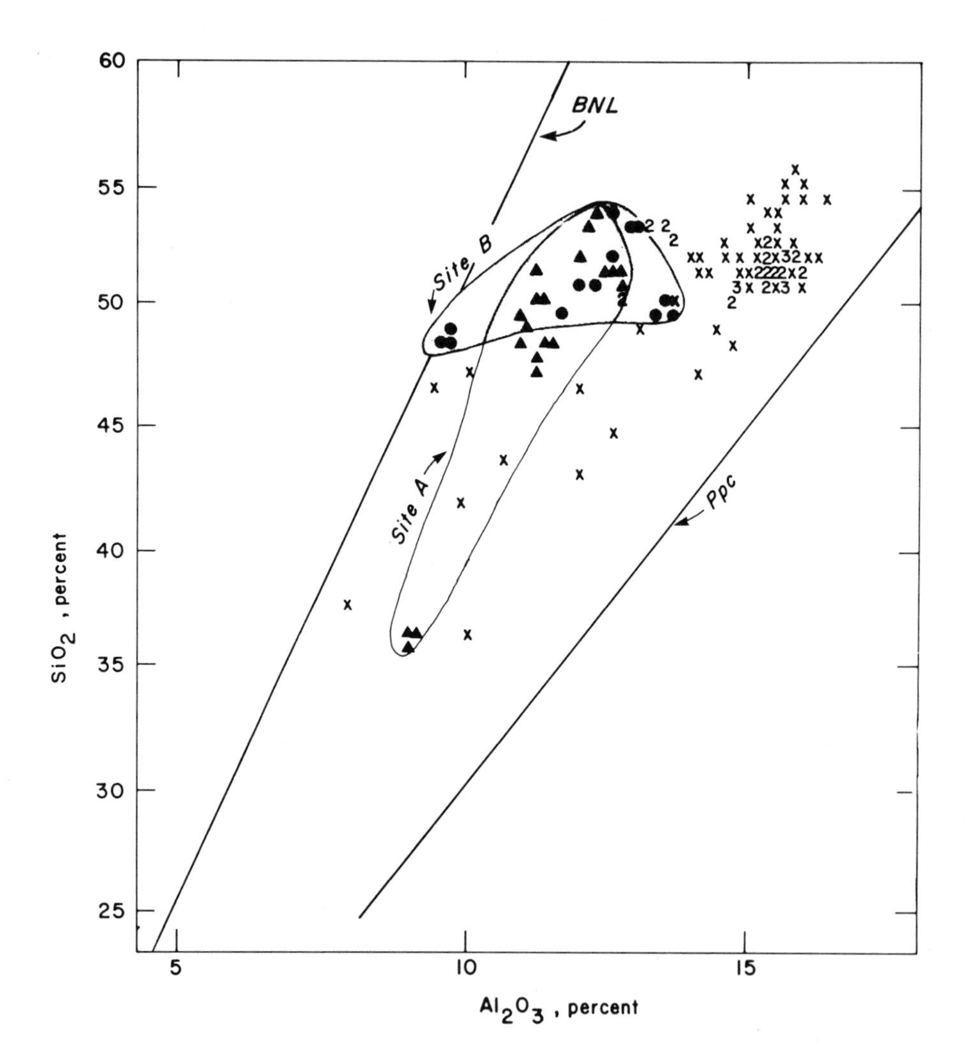

Fig. 2. SiO_2 versus Al_2O_3 for 118 sediment samples from DOMES Sites A (filled triangles), B (filled circles), and C (x's). Where two or more points overlap, a number designates the number of samples coinciding. Ppc is average Pacific pelagic clay, data from Table 2; BNL is benthic nepheloid layer, data from Baker et al., (this volume).

variations in local rates of supply relative to Mn, or heterogeneous distribution among more than one sediment fraction.

As a first approximation, therefore, 89.4, 91.5, and 98.8 percent of the sediments at Sites A, B, and C can be accounted for by a mixture of pelagic clay, $CaCO_3$, sea salts, and hydrogenous material (Table 4). The remainder, therefore, must represent biogenic SiO_2 and metalliferous sediment, which are treated below.

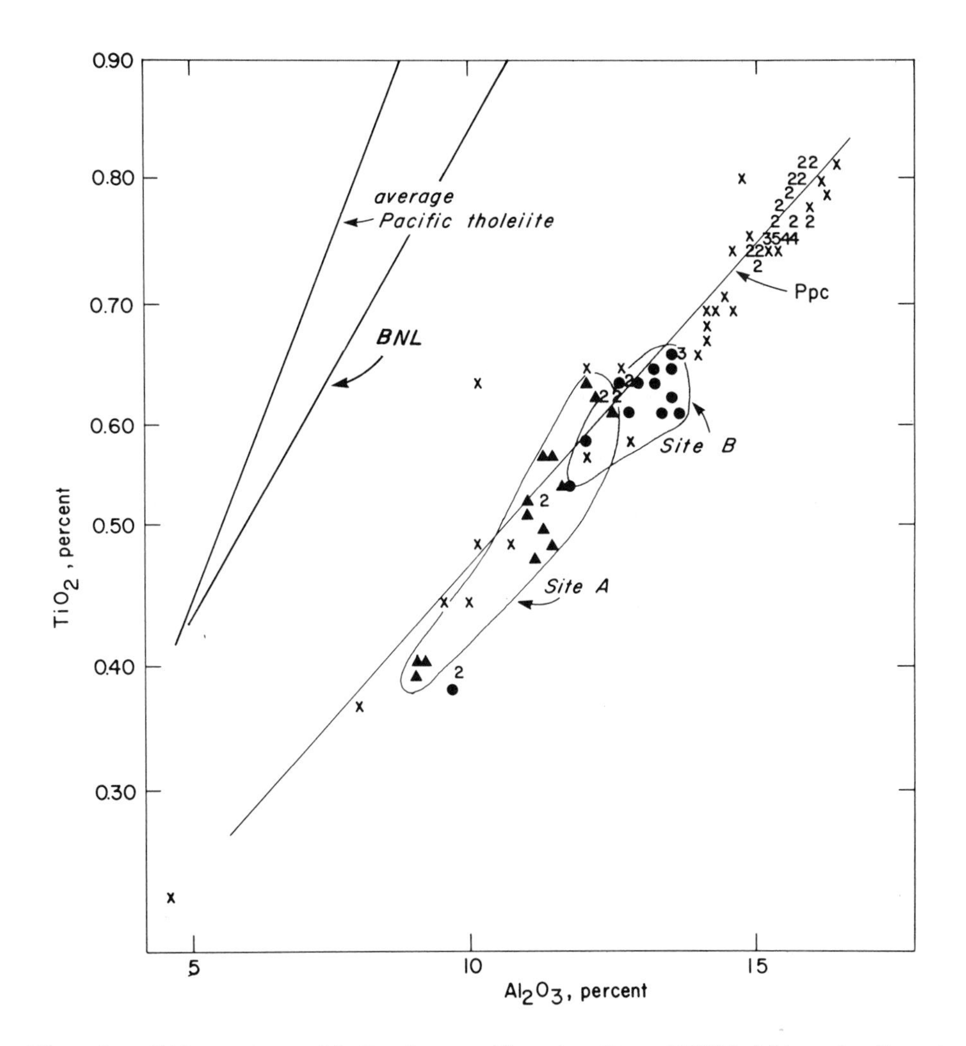

Fig. 3. TiO_2 versus Al_2O_3 for sediments from DOMES Sites A, B and C; symbols as in Figure 2. Data for average Pacific tholeiite taken from Engel et al.(1965).

Interelement relations

Interelement correlation coefficients (Table 5) and interelement plots (Figs. 2-6) can be used to compare individual samples with Ppc and BNL.

Al_2O_3, once again, is most useful for comparison. SiO_2, K_2O, and TiO_2 are strongly correlated with Al_2O_3 as might be expected (Table 5), but strong correlations of Fe_2O_3 and MgO with Al_2O_3, which would be expected if lithogenous material is homogeneous, were not found.

The plot of SiO_2 versus Al_2O_3 (Fig. 2) for all 118 analyses shows that Sites A and B differ markedly from Site C. All samples plot above the Ppc line and therefore have "excess SiO_2" (percent excess SiO_2 is equal to total SiO_2% - 3.3 Al_2O_3% where 3.3 is the SiO_2/Al_2O_3 ratio of Ppc). Although biogenic SiO_2 is virtually pure silica and probably accounts for most of this excess for Sites A and B, metalliferous sediment at Site C also contains considerable SiO_2 in excess of Al_2O_3 (Bischoff and Rosenbauer, 1977; Bischoff et al., 1979).

The plot of TiO_2 versus Al_2O_3 (Fig. 3) shows a very consistent relation, with ratios close to Ppc, and quite distinct from BNL.

This plot indicates that TiO_2 is almost exclusively and homogeneously distributed within the pelagic clay fraction at all three sites. Again, samples from Sites A and B tend to cluster and are distinct from Site C deposits.

A plot of Fe_2O_3 versus Al_2O_3 (Fig. 4) indicates that sediments from Sites A and B and for most samples from Site C are similar to Ppc. Excess Fe_2O_3, however, is present in about 16 samples from Site C. These samples are from cores described as containing varying proportions of metalliferous sediment (Bischoff et al., 1979). "Excess Fe_2O_3" is defined as total Fe_2O_3% - 0.46 Al_2O_3%, where .46 is the Fe_2O_3/Al_2O_3 ratio for Ppc. Values of excess Fe_2O_3 at Site C are as great as 7.3% and cannot be accounted for by, nor correlated with, hydrogenous MnO_2. Excess Fe_2O_3, therefore, to a first approximation, probably reflects the metalliferous sediment component.

K_2O versus Al_2O_3 (Fig. 5) displays a surprising scatter and shows a general enrichment of K_2O relative to Ppc for all samples from all three sites. Samples from Site C cluster along a trend parallel to Ppc ($r = 0.92$). Samples from Sites A and B, however, display considerable scatter (for A and B, $r = 0.78$) and have a trend quite different from C. Many of the samples from A and B have K_2O/Al_2O_3 ratios similar to BNL.

The plot of MgO versus Al_2O_3 (Fig. 6) shows that all samples from Sites A and B are moderately enriched in MgO. Most of the Site C samples, however, plot nearer the Ppc line, although some show extreme MgO enrichment. "Excess MgO", defined as total MgO%-0.2Al_2O_3%, has a correlation coefficient of -0.9 with Al_2O_3, a figure which indicates that the enrichment is greatest when the

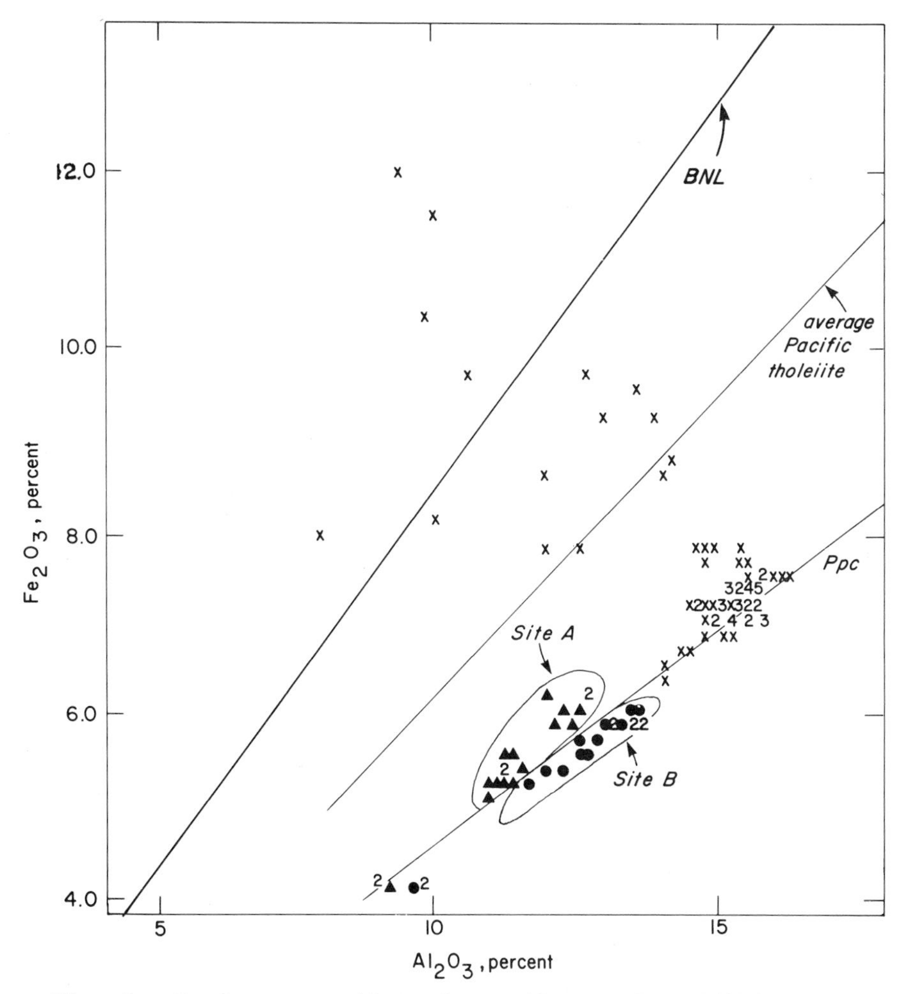

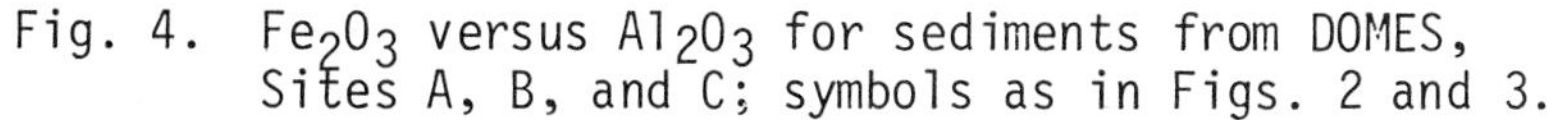
Fig. 4. Fe_2O_3 versus Al_2O_3 for sediments from DOMES, Sites A, B, and C; symbols as in Figs. 2 and 3.

content of pelagic clay is lowest. The strong correlation of excess MgO with excess Fe_2O_3, CaO, and Sr suggests that the enrichment may be in part associated with metalliferous sediment and in part with $CaCO_3$.

The strongest interelement relation is between CaO and CO_2 (r= .978). Most of the P_2O_5 appears to be present as apatite, because the correlation coefficient of CaO with 1.27 CO_2 + 1.33 P_2O_5 (the coefficients being the respective stoichiometric ratios of CO_2 and P_2O_5 in calcite and apatite) is .982.

Mo, Ni, Pb, and Co are very strongly correlated with MnO

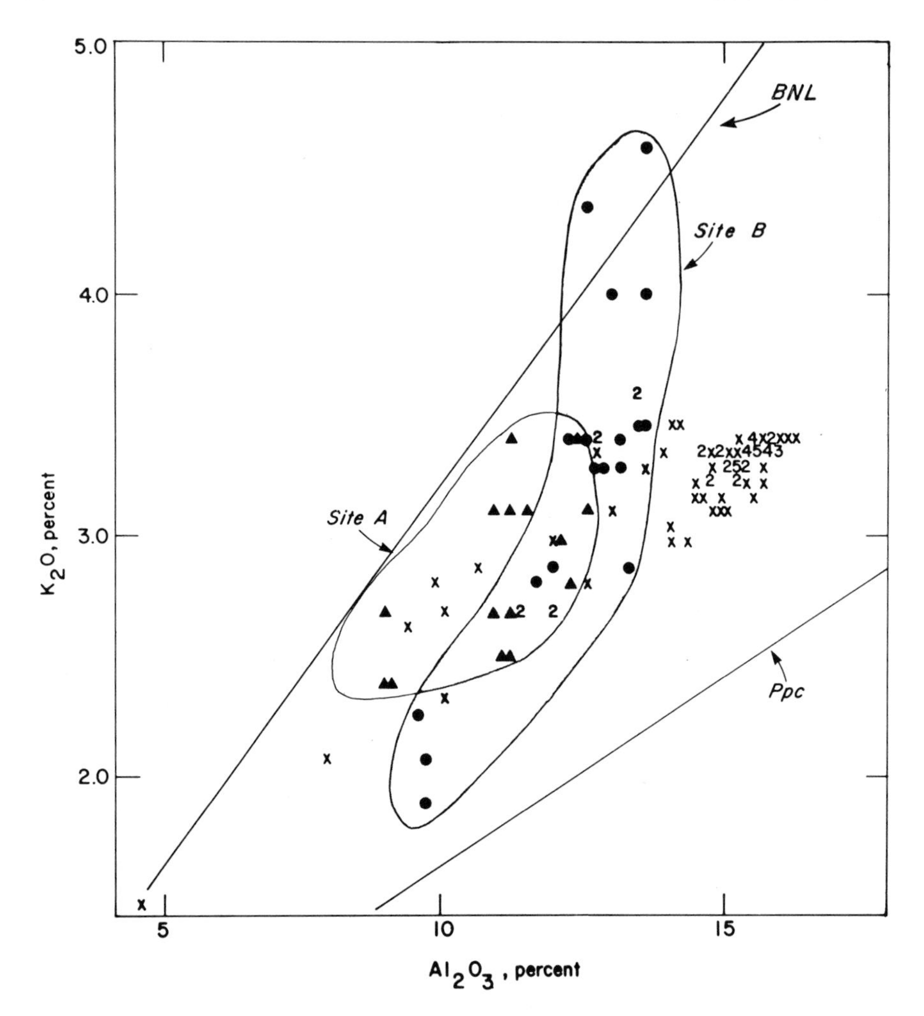

Fig. 5. K_2O versus Al_2O_3 for sediments from DOMES Sites A, B, and C; symbols as in Figure 2.

(Table 5). Their ratios to Mn (Table 3) are close to those in Pacific manganese nodules (Cronan and Tooms, 1969; Calvert and Price, 1977) suggesting the bulk of these elements resides in the hydrogenous fraction. The weaker correlation of Cu with MnO ($r = .69$) suggests either fractionation into other sediment components or inhomogeneous distribution in the hydrogenous fraction.

Distribution of the excess oxides

The preceding indicates that large amounts of SiO_2, Fe_2O_3, MgO, and K_2O cannot be accounted for by a single homogeneous pelagic-clay fraction. How else can these components be accounted for?

We begin with excess Fe_2O_3, significant amounts of which occur only in sediments from Site C. A plot of excess Fe_2O_3 versus MnO (Fig. 7) indicates that most of the excess Fe_2O_3 greater than about 1.5% is independent of MnO. Only four samples with excess

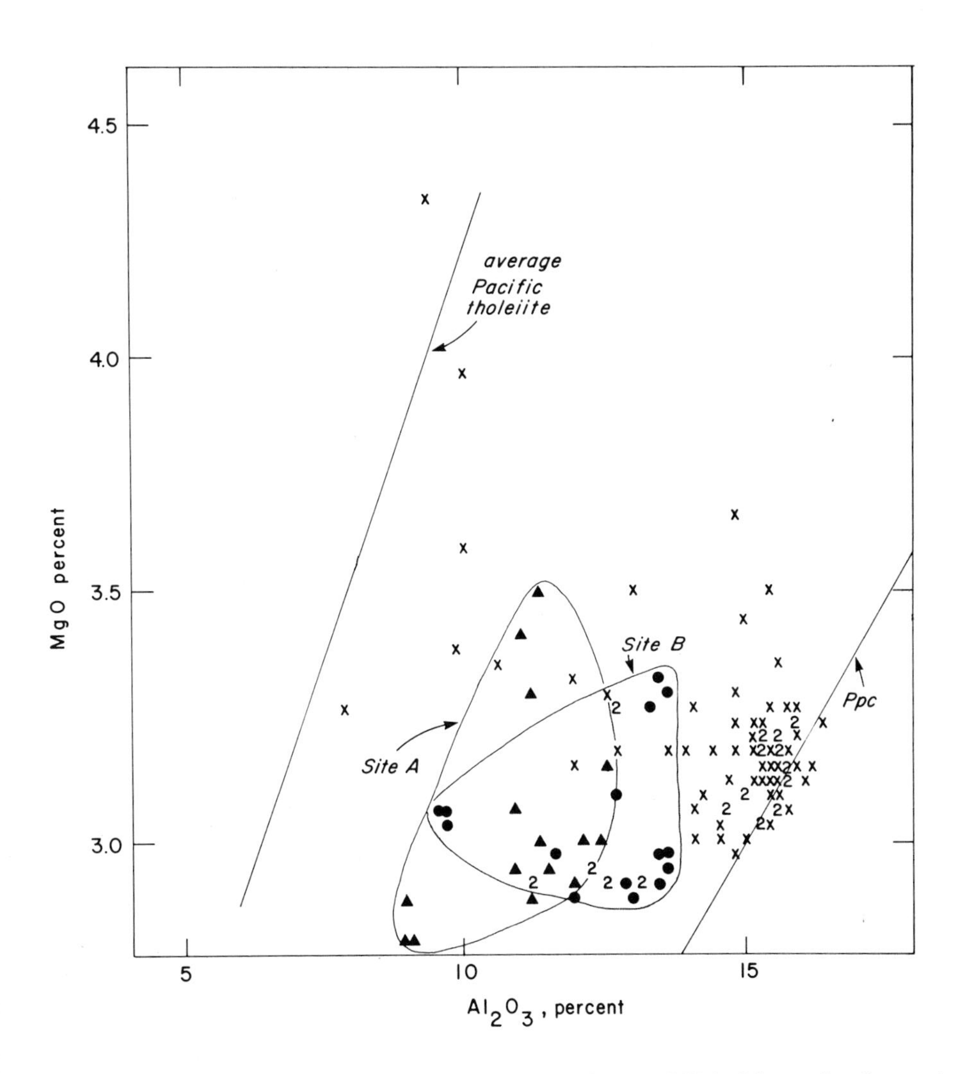

Fig. 6. MgO versus Al_2O_3 for sediments from DOMES Sites A, B, and C; symbols as in Figures 2 and 3.

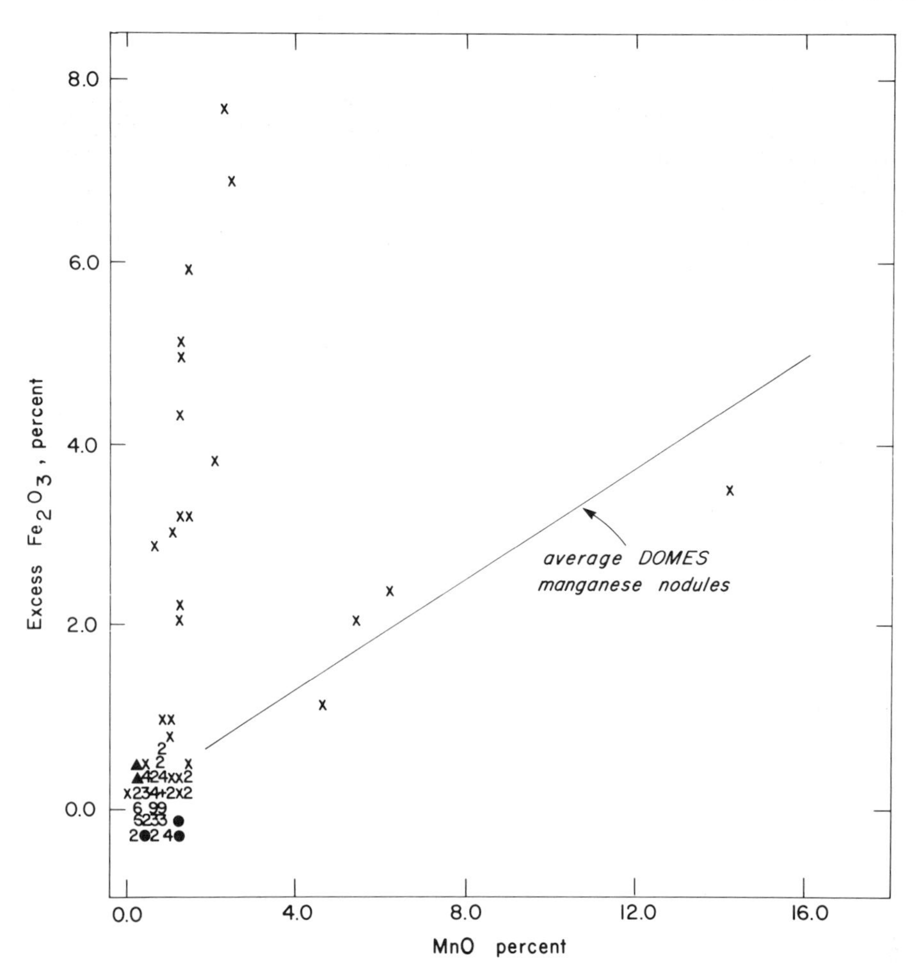

Fig. 7. Excess Fe_2O_3 versus MnO for sediments from DOMES Sites A, B, and C; symbols as in Fig. 2. Excess Fe_2O_3 = total Fe_2O_3 - 0.46 Al_2O_3, where .46 is the Fe_2O_3/Al_2O_3 ratio for Ppc

Fe_2O_3 can be related to high values of MnO, and these samples have Fe_2O_3/MnO ratios close to 2.5, the ratio for manganese nodules in the DOMES area (Piper et al., 1977; this volume). Excess Fe_2O_3 in these samples is presumably hydrogenous. Samples with excess Fe_2O_3 less than about 1% tend to scatter because such low values approach the analytical precision of the data. Thus, only values of excess Fe_2O_3 greater than 1% are considered significant.

On this basis, a group of 13 samples from Site C has excess

Fe_2O_3 values as great as 8% (Fig. 7), which cannot be explained as hydrogenous. These samples come from cores for which the visual descriptions suggest metalliferous sediment. Thus, "metalliferous" Fe_2O_3 = excess Fe_2O_3 - .32 MnO.

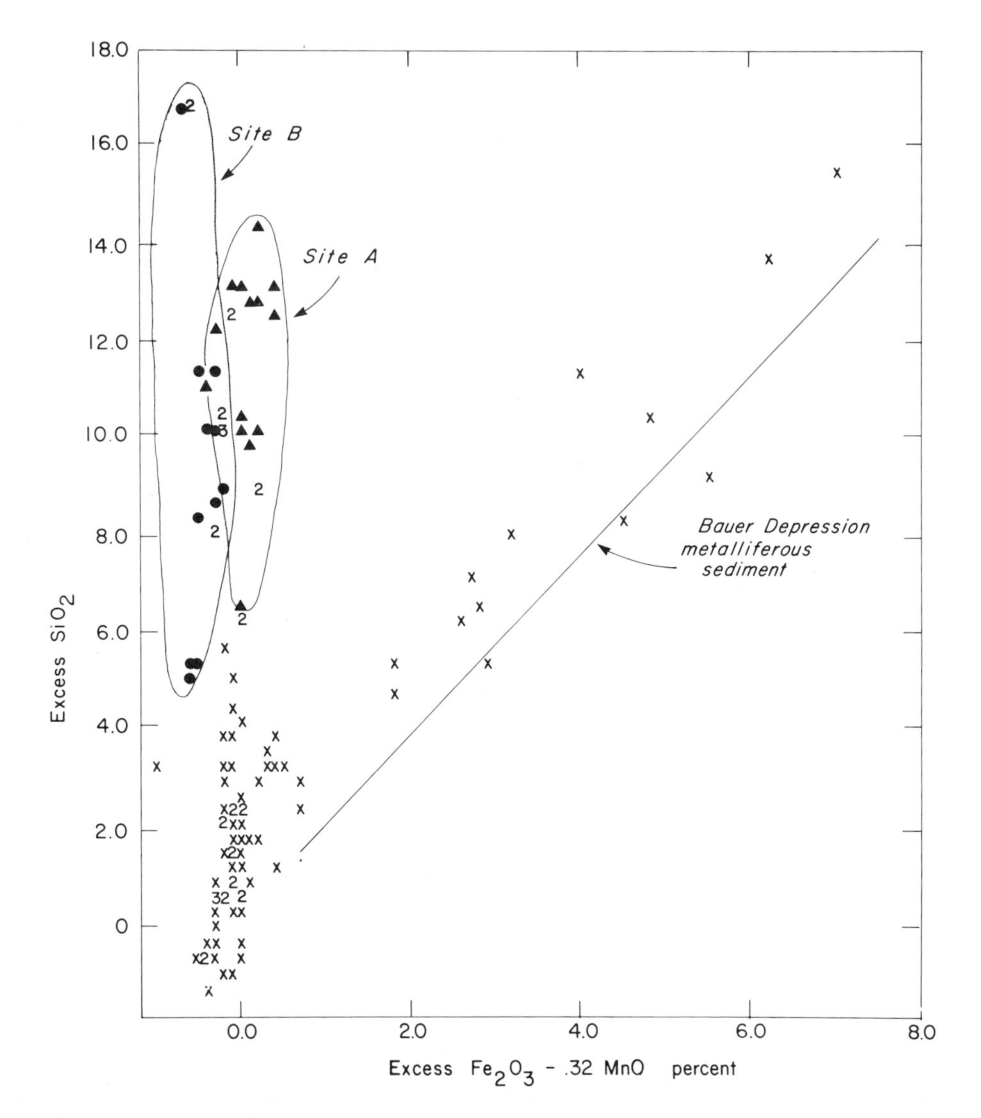

Fig. 8. Excess SiO_2 versus excess Fe_2O_3 - .32 MnO for sediments from DOMES Sites A, B, and C; symbols as in Figure 2. Excess SiO_2 = total SiO_2 - 3.3 Al_2O_3 where 3.3 is the SiO_2/Al_2O_3 ratio for Ppc. Data for average Bauer Depression metalliferous sediment from Sayles and Bischoff (1973)

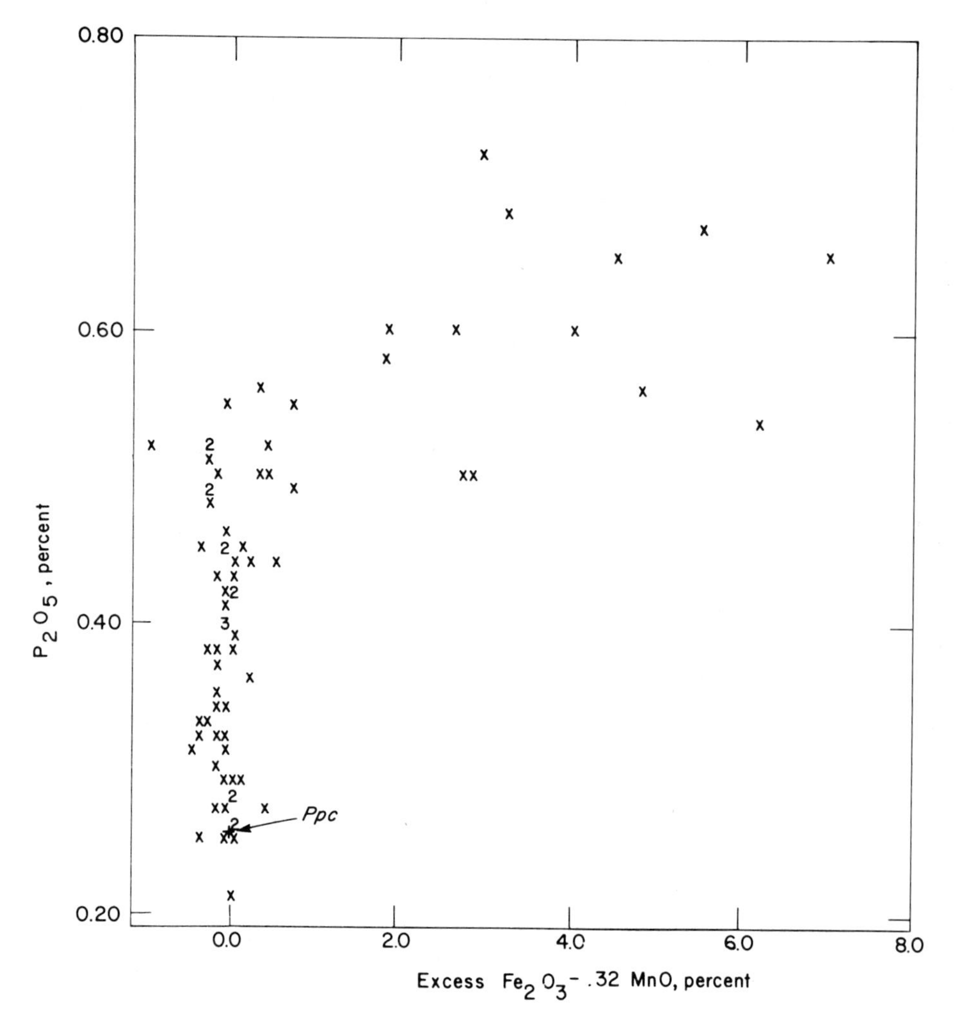

Fig. 9. P_2O_5 versus Fe_2O_3 - .32 MnO for sediment from DOMES Site C

A plot of excess SiO_2 versus "metalliferous" Fe_2O_3 shows two relations (Fig. 8). Samples from Sites A and B have excess SiO_2 ranging up to 17% but have no excess Fe_2O_3. In these samples biogenic silica apparently accounts for all the excess SiO_2. For Site C samples excess SiO_2 is strongly related to "metalliferous" Fe_2O_3 (for Site C samples, r = 0.9), and the points fall along the trend for metalliferous sediment of the Bauer Depression (Sayles and Bischoff, 1973).

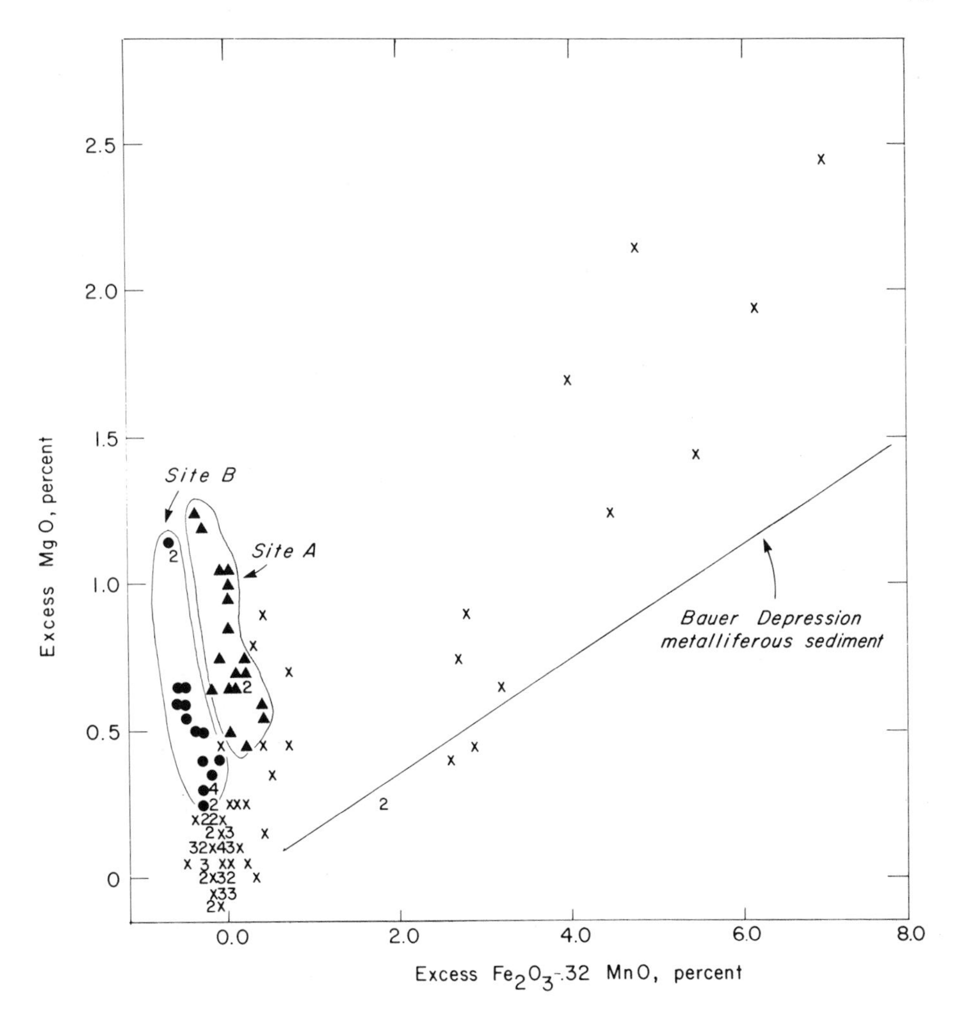

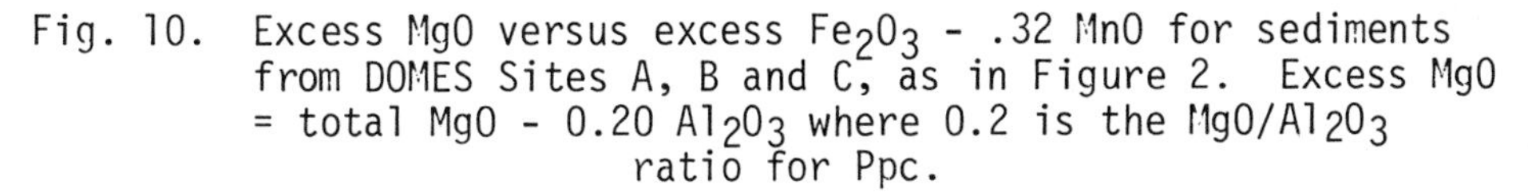
Fig. 10. Excess MgO versus excess Fe_2O_3 - .32 MnO for sediments from DOMES Sites A, B and C, as in Figure 2. Excess MgO = total MgO - 0.20 Al_2O_3 where 0.2 is the MgO/Al_2O_3 ratio for Ppc.

We conclude, therefore, that "metalliferous" Fe_2O_3 represents the metalliferous-sediment of the sediment in the same way that Al_2O_3 represents the pelagic-clay fraction.

We can then determine affinities of other components to metalliferous sediment by plotting against "metalliferous" Fe_2O_3. For example, some of the P_2O_5 found in Site C sediments, particularly the values in excess of 0.5%, is associated with "metalliferous"

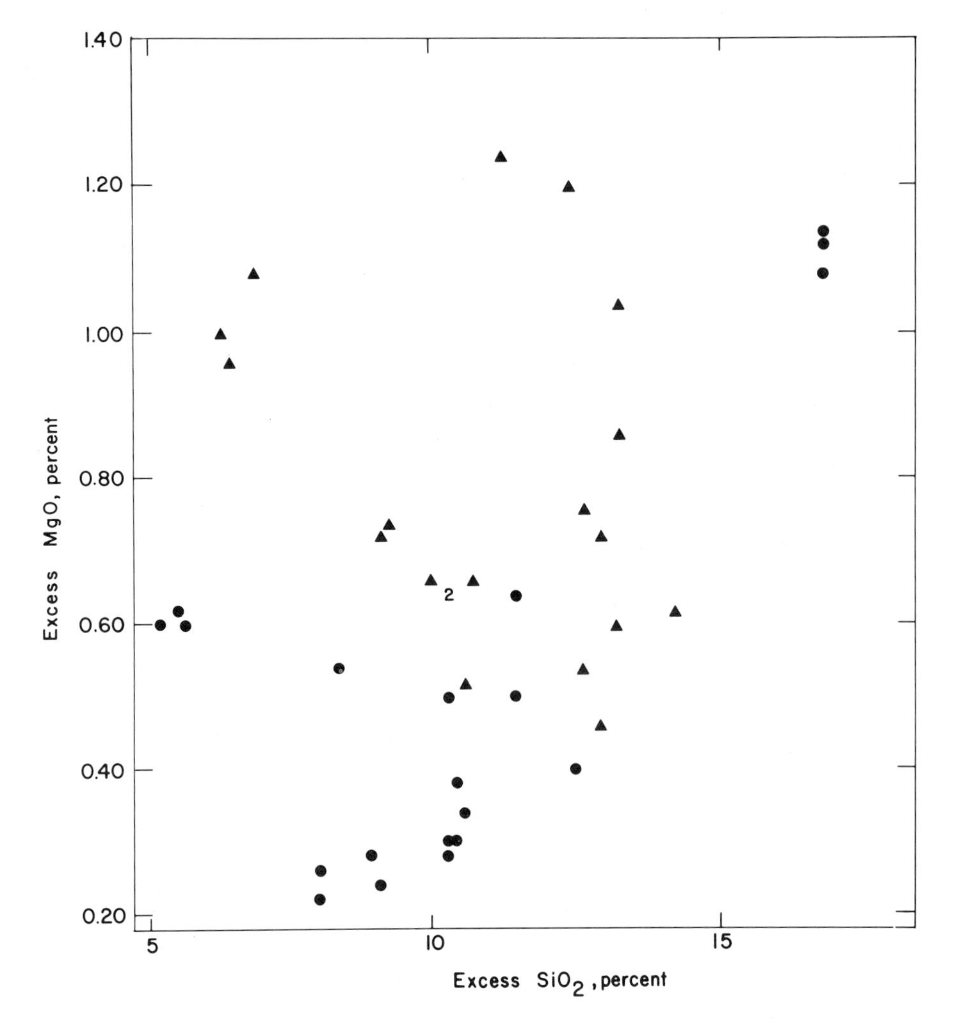

Fig. 11. Excess MgO versus excess SiO_2 for sediments from DOMES Sites A (filled triangles) and B (filled circles).

Fe_2O_3 (Fig. 9), a fact suggesting an affinity similar to that in metalliferous sediments from the southeast Pacific (Berner, 1973; Froelich et al., 1977).

It was only recently noted that MgO is an important component of metalliferous sediments (Bischoff and Rosenbauer, 1977). Thus, the strong correlation between MgO and "metalliferous" Fe_2O_3 in Site C sediments (Fig. 10, for Site C samples, r = .92) supports the inference that excess Fe_2O_3 represents a metalliferous sediment fraction and that it is indeed enriched in MgO. The Site C metalliferous sediment is somewhat richer in MgO than Bauer Depression deposits, however, (Sayles and Bischoff, 1973), with an average Fe_2O_3/MgO ratio of 3 compared to 5.9 for the Bauer Depression sediments. Further considerations of the metalliferous sediment are the subject of a separate paper (Bischoff et al., 1979).

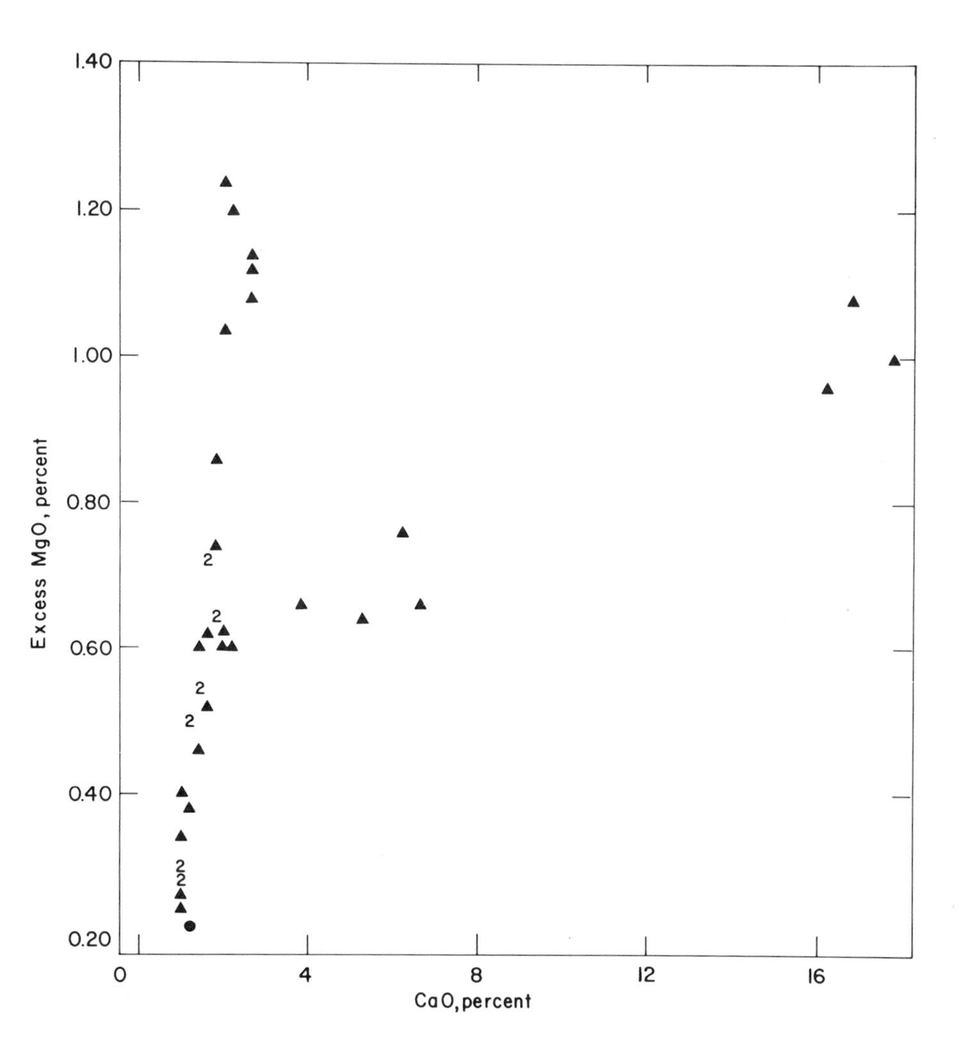

Fig. 12. Excess MgO versus CaO for sediments from DOMES Sites A and B; symbols as in Figure 11.

Sites A and B also contain significant excess MgO unrelated to metalliferous sediment (Fig. 10). Values average 0.5% and range up to 1.3%. Because there is neither identifiable metalliferous sediment at Sites A and B nor excess Fe_2O_3, the excess MgO must be associated with another sediment fraction. A plot of excess MgO versus excess SiO_2 for Sites A and B (Fig. 11) shows a weak but significant correlation (r = .31). All samples with high excess

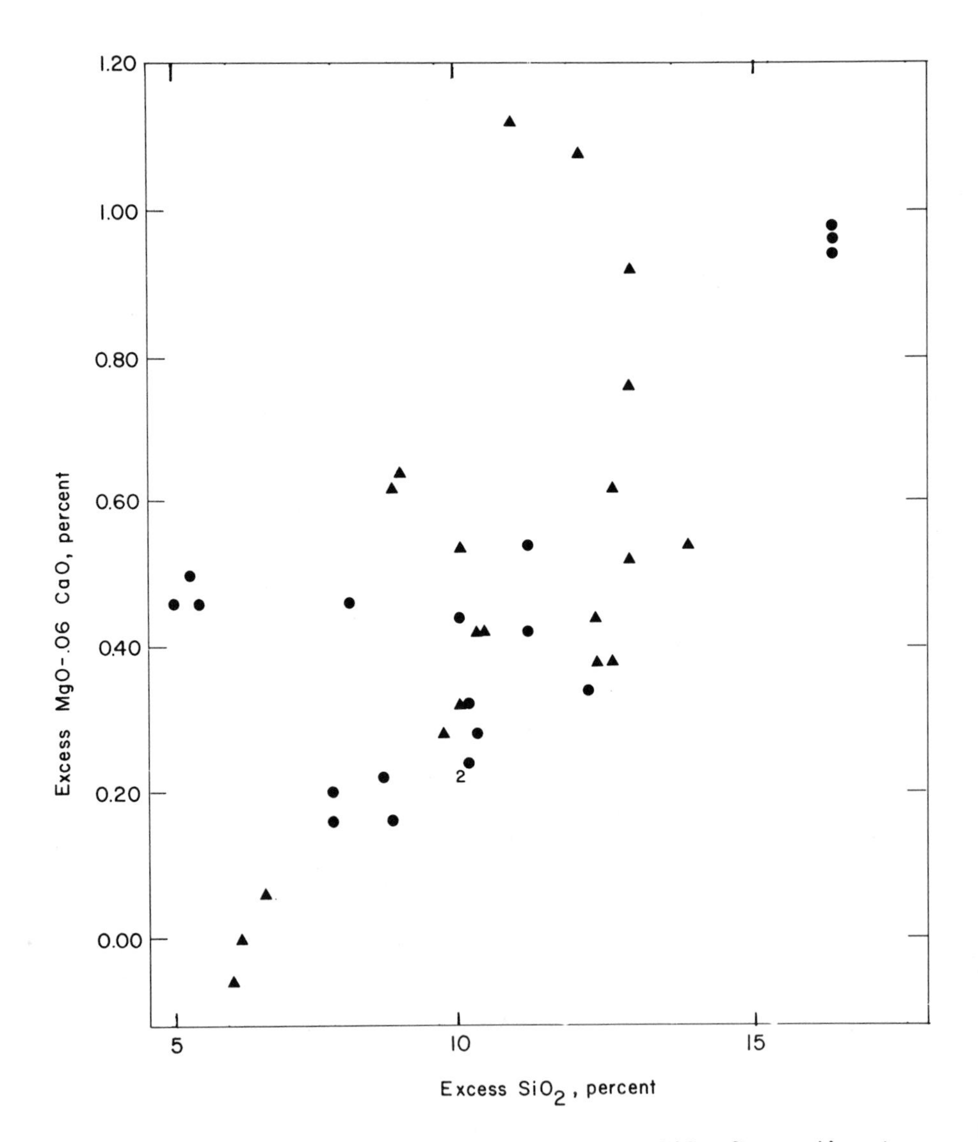

Fig. 13. Excess MgO - .06 CaO versus excess SiO_2 for sediments from DOMES Sites A and B; symbols as in Figure 11.

SiO_2 have high excess MgO as well, but some samples with high excess MgO have rather low SiO_2 contents. A plot of excess MgO against CaO (Fig. 12) suggests that high values of $CaCO_3$ are accompanied by significant quantities of excess MgO. The MgO/CaO ratios for these samples are close to 0.06, which is equivalent to

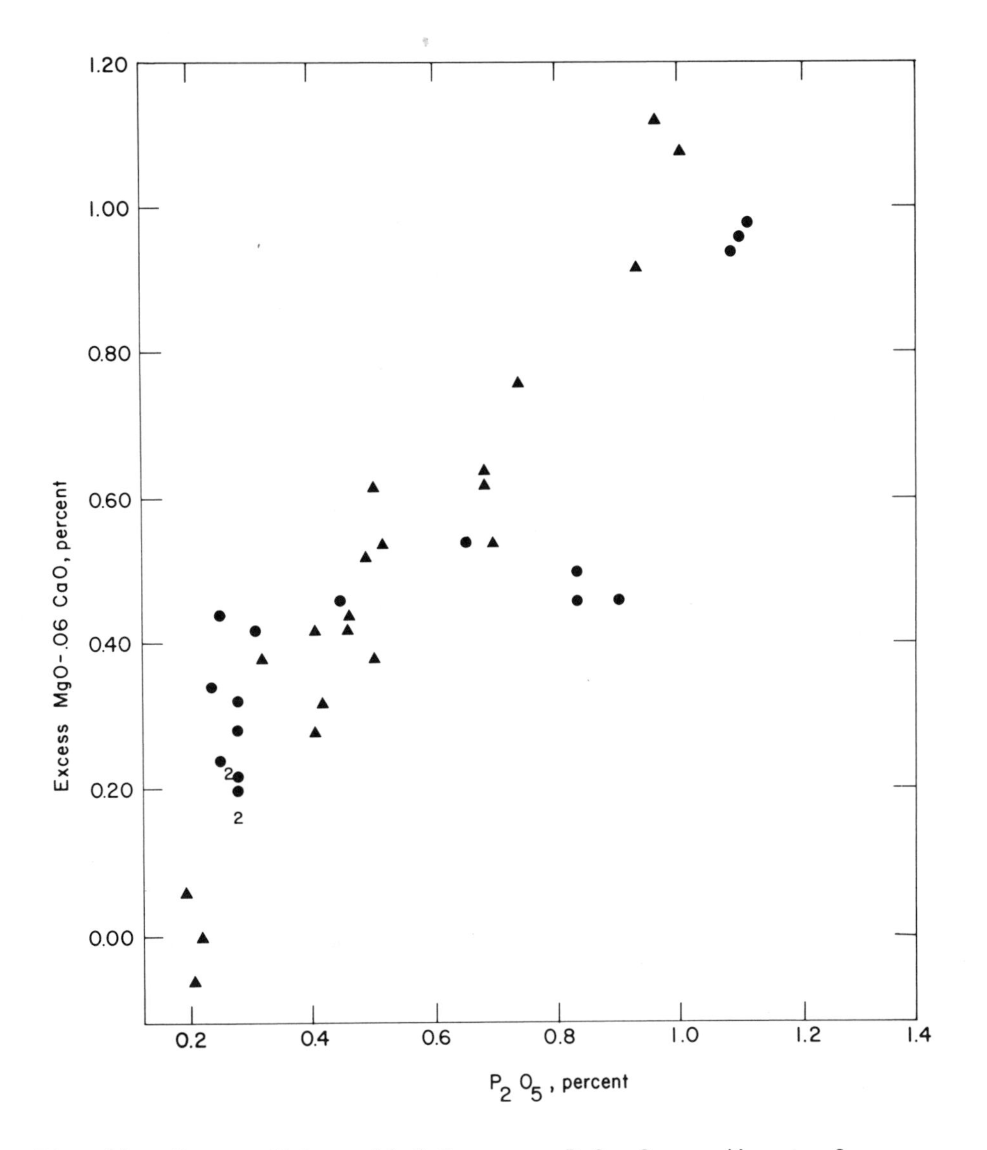

Fig. 14. Excess MgO - .06 CaO versus P_2O_5 for sediments from DOMES Sites A and B; symbols as in Figure 11.

a mole ratio of 0.07, implying that the $CaCO_3$ contains 6.6 mole percent $MgCO_3$, a very high value for pelagic calcite. Replotting excess SiO_2 versus excess MgO - 0.06 CaO, (to remove the carbonate contribution) improves the excess SiO_2 - MgO correlation to 0.63 (Fig. 13) and strengthens the implied association of these two components. In addition, carbonate-corrected excess MgO is correlated with P_2O_5 ($r = 0.89$, Fig. 14). This plot indicates that the mole ratio of Mg/P is very close to 1.0, which is a puzzling relation because the only phosphatic material observed is apatitic fish debris (Hein and Jones, 1977). There is no obvious genetic or mineralogic explanation for the affinity between Mg and P in the marine environment, or for a similarily strong correlation between excess MgO and MnO ($r = .82$). What do P_2O_5 and MnO have in common? Both MnO and P_2O_5 may be indicators of accumulation rate. Both accumulate extremely slowly, so that their absolute concentrations may be inversely proportional to sedimentation rate or degree of sediment reworking.

Reworking or recycling is simply the process of erosion and resedimentation of local older sediments. Micropaleontological data indicate that reworking in the DOMES sediments is highly variable from core to core (Quinterno and Theyer, this volume. The degree of reworking in a given sample is indicated by the percentage of older microfossils accompanying Quaternary microfossils. Some samples, for example, have 5% latest Quaternary radiolarians and 95% Miocene radiolarians, such as core 51-33 in Site B (Quinterno and Theyer, this volume). Thus, the sediment represented by this core is accumulating at present and, therefore, is classified as Quaternary, but the bulk of the sediment is not primary but is actually Miocene. Such sediment, is considerably more mature from a diagenetic and geochemical standpoint than Quaternary sediment represented by a nearby core, 54-51, which contains only 9% older microfossils (Quinterno and Theyer, this volume). Significantly, excess MgO averages 1.18 percent (39% of the total MgO) for samples from core 51-33 and 0.25% (8% of total MgO) for core 54-51.

The idea that P_2O_5 is related to sediment maturity can be tested by plotting P_2O_5 against percent reworked microfossils. Unfortunately, microfossil studies were not performed on the same samples for chemical analyses. Different intervals were chosen for biostratigraphic studies, although from the same cores sampled for chemical analysis; moreover, accuracy of visual estimates of the percentage of older microfossils will not be as precise as chemical data. An approach allowing semiquantitative comparison is to average both the chemical and microfossil results for each of the 20 cores from A and B. The resulting relation is highly suggestive, (Fig. 15). Cores with highest values of P_2O_5 also have the highest percentage of reworked microfossils and vice versa. Out of a total of ten cores, only one (46-1) does not plot on the trend, and the hypothesis that the P_2O_5 content is related to sediment maturity is supported.

Getting back to the question of excess MgO, it appears that excess MgO is also somehow related to sediment maturity. We have seen that excess MgO is related to excess SiO_2 (Fig. 13), which for Sites A and B represents biogenic silica. Reaction between sea water Mg^{+2} and biogenic silica to form a hydrated magnesium silicate (sepiolite-type) coating on the silica surface is thermodynamically favorable and has been suggested as a possible explanation of

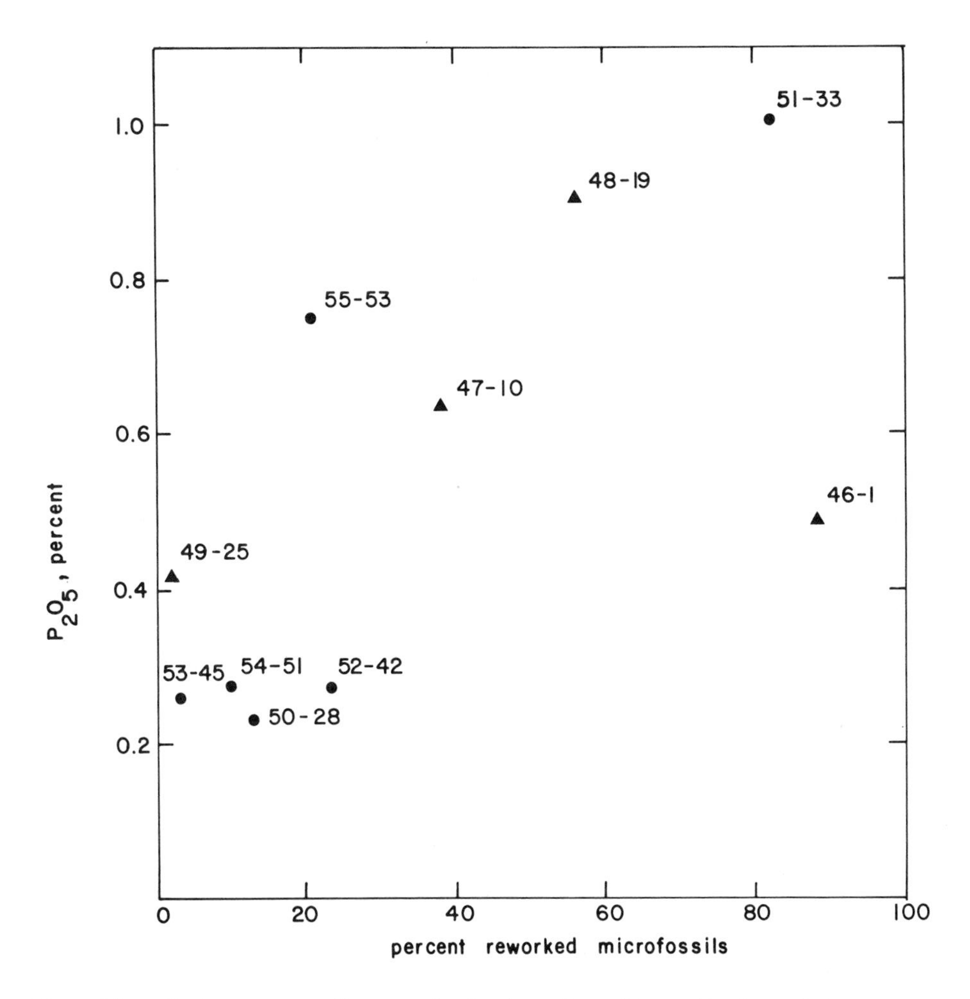

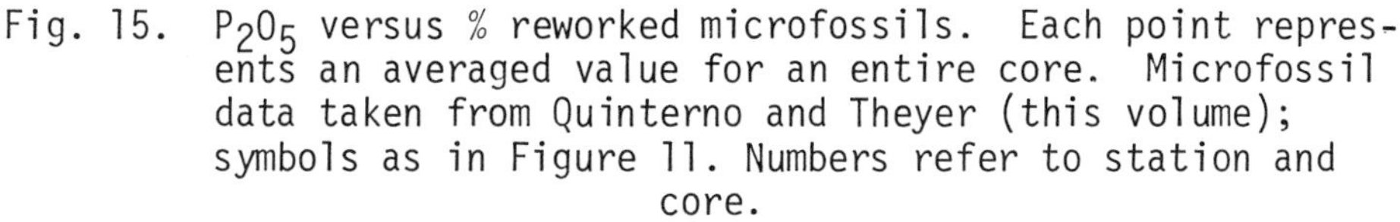
Fig. 15. P_2O_5 versus % reworked microfossils. Each point represents an averaged value for an entire core. Microfossil data taken from Quinterno and Theyer (this volume); symbols as in Figure 11. Numbers refer to station and core.

the low concentrations of dissolved SiO_2 in the pore waters of some siliceous marine sediments (Bischoff and Sayles, 1972). Thus, uptake of Mg^{+2} by biogenic SiO_2, if it occurs, must be a very slow process and the degree of reaction would be an indicator of diagenetic maturity. A plot of excess MgO versus the product (excess

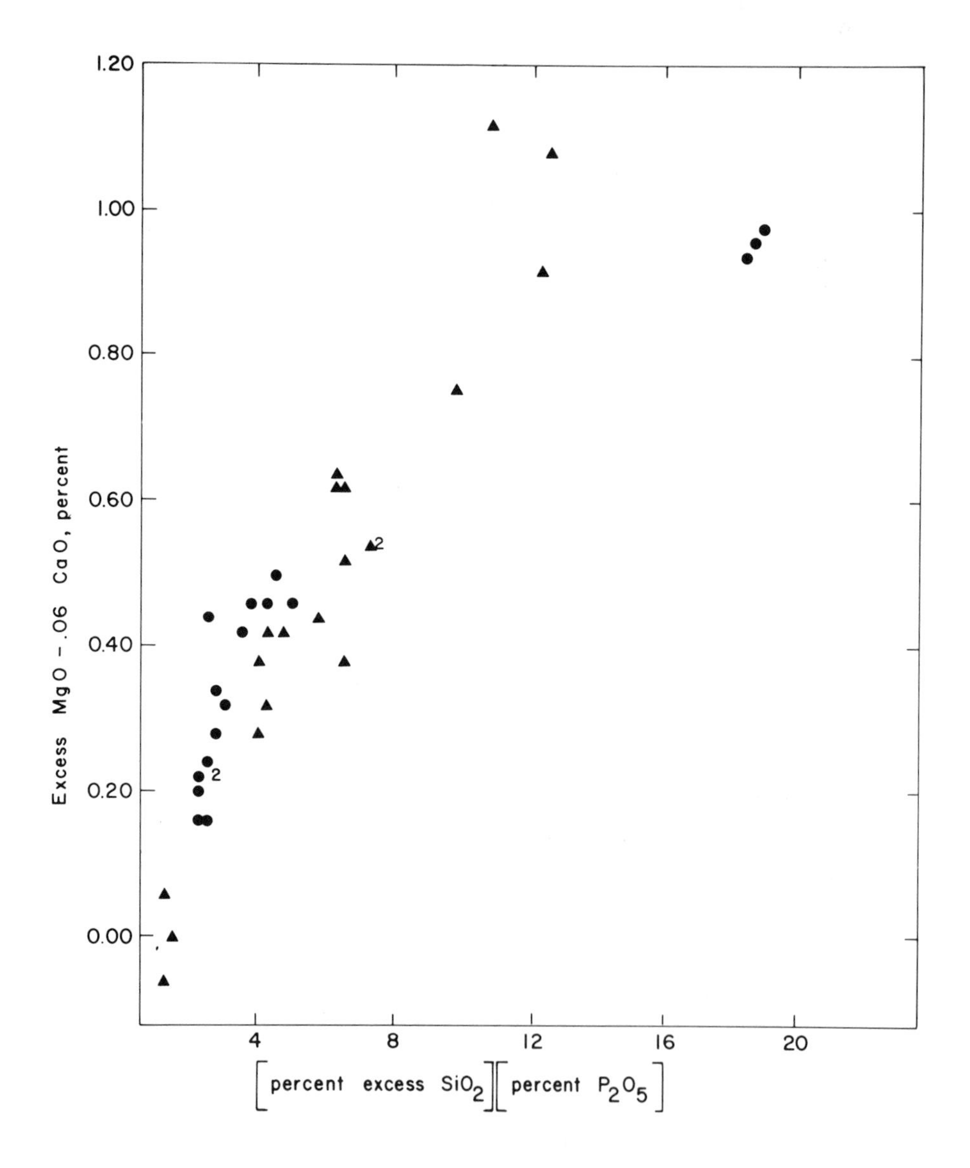

Fig. 16. Excess MgO - 0.06 CaO versus the product P_2O_5 times excess SiO_2 for sediments from Sites A and B; symbols as in Figure 11.

SiO_2 times P_2O_5) provides a test of this hypothesis. If the strong individual correlations of P_2O_5 and SiO_2 to excess MgO represent independent phenomena, then their product should not correlate with excess MgO. The correlation for the DOMES samples is enhanced, however, (Fig. 16), providing support for the concept of time-mediated silica uptake of Mg. Piper et al. (this volume) found that virtually all of the excess MgO from Sites A and B sediment was leachable by 25% acetic acid solution (pH = 4). This behavior is consistent with the excess MgO being bound as a hydrated magnesium silicate rather than an alumino-silicate.

The source of excess K_2O is another problem (Fig. 5). Excess K_2O for all three sites averages 0.98% of the bulk sediment, or 30% of the total K_2O. A part of this excess may be due to inaccurate analyses of K_2O for Ppc or to inhomogeneous distribution of K_2O in lithogenous material. Careful examination of the clay

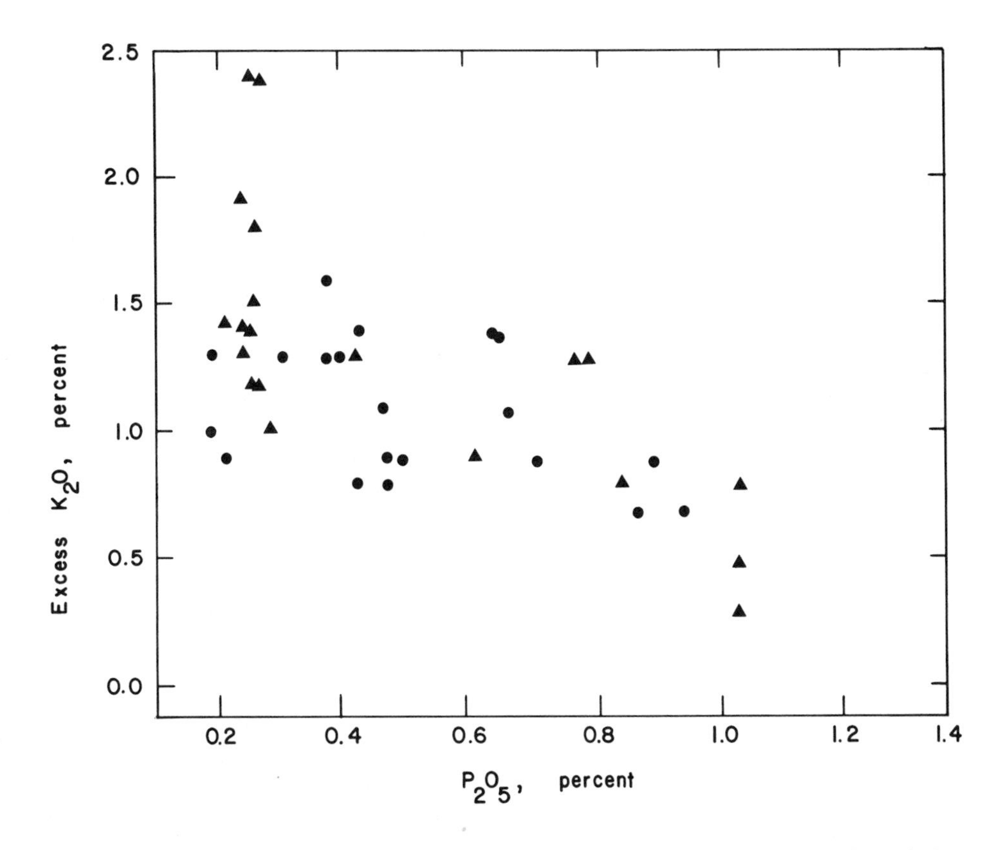

Fig. 17. Excess K_2O versus P_2O_5 for sediments from Sites A and B; symbols as in Figure 11.

mineralogical composition of the sediment presented by Hein *et al.* (1976) and Hein and Jones (1977) reveals no systematic mineralogic relation to excess K_2O.

Many samples from the three DOMES sites plot close to BNL on the K_2O-Al_2O_3 diagram (Fig. 5). This relation suggests that primary (as opposed to reworked) sediment might be richer in K_2O. This hypothesis is strengthened by the inverse relation between percent excess K_2O and percent P_2O_5 (Fig. 17, r = -0.6), for sediments from A and B (samples for Site C are excluded because the metalliferous sediment contributes an additional source of P_2O_5). Thus, much of the excess K_2O can be reasonably attributed to contribution of primarily relatively immature sediment fraction. Much of the K_2O in this fraction, therefore, is released during subsequent recycling and diagenesis. Experiments by Piper *et al.* (this volume) indicate that most of the excess K_2O is leachable by 25% acetic acid. The residues after leaching have K_2O/Al_2O_3 ratios very close to Ppc.

Distribution of Sediment Fractions

The various sediment components can now be defined geochemically. The amount of metalliferous sediment, for a first approximation, is considered proportional to excess Fe_2O_3% - 0.32 MnO% (see Fig. 7). Pure metalliferous sediment from the Bauer Depression appears to contain about 32% Fe_2O_3 (Bischoff and Rosenbauer, 1977). Therefore, the weight percent of Bauer-type metalliferous sediment in any sample is approximated by (3.13)[excess Fe_2O_3% - 0.32 MnO%]. Biogenic opal is essentially pure hydrated SiO_2, but we have seen that excess SiO_2 is associated with metalliferous sediment as well as with biogenic opal. Because the Fe_2O_3/SiO_2 ratio in our metalliferous component appears to be very constant (see Fig. 8), we take biogenic-silica to be excess SiO_2%-1.74 [excess Fe_2O_3%-0.32 MnO%]. Applying these expressions to the sediment compositions of Table 2 indicates that Sites A and B contain approximately 11% and Site C only about 1.8% biogenic SiO_2 (Table 4), whereas metalliferous sediment occurs only in Site C and accounts for only about 1.8%.

Summation of the contributions of the six sediment fractions yields totals that are close to 100% (Table 4). The close agreement supports the general validity of this simple geochemical approach to sediment classification. This classification requires that concentrations need only be known for Si, Al, Fe, Ca, Mn, and Na, a minimum suite of elements. That the technique does work depends on the fact that deep-sea sediments are slowly deposited and well mixed over limited geographic regions (Heath *et al.* 1970). The assessment of the local contribution of volcanic debris on the basis of geochemical data is difficult. CaO in excess over CO_2 and P_2O_5 is one possible criterion for unaltered volcanic debris, but during sea-floor alteration, the excess CaO may be lost. Excess

TiO_2 is also a possible index, somewhat more attractive because Ti is probably conserved during alteration. However, a use of Ti will not be highly sensitive to small percentages of volcanic debris. A mixture of 25% basalt and 75% pelagic clay will have an excess of only 0.025% TiO_2 out of a total TiO_2 content of 0.84%.

The variations of the sediment fractions for the entire suite of DOMES samples can be shown by triangular diagrams. The fractions of pelagic clay, biogenic ooze, and metalliferous sediment were calculated for each sample and normalized to 100% to allow construction of Figure 18. Before normalization, the sums of these three fractions averaged 88% with sea salts included and 92% with salts removed. Thus, normalization to 100% only negligibly changed the absolute percentages of the components. Samples from Sites A and B cluster along the pelagic clay-biogenic ooze axis and show little deviation from the averages shown in Table 4. Likewise, the majority of samples from Site C plot within a small group along the same axis but closer to the pelagic-clay end member. Twenty samples from C have significant fractions of metalliferous sediment with values ranging approximately to 30%.

Proportions of biogenic SiO_2 and pelagic clay calculated from the chemical data differ significantly from the visual estimates. Petrographic estimates (Cook et al. 1977) suggest that the average biogenic SiO_2 content for sediments of A and B is close to 50% rather than to the 10% estimated by the chemical model (Fig. 18 and Table 4). Such a discrepancy probably results in part from the difference in volume percent from weight percent for remains of siliceous organisms, and in part from the tendency for visual estimates to favor the large opal tests at the expense of the microscopically unresolvable clay particles.

Distribution of Minor Elements

Correlations can now be calculated not only between elements but also between elements and geochemically defined sediment components. Co, Cu, Mo, Ni, and Pb correlate strongly with MnO and, therefore, with the hydrogenous fraction (Table 5). CO_2, P_2O_5, and Sr and some excess MgO are correlated with CaO. Be and Cr appear to be concentrated in the pelagic component on the basis of correlations with Al_2O_3. Sc seems to favor the biogenic SiO_2 component ($r = 0.65$).

Strong correlations with metalliferous sediment at Site C were found for excess MgO ($r = 0.83$), excess SiO_2 ($r = 0.90$), and P_2O_5 ($r = 0.64$). The P_2O_5 affinity with metalliferous sediment has been found in metalliferous sediments from the southeast Pacific as well (Berner, 1973; Froelich et al. 1977). Thus P_2O has two affinities--one with CaO and one with metalliferous sediment.

Co, Cu, Mo, Ni, Pb, and Zn are not strongly fractionated into metalliferous sediment. This result was surprising because high concentrations of these metals were found in a single sample from Site C that is particularly enriched in metalliferous sediment

(Bischoff and Rosenbauer, 1977). Apparently this particular sample also had a large fraction of hydrogenous material so that the hydrogenous suite of metals was mistakenly assigned to the metalliferous fraction. The metalliferous sediment appears to be composed primarily of SiO_2, Fe_2O_3, MgO, and P_2O_5.

Strong correlations, and hence strong fractionation, with particular sediment components, was not found for the remaining elements.

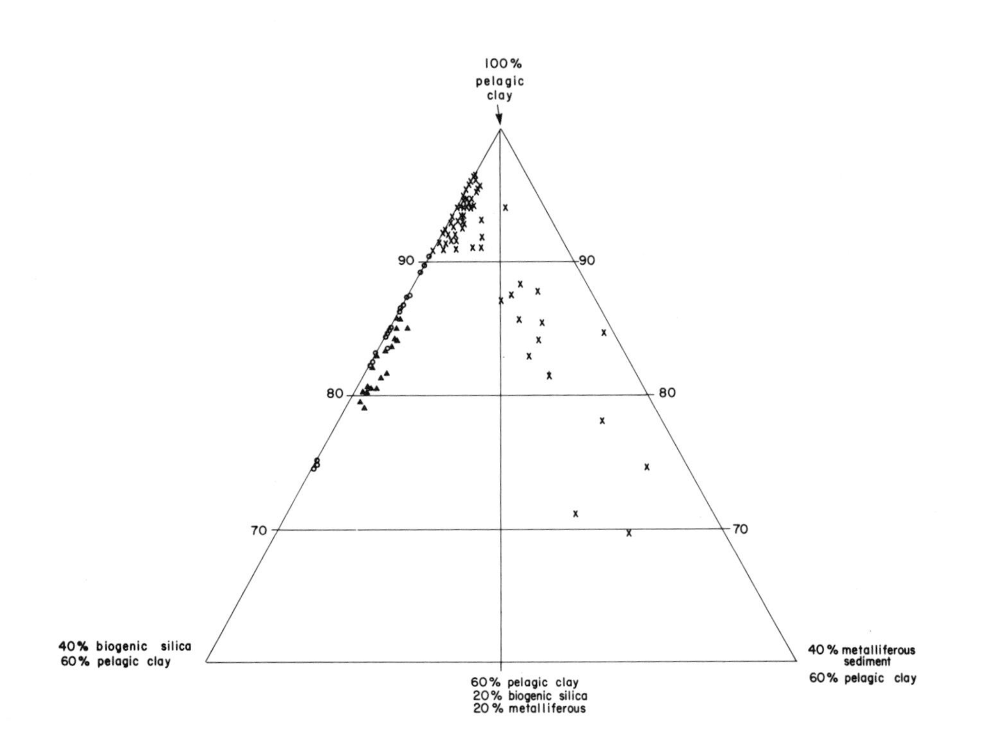

Fig. 18. Upper one third of a triangular diagram showing relative proportions of pelagic clay, biogenic silica and metalliferous sediment samples from DOMES Sites A (triangles), B (circles), and C (x's). Expression for calculating relative proportions are given in Table 4.

Alternative Partitioning Models

To test the simple partitioning model developed in the preceding sections, we subjected the data to multivariate analysis. In addition, we have attempted to partition Fe, Al, Si, Mn, Ni, Cu, Zn, and Ba between hypothesized detrital, metalliferous, hydrogenous, and biogenous sources using the model that Heath and Dymond (1977) developed for southeastern Pacific sediments.

The multivariate technique we used was Q-mode factor analysis (Imbrie and van Andel, 1964; Klovan and Imbrie, 1971). This technique describes each sample in terms of a small number of artificial variables (which can be though of as "ideal" end members) that distribute as equally as possible some specified fraction of the total variance in the data set. In order to give each element equal weight, the data were normalized so that each element had a mean of 0 and a variance of 1. A factor analysis of the complete data set was only partially successful. Even 10 factors accounted for less than 92 percent of the total variance, and less than half the variance of certain samples. In most sets of geologic data, a small number of factors accounts for most of the information in the data, and additional factors explain only a small percentage of residual noise. For the DOMES data, however, even the eighth factor still accounted for 4 percent of the total variance. The first two factors, with high loadings in Si, Al, Fe, Ti, Be, Cr, and Pb in the first case, and Fe, Mg, P, Mn, Ba, Co, Cu, Mo, Ni, Pb and Zn in the second, are geologically reasonable (continental detritus and metalliferous material), but later factors are not. We attribute this to analytical errors in the optical spectrographic data that must be comparable in magnitude to some of the real variations in the DOMES samples.

Factor analyses of only major elements at each of the sites gave cleaner patterns. In Site A, 4 factors accounted for 94% of the variance; in Site B, 94%; and in Site C, 88%. In each case, one of the factors, which is dominated by Na, is clearly attributed to sea salt. The other three factors include a detrital member (heavily loaded in Al, Fe or Mg, K, Ti, and Si) and a smectite-authigenic member (heavily loaded in Mg and Fe, with lesser Mn) in all areas. In addition, Site C has a metalliferous factor (high Fe, P, and K loadings), and Sites A and B have biogenic factors (high Ca, Si, K, and P loadings). These groupings agree well with the partitioning model.

Finally, we factored just the major elements (excluding Na) for all the DOMES samples. The results are summarized in Figure 19. This strictly objective partitioning, which allows elements to be shared between factors, can be compared with our simpler, more subjective partitioning scheme (Fig. 18). The plot shows that sediments from Sites A and B are very similar. Site C samples, however, occupy a totally distinct field and are depleted in biogenic material (F3) relative to Sites A and B. In addition, the Site C samples include the most metalliferous as well as the most terrigenous sediments of the entire DOMES suite.

The partitioning model of Heath and Dymond (1977) assumes that all Fe and Al are either hydrothermal or detrital and that all Ni is detrital, hydrothermal, or hydrogenous (authigenic). By assuming that all other elements have constant ratios to Al, Fe, and Ni in detrital, hydrothermal, and hydrogenous material, respectively (Table 6), we can partition each element in each sample according to the four hypothesized sources (the portion not accounted for by the three preceding sources is assumed to be biogenous).

For the DOMES samples, we have used a detrital composition that

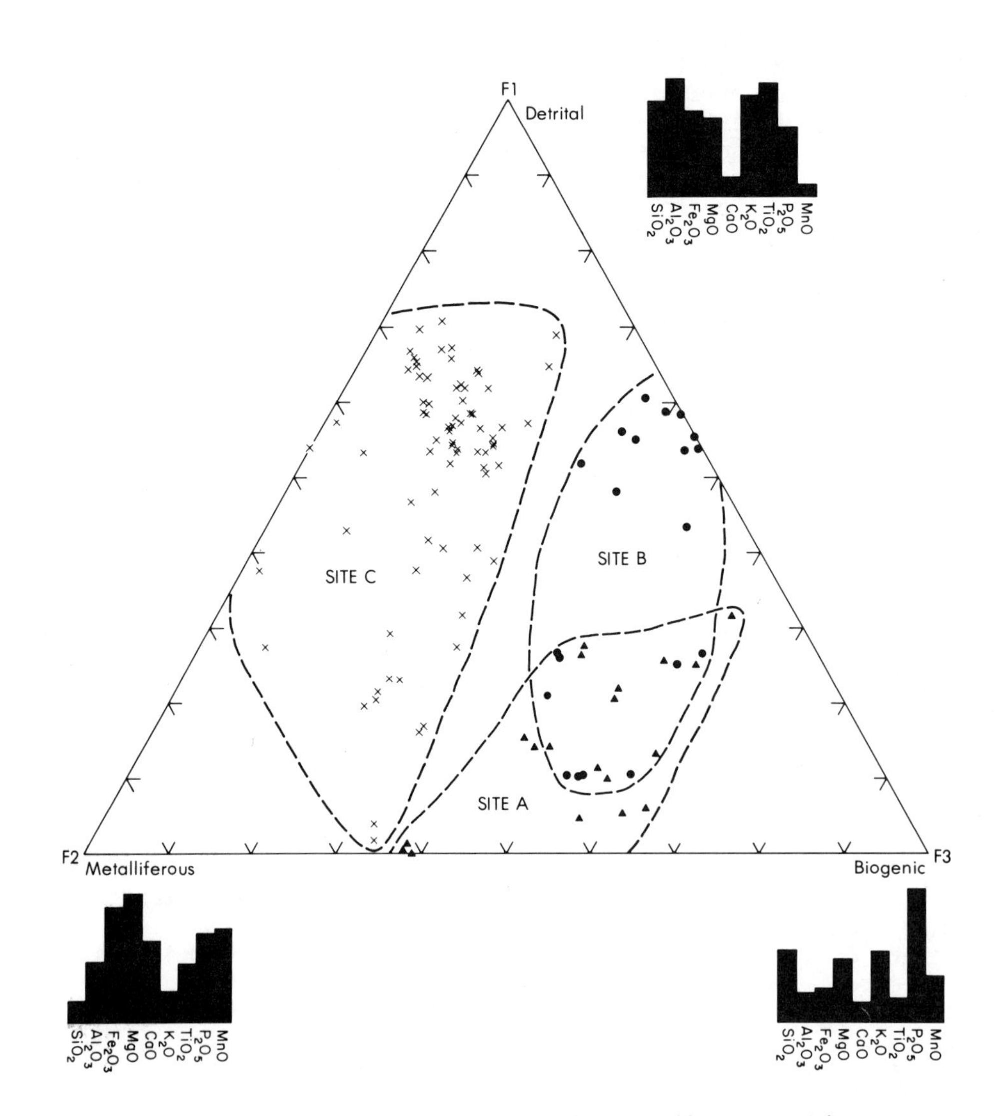

Fig. 19. Q-mode factor loadings for DOMES sediments. Histograms show the relative importance ["factor scores"] of oxides used in the analyses for each end member or factor.

is more granitic than Heath and Dymond's (1977) andesitic-basaltic material. This change is reasonable in view of the likely sources of detritus in the two areas. It is mandated by the low Fe/Al ratios in DOMES samples (lower than Heath and Dymond's detrital ratio of 0.75 in many cases). If the ratios of Table 6 were to be applied to the southeast Pacific data, they would make the sediments, which already are dominated by a metalliferous component, look even more hydrothermal.

The results of the partitioning exercise are summarized in Figure 20 and Table 7. The most striking differences between the three sites are: 1) Si in Sites A and B includes twice the proportion of biogenic input as in Site C. 2) Fe, Cu, Zn, and to a lesser extent Ni, show markedly more hydrothermal influence in Site C than in Sites A and B. 3) Site B is more "detrital" in aspect than Sites A and C, particularly in Zn, Ni, Cu, and Mn. 4) In contrast to the Si partitioning, biogenic Zn is much more important in Site A than it is in Sites B and C. 5) The relative importance of hydrogenous contributions are quite uniform from area to area, a characteristic suggesting that the authigenic reactions responsible for the formation of metal-rich nodules vary little across the DOMES region. It is noteworthy that Zn, which we could not assign to any one of our sediment components, shows the greatest site to site variation in the Heath and Dymond model.

Table 6

Elemental Ratios Used to Partition DOMES Sediments According to Heath and Dymond's (1977) Model

	Detrital $(\text{Element/Al})_D$	Hydrothermal $(\text{Element/Fe})_H$	Authigenic $(\text{Element/Ni})_A$
Fe (%)	0.50	1.00	0.0
Mn (%)	0.0073	0.102	0.0015
Si (%)	2.70	0.100	0.0
Al (%)	1.00	0.0058	0.94
Mg (%)	0.095	0.0	0.0
Cu (ppm)	6.70	46.0	0.11
Ni (ppm)	1.90	15.5	1.00
Zn (ppm)	8.00	16.0	0.15
Ba (ppm)	70.00	60.0	10.00

D - Refers to detrital fraction

H - Refers to hydrothermal fraction

A - Refers to authigenic or hydrogenous fraction

Table 7

Partitioning of DOMES Sediments by Percentages of Each Element According to Heath and Dymond's (1977) Model. Values in brackets are standard deviations.

		A	B	C
Al	Detrital	100 (0)	100 (0)	100 (0)
	Hydrothermal	0	0	0
Si	Detrital	72 (3)	74 (7)	87 (6)
	Hydrothermal	<1	<1	1
	Biogenous	27 (3)	25 (7)	12 (7)
Mg	Detrital	31 (3)	35 (4)	38 (6)
	Biogenic	69	65	62
Fe	Detrital	85 (2)	91 (2)	78 (16)
	Hydrothermal	15	9	22
Mn	Detrital	11 (5)	15 (11)	9 (5)
	Hydrothermal	15 (7)	11 (9)	15 (9)
	Hydrogenous	74	74	76
Cu	Detrital	10 (3)	22 (7)	13 (6)
	Hydrothermal	6 (2)	8 (3)	12 (10)
	Hydrogenous	4	4	5
	Biogenic	80 (5)	66 (9)	70 (11)
Ni	Detrital	7 (3)	16 (9)	7 (4)
	Hydrothermal	5	7	9
	Hydrogenous	88 (5)	77 (13)	84 (13)
Zn	Detrital	21 (8)	57 (11)	43 (14)
	Hydrothermal	4	5	11
	Hydrogenous	11 (7)	14 (9)	18 (8)
	Biogenic	64 (13)	24 (8)	28 (16)
Ba	Detrital	15 (3)	16 (5)	16 (6)
	Hydrothermal	1	1	2
	Hydrogenous	51 (18)	29 (18)	47 (22)
	Biogenic	33 (18)	54 (17)	36 (26)

Summary and Conclusions

1. The relative amounts of the deep-sea sediment components, pelagic clay, biogenic SiO_2, biogenic $CaCO_3$, sea salts, hydrogenous metals, metalliferous sediments, and, to some extent, local volcanic material, can be estimated from the concentrations of Si, Al, Fe, Ca, CO_2, Mn, Ti, and Na in the bulk sediment.

2. The partitioning model implies that average sediment at Sites A and B contains about 11% weight percent biogenic SiO_2 and 70% pelagic clay, whereas sediment at Site C contains less than 2% biogenic SiO_2, 87% pelagic clay, and 1.8% metalliferous sediment. Petrographic determinations of 25 to 60% biogenic opal for Sites A and B appear to be overestimates.

3. Most of the P_2O_5 and MnO in DOMES sediments are in fish debris and hydrogenous phases, respectively. Their intercorrelation and the correlation of P_2O_5 to degree of sediment reworking suggest that their contents can be used as an indicator of relative sedimentation rates or degree of sediment recycling.

4. Interelement correlations indicate that Mo, Ni, and Pb are strongly fractionated into the hydrogenous fraction. The composition of the hydrogenous fraction shows little variation and is comparable with the composition of DOMES manganese nodules.

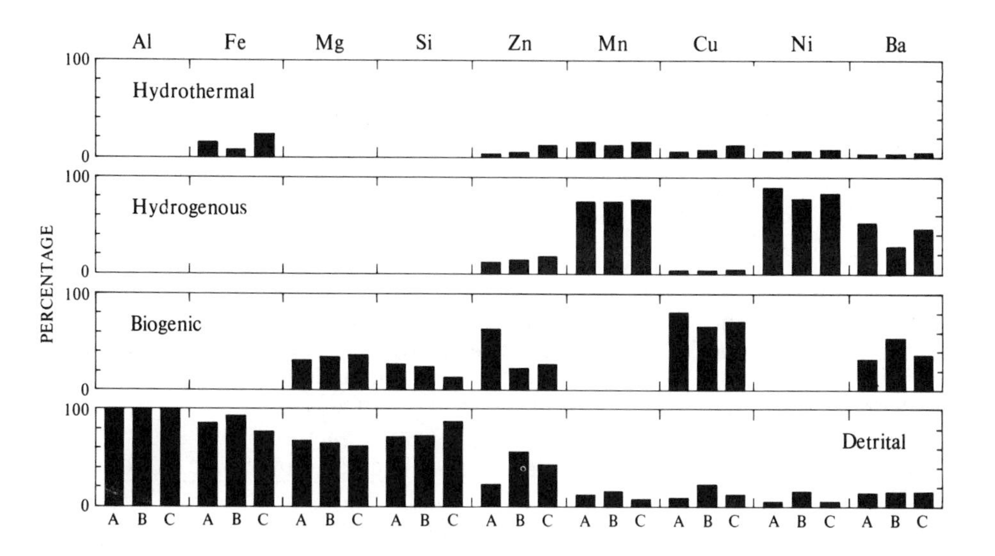

Fig. 20. Partitioning of selected elements in DOMES samples according to Heath and Dymond's (1977) model. Values are percentages of each element in each sample, averaged by area (see Table 7).

5. The Fe_2O_3 in DOMES sediments is distributed among pelagic clay, hydrogenous material, and metalliferous sediment. Metalliferous Fe_2O_3 if present, can be recognized at concentrations of a few percent by subtracting the pelagic and hydrogenous contributions from the total Fe_2O_3.

6. Several cores at Site C have large components of metalliferous sediment. Fe_2O_3, SiO_2, and MgO, in excess of the pelagic clay contributions, are present in remarkably uniform proportions in the metalliferous sediment fraction.

7. The pelagic clay fraction at all three DOMES sites contains more K_2O than Ppc. Part of the difference may be due to underestimates of K_2O in Ppc. However, primary sediment, as opposed to reworked sediment, has distinctly higher contents of excess K_2O.

8. MgO in excess of Ppc at Sites A and B represents Mg uptake by very slowly deposited biogenic SiO_2.

9. Factor analysis and elemental partitioning by the Heath and Dymond model yield results in good agreement with the partitioning scheme developed here. These techniques provide additional insight to the origin of elements that are not fractionated strongly into one or two of the basic sediment components.

Acknowledgments

Chemical analyses were performed by the U.S. Geological Survey. We are grateful to D. Piper, R. Rosenbauer, G. Moore, V. Kennedy, and to V. Marchig for helpful discussions and critical review of the manuscript.

References

Baker, E.T., Feeley, R.A. and Takahashi, K. (1979). Chemical composition, size distribution and particle morphology at DOMES Sites A, B, and C: Relationships with local sediment composition, In Bischoff, J.L. and Piper, D.Z. (eds.), Marine Geology and Oceanography of the Pacific Manganese Nodule Province, Plenum Publ. Corp., New York, (this volume).

Berner, R. (1973) Phosphate removal from sea water by absorption on volcanogenic ferric oxides. Earth and Planet. Sci. Lett. 18, 77-86.

Bischoff, J.L. and Sayles, F.L. (1972) Pore fluid and mineralogical studies of recent marine sediments: Bauer Depression region of the East Pacific Rise, Jour. Sed. Petrol. 42, 711-724.

Bischoff, J.L. (compiler) (1976) Deep Ocean Mining Environmental Study, N.E. Pacific Manganese Nodule Province, Site C, Geology and Geochemistry, U.S. Geol. Survey Open-file Report 76-548, 275 p.

Bischoff, J.L. and Rosenbauer, R.J. (1977) Recent metalliferous sediment in the North Pacific Manganese Nodule Area, Earth and Planet. Sci. Lett. 33, 379-388.

Bischoff, J.L., Piper, D.Z. and Quinterno, P. (1979) Nature and origin of metalliferous sediment in DOMES Site C, Pacific Manganese Nodule Province, In Lalou, C. (ed.) "International Colloquium on the Genesis of Manganese Nodules", Colloques Internationaux du C.N.R.S., Paris, pp.

Calvert, S.E. and Price, N.B. (1977) Geochemical variation in ferromanganese nodules and associated sediments from the Pacific Ocean, Marine Chem.5, 43-74.

Cook, H.E., Piper, D.Z. and Gardner, J.V. (1977) Geological and oceanographic framework of DOMES Sites A, B and C: Central Equatorial Pacific, In Piper, D.Z. (compiler), Deep Ocean Environmental Study: Geology and Geochemistry of DOMES Sites A, B, and C, Equatorial North Pacific, U.S. Geol. Survey Open-file Report 77-778, 15-81.

Cronan, D.S. (1969) Average abundances of Mn, Fe, Ni, Co, Cu, Pb, Mo, V, Cr, Ti and P in Pacific pelagic clays, Geochim. Cosmochim Acta 33, 1562-1565.

Cronan, D.S. and Tooms (1969) The Geochemistry of manganese nodules and associated pelagic deposits from the Pacific and Indian Oceans, Deep Sea Research 16, 335-339.

El Wakeel, S.K. and Riley, J.R. (1961) Chemical and mineralogical studies of deep sea sediments, Geochim. Cosmochim. Acta 25, no. 2, 110-146.

Engel, A.E., Engel, C.G. and Havens, R.G. (1965) Chemical characteristics of oceanic basalts and the upper mantle, Bull. Geol. Soc. Amer. 76, p. 719.

Froelich, P.N., Bender, M.L. and Heath, G.R. (1977) Phosphorus accumulation rates in metalliferous sediments on the East Pacific Rise, Earth and Planet. Sci. Lett. 34, 351-359.

Goldberg, E.D. and Arrhenius, G.O.S. (1958) Chemistry of Pacific pelagic sediments, Geochim. Cosmochim. Acta 13, 153-212.

Heath, G.R., Moore, T.C., Somayajulu, B.L.K. and Cronan, D.S. (1970) Sediment budget in a deep-sea core from the Central Equatorial Pacific, Jour. Mar. Res. 28, 225-234.

Heath, G.R. and Dymond, J. (1977) Genesis and transformation of metalliferous sediments from the East Pacific Rise, Bauer Deep and Central Basin, Northwest Nazca Plate, Bull. Geol. Soc. Amer. 88, 723-733.

Hein, J.R., Gutmacher, C., and Miller, J. (1976) DOMES Area C: General statement about mineralogy, diagenesis and sediment classification, in Bischoff, J.L. (compiler), Deep Ocean Mining Environmental Study, N.E. Pacific Nodule Province, Site C, Geology and Geochemistry, U.S. Geol. Survey Open-file Report 76-548, 275 pp (see 63-80).

Hein, J.R. and Jones, M.G. (1977) Mineralogy and diagenesis of DOMES Sites A and B, In Piper, D.Z. (compiler), Deep Ocean Mining Environmental Study: Geology and geochemistry of DOMES Sites A, B, and C, Equatorial North Pacific, U.S. Geol. Survey, Open-file Report 77-778, 527 pp. (see 146-178).

Imbrie, J. and van Andel, Tj. H. (1964) Vector analysis of heavy mineral data, Bull. Geol. Soc. Amer. 75, 1131-1156.

Klovan, J.E. and Imbrie, J. (1971) Algorithm and FORTRAN IV Program for largescale Q-mode factor analysis and calculations of factor scores, Mathematical Geol. 3, 61-77.

Krishnaswami, S. (1976) Authigenic transition elements in Pacific pelagic clays, Geochim. Cosmochim. Acta 40, 425-434.

Landergren, S. (1964) On the geochemistry of deep-sea sediments. Reports of the Swedish deep-sea expedition, v. X, Spec. Invest. no. 5, 61-148.

Piper, D.Z., Cannon, W.F., and Leong, K. (1977) Composition and abundances of ferromanganese nodules at DOMES Sites A, B, and C: Relationship with bathymetry and stratigraphy, In Piper, D.Z. (compiler), Deep Ocean Environmental Study: Geology and Geochemistry of DOMES Sites A, B, and C, Equatorial North Pacific, U.S. Geol. Survey Open-file Report 77-778, 217-266.

Piper, D.Z. (compiler) (1977) Deep Ocean Environmental Study: Geology and Geochemistry of DOMES Sites A, B, and C, Equatorial North Pacific, U.S. Geol. Survey Open-file Report 77-778, 527 pp.

Piper, D.Z., Cannon, W.F., and Leong, K. (this volume) Ferromanganese nodules at DOMES Sites A, B, and C: Relationship with bathymetry and stratigraphy, In Bischoff, J.L. and Piper, D.Z. (eds.), Oceanography of the Equatorial Pacific Manganese Nodule Province.

Piper, D.Z., Cook, H.E., and Gardner, J.V. (this volume). Acoustic and litho-stratigraphy of the Central Equatorial Pacific, Domes Sites A, B, and C.

Quinterno, P. and Theyer, F. (this volume). Biostratigraphy of the Equatorial North Pacific, DOMES Sites A, B, and C.

Sayles, F.L., and Bischoff, J.L. (1973) Ferromanganese sediments in the Equatorial East Pacific, Earth and Planet. Sci. Lett. 19 , 330-336.

Sayles, F.L. and Mangelsdorf, P.C. (1977) The equilibrium of clay minerals with sea water: exchange reactions, Geochim. Cosmochim. Acta 41, 951-960.

Turekian, K.K. and Wedepohl, K.H. (1961) Distribution of the elements in some major units of the earth's crust, Bull. Geol. Soc. Amer. 72, 175-192.

MANGANESE NODULE AND SURFACE SEDIMENT COMPOSITIONS: DOMES SITES A, B, AND C

David Z. Piper and Kam Leong

U.S. Geological Survey
Menlo Park, California

and

William F. Cannon

U.S. Geological Survey
Reston, Virginia

Abstract

Manganese nodules and associated sediment from the equatorial North Pacific have been examined to ascertain relations between nodule abundance, texture, and composition, and sediment composition, lithology and acoustic stratigraphy. Nodules are most abundant in areas where the uppermost acoustically transparent layer of sediment is less than approximately 15 m thick. Abundance drops off sharply in areas where this acoustic unit is thicker than 20 m.

Nodules have average Zn/Mn and Co/Mn ratios which are approximately the same as these ratios for the component of sediment soluble in hydroxylamine hydrochloride-acetic acid. The relations between Cu and Mn and Ni and Mn in the sediment are also approximately the same as in nodules, but only for those nodules which have a smooth surface texture. Granular nodules have relatively more Ni and Cu. Antimony in nodules, similar to Ni, Cu, and Zn, is strongly correlated with Mn, whereas Co, Hf, Th, and U are strongly correlated with Fe.

The insoluble component of sediment has a uniform composition which is similar to the composition of terrigenous shale. This similarity holds for the major oxides as well as for Co, Cr, Zn, Hf, Sb, Th, U, Sc, Cs, Rb, and Ta.

Introduction

The fine fraction of pelagic sediment consists of a mixture of sedimentary components, classified according to their origin as hydrogenous, biogenous, lithogenous, and cosmogenous (Goldberg, 1963). These same components make up ferromanganese nodules, but their relative concentrations are quite different in sediments from their concentrations in nodules. Sediments consist largely of lithogenic and biogenic components, whereas nodules usually contain between 10 and 50% hydrated oxides each of Fe and Mn (hydrogenous component) plus approximately 30% ± 20% of aluminosilicates (lithogenous component) and opaline silica (biogenous component) (Riley and Sinhaseni, 1958; Calvert and Price, 1977).

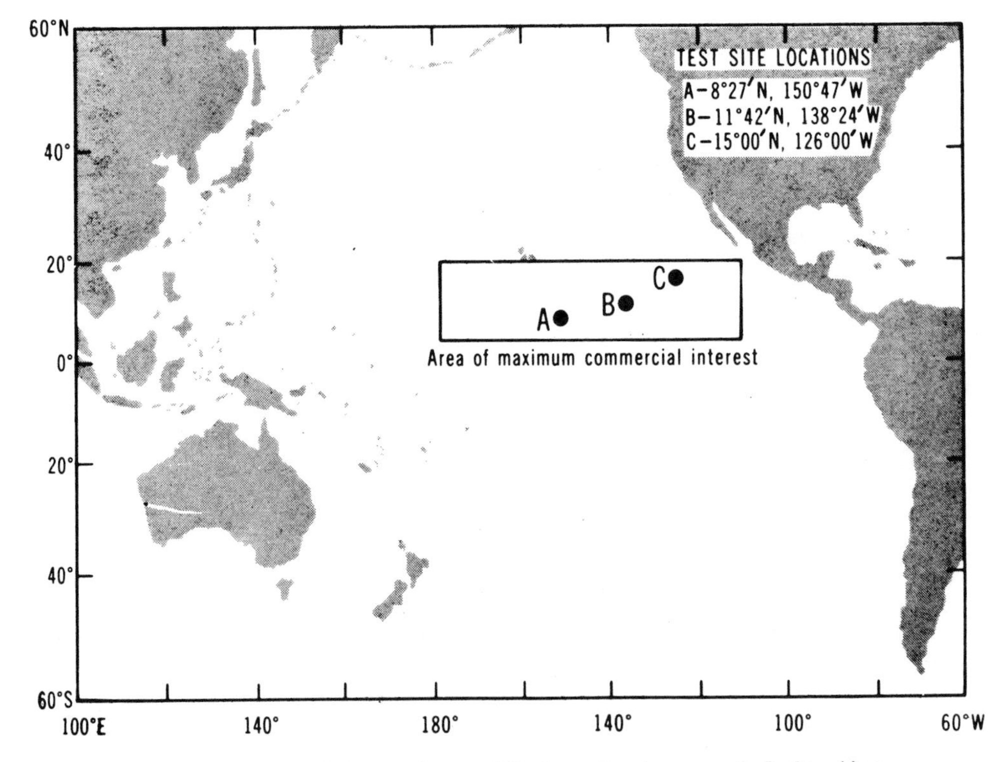

Fig. 1. Location of Deep Ocean Mining Environmental Studies (DOMES) Project area and of DOMES Sites A, B, and C.

Nodules from the equatorial North Pacific, an area that includes the three DOMES sites (Fig. 1), contain relatively high concentrations of Ni and Cu (Horn et al., 1973) within the hydrogenous component, each in the range of 1 to 2%. Other elements are present in one or more nodule components in concentrations of 10's of ppm up to several ppt. The more abundant elements include Co, Zn, Ba, Mo, Pb, and the light rare earth elements.

As nodules are composed largely of the hydrogenous component, it may be expected that nodule composition is related to this component in the fine fraction of sediment. Attempts to relate the composition of nodules to the composition of various components in sediments, however, have met with limited success (Calvert _et al._, 1978; Skornyakova, this volume).

In our investigation of the three DOMES sites, we attempted to separate the hydrogenous component in the fine-grain fraction of sediment by a chemical leaching technique (Chester and Hughes, 1967). The aim of this analysis was to ascertain the relations between the composition of this component of sediment with the composition of this component in nodules. Also, the abundance, size, and texture of nodules at each DOMES site were measured to ascertain possible relations with sediment composition, age, lithology, and acoustic stratigraphy.

Techniques

Nodule abundances (A, in kg/m^2) were estimated from photographs of box cores by determining nodule coverage (P) and average maximum nodule diameter (D), and using the following equation: A= 19.5 DP/100 (This equation assumes that the vertical nodule axis is 0.57 times D). In the case of 15 box cores, these estimates have been compared to actual nodule weights. With the exception of two box cores, agreement of the two techniques was within 25%. In one core, the surface sediment appeared to be strongly disturbed, which appeared to account for many buried nodules in the upper few centimeters of sediment. In the other core nodule coverage was extremely low, approximately 5%, and the measured weight was three times the value calculated from the photograph.

Bottom photographs also were used to estimate nodule size and abundance. Both estimates are dependent upon an accurate scale. Thus, only those frames that included a compass resting on the bottom give accurate values; these represent only about 200 of 13,000 frames scanned. The values obtained from these photographs and nearby box cores agree to within approximately 25%, but estimates of nodule coverage and abundance are consistently greater from bottom photographs than from box-core photographs.

Nodules selected for bulk analysis were taken from the dominant size class of each box core when possible, after sorting nodules into the following groups based on their maximum dimension: less than 2 cm, 2-4 cm, 4-6 cm, 6-8 cm, and greater than 8 cm. A slice taken through the nodule center, or the entire nodule in the case of small nodules, was used for bulk chemical and mineralogical analysis. Samples were washed free of sediment and sea salt, dried at 110°C, and powdered. Mineralogy was determined by X-ray diffraction on an aliquot that had been dried at room temperature. Six elements (Ni, Cu, Co, Zn, Fe, and Mn) were determined by atomic absorption spectrophotometry (AA) on the fraction soluble in 6N HCl and dried at 110°C. Nine elements (Fe, Co, Zn, Ba, Hf, Sb, Th, U and Sc) were measured on the bulk nodule sample by neutron activation analysis (NAA). Cobalt and zinc values, measured by both techniques, agreed to within 15%. Iron values measured by AA were 20% higher, on the average, than NAA values. We are unable to account for this difference. Iron values obtained by both techniques for the nodule standard GRLD-126 agree to within 10% of average values obtained by other laboratories (J. Greenslate, 1977, personal communication). We have reported the iron values obtained by AA. These values may be high, but the error is systematic and should not affect our conclusions.

Minor-element composition was measured by NAA of bulk sediment and of the residue of samples leached with hydroxylamine hydrochloride-acetic acid (Chester and Hughes, 1967). Six elements (Fe, Mn, Ni, Cu, Co and Zn) were measured on the acid solutions by AA. Leachable Fe, Co, and Zn, measured by AA, agreed to within 15% at 2σ with NAA measurements of losses of these elements from sediment.

Major-oxide composition of leached and unleached sediment was measured by X-ray fluorescence. Bischoff et al. (this volume) report on the latter. The fraction of sediment soluble in hydroxylamine hydrochloride-acetic acid was determined by weight loss.

Stratigraphy and Lithology

The acoustic stratigraphy for DOMES Sites A, B, and C has been presented by Piper et al. (this volume). Briefly, three acoustic units are recognized at B and C. Unit 1, the uppermost unit, is a transparent layer of highly variable thickness (Fig. 2) with a maximum at Site C of approximately 25 m. Box cores taken in areas where this sediment is at the surface at Site C contain a sparse, poorly preserved assemblage of Quaternary radiolarians and coccolithophorids. At Site B the unit contains a rich assemblage of Quaternary radiolarians mixed with older Tertiary species (Quinterno and Theyer, this volume). Unit 2 has a rather uniform thickness of 25 to 30 m. It consists of a series of acoustically opaque layers that crop out on the seafloor in many areas. Several cores taken at Site C, where the contact between this and the underlying unit appears to be at the surface, contain unmixed early Miocene flora and fauna. Unit 3 is acoustically much more transparent than unit 2. It has a thickness of approximately 35 m at Site C above a poorly defined acoustic basement. This unit is clearly recognized above a strongly reflecting basement at Site B, where its thickness is approximately 55 m.

At Site A the acoustic records are similar to those at Sites

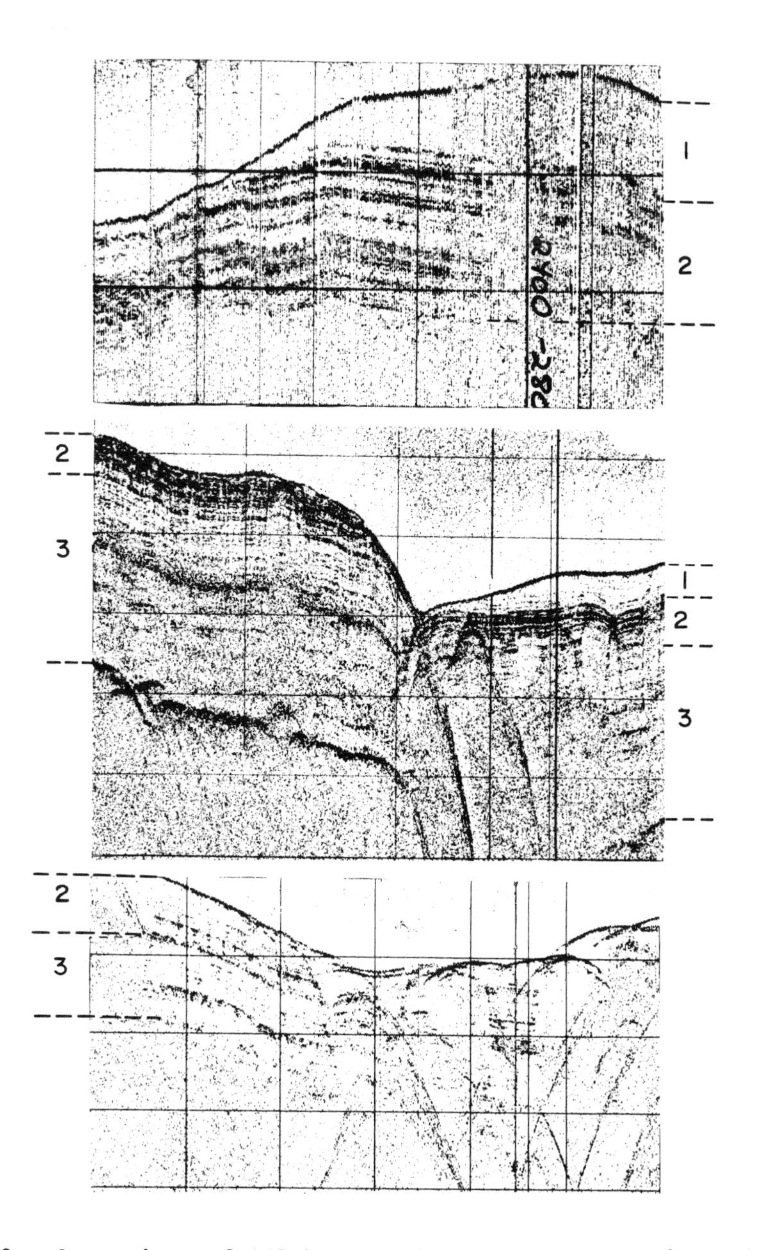

Fig. 2. Comparison of 3.5-kHz profiles from Sites A (upper), B (middle), and C (lower). Acoustic units 1 and 2 are possibly representative of the Clipperton Formation whereas acoustic unit 3 may be representative of the Marquesas Formation (Piper et al. this volume). The distance between horizontal lines equals 36.6 m (20 f) in seawater.

B and C, but units 1 and 2 appear to be significantly thicker than at Sites B and C (Fig. 2). The relief of the top of unit 2 often differs strongly with that of the seafloor, perhaps more so than at Sites B and C. No strong acoustic basement was observed in the records making it impossible to identify acoustic units lower than unit 2.

Box cores from Sites A and B usually exhibited moderate bedding and strongly mottled contacts (Cook et al., 1977). Although moderate bedding was also observed in a few cores from Site C, most Site C box cores exhibited massive bedding.

Sediment Composition

Surface sediment at the three DOMES sites is classified as radiolarian-diatom-bearing pelagic clay to pelagic clay. It averages 85 to 95% lithogenous material, less than 15% biogenic silica, and minor amounts of amorphous metal oxides. Biogenic $CaCO_3$ was observed in only a few of the samples. Several Site C cores contain an Fe-smectite (Bischoff et al.1979a) of possible hydrothermal origin. The lithogenic material consists of a mixture of clay minerals dominated by smectite and illite (Hein et al. this volume) plus a minor amount of phillipsite, and trace amounts of quartz, feldspar, and volcanic debris (Cook et al., 1977; Hein et al. this volume).

The bulk sediment was leached with hydroxylamine hydrochloride-acetic acid (HHA). This solution removes amorphous metal oxides, $CaCO_3$, and adsorbed ions and attacks weakly any crystalline phases other than $CaCO_3$ (Chester and Hughes, 1967). There is no significant change in mineralogy from the leaching treatment. The composition of the leachate is listed in Table 1. In two samples in which $CaCO_3$ was not observed in smear slides, the leached component contained less than 1% of the total Si in the bulk sediment. The major element was Na, with lesser but approximately equal amounts of Mn, Fe, K, Ca, and Mg.

Insoluble Sediment Fraction (ISF)

This fraction of sediment is uniform in composition. At Site C the major oxides exhibit less than 5% variation at 2σ (Table 2) and the minor elements less than 10%. At Sites A and B variations are only slightly greater.

Comparison of the composition of the ISF fraction (Table 2) with that of bulk sediment (Table 3; Bischoff et al., Table 5, this volume) shows that almost all of the SiO_2, Al_2O_3, TiO_2, Ba, Cr, Hf, Sb, and Ta are present in the ISF. Approximately 75 to 95% of the Th, Sc, Cs, and Rb, 25 to 75% of the K_2O, MgO, CaO, and Na_2O, and less than 25% of the P_2O_5 and MnO are present in the ISF. Comparison of the composition of the solution obtained by leaching the sediment with HHA with that of bulk sediment shows that 85 to 95%

Table 1. Composition of the fraction of sediment soluble in hydroxylamine and hydrochloride-acetic acid for two samples from Core 22-28 (Site C). The samples were washed in distilled water and dried at 110°C prior to leaching.

Elemental composition expressed as % of bulk sediment

	Si	Ca	Mg	Na	K	Fe	Mn	Total*
22-28-1	.22	.492	.309	1.25	.655	.371	.651	4.66
22-28-8	.19	.487	.291	1.20	.656	.452	.651	4.60

Oxide composition expressed as % of bulk sediment

	SiO_2	CaO	MgO	Na_2O	K_2O	Fe_2O_3	MnO	Sum	Total+
22-28-1	.47	.68	.40	1.69	.79	.53	.84	5.02	5.59
22-28-8	.41	.68	.38	1.62	.79	.64	.84	5.02	5.53

*The total represents percent weight loss.

+The total represents an adjustment of the weight loss, assuming that all Na, K, Ca, and Mg were present as exchangeable ions and Si, Fe, and Mn were present as oxides. The adjustment, then, is H_2O added corresponding to the amount of Na, K, Mg, and Ca measured in the solution. No adjustment was made to compensate for hydration of Fe_2O_3 and MnO in the sample.

of the Fe_2O_3 and Zn, 25 to 50% of the Cu, and 10 to 30% of the Co, Ni, and Mn are present in the ISF.

The bulk concentrations of the minor elements (Table 3) which occur predominantly in the ISF, exhibit strong interelement correlations (Table 4). The exceptions are Ba and Sc which do not correlate strongly with any elements, and Sb and Zn which correlate with each other and with Co. The relation to Co is somewhat surprising in that most of the Sb and Zn are present in the ISF but only about 15% of the total Co occurs in this fraction.

The average composition of the ISF is similar to that of shale (Turekian and Wedepohl, 1961); the only exceptions are CaO and Ba. The tenfold greater concentration of Ba in pelagic sediment probably is due to the occurrence of barite, a common minor mineral of equatorial sediments. The difference in CaO values may be due to the presence of $CaCO_3$ in shale. The otherwise close similarity suggests that the ISF has a terrigenous source. The average composition for Site C sediment may represent a close approximation to that of terrigenous material entering the deep ocean by riverine and aeolian transport. This sediment fraction at Sites A and B is diluted 10 to 15% by biogenic silica (Bischoff et al. this volume), which accounts for most of the difference of the major oxide concentrations with those for Site C sediment.

Table 2. The average elemental composition of the insoluble sediment fraction from the three DOMES sites and of shale (Turekian and Wedepohl, 1961). Concentrations are in ppm except where noted otherwise. The values in parentheses represent 2σ for A and C and the range of 4 analyses for B. The minor elements were measured by NAA and the major oxides by XRF.

Element	Site A	Site B	Site C	Shale
Ba (%)	.474 (.05)	.540 (.13)	.583 (.12)	.058
Co	16.1 (.6)	17.8 (.8)	14.9 (1.3)	19
Cr	62.8 (6)	60 (8)	73.8 (3.5)	90
Cs	8.6 (.8)	9.5 (1)	11.2 (.8)	5
Hf	4.2 (.3)	4.4 (.1)	4.7 (.3)	2.8
Rb	90 (8)	102 (8)	129 (12)	140
Sb	2.1 (.2)	2.1 (.6)	2.8 (.6)	1.5
Ta	0.9 (.3)	1.0 (.1)	1.2 (.1)	0.8
Th	13.3 (1)	14.0 (.8)	15.6 (1)	12
U	1.5 (.3)	2.2 (.3)	2.2 (1.1)	3.7
Zn	158 (11)	165 (12)	163 (15)	95
Sc	29.6 (1.1)	28.5 (1.4)	25.3 (.6)	13
SiO_2	61.80 (.56)	62.08 (.62)	58.41 (.40)	58.43
Al_2O_3	13.66 (.31)	13.64 (.93)	16.36 (.31)	15.15
Fe_2O_3	6.93 (.19)	6.42 (.46)	7.48 (.32)	6.75
MgO	2.36 (.05)	2.27 (.11)	2.55 (.05)	2.49
CaO	0.79 (.03)	0.68 (.08)	0.69 (.01)	3.09
Na_2O	1.11 (.06)	1.19 (.04)	1.39 (.07)	1.29
K_2O	2.26 (.02)	2.45 (.31)	3.14 (.06)	3.20
TiO_2	0.78 (.03)	.75 (.05)	0.87 (.01)	0.77
P_2O_5	0.10 (.01)	.09 (.02)	0.09 (.01)	0.16
MnO	.059 (.001)	.054 (.004)	0.060 (.002)	0.11
LOI	9.38	9.12	8.82	
TOTAL	100.23	99.88	101.09	

Table 3. Bulk composition of sediment from DOMES Sites A, B, and C. Sample numbers give station-box core interval, where interval corresponds to 2 cm. For example, 1 = 0 to 2 cm, 2 = 2 to 4 cm, 3 = 4 to 6 cm, etc. Elemental concentrations are given in ppm, except for Fe, which is in percent.

Sample	Fe	Ba	Co	Cr	Cs	Hf	Rb	Sb	Ta	Th	U	Zn	Sc
SITE A													
46-1-1	4.47	3730	79.5	58.5	7.5	3.7	82	2.1	0.79	13.9	1.7	154	26.95
46-1-15	4.77	4632	131.5	62.1	7.8	4.1	92	2.3	0.82	14.1	2.2	183	29.27
46-1-19	4.96	5977	112.5	65.1	7.2	4.4	84	3.4	0.85	13.7	2.3	210	31.57
47-10-1	4.25	3874	74.8	55.4	6.6	3.7	78	2.0	0.77	15.8	1.8	166	28.31
47-10-9	4.52	4931	110.7	54.8	7.1	4.1	87	2.0	0.76	15.2	2.0	171	30.32
47-10-15	4.90	5717	196.3	49.6	6.5	4.8	84	3.4	0.93	13.6	2.3	231	35.63
47-10-20	4.62	5480	181.2	41.4	5.8	4.8	72	5.6	0.66	10.6	2.6	279	37.36
50-28-1	3.91	4988	70.3	53.5	6.9	3.3	59	1.6	0.80	10.4	1.6	165	24.11
50-28-4	3.65	3978	71.1	48.8	6.5	3.0	79	1.5	0.70	9.9	1.5	141	22.10
50-28-10	3.15	3175	19.9	42.0	5.3	3.0	60	1.1	0.57	8.4	1.2	108	18.53
SITE B													
51-33-1	4.33	6496	84.0	52.6	7.5	3.8	94	1.9	0.79	12.5	1.9	203	28.27
51-33-5	3.70	12773	55.6	26.4	4.3	3.7	75	1.5	0.58	6.8	1.7	227	30.22
53-45-1	4.32	4878	75.3	58.8	8.2	3.9	101	1.9	0.95	13.5	2.0	168	25.28
54-51-1	4.35	5043	76.9	58.0	8.5	4.5	105	2.4	0.83	12.3	1.8	168	25.33
54-51-6	4.56	4607	52.0	61.3	9.1	4.3	131	1.7	0.98	14.0	2.2	176	26.26
54-51-13	4.65	5636	81.4	61.5	9.6	4.6	108	2.4	1.00	14.4	2.4	165	26.52
54-51-15	4.67	5417	103.5	57.9	9.5	4.6	106	2.0	1 13	14.9	2.3	171	26.17

Table 3. (Cont'd).

							SITE C							
2-7-1	4.71	4723	294.8	61.2	8.3	3.8	95	3.8	0.80	13.4	2.0	223	21.18
2-7-8	4.96	4305	79.9	69.3	9.7	4.2	115	2.2	0.96	14.4	2.0	156	22.80
2-7-16	5.25	5217	93.0	71.3	10.6	4.2	114	2.5	1.11	15.6	1.9	171	24.32
4-9-1	4.82	5336	90.1	68.9	10.0	4.0	103	2.5	0.89	13.9	2.2	167	22.64
4-9-14	5.09	5229	87.7	69.5	10.4	4.1	114	2.5	0.94	14.4	2.2	156	23.21
7-12-1	4.87	5133	88.7	69.2	10.0	4.0	107	2.4	1.05	13.5	2.2	166	22.42
15-21-1	5.21	5510	93.6	66.3	10.0	4.1	102	2.7	1.00	13.9	1.9	181	23.68
15-21-3	4.95	5498	91.8	67.2	9.4	3.8	107	2.7	0.88	13.8	1.9	163	22.34
15-21-5	5.07	5437	90.9	70.4	10.2	4.3	102	2.4	1.13	14.6	2.4	166	22.92
15-21-7	5.13	4762	84.0	67.3	10.1	4.0	115	2.3	0.99	14.8	2.0	167	22.54
15-21-8	5.75	5481	84.4	59.5	9.4	4.3	107	3.1	0.93	13.7	1.9	184	21.83
15-21-9	7.13	6803	54.2	51.6	8.0	4.3	97	2.4	0.90	12.3	2.1	203	20.67
15-21-10	8.86	9716	57.8	28.8	4.4	3.2	70	4.2	0.64	8.3	1.8	271	14.99
15-21-12	9.67	10795	55.4	19.4	3.2	3.3	60	4.5	0.54	7.2	2.0	284	13.01
18-24-1	4.71	5678	91.5	70.3	9.8	3.9	111	2.6	0.96	13.8	2.3	156	21.77
18-24-7	4.33	4441	72.3	60.3	8.2	3.6	104	1.6	0.89	12.3	1.9	134	19.65
18-24-16	5.10	6066	86.8	69.0	10.2	4.3	119	2.4	1.08	15.1	2.1	177	23.23
22-28-1	5.04	5306	98.9	70.9	9.7	4.0	126	2.8	0.98	14.5	2.6	176	23.15
22-28-4	5.09	4693	79.7	73.1	10.0	4.4	129	2.2	1.02	14.8	2.4	170	23.59
22-28-8	5.02	4789	79.0	70.0	10.0	4.3	112	2.8	1.02	14.6	2.1	181	23.13
22-28-15	5.37	5809	83.2	70.9	10.1	4.4	125	2.3	1.04	15.5	2.4	176	24.22

Table 4. Correlation coefficients of elemental concentrations of bulk sediment (Table 3).

	Ba	Co	Cr	Cs	Hf	Rb	Sb	Ta	Th	U	Zn	Sc
Fe	.03	.30	.72	.74	.65	.69	.51	.70	.75	.58	.26	.10
Ba		00	.40	.24	.14	.06	.06	.14	.39	.10	.47	.32
Co			.03	.1	.28	.1	.75	.10	.14	.24	.71	.26
Cr				.92	.30	.76	.06	.81	.81	.46	.32	.45
Cs					.39	.85	.01	.89	.74	.47	.30	.49
Hf						.46	.53	.55	.51	.77	.14	.49
Rb							.03	.78	.66	.52	.14	.36
Sb								.04	.14	.46	.78	.37
Ta									.75	.56	.20	.26
Th										.52	.10	.10
U											.32	.48
Zn												.62

Variations in the ISF between sites, exclusive of variations in the concentration of biogenic SiO_2, are seen by normalizing to Al_2O_3 (Table 5). Fe_2O_3, MgO, CaO, and P_2O_5 increase and K_2O decreases from Site C to B to A. The increase in P_2O_5 may be related to sediment maturity (Bischoff et al. this volume), where maturity is estimated by the relative amount of Tertiary radiolarians mixed with Quaternary species. The mineralogy exhibits a similar trend. Site A sediment has the highest relative concentrations of Tertiary radiolarians (Quinterno and Theyer, this volume), is considered to be the most mature, and has a relatively high smectite concentration (Fig. 3). The trends in major oxide composition are, thus, apparently controlled primarily by this trend in mineralogy of the clays. It is not possible to ascertain the extent to which these trends may be related to sediment diagenesis and to variations in sediment source between the Tertiary and Quaternary.

Soluble Sediment Fraction (SSF)

The soluble sediment fraction consists of amorphous metal oxides (mostly hydrated oxides of Fe and Mn), adsorbed ions (mostly Na^+, K^+, Ca^{++}, and Mg^{++}), $CaCO_3$, micro-nodules and fragments of larger nodules. The nodules are thought to have contributed to the relatively high metal values of the surface sample from core 2-7 (Table 6). Metal ratios for the SSF of this sample are close

Table 5. Major oxide ratios of the lithogenic component of sediment, i.e., the fraction of sediment insoluble in hydroxylamine hydrochloride-acetic acid.

	Site A	Site B	Site C
Fe_2O_3/Al_2O_3	.517	.471	.457
MgO/Al_2O_3	.173	.166	.156
CaO/Al_2O_3	.058	.050	.042
Na_2O/Al_2O_3	.0813	.0872	.0850
K_2O/Al_2O_3	.165	.180	.192
TiO_2/Al_2O_3	.057	.055	.053
P_2O_5/Al_2O_3	.0073	.0066	.0055
MnO/Al_2O_3	.0043	.0040	.0037
SiO_2/Al_2O_3	4.52	4.55	3.57

to those for Site C nodules.

Concentrations of Mn, Cu, Ni and Co in the SSF average 0.75%, 200 ppm, 185 ppm, and 75 ppm respectively. The average value for Mn is considerably higher than that reported by Calvert et al. (1978) for the Wahine area, approximately 2° west of Site A, but the values for surface sediment at Site A (Table 6) are similar to the range of concentrations measured by them. The three minor metals are strongly correlated with Mn (Table 7). Except for Cu, none of these metals correlates with Fe and, in the case of Cu, the correlation coefficient is only 0.37.

Zinc is similar to Fe in that only about 10% of its total is in the SSF. However, it shows a strong correlation with Mn (Table 7), similar to Cu, Ni, and Co.

Manganese and its related metals (Cu, Ni, Co, and Zn) exhibit a wide range of concentrations in the SSF. Mn varies from 0.05 to 2.41% of the total sediment. The lowest values are associated with carbonate ooze in core 50-28.

The strong correlations between Mn and Cu, Ni, Co and Zn (Table 7) suggest that these metals may be bound largely in a single phase, possibly micronodules. Although micronodules were observed in several cores, no analysis has been made of them and

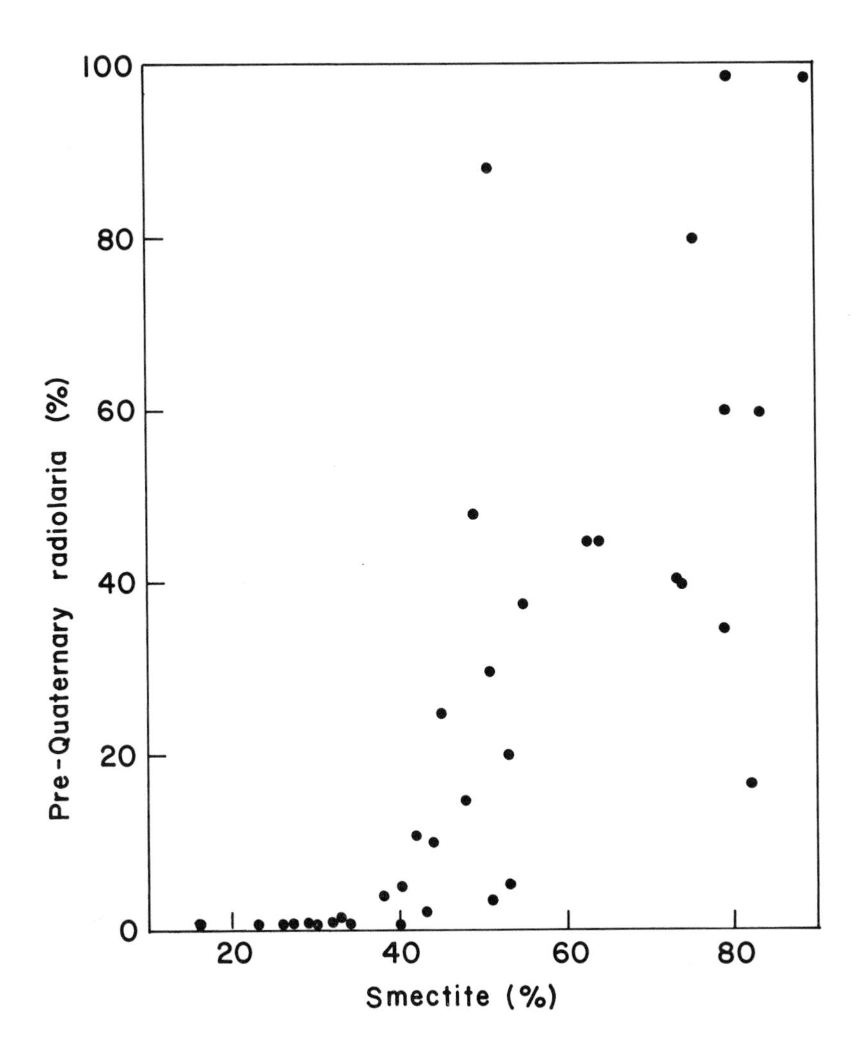

Fig. 3. Abundance of smectite in DOLES box cores (Hein *et al.*, this volume) versus relative abundance of Pre-Quaternary radiolarians (Quinterno and Theyer, this volume). Individual points represent core intervals for which neither parameter varied by more than 10%, regardless of the length of the interval. The correlation coefficient is 0.79.

Table 6. The concentration and composition of the hydrogenous component of sediment, i.e., the component of sediment soluble in hydroxylamine hydrochloride-acetic acid. Concentrations of Mn, Fe, and Total Soluble are in percent of bulk sediment; Cu, Ni, Co, and Zn are in ppm.

Cruise	Sample	Cu	Ni	Co	Zn	Mn	Fe	Soluble
				SITE A				
RP-8-OC-75	46-1-1	185	109	65.5	25.8	0.38	0.35	8.06
"	46-1-15	186	182	130	36.9	0.73	0.31	6.82
"	46-1-19	249	387	108	58.4	1.33	0.34	7.25
"	47-10-1	149	110	59.7	26.9	0.35	0.29	9.38
"	47-10-9	128	93.8	104	23.9	0.27	0.28	5.21
"	47-10-15	337	375	164	45.9	1.33	0.28	6.21
"	47-10-20	383	708	162	90.8	2.41	0.27	10.22
"	48-19-2	206	201	97.4	29.5	0.54	0.34	-
"	48-19-8	152	170	104	30.5	0.54	0.30	-
"	50-28-1	174	167	75.9	26.0	0.40	0.38	17.07
"	50-28-2	178	140	69.0	26.0	0.37	0.31	-
"	50-28-4	171	144	71.4	25.8	0.38	0.33	23.65
"	50-28-10	59.4	14.8	15.8	13.4	0.05	0.22	-
"	50-28-11	51.5	12.4	7.93	11.9	0.05	0.25	37.57
"	50-28-13	61.8	14.8	7.7	10.9	-	0.19	-
				SITE B				
"	51-33-1	306	305	85.3	44.6	0.66	0.36	8.37
"	51-33-5	261	274	55.7	57.7	0.57	0.37	11.44
"	51-33-6	289	359	73.7	62.3	0.85	0.33	-
"	53-45-1	164	115	69.9	20.0	0.45	0.38	8.13
"	54-51-1	195	157	75.7	25.9	0.36	0.37	8.35
"	54-51-2	169	84.6	66.7	23.4	0.24	0.31	-
"	54-51-6	107	59.7	37.8	15.9	0.16	0.27	7.51
"	54-51-13	105	27.3	83.5	16.9	0.23	0.27	5.52
"	54-51-15	104	69.7	108	11.9	0.13	0.31	6.44

Table 6. (Cont'd).

Cruise	Sample	Cu	Ni	Co	Zn	Mn	Fe	Soluble
				SITE C				
RP-8-OC-76	2-7-1	880	1047	319	102	2.07	0.56	11.43
"	2-7-8	145	141	63.7	25.9	0.44	0.26	7.41
"	2-7-16	157	139	81.4	23.8	0.52	0.31	7.72
"	4-9-1	197	161	83.5	33.8	0.52	0.45	4.64
"	4-9-14	154	134	81.8	20.9	0.52	0.37	5.25
"	7-12-1	194	147	81.8	21.9	0.50	0.38	4.22
"	15-21-1	218	181	83.4	27.7	0.64	0.52	5.70
"	15-21-3	219	179	87.5	29.8	0.63	0.51	4.52
"	15-21-5	186	161	79.3	28.8	0.56	0.36	3.81
"	15-21-7	175	149	69.5	31.8	0.55	0.40	4.73
"	15-21-9	359	383	53.6	85.3	1.98	1.17	14.70
"	15-21-10	345	349	53.8	85.6	2.04	1.31	16.36
"	15-21-12	340	330	38.0	92.9	2.20	1.52	17.08
"	15-21-13	338	308	42.0	88.9	2.22	1.47	16.08
"	18-24-1	185	165	80.0	27.8	0.55	0.38	12.49
"	18-24-2	199	189	90.0	29.8	0.52	0.40	-
"	18-24-7	138	93.7	65.8	30.9	0.41	0.37	19.22
"	18-24-16	169	109	79.6	32.8	0.52	0.38	4.53
"	22-28-1	272	228	95.3	44.7	0.65	0.45	4.66
"	22-28-4	139	102	65.7	27.9	0.38	0.37	3.42
"	22-28-8	246	209	71.5	33.8	0.65	0.37	4.60
"	22-28-15	158	112	69.8	23.9	0.44	0.34	4.31

Table 7. Correlation coefficients of elements in the hydrogenous component of sediment (the sediment fraction soluble in hydroxylamine hydrochloride-acetic acid, Table 6).

	Ni	Co	Zn	Mn	Fe
Cu	.91	.69	.87	.83	.37
Ni		.69	.94	.95	.14
Co			.58	.73	.20
Zn				.88	.14
Mn					.05

no measurement made of their abundance. In smear slides, however, they were infrequently observed, suggesting that their contribution of metals to the SSF component may be minor. Certainly, the phase in which these metals and Fe are bound needs to be identified.

Manganese Nodules

Size, Abundance, and Texture

Box cores from the three DOMES sites contain nodules of two types of populations (Fig. 4), based on surface texture and average maximum nodule diameter (MND). One population consists of K-type nodules, as defined by Meyer (1973). Individual nodules are small, ellipsoidal, have a smooth surface texture, and are commonly intergrown forming polynodules. The population has a well defined modal size that is usually between 1.5 and 3 cm (Fig. 5).

The second population consists of B-type (Fig. 4) and E/S-type nodules. The B-type nodules usually have a granular surface texture on the side in contact with the sediment and a smooth surface in contact with the overlying seawater. The surface is commonly botryoidal. The modal size of the population is greater than 4.5 cm, although it is commonly poorly defined (Fig. 5). These nodules appear to form a series with E/S-type nodules, which are smaller. E/S-type nodules, which have a MND less than approximately 4 cm, often have a granular texture over the entire surface. Nodules with a MND less than 1.5 cm and polynodules seldom occur with B- and E/S-type nodules.

At Site C most of the box cores contained B- and E/S-type nodules, i.e., nodules with a granular surface texture (Table 8). Only four box cores contained smooth K-type nodules (Table 8). Approximately 20% of the box cores at Site B had no nodules and, of

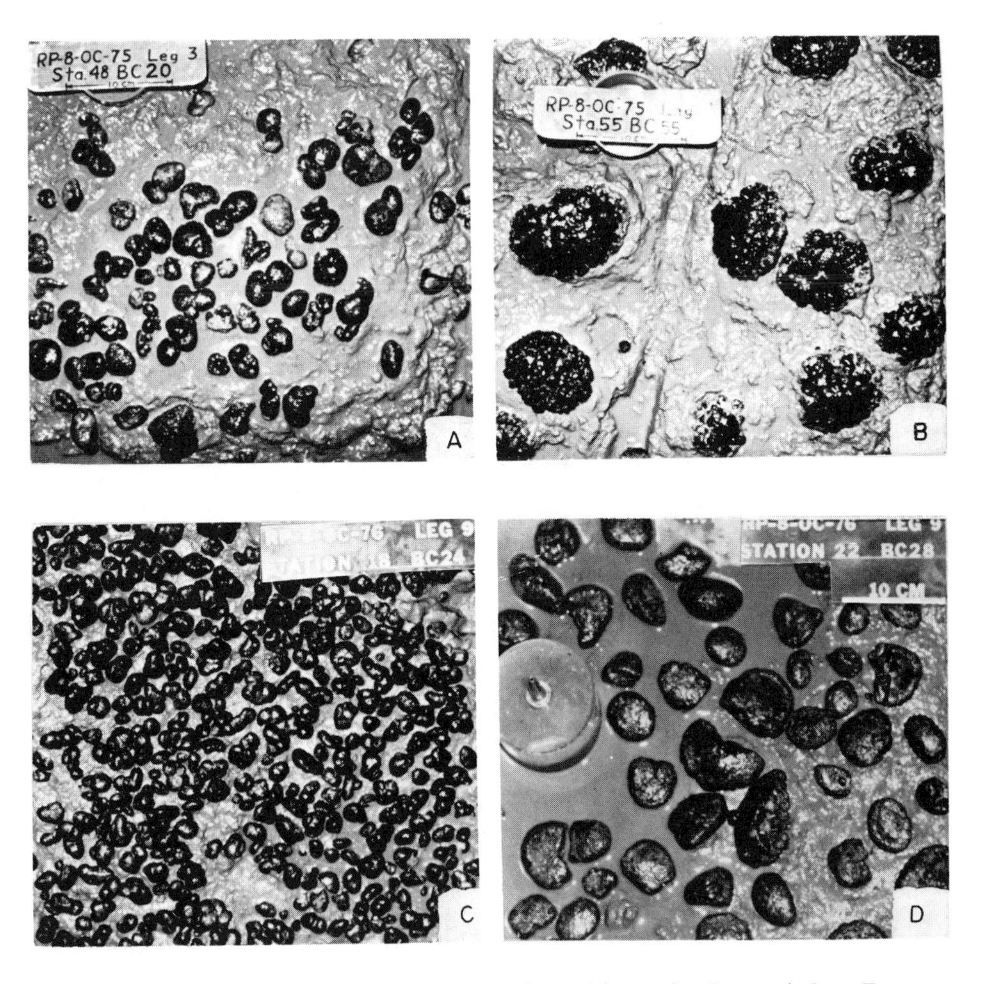

Fig. 4. Photographs of box cores from Sites A, B, and C. Two distinct nodule populations are present. One (A and C) is composed mainly of small nodules, but with considerable variation. It includes 48-20 (Site A) and 18-24 (Site C). The other population (B and D) consists predominantly of large nodules. Representative box cores of this group include 55-55 (Site B) and 22-28 (Site C). Photographs were provided by R. Sorem.

those with nodules, all appear from photographs of box cores to have a population of B- and E/S-type nodules. At Site A, 18 of the 26 box cores had smooth K-type nodules.

At Site C, a 225-km^2 area (Fig. 6) was surveyed in detail on R/V OCEANOGRAPHER cruise RP-8-OC-76. The area consists of three abyssal hills separated by north-south-trending channels (Fig. 6),

Table 8. Abundance, size and surface texture of nodules in box cores from DOMES Sites A, B, and C. The M.W. Series of samples were collected by the R/V MOANA WAVE.

Station	Box Core	Coverage (%)	Average Diameter (cm)	Concen-tration (kg/m²)	Texture*	Station	Box Core	Coverage (%)	Average Diameter (cm)	Concen-tration (kg/m²)	Texture*
SITE A											
D. 46	1	10	2.2	1.8	S	D. 48	19	1	4.6	0.4	S
D. 46	4	30	2.7	6.5	S	D. 48	20	30	2.3	5.5	S
D. 46	6	25	1.7	3.4	S	D. 48	22	1	5.7	0.5	G
D. 46	7	50	1.5	6.0	S	D. 48	23	1	-	0.0	
D. 46	8	55	1.6	7.0	S	D. 49	24	1	2.4	0.2	S
D. 46	9	25	2.7	5.4	S	D. 49	25	45	2.5	9.0	S
D. 47	10	0	-	0.0		D. 49	26	40	3.2	10.2	S
D. 47	11	5	5.4	2.2	C	D. 49	27	25	2.8	5.6	C
D. 47	12	20	4.3	6.9	C	D. 50	28	55	2.2	9.7	S
D. 47	13	7	3.3	1.8	G	D. 50	29	55	2.6	11.4	S
D. 47	14	4	4.8	1.5	G	D. 50	30	30	3.1	7.4	S
D. 47	15	4	5.7	1.8	G	D. 50	31	10	3.1	2.5	S
D. 47	16	15	3.4	4.1	G	D. 50	32	30	1.9	4.6	S
D. 47	18	15	3.1	3.7	S	Average:		21	3.0	4.3	S
SITE B											
D. 51	33	2	6.1	1.0	G	D. 55	53	5	1.6	0.6	C
D. 51	34	7	7.1	4.0	G	D. 55	54	10	5.6	4.5	C
D. 51	35	0	-	0.0		D. 55	55	22	7.6	13.4	C
D. 51	36	1	2.0	0.2	G	D. 55	56	35	6.2	17.4	C
D. 52	37	1	2.0	0.2		M.W. 16	27	-	-	0.1	
D. 52	39	5	9.0	3.6	G	M.W. 16	28	-	-	6.4	
D. 52	40	3	4.1	1.0	G	M.W. 16	29	-	-	12.1	
D. 52	41	0	-	0.0		M.W. 16	30	-	-	5.6	
D. 52	42	0	-	0.0		M.W. 16	31	-	-	6.9	
D. 53	43	0	-	0.0		M.W. 16	33	-	-	3.7	
D. 53	44	.5	.7	0.0		M.W. 16	34	-	-	12.7	
D. 53	45	3	8.0	1.9	G	M.W. 16	35	-	-	2.0	
D. 53	46	0	-	0.0		M.W. 16	36	-	-	0.6	
D. 54	47	37	4.8	14.2	G	M.W. 16	37	-	-	1.1	
D. 54	48	5	5.8	2.3	C	M.W. 8	30	-	-	21.2	
D. 54	51	35	5.6	15.7	C	M.W. 8	32	-	-	15.0	
D. 54	52	25	6.9	13.8	C	M.W. 8	33	-	-	5.6	
						Average:	-	10	5.2	5.5	

Table 8. (Cont'd).

SITE C

D. 3	5	50	2.5	10.0	
D. 6	19	60	5.4	25.9	
D. 9	21	60	1.8	8.6	
D. 20	27	62	1.5	7.4	
D. 24B	29	62	4.3	21.3	
D. 18	32	10	1.7	1.4	
D. 18	33	47	1.3	4.9	
D. 18	35	65	1.5	7.8	
D. 18	37	37	2.6	7.7	
D. 14	38	57	1.4	6.3	
D. 14	39	47	2.4	9.0	
D. 16B	43	62	2.7	13.4	
D. 16B	45	30	2.2	5.3	
D. 16B	48	5	1.9	0.8	
D. 16B	49	35	2.3	6.4	
D. 15B	50	50	2.1	8.4	
D. 27	52	40	1.9	6.1	
D. 11	53	55	1.2	5.3	
D. 57	59	0	-	0.0	
D. 58	60	60	1.6	7.7	
D. 1	6	20	3.1	5.1	C
D. 2	7	60	2.9	13.9	C
D. 3	8	10	2.3	1.8	G
D. 4	9	45	2.6	9.4	S
D. 5	10	55	2.5	11.0	C
D. 6	11	40	3.0	9.6	C
D. 7	12	45	2.5	9.0	G
D. 8	13	35	2.4	6.7	C
D. 10	15	20	4.7	7.5	C
D. 11	16	30	7.2	17.5	G
D. 12	18	50	2.7	10.8	C
D. 13	19	55	2.8	12.3	G
D. 14	20	50	5.1	20.4	C
D. 15	21	60	2.7	13.0	C
D. 16	22	70	2.1	12.8	S
D. 17	23	30	2.7	6.5	S
D. 18	24	60	2.4	11.5	S
D. 21	27	60	3.4	16.3	C
D. 22	28	45	5.2	18.7	C
M.W. 13A	2	-	-	4.2	
M.W. 13A	3	-	-	3.5	
M.W. 13A	4	-	-	1.9	
M.W. 13B	5	-	-	20.0	
M.W. 13B	6	-	-	0.1	
M.W. 13B	7	-	-	15.0	
M.W. 13B	8	-	-	11.9	
M.W. 13B	9	-	-	15.0	
M.W. 13B	10	-	-	0.6	
M.W. 13B	11	-	-	30.0	
M.W. 13B	12	-	-	17.5	
Average:		43	2.6	9.2	

* "S" refers to smooth surface texture, "G" to granular, and "C" to combination of smooth top and granular underside

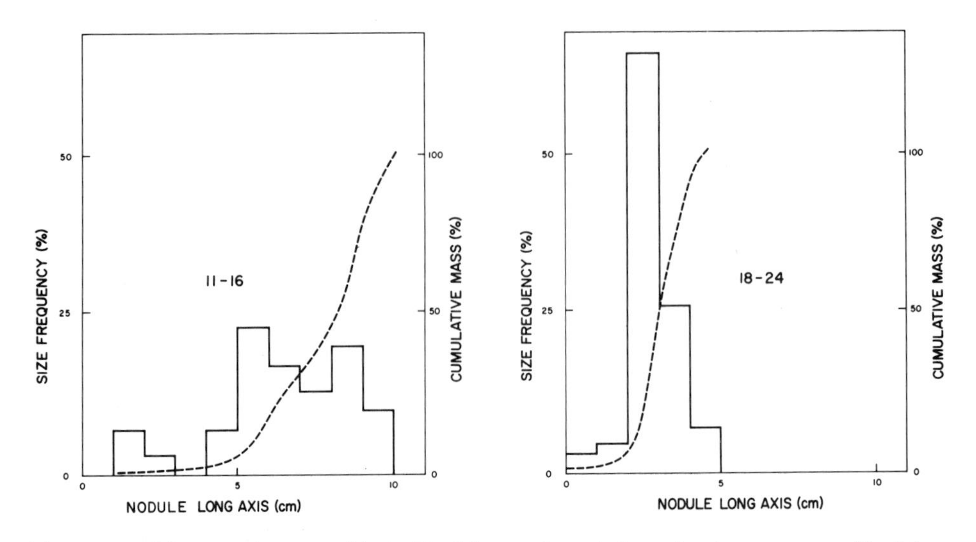

Fig. 5. Size and mass distribution of nodules in box cores 11-16 and 18-24 from Site C.

with maximum relief of approximately 300 m. The region was divided into five areas on the basis of bathymetry, acoustic stratigraphy (3.5 kHz), and nodule population.

Area 1 is the westernmost abyssal hill (Fig. 6). Acoustic unit 1 (Fig. 2) is apparently present as a thin veneer of sediment less than 10 m thick. E/S-type nodules are present and tend to be small. Seafloor coverage is uniformly high (Fig. 7). The easternmost abyssal hill appears to be similar, but few 3.5-kHz records and box cores were collected from this area.

Area 2 is the channel to the west of the central abyssal hill. Acoustic unit 2 appears to be present throughout much of the area (Piper *et al.* this volume). The upper transparent layer of sediment, however, is present on several profiles as a sediment wedge, that pinches out toward the center of the channel. Nodule coverage within the channel is variable (Fig. 8); nodule barren areas give way abruptly to areas in which coverage is as great as 70%. This is the only area within the 225-km^2 region for which box cores and bottom photographs showed that the nodule population consisted of B-type nodules (Fig. 7).

The channel to the east of the central abyssal hill, area 5a, is similar to this western channel, except possibly for size of nodules. Box core 7-12, located in the central part of this channel, recovered small nodules. One box core (57-58) taken just to the north of this area on an earlier cruise and apparently located within this channel, however, contained B-type nodules (Sorem et al. this volume).

Area 3 is the western part of the central abyssal hill (Fig. 6). Acoustic unit 2 and possibly unit 3 are at or very near the

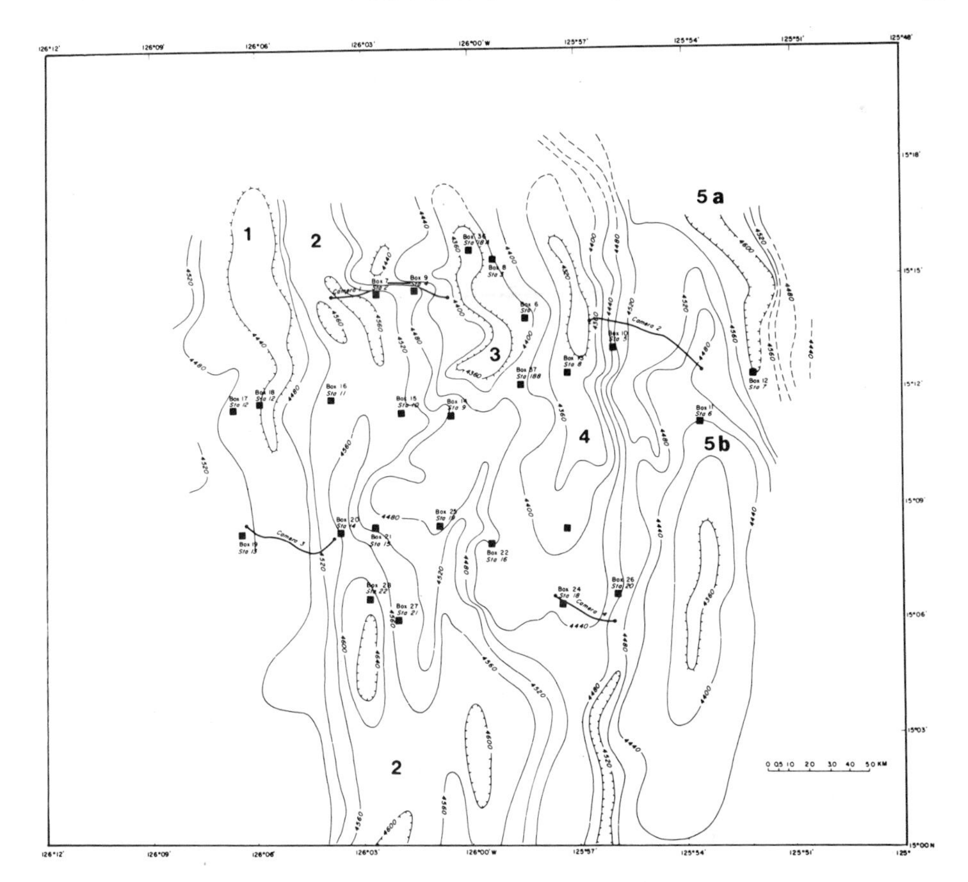

Fig. 6. Bathymetry, box core and camera flight locations for NOAA cruise RP-8-OC-76, Site C. The large numerals designate the five areas that are distinguishable due to nodule population, acoustic stratigraphy and bathymetry (Table 9).

surface. Nodules are small and have granular surface textures. Nodule coverage in box cores was small, but in bottom photographs coverage appears to be equal to that of the other areas.

Area 4 is the eastern half of the central abyssal hill. The transparent layer of sediment was clearly observed in the 3.5-kHz record. Its thickness is quite variable, with a maximum of approximately 20 m. Nodule coverage is high in all box cores and bottom photographs. K-type nodules were recovered in the box cores from this area.

Seafloor coverage of surface nodules throughout Site C is uniformly high (Table 9). The greatest nodule abundance, estimated from bottom photographs and from box cores, is in the western channel, where B-type nodules occur. The average abundance of nodules at Site C is 9.2 kg/m^2 (Table 8). Abundance measurements for box cores and bottom photographs exhibit a normal

Table 9. Nodule abundance in five areas within Site C, as shown in Fig. 6 and metal reserves (kg/m^2) of Fe, Mn, Cu, Ni, Co, and Zn. Nodule concentrations have been estimated form bottom photographs (column 5) and from nodule weights for box cores (column 7). For 19 box cores, nodule weights were also calculated from photographs (column 8). In calculating the average nodule concentrations for each area (column 9), the average of 200 bottom photographs was given equal weight to that of one box core. The average value listed in column 9 was lowered by 20% to give a nodule concentration on a dry-weight and salt-free basis. Metal concentrations used to estimate resource are listed in Table 10.

	Nodule Concentration							
	Camera Flights				Box Cores			
Nodule Area	Flight	Frame	Coverage Ave. (Range)	Conc. kg/m^2	No.	Weight kg/m^2	Photo kg/m^2	Average kg/m^2
	3	0-200	50 (10-65)	11	17	-	6	
1.	3	200-400	55 (10-65)	13	18	11	12	10
	3	400-600	60 (55-65)	16	19	10	-	
					7	15	14	
	1	0-200	25 (0-70)	15	15	7	8	
2.	1	200-400	70 (60-70)	29	16	19	18	14
	3	600-800	55 (10-65)	17	20	20	20	
					27	17	16	
					28	14	19	
					6	7	5	
3.	1	400-600	70 (65-70)	15	8	3	12	8.2
	1	600-800	65 (35-70)	16	9	7	9	
					21	13	13	
	2	0-200	55 (45-60)	10	10	11	11	
	4	0-200	60 (55-70)	12	22	14	13	9.1
4.	4	200-400	60 (5-70)	12	23	-	6	
	4	400-600	60 (55-65)	12	24	11	12	
	4	600-800	60 (55-65)	11	13	9	12	
	2	200-400	40 (5-65)	8	11	-	10	
5.	2	400-600	40 (5-65)	8	12	8	9	8.6
	2	600-800	45 (10-70)	15				

Table 9. (Cont'd).

Metal Resource (kg/m²)						Area Description
Fe	Mn	Cu	Ni	Co	Zn	
.92	2.29	.082	.110	.024	.0094	Western abyssal hill. Small nodules predominate. Surface sediment possibly Unit 1.
1.03	3.76	.15	.18	.035	.020	Western channel. Large nodules. Surface sediment mostly Unit 2 and Unit 1.
0.71	2.09	.085	.10	.018	.0091	Western part of central abyssal hill. Nodules small to large. Surface sediment mostly Units 2 and 3.
1.10	2.04	.056	.090	.030	.007	Eastern part of central abyssal hill. Small nodules. Surface sediment Unit 1.
0.84	2.26	.079	.10	.025	.010	Eastern portion of area. Nodules variable. Surface sediment Unit 2 in channel and possibly Unit 1 on abyssal hill.

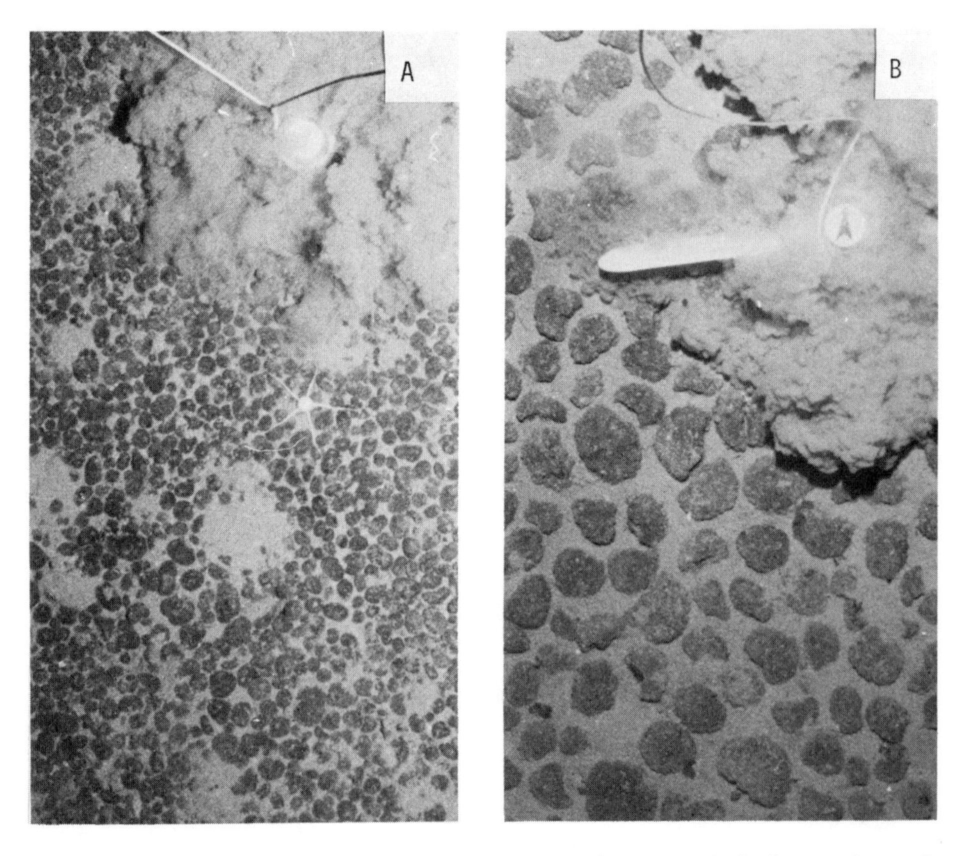

Fig. 7. Bottom photographs of Site C. Photograph A is a view of Area 1 (Fig. 6) taken on camera flight 3. The barren areas are mounds. Photograph B is a view of Area 2 (Fig. 6) frame 226, flight 1. The compass face is 5.7 cm in diameter.

distribution (Fig. 9). The curve based on measurements of bottom photographs, however, is shifted in the direction of greater abundance.

At Site B, average nodule coverage is 5.5 kg/m^2 and the frequency curve exhibits a log-normal distribution with a secondary mode at approximately 13 kg/m^2 (Fig. 9). Similar to Site C, the curve based on bottom photographs is shifted in the direction of greater abundance. Although roughly 40% of the area has few or no nodules, about 30% of the area has a nodule abundance greater than 10 kg/m^2.

The nodule coverage at Site A is 4.2 kg/m^2, and the frequency curve based on box cores exhibits a log-normal distribution. The curve based on bottom photographs, however, differs from this curve (Fig. 9).

Mizuno _et al_. (1976) have shown that nodule abundance may be

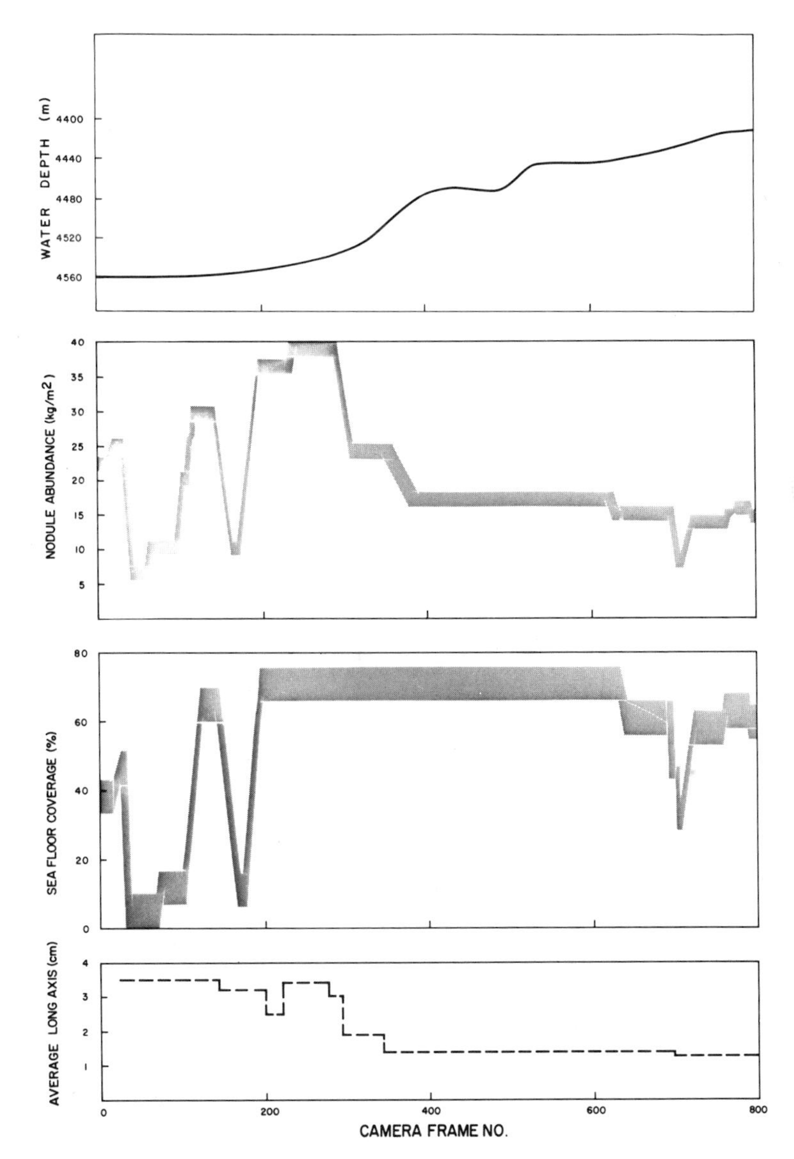

Fig. 8. Relation between bathymetry, seafloor coverage of nodules, average nodule long axis, and nodule abundance for camera flight 1 (Fig. 6). Basically, two nodule populations are observed. One has an average long axis of approximately 3.5 cm, camera frames 0 through 280. The other population has an average long axis of 1.3 cm, camera frames 350 through 800.

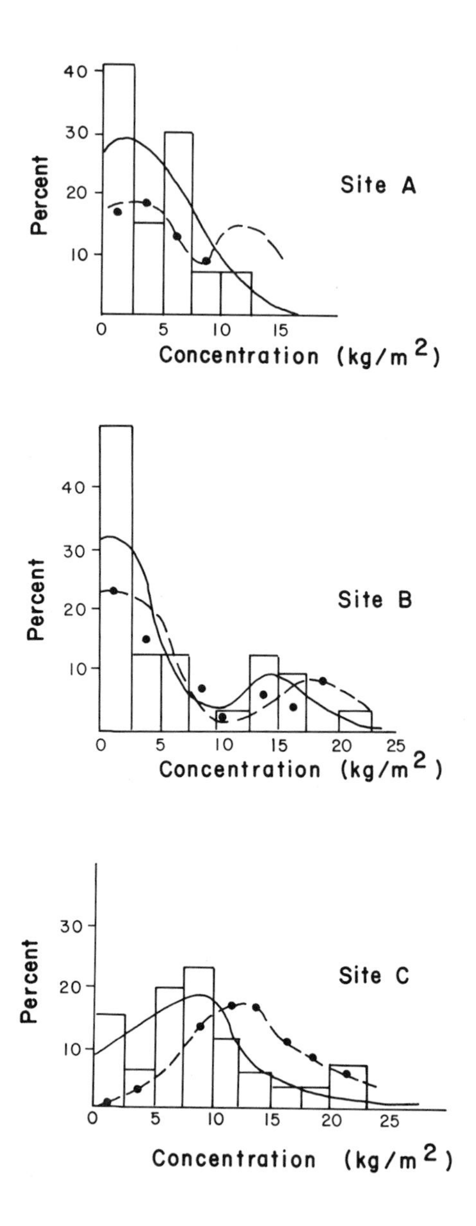

Fig. 9. Nodule abundance at Sites A, B, and C versus frequency of occurrence, as estimated from box cores (block diagram and solid curve) and from bottom photographs (dots and broken curve). The two curves represent floating-point averages of three points.

related to the distribution of the uppermost acoustically transparent sediment layer. They reported that in the Central Pacific Basin nodules were most frequently observed where this layer of sediment was less than 20 m thick. Nodular pavement was occasionally observed where the transparent layer was absent. Our data tend to support their observation. Nodules are most abundant at Site C where acoustic unit 1, the transparent layer, is very thin or absent (Piper et al. this volume). Nodules are least abundant at Site A where this acoustic unit is thickest.

The data for individual stations at Site B suggest that this relationship may hold for an area as small as individual sites. At Stations 52 and 53 acoustic unit 1 has its greatest distribution and thickness, averaging about 30 m (Piper et al. this volume). Nodules averaged about 1.5 per box core for 10 box cores (Sorem et al. this volume). At Stations 54 and 55, unit 1 is absent over much of the seafloor and, where present, it usually is less than 15 m thick. Nodules averaged approximately 22 per box core and the smallest number observed in any box core was 5.

Nodule texture and size are highly variable over distances of only a few hundred meters. At Site C there is a possible relation between nodule size and bathymetry, (Fig. 8). Large B-type nodules apparently are restricted to the channels between abyssal hills (Figs. 4,7,8). The K-type nodules appeared to be restricted to area four (Fig. 6), the area where the uppermost transparent layer of sediment, acoustic unit 1, is observed as having a continuous areal distribution. No such relations, however, appear to hold for Sites A and B.

Mineralogy

The minerals todorokite and δ-MnO_2 are present in all nodules, as deduced from X-ray diffraction peaks at 9.6, 2.44, and 1.44 Å. The convention used for their identification is that suggested by Burns and Burns (1977). In most samples, a peak at 7 Å was also observed, suggesting the presence of birnessite. In those samples that exhibited a rather intense 7 Å reflection, however, other reflections corresponding to the weaker reflections of phillipsite were always present.

Chemical Composition

Hydrated oxides of Mn and Fe constitute the major fraction of the hydrogenous component in nodules. The lack of major element analyses, particularly for Al and Si, precludes, however, our ascertaining quantitatively the concentrations of biogenic and lithogenic material. Comparison of the composition of the ISF (Table 2), dominantly lithogenic material, with that of nodules (Table 10) shows that within nodules essentially all of the Co, Zn, Sb, Ni, and Cu and most of the Th, Hf, and U are associated with Mn and Fe oxides. Calvert and Price (1977) arrived at this same conclusion for nodules for which they had major element analyses. Interelement correlations of the minor elements with Mn and Fe (Table 11)

Table 10. Bulk compositions of nodules from DOMES Sites A, B, and C Concentrations are in ppm except where noted otherwise. The averages for each site enclosed in parentheses include analyses reported by Mouget _et al_. (1976).

Station-Box Core	Lat. (North)	Long. (West)	Depth (Meters)	Fe (%)	Mn (%)	Cu (%)	Ni (%)
46-3	9°19.0'	150°52.5'	4900	11.6	19.3	0.49	0.77
46-3	9°19.0'	150°52.5'	4900	9.7	25.6	0.71	1.12
46-7	9°19.6'	150°50.7'	5116	6.7	22.3	0.77	1.24
46-7	9°19.6'	150°50.7'	5116	6.3	12.1	0.40	0.61
47-12	9°03.5'	151°11.1'	5005	4.1	26.8	1.39	1.59
47-12	9°03.5'	151°11.1'	5005	5.9	25.6	1.35	1.37
47-13	9°02.3'	151°11.2'	5039	5.4	24.3	1.34	1.38
47-13	9°02.3'	151°11.2'	5039	4.0	26.2	1.57	1.54
48-21	8°17.6'	151°13.2'	5097	5.6	22.0	1.17	1.17
49-25	8°28.5'	150°44.5'	5005	7.0	25.2	0.99	1.29
49-27	8°30.2'	150°47.8'	5039	5.8	26.7	1.26	1.46
50-28	8°41.7'	150°18.7'	5009	10.7	22.8	0.56	0.95
50-29	8°43.8'	150°18.7'	5033	11.6	24.1	0.58	0.95
50-29	8°43.8'	150°18.7'	5033	11.9	22.2	0.59	0.65
50-30	8°41.1'	150°15.1'	4928	10.9	23.4	0.59	0.98
50-30	8°41.1'	150°15.1'	4928	11.3	22.0	0.57	0.98
50-32	8°43.1'	150°14.1'	4948	10.4	24.7	0.64	1.05
Average (Site A)				8.17 (9.2)	23.3 (20.7)	0.85 (0.79)	1.12 (1.00)
51-34	11°42.4'	139°10.8'	4925	2.9	26.8	1.27	1.23
54-47	12°10.4'	137°44.1'	4930	4.3	26.4	1.26	1.51
54-49	12°12.0'	137°44.7'	4942	3.0	26.9	1.31	1.52
54-50	12°08.5'	137°46.0'	4978	5.6	25.6	1.21	1.41
Average (Site B)				4.0 (5.5)	26.4 (28.1)	1.26 (1.20)	1.42 (1.54)
1-6	15°13.9'	125°58.4'	4380	8.7	28.9	1.18	1.36
2-7	15°14.5'	126°02.6'	4359	8.3	27.5	1.08	1.26
3-8	15°15.4'	126°59.3'	4363	8.5	28.8	1.27	1.29
4-9	15°14.6'	126°01.6'	4490	9.1	16.2	0.53	0.82
5-10	15°13.0'	125°55.9'	4465	9.5	26.7	0.99	1.03
6-11	15°10.9'	125°53.6'	4442	8.6	26.8	1.05	1.38
7-12	15°12.3'	125°52.1'	4601	10.1	25.5	0.86	1.12
8-13	15°12.4'	125°57.2'	4367	11.3	25.9	0.81	1.14
9-14	15°11.3'	126°01.5'	4444				
10-15	15°11.4'	126°02.2'	4516	8.9	27.2	1.05	1.21
11-16	15°11.7'	126°03.9'	4524	7.4	27.3	1.24	1.30
12-18	15°11.6'	126 05.9'	4459	9.1	24.3	0.89	1.21
13-19	15°08.2'	126 06.4'	4481	9.4	21.5	0.74	0.95
14-20	15°08.2'	126 03.7'	4586	7.7	25.4	1.05	1.31
15-21	15°08.4'	126 02.7'	4501	8.6	24.7	0.95	1.26
16-22	15°07.9'	125 59.4'	4442	10.6	21.8	0.61	1.04
17-23	15°08.3'	125 57.3'	4376	14.2	21.4	0.63	0.88
18-24	15°06.3'	125 57.4'	4426	11.7	24.1	0.61	1.05
19-25	15°08.4'	126 00.9'	4465	9.3	26.8	1.17	1.31
21-27	15°05.9'	126 02.1'	4522	6.9	26.0	1.15	1.33
22-28	15°06.4'	126 02.8'	4641	6.8	26.9	1.06	1.34
15-50	15 45.6'	126 00.5'	4552	5.6	26.5	1.08	1.19
16-49	15 46.7'	126 11.2'	4552				
57-57	15 17.8'	125 55.3'	4520	7.41	25.3	.93	1.35
58-60	15 17.8'	125 28.4'	4375	7.92	24.5	0.87	1.20
58-61	15 20.6'	125 26.8'	4414	5.83	28.0	1.25	1.33
Average (Site C)				8.8 (8.5)	25.4 (27.0)	0.97 (0.99)	1.20 (1.48)

Table 10. (Cont'd).

Co	Zn	Ba	Hf	Sb	Th	U	Sc
2680	630	1455	10.7	23.3	25.2	5.8	16.87
3360	900	-	-	-	-	-	-
2140	920	1240	7.9	27.4	22.9	4.0	13.95
1360	620	1582	-	14.5	22.5	3.1	14.88
1710	1200	1777	4.6	27.4	9.0	1.1	11.54
1660	1700	2478	4.2	34.3	9.2	3.8	11.56
1440	1200	1617	6.0	21.9	10.1	2.8	18.14
1730	1200	1769	3.7	27.4	9.3	3.6	10.47
1490	920	2151	4.9	35.6	10.0	5.2	10.92
2300	980	2010	6.9	28.3	17.7	5.0	10.26
1950	1200	2729	6.3	30.1	12.2	4.4	12.37
3140	760	1919	12.2	34.2	31.8	8.6	12.77
3580	830	1587	11.1	33.4	31.7	8.6	13.12
3620	700	1745	11.3	27.8	34.3	6.2	12.57
3310	780	1823	11.2	31.9	25.6	7.2	12.66
3170	740	1452	10.7	31.2	29.4	7.8	12.33
3570	820	2396	14.2	32.9	30.1	6.9	12.42
2480	950						
(2400)	-						
1450	1100	1756	2.7	43.9	9.4	1.2	7.40
2970	1100	1771	3.6	35.8	18.5	3.6	8.63
1780	1500	1755	3.9	36.2	13.6	3.1	8.88
2040	1300	1642	4.4	43.4	16.0	3.4	9.53
2060	1250						
(2900)	-						
2550	1600	1894	2.8	42.6	13.2	3.0	9.08
2160	1500	2593	4.2	40.7	17.0	3.1	10.26
2050	1700	2144	2.7	47.4	12.4	4.3	9.33
2050	690	1878	5.2	18.2	19.3	3.2	13.95
2720	1400	1269	4.0	34.8	18.8	4.5	9.32
2960	1200	2421	4.9	37.1	16.7	4.1	9.87
3060	1100	1886	5.3	41.4	24.2	5.3	9.27
2960	1300	1885	5.9	36.1	23.8	6.2	9.67
2502	652	1963	7.3	22.0	29.4	5.2	12.24
2370	1400	2258	2.6	45.5	11.6	1.0	8.96
2730	1400	2345	3.4	43.5	15.8	2.8	9.08
2640	1000	2115	5.6	34.1	19.0	3.7	11.97
2250	940	2118	6.3	32.6	24.2	5.6	14.75
3090	950	2947	3.7	33.4	13.9	4.3	9.91
2140	1200	2800	4.5	34.5	17.0	3.9	11.18
3420	770	1686	8.4	31.0	35.1	5.3	11.16
4310	640						
2190	920	1904	8.9	32.8	38.4	6.6	11.02
2320	1500						
2430	1600	1626	4.6	24.7	21.0	3.4	15.34
2380	1500	2676	4.1	36.0	13.8	4.1	10.08
2380	1100	1576	3.9	35.4	17.3	5.1	7.79
2460	1300	1279	3.6	40.7	20.0	6.3	7.32
2640							
(2700)							

Table 11. Correlation coefficients of metal concentrations in DOMES nodules.

	Mn	Cu	Ni	Co	Zn	Ba	Hf	Sb	Th	U	Sc
Fe	-.30	-.77	-.69	-.76	-.44	00	.65	00	.78	.71	.24
Mn		.72	.76	00	.76	.30	-.57	.73	-.42	-.24	-.63
Cu			.89	-.56	.77	.24	-.78	.39	-.85	-.65	-.44
Ni				.41	.72	.33	-.66	.39	-.70	-.57	-.42
Co					-.40	00	.10	.14	.67	.56	.24
Zn						.24	-.73	.58	.64	-.50	-.45
Ba							-.20	.24	-.30	-.17	-.10
Hf								-.44	.79	.60	.55
Sb									-.26	-.14	-.76
Th										.75	.30
U											.20

also support this conclusion.

Barium may be present as barite, as this mineral has been identified in nodules (Dugolinsky, 1976). Calvert and Price (1977) observed a strong correlation between Ba and Mn and suggested that much of the Ba may be in the oxide phase. However, DOMES nodules show a weak correlation (Table 11). Scandium may be restricted to the lithogenous component of nodules, in which it is approximately threefold more concentrated than in nodules.

The composition of nodules from all three DOMES sites are quite similar . The relation between Ni plus Cu and the Mn/Mn + Fe ratio (Fig. 10) suggests only very subtle differences between the three sites. This relation for Site A nodules has a correlation coefficient of 0.95. Nodules from Site C fall along the curve, but on the high Ni + Cu side. Site B nodules plot on the low Ni + Cu side, but only four nodules from Site B were analyzed. The relation between Zn and Mn/Mn + Fe (Fig. 10), however, clearly distinguishes Site A and Site C nodules: Site C nodules have a relatively higher Zn concentration and show a poor correlation of 0.59. Nodules from the Wahine area (Calvert *et al.*, 1978) tend to fall on the low minor-element side of the curves of Site A nodules (Fig. 10).

Calvert and Price (1977) have used plots of Ni versus Mn and Cu versus Mn to show that nodules from the pelagic environment of the Pacific Ocean belong to one of two populations. One population contains δ-MnO_2 and the other todorokite plus δ-MnO_2.

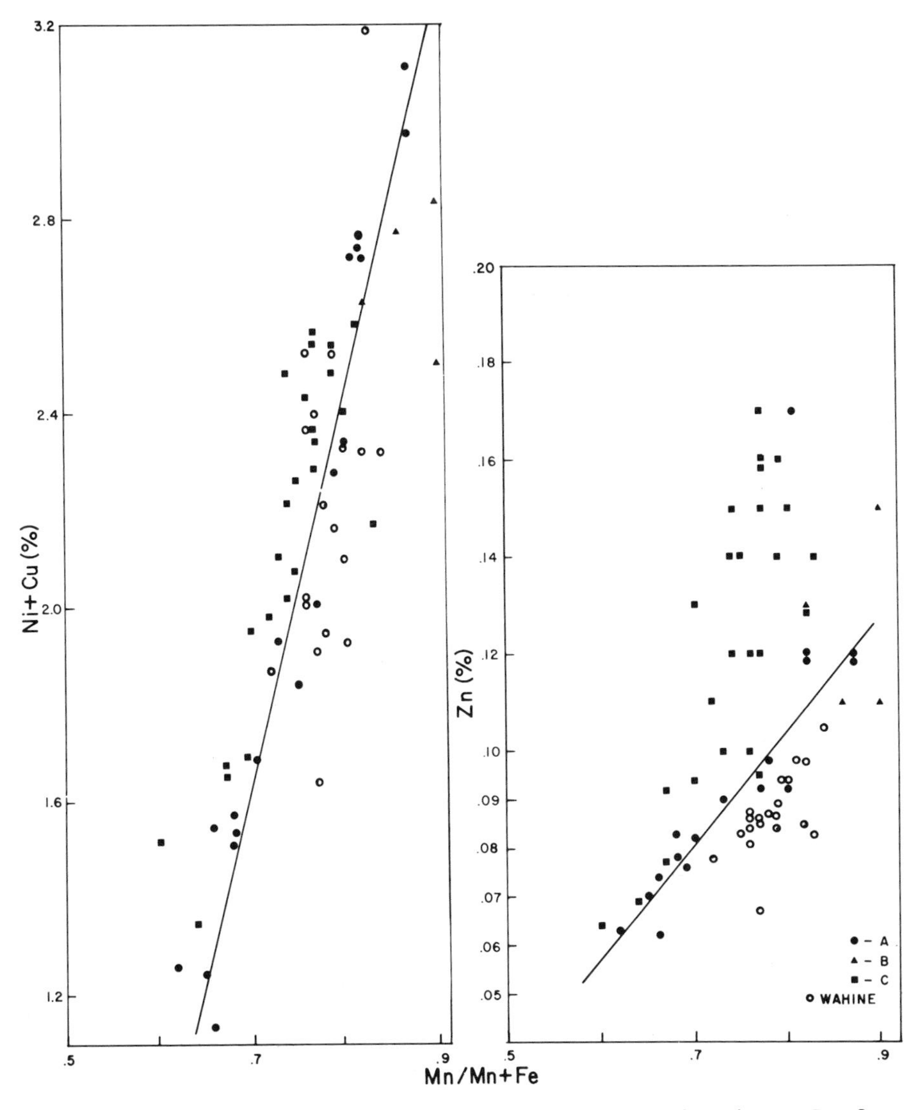

Fig. 10. Relations between Zn and Ni plus Cu and Mn/Mn + Fe for nodules from Sites A, B, and C. The curves are a linear regression by method of least squares using only the data for Site A nodules. The equation for the Zn curve is y = .24x - .087, σ = 0.95; the equation for the Ni plus Cu curve is y = 7.88 x -3.85, σ = 0.97. The Wahine samples are from 8°20'N, 153°W (Calvert et al. 1978).

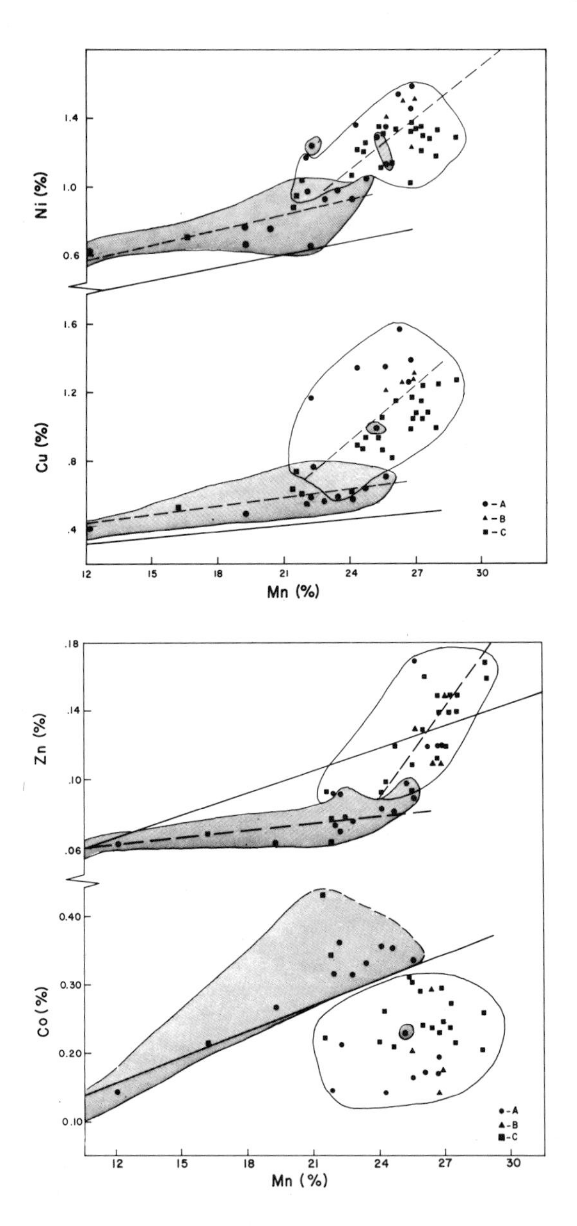

Fig. 11. Relation between Mn and Cu, Ni, Zn, and Co for nodules from Sites A, B, and C (Table 10). The shaded areas include all K-type nodules. The unshaded areas include all B- and E/S-type nodules. The broken lines are curves of best fit, for the two populations, determined by method of least squares. The solid lines are the extrapolated curves of elemental relations within the SSF.

Although both populations exhibit strong interelement correlations, the slopes of the curves for the δ-MnO_2 population are significantly less and their curves extrapolate to the origin and approximately through the field for the composition of bulk sediment.

The DOMES nodules at first might not appear to fall into these two groups, principally because they all contain todorokite. Plots of Ni versus Mn and Cu versus Mn (Fig. 11), however, show that nodules with relatively low Ni and Cu fall along curves that parallel the curves for δ-MnO_2 nodules. The remaining nodules scatter about a curve that parallels the curves for todorokite-bearing nodules.

Although all DOMES nodules contain todorokite, they can be divided into two populations on the basis of surface textures. The smooth, K-type nodules from DOMES cores have a composition similar to the δ-MnO_2 nodules analyzed by Calvert and Price (1977). The second population consists of B- and E/S-type nodules and they have relatively high Ni and Cu values, similar to nodules analyzed by Calvert and Price which contain todorokite. The compositions of these two populations have very limited overlap (Fig. 11).

The curves of Ni versus Mn and Cu versus Mn for the SSF, when extrapolated into the compositional field of nodules, define exceedingly well the composition of smooth nodules. The sediment curves, however, are displaced slightly in the direction of low Ni and Cu values (Fig. 11).

Several investigators have suggested that K-type nodules grow by incorporation of metals from seawater, whereas B- and E/S-type nodules derive an additional metal component from interstitial water during sediment diagenesis. For example, Marchig and Gundlach (1978) observed a decrease in micronodule abundance with depth, in box cores, accompanied by a slight decrease in their Ni and Cu concentrations relative to Mn. Thus, one possible explanation for the relation between the composition of the SSF and macronodules analyzed in this study is that both the K-type nodules and the SSF derive metals from seawater. During sediment diagenesis, Ni and Cu go into solution more so than Mn. These metals are then incorporated into B- and E/S-type nodules.

This interpretation suggests that the SSF may represent the residue resulting from sediment diagenesis. The distribution of Sb between nodules and sediment tends to support this interpretation. Its concentration in nodules (Table 10) is approximately 30 times its concentration in bulk sediment (Table 3), similar to the partitioning of Mn. Unlike Mn, however, essentially all Sb present in sediments is in the ISF.

The relation between Zn and Mn (Fig. 11) in nodules is similar to that between Cu and Mn (Fig. 11). The curve defining their distribution in the SSF, however, passes through the compositional fields for both types of nodules, rather than paralleling the curve that defines the composition of K-type nodules.

Within the SSF the concentration of Co correlates strongly with Mn (Table 7) but it correlates with Fe in nodules (Table 11).

When the nodule values are plotted against Mn, however, the two populations based on textural differences exhibit separate ranges of composition, where the K-type nodules are relatively enriched in Co. The curve which defines their relation in the SSF lies between the two nodule fields (Fig. 11).

Mineral Resource

The composition and abundance of nodules at Site C are known well enough to estimate the variations in resource for a small area. Throughout the area shown in Fig. 6, abundance is quite high (Table 9). The somewhat greater nodule abundance in area 2 accounts for the greater metal resource of this area over that of areas 1, 3, and 5. The low Ni, Cu, and Zn resources of area 4, however, reflect the low grade of K-type nodules of this area, rather than low nodule abundance.

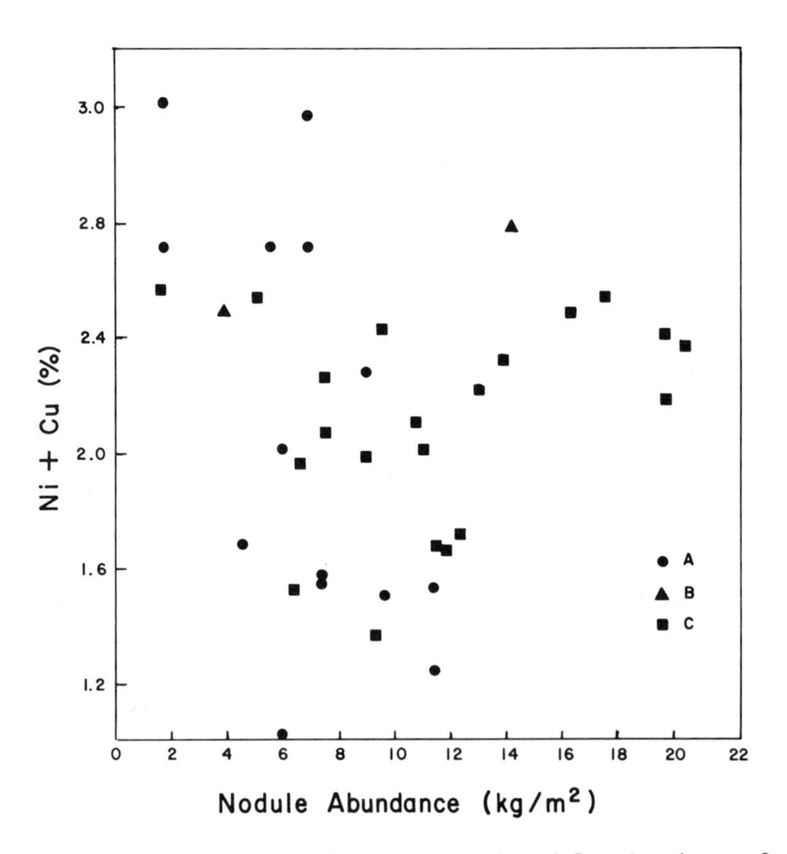

Fig. 12. Relation between Ni plus Cu and nodule abundance for DOMES Sites A, B, and C.

No relationship between nodule grade and abundance was observed for area C (Fig. 12). For the box cores from Site A, however, a possible negative correlation is suggested, as observed by Menard and Frazer (1978) for nodules from throughout the Pacific. However, few data are available for Site A and even fewer for Site B.

Conclusions

1. The major- and minor-element compositions of the lithogenous component of sediment at DOMES Sites A, B, and C are extremely uniform. The main difference is slight dilution of this component at Sites A and B with biogenic silica. The composition of this sedimentary component is nearly identical to that of terrigenous shale. Several elements have been measured (Cr, Cs, Hf, Rb, Sb, Ta, and Sc) for which few analyses have been published previously.

2. The fraction of sediment soluble in hydroxylamine hydrochloride-acetic acid, the hydrogenous component of sediment, consists mostly of adsorbed ions of Na^+, K^+, Ca^{++}, and Mg^{++}, amorphous oxyhydrates of Fe and Mn, and minor amounts of Co, Cu, Ni, and Zn. The last four metals correlate strongly with Mn.

3. Nodules can be divided into two populations by surface texture, size, and frequency of polynodules. K-type nodules (smooth surface texture, maximum dimension usually less than 3 cm, polynodules abundant) have low Ni, Cu, and Zn and high Co concentrations relative to B- and E/S-type nodules (granular surface texture, maximum diameter usually greater than 5 cm for B-type nodules but less for E/S-type nodules, polynodules rare).

4. The relations between Ni and Mn and between Cu and Mn in K-type nodules is similar to their relation in the hydrogenous component of sediment, but with the nodules slightly enriched in Ni and Cu relative to Mn. B- and E/S-type nodules are strongly enriched in Ni and Cu. One possible interpretation for the marked enrichment of Ni and Cu in B-type and E/S-type nodules is that the composition of K-type nodules resembles the initial composition of the hydrogenous component of sediment. During sediment diagenesis this component within the fine-grained fraction of sediment loses Ni and Cu relative Mn, and B- and E/S-type nodules take up Ni and Cu. The distributions of Co and Zn between nodules and the hydrogenous component of the sediment, however, can not be accounted for by this explanation.

5. The abundance of nodules at the DOMES sites appears to be related to the thickness of the uppermost acoustically transparent layer. Nodules are abundant (greater than approximately 5 kg/m^2) where this layer of sediment is less than approximately 15 m thick and significantly less abundant where it is thicker than 20 m.

Acknowledgments

We would like to thank the ship's crew and officers of the R/V OCEANOGRAPHER for their full cooperation and assistance during the cruises devoted to geologic studies. Jeanne Henning patiently edited and typed the manuscript. J. Bischoff and F. Manheim critically reviewed a preliminary draft and made numerous helpful suggestions. Any mistakes or misinterpretations, however, are the responsibility of the authors.

References

Bischoff, J.L., Piper, D.Z., and Quinterno, P., (1979a) Nature and origin of metalliferous sediment in DOMES Site C, Pacific manganese nodule province, In Sur la Genese des Nodules de Manganese (editor, C. Lalou) (in press).

Bischoff, J.L., Heath, G.R., and Leinen, M., (1979b) Geochemistry of deep-sea sediments from the Pacific manganese nodule province: DOMES Sites A, B, and C: (this volume).

Burns, R.G. and Burns, V.M.,(1977) Mineralogy. In Marine Manganese Deposits, (editor, G.P. Glasby), Elsevier Ocean. Ser. 185-248.

Calvert, S.E. and Price, N.B. (1977) Geochemical variation in ferromanganese nodules and associated sediments from the Pacific Ocean, Mar. Chem. 5, 43-74.

Calvert, S.E., Price, N.B., Heath, G.R., and Moore, T.C., Jr. (1978) Relationship between ferromanganese nodule compositions and sedimentation in a small survey area of the equatorial Pacific, J. Mar. Res. 35, 161-183.

Chester, R. and Hughes, M.J. (1967) A chemical technique for the separation of ferromanganese minerals, carbonate minerals, and adsorbed trace elements from pelagic sediment, Chem. Geol. 2, 249-262.

Cook, H.E., Piper, D.Z. and Gardner, J.V., (1977) Geologic and oceanographic framework of DOMES Sites A, B, and C: Central Equatorial Pacific, In Deep Ocean Environmental Study: Geology and Geochemistry of DOMES Sites A, B, and C, Equatorial North Pacific, (compiler, D. Piper), U.S. Geol. Survey Open-file Rept. 77-778, 15-81.

Dugolinsky, B.K. (1976) Chemistry and morphology of deep-sea manganese nodules and the significance of associated encrusting protozoans, PhD dissertation, Dept. Ocean., Univ. Hawaii, 228 p.

Goldberg, E.D., (1963) Mineralogy and chemistry of marine sedimentation, In Submarine Geology (F.P. Shepard), New York, Harper and Row, 436-466.

Hein, J.R., Ross, C.R., Alexander, E. and Yeh, H-W (1979) Mineralogy and diagenesis of surface sediment from DOMES Areas A, B, and C: (this volume).

Horn, D.R., Delach, M.N. and Horn, B.M., (1973) Metal content of

ferromanganese deposits of the oceans, IDOE Tech. Report No. 3, NSF, Washington, D.C., 55p.
Marchig, V., and Gundlach, H., (1978) Diagenetic changes in the radiolarian oozes of the Central Pacific and their influence on the growth of manganese nodules, (in press).
Menard, H.W., and Frazer, J.Z., (1978) Manganese nodules on the sea floor: inverse correlation between grade and abundance, Science 199, 969-971.
Meyer, K., (1973) Surface sediment and manganese nodule facies, encountered by R/V VALDIVIA cruises 1972/1973: In The Origin and Distribution of Manganese Nodules in the Pacific and Prospects for Exploration, (editor, M. Morgenstein), Hawaii Inst. Geophys., Honolulu, 125-135.
Mizuno, A., Chuijo, J. and Yamakade, N., (1976) Prospecting of manganese deposits of the eastern central Pacific Basin, 3rd Intern. Devel. Conf. , Tokyo 5, 322-335.
Monget, J.M., Murray, J.W. and Mascle, J., (1976) A world-wide compilation of published, multicomponent/analysis of ferromanganese concretions: NSF-IDOE Tech. Report 12, 127 pp.
Piper, D.Z., Cook, H.E., and Gardner, J.V., (1979) Lithic stratigraphy and acoustic stratigraphy of the equatorial North Pacific, DOMES Sites A, B, and C (this volume).
Quinterno, P., and Theyer, F., (1979) Biostratigraphy of the equatorial North Pacific DOMES Sites A, B, and C (this volume).
Riley, J.P. and Sinhaseni, P., (1958) Chemical composition of three manganese nodules from the Pacific Ocean, J. Mar. Res. 17, 466-482.
Skornyakova, N.S., (1979) Zonal regularities in occurrence, morphology and chemistry of the Pacific Ocean manganese nodules (this volume).
Sorem, R.K., Reinhart, W.R., Fewkes, R.H., and McFarland, (1979) Occurrence and character of manganese nodules in DOMES Sites A, B, and C east equatorial Pacific Ocean (this volume).
Turekian, K.K., and Wedepohl, K.H., (1961) Distribution of the elements in some major units of the earth's crust: Geol. Soc. Amer. Bull. 72, 175-192.

OCCURRENCE AND CHARACTER OF MANGANESE NODULES IN DOMES SITES A, B, AND C, EAST EQUATORIAL PACIFIC OCEAN

R.K. Sorem, W.R. Reinhart, R.H. Fewkes, and
W.D. McFarland
Department of Geology
Washington State University
Pullman, Washington 99164

Abstract

Study of the occurrence and physical and chemical character of manganese nodules collected in box cores in DOMES Sites A, B, and C indicates that virtually all nodules lie at the sediment-water interface. Buried nodules are rare. Nodules seem to be smaller and have smoother surfaces at Site A than at Sites B and C. Site C nodules are predominantly 2 to 6 cm in size and have smooth tops and granular bottoms.

Nodules from all sites show non-uniform contents of Mn, Fe, Ni and Cu within individual box cores. The greatest variations were found in box cores from Sites A and C. Mn, Ni, and Cu are generally covariant whereas Fe content is antithetic to Mn content. Large variations in metal content are generally due to differences in oxide mineralogy but some variation is related to clay content. Large nodules tend to be relatively low in Mn, Ni, and Cu.

Site B nodules have the highest mean values for Mn, Ni, and Cu and the lowest mean Fe content. Site C nodules and Site A nodules have the same Mn content, but Site C nodules are richer in Ni and poorer in Fe than Site A nodules.

Introduction

In the past 25 years, several authors have described a general geographic pattern of variations in bulk chemical composition of manganese nodules in the central Pacific Ocean (Arrhenius, 1963; Mero, 1965; Horn *et al.*, 1972; and others). There is little doubt that extensive deposits of manganese nodules relatively rich in nickel and copper exist in a broad belt between Hawaii and Baja

California, Mexico.

Although there is great interest in the possible resource value of manganese nodules, it is rarely known if the individual nodules or pieces of nodules analyzed are representative samples of even small deposits. An analyzed nodule may or may not be an "average" nodule, typical of a large population at a bottom location. The simplest solution to this problem is to analyze several additional nodules from a site of interest. Then, knowing the mean metal content and standard deviation, an intelligent estimate of nodule grade can be made. However, where additional nodules for some reason cannot be analyzed, other criteria to evaluate the significance of a single analysis must be sought.

The two chief objectives of the research reported here are to document the natural chemical variations in nodules in part of the central East Pacific Ocean and to seek relationships between chemical composition and other factors.

Shipboard Operations and Observations

General

In 1975 and 1976, box core sampling, bottom photography, and bathymetric surveys were carried out during three cruises of NOAA Ship OCEANOGRAPHER to Sites A, B, and C of DOMES.

Site A is centered around 8°27'N, 150°47'W; Site B, 11°42'N, 138°24'W; and Site C, 15°00'N, 126°00'W.

Bottom Photographs

More than 10,000 bottom photographs were obtained during the 1975 and 1976 cruises. In many photographs, not only nodule abundance and distribution but also nodule type can be recognized. In some, features caused by benthic organisms are clearly seen as are many details of the upper surface of the sediment-water interface. In 1976, stereoscopic pairs of photographs were obtained in an attempt to document the small-scale relief of the sea floor deposits (Fig. 1).

A study of several dozens of the photographs, coordinated with study and sampling of "undisturbed" box cores, points to the widespread presence of a "sediment-water interface boundary layer" (Sorem *et al.*, 1978a, 1978b). This diffuse layer, which may attain a thickness of several centimeters in some places, may have a profound influence on the mineralogy and chemical composition of the layers now forming on nodules.

Box Cores: Surface Features

85 box cores were recovered during Leg 3 of RP8-OC-75 and Leg 9 of RP8-OC-76 and were documented by R.K. Sorem, R.H. Fewkes, and D.L. Banning at sea; data are given in the Appendix.

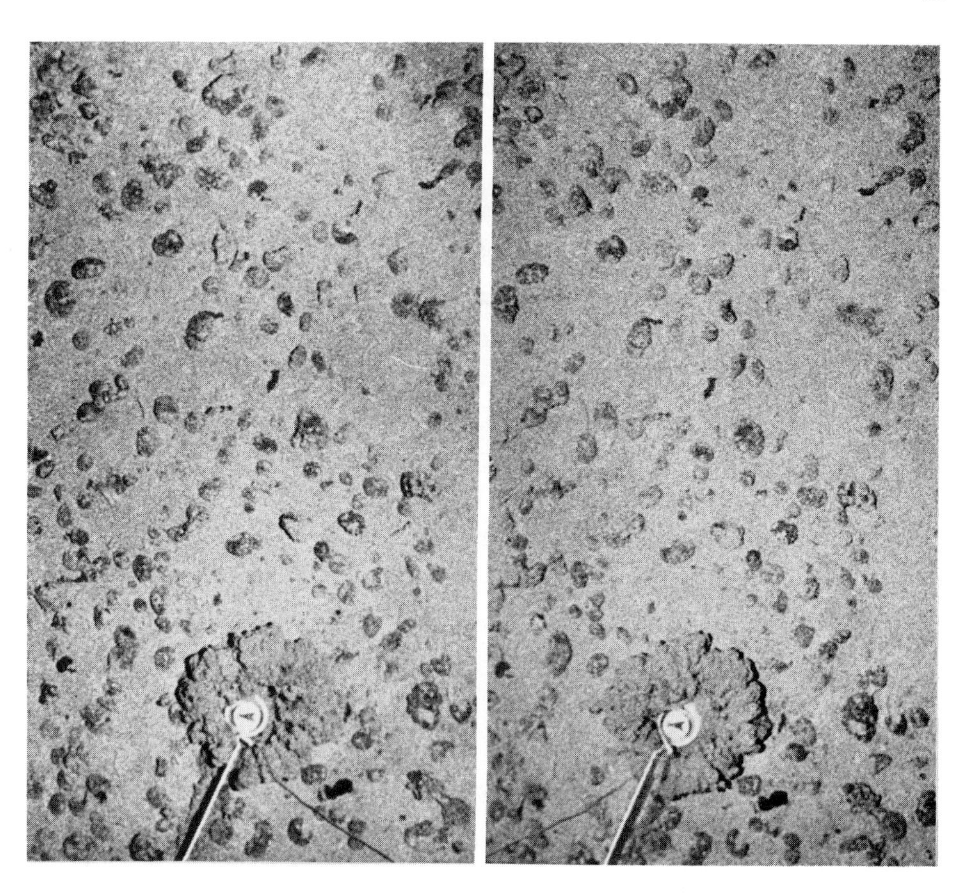

A B

Fig. 1. Pair of stereoscopic photographs of the sea bed. The compass head is 8 cm in diameter. Note the varying degree to which nodules are covered with sediment (or sediment-laden water). The easily dispersed nature of the sediment is shown by the mud cloud produced by compass impact. The flash lamp was mounted between the two cameras and the compass was suspended below the camera on a line about 2.4 m in length. Approximate location: Lat. 15°N, Long. 126°W, depth, ca. 4580 m. Photographs courtesty of NOAA (DOMES Project, RP8-OC-76, Leg 9).

The sediment in many of the box cores was free from serious disturbance, as shown by the lack of physical breakup and distortion and by the preservation of delicate organic features on many manganese nodules on the top surface (Figures 2 and 3). The top surfaces of typical cores are illustrated in Figure 4.

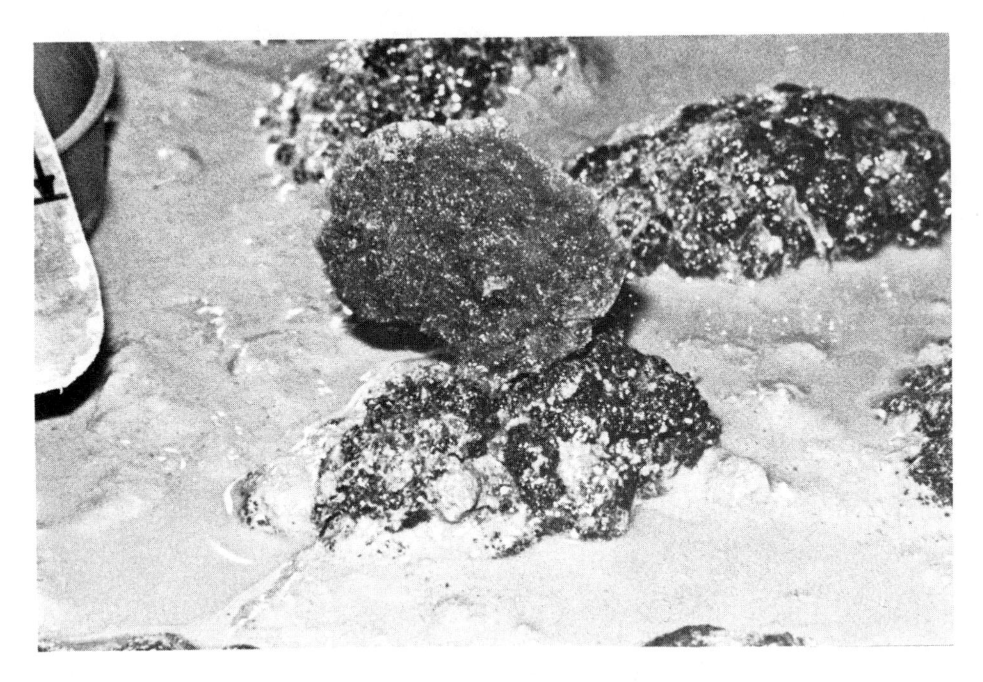

Fig. 2. Delicate fan-like sponge in growth position on manganese nodule in box core 56 (Sta. 55, RP8-OC-75). Similar organisms are seen on nodules in some bottom photographs. The position of the nodule shown was not disturbed by box coring. Nodule is about 6 cm long.

Nodules from undisturbed box cores were collected and cataloged individually. Because of the established importance of the study of top and bottom surfaces in problems of nodule origin, (Raab, 1972; Fewkes, 1975; Sorem, 1975; Fewkes, 1976; Price and Calvert, 1977; and others) many nodules were stored with their orientation at the time of recovery marked for future reference.

Box Cores: Cross Sections

The sediment was sampled in several ways. Before removal of nodules from the top surface, several cylindrical sub-coring tubes (diameter 3 cm to 10 cm) were driven into most of the box cores (for details see Bischoff, 1976). After nodule removal, the side of the box was removed, the sediment cross section was photographed and/or sketched, and additional sediment samples were removed.

Schematic diagrams of the major features of the sediment sections are reproduced in the Appendix.

The sedimentary section sampled by most box cores showed a variety of features. Few cores were identical in cross-section.

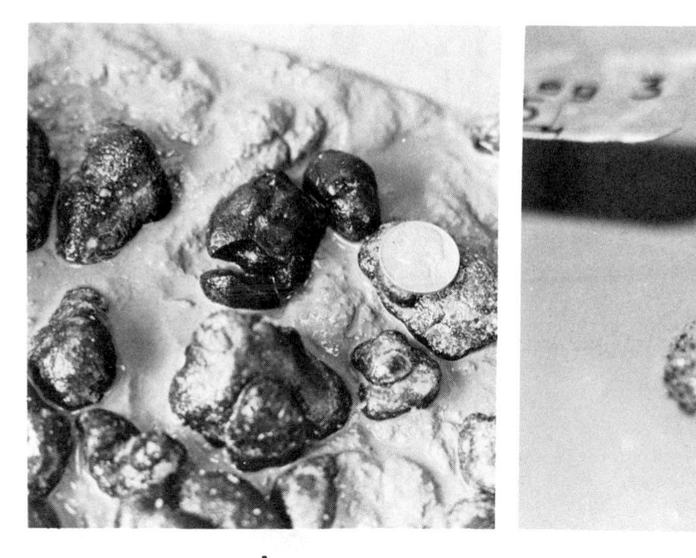

Fig. 3. Delicate features preserved in box cores.
A. Nodule broken by auto-brecciation resting in natural position. Mn oxides on fracture prove that break predates collection. Box core 23 (Sta. 49). Coin is 2.1 cm in diameter.
B. Delicate agglutinated foraminifera in growth position on manganese nodule. Box core 15 (Sta. 47). Nodule width ca. 5.5 cm.

The only uniformity over the entire region seems to be the presence of an upper mud unit which is even-textured, soft, and without color banding or pattern. This surface mud generally ranges in thickness from 10 to 20 cm and rests on firmer clay at a contact which may be either gradational or distinct. Except for the few cores in areas where the underlying clay was especially dark, the mud unit is tan or light brown in color. Good examples of the contrast between the mud unit and clay below can be seen in Fig. 5. Where very dark brown clay underlies the mud, the mud tends to have a grayish-brown or grayish-tan color.

Color mottling was found in many cores at depth. The most common clay types were given the simple field term buff clay and tan clay, and each of these units was found in places to contain irregular patches of the other. In many cores, the dominant unit in such a mottled zone just below the upper mud was found to show less and less mottling as depth increased. In several box cores the entire clay section exposed was mottled (Fig. 5).

Buried Nodules

After work on the side of each box core was completed, the entire volume of sediment was searched by hand for buried nodules.

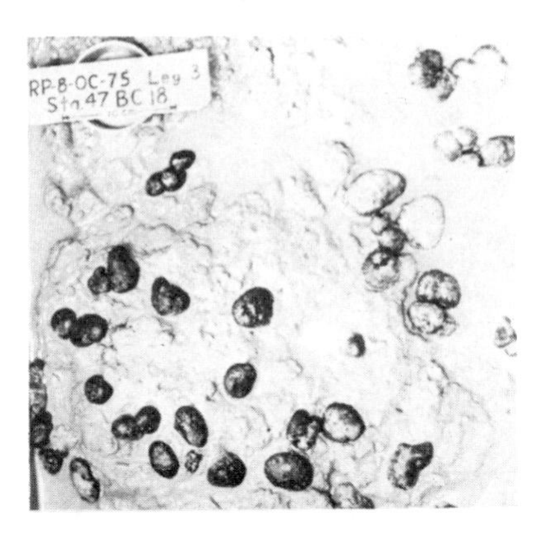

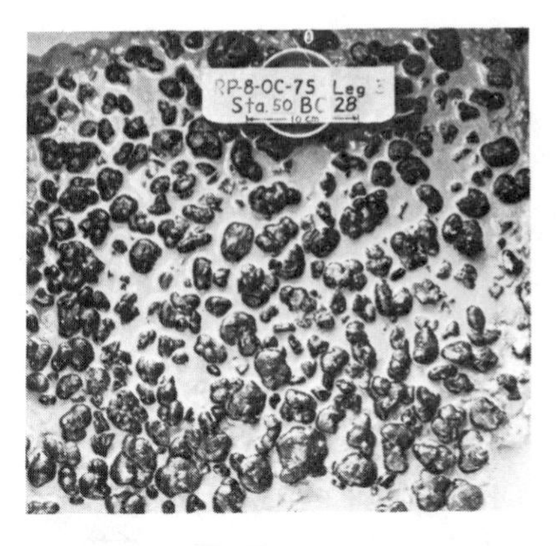

Top Surfaces of Selected Box Cores, Site A

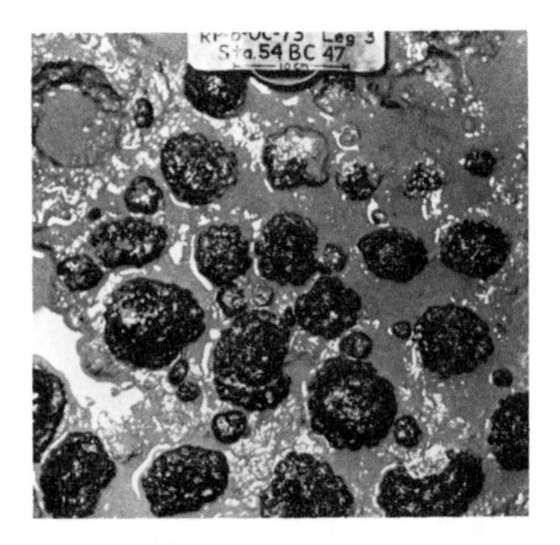

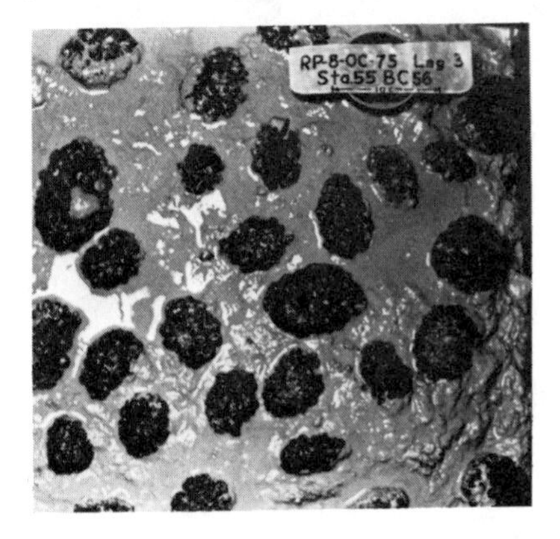

Top Surfaces of Selected Box Cores, Site B

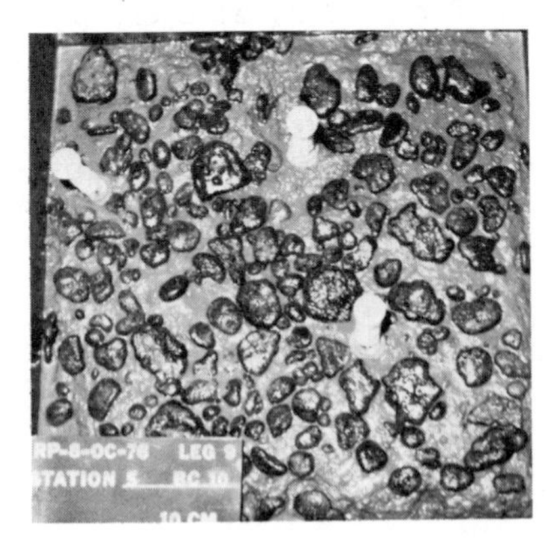

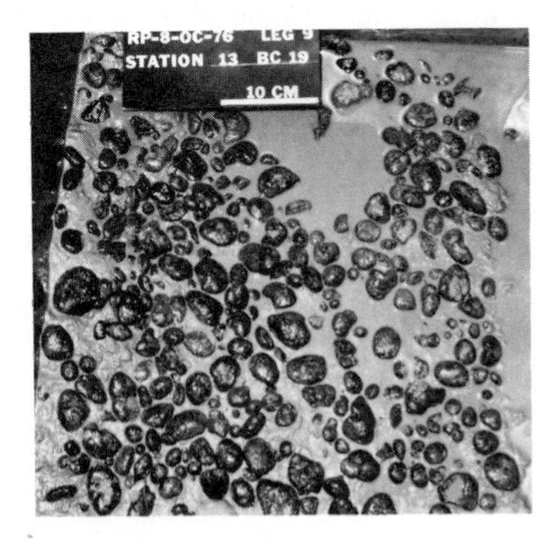

Top Surfaces of Selected Box Cores, Site C

Fig. 4. Shipboard photographs of box cores.

In only ten were buried nodules found (see Appendix). Most of these nodules were found at depths between 20 and 30 cm, and only rarely was more than one buried nodule found in a single core. Buried nodules were found in only one core from Site B and one core from Site C; the rest were discovered in Site A.

The buried nodules are generally similar in size and shape to the nodules at the surface of the core in which they were found. They commonly have a very fragile granular outer surface which is more delicate than that of most surface nodules. There is no megascopic evidence such as staining in the clay to suggest that partial dissolution of the oxides has occurred.

External Physical Characteristics of Nodules

Nodule Size

The size range of closely associated nodules as well as the predominant nodule size in bottom sites sampled or photographed has rarely been accurately documented. Recently several reports have provided partial data in both the eastern Central Pacific Ocean (Andrews *et al.*, 1974; Margolis *et al.*, 1975) and in the southwestern Pacific basin (Bäcker *et al.*, 1976), but site collections were made largely by free fall grab samplers or pipe dredges, both of which tend to yield incomplete samples. The large size box cores (surface area 2500 cm^2) used during the DOMES project, in contrast, provide a means of recovering all nodules at a site and therefore are of special value in size analyses.

Size distribution: RP8-OC-75, Leg 3. Ideally, the size distribution of nodules in a box core should be determined by measuring each nodule individually, but this was not possible with cores of Leg 3 of RP8-OC-75 because the nodules were required for biological studies. It is possible to approximate the size distribution from the photographs (i.e., Fig. 4) and the data on nodule abundance and size range in Table 1. In general, small nodules (ca. 1-2 cm) were dominant in box cores from two of the five stations at Site A. At the other stations, nodules were mostly larger but were not as numerous. At Site B, nodules in box cores from all six stations occupied were relatively large and sparse. At the two stations in Site C, nodules were small and numerous in two box cores and larger and fewer in two others.

The determination of both nodule size and abundance from photographs produces results of doubtful accuracy, however, even if carefully done. Some nodules may be partly covered by mud, and others may be completely hidden, especially those of small size. Another problem is that the view is only two-dimensional and there is no way to know if the maximum dimension is exposed. Therefore, data in Table 1 is considered semi-quantitative.

Size distribution: RP8-OC-76, Leg 9 (Site C). Fortunately, a

A B

Fig. 5. Features of sediment sections in box cores.

A. Typical section of uniform mud layer grading downward through mottled zone to light tan clay with minimal mottling. Tube at left shows thin mud zone in surface channel of box core 9 (Sta. 46).

B. Prominent mottling in lower part of box core 56 (Sta. 55). Note circular cross-sections of filled "burrows". Buff and brown clay.

much more complete and accurate record of nodule size distribution was possible on Leg 9, RP8-OC-76. Virtually all nodules were returned to the laboratory intact, and accurate population counts, size measurements, and weights were obtained. Box core locations, ship's tracks during bottom photography runs, and results of the detailed bathymetric surveys are shown in Figure 6.

Nodules were classified by size in two different ways. All nodules available from each core were first roughly categorized according to maximum dimension by use of a template with circular holes of 2, 4, 6, and 8 cm diameter. Nodules in each of the resulting five size classes were photographed and weighed, permitting the calculation of size distribution in terms of percent of the total weight of nodules in the box core (Table 2).

If the size and weight data in Table 2 are compared with box core locations shown in Figure 6, it is found that a correlation appears to exist between topography and nodule size distribution. For example, large nodules are more common in and near depressions.

Fig. 5C. Unusual mottling on both horizontal and vertical planes, with well-defined patterns in both very dark brown and buff clay. Note upper mud layer with uniform gray color preserved in core tube. Box core 60 (Sta. 58).

External Textures

Very small structures superimposed on the outer surface of nodules have been referred to as external textural features (Sorem and Banning, 1976). They are often ignored in nodule investigations but may be of great importance in studying nodule origin as well as in evaluating probable behavior of nodules during large scale dredging. The simplest classification, based upon megascopic examination, distinguishes nodules with smooth, granular, and combination texture (Fig. 7). A nodule which appears to have a smooth surface feels smooth and firm to the hand, whereas one with a granular texture feels rough and commonly leaves a gritty powder on the hand when touched. Nodules with combination texture generally have a smooth upper surface and a granular bottom surface, as pointed out by Raab (1972) and Raab and Meylan (1977). Binocular microscopic examination reveals that the surface of a smooth nodule consists of innumerable closely packed microbotryoids of hemispherical shape, whereas a granular or gritty surface is made up

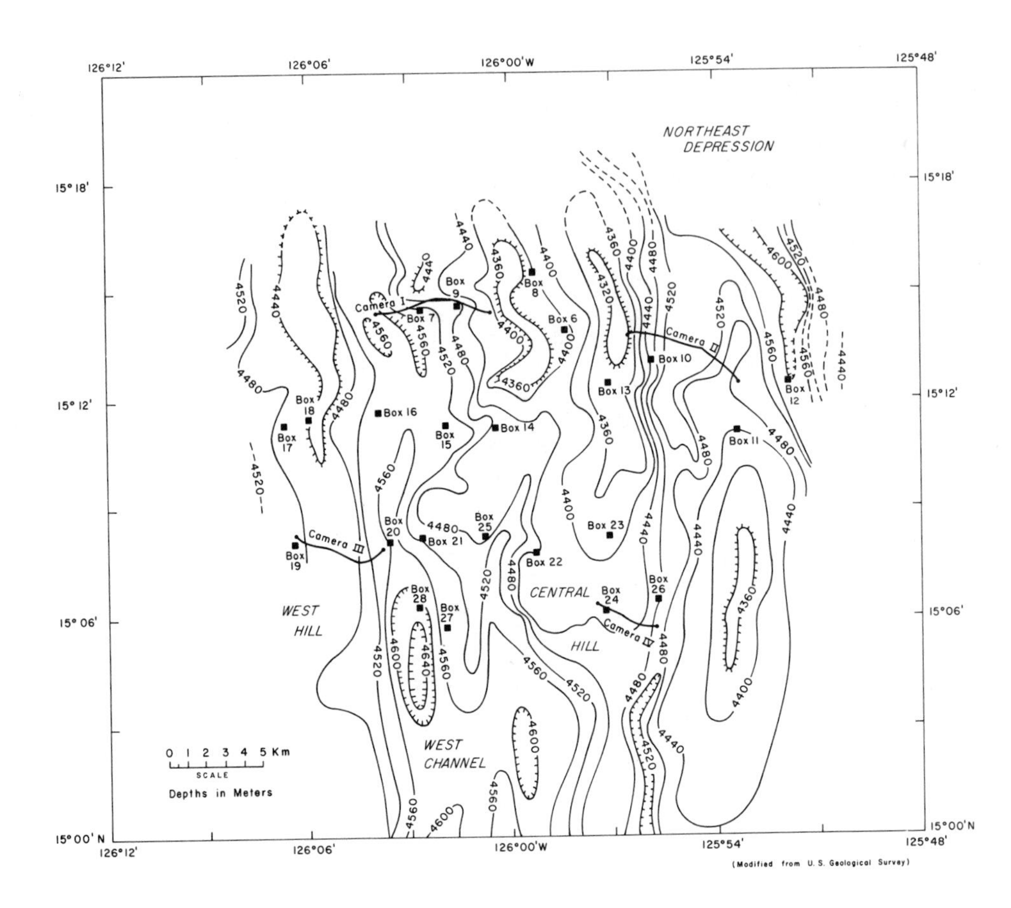

Fig. 6. Base map, part of DOMES Site C. Bathymetry modified from Piper (1977). Locations are shown for box cores and deep sea camera runs on Leg 9, RP8-OC-76.

of fragile, elongate dendritic growths which call to mind a top view of a miniature boxwood hedge (Figure 8). Cross sections which show the internal structures may be seen in Figures 9, 10, and 11.

The genetic significance of surface textures has been discussed in detail recently by Fewkes (1975), Sorem (1975), Fewkes (1976), and Sorem et al., (1978 a, b). In essence, there is good evidence from micro-X-ray diffraction research and X-ray macroprobe and electron microprobe analyses of oriented nodules that dense iron-rich oxides which are amorphous to X-rays, form on nodule surfaces exposed to sea water virtually free of solids, whereas todorokite and birnessite, crystalline hydrous oxides rich in manganese (and commonly nickel and copper as well), are deposited as complex dendritic crusts where nodule surfaces are

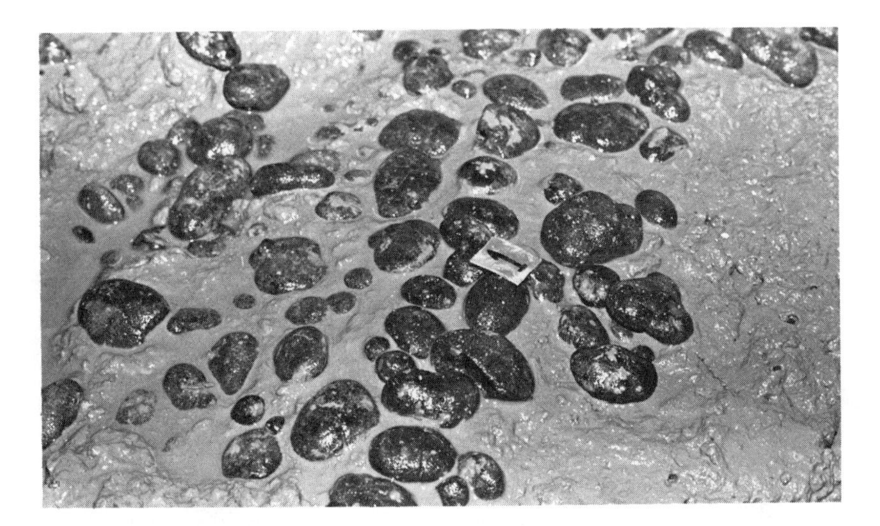

A

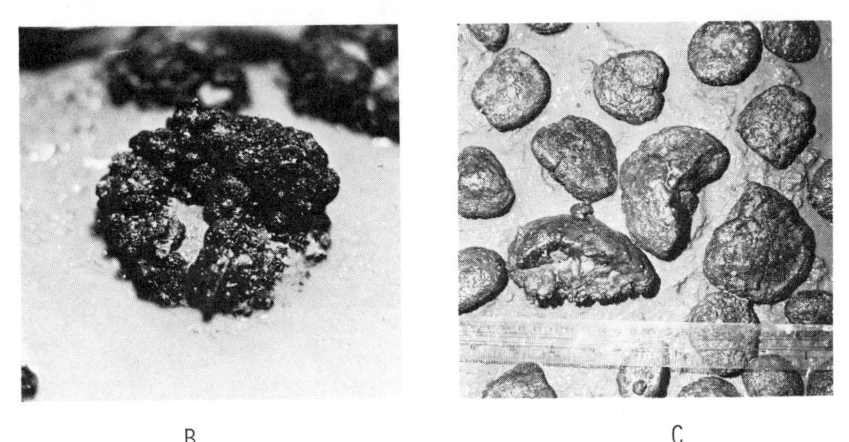

B C

Fig. 7. Examples of nodules with different surface textures. All nodules lie in natural position on undisturbed box core surfaces.
A. Small discoid manganese nodules with smooth surface texture. The number "1" is about 1.5 cm long. Box core 57 (Sta. 57) RP8-OC-75.
B. Large spheroidal nodule, with abundant large botryoids covered with coarse dendritic oxides: a coarse granular surface. Diameter ca. 10 cm. Box core 34 (Sta. 51) RP8-OC-75.
C. Discoid nodules with combination surface texture. In the large nodule at lower left, the coarse granular texture of the bottom can be seen. The upper surface of all of the nodules is smooth. Where smooth and granular surfaces meet, a coarse granular "equatorial rim" commonly develops, as in the broken nodule at lower right. Box core 28 (Sta. 22) RP8-OC-76, Leg 9 (do not confuse with box core 28 (Sta. 50)). Width of field is 15 cm.

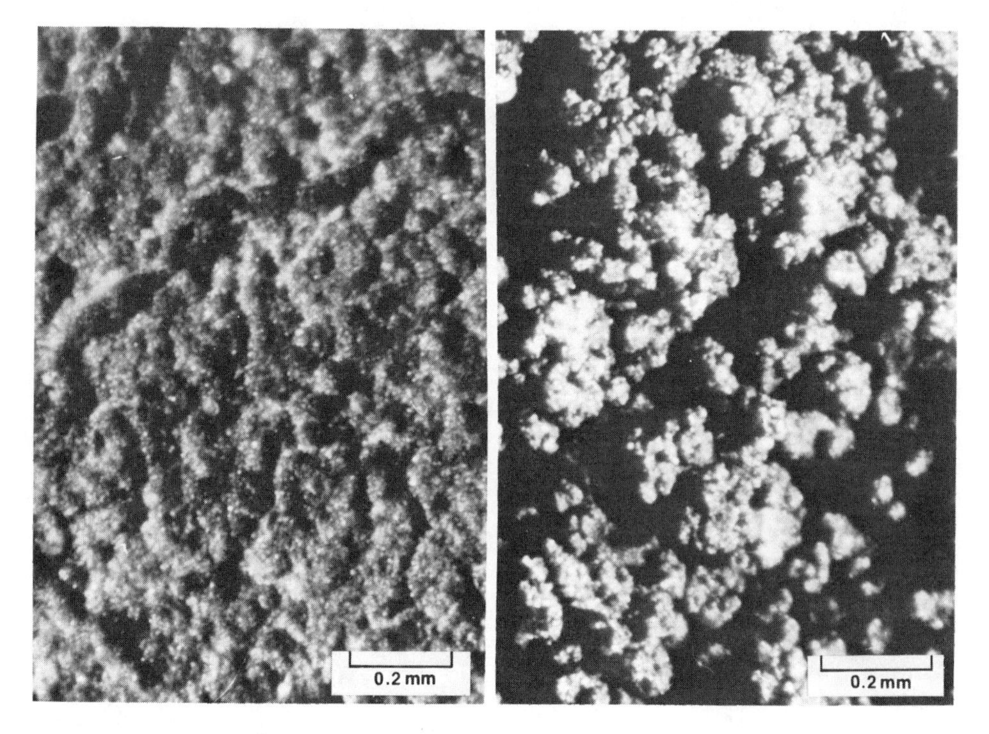

A B

Fig. 8. Microfeatures of smooth and granular nodule surfaces.

A. "Smooth" top of nodule NP 60A-10 (Sta. 15B). Minute hemispherical microbotryoids are so densely packed as to leave little interstitial space. See also Figure 10.

B. Bottom surface of same nodule at same magnification as A. Note great height to width ratio compared to forms in A and finely irregular surface of the individual forms. Extreme fragility of these forms is suggested by the large amount of void space and is emphasized in the cross-sectional view of same features, shown in Figure 9. See also Figure 11.

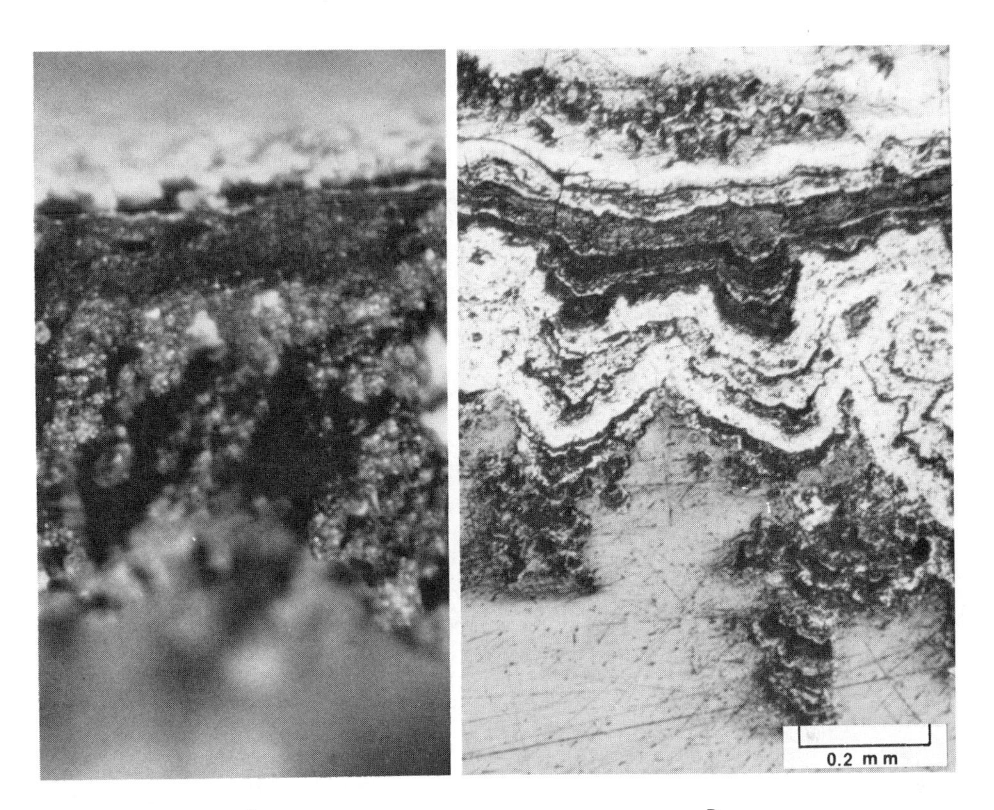

A B

Fig. 9. Internal microstructure at bottom of nodule NP 60A-10.

A. Side view of bottom surface features of nodule NP 60A-10 shown in Fig. 8B, at same magnification, as seen on fracture surface. Interior of nodule is toward top of page. Note great porosity and delicate structure of the dendritic microbotryoids. A polished section across these forms in the same nodule is shown in B.

B. Photomicrograph of polished section of same features as shown in A (above). Vertical illumination. Note the intricate structure of these bottom layers, compared to the structure of the top layers shown in Figure 10. This type of growth gives rise to the dendritic forms shown in Figure 8B. Most of the oxides shown (white and gray) are crystalline todorokite, a hydrous impure manganese oxide relatively rich in Ni and Cu.

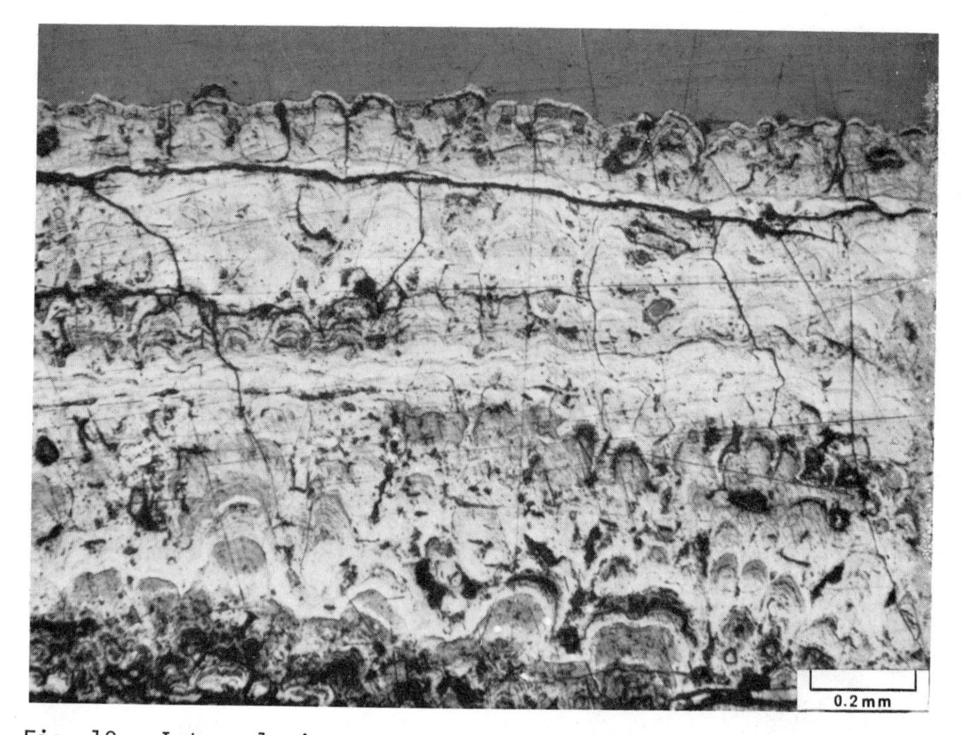

Fig. 10. Internal microstructure at top of nodule NP- 60A-10. Photomicrograph of part of polished section of same nodule as that shown in Figure 8A. Vertical illumination. Plane-polarized light. At the top just below the plastic mounting medium (even gray) the internal structure of the hemispherical microbotryoids on the upper surface of the nodule can be seen. The layers of oxides (light gray) which form the structure are very dense and relatively uniform in character passing downward toward the center of the nodule. Optical character of these oxides is that of Fe-rich X-ray amorphous nodule material.

in contact with sediment-laden water (or watery sediment) at and below the sediment-water interface.

The composition of layers in manganese nodules from this region varies on a microscale, and the outermost layers and the surface texture do not necessarily relate to the bulk composition of the nodules. However, there are no data published thus far which examine the possibility that there is a relationship of some kind between outer surface texture and bulk nodule composition. This report, therefore, explores that possibility.

An important consideration when evaluating surface texture of associated nodules is the relationship between size and texture. Not all nodules in a box core necessarily have the same texture,

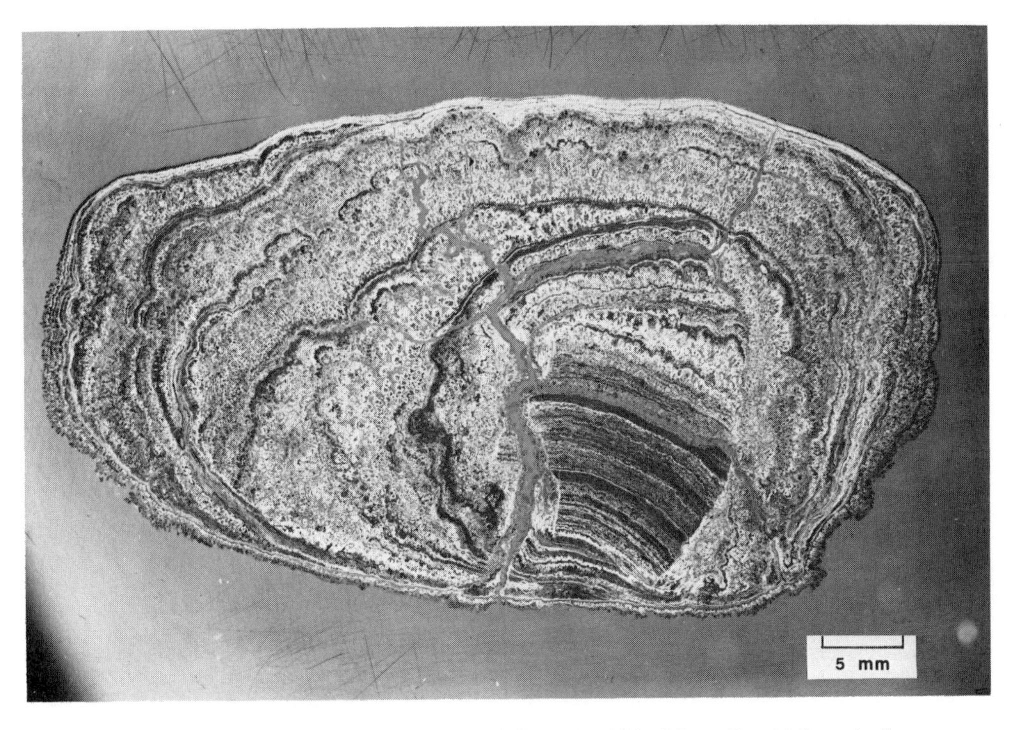

Fig. 11. Polished section of nodule NP 60A-10. In this whole cross-section, viewed in vertical illumination, the internal structure of the nodule is shown in natural orientation. The top and bottom photographs shown in Figures 9 and 10 are of the same nodule. The morphology of the nodule was termed "irregular discoid". Note that the crustal zone over much of the top is clearly more dense and highly reflective than along much of the bottom. The overall growth structure consists of wedging zones of oxides rich in clay around a core which is a piece of an older nodule. The older nodule evidently was much different in internal structure. A complex growth history is suggested.

but where differences exist a pattern can be recognized. For example, where a granular surface texture prevails in one box core, nodules above a certain size (3 to 4 cm) tend to have a combination texture, presumably indicating the attainment of a size where deposition on the top surface from sea water began. In box cores where large nodules have a smooth surface, however, nodules of all sizes are generally smooth.

The most prominent surface textures observed in selected nodules from each box core collected during RP8-OC-75 are recorded in Table 1. If the small group of nodules examined from each box core is reasonably representative, as we believe, the table shows that

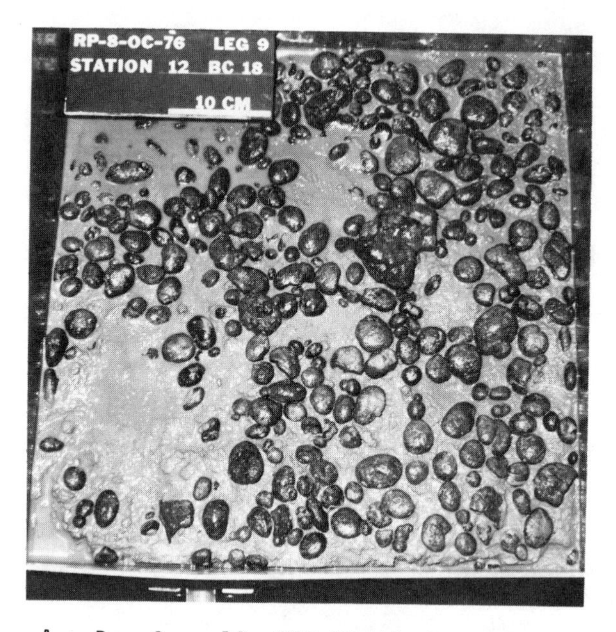

A. Box Core 18, RP8-OC-76, Top View

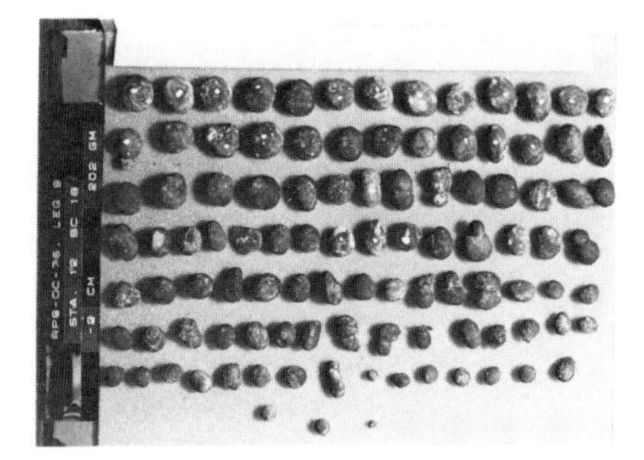

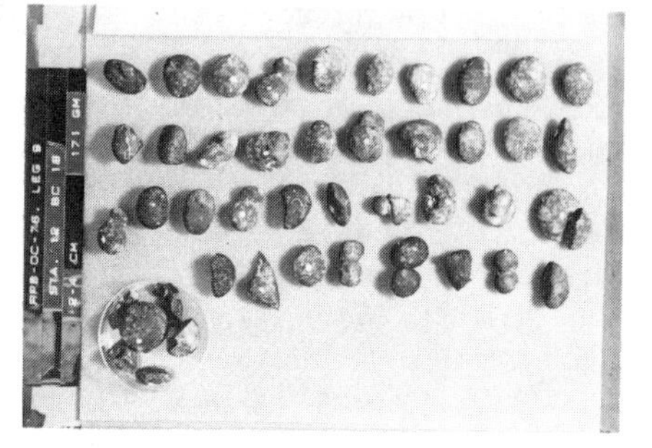

B. Nodules < 2 cm

C. Nodules 2 - 4 cm

Fig. 12. Manganese nodules, box core 18, RP8-OC-76, Leg 9.
A. Box core photograph showing top surface immediately after recovery. Nodules are obviously non-uniform in size and morphology. Other photographs show nodules, after removal, grouped by size.
B. Nodules smaller than 2 cm in length.
C. - G. Nodules in 2 to 4 cm size class.
H. Nodules in 4 to 6 cm size class.
I. Nodules in 6 to 8 size class.
J. Nodules larger than 8 cm.

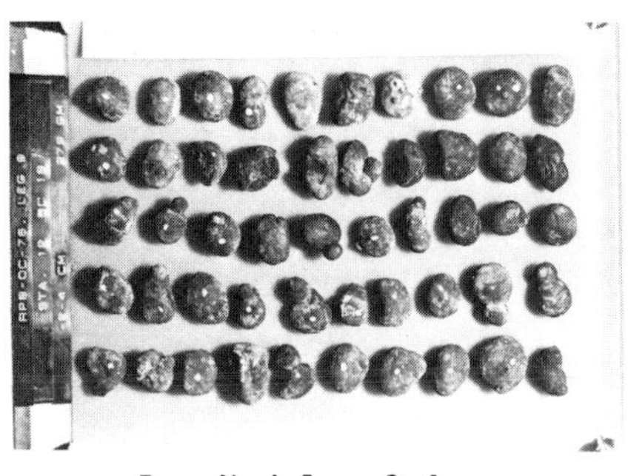

D. Nodules 2-4 cm

E. Nodules 2-4 cm

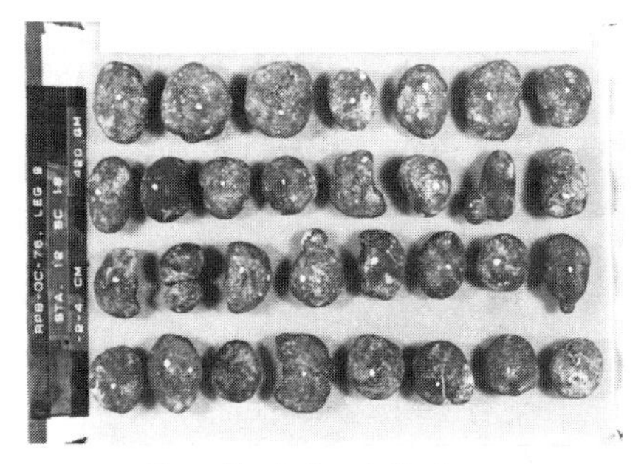

F. Nodules 2-4 cm

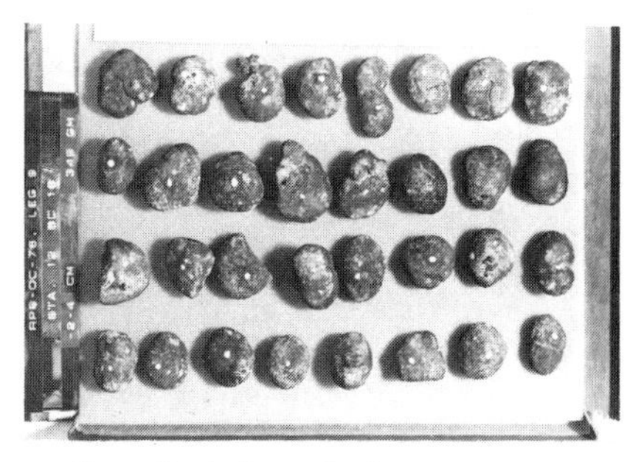

G. Nodules 2-4 cm

H. Nodules 4-6 cm

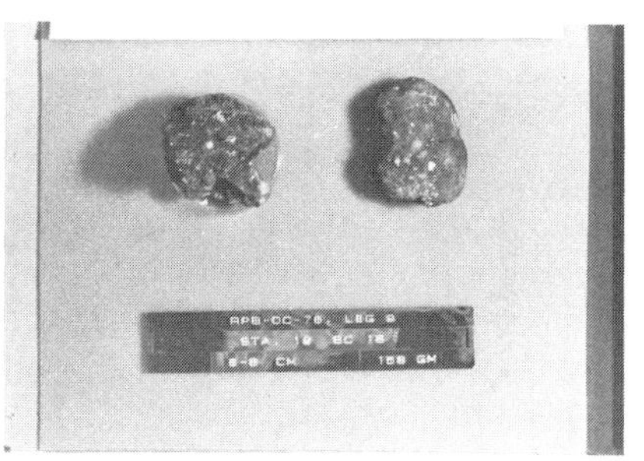

I. Nodules 6-8 cm

J. Nodules > 8 cm

Fig. 12, Continued

nodules at the three DOMES sites are basically different. Of the 32 box cores recovered at Site A, smooth-surfaced nodules predominated in 22, whereas in 8 cores most nodules had a granular surface. At Site B, the majority of nodules in 11 of the 19 nodule-bearing cores had a granular texture, and the rest of the cores contained nodules of combination texture. In four box cores from Site C, two contained smooth nodules and two combination-texture nodules.

Nodule Shape or Morphology

Most DOMES nodules can be classed as spheroidal, ellipsoidal, or discoidal, depending upon whether a nodule has orthogonal axes all of equal length, two short and one long, or two long and one short, respectively. Some nodules have two or more flat surfaces which intersect, resulting in angular shape. Obviously, boundaries between the classes are gradational, and some hybrid terms may be useful. A few nodules are best termed botryoidal, slabby, or conglomeratic.

Many nodules, especially those with a discoidal shape, tend to show a girdle or equatorial rim (Raab, 1972), along which botryoidal forms may have developed even if they are absent elsewhere. Whereas the other shapes noted above may be found in nodules of any size, a pronounced equatorial rim is rarely found in nodules less than 3 or 4 cm in maximum dimension.

Texture and Morphology Relationships in Site C Nodules

By far the most common surface texture is the combination type (smooth top, granular bottom). In five box cores (6, 8, 12, 16, and 19), all nodules, even the largest, showed only granular textures. In one box core, 22, all nodules were smooth. Nodule morphology varied somewhat with size, as expected, with ellipsoidal and discoidal shapes the most common. Nodules in several box cores (13, 20, 22, 23, 24, 25, 27, and 28) were in part angular or irregular.

These observations indicate that nodule surface textures and morphology, like size distribution, vary somewhat among the RP8-OC-76 box cores. Nevertheless, it seems likely that discoidal and ellipsoidal nodules predominate in the area. In all probability, this is true of a considerably larger area in Site C as well.

Analytical Procedure

Nodules were analyzed for Mn, Fe, Ni, Cu, Co and Zn. Samples were selected from each box core on the basis of physical characteristics, to represent the dominant types and sizes present. Preference was given to whole nodules from which a polished section could be prepared, but otherwise the choice of nodules within physical groups was random. Most cores were represented by five or more nodules. In all, 456 nodules were selected for analysis.

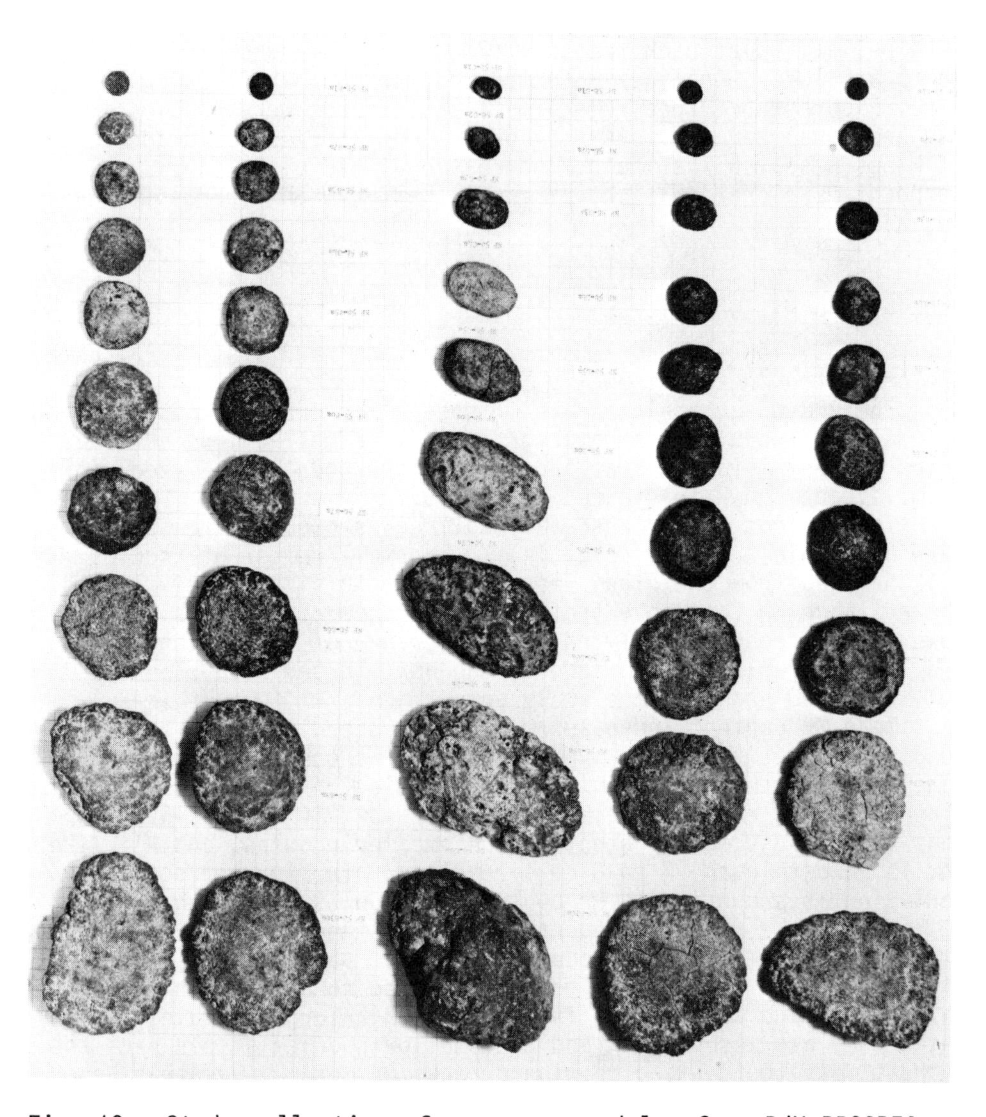

Fig. 13. Study collection of manganese nodules from R/V PROSPECTOR at Washington State University (Sorem and Fewkes, 1976). Heavy grid lines are 1 cm apart. Nodules in two vertical rows at left are spheroidal-ellipsoidal in morphology. Center row nodules are elongate-ellipsoidal, and those in two rows at right are discoidal. Nodules are representative of types in one-ton dredge haul near DOMES RP8-OC-76 Site C area.

After the documentation, each nodule was cut into symmetrical halves, where size permitted. One half was pulverized and the other was used to prepare a polished section. The powdered samples were pelletized at high pressure (2000 kg/cm^2) and dried for 4 hours at 110°C. Analyses were made by use of a Phillips PW 1410-10 X-ray Fluorescence Spectrometer. The standards were nodules which had been analyzed previously by other laboratories (Colorado School of Mines Research Institute, International Nickel Company, and Kennecott Exploration). Precision is better than 5 percent (one sigma).

Nodule Composition, Site C

Composition and nodule size. Analytical and statistical data are presented in the Appendix.

The RP8-OC-76 analyses are of special interest because of the abundance of nodule size data for the entire collection. Relationships between metal content and size in the nodules analyzed, for example, are shown in Figures 14 and 15.

The various size classes are not represented by the same numbers of nodule analyses (Fig. 14). The small number of nodules (6) analyzed from the class less than 2 cm in size clearly limits the reliance on the concentration values shown for that size class. However, the nodules analyzed showed a range of values for Mn generally comparable in magnitude to the next larger size class (2-4 cm), where 60 nodules were analyzed.

Mean Mn content increases slightly proceeding from nodules 6 to 8 cm in size to smaller nodules. Mean Fe content shows the same trend, with the lowest value in nodules of the greatest size (larger than 8 cm). Nickel and copper, shown in Figure 15, also show the highest mean content in the smaller nodules, and Ni, like Mn, is lowest in the 6 to 8 cm size class. The largest degree of variation is shown in the Mn content of nodules larger than 4 cm. Not only is the range very great, but the standard deviations are also extreme in the large nodules. In this part of Site C, nodules less than 4 cm in size may be expected to be more uniform in Mn, Fe, Ni, and Cu content, in general, than larger nodules with which they are associated, and the smaller nodules should have the highest metal content. The widest range of metal concentrations should be found in the 4 to 8 cm size range. Finally, whatever the nodule size, Ni content can always be expected to exceed Cu content.

In view of the differences in mean concentrations in nodules of different size, the question of calculation of the best average composition for nodules in individual box cores must be considered. The simple arithmetic mean values are given in Table 3. Most of the box cores for RP8-OC-76, however, contained different proportions, by weight, of nodules of different size, as shown in Table 2. We therefore calculated the weighted average concentrations for each box core, taking into account the fraction of the total

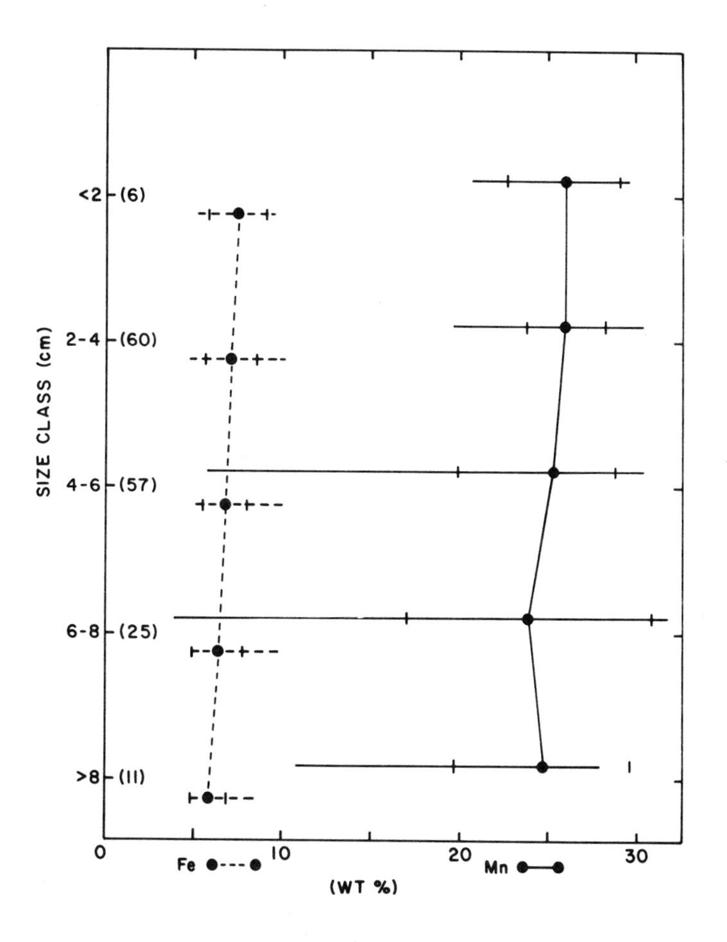

Fig. 14. Relationships between bulk Mn and Fe content and nodule size, analyzed nodules from RP8-OC-76 box cores. Horizontal lines show range of analyses, with vertical ticks at first standard deviation value. Solid circles joined by lines show mean values. Number of nodules analyzed in each size class shown in parentheses.

nodule weight in the box core represented by each size class. Thus the average composition of nodules in a size class which made up a small part of the total nodule weight in the box core has less influence on the final result than that of nodules from size classes which made up larger parts of the total mass.

The weighted averages calculated for RP8-OC-76 box cores are shown in Table 4. Most of the metal concentrations differ little from the mean values reported in Table 3, indicating that the nodules of all sizes were fairly uniform in individual box cores.

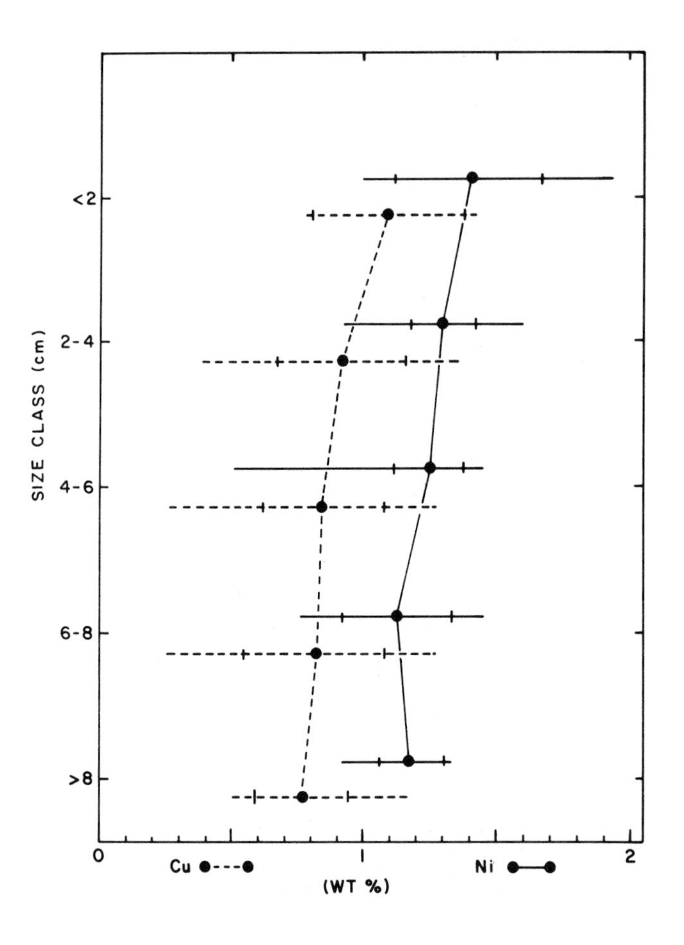

Fig. 15. Relationships between bulk Ni and Cu content and nodule size, RP8-OC-76. See Figure 14 caption for explanation of symbols.

However, in three box cores, 19, 23, and 27, the difference between nodules was appreciable. The value of Mn increased by 8.1 percent of the mean in box core 19 and by 18 percent in box core 23. In box core 27, Mn dropped by 7.4 percent of the mean value. The only remarkable change in copper or nickel was in box core 19, where the weighted average was 12 percent greater than the mean value of 0.92 percent Cu. For Ni and Cu values as a whole, some increased and some decreased slightly.

TABLE 1

Box Core Observations
RP8-OC-75, Leg 3

Ship Sta.	Box Core	Approx. Core Thickness (cm)	Observations on top of core: Nodules Exposed (Approx)	Nodule Long Axis (cm) Max.	Nodule Long Axis (cm) Min.	Nodule Surface Texture (Megascopic) Selected Specimens)
Site A						
46	1	33	120	4.5	0.5	Smooth
	2	40	>15	5.0	1.0	*Combination
	3	18	> 25	2.0	< 1.0	Smooth
	4	32	> 100	5.0	< 1.0	Smooth
	5	25	> 10	2.1	1.0	Granular
	6	28	250	8.8	0.2	Smooth
	7	20	> 100	5.0	0.5	Smooth
	8	30	> 100	5.0	0.6	Smooth
	9	40	> 100	6.0	0.4	Smooth
47	10	45	1	2.0		Granular
	11	30	4	6.4	3.9	Combination
	12	30	16	5.0	1.2	Granular
	13	25	9	5.1	2.0	Granular
	14	40	7	7.9	3.8	Smooth
	15	40	10	8.1	4.2	Smooth
	16	40	32	13	0.4	Granular
	17	30	No photograph			Granular
	18	50	47	5.0	1.8	Smooth
48	19	45	4	6.0	2.5	Smooth
			100	6.1	1.5	Smooth
	21	30	No photograph			Granular
	22	50	2	6.4	3.2	Smooth

Ship Sta.	Box Core	Approx. cm.	Nodules Exposed	Nodule Long Axis (cm) Max.	Min.	Nodule Surface Texture (Megascopic)
	23	50	2	2.0		Granular
49	24	40	3	6.0	1.0	Smooth
	25	48	140	5.5	1.0	Smooth
	26	48	> 20	6.0	1.0	Smooth
	27	48	79	8.1	0.4	Smooth
50	28	50	230	5.2	0.6	Smooth
	29	50	200	7.2	0.4	Smooth
	30	42	100	7.5	0.5	Smooth
	31	47	125	10.5	1.6	Smooth Slabs
	32	49	134	4.4	0.3	Smooth
Site B						
51	33	28	4	9.0	1.0	Granular
	34	50	12	9.5	1.8	Smooth
	35	53	0	0	0	
	36	45	1	1.8		Granular
52	37	47	0			
	38	40	3	3.0	2.0	Granular
	39	46	4	11.0	9.4	Granular
	40	52	4	10.0	3.9	Granular
	41	48	0			
	42	50	0			
53	43	45	0			Granular
	44	50	1	2.0		Granular (?)
	45	45	3	9.4	2.0	Granular
	46	52	1	2.5		Granular (?)
54	47	50	45	8.6	0.9	Granular
	48	40	5	8.0	1.2	Combination
	49	> 20	> 5	3.0	1.0	Granular
	50	> 25	20	2.6	1.0	Granular

Ship Sta.	Box Core	Approx. cm.	Nodules Exposed	Nodule Long Axis (cm)		Nodule Surface Texture
	51	50	35	8.0	1.2	Combination
	52	50	15	9.0	1.0	Combination
55	53	36	34	5.0	0.4	Combination
	54	40	6	8.0	2.6	Combination
	55	52	18	11.8	1.0	Combination
	56	50	30	10.5	4.6	Combination
Site C						
57	57	50	215	4.1	0.4	Smooth
	58	45	22	9.0	1.6	Combinations
	59	47	0			
58	60	47	315	4.6	0.2	Smooth
61	Partial	2	3.0			Combination
62			0			Rock

* Combination classification indicates that large nodules showed this texture, whereas smaller nodules in the same box core were granular (see text).

Variations within box cores. Although useful in comparing box cores, mean values of metal content of nodules do not give any indication of the degree to which nodules vary within each core.

In Figure 16 data are presented for Mn and Fe content of the nodules analyzed. Except in a few box cores, the range and standard deviation values are of the same general magnitude. For Mn, box cores 6, 7, 15, 24, and 28 show unusually small variations, whereas box cores 9, 23, and 27 show very wide variations in nodule composition. In box cores 23 and 27, the large standard deviation values indicate that the entire group of nodules showed little uniformity. The low mean values for these box cores show that the nodules, as a group, were also unusually low in Mn content.

Variations in Fe content also differ from one core to another, but the degree of variation is not related to variations in Mn. Of the 8 box cores with the lowest Fe range, only two (6 and 7) had a very low Mn range. Maximum variations in Fe content also seem to be independent of Mn variations. The only general pattern

TABLE 2

Distribution of Nodules by Size and Weight in RP8-OC-76 Leg 9 Box Cores

Box Core No.	Total Weight (gm)	Total Number Nodules	Number and Weight of Nodules in Each Size Class (Length) < 2 cm No.	Gm	2-4 cm No.	Gm	4-6 cm No.	Gm	6-8 cm No.	Gm	> 8 cm No.	Gm
6	1180	67	14	33	33	362	17	588	3	197		
7	2059	147	30	73	87	954	28	938	2	94		
8	431	111	85	94	24	218	1	60	1	59		
9	1568	227	59	98	153	1205	14	231	1	34		
10	2151	234	72	115	122	896	33	709	7	431		
11	2976	155	3	3	118	1831	32	1022	2	120		
12	2316	170	27	108	134	1244	4	105	4	476	1	383
13	2044	370	173	291	174	1146	23	607				
14*	325	17	3	9	12	246	2	70				
15	1563	24	2	6	4	39	10	545	7	809	1	164
16	3327	30	2	7	1	6	7	384	13	1336	7	1593
17												
18	2755	321	103	202	192	1563	23	575	2	158	1	257
19	2460	307	111	228	165	1522	31	710				
20	3668	81			29	416	31	1247	18	1428	3	577
21	2052	261	67	95	175	1532	18	425				
22	2226	398	121	249	267	1818	10	159				
23	3922	411	189	291	208	1176	4	101	9	714	1	1640
24	2480	381	94	196	275	2089	12	195				
25*	833	92	18	56	69	658	5	119				
27	3870	260	91	151	126	983	32	962	9	843	2	931
28	2828	49			8	97	26	1129	13	1246	2	356

* Recovery probably incomplete.

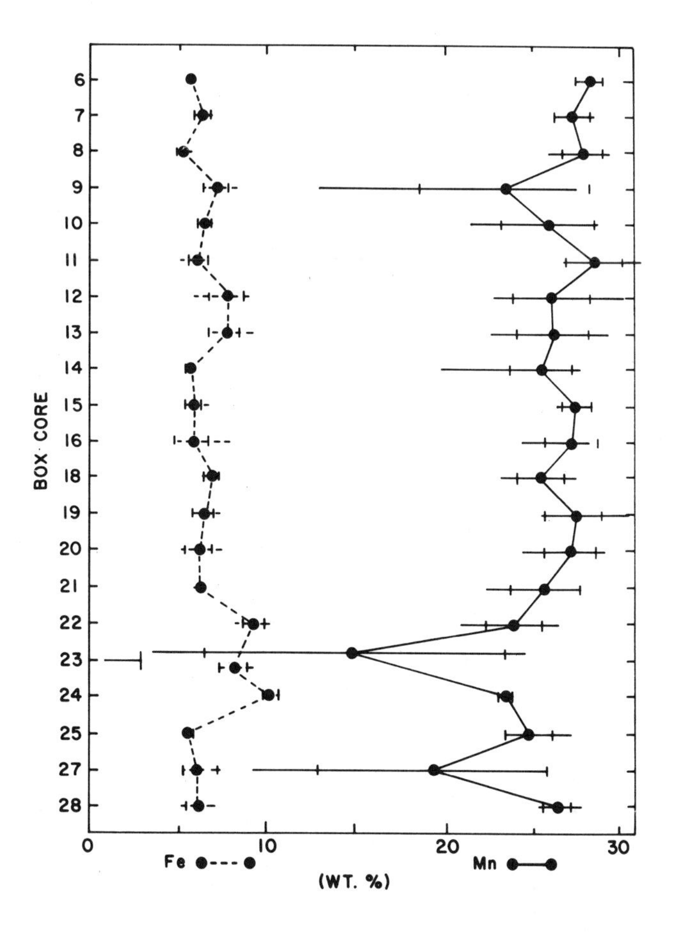

Fig. 16. Summary of analyses for Mn and Fe, RP8-OC-76 nodules. Range of analyses in each box core is shown by horizontal line, with vertical ticks at value of first standard deviation. Box core means are shown by solid circles.

suggested is in the mean metal concentrations, which are joined for each metal by a line.

In Figure 17 data for Ni and Cu are plotted in the same manner, and the mean Mn content of the nodules in each box core is shown for reference. By comparing the mean metal values for Mn, Ni, and Cu, it is apparent that a general sympathetic relationship exists between Mn and Cu content, whereas a similar pattern for Mn and Ni is somewhat less well developed. The

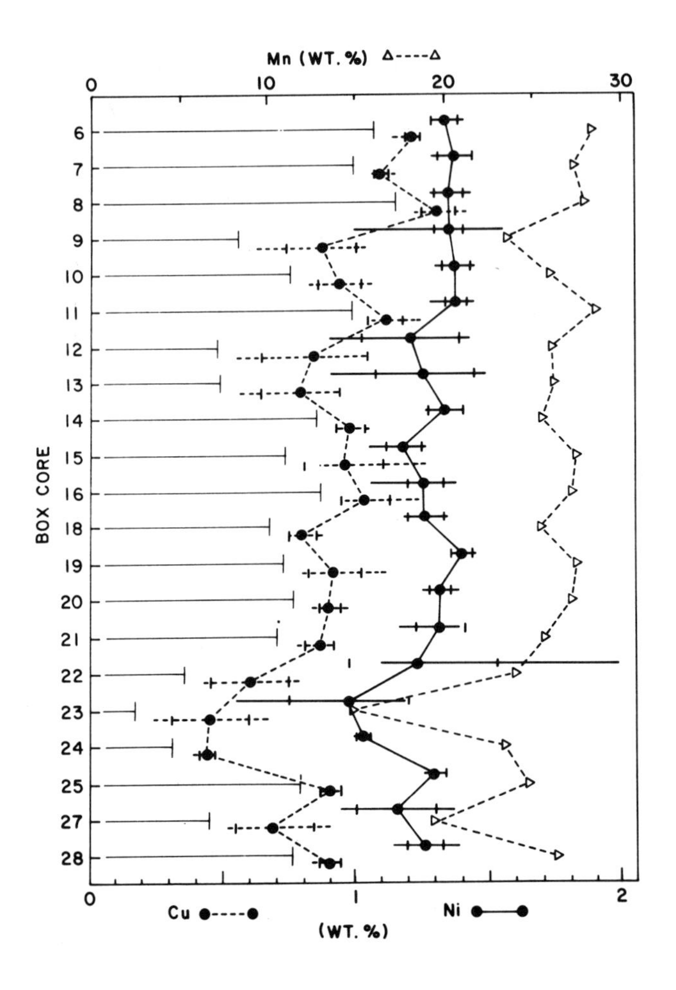

Fig. 17. Summary of analyses for Ni and Cu, RP8-OC-76 nodules. See Fig. 16 caption for explanation of symbols. Mean values for Mn (from Fig. 16) are plotted for reference. Note scale for Mn at top of diagram.

correlation coefficients bear out this observation (Mn:Cu = 0.69, Mn:Ni = 0.64). With regard to intra-box core metal variations, the range of Ni and Cu content clearly is small in some box cores and large in others. Among the box cores with the largest ranges are those which varied widely in Mn content (box cores 9, 23, and 27), but in addition large ranges are found in box cores 12, 13, and 22. In these box cores, as well as in those with a smaller range of composition, Ni and Cu variations show comparable ranges,

TABLE 3

Mean Metal Content of Nodules in RP8-OC-76 Box Cores
(in Weight Percent)

Box Core Number	Mn	Fe	Ni	Cu	Co	Zn
6 (6)*	28.31	5.74	1.34	1.22	0.24	0.23
7 (7)	27.44	6.45	1.38	1.10	0.29	0.18
8 (8)	28.08	5.35	1.36	1.32	0.25	0.24
9 (7)	23.55	7.32	1.36	0.88	0.29	0.14
10 (7)	25.95	6.66	1.38	0.95	0.29	0.17
11 (7)	28.65	6.24	1.39	1.12	0.30	0.20
12 (9)	26.31	7.87	1.21	0.85	0.31	0.15
13 (8)	26.19	7.76	1.27	0.80	0.30	0.14
14 (4)	25.53	5.77	1.34	0.99	0.22	0.23
15 (8)	27.54	6.05	1.18	0.96	0.24	0.18
16 (8)	27.13	5.89	1.26	1.04	0.29	0.15
18 (8)	25.54	7.02	1.26	0.80	0.28	0.16
19 (6)	27.46	6.51	1.40	0.92	0.28	0.19
20 (11)	27.16	6.20	1.32	0.91	0.29	0.16
21 (6)	25.82	6.30	1.32	0.86	0.24	0.18
22 (9)	24.07	9.39	1.25	0.60	0.37	0.09
23 (8)	14.97	8.21	0.97	0.45	0.29	0.10
24 (6)	23.34	10.11	1.04	0.44	0.36	0.11
25 (6)	24.83	5.65	1.30	0.90	0.23	0.22
27 (10)	19.36	6.21	1.17	0.69	0.24	0.15
28 (10)	26.30	6.13	1.26	0.91	0.27	0.17
Mean	25.41	6.80	1.27	0.89	0.28	0.17
Maximum	28.65	10.11	1.40	1.32	0.37	0.24
Minimum	14.97	5.35	0.97	0.44	0.22	0.09

*Number of nodules analyzed

TABLE 4

Weighted Average Metal Content of Box Cores RP8-OC-76 (in Weight Percent)*

Box Core Number	Mn	Fe	Ni	Cu	Co	Zn
6	28.70	5.74	1.35	1.23	0.23	0.23
7	27.50	6.40	1.39	1.11	0.29	0.18
8	28.04	5.34	1.36	1.32	0.25	0.24
9	24.58	7.29	1.41	0.91	0.30	0.14
10	25.84	6.67	1.37	0.94	0.29	0.17
11	28.22	6.41	1.40	1.10	0.32	0.20
12	25.67	8.09	1.16	0.89	0.31	0.14
13	26.52	7.68	1.26	0.77	0.30	0.15
14	25.23	5.70	1.33	0.98	0.22	0.23
15	27.05	5.83	1.15	0.88	0.23	0.18
16	27.70	5.64	1.22	0.99	0.28	0.16
17	.	.	.	.	.	.
18	25.32	7.22	1.30	0.78	0.29	0.16
19	29.05	7.03	1.41	1.03	0.30	0.16
20	27.66	5.97	1.33	0.93	0.27	0.16
21	26.68	6.41	1.34	0.87	0.24	0.18
22	24.45	9.46	1.20	0.63	0.36	0.07
23	17.69	8.35	1.01	0.48	0.33	0.10
24	23.42	10.06	1.04	0.43	0.36	0.11
25	24.66	5.64	1.30	0.89	0.23	0.21
26	.	.	.	.	.	.
27	17.92	6.21	1.12	0.66	0.24	0.13
28	26.36	6.35	1.28	0.91	0.27	0.16

* Weighted average metal content based upon nodule samples analyzed from different size classes. In general, analyses represent all major size classes found in each box core. See text for explanation.

in terms of percent deviation from the mean.

A concise comparison of the magnitude of variations in concentrations of the different metals is difficult using either the numerical data in the Appendix or the graphs in Figures 16 and 17, chiefly because the concentration levels of the metals are different. Comparison is facilitated, however, if the metal concentrations are examined in terms of the coefficient of variation, CV (Steel and Torrie, 1960). The coefficient of variation for each metal in a box core is calculated by dividing the first standard deviation value for the metal analysis by the mean metal content and multiplying by 100. The result, expressed as a percent, provides a uniform numerical basis for expressing variations in metal concentrations. Magnitude of variation is directly proportional to the CV number. Values of CV for the RP8-OC-76 box cores are listed in Table 5.

The usefulness of CV values can be seen by comparing values for some of the box cores just discussed. For example, the large standard deviations in analyses for Mn in box cores 9, 23, and 27 result in the three highest CV values in the table. In contrast, box cores 6, 7, 15, 24, and 28 have the lowest Mn variations and the lowest values of CV. A major advantage of the use of the coefficient of variation, however, is in comparing variations between metals. As an example, it can be seen that box cores 6 and 15, although nearly the same in Mn variation, show widely different variations in Fe, Ni, Cu, Co, and Zn. Similarily, it is possible to say that the variability of Ni in box core 15 is far less than the variability of the other metals, except Mn. Although there are notable exceptions, the average variability of Ni is less than that of Cu, as shown by the mean CV values for all 21 box cores (Table 5). It is interesting to note, in fact, that only Fe shows a lower mean CV value than Ni.

Causes of chemical variations in nodules. The data show that mean metal content and variability in composition are both related to nodule size in the RP8-OC-76 collection. The relationships suggest that differences in average nodule composition among the box cores may be explained by the fact that the size distribution of nodules analyzed was different from one box core to another. Thus it would be expected that in box cores where a number of nodules in the 4 to 8 cm size range were analyzed the mean metal content would be somewhat lower and more variable than in cores where the nodules analyzed were only of smaller size. The data at hand do not, however, support this proposition.

The size classes of all nodules analyzed are shown in Table 6 together with CV values for Mn, Fe, Ni, and Cu from Table 5. It can be seen that several box cores with generally high metal content and low CV values (e.g., 6, 7, 15, 19, and 28) are represented by analyses of about the same number of nodules in the 4 to 8 cm size range as other box cores where metal content is lower and more variable (e.g., 9, 23, and 27). In box cores 27 and 28, the number and size distribution of nodules analyzed was, in fact, identical. Further examination of Tables 4 and 6 produces no suggestion that nodule size is a significant factor in determining mean metal content and compositional variation range within individual box cores.

TABLE 5

Coefficient of Variation Values for Major Metal Content of Nodules in RP8-OC-76 Box Cores (Percent)

Box Core Number	Mn	Fe	Ni	Cu	Co	Zn
6	2.43	2.53	3.50	2.80	3.15	9.69
7	3.47	5.35	4.33	2.90	4.83	8.54
8	4.42	4.62	3.32	4.62	8.76	7.16
9	20.70	9.10	13.38	14.89	7.45	20.14
10	9.70	4.13	3.83	8.69	4.14	12.14
11	5.55	7.99	3.07	5.73	14.23	15.10
12	8.42	12.14	15.35	24.06	9.81	22.15
13	7.57	10.61	14.20	16.88	11.25	32.63
14	14.93	3.28	4.50	4.66	3.73	4.30
15	2.97	7.99	5.55	14.89	10.94	17.53
16	4.92	17.22	4.48	8.84	11.18	19.33
18	5.04	5.32	4.73	5.97	6.72	6.13
19	5.87	8.46	2.78	11.48	6.32	16.10
20	5.38	11.72	3.14	4.27	15.69	8.59
21	8.10	4.41	6.64	5.36	4.05	2.20
22	6.92	6.81	22.13	23.38	5.14	36.36
23	56.12	8.63	22.56	33.55	24.96	21.78
24	1.60	2.38	2.36	7.52	3.61	3.69
25	5.82	2.81	3.31	4.54	4.44	6.51
27	33.47	10.83	13.30	20.90	18.24	29.86
28	2.79	11.28	5.51	3.98	14.79	10.59
Mean	10.29	7.51	7.71	10.95	9.21	14.79

TABLE 6

Coefficients of Variation and Sizes of Nodules
Analyzed from RP8-OC-76 Box Cores

Box Core Number	Mn	Fe	Ni	Cu	Number of Nodules Analyzed by Size Class* 1	2	3	4	5
6	2.43	2.53	3.50	2.80		3	2	1	
7	3.47	5.35	4.33	2.90		3	3	1	
8	4.42	4.62	3.32	4.62	2	4	1	1	
9	20.70	9.10	13.38	14.89		4	2	1	
10	9.70	4.13	3.83	8.69		3	3	1	
11	5.55	7.99	3.07	5.73		3	3	1	
12	8.42	12.14	15.35	24.06		2	4	2	1
13	7.57	10.61	14.20	16.88	2	3	2	1	
14	14.93	3.28	4.50	4.66		3	1		
15	2.97	7.99	5.55	14.89		2	3	3	
16	4.92	17.22	4.48	8.84		1	2	1	4
18	5.04	5.32	4.73	5.97		1	5	1	1
19	5.87	8.46	2.78	11.48		3	3		
20	5.38	11.72	3.14	4.27		3	3	2	3
21	8.10	4.41	6.64	5.36		3	3		
22	6.92	6.81	22.13	23.38	2	3	4		
23	56.12	8.63	22.56	33.55		3	2	3	
24	1.60	2.38	2.36	7.52		4	2		
25	5.82	2.81	3.31	4.54		3	3		
27	33.47	10.83	13.30	20.90		3	3	3	1
28	2.79	11.28	5.51	3.98		3	3	3	1
Mean	10.29	7.51	7.71	10.95					

*Size Classes (maximum length)
1 <2 cm
2 2-4 cm
3 4-6 cm
4 6-8 cm
5 >8 cm

TABLE 7

Summary of Microprobe Analyses *
(Weight percent)

Element	Crystalline Oxides Min.	Crystalline Oxides Avg.	Crystalline Oxides Max.	Amorphous Oxides Min.	Amorphous Oxides Avg.	Amorphous Oxides Max.
Mn	23.46	35.81	47.49	12.31	19.42	27.60
Fe	0.12	1.28	4.36	10.70	16.94	27.12
Ni	0.05	1.74	3.67	0.02	0.27	0.83
Cu	0.13	1.25	2.95	0.01	0.16	0.41
Co	0.00	0.07	0.43	0.08	0.31	0.74
Zn	0.01	0.23	0.77	0.00	0.01	0.12

* Summary based upon 118 spot analyses on six Pacific Ocean nodules

Analyst: D.L. Banning, Washington State University.

Thus, it appears that nodule composition must be a function of more basic attributes of nodules than size. The two most important factors appear to be composition of the oxide components and the proportions of oxides and non-oxide materials in a nodule. Although it has long been known that the oxides in individual nodules are not homogeneous (Burns and Fuerstenau, 1966; Sorem, 1967; Foster, 1970; Carr, 1970), Banning (1975) was the first to thoroughly document the virtual restriction of high concentrations of Ni and Cu to the Mn-rich crystalline oxides (todorokite and birnessite) and the preferential concentration of Co in the Fe-rich X-ray amorphous oxides. Banning stated that the ratio of crystalline to amorphous oxides appears to be the main factor in controlling the bulk chemical composition of a nodule. A summary of the results of Banning's microprobe analyses is given in Table 7.

We have examined polished sections of many of the analyzed nodules in the RP8-OC-76 collection with the ore microscope (Figs. 18 and 19), and proportions of the two different types of oxide clearly differ from nodule to nodule. There is no doubt that oxide proportions could play an important role in determining the bulk metal content of these nodules. Furthermore, the bulk composition is affected not only by the proportion of the oxides but also by compositional variations in each type of oxide.

In the nodules studied by Banning, clay and other non-oxide material was sparse and had little effect on the bulk nodule composition. In some of the RP8-OC-76 nodules, however, clay is abundant and has an important effect on nodule bulk composition

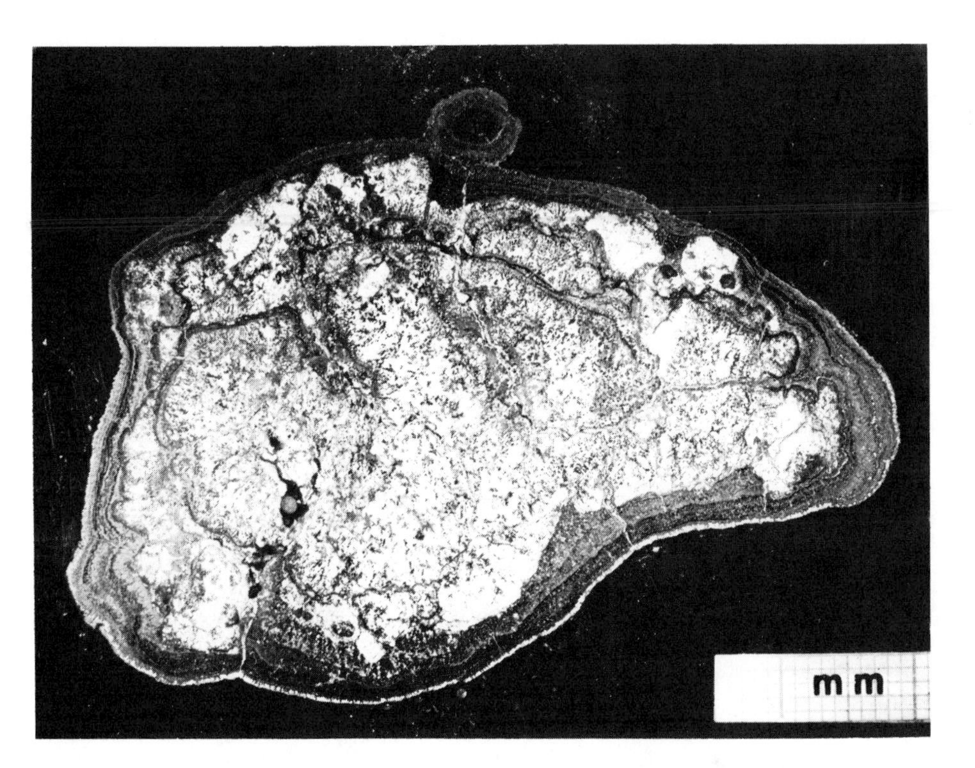

Fig. 18. Polished section of clay-rich manganese nodule from box core 23, RP8-OC-76. Oblique illumination has been used to emphasize the contrast between the clay core and the oxide crustal zone. The core shows intensive dendritic impregnation with oxides. Bulk chemical composition (weight percent) given below (for matching half of nodule) reflects clay dilution.

Mn	Fe	Ni	Cu	Mn + Fe	Mn/Fe
15.13	7.30	0.98	0.54	22.43	2.07

(Reinhart *et al.*, 1978). The diluent effect of a clay nodule core, for example, is illustrated by sections and analyses of two nodules from box core 23 in Figures 18 and 19.

Analyses of nodules rich in non-oxide material, like that shown in Figure 18, generally reveal lower percentages of all four major metals (Mn, Fe, Ni, and Cu) than do nodules with little clay or rock. In the RP8-OC-76 collection, examination of polished sections shows that most nodules with combined Mn and Fe content of

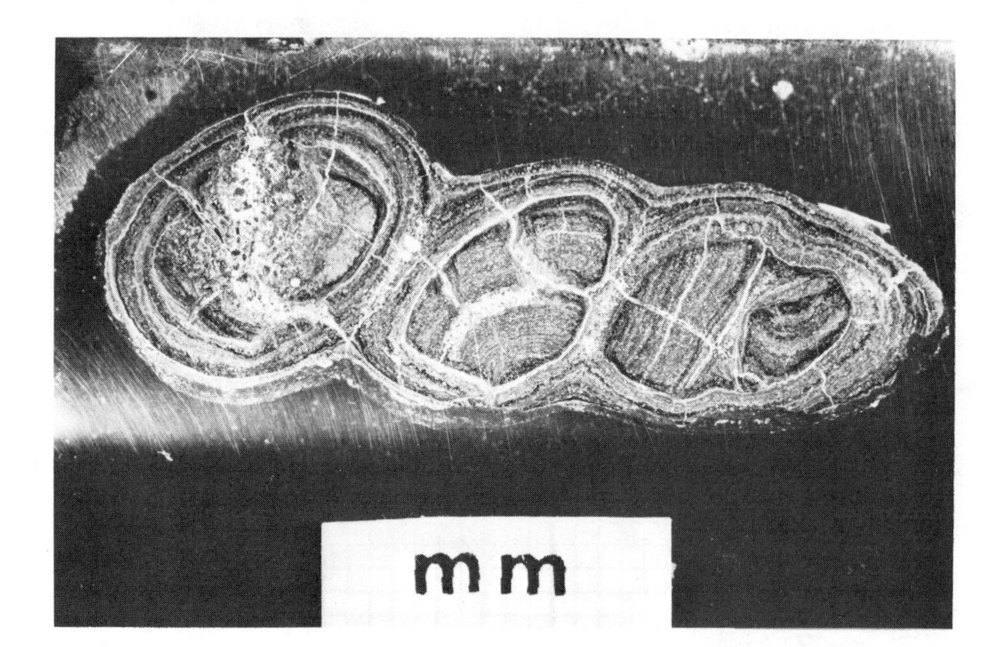

Fig. 19. Polished section of nodule very poor in clay from same box core as nodule in Figure 17 (box core 23, RP8-OC-76). Oblique illumination. The oxides are gray to black, and the thin white seams are clay-filled veinlets and partings between oxide layers. Compare with section shown in Figure 17. This nodule contains very little clay (about 5 percent). The low nickel and copper content shown by the bulk analysis (below) is due to presence of amorphous oxides, not clay, as shown by the Fe content and Fe/Mn ratio.

Mn	Fe	Ni	Cu	Mn + Fe	Mn/Fe
24.50	9.31	1.16	0.55	33.85	2.63

less than about 30 percent and a ratio of Mn/Fe less than about 3 contain appreciable silicates in the core. The resulting dilution also affects Ni and Cu content, but low Ni and Cu content alone is not a reliable indicator of non-oxide impurities. Low content of these metals can also be due to an unusually large proportion of Fe-rich amorphous oxides in the sample. In most samples of this kind (Mn + Fe) content exceeds 30 percent and Fe content is relatively high (more than 7 percent). Mn content may be moderate

(24 percent) to high (29 percent). Low Mn values are not found because the amorphous oxides contain appreciable Mn (Table 7). In this region, low bulk Mn content is only found in nodules where both oxide components are of minor abundance (Fig. 18).

Their relationships are best illustrated with single nodules, as in Figures 18 and 19, but the criteria can also be used for interpretation of average composition of box cores. The basic data needed, which are based upon the weighted average box core analyses listed in Table 4, are listed in Table 8. To permit a concurrent evaluation of the spatial distribution of the box cores, the values for weight percent of Ni and Cu, combined Mn and Fe, and Mn/Fe ratio are plotted in Figure 20.

To test the method described above, only the box cores with the lowest content of Ni and Cu will be considered. Box cores 22, 23, 24, and 27 each have a combined content of Ni and Cu less than 2 percent by weight. Considering the close proximity of the sites,

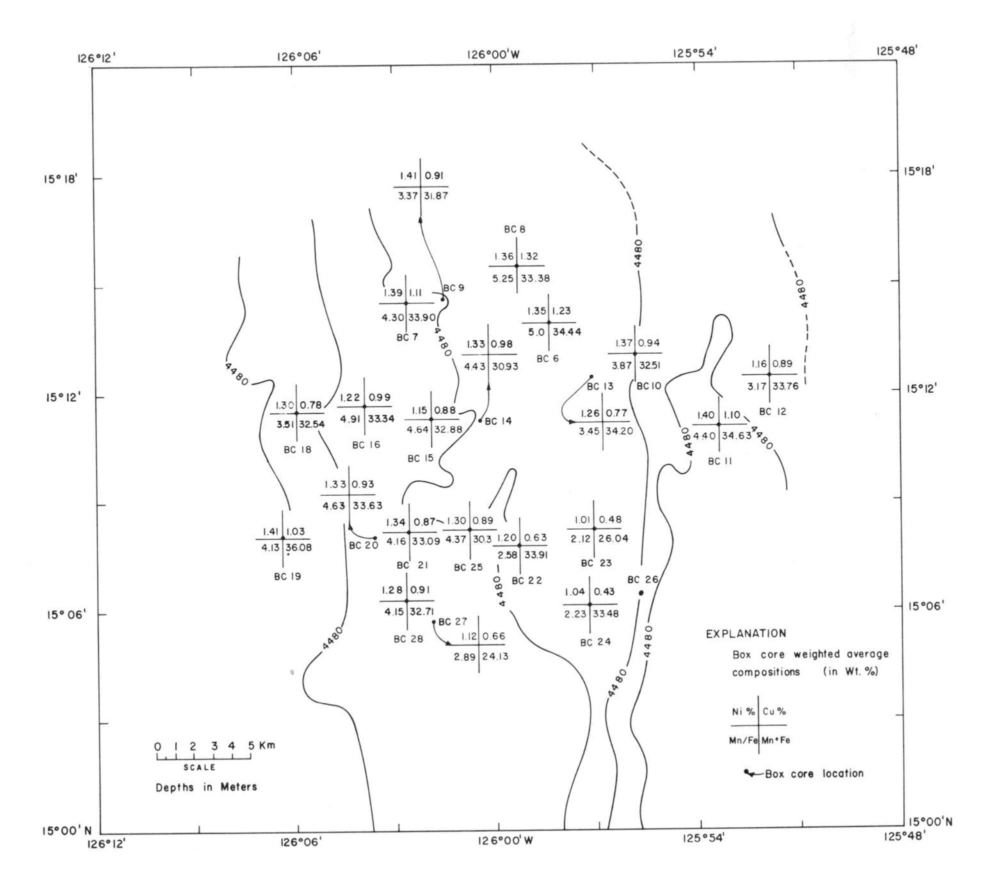

Fig. 20. Selected chemical data for RP8-OC-76 box cores. The metal values shown are weighted average concentrations (Table 5). Base map adapted from Fig. 6.

it might be expected that nodules from box cores 23 and 24 are low in (Ni + Cu) for the same reason. The (Mn + Fe) content is very different, however. The value for box core 23 is 26.04, more than 7 percent below that of box core 24. This suggests strongly that non-oxide dilution is responsible for the low (Ni + Cu) content. The low Mn/Fe ratio (2:12) suggests that relatively abundant amorphous oxides are also a factor. In box core 24, the high (Mn + Fe) content (33.48 percent) together with the low Mn/Fe ratio (2.23) indicates that non-oxides are sparse but that Fe-rich amorphous oxides are abundant, resulting in a low bulk (Ni + Cu) content.

By the same line of reasoning, box cores 22 and 27 differ in the same general way. Box core 27 nodules must contain abundant non-oxide material, because the (Mn + Fe) value is only about 24 percent. The Mn/Fe ratio (2.89) also suggests that Fe-rich oxides are relatively abundant. In box core 22, the high (Mn + Fe) content (33.91 percent) indicates minor non-oxides, but the low Mn/Fe ratio (2.59) shows that the proportion of amorphous oxides is greater than in BC 27.

Further examination of the data in Table 8 and Figure 20 suggests that the average content of Ni and Cu in the other box cores varies primarily with differences in oxide composition, with little dilution by non-oxides. Possible exceptions are box cores 14 and 25, where (Mn + Fe) content only slightly exceeds 30 percent. In both, the Mn/Fe ratio is more than 4.3, however, indicating that little Fe is likely to be present in a component that does not also contain Mn. It is therefore concluded that 17 of the 21 box cores owe differences in mean metal concentration chiefly to non-uniform metal content of the oxide portions of the nodules. Considering Banning's data (Table 7), the variety of bulk chemical compositions possible is thus virtually limitless.

Considering the weighted average metal concentrations of Ni and Cu plotted in Figure 20 and the depth contours (see also Fig. 5), it is clear that most of the box cores rich in Ni were recovered from relatively high elevations. Of the 7 box cores with a mean Ni content of 1.35 percent or more, only one (BC 7) is at a depth greater than 4500 meters. The 5 highest mean values for Cu (1.10 percent and greater) are also found in this group of box cores. Reasonably high Ni and Cu values are not confined to the high elevations, however, as shown by the data for two of the deepest sites, BC 20 and BC 28. Similarily, Ni and Cu content is not high at all sites sampled at higher elevations. For example, the three lowest mean values for Cu and the two lowest Ni values were found in box cores 22, 23, and 24, high in the south part of the area. In summary, a pattern in which the highest concentrations of Ni and Cu exist in nodules at high elevations is suggested, but exceptions are numerous. There is also a trend of alignment of the richest nodules in a northwesterly direction in the north part of the area.

These observations suggest that Ni and Mn content of the nodules in the area may be related to bottom topography or associated factors such as sediment type or bottom water currents. Relationships cannot be evaluated until more data are available.

The nodule size distribution and total mass of nodules in each box core were also examined for possible correlations with average box core composition. With regard to size distribution, the data in Tables 4 and 6 reveal no obvious relationship between the sizes of nodules analyzed and weighted average concentrations and CV values for Mn, Fe, Ni and Cu. The question still remains, however, as to whether overall size distribution is related to mean metal content of nodules from individual box cores.

To study this problem, the overall size distribution in each box core is shown graphically in Figure 21. The weighted average content of Ni and Cu of each box core may be seen in Figure 20,

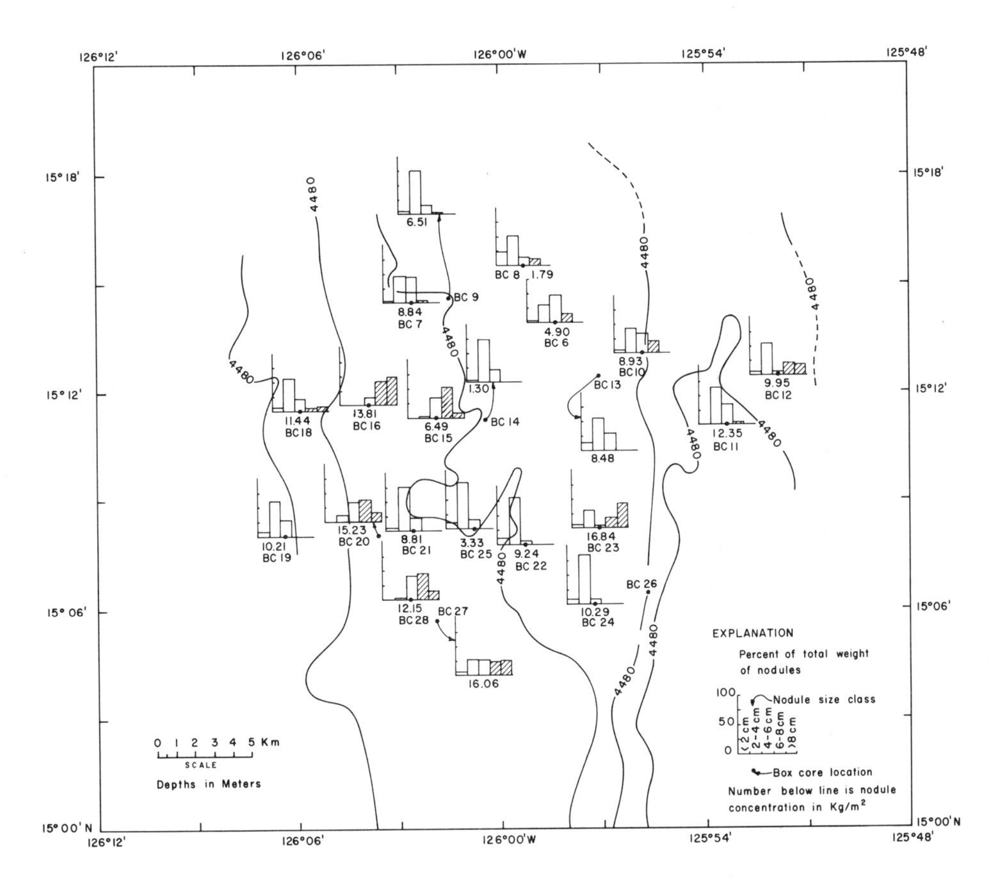

Fig. 21. Size distribution of nodules in RP8-OC-76 box cores and calculated nodule concentration for each box core. Size classes 6 to 8 cm and greater than 8 cm are cross-hatched for emphasis. Base map same as Fig. 20 (adapted from Fig. 6).

which is plotted on the same base. The wide variations in size distribution from one box core to another recorded in Table 2 are easily seen, as is the predominance of nodules in the range from 2 to 6 cm. There is only a weak indication that size distribution is related to mean metal content. Of the 8 box cores with the highest (Ni + Cu) content (2.31 to 2.68 weight percent; box cores 6, 7, 8, 9, 10, 11, 14 and 19) all but two contain at least a few nodules larger than 6 cm but none has nodules larger than 8 cm. In contrast, the 8 box cores which do contain nodules larger than 8 cm are all relatively low in (Ni + Cu) content, only three (BC 16, 20, and 28) having mean values of 2.19 percent or more. Thus, it appears that the presence of nodules larger than 8 cm at a sample site in this region could signal relatively low mean (Ni + Cu) content. The same cannot be said for 6 to 8 cm nodules, which are more abundant in box cores with high (Ni + Cu) content than low.

Finally, the possibility of a correlation between average nodule composition and total nodule mass in each box core was investigated. Nodule mass is represented in Figure 21 as <u>nodule concentration</u>, in kilograms per square meter, based upon the laboratory weight of nodules recovered and the box core sample area of 2500 cm^2, with a correction for box core area removed during shipboard sub-coring. Concentrations range from 1.30 to 16.84 kg/m^2.

The relationship between nodule concentration and mean (Ni + Cu) content is shown in Table 9 which should be used in conjunction with Figure 21. Data for Table 9 are taken from Table 8.

In the table it is apparent that the 8 box cores with the highest mean metal values (2.31 - 2.68 percent (Ni + Cu) include 3 with low nodule concentrations, 3 with medium, and 2 with high nodule concentrations. Box cores with a middle range metal content (2.19 - 2.26) include one each of low and medium nodule concentrations and 3 with high concentrations. Among the 8 box cores with lower metal content (1.49 - 2.08), 4 have medium and 4 have high nodule concentrations.

Thus, it is clear that the data indicate that a high mean (Ni + Cu) content may be expected in deposits of any concentration in this area. However, high average content of these metals is more likely to be found in box cores with low and medium than high nodule concentrations. The degree to which this inverse relationship is developed in the RP8-OC-76 collection is expressed statistically (after omitting data for box cores 14 and 25, where recovery was probably partial) by the correlation coefficient of -0.48 between nodule concentration and (Ni + Cu) content. This correlation is similar to that reported recently for Pacific Ocean nodules in general (Menard and Frazer, 1978).

Given these relationships between (Ni + Cu) content, overall size distribution, and nodule concentrations, together with the observations made earlier on the general topographic distribution of box cores of different mean metal content, it might be expected that a topographic pattern exists in the variation of nodule concentrations. Using the three-fold classification already set up and

TABLE 8

Box Core Metal Data Based upon Weighted Averages of Nodule Analyses

RP8-OC-76 Collection, DOMES Site C

Box Number	Metal Content (Weight Percent) Ni	Cu	Ni + Cu	Mn + Fe	Ratio Mn/Fe	Nodule Conc. (kg/m^2)*
6	1.35	1.23	2.58	34.44	5.0	4.90
7	1.39	1.11	2.50	33.90	4.30	8.84
8	1.36	1.32	2.68	33.38	5.25	1.79
9	1.41	0.91	2.32	31.87	3.37	6.51
10	1.37	0.94	2.31	32.51	3.87	8.93
11	1.40	1.10	2.50	34.63	4.40	12.35
12	1.16	0.89	2.05	33.76	3.17	9.95
13	1.26	0.77	2.03	34.20	3.45	8.48
14	1.33	0.98	2.31	30.93	4.43	1.30
15	1.15	0.88	2.03	32.88	4.64	6.49
16	1.22	0.99	2.21	33.34	4.91	3.81
18	1.30	0.78	2.08	32.54	3.51	11.44
19	1.41	1.03	2.44	36.08	4.13	10.21
20	1.33	0.93	2.26	33.63	4.63	15.23
21	1.34	0.87	2.21	33.09	4.16	8.81
22	1.20	0.63	1.83	33.91	2.58	9.24
23**	1.01	0.48	1.49	26.04	2.12	16.84
24	1.04	0.43	1.47	33.48	2.23	10.29
25	1.30	0.89	2.19	30.30	4.37	3.33
27	1.12	0.66	1.78	24.13	2.89	16.06
28	1.28	0.91	2.19	32.71	4.15	12.15

* Calculated from total nodule weight (wet) in 2500 cm^2 box cores. For dry concentration subtract 30 percent of wet concentration. Metal content is given in weight percent dry.

** In box core 23, 49 percent of total nodule weight was in nodule sizes not represented by analyzed samples. Weighted averages shown here are based only upon nodule sizes analyzed.

TABLE 9

Nodule Concentration and Metal Content, RP-8-OC-76 Box Cores

Nodule Concentration (kg/m^2)	Weighted Average (Ni + Cu) Content (Weight Percent)			Total Box Cores
	1.49-2.08	2.19-2.26	2.31-2.68	
< 6		BC 25	BC 6,8,14	4
6 - 10	BC 12,13,15,22	BC 21	BC 7,9,10	8
> 10	BC 18,23,24,27	BC 16,20,28	BC 11,19	9
Total Box Cores	8	5	8	

the values shown in Figure 21, however, only a very general pattern can be seen. The four sites where nodule concentration is less than 6 kg are all in high areas (depth less than 4480 meters). The eight box cores in the intermediate range (6 to 10 kg/m^2) are divided equally between sites above and below 4480 meters depth, and 5 of the 9 box core sites where nodule concentration is greater than 10 kg/m^2 are deeper than 4480. Thus, the sites with medium or high concentrations are almost randomly distributed in this classification.

There is preferred distribution of sites with high nodule concentration with respect to topographic features, however, perhaps better seen by reference to Figure 6. Four of the six sites with highest nodule concentrations are located in the West Channel, which Piper et al. (1977) reported to have the highest nodule concentrations in the RP8-OC-76 area. Their conclusion that the Channel also represents the greatest resource of Mn, Ni, and Cu in the area is supported by our chemical data (Fig. 20), but their chemical analyses are not everywhere in agreement with ours. The chief reason probably is that we analyzed a far greater number of nodules from the same box cores and have plotted mean values on our maps. As an example, Piper et al. (1977, p. 230) report Ni content of 0.82 and 1.03 percent for nodules from box cores 9 and 10, respectively, whereas our weighted average Ni percentages are 1.41 for box core 9 and 1.37 for box core 10.

Again, with regard to relationships between topography and nodule concentration, it should be pointed out that our findings and those of Piper et al. (1977) agree in general and are not in agreement with a general statement made by Menard and Frazer (1978). Menard and Frazer state that "This negative correlation is apparently related to local topography", and they cite as support the observations of others that ". . .nodules are usually more abundant on hilltops and slopes than in the valleys and at the base of hills". We wish to emphasize not only that our nodule concentration measurements show the opposite but that our nodule analyses show that the highest (Ni + Cu) values are found on the hills.

It is appropriate, moreover, to point out that the terms "nodule abundance" and "nodule concentration" are often confused or used interchangeably in the literature. To avoid confusion, we strongly urge the uniform use of these or other terms according to rigid definitions, such as those proposed recently by Fewkes et al. (1979). Abundance is equivalent to population density, which means the proportion of the sea floor covered by nodules and is expressed as a percent. Nodule concentration is the total weight of nodules per unit area of the sea floor (kg/m^2). Conversion of population density determinations to nodule concentrations with acceptable accuracy requires a detailed knowledge of nodule sizes in the field being evaluated so that realistic weights can be assigned to the nodules.

The ultimate causes of the distribution of nodules of different size, composition, and abundance in the area explored may be mostly factors in the physical environment. The observed trends such as the restriction of sites of relatively small high-grade nodules, in low to moderate concentrations, chiefly to the highest part of the area, suggest some kind of topographic control, as does the distribution of sites with larger nodules to the south and in the depressions. The effect of the topography on bottom currents may be a factor. The direction of bottom currents is known only from the behavior of the compass and vane during bottom photography (with the ship not under way), but the suggested prevalence of current flow from east to west may have influenced the trends just described. Although slopes are very gentle, it seems reasonable to expect that thickness and type of surface sediment (equivalent to the top mud unit in box cores) would also be largely determined by the current and the topography. The long term drifting of sediment over the high area and settling in the depressions or low slopes might well lead to the nodule distribution which exists.

Site A Nodule Composition

Collection data for RP8-OC-75 box cores from Site A are presented in the Appendix. Some of the box core top surfaces are illustrated in Figure 4, and basic nodule data are presented in Table 1.

The station summary data (Table 10) show that mean metal content of the box cores is remarkably uniform within a station area, but the mean metal content at each station differs considerably from one station to the next. Among the five stations at Site A, for example, mean Mn content ranges from 23 to nearly 34 weight percent. Almost identical low mean Mn content (23 percent) is found in nodules at both stations 46 and 50. An intermediate Mn content (28 percent) is found at Station 49, and high values,

TABLE 10

Summary of Major Metal Content of Nodules from Site A-RP8-OC-75 (in Weight Percent)

Sta.	Box Core No.	Anal.	Mn Mean	Mn Max.	Mn Min.	Fe Mean	Fe Max.	Fe Min.	Ni Mean	Ni Max.	Ni Min.	Cu Mean	Cu Max.	Cu Min.
46	1	(8)	24.79	27.50	18.60	8.66	9.36	8.11	1.51	1.93	1.37	1.10	1.15	1.01
	3	(10)	23.08	26.00	15.30	12.08	13.26	9.45	0.79	0.95	0.52	0.56	0.65	0.50
	4	(10)	20.89	24.60	17.90	11.47	13.57	8.40	0.81	0.89	0.65	0.55	0.70	0.50
	6	(9)	21.51	28.50	12.80	10.60	11.82	9.42	1.27	1.55	1.03	0.86	1.02	0.69
	8	(8)	26.85	31.40	20.50	9.08	10.25	7.09	1.34	1.70	1.08	0.91	1.17	0.77
	9	(9)	22.70	27.30	11.00	12.76	14.70	9.71	0.86	1.02	0.60	0.64	0.73	0.51
Station Summary		(54)	23.16	31.40	11.00	10.88	14.70	7.09	1.06	1.93	0.52	0.75	1.17	0.50
47	11	(4)	35.45	35.90	35.00	4.24	4.61	3.60	1.55	1.72	1.22	1.61	1.67	1.49
	14	(5)	32.72	34.60	31.40	5.40	6.18	4.28	1.56	1.67	1.41	1.49	1.55	1.42
	15	(7)	33.34	35.20	28.80	5.01	5.80	4.54	1.68	1.82	1.29	1.55	1.65	1.44
Station Summary		(16)	33.67	35.90	28.80	4.94	6.18	3.60	1.61	1.82	1.22	1.55	1.67	1.42
48	19	(10)	29.26	32.50	27.30	8.25	9.42	6.16	1.28	1.43	1.08	1.09	1.59	0.97
	20	(10)	29.66	31.20	24.90	9.39	10.19	6.93	1.32	1.43	1.17	1.06	1.10	0.97
	22	(9)	33.39	35.70	29.10	5.46	6.24	3.75	1.57	1.72	1.42	1.64	1.71	1.52
Station Summary		(61)	30.88	35.70	24.90	7.62	10.19	3.75	1.40	1.72	1.08	1.28	1.71	0.97
49	24	(1)	7.30	7.30	7.30	5.62	5.62	5.62	1.08	1.08	1.08	0.83	0.83	0.83
	25	(10)	26.55	30.40	21.90	10.93	11.65	8.05	1.07	1.23	0.62	0.79	0.88	0.62
	26	(10)	29.56	31.20	27.70	8.21	8.69	7.48	1.35	1.46	1.26	1.04	1.11	0.86
	27	(10)	30.49	33.30	27.40	6.17	8.91	4.47	1.43	1.59	1.21	1.23	1.37	1.02
Station Summary		(31)	28.17	33.30	7.30	8.35	11.65	4.47	1.28	1.59	0.62	1.01	1.37	0.62
50	28	(10)	23.92	27.30	20.40	12.31	14.52	10.73	0.94	1.41	0.75	0.66	0.81	0.54
	29	(10)	23.23	29.60	16.80	11.20	13.32	8.34	0.91	1.41	0.73	0.61	1.03	0.29
	31	(10)	20.80	28.10	8.50	11.80	14.05	10.09	0.97	1.21	0.58	0.58	0.77	0.33
	32	(10)	25.10	28.00	20.80	12.74	14.55	10.59	0.99	1.17	0.81	0.69	1.19	0.33
Station Summary		(40)	23.26	29.60	8.50	12.01	14.55	8.34	0.95	1.41	0.58	0.64	1.19	0.29
Site A Summary		(170)	26.37	35.90	7.30	9.60	14.70	3.60	1.18	1.93	0.52	0.93	1.71	0.29

greater than 30 percent, are present at Stations 47 and 48. Considering all the analyses, Ni and Cu content show a generally positive correlation with Mn content (Mn:Ni, r = 0.69; Mn:Cu, r = 0.73); Fe varies in a more random fashion and is antithetic to Mn (r = 0.62).

The low mean values for Mn, Ni, and Cu at Station 46 reflect not only a general lack of nodules high in metal content but also the particularly low Ni and Cu content of box cores 3, 4, and 9. These same box cores have a relatively high Fe content. At Station 50 the situation is somewhat different, for all of the box cores are low in Ni and Cu content and relatively rich in Fe.

The intermediate metal content at Station 49 is due primarily to the presence of Fe-rich nodules in box core 25, with their correspondingly low Ni and Cu content. The one nodule from box core 24 is a puzzle, with very low Mn, normal Fe, and appreciable Ni and Cu content. The specimen has the appearance of oxide-stained white clay. No error can be found in the analysis.

Box cores from Stations 47 and 48 are all characterized by high Mn and Ni content, but Cu content is more variable. The mean Cu content is lowest in the box cores with highest Fe content (19 and 20). The large number of nodules analyzed from these box cores reduces the Station 48 mean to 1.28 percent Cu, although the mean in the other box core (22) is much higher (1.64 percent Cu).

In the Site Summary (Table 10) the values for mean metal content of the 170 Site A nodules analyzed are 26.37 percent Mn, 9.60 percent Fe, 1.18 percent Ni, and 0.93 percent Cu. The relatively low average values for Mn, Ni, and Cu are due primarily to the large number of nodules (94) analyzed from the low-grade box cores at Stations 46 and 50.

Chemical Composition of Nodules, Site B

The analyses of 79 nodules from Site B box cores are reported in the Appendix.

Although the number of analyses is relatively small, our data (Table 11) indicate that Site B is a unique area chemically compared to Sites A and C. Metal content is uniformly high, to judge from mean values for Mn, Ni, and Cu in box cores. Fe content is very low. The Station Summary values (Table 11) for mean metal content show only small variations from station to station over the entire area. Unfortunately the number of nodules available from three stations (Stations 51, 52, and 53) was very small, but the analyses are in the same general range. For the present, they are assumed to give a good indication of approximate box core compositions. The mean Mn content is less than 30 percent in only one box core (52), where 3 of the 10 nodules analyzed contained only 19.9 to 26.6 percent Mn (see Appendix).

There is some uncertainty about the analyses of Site B nodules. The unusually high metal concentrations may be partially due to the storage and shipping conditions of the nodules (these nodules were

TABLE 11

Summary of Major Metal Content of Nodules from Site B-RP8-OC-75

(in Weight Percent)

Sta.	Box Core No.	Anal.	Mn Mean	Mn Max.	Mn Min.	Fe Mean	Fe Max.	Fe Min.	Ni Mean	Ni Max.	Ni Min.	Cu Mean	Cu Max.	Cu Min.
51	33	(2)	30.30	30.70	29.90	5.08	5.63	4.52	1.87	1.89	1.84	1.54	1.60	1.41
	34	(4)	33.68	37.20	31.70	5.01	5.56	4.41	1.54	1.66	1.39	1.14	1.28	1.01
Station Summary		(6)	32.55	37.20	29.90	5.03	5.63	4.41	1.65	1.89	1.39	1.27	1.60	1.01
52	40	(3)	34.87	35.70	34.20	5.33	5.65	5.07	1.62	1.63	1.61	1.16	1.21	1.10
53	45	(1)	33.40	33.40	33.40	6.32	6.32	6.32	1.64	1.64	1.64	1.27	1.27	1.27
54	48	(10)	32.52	34.00	30.80	5.93	9.71	4.40	1.51	1.74	1.29	1.24	1.54	1.03
	51	(10)	32.47	34.10	31.20	5.71	5.99	5.36	1.63	1.77	1.54	1.23	1.38	1.13
	52	(10)	29.62	33.80	19.90	5.47	5.79	5.09	1.52	1.66	1.16	1.20	1.30	0.97
Station Summary		(30)	31.54	34.10	19.90	5.70	9.71	4.40	1.55	1.77	1.16	1.22	1.54	0.97
55	53	(10)	32.24	36.30	26.50	4.37	7.54	3.24	1.59	1.75	1.12	1.47	1.61	0.96
	54	(10)	32.73	35.80	23.90	5.42	7.61	4.79	1.63	1.76	1.53	1.20	1.41	0.81
	55	(9)	33.47	34.80	32.30	5.05	5.34	4.77	1.65	1.71	1.56	1.28	1.42	1.21
	56	(10)	32.33	34.90	25.80	5.04	5.58	4.58	1.64	1.74	1.54	1.25	1.34	1.11
Station Summary		(39)	32.69	36.30	23.90	4.97	7.61	3.24	1.63	1.76	1.12	1.30	1.61	0.81
Site B Summary		(79)	32.32	37.20	19.90	5.28	9.71	3.24	1.60	1.89	1.12	1.26	1.61	0.81

first collected for biological studies by Dr. A. Paul and were stored in a weak solution of formalin and sea water). The effect may be negligible, however, to judge from the analyses of Site A nodules which had been treated in the same manner. It is possible to check the Site A analyses, for the nodules analyzed from two box cores (BC 3, Sta. 46; BC 14, Sta. 47) were not so treated and the results agree well with those from treated nodules from nearby box cores (BC 1 and 4, Sta. 46; BC 15, Sta. 47).

It has not been possible to check the Site B nodules in a similar way, but caution is needed because the Mn, Ni, and Cu values, in particular, are higher in our analyses than in the few analyses available for comparison (Horn et al., 1972; Frazer and Arrhenius, 1972; Piper et al., 1977). Therefore, we suggest that the Site B analyses in this report be considered possibly high by 5 to 10 percent of the amount present for Mn, Ni, and Cu. If the Site B mean percentages are reduced by 10 percent, Mn changes from 32.32 to 29.09, Ni from 1.60 to 1.44, and Cu from 1.26 to 1.13. This tentative "correction" will not affect the discussion of the differences and similarities of the three sites (A, B, and C) significantly.

In terms of mean metal content of nodules, Sites A, B, and C show a number of clearly different characteristics. To summarize the data presented in Tables 3, 10, and 11 for discussion, values for the mean metal content of nodules and the first standard deviation at each site are listed in Table 12.

Site A nodules are notable for relatively low Mn and high Fe. The standard deviation values for both metals show that appreciable variation is common from nodule to nodule. In addition, Ni

TABLE 12

Comparison of Metal Content of Manganese Nodules-Sites A, B, and C

	Site A		Site B		Site C	
Metal	Mean Weight Percent	Standard Deviation	Mean Weight Percent	Standard Deviation	Mean Weight Percent	Standard Deviation
Mn	26.37	5.41	32.32	2.81	26.29	4.64
Fe	9.60	2.81	5.28	0.87	7.01	1.34
Ni	1.18	0.31	1.60	0.13	1.30	0.16
Cu	0.93	0.36	1.26	0.16	0.93	0.24
Co	0.25	0.06	0.22	0.03	0.28	0.04
Zn	0.09	0.04	0.14	0.03	0.15	0.05

and Cu content is typically low, and the standard deviation for these metals indicates even greater variations. The mean Co content is relatively high, with a moderate standard deviation value. The Zn content is very low and has a wide range.

By comparison, nodules from Site B are much richer in Mn, Ni, and Cu and contain less Fe. In contrast, the standard deviation values are very low, especially for Mn and Ni. Standard deviation for the other metals is also less than in the counterpart analyses of Site A nodules. Co and Zn were not greatly different in mean concentration from Site A.

The Site C data (Table 12) include analyses of nodules from the RP8-OC-75 and RP8-OC-75 box cores as well as those from RP8-OC-76 presented above. They show somewhat higher values than those for RP8-OC-76, only given in Table 3.

It is immediately apparent that the mean Mn content of Site C nodules is virtually the same as that for Site A nodules. The standard deviation is somewhat less. Ni content is intermediate between Sites A and B, but Cu content is the same as that for Site A nodules. Co and Zn are very slightly higher than at Sites A and B. Mean Fe content is also intermediate, as is the standard deviation value. The Mn and Fe content, taken together, suggest that the Site C nodules are, on the average, richer in non-oxide material (clay) than those from Site A. In both sites, Mn and Fe is greater than 33 percent, however, indicating that clay dilution of metal content is not extensive at either site. The higher Fe content at Site A suggests that amorphous oxides were relatively abundant in the nodules analyzed.

The apparent differences in chemical composition of nodules from the three sites suggest that all factors which control nodule metal content have not yet been considered, although there is the strong indication that, at Site C, significant relationships exist between metal content and nodule size, nodule concentration, and sea floor conditions (topography, sediment type, or currents, for example). None of these seems adequate to explain the differences from site to site. There is the definite possibility that the regional chemical differences are related to recognizable differences in nodule physical character. For example, Site C nodules seem to be largely asymmetrical and granular in surface textures, whereas Site B nodules are more commonly large and spheroidal, with botryoidal and granular outer surfaces. Site A box cores show widespread distribution of many relatively small and smooth-surfaced nodules. It remains to be determined, however, whether these characteristics are indicative of certain ranges of metal content or are instead simply accidental differences.

Assuming for the moment that differences in nodule physical character are regional in pattern, it seems likely that the same conditions which affect nodule metal content influence nodule shape, size, and texture. Following this line of reasoning, the key to solving the mystery of regional variations in nodule composition probably lies in the study of the relationships between nodules of different physical as well as chemical type and large scale geological features such as broad sedimentation patterns, crustal structure, hydrothermal activity, and current patterns.

Summary and Conclusions

Box cores of exceptionally good quality from the recent DOMES expeditions in the Horn Region of the Pacific Ocean have provided much new information on manganese nodule deposits. The study of whole box cores at sea showed that nodules vary greatly in abundance, size, shape, and texture, over both large and small distances, but that certain similarities exist among nodules within each two-degree DOMES site. The nodule deposits lie almost entirely at the sediment-water interface, with a negligible proportion of nodules buried in the upper 40-50 cm of the subjacent sediment. Most nodules in the size range most commonly found (2 to 6 cm) rest in a very spongey and mobile sediment "soup"; probably only large nodules are actually imbedded in the bottom sediment. This situation probably varies, however, depending upon the physical properties of the underlying sediment and topography.

Chemical analyses of more than 450 manganese nodules from DOMES Sites A, B, and C demonstrate wide variations in bulk composition. At each site nodules show characteristic mean values for Mn, Fe, Ni, and Cu, but within each site nodule composition varies considerably. The range in content of the metals among nodules in a single box core commonly exceeds the range of mean values from one DOMES site to the next several-fold. This is true for some box cores even if standard deviation is considered instead of the entire numerical range of analyses.

Whether it is required to find the mean metal content of a box core, of a station several kilometers wide, or of a site two degrees of latitude and longitude in size, the implications are the same. It is possible to find the mean metal content of the area without analyzing a number of nodules from each sampling point, such as a box core location. Only after the variations of metal content from nodule to nodule are determined can the mean values be calculated. The arithmetic mean probably is representative of a box core even where nodules of different size are present in different proportions, since metal content does not seem to vary greatly with nodule size in most box cores. In box cores where nodule size is related to metal content, nodules from all size groups should be analyzed and a weighted average calculated.

The number of nodules to be analyzed to arrive at reasonably representative mean values for a box core cannot be arbitrarily set but depends upon the demonstrated variability of the group in question. In the Site C area, for example, nodule variability is great and mean values should be based upon a larger number of analyses than would be required at typical localities in Site B, where nodule composition is more uniform. Fewkes et al., 1979, suggest that an average of 8 nodules per box core should be analyzed to predict mean Ni and Cu content of a box core.

Variability of metal content in nodules at a given locality can generally be traced to two factors, which may operate together or separately. Low content of Mn, Ni, and Cu may be due to simple dilution of the sample by non-oxide material, such as clay, or by oxides with low Mn, Ni and Cu. An examination of the analyses and an elementary study of polished sections of representative nodules can quickly establish which is the case in a problem area.

We hope that the data and the ideas presented in this report will stimulate more research into the relationships between metal content and nodule size, nodule concentration, and bottom features such as topography, and sediment type, and basement structure. A better understanding of these relationships is clearly essential for progress in solving problems of nodule origin as well as in planning the possible economic development of nodule deposits in the region. Both the scientist interested in nodule origin and the economic geologist searching for a mine site must take into account the physical and chemical features characteristic of nodules as a group and individually. Hypotheses of origin must account not only for the vast amounts of metals incorporated in nodules, for example, but also for the concentration of nodules of different composition locally and regionally at the present water-sediment interface. For economic development, the hopeful miner must also look farther than the simple discovery of deposits with the highest metal content. If he realizes, for instance, that some nodules are low grade because of clay content, he can then weigh the possibility of physical beneficiation of abundant low grade nodules against the expensive search for the more elusive high grade deposits.

Acknowledgments

We wish to acknowledge, with thanks, the important contributions to our work made by the officers and crew of OCEANOGRAPHER. We are especially indebted to Barrett Erickson, chief scientist, for his willingness to apply his many skills to help solve problems for us and for the overall success of the work at sea. In 1975, on Leg 3 of RP8-OC-75, D.L. Banning was a valuable assistant.

We also wish to acknowledge financial support from the DOMES Project, the U.S. Geological Survey, Ocean Management, Inc., and the U.S. Bureau of Mines for various phases of the work reported here.

References

Andrews, J.E., Callender, E., Bowser, C.J., Mero, J.L., Gauthier, M., Meylan, M.A., Craig, J.D., Binder, K., Volk, P., Chane, A., and Bachman, W. (1974), Ferromanganese Deposits of the Ocean Floor, Crust Report Mn-74-01, Hawaii Instit. Geophys, Rept. HIG-74-9, 194 p.

Arrhenius, G., (1963) Pelagic sediments, In The Sea: Ideas and observations on progress in the study of the seas, M.N. Hill, Ed., Interscience, New York, v. 3, 655-727.

Bäcker, H., Glasby, G.P., and Meylan, M.A., (1976) Manganese nodules from the southwestern Pacific Basin: Oceanographic Field Report No. 6, New Zealand Oceanographic Institute, D.S.I.R., Wellington, 88 p.

Banning, D.L., (1975) Copper-nickel-cobalt associations in marine manganese nodules (abstract): Am. Geophys. Union Trans., v. 56, no. 12, p. 1000.

Bischoff, J.L. (ed) (1976) Deep Ocean Mining Environmental Study, N.E. Pacific Nodule Province, Site C, Geology and Geochemistry, U.S. Geol. Survey Open-file Report no. 76-548, 275 p.

Burns, R.G. and Fuerstenau, D.W., (1966) Electron-probe determination of inter-element relationships in manganese nodules: Am. Mineralogist v. 51, 895-902.

Carr, G.A., (1970) Marine manganese nodules: identification and occurrences of minerals, MS Thesis, Washington State Univ., Pullman, 101 p. (unpublished).

Fewkes, R.H., (1975) Correlation between morphology and mineralogy of marine manganese nodules -- possible economic significance (abs.): Am. Geophys. Union Trans., v. 56, no. 12, p. 1000.

Fewkes, R.H., (1976) The origin of marine manganese nodules as determined by textural and mineralogical analysis: Ph.D. Thesis (unpubl.) Washington State Univ., Pullman, Wa, 169 p.

Fewkes, R.H. and Sorem, R.K., (1976) Todorokite and birnessite in MOANA WAVE nodules from the northeastern Pacific Ocean 1974 (abstr): Second Internat. Sympos. on Manganese, 25th Internatl. Geol. Congress, Sydney, August 1976.

Fewkes, R.H., McFarland, W.D., Reinhart, W.R., and Sorem, R.K., (1979) Development of a reliable method for evolution of deep sea manganese nodule deposits, Final Report, Contract No. GO 274013-MAS, U.S. Bur. of Mines, Spokane, 91 p.

Foster, A.R. (1970) Marine manganese nodules: nature and origin of internal features, MS Thesis, Washington State Univ., Pullman, Wa, 131 p. (unpubl).

Frazer, J.Z. and Arrhenius, G., (1972) World-wide distribution of ferromanganese nodules and element concentration in selected Pacific nodules, Tech. Rep. Off. Int. Decade Ocean Explor. 2, 51 p.

Horn, D.R., Horn, B.M., and Delach, M.N., (1972) Ferromanganese deposits of the North Pacific, Tech. Rep. Off. Int. Decade

Ocean Explor. 1, 78 p.
Margolis, S.V., Bowser, C.J., Murray, J., Mero, J.L., Hardy, W., Dudley, W.C., Dugolinsky, B.K., Binder, K., Hall, R., and Boatman, C., (1975) Ferromanganese Deposits of the Ocean Floor, Cruise Report Mn-74-02, Hawaii Instit. Geophys. Report HIG-75-17, 121 p.
Menard, H.W. and Frazer, J.Z., (1978) Manganese Nodules on the Sea Floor: Inverse Correlation Between Grade and Abundance, Science, v. 199, 969-971.
Mero, J.L., (1965) The Mineral Resources of the Sea, Elsevier, Amsterdam, 312 p.
Piper, D.Z. (compiler) (1977) Deep Ocean Environmental Study: Geology and Geochemistry of DOMES Sites A, B, and C, Equatorial North Pacific, U.S. Geol. Survey Open-file Report No. 77-778, 527 p.
Price, N.B. and Calvert, S.E., (1977) Geochemical variation in ferromanganese nodules and associated sediments from the Pacific Ocean, Marine Chemistry 5, 43-74.
Raab, W.J., (1972) Physical and chemical features of Pacific deep sea manganese nodules and their implication to the genesis of nodules, In Ferromanganese Deposits on the Ocean Floor, D.R. Horn (ed.), National Sci. Found., Washington D.C., 31-49.
Raab, W.J. and Meylan, M.A., (1977) Morphology of manganese nodules, In Marine Manganese Deposits, G. Glasby (ed.), Elsevier, 109-146.
Reinhart, W.R., Sorem, R.K., and Fewkes, R.H., (1978) Metal variations in Pacific manganese nodules from a potential deep sea mining area (abstr).: IAGOD Internat. Sym., Snowbird, Utah, August 1978.
Sorem, R.K., (1967) Manganese nodules: nature and significance of internal structures: Econ. Geol. v. 62, 141-147.
Sorem, R.K., (1975) Rest-position relationships of manganese nodules and subjacent sediments as a key to nodule origin (abstr).: Am. Geophys. Union Trans. v. 56, no. 12, p. 999.
Sorem, R.K., and Fewkes, R.H., (1976) Variations in manganese nodules in a one-ton dredge haul: an unsolicited report submitted to Deep Sea Ventures, Inc., 28 May 1976.
Sorem, R.K., and Banning, D.L., (1976) Microfeatures of typical manganese nodules from six box cores from NOAA cruise RP8-OC-75: In Geology and Geochemistry of Site C, Deep Ocean Environmental Study, NE Pacific Nodule Province, U.S. Geol. Survey Open-file Report no. 76-548, 167-216.
Sorem, R.K., McFarland, W.D., and Fewkes, R.H., (1978a) Nature and significance of a diffuse sediment-water interface "boundary layer" in East Pacific manganese nodule deposits (abstr).: IAGOD Internat. Sym., Snowbird, Utah August 1978.
Sorem, R.K., Fewkes, R.H., McFarland, W.D., and Reinhart, W.R., (1978b) Physical aspects of the growth environment of manganese nodules in the Horn Region, east equatorial Pacific

Ocean: Invited paper, Colloque International No. 238, Sur la Genese des Nodules de Manganese, Gif-sur-Yvette, France, 25-30 Sept. 1978, 56 p.
Steel, R.G.D., and Torrie, J.H., (1960) Principles and procedures of statistics, McGraw Hill, New York, p. 20.

GEOLOGICAL INVESTIGATION OF THE EQUATORIAL NORTH PACIFIC SEAFLOOR: A DISCUSSION OF SEDIMENT REDISTRIBUTION

James D. Craig

Department of Oceanography
Hawaii Institute of Geophysics, University of Hawaii
Honolulu, Hawaii 96822

Abstract

Detailed investigations in the equatorial North Pacific (9°05'N, 148°45'W) are used to describe geologic evidence of, and processes involved in, sediment erosion/redistribution on the abyssal seafloor. Geophysical surveys indicate that sediment thickness typically varies by a factor of two over distances of 10 km, and that sediment thicknesses are inversely correlated to bathymetric slopes. Although most of the sediment redistribution occurred prior to late Miocene time (Line Islands Formation), variation in Quaternary sediment thickness and admixture of reworked Tertiary microfossils indicates that processes are still active. Grain-size analysis, employing a rapid photometric centifuge technique, shows, however, no evidence of grain-size sorting between net erosional and depositional bathymetric settings. Distribution of manganese nodules, which are divided into two distinct facies based on size, morphology, and chemical composition, appears unrelated to sedimentation patterns and is primarily influenced by the proximity of a small seamount. Collection of undisturbed sediment samples by a spade corer and shipboard measurement of shear strength parameters are used as a basis for evaluating seabed erodibility. A speculative sedimentary model involves erosion of bioturbated "mixed layer" sediment by anomalously strong bottom currents, and downslope transport by low-density turbid or "lutite" suspensions.

Introduction

Manganese nodule deposits on the seafloor of the equatorial North Pacific have captured the interest of several international mining consortia in recent years and prompted a surge in scientific effort to map their distribution and determine their origin. Deep-sea surveys have typically encountered a variability in seafloor abundance and chemical content of nodules over distances of a few km or less (Piper *et al.*, 1977; Craig, in press). This variability affects both scientific studies of nodule origin and economic evaluation of deep ocean mining sites. Sedimentation has often been cited as a major factor in nodule growth and distribution because nodules characteristically occur on the seafloor surface (Moore, 1970; Horn *et al.*, 1973a; Mizuno and Chujo, 1975; Cook *et al.*, 1977), however, localized sedimentary processes have been overlooked because of previous misconceptions that the deep sea acts as a uniform depositional basin. Detailed investigations in the equatorial North Pacific have reported evidence of active sedimentary erosion/redistribution processes over localized scales in the form of outcropping Tertiary strata, major hiatuses, mixed micro-fossil assemblages, and variations in stratigraphic thicknesses (Luyendyk, 1970; Moore, 1970; Johnson and Johnson, 1970; Johnson, 1972; Mudie *et al.*, 1972; Andrews *et al.*, 1974; van Andel *et al.*, 1975; Piper *et al.*, 1977). These anomalous sedimentary features, which have generally been attributed to sediment erosion and lateral transport controlled by bottom currents and bathymetric relief (van Andel *et al*, 1975; Piper *et al.*, 1977), might be related to the physical and chemical variability of nodule deposits over similar seafloor distances. Despite the lack of *in situ* data at present, semi-quantitative models, which are based on both geotechnical studies of surface sediment and results of laboratory experiments using seawater flumes, can expand our understanding of abyssal sedimentary processes in relation to nodule occurrence, and indicate critical objectives for future research. The purpose of this paper is to (1) describe the geologic setting and distribution of manganese nodules in an area where lateral redistribution of sediment has occurred; (2) present the results of shipboard and shore laboratory geotechnical tests relative to sediment strength and erodibility; and (3) provide a preliminary review of erosional and depositional dynamics affecting the benthic boundary in the equatorial North Pacific region.

Data Collection

Detailed sampling of nodules and sediments in an area (approx. 28 x 28 km) was made over a period of 19 days during cruise VA 13/1 of the R/V VALDIVIA (Fig. 1). Survey work was initiated by 12 kHz echo-sounding profiles on 1.85 km grid, which expanded and modified an earlier VALDIVIA survey made during cruise VA-08 (Beiersdorf and Wolfart, 1974). Navigation throughout survey operations was by VALDIVIA's INDAS system, which produced continuous navigation updates with mean errors of less than 1 km between satellite fixes and dead reckoning positions. Nodule samples and bottom photographs were collected by Preussag freefall devices deployed at

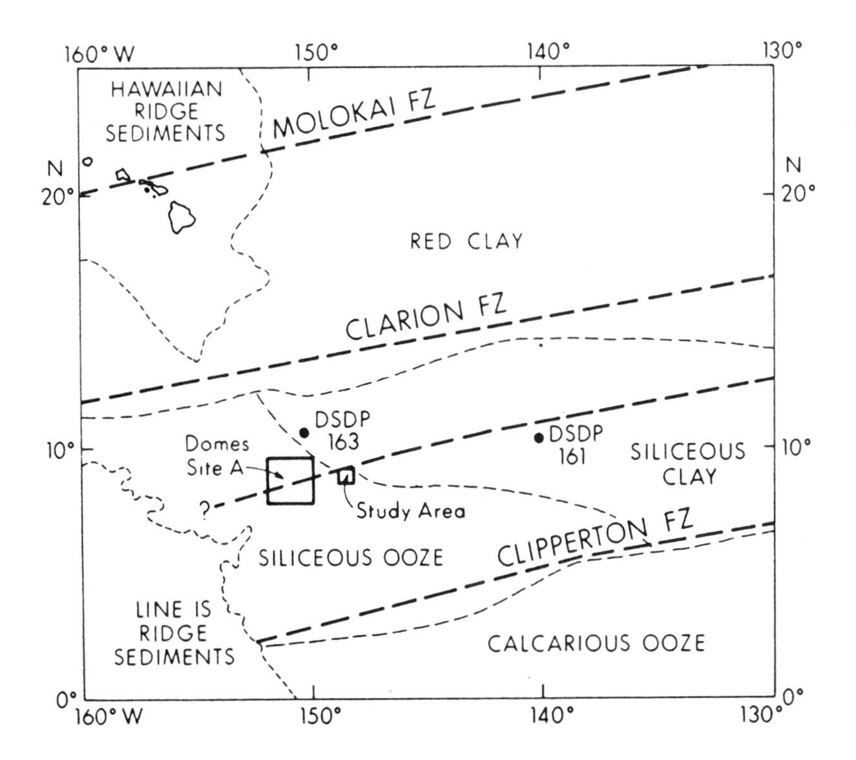

Fig. 1. Geography of the equatorial North Pacific Ocean. Locations of tectonic features and DSDP sites are from van Andel and Heath (1973). Note the postulated trend of the "intermediate" fracture zone through DOMES Site A and the study area. Distribution of surface sediment is from Horn et al. (1973b).

intervals of 200-500 m in circular patterns and on long traverses. Sediment samples were obtained by spade cores (20 x 30 x 50 cm), long box cores (30 x 30 cm x 12 m) and piston cores (9 cm x 10 m). Piston cores and long box cores were released 8 m above the seafloor and allowed to freefall into the sediment, thus producing severe disturbance of the surface layer. In contrast, spade cores were set gently on the seafloor and driven into the sediment by a 200 kg weight, thereby collecting fairly undisturbed samples. The sediment surface was photographed after spade cores were retrieved; nodules on the surface were removed, separated by size and weighed. Sections of piston core liner were then carefully inserted to provide subsamples for archive, geotechnical tests, and studies by other investigators. Subcores to be used for shore laboratory geotechnical tests were capped, heat sealed in plastic, and placed in refrigerated storage.

Geologic Setting

Regional Bathymetry and Stratigraphy

The study area is located in the western portion of the structural block bounded on the north and south by the Clarion and Clipperton fracture zones, respectively (Fig. 1). Westward seafloor spreading from the East Pacific Rise has controlled the development of north-south elongated abyssal hills of 50-200 m in relief which cover most of the north Pacific region (Luyendyk, 1970). Although the initial formation of abyssal hills is controlled by block-faulting and volcanism on the flanks of spreading ridges (Luyendyk, 1970), evidence for tectonic activity at considerable distances away from the EPR crest is seen in irregularities of micro-topography on slopes of some abyssal hills (van Andel *et al*., 1973). Variation in the characteristic north-south orientation of abyssal hills has also been interpreted as the effect of later tectonism involving transcurrent faulting along fracture zones (Andrews, 1971). A zone of volcanic peaks, large normal faults, and anomalously thin sedimentary sections, recognized between DSDP Sites 161 and 163 (van Andel *et al*., 1973), was attributed to an "intermediate" fracture zone predicted by Sclater *et al*. (1971) and Herron (1972). Figure 1 shows the postulated structural trend of this "intermediate" fracture zone through the study area and DOMES Site A.

The generalized patterns of regional sedimentation based on LDGO piston cores (Horn *et al*., 1973b) is also shown in Fig. 1. Stratigraphic sections for DSDP Site 163 (van Andel *et al*., 1973) and DOMES Site A (Piper *et al*., 1977) have been correlated to the following lithostratigraphic units: an uppermost unit, the Clipperton Formation, which is an acoustically transparent layer of variable thickness (0-30 m) composed of brown zeolitic clay with varying proportions of Quaternary and reworked Tertiary radiolaria, and a lower unit, the Line Islands Formation, which is an acoustically stratified layer of late Eocene to early Miocene age composed of clayey radiolarian ooze and thin chert beds. At DSDP Site 163, 28 m of zeolitic radiolarian clay of Oligocene and (?)Quaternary age (Clipperton Formation) overlies a radiolarian ooze and siliceous clay of Eocene age (Line Islands Formation) (van Andel *et al*., 1973). Theyer (1977) reported a consistent Quaternary age for all box core samples from DOMES Site A but noted a significant (20-80%) admixture of early Miocene and middle-late Eocene radiolaria taxa. Acoustic basement is composed of extrusive basalt with an approximate age of 65-70 Myr in the study area (van Andel and Heath, 1973).

Geology of Study Area

The bathymetry of the study area (Fig. 2) is dominated by a seamount named Gundlach Höhe (peak) rising 650 m above a 5200 m-deep central plain. The oval shape and steep basaltic slopes of this peak and others in the study area suggest they are of volca-

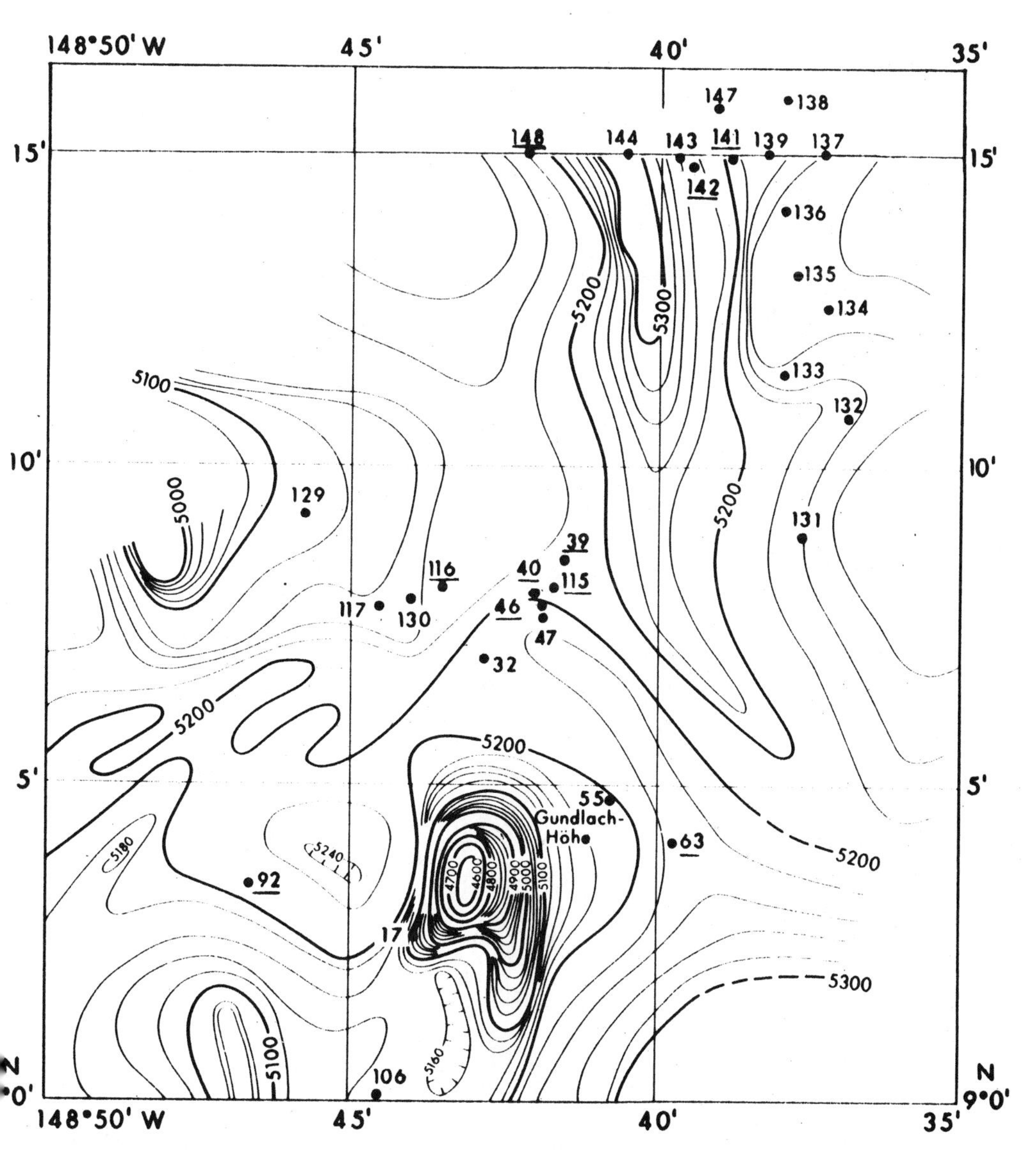

Fig. 2. Bathymetry of the study area compiled from Beiersdorf et al. (1976) (cruise VA 13/1) and Beiersdorf and Wolfart (1974) (cruise VA-08). Contours are given in uncorrected meters with a 20 m contour interval. Survey tracks were oriented N-S and E-W on a regular 1 n. mile grid. Location of coring stations are shown as labeled dots. Stations 32, 47, 106, and 130 are long box cores (6 m). Remaining stations are spade cores. Grain-size analyses were made on samples from underlined stations.

nic origin. North-south elongated abyssal hills in the northern portion of the area plunge southward into the central plain and are separated by a steep-walled valley which slopes northward. Slopes less than 1° cover most of the area with exception of the volcanic peaks (10°-15°) and the abyssal hills (1°-7°).

Reflection profiles show several acoustic facies which are correlated by Dürbaum and Schlüter (1974) to lithostratigraphic units at DSDP Site 163. An uppermost acoustic facies, designated as Unit 1 and correlated to the Clipperton Formation, is subdivided into an acoustically "diffuse" unit, which occurs on slopes of abyssal hills, and a "transparent" unit, which occurs in the steep-walled valley. A stratified or "ribboned" facies, designated as Unit 2 and correlated to the Line Islands Formation, underlies Unit 1 and outcrops on some steep abyssal hill slopes. An isopach map produced by Dürbaum and Schlüter (1974) (Fig. 3) shows that the thickest sedimentary sections (175 m) occur on the central plain, intermediate thicknesses (125-150 m) occur on abyssal hill crests, and a thin (100 m) section occurs in the steep-walled valley. As shown by reflection records and bottom photographs, sediment is generally absent from Gundlach Peak at depths less than 5000 m. Biostratigraphic dating by Beiersdorf and Wolfart (1974) (Fig. 3) indicates a general Quaternary age for surface sediment, average Quaternary sedimentation rates of 1.5-3 m/Myr, and a typical admixture of Eocene and Miocene radiolaria similar to surface sediments in DOMES Site A (Theyer, 1977).

Redistribution of sediment is indicated both by variable sediment thicknesses (Fig. 3) and by the corroded and/or fragmented appearance of Tertiary microfossils mixed with Quaternary taxa. Beiersdorf and Wolfart (1974) suggest that the highly corroded appearance of Eocene radiolaria, in comparison to Upper Miocene and younger taxa, indicates that this redistribution took place prior to the late Miocene (10 Myr B.P.). This interpretation agrees with van Andel _et al_. (1975) who define regional erosional episodes in late Eocene/Oligocene (38-40 Myr B.P.) and in Middle Miocene (10-12 Myr B.P.) associated with changes in deep circulation and production of Antarctic Bottom Water. Although extensive sediment erosion/redistribution during Tertiary episodes largely accounts for the variations in sediment thicknesses in this area, variable thicknesses of Quaternary strata (Fig. 3) suggest more recent episodes of sediment redistribution, possibly associated with Pleistocene climatic changes as suggested by Johnson (1972).

The distribution of manganese nodules is of interest both to characterize micro-topographic roughness of the seabed, subsequently referred to as bottom roughness (k_s), and to determine relationships of nodule distribution to sediment accumulation. Two distinct nodule facies were recognized by Friedrich and Plüger (1974), and the facies map (Fig. 4) has been modified slightly from their observations. A symmetrical distribution of intergrown

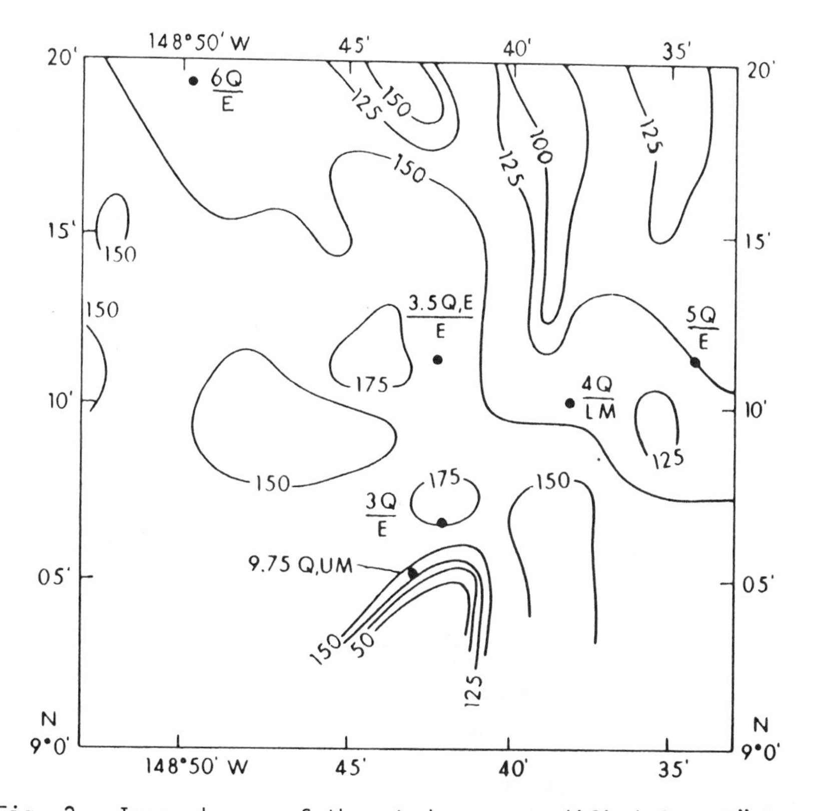

Fig. 3. Isopach map of the study area modified from Dürbaum and Schlüter (1974) (cruise VA-08). Reflection profiles were recorded using a 5000 cm^3 air cannon and a 100 hydrophone eel system towed at 6 knots. Contours are sediment thickness to acoustic basement assuming an average sound velocity of 1600 m/sec. Labeled points are piston core locations (Beiersdorf and Wolfart, 1974). Letters refer to sediment age based on biostratigraphy; (Q) Quaternary; (UM) upper Miocene; (LM) lower Miocene; (E) Eocene. Numbers refer to thickness of strata containing Quaternary radiolaria. For example, 3.5 Q,E/E means 3.5 m of mixed Quaternary and Eocene age overlie Eocene strata.

or "poly" nodules (Facies A) (Meyer, 1973) of small size, variable seafloor density, low Mn/Fe ratios (ave. 2.4), and low Cu + Ni contents (ave. 1.4%) surrounds the ferromanganese-encrusted outcrops on Gundlach Peak. Facies C, which occurs on the abyssal hills, consists of larger "B type" discoidal nodules (Meyer, 1973) of higher Mn/Fe ratios (ave. 4.8) and Cu + Ni content (ave. 2.50%). These two populations are mixed on the central plain surrounding Gundlach Peak. Nodule coverage, mapped during four TV transects and by freefall bottom photos, varies from 10 to 50% of the sediment surface, and the highest (30 kg/m) and most variable densities (ave. 7.5 ± 7.0 kg/m) occur near Gundlach Peak (Facies A). More uniform densities of large nodules (ave. 10.3 ± 3.5 kg/m) cover the abyssal hills (Facies C). The percentage of barren areas, defined by freefall grabs that returned without nodules, decreases from Facies A (approx. 20%) to Facies C (approx. 5%). Bottom roughness created by partially buried nodules is estimated to range from 1-2 cm on the central plain to 2-3 cm on the abyssal hills (Fig. 4).

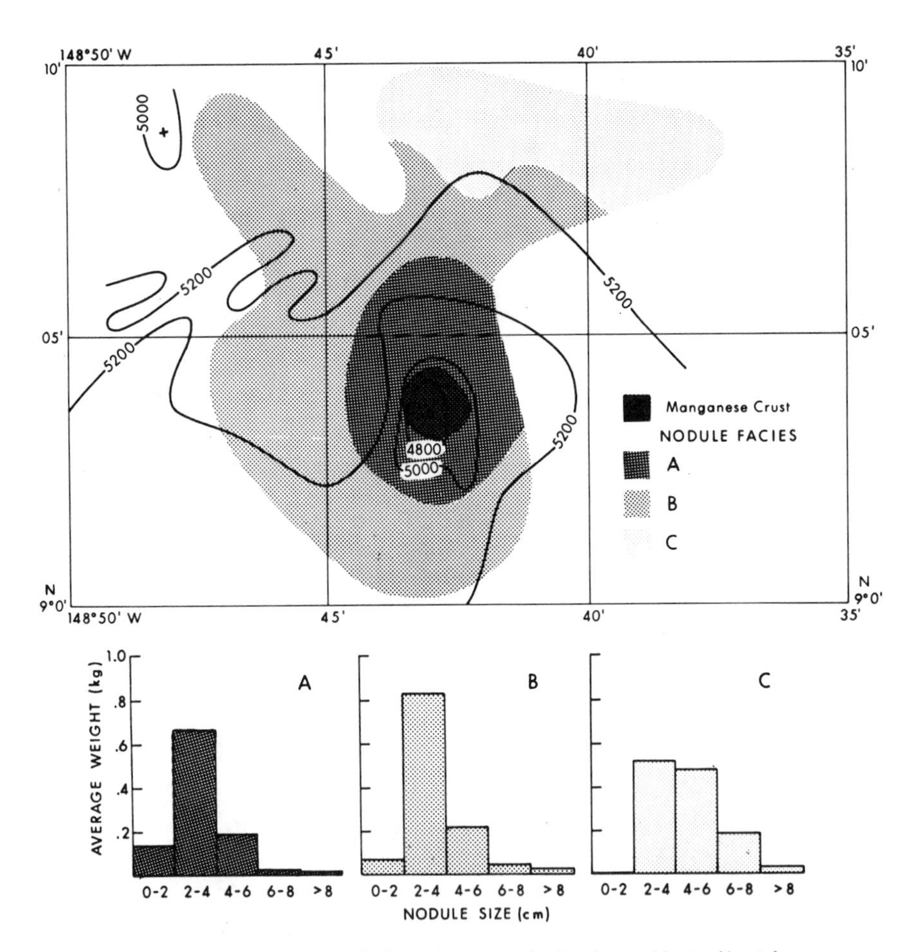

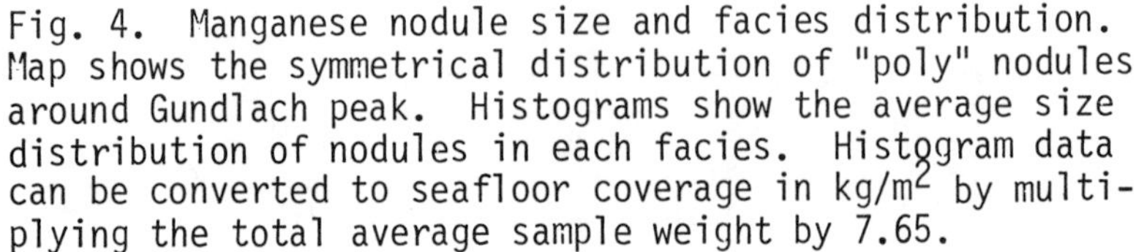

Fig. 4. Manganese nodule size and facies distribution. Map shows the symmetrical distribution of "poly" nodules around Gundlach peak. Histograms show the average size distribution of nodules in each facies. Histogram data can be converted to seafloor coverage in kg/m^2 by multiplying the total average sample weight by 7.65.

Geotechnical Properties and Techniques

Geotechnical properties which characterize the nature of the seabed interface are of primary interest to studies of sediment redistribution. Although *in situ* geotechnical tests have been made in a few deep sea settings, the majority of data are based on sediment samples collected by a variety of coring devices. Models of sedimentary processes based on these data have often been questioned due to sample disturbance during coring and variable changes in geotechnical properties during storage prior to testing. These uncertainties were minimized in the present study by the use of spade corers which collect relatively undisturbed samples of the sediment-water interface (Hagerty, 1974; Sorem, 1977), and by conducting tests of sensitive geotechnical properties 1-2 hours after retrieval of the sample.

Spade core samples used for this study were collected in a variety of locations shown in Fig. 2. Surface sediments (upper 50 cm) are similar in appearance throughout the study area, and are comparable to sediment described for DOMES Site A (Cook *et al.*, 1977). The sediment surface (0-2 cm) was typically covered by a variety of manganese nodules imbedded in a low strength/high water content layer of "soupy" consistency (Fig. 5). A homogeneous, dark yellowish brown (10 YR 4/4) layer, at depths less than 5 to 15 cm, corresponds to the intensely bioturbated "mixed layer" of Berger and Heath (1968) and Guinasso and Schink (1975). The underlying "historic layer" (Berger and Heath, 1968) has a mottled appearance of dark yellowish brown (10 YR 4/4) mass with light yellowish brown (10 YR 7/4) burrows of 1 to 2 cm in diameter. Sediment lithology, determined by smear slides, ranged from a clayey radiolarian ooze to a radiolarian-rich clay, bearing (2-10%) other siliceous fossils (sponge spicules and diatoms), micronodules and red semi-opaque globules (termed RSO in DSDP reports).

Shear Strength

Undisturbed sediment shear strength was measured before taking subcores using a Lockheed hand-held vane (2.5 x 2.5 cm vane; approx. 180°/min rotation rate) (Simpson *et al.*, 1977) and on sediment subcores (using pre-split core liner) with a Haake motorized shear vane (1 x 1 cm; 360°/min) and a Geonor fallcone penetrometer. For remolded shear strength tests using the shear vane systems, sediment was mechanically stirred in place after failure of the original structure and immediately re-tested at rotation speeds of 270°/min (Lockheed hand-held vane) and 360°/min (Haake vane). For remolded tests using the Geonor fallcone, 5 cm portions of the split subcore were stirred in a cup and re-tested using the suggested cone (10 g/60°) and Geonor calibration tables. Laboratory tests were conducted about two months following cruise VA 13/1 by Lockheed Ocean Laboratory using a motorized Diversified Marine

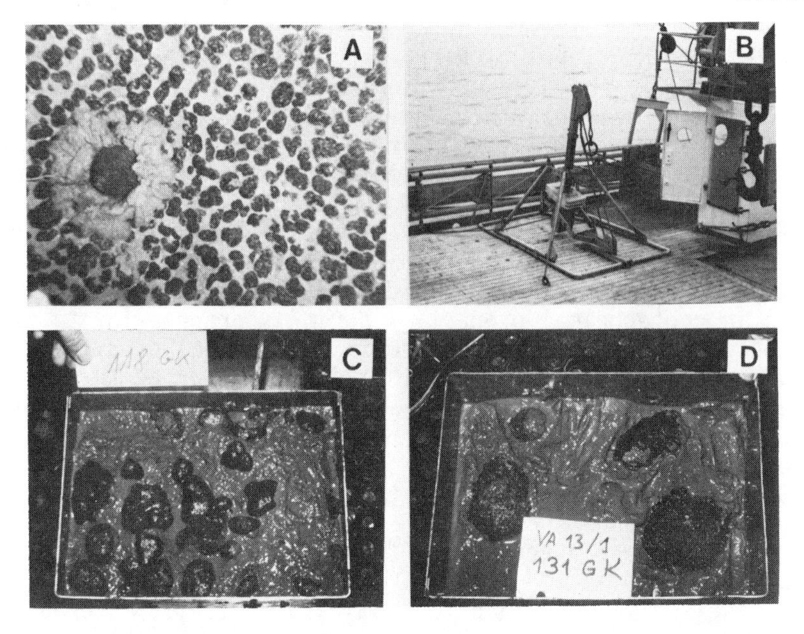

Fig. 5. Photo documentation. (A) Bottom photo taken by Preussag freefall grab/camera showing the variety of nodule sizes and types on the seabed surface. Sediment cloud produced by trigger weight (7 cm diam.) indicates degree of disturbance of the surface layer caused by freefalling sampling devices. (B) Spade corer used in collecting undisturbed samples of the seabed surface. (C) Surface of recovered spade core sample showing a variety of "poly" nodules (Facies A) partially buried in very soupy sediment. (D) Surface of spade core sample containing large "B-type" nodules (Facies C).

Corporation (DMC) vane system (1.25 x 1.25 cm, 48°/min) (Simpson et al., 1977). All shear strength tests were presumably conducted under undrained conditions due to the rapid rates of applied shear strength and saturated nature of the sediment (Monney, 1974), and shear strength is thus equivalent to cohesion (Hirst and Richards, 1975). Several investigators (see Monney, 1974; Perlow and Richard, 1977) have reported moderate differences in undisturbed shear strengths (20 to 30%) due to variations in angular shear velocities between 6°/min and 90°/min. Since both vane size and rotation rate determine the angular shear velocity, the vane rotation speeds used in this study are more appropriately reported as follows: Lockheed hand-held vane, 0.65 mm/sec; Lockheed lab-vane, 0.087 mm/sec; Haake vane, 0.52 mm/sec. As noted by one reviewer (A. Richards), the shear strength values obtained by these instruments may not be strictly comparable because of the wide range of angular velocities.

The results of shear strength tests by these techniques are shown in Fig. 6. Profiles of shear strength by all test instruments show a rapid increase in original or undisturbed shear strength (Tmax) through the mixed layer, with average values of 1.0 kPa at 2.5 cm and 5.0 kPa at 15 cm. Average Tmax values increase gradually to about 7.5 kPa between 15 and 40 cm depths. Remolded shear strengths (Trem) gradually increase from 0.25 to 1.0 kPa over the 2.5 to 40 cm depth interval, and the Trem values

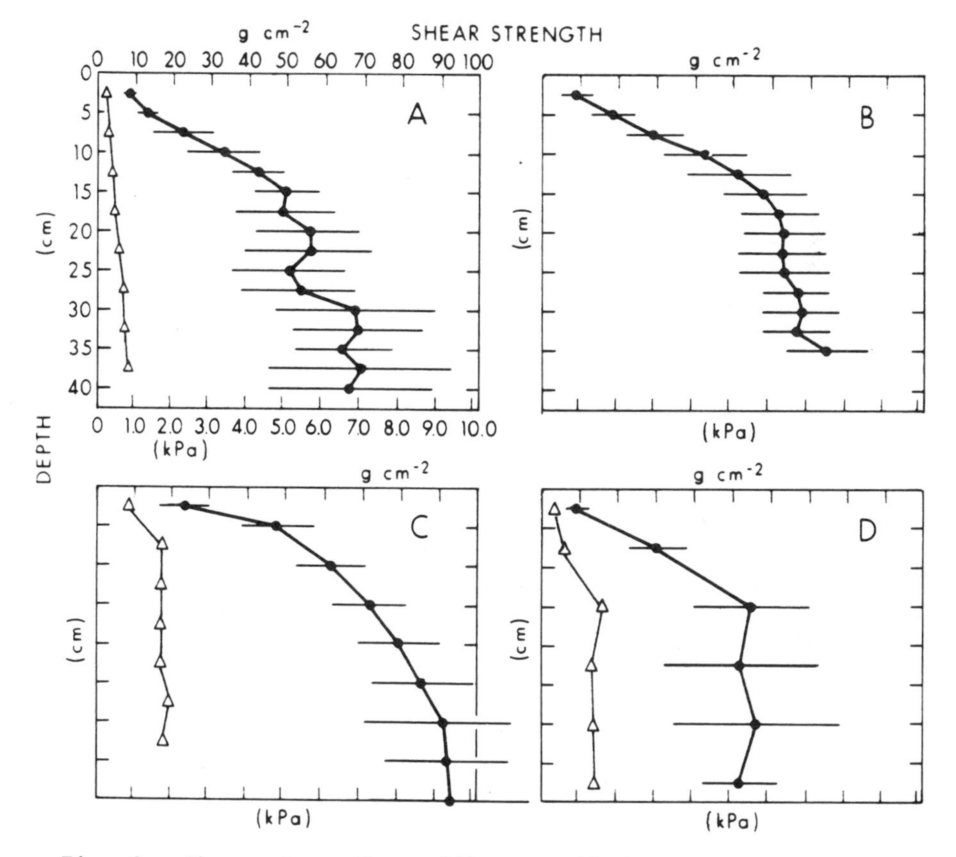

Fig. 6. Shear strength profiles compiled from spade core samples. (6a,b,d) are from tests shortly after retrieval (1-2 hr). (6c) was from tests about 2 months later in a shore laboratory. Dot profile is original or maximum shear strength (Tmax), with error bars (1σ). Triangle profile is remolded shear strength (Trem). Depth (cm) and shear strength (kPa, g/cm^2) scales given in (6a) apply to the other graphs. (6a) Geonor fallcone profile; (6b) Haake shear vane profile; (6c) Shore laboratory profile using a motorized shear vane; (6d) Lockheed hand-held shear vane profile.

measured by the Geonor fallcone (Fig. 6a) are generally lower (approx.50%) than the values measured by the shear vanes. This discrepancy is probably caused by sample preparation, that is, the fallcone samples are more thoroughly remolded prior to testing. The large standard deviations of Tmax below the mixed layer (15-40 cm depth) are largely caused by scattered burrows in which Tmax is considerably lower (50-75%) than in the surrounding sediment. This benthic burrowing accounts for much of the small-scale variability in shear strength of deep-sea cores (Hagerty, 1974; Simpson *et al*., 1977). Discrepancies in average Tmax values at various core depths are probably due to differences in angular shear velocity which was previously discussed. The shore laboratory profile (Fig. 6c) has consistently higher Tmax values and smaller standard deviations than the shipboard profiles. Shipboard profiles is generally favored over shore-lab testing because the natural structure of the sediment is not appreciably altered during spade coring operations and shipboard tests eliminate physical changes caused by dehydration, drainage, and vibration during shipment and storage. Another factor favoring shipboard testing is the noticeable strength "regain" in both burrows and the mixed layer with increasing time. Shipboard tests of strength "regain" for remolded sediments showed a rapid strength increase during the first twelve hours (approx.100-200%), followed by a gradual increase in strength to about half of the initial Tmax value in periods of 1 to 2 days. Future geotechnical studies should use the standard angular shear velocity (0.15 mm/sec) suggested by Perlow and Richards (1977) in order to eliminate uncertainties created by differences in vane size and rotation rate.

A comparison of shear strength profiles to visual core descriptions in three bathymetric settings is shown in Fig. 7. This figure illustrates the geotechnical variability of sediments over small-distance scales as well as the variation in seafloor density of manganese nodules. The coverage of manganese nodules at the seafloor surface typically varies by a factor of 2 to 3 in the three localities shown, with densities generally ranging from about 5 to 15 kg/m^2. The stations shown in Fig. 7a are clustered in the central plain area and exhibit a uniform mixed layer 7.5 to 10 cm thick. The shear strength profiles of Fig. 7a differ from profiles on a gentle abyssal hill slope (Fig. 7b) and steep abyssal hill slope (Fig. 7c) in that they have generally lower shear strengths at the base of the mixed layer (< 2.0 kPa) and show an abrupt increase in shear strength below about 10 cm. In contrast, the other profile sets show a rapid increase in shear strength beginning near the sediment surface (2.5 cm depth) and continuing to the base of the mixed layer (10-15 cm depth) where values typically exceed 4.0 kPa. Only station 139 (Fig. 7c), located on a steep slope, shows clear evidence of diagenetic recrystallization in surface sediments and contains no manganese nodules. The occurrence of a uniform, low-strength, mixed layer in the central plain stations and the lack of a similar layer on the slopes of the abyssal hill could be the result of erosion of the bioturbated layer on the slopes and redistribution of sediment into the central plain, as is also suggested by the variation of total sediment thickness shown in Fig. 3, or it could be caused by more intensive bioturbation in the locality sampled by the cores of Fig. 7a.

The profiles shown in Fig. 7 illustrate the interaction of two opposing forces, benthic bioturbation and consolidation, which create gradients of shear strength with depth. Due to very low accumulation rates in abyssal areas, most deep-sea sediments are "overconsolidated", that is, they behave as if they had been consolidated under an effective load greater than the present sediment overburden pressure. Consolidation tests conducted on siliceous ooze by Noorany (1972) indicate that a preconsolidation

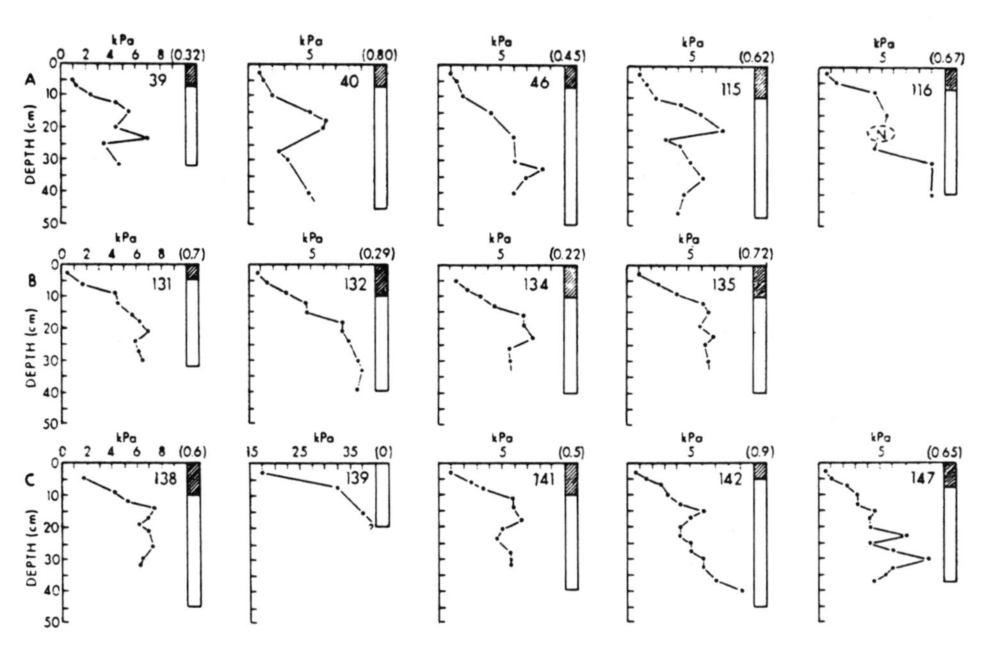

Fig. 7. Shear strength profile from three different bathymetric settings. Maximum shear strength values are given as dots along profiles. Stations numbers are given in upper right corner; locations are shown in Fig. 2. Weight (kg) of manganese nodules on core surface is given in parentheses. Visual core description is illustrated on right side of each profile, with shaded area representing the bioturbated "mixed layer". (7a) Stations clustered on the central plain. (7b) Stations located on a low-angle abyssal hill slope. (7c) Stations clustered on a steep slope adjacent to the northern valley.

pressure equivalent to several meters of overburden occurs in sediments from a depth of about 20 cm. This overconsolidated character is probably caused by the combined effects of prolonged secondary compression and some early diagenetic bonding (Noorany, 1972), and it would produce an erosion-resistant bed if deposit-feeding organisms did not modify this condition near the sediment surface. Although decreases in animal abundance and metabolism in abyssal areas suggests a decrease in the importance of bioturbation, biological mixing at rates several orders of magnitude higher than sedimentation rates (Berger and Heath, 1968; Guinasso and Schink, 1975) can provide an effective process to lower sediment strength. At present, knowledge of the behavior, vertical and horizontal distribution, and metabolism of benthic organisms is an important gap in models of abyssal sedimentary processes.

Other Index Properties

Figure 8 provides a relationship between shear strength (or cohesion) and water content. These data can be used to model sediment erosion on the basis of laboratory studies where the physical state of the erosional surface is reported in parameters of void ratio, porosity, or water content (Einsele et al., 1974; Lonsdale and Southard, 1974). In Fig. 8, the water content (w%), given in percent dry weight, is corrected for residual salt weight by assuming a pore-water salinity of 35°/₀₀. The UH subsamples, which were taken from the split subcores, individually sorted in heat-sealed plastic bags, and refrigerated, show a wider range of water contents (w% = 200-350%) than the Lockheed samples left intact in their subcores and stored in refrigerated seawater (w% = 250-350%). Although the Lockheed measurements, which were made about two months after sample collection, have generally higher Trem values than the fallcone values obtained at sea, both data sets confirm a trend of increasing Trem with decreasing water content typical of rheological materials. The range of water contents for both data sets is in agreement with previously reported values for siliceous sediments in this region (Horn et al., 1973b; Richards, 1977a), and is substantially higher (approx. 2x) than the Atterberg Liquid Limit (Fig. 8). The liquid limit (LL) is the water content marking the transition between liquid and plastic consistency states of the sediment, and above this liquid limit the deformation of remolded sediment acts as a slurry-like flow. The liquid limit test, which is analogous to a shear test, was determined using the Geonor fallcone method (Karlsson, 1961) because of its speed and accuracy compared to the ASTM technique. Liquid limits for DOMES Site A sediment (ASTM method) are slightly higher (183 ± 29; Richards, 1977b) than these results (146 ± 15; Fig. 8) due probably to differences in the definition of shear strength at the liquid limit. Karlsson (1961) found a shear strength of 3.0 kPa at the liquid limit, while other investigators have generally defined the transition between liquid and plastic

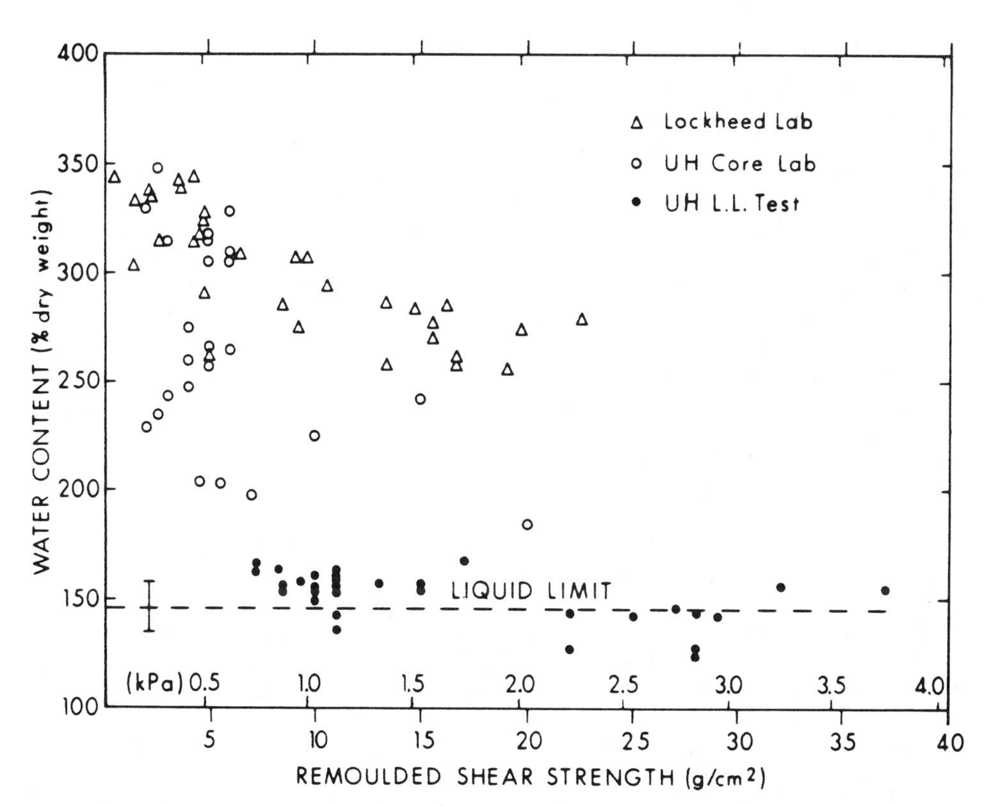

Fig. 8. Diagram of water content (% dry weight) and remolded shear strength. All water contents were determined about 1 to 2 months following cruise VA 13/1 in shore laboratories. UH data (Trem) measured at sea using a Geonor fallcone. Lockheed data (Trem) measured in shore lab using a motorized shear vane. Atterberg liquid limit determined by a fallcone method (Karlsson, 1961); error bar is 1σ.

consistency states to lie in the range of 2.0 to 3.0 kPa (see Mitchell, 1976).

Sensitivity (St) is an index property which describes the thixotropic nature of fine-grained sediment. This parameter is defined as the ratio of maximum to remolded shear strength (St = Tmax/Trem) at constant water content. The sediment studied typically loses 80 to 90% of its original shear strength upon remolding (St = 5-10), and it is described as being very sensitive to slightly quick (Rosenquist, 1953).

Grain-Size Analysis

Textural analysis has long been fundamental to studies of sedimentation processes, although the scarcity of these data for fine-grained sediments reflects the difficulty in analytical technique. The results of fine-grain analyses by standard pipette methods which use dispersive agents and distilled water are difficult to interpret relative to natural depositional behavior in flocculated states. Previous investigations of fine-grained sediment in deep-sea regions have correlated trends in grain-size parameters to segregation by settling velocity during transport (van Andel, 1973), by erosional winnowing of finer fractions by bottom currents (Huang and Watkins, 1977), and by dissolution of siliceous microfossils in the surface mixed layer (Johnson, 1974).

The "photometric centrifuge" technique (Woodruff, 1972) used in this study utilizes the principle of Stokes' Law, photo-electric measurement of transmitted light through a suspension, and centrifugal force to accelerate settling. This method eliminates both the lengthy settling period and the residual salt correction inherent in pipette analysis, and produces a continuous plot representing grain-size distribution from 62.5 to 4 μm (4-8ϕ) in less than 5 min. Grain-size analyses were conducted on archive subcores stored in refrigeration for about one year, during which time water contents decreased to 50% of their original value (Fig. 8). As for shear strength measurements, the best estimate of natural particle size distribution is probably derived from fresh samples of sediment at natural water contents using seawater as a suspensate. Samples were prepared for the photometric centrifuge by mechanically dispersing the sediment in tap water and wet sieving to separate the > 62.5 μm (4ϕ) fraction. The suspension containing the fine fraction (<4ϕ) was allowed to settle for 7 to 10 days before decanting the excess tap water and returning the samples to refrigerated storage. Three sets of photometric analyses were made for each sample at suspension concentrations of about 1.0 g/L in millipore filtered (0.4 μm) seawater at 25°C. Moment measures for three depth intervals (1-5 cm, 15-20 cm, and 35-40 cm) were calculated on the basis of ϕ16, ϕ50, and ϕ84 cumulative percentile averages for each set.

All sample intervals contained a low, but constant, >4ϕ fraction (2.5-7.0 wt%) dominated by radiolarians (75-90%), the majority of which were abraded and corroded, indicating reworking. Other biogenic siliceous debris accounted for less than 10% of the >4ϕ fraction. Micronodules and RSO's were a consistent component (2-5%) that occasionally became a dominant fraction in "Fe-rich" layers (Station 63, 35-40 cm). Trace amounts (< 2%) of glass shards, zeolites, and fish teeth were also identified.

Table 1 summarizes the results of the grain-size analyses listed in Appendix A. The mean grain size ranges from 6.15 to 7.10ϕ (14.1-7.3 μm), similar to the size range reported for DOMES Site A sediments (Sallenger _et al_., 1977) and considerably larger than the values (8-9ϕ; 4-2 μm) reported by Horn _et al_. (1974) and Richards (1977b). The effect of particle flocculation in the seawater suspensate probably accounts for the larger mean grain-size in this analysis. A characteristic decrease in both mean size

TABLE 1

Photometric centrifuge grain-size analysis

DEPTH (cm)	$\bar{X}_{M\phi}$	$\delta_{M\phi}$	$\bar{X}_{\sigma\phi}$	$\delta_{\sigma\phi}$	$\bar{X}_{sk\phi}$	$\delta_{sk\phi}$	$\bar{X}_{4\phi}$	$\delta_{4\phi}$	
0- 5	6.46	.25	.77	.11	-.38	.09	5.54	.55	
15-20	6.77	.20	.68	.11	-.29	.08	5.24	1.09	A
> 30	6.70	.23	.68	.12	-.31	.15	3.79	1.76	
0- 5	6.53	.30	.74	.15	-.29	.09	5.30	.61	
15-20	6.82	.17	.58	.17	-.30	.10	5.26	1.42	B
> 30	6.75	.20	.70	.15	-.32	.19	3.86	2.22	
0- 5	6.46	.26	.79	.05	-.39	.05	5.87	.55	
15-20	6.71	.25	.73	.06	-.33	.06	5.57	.21	C
> 30	6.69	.27	.70	.14	-.25	.11	4.30	.44	

Symbols: $\bar{X}$ (mean); δ (standard deviation); mean ($M\phi$)=ϕ16+84/2; Sorting (σ)=ϕ84-ϕ16/2; Skewness (sk)=$M\phi$-ϕ50/σ; 4ϕ = weight % greater than 62.5 μm; (A) all cores (10); (B) abyssal plain cores (39, 40, 46, 115, 116); (C) abyssal hill slope cores (141, 142, 148)

($\bar{\chi}m\phi$) and coarse fraction ($\bar{\chi}4\phi$) below the surface interval (0-5 cm) is best explained by dissolution of siliceous debris in the mixed layer (Johnson, 1974; Schink and Guinasso, 1977) since the uniformity in sorting and skewness parameters do not indicate surface erosion effects (Huang and Watkins, 1977). Both sorting coefficients ($\bar{\chi}_\sigma$ = 0.58-0.79) and skewness values ($\chi_{Sk\phi}$ = -0.25 to -0.39) are typical of moderately well-sorted, fine-grained, deep-sea sediments (Huang and Watkins, 1977). Grain-size parameters in Table 1 show no significant difference (at 90% confidence limits) in moment measures between the abyssal plain (net depositional site) and the abyssal hill slopes (net erosional sites), although compared to all depth intervals, the abyssal plain samples have a slightly finer mean size, a lower weight percent of >4ϕ fraction, and a generally larger standard deviation than the abyssal hill samples.

Several explanations can account for the observed uniformity in grain-size parameters. Surface erosion and sediment redistribution was not active during the time interval represented by these shallow cores (0.4-1.2 Myr, using accumulation rates for Site A; Theyer, 1977). In other deep-sea areas, Huang and Watkins (1977) and Johnson et al. (1977) report significant erosional activity during the lower Matuyama epoch (1.7-2.4 Myr B.P.) and uniform depositional conditions during the Brunhes epoch (0.7 Myr to present). Sediment erosion may occur as a "mass erosion" of sediment chunks over rough seabeds (Lonsdale and Southard, 1974) rather than by "surface erosion" of particles, whereby selective winnowing of the fine fraction is not involved. Localized transport mechanisms (e.g., lutite flow or seabed creep) may not allow size-selective deposition as modeled by van Andel (1973) and McCave and Swift (1976).

Discussion

Although much of the bathymetry in the study area is of low relief (< 100 m), sediment thickness varies considerably between a plain (175 m) and the slopes of the abyssal hills (100 m), the two features separated by less than 10 km. Sediment thickness is inversely correlated to bathymetric slope, that is, the thinnest sections (< 100 m) occur on steep slopes, and thick sections (175 m) occur primarily in flat, low-lying areas (Figs. 2 and 3). The variation in thickness of an acoustically stratified unit, which correlates with the Line Islands Formation, indicates that most of the sediment redistribution occurred prior to late Miocene time. This redistribution was probably associated with the regional erosional activity in Eocene/Oligocene and Middle Miocene time suggested by van Andel et al. (1975). Thicknesses of Quaternary sediment also typically vary by a factor of 2 and contain significant portions of reworked Tertiary radiolaria, which indicates continued redistribution of sediment. Grain-size analyses of surficial sediments (upper 40 cm) show, however, no clear evidence of size-sorting between net erosional and net depositional areas. Characteristics of the benthic boundary are uniform throughout the study area with respect to bottom roughness and geotechnical properties of the bioturbated "mixed layer". Since these geologic characteristics are typical of many local areas in the equatorial North Pacific, a general discussion of sediment erosion/redistribution processes is relevant to studies of manganese nodule deposits which may be affected by these processes because of their occurrence at the sediment/water interface.

Benthic Boundary Layer Relationships

On a fundamental level, sediment erosion and deposition involves the interaction of shear stress along the seabed, exerted by bottom water flow and gravitational force, and the shear strength of the seabed or weight of particles in suspension to overcome this stress. As water flows past the seabed, there is a frictional drag which causes a decrease in velocity and an increase in turbulence relative to the free stream or mean flow ($\bar{U}$). This region, termed the benthic boundary layer (BBL), varies from 5-30 m thick in the deep sea (Wimbush and Munk, 1970). Near the boundary (within approx. 1 m), the current speed decreases logarithmically from a mean flow ($\bar{U}$) to a friction velocity (U_*) at the seabed (Wimbush and Munk, 1970). The shear stress exerted on the seabed by the flow is given by the equation $\tau_0 = \rho_{sw} U_*^2$, where $\rho_{sw} = 1.028$ g/cm^3. The constant logarithmic decrease in current velocity allows an accurate estimation of friction velocity to be made from current measurements one meter above the bottom (U_{100}). Empirical "law of the wall" equations (Monin and Yaglom, 1971; Komar, 1976) or proportionality constants, termed "drag coefficients" (Sternberg, 1972), yield relationships of $U_*/U_{100} = 0.054$ (transitional to rough flow) and $U_*/U_{100} = 0.038$ (smooth flow). Hydrodynamic conditions in the BBL are described as being "smooth", "transitional", or "rough" on the basis of bottom roughness (k_s) and frictional velocity (U_*) (Komar, 1976). Under smooth flow conditions, a thin viscous sublayer (2 cm thick at $\bar{U}$ = 3 cm/sec) partially restricts turbulent eddies from reaching the seabed. Along a transitional to rough boundary, where bottom roughness exceeds the viscous sublayer thickness, turbulent eddies extend to the seabed and increase bottom shear stress. In the equatorial North Pacific region, typical bottom roughness of 2 to 3 cm created by partially buried manganese nodules causes transitional flow conditions at bottom current speeds of $\bar{U}$ = 0.7-10.2 cm/sec.

Seafloor Erosion

Sediment erosion can be divided into general categories of

erosion by bottom currents, and downslope mass movement due to gravity. These processes are probably confined to the sediment surface because of the rapid increase in shear strength (τ_y = cohesion) through the mixed layer.

Laboratory studies have greatly contributed to the evaluation of critical erosion velocity (U_{*crit}) relative to geotechnical properties of sediment surfaces. Einsele et al. (1974), working with freshly deposited illite mud in a seawater flume, derived the relationship ($\tau_{ocrit} \simeq \tau_y \times 10^{-2}$), where erosion of a smooth sediment surface (w% = 400%, τ_y = 0.04 kPa) occurred at U_{*crit} = 1.94 ($\bar{U}$ = 51 cm/sec). Lonsdale and Southard (1974), working with redeposited North Pacific brown clay, found a similar range of erosion velocities of U_{100} = 53-43 cm/sec (w% = 362-411%) for smooth beds and velocities of U_{100} = 48.2-32.2 cm/sec (w% = 364-423%) for erosion of a rough bed containing manganese nodules (k_s = 2 cm). At higher water contents (w% > 600%), Lonsdale and Southard (1974) reported erosion of a rough bed at velocities as low as U_{100} = 12.3 cm/sec. The correlation between water content and U_{*crit} was also recognized by other investigators (Einstein and Krone, 1962; Gust and Walger, 1976) who found U_{*crit} of 0.7 to 1.1 cm/sec (U_{100} = 18.4-28.9 cm/sec) at w% ≃ 600% for smooth beds.

Comparison of these results to seafloor erosion in the study area is difficult because of the lack of geotechnical measurements at the sediment/water interface (0-2 cm), and the scarcity of long-term bottom current measurements in this region. Using the data in Fig. 8 for geotechnical properties at 2.5 cm depth (τ_y = 0.1 kPa, w% = 350%) and these laboratory flume results, we derive an erosional velocity of $\bar{U}$ = 57.7 cm/sec. However, assuming that water contents are higher at the sediment/water interface (w% = 400-600%), mean current velocities of $\bar{U}$ = 18.5-37.0 cm/sec may be sufficient to initiate erosion over rough bottoms. These current speeds are higher than values measured in most deep-sea regions (Heezen and Hollister, 1964; Lonsdale and Spiess, 1977). Bottom current measurements in the equatorial North Pacific region by Johnson (1972), Amos et al. (1977), and Hayes (1977) defined a variable direction mean flow of 2 to 3 cm/sec with short-term fluctuations in the range of 10 to 25 cm/sec related to tidal frequencies. The spacial/temporal extent of so-called "benthic storms", which may produce current velocities in the erosional range, are undocumented at present.

Downslope mass movement has generally been discounted due to the stable, low-relief character of most abyssal regions; however, the possibility of surface creep has often been mentioned in the literature. Mass movement (slumping) involving the slow progressive failure of thick sediment sections under "drained conditions" is not favored where the slope angle (θ) is less than the angle of internal friction (ϕ) (Taylor, 1958). Slumping is therefore not probable on abyssal hill slopes where $\theta < 3°$ and $\phi = 33°$ (Noorany, 1972). However, under "undrained conditions" internal strength is independent of ϕ and can be equated to τ_y. If we use the formula for downslope shear stress exerted on a homogeneous layer on an infinite slope ($\tau_g = \rho_{sat} z \sin \theta$; Taylor, 1958) and assumed properties of the sediment surface ($z = 2$ cm, $\rho_{sat} = 1.1$ g/cm^3, $\tau_y = 0.04$-0.1 kPa), we derive critical slope angle of $\theta = 10.5$ to $27°$. This critical slope range is similar to the results of slide experiments by Einsele et al. (1974) who found $\theta_{crit} = 17.3$-$22.0°$ for illite mud (w% = 280-485%). The erosional mechanism of "surface creep" would therefore only be possible on anomalously steep abyssal hill slopes or on seamounts (Stanley and Taylor, 1977).

Deposition

Deposition of sediment particles occurs when the settling force, related to particle weight and settling velocity, overcomes the shear stress exerted by currents along the seabed (τ_0). Since flocculation dynamics which control settling velocity are difficult to monitor in natural settings, many depositional models use a limiting shear stress parameter (τ_1) to define depositional conditions, where $\tau_0 < \tau_1$. Experimental deposition of clay suspensions in seawater yields values of $\tau_1 < 0.06$ Pa, corresponding to $U_* = 0.76$ cm/sec and $U_{100} = 19$ cm/sec in smooth flows (Einstein and Krone, 1962). Calculations by McCave and Swift (1976), which include the effect of decreasing wet-bulk density with increasing floc size, yield values of $\tau_1 = 0.01$-0.18 Pa (8-6 ϕ), corresponding to $U_* = 0.31$-0.41 cm/sec and $U_{100} = 7.8$-10.5 cm/sec. Without comparing the relative merits of these results, it can be concluded that bottom currents higher than typical mean flow velocities in this abyssal region ($\bar{U} = 2$-3 cm/sec) would be required to preclude deposition of fine-grained suspended sediment.

Equations relating settling parameters to depositional rates for dilute suspensions (< 300 mg/l) are given by Krone (1976), McCave and Swift (1976) and Owen (1977). Depositional models commonly eliminate the problem of time-varying sediment concentration (C_0), caused by vertical turbulent mixing, by assuming an entrapment of suspended particles in the viscous sublayer (McCave, 1970; McCave and Swift, 1976). These models cannot be strictly applied to the study area where the viscous sublayer is theoretically absent under prevailing transitional flow conditions. Problems are encountered in modeling the deposition of dense suspensions (0.3 g/l $< C_0 < 170$ g/l) where turbulent drag reduction caused by suspended cohesive sediment modifies the BBL velocity structure (Gust and Walger, 1976). In this poorly documented realm of "hindered settling" or "two phase flow", interfloc forces become dominant and cause the suspension to behave like a gel. Deposition of these "fluid mud" layers takes place in short periods of time (2-10 hr) even at high current velocities (approx 100 cm/sec) (Einstein and Krone, 1962).

Transport Processes

Sediment transport is difficult to observe in natural settings or study in laboratory flumes, and therefore is poorly understood at present. Since, however, sediment redistribution from bathymetric highs to low-lying depositional sites is a common characteristic of this abyssal region, and both net erosional and depositional settings are similar with respect to hydrodynamic conditions, gravity is suspected as the major driving force. As previously discussed, simple depositional equations are difficult to apply to sedimentary processes involving dense suspensions and "two phase flow" characteristics. However, models which treat these suspensions as low-density turbidity flows, combining parameters of slope angle, gravity, particle size, and suspension concentration, may provide the best explanation for the inverse correlation of sediment thickness to bathymetric slope. A turbid or "lutite" flow will move downslope, driven by gravity, until the so-called "autosuspension limit" is reached. Sediment suspended in the lutite flow would then be deposited rapidly in the low-lying site.

Observations of sediment redistribution by turbid layer transport can be compared to the sedimentation and grain-size parameters in the study area. Komar *et al.* (1974) reported the movement of a 2 m-thick turbid layer at speeds of 3 to 5 cm/sec on the Oregon shelf. The turbid layer flowed down a slope of θ = 0.13° driven by an excess density of suspended sediments of 15 to 40 g/L. A turbid or "fluid mud" layer with properties similar to those investigated by Einstein and Krone (1962) (τ_y = 0.1-0.5 Pa, C_0 = 20-100 g/l) is theoretically unstable on slopes greater than 0.05 to 0.26° (using $\tau_g = \rho_{sat}\, z \sin\theta$; where ρ = 1.1 g/cm^3, z = 100 cm). Fluid mud suspensions in this excess density range could be produced by eroding the seabed to a depth of 2 cm and mixing the eroded sediment to heights of 1.0 m in the BBL.

Conclusion

In the study area, the variation of sediment thickness indicates that sediment redistribution was most active prior to late Miocene time, and this process generally resulted in the net transport of sediment from the slopes of abyssal hills to a low-lying plain. Quaternary surface sediment has uniform geotechnical properties and grain-size distribution throughout this locality, despite stratigraphic evidence that sediment redistribution continues. The morphology, chemistry, and seafloor density of manganese nodules lying on this Quaternary surface are apparently unaffected by continued sediment redistribution, and nodule facies are primarily influenced by the proximity of a small seamount of unknown age.

A speculative description of sediment erosion/redistribution is based on the geotechnical properties of sediment and published laboratory studies. Sediment erosion by bottom currents is suggested because downslope slumping is not probable on the low-angle slopes covering most of this locality. On a rough seabed containing manganese nodules and which is actively bioturbated, bottom current speeds approximately one order of magnitude higher than normal mean current speed would be required to initiate erosion. Erosional episodes would then probably be infrequent, associated with major global climatic change or postulated "benthic storms", and they would rapidly erode portions of the uppermost bioturbated "mixed layer". Rapid erosion during these episodes would create dense suspensions of fine-grained sediment which might effectively dampen turbulence in the benthic boundary layer. "Two phase flow" conditions would promote transport of sediment down slopes steeper than 1/500 by low-velocity lutite flow. Rapid redeposition of suspended material in flat, low-lying areas would preclude any effective grain-size sorting. Manganese nodules would not be displaced horizontally during these episodes, although nodules in depositional areas would probably be covered by a thin blanket of sediment, as would benthic organisms. Nodules lying under redeposited sediment might return to the sediment surface by a combination of biological activity and scouring by normal bottom currents.

Evidence of sediment erosion/redistribution on the equatorial North Pacific seafloor is widely reported; however, models which describe this process are scarce because of the lack of observations in the benthic boundary layer. Fundamental data such as geotechnical properties of the sediment/water interface, measurements of both fine-scale turbulence and long-term bottom current characteristics, distribution and bioturbation rates of benthic organisms, and detailed stratigraphy, would both significantly improve our knowledge of the deep-sea environment and perhaps explain the enigmatic variability of manganese nodules.

Acknowledgments

This study was supported by NSF Grant IDO-12953 as part of the Inter-University Ferromanganese Research Program, James E. Andrews, principal investigator. I appreciate the opportunity for participation in research activities of R/V VALDIVIA supported by the Federal Republic of Germany, and sincerely thank H. Beiersdorf (chief scientist) and Captain Ehle for their invaluable assistance during cruise VA 13/1. I also wish to acknowledge the cooperative exchange of ideas, geotechnical data, and testing instruments by the Clausthal research group (P. Halbach, M. Özkara, and E. Rehm) and by Lockheed Ocean Lab (F. Simpson) which helped greatly in this study. I also thank Stanley Margolis, David Hurd, Maury Meylan, James Gardner, Adrian Richards, and James Bischoff for their crittical review of the manuscript. I am sincerely grateful to Lora Hingston, Rita Pujalet and Bonnie Kempner for typing, editing, and layout of this manuscript.

References

Amos, A.F., O.A. Roels, C. Garside, T.C. Malone, and A.Z. Paul, (1977) Environmental aspects of nodule mining, In: Glasby, G.P., ed., Marine Manganese Deposits, Elsevier, New York, 391.

Andrews, J.E., (1971) Abyssal hills as evidence of transcurrent faulting on North Pacific Fracture Zones, Geol. Soc. Am. Bull. 82, 465.

Andrews, J.E., et al., (1974) Ferromanganese Deposits on the Ocean Floor, Cruise Report Mn 74-01, R/V MOANA WAVE, Haw. Inst. Geophys. Publ. HIG-74-9.

Beiersdorf, H., and R. Wolfart, (1974) Sedimentological-biostratigraphical investigations on sediments from the Central Pacific Ocean, Meerestechnik 5, 192.

Beiersdorf, H., et al., (1976) Manganknollen-wissenschaftsfahrt VA 13/1, Zentralen Pazifischer Ozean, Bundesanstalt für Geowissenschaften und Rohstoffe, Hannover, Fed. Rep. Germany.

Berger, W.H., and G.R. Heath, (1968) Vertical mixing in pelagic sediments, J. Mar. Res. 26, 134.

Cook, H.E., D.Z. Piper, and J.V. Gardner, (1977) Geologic and oceanographic framework of DOMES Sites A, B, and C: Central Equatorial Pacific, In: Piper, D.Z., ed., Deep Ocean Environmental Study: Geology and Geochemistry of DOMES Sites A, B, and C, Equatorial North Pacific, USGS Open-File Rept. 77-778, 15, Menlo Park.

Craig, J.D., (in press) The relationship between bathymetry and ferromanganese deposits in the North Equatorial Pacific, Mar. Geol.

Dürbaum, H.J., and H.V. Schlüter, (1974) Possibilities of reflection seismics for the exploration of manganese nodules, Meerestechnik 5, 188.

Einsele, G., R. Overbeck, H.V. Schwartz, and G. Unsöld, (1974) Mass physical properties, sliding, and erodibility of experimentally deposited and differentially consolidated clayey muds, Sedimentol. 21, 339.

Einstein, H.A., and R.B. Krone, (1962) Experiments to determine modes of cohesive sediment transport in salt water, J. Geophys. Res. 67, 1451.

Friedrich, G., and W. Plüger, (1974) The distribution of manganese, iron, cobalt, nickel, copper and zinc in manganese nodules from different fields, Meerestechnik 5, 203.

Guinasso, N.L., and D.R. Schink, (1975) Quantitative estimates of biological mixing rates in abyssal sediments, J. Geophys. Res. 80, 2032.

Gust, G., and E. Walger, (1976) The influence of suspended cohesive sediments on boundary-layer structure and erosive activity of turbulent seawater flow, Mar. Geol. 22, 189.

Hagerty, R., (1974) Usefulness of spade cores for geotechnical studies and some results from the northeast Pacific, In: Inderbitzen, A.L., ed., Deep Sea Sediments: Mechanical and Physical Properties, Plenum Press, New York, 169.

Hayes, S.P., (1977) Study of the benthic boundary layer in the North Equatorial Pacific, Off. Tech. Conf., 387.

Heezen, B.C., and C.D. Hollister, (1964) Deep-sea current evidence from abyssal sediments, Mar. Geol. 2, 141.

Herron, E.M., (1972) Seafloor spreading and the Cenozoic history of the East Central Pacific, Geol. Soc. Am. Bull. 83, 1671.

Hirst, T.J., and A.F. Richards, (1975) Analysis of deep-sea nodule mining - seafloor interaction, Off. Tech. Conf., paper no. OTC 2241, Dallas.

Horn, D.R., B.M. Horn, and M.N. Delach, (1973a) Factors Which Control the Distribution of Ferromanganese Deposits and Proposed Research Vessel's Track, North Pacific, IDOE-NSF, Tech. Rept. 8, NSF GX 33616.

Horn, D.R., B.M. Horn, and M.N. Delach, (1973b) Copper and nickel content of ocean ferromanganese deposits and their relation to properties of the substrate, In: Morgenstein, M., ed., The Origin and Distribution of Manganese Nodules in the Pacific and Prospects for Exploration, Haw. Inst. Geophys., Honolulu, 77.

Horn, D.R., B.M. Horn, and M.N. Delach, (1974) Physical properties of sedimentary provinces, North Pacific and North Atlantic Oceans, In: Inderbitzen, A.L., ed., Deep Sea Sediments: Mechanical and Physical Properties, Plenum Press, New York, 417.

Huang, T.C., and N.D. Watkins, (1977) Contrasts between the Bruhnes and Matuyama sedimentary records of bottom water activity in the South Pacific, Mar. Geol. 23, 113.

Johnson, D.A., and T.C. Johnson, (1970) Sediment redistribution by bottom currents in the Central Pacific, Deep Sea Res. 17, 157.

Johnson, D.A., (1972) Ocean floor erosion in the equatorial Pacific, Geol. Soc. Am. Bull. 83, 3121.

Johnson, D.A., (1974) Dissolution of siliceous microfossils in surface sediments of the eastern tropical Pacific, Deep Sea Res. 21, 841.

Johnson, D.A., M. Ledbetter, and L.H. Burkle, (1977) Vema Channel paleoceanography: Pleistocene dissolution cycles and episodic bottom water flow, Mar. Geol. 23, 1.

Karlsson, R., (1961) Suggested improvements in the Liquid Limit Test with references to flow properties of remoulded clays, 5th Internat. Cong. Soil Mech. and Found. Eng., Paris, 171.

Komar, P.D., L.D. Kulm, and J.C. Harlett, (1974) Observations and analysis of bottom turbid layers on the Oregon continental shelf, J. Geol. 82, 104.

Komar, P.D., (1976) Boundary layer flow under steady unidirectional currents, In: Stanley, D.J. and J.P. Swift, eds., Marine Sediment Transport and Environmental Management, Wiley, New York, 91.

Krone, R.B., (1976) Engineering interest in the benthic boundary layer, In: McCave, I.N., ed., The Benthic Boundary Layer, Plenum Press, New York, 143.

Lonsdale, P., and J.B. Southard, (1974) Experimental erosion of North Pacific Red Clay, Mar. Geol. 17, M51.

Lonsdale, P., and F.N. Speiss, (1977) Abyssal bedforms explored with a deeply-towed instrument package, Mar. Geol. 23, 57.

Luyendyk, B.P., (1970) Origin and history of abyssal hills in the northeastern Pacific Ocean, Geol. Soc. Am. Bull. 81, 2237.

McCave, I.N., (1970) Deposition of fine-grained suspended sediments from tidal currents, J. Geophys. Res. 75, 4151.

McCave, I.N., and S.A. Swift, (1976) A physical model for the rate of deposition of fine-grained sediments in the deep sea, Geol. Soc. Am. Bull. 87, 541.

Meyer, K., (1973) Surface sediment and manganese nodule facies encountered on R/V VALDIVIA cruises 1972/73, Meerestechnik 4, 196.

Mitchell, J.K. (1976) Fundamentals of Soil Behavior, Wiley, New York.

Mizuno, A., and J. Chujo, eds., (1975) Deep Sea Mineral Resources Investigation in the Eastern Central Pacific Basin, August-October 1974 (GH74-5 Cruise), Geol. Surv. Japan Cruise Rept. 4.

Monin, A.S., and A.M. Yaglom, (1971) Statistical Fluid Mechanics: Mechanics of Turbulence, M.I.T. Press, Cambridge, Massachusetts.

Monney, N.T., (1974) An analysis of the vane shear test at varying rates of shear, In: Inderbitzen, A.L., ed., Deep Sea Sediments: Mechanical and Physical Properties, Plenum Press, New York, 151.

Moore, T.C., (1970) Abyssal hills in the central equatorial Pacific: sedimentation and stratigraphy, Deep-Sea Res. 17, 573.

Mudie, J.D., J. A. Grow, and J.S. Bessey, (1972) A near-bottom survey of lineated abyssal hills in the equatorial Pacific, Mar. Geophys. Res. 1, 397.

Noorany, I., (1972) Engineering properties of submarine clays from the Pacific, Proceed. First Internat. Conf. on Port and Ocean Eng. under Arctic Cond., Trondheim, Norway, 255.

Owen, M.W., (1977) Problems in the modeling of transport, erosion, and deposition of cohesive sediments, In: Goldberg, E.D. et al., eds., The Sea, Vol. 6, Wiley, New York, 515.

Perlow, M. Jr., and A.F. Richards, (1977) Influence of shear velocity on vane stress strength, J. Geotech. Eng. Div. GTI, 19.

Piper, D.Z., W. Cannon, and K. Leong, (1977) Composition and abundance of ferromanganese nodules at DOMES Sites A, B, and C: relationship with bathymetry and stratigraphy, In: Piper, D.Z., ed., Deep Ocean Environmental Study: Geology and Geochemistry of DOMES Sites A, B, and C, Equatorial North Pacific, USGS Open-File Rept. 77-778, Menlo Park, 217.

Richards, A.F., (1977a) Marine geotechnical data compilation, presentation and synthesis for the pelagic nodule area, Central Northeast Pacific Ocean, Final Report USGS Grant 14-08-001-G-258, Marine Geotechnical Lab, Lehigh Univ.

Richards, A.F., (1977b) Geotechnical testing of U.S. Geological Survey cores collected from the pelagic nodule area, Central Northeast Pacific Ocean, In: Piper, D.Z., ed., Deep Ocean Environmental Study: Geology and Geochemistry of DOMES Sites A, B, and C, Equatorial North Pacific, USGS Open-File Rept. 77-778, Menlo Park, 267.

Rosenquist, I.Th., (1953) Considerations on the sensitivity of Norwegian quick clays, Geotechnique 3, 195.

Sallenger, A.H., D.Z. Piper, and M.L. Ibsen, (1977) Setting velocities of sediment in seawater for DOMES Sites A, B, and C, In: Piper, D.Z., ed., Deep Ocean Environmental Study: Geology and Geochemistry of DOMES Sites A, B, and C, Equatorial North Pacific, USGS Open-File Rept. 77-778, Menlo Park, 518.

Schink, D.R., and Guinasso, N.L., (1977) Effects of bioturbation on sediment-seawater interaction, Mar. Geol. 23, 133.

Sclater, J.G., R.N. Anderson, and M.L. Bell, (1971) Elevation of ridges and evolution of the central eastern Pacific, J. Geophys. Res. 76, 7888.

Simpson, F., C.S. Wallin, H.M. Crawford, and P.C. Springer, (1977) Geotechnical properties of NE Pacific sediments from DOMES Sites A, B, and C, In: Piper, D.Z., ed., Deep Ocean Environmental Study: Geology and Geochemistry of DOMES Sites A, B, and C, Equatorial North Pacific, USGS Open-File Rept. 77-788, Menlo Park, 351.

Sorem, R.K., (1977) Shipboard documentation of samples and preliminary results of manganese nodule textured studies: cruise RP-8-OC-75, In: Piper, D.Z., ed., Deep Ocean Environmental Study: Geology and Geochemistry of DOMES Sites A, B, and C, Equatorial North Pacific, USGS Open-File Rept. 77-788, Menlo Park, 82.

Stanley, D.J., and P.T. Taylor, (1977) Sediment transport down a seamount flank by a combined current and gravity process, Mar. Geol. 23, 78.

Sternberg, R.W., (1972) Predicting initial motion and bedload transport of sediment particles in the shallow marine environment, In: Swift, D.J.P., D.B. Cuane, and O.H. Pilkey, eds., Shelf Sediment Transport, Dowden, Hutchinson and Ross, Stroudsburg, PA, 61.

Taylor, D.W., (1958) Fundamentals of Soil Mechanics, Wiley, New York.

Theyer, F., (1977) Micropaleontological dating of DOMES project box cores from test areas A and B, Tropical Pacific, In: Piper, D.Z., ed., Deep Ocean Environmental Study: Geology and Geochemistry of DOMES Sites A, B, and C, Equatorial North Pacific, USGS Open-File Rept. 77-778, Menlo Park, 179.

van Andel, Tj.H. et al., (1973) Initial Reports of the Deep Sea Drilling Project, Vol. 16, U.S. Gov't. Print. Off., Washington.

van Andel, Tj.H., (1973) Texture and dispersal of sediments in the Panama Basin, J. Geol. 81, 434.

van Andel, Tj.H., and G.R. Heath, (1973) Geological results, central equatorial Pacific - West East Pacific Rise, In: van Andel, Tj. et al., Initial Reports of the Deep Sea Drilling Project, Vol. 16, U.S. Gov't. Print. Off., Washington, D.C., 937.

van Andel, Tj.H., G.R. Heath, and T.C. Moore Jr., (1975) Cenozoic History and Paleoceanography of the Central Equatorial Pacific Ocean, Geol. Soc. Am. Mem. 143.

Wimbush, M., and W. Munk, (1970) The benthic boundary layer, In: Maxwell, A.E., ed., The Sea, Vol. 4 (1), Wiley, New York, 731.

Woodruff, J.L., (1972) A photometric centrifuge for rapid size analysis of fine sediment, Unpubl. M.S. thesis, Univ. of Hawaii, Honolulu.

SEDIMENTATION, HIATUSES, AND DEVELOPMENT OF MANGANESE NODULES: VALDIVIA SITE VA-13/2, NORTHERN CENTRAL PACIFIC

Ulrich von Stackelberg

Bundesanstalt für Geowissenschaften und Rohstoffe

Hannover, Federal Republic of Germany

Abstract

An area 20 x 20 nautical miles, within the abyssal hill region of the northern Central Pacific, has been sampled intensively by R/V VALDIVIA. The sediments represent four facies characterized by variations in color, structure and composition. A marked unconformity between facies II and III was dated by radiolaria and coccoliths as early Miocene to late Pliocene and is explained by an intensification of AABW circulation. We note a maximum of reworking, dissolution of siliceous organisms and growth of authigenic minerals above and below this unconformity. Dissolution of siliceous organisms seems to control the growth of minerals such as zeolites and Fe-Mn-oxides. Fe-Mn-oxides were precipitated both as micro and macro nodules. While the micro nodules remained in place, most of the macronodules were pushed upwards by benthic organisms and were kept at the sediment surface. Thus, the surface nodules appear to be "foreigners" in the recent sediments and represent the conditions of a past environment.

Introduction

The investigations of R/V VALDIVIA cruise VA-13/2 from February 9 to March 10, 1976 concentrated on an area 20 x 20 nautical miles (centered at 9°25'N, 146°00'W; Fig. 1). This area is situated within the manganese nodule belt between the Clarion and Clipperton fracture zones at the northern edge of the equatorial sedimentary lense.

The aim of the present study is to reconstruct the sedimentary processes from early Miocene to Recent, based on sedimentological

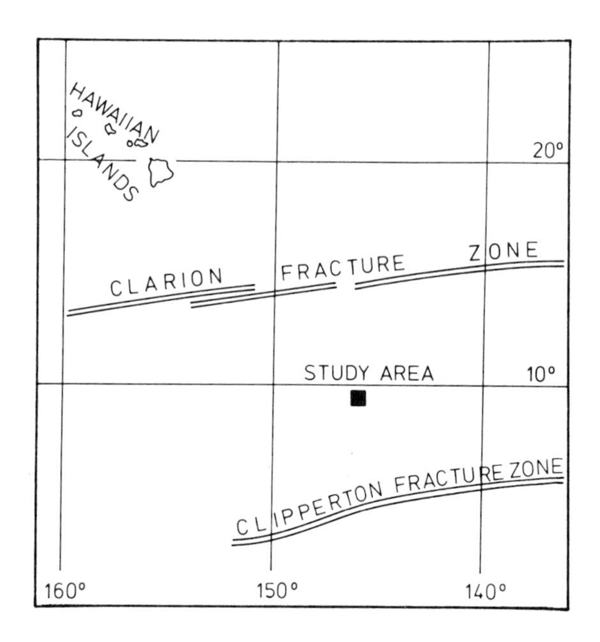

Fig. 1. Location of study area, Site VA-13/2

analyses of core materials. The information should contribute to solving the problem of the genesis of manganese nodules.

The R/V VALDIVIA visited the area on earlier cruise VA-08 in 1974. The results of this cruise (Beiersdorf and Wolfart, 1974; Dürbaum and Schlüter, 1974; Friedrich and Plüger, 1974) were fundamental to the investigations of the cruise VA-13/2.

Morphology and Oceanography

The study area is characterized by abyssal hill topography. It lies between 5400 and 4740 m water depth (Fig. 2), with three submarine hills (N-, W- and E-hill), with predominantly N-S orientation. The morphology is determined by volcanism and fault tectonics. TV-recordings revealed that the morphology is particularly rough around the submarine hills. Steps with tens of meters relief have been observed here. Sediments of the equatorial zone of the Central Pacific Ocean are strongly influenced by high productivity of the surface waters. This productivity has an effect on the study area which lies on the northern margin of this zone. Another factor influencing sedimentation is Antarctic Bottom Water (AABW). Following the outline map of Gordon and Gerard (1970) AABW flows northeastward through the Line Island chain, in a broad front over the study area. This results in an increase of the

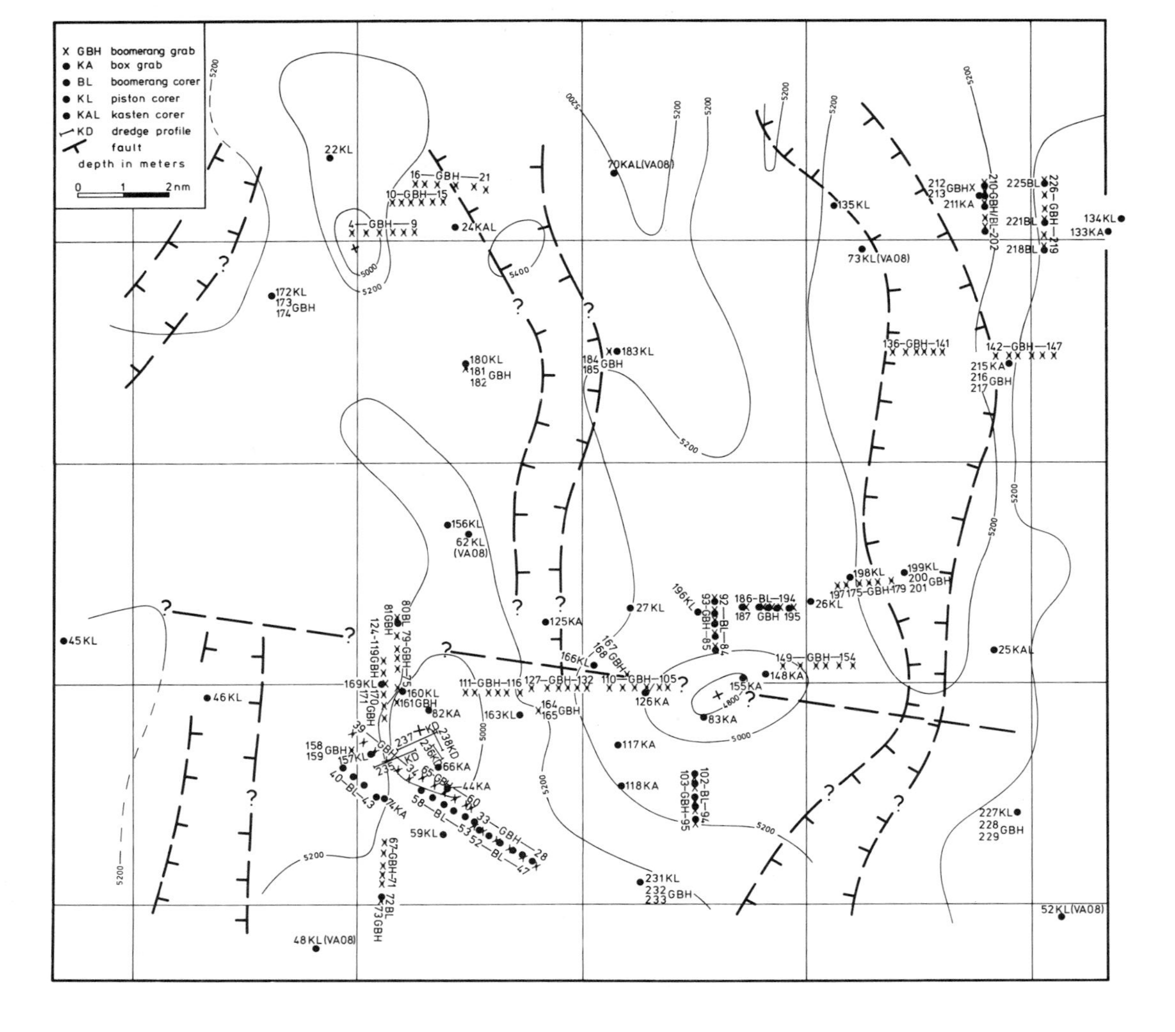

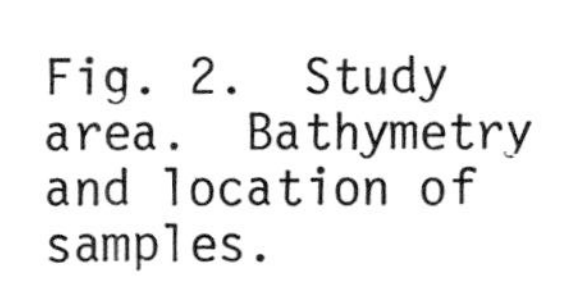
Fig. 2. Study area. Bathymetry and location of samples.

current velocities and contents of oxygen and in a lowering of the temperatures (Ryan and Heezen, 1976). Bottom current measurements are not available for the study area. Measurements from neighboring areas, however, have yielded values which normally lie below 10 cm/sec (Johnson, 1972; Baker and Feely, 1978). Johnson (1972) concludes that erosion and sediment reworking are not taking place at present. During former periods, however, the intensity of AABW may have been much greater.

Methods of Investigation

Narrow beam echosounder records from 282 nautical miles of track lines were evaluated for the bottom topography of the study area. Video tape records of 44 nautical miles of deeptowed TV were studied.

From 60 locations 79 cores were taken (total core length 255 m). They include one kasten core (KAL; 575 cm core length), 20 piston cores (KL; 198 m cores, 14 m max. core length), 20 gravity cores (SL) associated with piston cores (15 m cores, 106 cm max. core length, and 38 boomerang cores (BL; 37 m cores, 105 cm max. core length). Only kasten and piston cores will be described in detail. The data of five cores (48-, 52-, 62-, 73-KL and 70-KAL) from cruise VA-08 (Beiersdorf and Wolfart, 1974) are also included.

The sediment colors were determined according to the Munsell soil Color Chart. To observe fine scale sedimentary structures, 252 radiographs (1:1 scale) were made from 5 mm thick slices of sediment. These radiographs represent altogether 50.4 m core length.

Two-cm thick samples were taken for sedimentological and biostratigraphical studies at intervals of 50 cm or less. Two hundred and ninety one samples were wet sieved into size fractions of < 32 μ, 32-63 μ, and > 63 μ. The composition of the two coarser fractions was estimated by microscope. Up to 15 single components were determined which, combined in 8 groups, are presented below. In the following, only the > 63μ fraction will be dealt with because the composition of the two coarser fractions is very similar and the quality of estimation declines with decreasing grain size. Sixty smear slides of the total samples were examined to study the finer size fraction.

Sediments

Stratigraphy

Based on lithologic investigations the sediments can be divided into four facies: Facies I is a light-brown siliceous mud containing varying quantities of manganese nodules. This facies is mainly near the surface. Facies II is a dark-brown zeolitic mud with a high content of micro nodules. Facies III is a reddish-brown zeolitic mud with similar content of micro nodules. Facies IV shows alternation in colors between light-brown and reddish-brown. The reddish-brown parts resemble facies type III. The light-brown sediments consist of radiolarian-bearing nanno ooze.

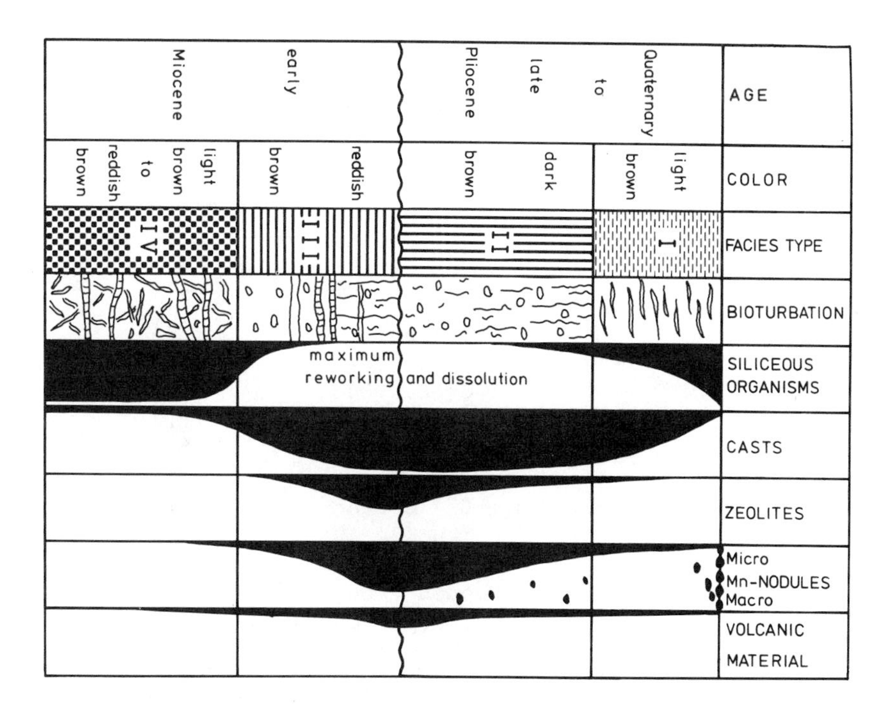

Fig. 3. Schematic core diagram (for explanation see Fig. 9)

The sediments of facies I to III sporadically contain nannofossils of Cretaceous and Tertiary age which are probably redeposited (Tables 1 and 2). Two samples from facies IV yielded a rich nannoplankton flora of early Miocene age and is considered to be mainly autochthonous. Foraminifera are extremely rare and of little stratigraphic value. Wolfart (1977) describes that in surface sediment radiolaria of late Pliocene to Quaternary age are predominant; in facies type IV they are mainly of late Oligocene to early Miocene age. Admixing of reworked late Eocene to early Oligocene radiolaria is observed in almost all samples, particularly in facies II and III. Pliocene-Quaternary radiolaria are present from the top of the cores down to varying depths in facies I and II. In core 169-KL, Pliocene-Quaternary radiolaria were encountered down to 502 cm, whereas in cores from neighboring areas they do not extend as deeply (157-KL, 337 cm; 46-KL, 113 cm).

It is concluded that facies I and II are Pliocene-Quaternary,

TABLE 1

Nanno-fossils

core	cm	nanno fossils	facies
134-KL	750	none	II
156-KL	750	1 Cretac.Cocc.	II
157-KL	550	none	I
157-KL	820	1 Cretac.Cocc.	II
169-KL	0-2	none	I
169-KL	850	3 Cretac.Cocc.	II
183-KL	40	none	I
"	400	none	II
"	650	1 Cretac.Cocc.	II
"	820	none	III
"	1150	none	III
198-KL	0-2	none	I
"	450	1 (?) Cretac.C.	III
"	800	none	III
"	1210	abundant;early Miocene;Zone NN1(upper part) to NN2	IV
227-KL	800	1 Tertiary Cocc.	II
"	1100	1 Cretac.Cocc.	III
231-KL	550	abundant;middle early Miocene (Zone NN2)	IV

even though the in-situ fauna has been more or less completely dissolved and the reworked fauna would imply an older age. Late Oligocene to early Miocene radiolaria are found in the sediments below the boundary between II and III. These are also intermixed with older radiolaria.

The results of the nannoplankton and radiolarian investigations suggest that the boundary between facies II and III represents a hiatus between early Miocene and late Pliocene. The poor preservation of microfossils around the unconformity precludes a more definite dating of the hiatus range. Only a few cores could be dated. Therefore, core to core correlation is mainly based on lithostratigraphy.

Sedimentary structures

Bioturbation structures are visible macroscopically due to contrasts in colors. Finer structures are visible by X-radiography. Differences in density are apparent, and are due in part to authigenic Fe-Mn oxides in tube-like shape. Fe-Mn oxide grains are also irregularly distributed or are in chains within the burrows. These probably represent fecal pellets.

TABLE 2

Selected Results from Smear Slide Investigations

core	cm	nanno fossils	diatoms	zeolites	barites	volcanic glass	facies
26-KL	0-2	-	+	+	+	+	I
"	50	-	-	++	+	-	II
"	90	+	+	+	+	-	
"	120	+	-	++	+	-	
"	170	+	-	+	+	-	III
"	190	-	-	+	+	-	
"	400	+	-	+	-	+	
"	600	-	-	-	-	+	
"	790	90	-	-	-	-	IV
27-KL	20	10	-	+	-	-	I
"	100	70	-	-	-	-	
"	200	70	-	-	-	+	IV
"	600	80	-	-	-	-	
45-KL	0-2	-	+	+	-	-	I
"	50	-	-	+	-	-	
"	200	-	-	-	-	-	III
"	400	-	+	-	-	-	
"	600	60	+	-	-	-	
"	800	20	+	-	-	-	IV

TABLE 2

(Continued)

core	cm	nanno fossils	diatoms	zeolites	barites	volcanic glass	facies
59-KL	810	10	–	–	+	–	
"	840	+	–	–	–	–	IV
"	860	+	–	–	–	–	
"	890	–	–	–	+	–	
163-KL	0-2	–	–	–	–	–	I
"	20	–	–	–	+	–	
"	100	+	+	–	–	–	III
"	350	10	–	–	–	–	
"	700	40	–	–	–	–	IV
"	950	30	–	–	–	–	
169-KL	0-2	–	+	+	–	–	
"	120	–	+	+	–	–	
"	404	–	+	+	–	–	I
"	650	–	–	+	–	–	
"	740	–	–	++	–	–	II
180-KL	50	–	–	+	–	–	
"	100	+	–	+	+	–	I
"	200	–	–	+	+	–	
"	400	–	–	++	+	–	II
"	500	–	–	++	++	–	
"	950	–	–	++	–	–	III
"	1100	–	–	++	–	–	
183-KL	0-2	–	+	+	+	+	I
"	115	+	–	++	+	–	
"	250	–	–	++	+	–	II
183-KL	450	–	–	+	+	–	II
"	650	–	–	+	+	–	
"	815	+	–	+++	+	–	
"	1250	–	–	+	–	–	III

TABLE 2

(Continued)

core	cm	nanno fossils	diatoms	zeolites	barites	volcanic glass	facies
198-KL	150	-	-	+	-	-	III
199-KL	50	+	-	+	-	-	I
"	100	-	-	++	-	-	
"	200	+	-	++	-	-	II
"	300	-	-	+++	-	-	
"	600	-	-	++	-	-	
"	1000	+	-	+	-	-	III
"	1250	-	+	+	-	-	
"	1350	-	+	+	-	-	
227-KL	800	-	-	++	-	-	II
"	900	+	-	+	+	-	
"	1000	+	-	-	+	-	III

- = absent; + = rare; ++ = common; +++ = abundant;
numbers (nanno fossils) = estimated percentage

Facies I is characterized by indistinct horizontal burrowing. In facies II light-colored, more or less sharply bounded ring structures are frequent. In addition, one observes a multitude of fine (< 1 mm diameter), mostly vertical, burrows in the radiographs (Fig. 4). In some cores, especially in the lower half of facies II, lenticular bioturbation structures which have been cut by vertical burrows are faintly visible.

Facies III also shows big light-colored ring structures (Fig. 4). In several cores horizontal planar burrows with internal structures typical of zoophycus, were observed (Fig. 4). In cores with well-developed contacts between facies II and III intensive vertical bioturbation extends downward from the boundary (Fig. 4). The degree of bioturbation decreases downward. The burrows have diameters between 0.5 and 5 mm, and appear darker in the radiographs, implying a greater density. Most of the sediments of facies III show very little distinctive bioturbation. Facies IV shows the most intensive bioturbation with irregularly oriented

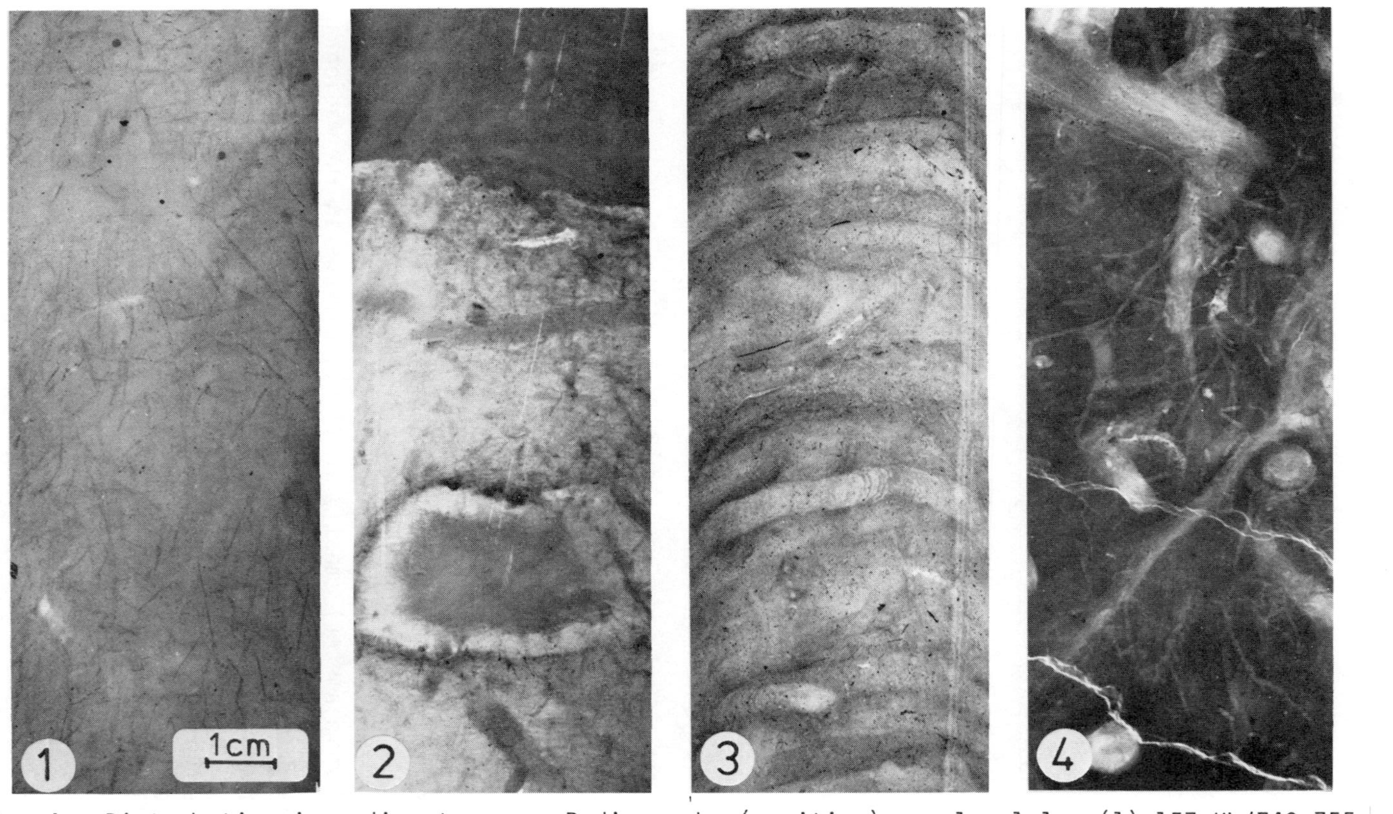

Fig. 4. Bioturbation in sediment cores: Radiographs (positive), scale, 1:1. (1) 157-KL/743-755 cm, facies II; (2) 227-KL/844-856 cm, boundary between facies II and III; (3) 156-KL/794-806 cm, facies III; (4) 27-KL/229-241 cm, facies IV.

burrows (Fig. 4). Zoophycus are quite common. Also, burrows with greater diameters (1 cm) and numerous finer ones (1 mm) are also present. Studies of burrow structures in DSDP cores from the Pacific (Chamberlain, 1975) have revealed that ring-shaped burrow sections are predominant in radiolarian oozes. Bioturbation is particularly intensive in Miocene calcareous nanno oozes. Here zoophycus traces were frequently observed, similar to our observations.

Grain-size distribution

Out of the three fractions (< 32μ, 32-63μ, >63μ) the finest size range forms the major part of the sediments (up to 98%). Generally, coarser fractions contribute only a very small proportion of the total weight (Figs. 5-8) Normally, facies I and IV and the lower part of III are coarser than II and the upper part of III.

Biogenic components

Calcareous organisms. The carbonate content of facies I to III is generally very low (Table 3). It varies between 1.2% (facies I) and 4.9% (facies III). Facies IV is essentially a calcareous ooze with a carbonate content up to 69.6%. Foraminifera are rare, even in calcareous ooze cores. Coccoliths range from rare in facies I to III to frequent in facies IV (Tables 2 and 3).

Siliceous organisms. Paralleling the decreasing quantity of siliceous tests from the surface downward (Figs. 4 and 9), is a deterioration in the preservation state of the radiolaria. This extends to nearly complete dissolution of the tests in facies II and III. The preservation state of radiolaria is of the same high order in the carbonate-rich sediments of facies IV as in the surface sediments. Diatoms are normally found in the upper part of facies I. In some cores they were also identified in facies IV (26-, 45-, 199-, 231-KL). Simultaneous to the decrease of the Pliocene-Quaternary radiolarian fauna with depth and a proportionate increase in reworked fauna there is normally a general worsening in the state of preservation of the radiolarian tests. Diatoms are generally limited to the top decimeter of sediments and are gone below this layer. In core 169-KL they can, however, be observed down to a depth of 490 cm. This core recovered 724 cm of facies I, which is a maximum for all cores. Apparently, the silica dissolution was impeded because of a comparatively high sedimentation rate.

In some cores the ready supply of reworked material and selective dissolution resulted in the presence of a nearly pure late Eocene to early Oligocene fauna. The fauna of younger age has been almost completely dissolved. According to Heath (1974), Eocene radiolaria are more resistant to dissolution than Quaternary ones.

Up to several meters below the early Miocene to late Pliocene hiatus (i.e., in facies III) siliceous organisms are almost completely dissolved and reworked. In core 199-KL this zone of dis-

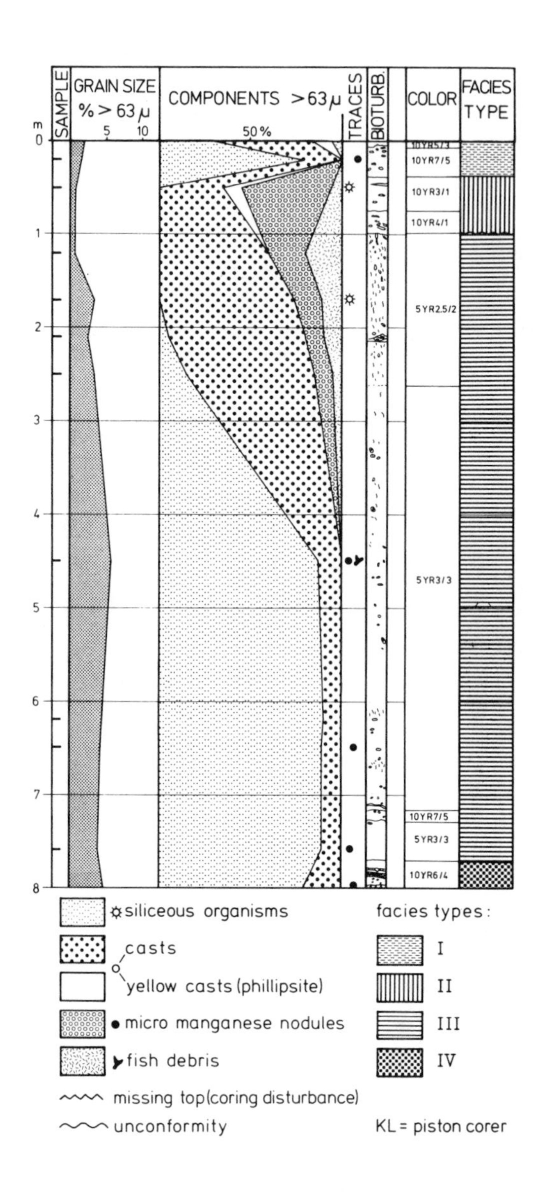

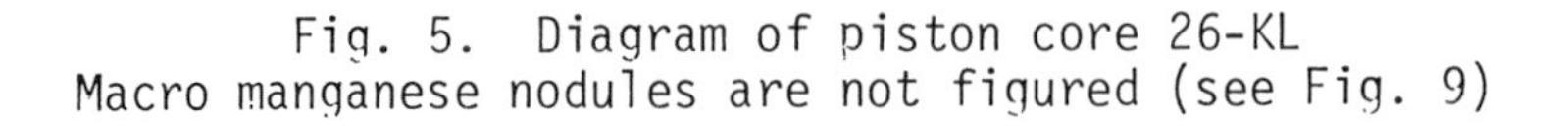
Fig. 5. Diagram of piston core 26-KL
Macro manganese nodules are not figured (see Fig. 9)

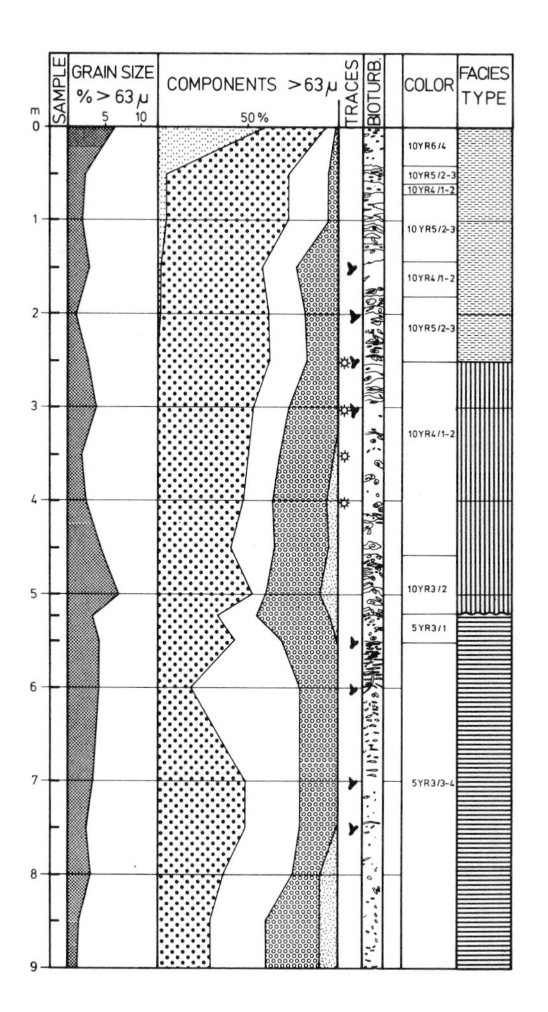

Fig. 6. Diagram of piston core 46-KL
For explanation, see Fig. 5

solution extends to 882 cm (218-1100 cm). Because of the thickness of this zone, dissolution from the surface of the unconformity towards greater depth at a later time is not probable. As in facies II and I, dissolution could have taken place in the near-surface zone during, or just after, the sediment accumulated.

What is the cause of this extensive dissolution of opaline

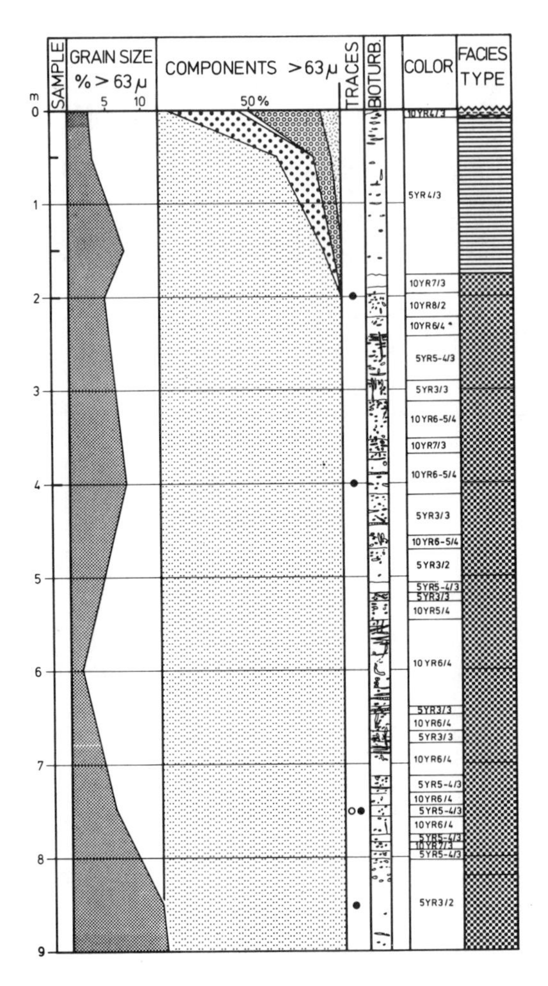

Fig. 7. Diagram of piston core 59-KL
For explanation, see Fig. 5

silica? Sea-water is undersaturated relative to silica at the surface as well as at depth. AABW contains even less dissolved silica. Therefore most of the siliceous skeletons produced in the photic zone are dissolved before settling to the sea-floor and only 2% of the tests enter the fossil record (Heath, 1974). Heath stresses the importance of the turbulence of the bottom waters for the dissolution of opaline skeletons. The stirring of the sediments by currents and their transport as suspension enhances the dissolving process. Without doubt, the time factor is also of great importance. Intensified AABW flow is therefore the preferred mechanism to explain gaps in sedimentation and the reworking and dissolution of siliceous organic remains.

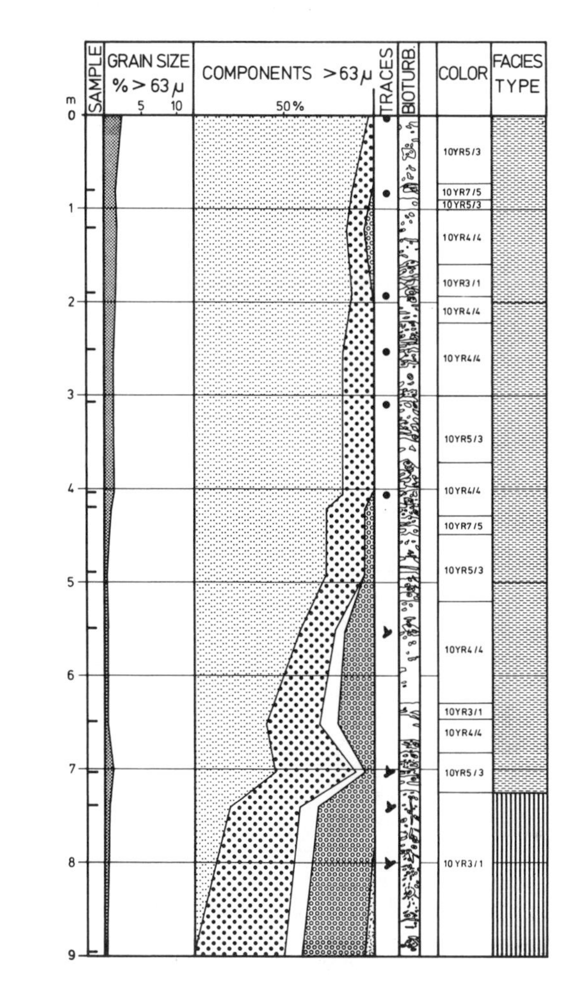

Fig. 8. Diagram of piston core 169-KL
For explanation, see Fig. 5

Casts

Spherical to irregularly shaped aggregates of fine-grained, mostly white, sedimentary material are found in the coarse fractions of all cores (Fig. 9). Some are colored yellow and have been recrystallized to a great extent into phillipsite. Frequently, these aggregates are found inside radiolarian tests in samples with variably preserved siliceous organic remains. All transitional stages of the dissolution of siliceous skeletons are present. Remains of radiolarian tests were found adhering to the surface of these internal casts. Sometimes, only impressions of the skeletal patterns are discernible on the surface of these casts.

TABLE 3

Calcium Carbonate

core	cm	$\%CaCO_3$	facies type
26-KL	0	1,84	I
"	50	3,25	II
"	120	3,81	III
"	450	3,71	III
"	800	69,67	IV
27-KL	100	49,07	IV
"	650	61,00	IV
45-KL	0	2,41	I
"	20	3,50	III
"	200	2,62	III
"	600	61,87	IV
169-KL	50	1,25	I
"	650	1,62	II
"	740	2,00	II
198-KL	100	4,53	III
"	350	4,90	III
"	950	3,93	III
"	1240	62,31	IV

Increased crystallization of clay minerals in the surface part of these internal casts is observable in thin sections.

X-ray investigations of the > 63μ fraction of sample 183-KL, 815 cm, in which the whitish internal casts are predominant and the yellow ones are missing, revealed smectite as the main crystalline component (Table 4). Increase in smectite content supports the assumption of an increase in the crystallinity of clay minerals in the internal casts.

The major part of the aggregates consists of radiolarian casts but some may be burrow-fillings or fecal pellets. The distribution of casts in coarse fractions negatively correlates with siliceous tests (Fig. 9). The formation of internal casts appears to be closely related to the dissolution of opaline silica.

Fish remains. Fish remains are very common in the coarse fraction. They consist partly of small, sharp-pointed teeth and apatite flakes and fragments which cannot be definitely identified. Their concentration in zones of increased silica dissolution is conspicuous. They are very resistant to dissolution and are concentrated in the solution zones.

TABLE 4

X-Ray Analysis

core no.	depth cm	size fraction	main components	common components	traces	facies
183-KL	0-2	$< 32\ \mu$	amorph.	phillipsite (muscovite) illite smectite quartz		I
183-KL	115	$<32\mu$	amorph.	phillipsite smectite	quartz illite	II
183-KL	815	$<32\mu$	amorph. phillipsite smectite		quartz illite calcite	III
183-KL	815	$>63\mu$	amorph. smectite	phillipsite	quartz mixed-layer ?	

Volcanogenic components

Materials of volcanic origin occur as a minor part in the cores studied. Small bits of pumice or of colorless volcanic glass were found at a few places in the coarse fraction (26-KL, 450 and 733 cm; 27-KL, 0 and 200 cm; 163-KL, 353 cm) and in smear slides (Table 3). Feldspars, magnetite, hornblende, and biotite are more frequent. In order to make more precise observations, 131 samples were taken from 10 cores (22-SL; 25-KAL; 26-, 46-, 135-, 157-, 166-, 180-, 183-, 227-KL) and the fraction $> 63\mu$ was completely checked for volcanic materials. The maximum number of grains detected per sample were for feldspar 40, and for magnetite 9. The sharpest maximum for feldspar content was observed at the boundary of facies II to III. Generally, the feldspar content is very low for the upper part of facies I. No feldspars were observed in facies IV. Magnetite content is positively correlated with the feldspar in most cores.

Feldspar, magnetite, hornblende and biotite may be the weathered products of submarine volcanic rocks. Their wide scatter over the complete zone of reworking at the hiatus (Fig. 4) suggests that these minerals were reworked from volcanic debris and cannot be connected directly with volcanic activity.

Authigenic minerals

Micro nodules. Dark-brown, spherical to irregularly

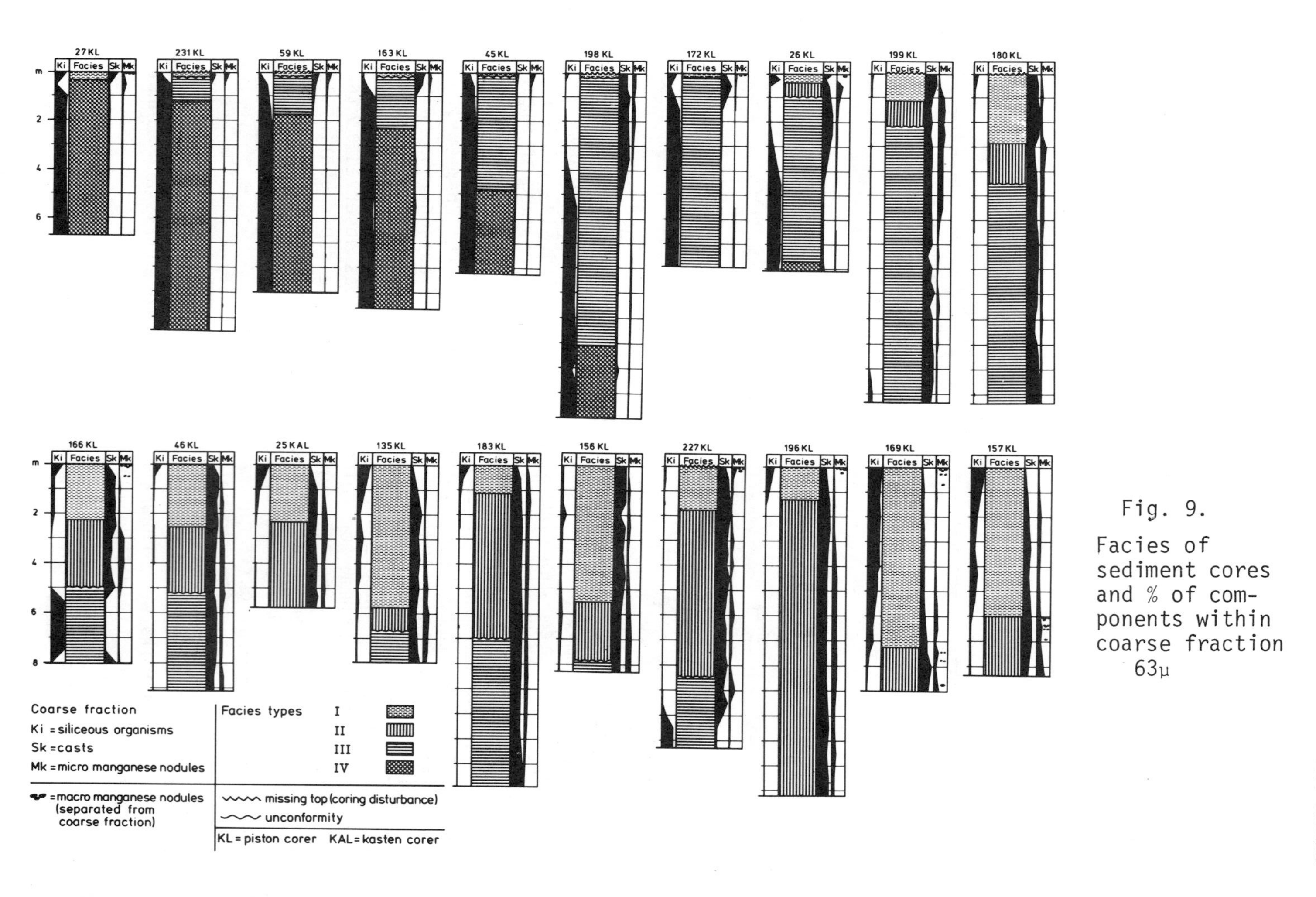

Fig. 9. Facies of sediment cores and % of components within coarse fraction 63μ

formed particles composed of Mn-Fe oxides were found in the coarse fractions. In individual cases foraminifera, radiolaria, casts, feldspar, and magnetite crystals are coated by an oxide layer. Generally, however, no foreign material can be observed at the center of these micro manganese nodules. Frequently hollow spherules made of Mn-Fe oxides, or their fragments, are present which indicate the total replacement of the radiolaria.

Micro nodules occur mainly in facies II and III, where they are particularly concentrated near the hiatus (Fig. 9). The light-colored sediment of facies IV have a smaller content of micro nodules than the darker ones. Micro nodules are not very abundant in the surface sediments. Exceptions are cores with greatly reduced Pliocene-Quaternary sediment thickness (e.g. 59-KL, 231-KL and 227-KL) or with abraded material from macronodules (e.g. 227-KL).

The frequent occurrence of micronodules in and around the burrows and biogenic remains suggests a promotion of oxide precipitation by organic matter. The generally positive relationship with internal casts or the negative relationship with siliceous tests is characteristic for the distribution of the micro nodules. According to Margolis (1973) the formation of micro nodules is supported particularly by low rates of sedimentation and oxygen-rich bottom currents with intensified reworking. Decay of organic substances in areas containing biogenic remains influences the precipitation. Greenslate (1974) describes the formation of micro manganese nodules in and around the remains of siliceous organisms. Thiel (1978) reports that agglutinating foraminifera attached to macronodules collect micro nodules thus causing the nodules to grow.

An increased formation of micro nodules was also discovered (von Stackelberg et al., 1976) in sediments on the eastern flank of Great Meteor Seamount (North Atlantic) in areas of increased input of foreign material. The relatively quick burial of the in-situ fauna containing organic matter by older, reworked sedimentary material poor in organic carbon, stopped the decay at the surface of the sea-floor. The decay of organic carbon started under low sediment burial and under conditions still rich in oxygen. Mobilization and reprecipitation of Mn was promoted in such microenvironments. A reductive mobilization of manganese is restricted to the top 20 cm of sediment in Central Pacific (Schnier et al., 1978). Erosion and resedimentation in the study area may influence the formation of micro nodules in a similar way as around Great Meteor Seamount.

Macro nodules. Macro nodules from the study area recovered by boomerang grabs and dredges were studied and described by Friedrich et al. (1976). Nodules were also encountered in 16 boomerang cores and 11 piston and gravity cores. In 25 cases they occurred on or near the upper surface of the cores, and in 14 cases at depths below 20 cm. Particularly conspicuous is the accumulation of nodules in facies II, in the cores 157- and 169-KL (Fig. 9).

The occurrence of nodules in cores is limited to the vicinity of submarine hills and elevations; here the surface density of nodules is very high. Generally, they are polynodules with more or less smooth surfaces. Only in cores with very little Pliocene-Quaternary sediment thicknesses are the nodule surfaces rough. They are similar to mono-nodules typical of the basins.

Sections of nodules show the well-known concentric layered structure. Several phases of growth, with different internal structures, can be distinguished (Heye, 1975).

Generally, the sea-floor surface nodules are multilayered. Frequently, there are inclusions in the center part of the nodules: biogenic remains, feldspar crystals, fragments of basalt, and white sediment pieces of up to 1.5 cm thickness. The white sediment consists mainly of clay minerals. Occasionally, fish teeth and radiolaria are found in the white sediment. Nannofossils are absent. Several relatively thick zones, distinctly separated from each other and with differing growth structures, surround this center.

The nodules buried deeper in the core have, in contrast, mostly a two-layered constitution. The nucleus is similar in structure and composition to that of the surface nodules. Mn-oxide layers, generally only a few mm thick, surround this center in close succession. They form the outer layer which is appreciably different from the outer zones of the surface nodules.

From the similarity of the nuclei, we conclude that the surface nodules and the buried nodules are genetically similar. The initial growth stage for both was the same; their further growth took place in different ways. The growth of the buried nodules was stopped by burial, whereas the surface nodules which started their growth at the same level, continued to grow until today, since they were constantly kept at the surface of the sediment, perhaps by the activity of benthic animals. Observations made with TV-recordings during cruise VA-13/2 and studies by Paul (1976) suggest that benthic organisms move the nodules and free them from the sediment cover. Because they are overgrown by sessile benthos (Dugolinsky *et al.*, 1977), the nodules are probably a favored target for these browsers. The intensively burrowed nature of the sediment in the nodule-rich area (Fig. 3) is further evidence of this mechanism.

Buried nodules occur mostly in facies II. However, in one core (221-BL) manganese nodules were found in facies III, 50 cm below the unconformity. White sediment in the nucleus of many nodules does not contain nannofossils, but frequently includes radiolaria of unknown age. This suggests that this material did not originate from the light-colored facies IV. It rather, stems from the type present at the boundary between facies II and III, before the radiolaria were completely dissolved and before the formation of micro nodules imparted the brownish coloring to the sediment. The oxide coating may have protected the radiolaria from dissolution and prevented the formation of micro nodules.

These observations suggest that both the buried nodules, as well as those from the surface, had their origin and main growth predominantly in facies II, especially near the unconformity at the base of II. The sedimentary conditions near the hiatus level were suitable in several ways for the formation of macro nodules. Oxygen-rich, turbulent bottom waters created suitable conditions for the precipitation of manganese-iron oxides. Erosion and reworking provided an increase in the supply of coarse particles such as basalt fragments, feldspar, and remains of organisms to act as nuclei for growth, and an increase in dissolved metallic ions from weathered biogenic and volcanic materials. As the nodules were kept free from sediments for a long time, they could grow without hindrance. In contrast to the micro nodules, benthic organisms kept the macro nodules moving and at the surface. The macro nodules got buried only at those places where the sedimentation rate was higher than the rate of keeping them free by the macro benthos. Thus, surface nodules are representative of deeper horizons, and formed under ancient environmental conditions totally different from their present surroundings. The occurrence of manganese nodules of very different sizes in the same surface-assemblage could be explained if each size originates from a hiatus level of different age.

The nodules of the study area may have begun growth between early Miocene and late Pliocene. This agrees with growth rates measured by the isotope method. Particularly low growth rates are typical for the iron-rich nodules from the Pacific (Heye and Marchig, 1977). As nodules from the slopes are richer in iron than those from the basins (Friedrich and Pluger, 1974) we conclude that the former grew more slowly than the later ones. According to Piper and Williamson (1977) and Calvert et al. (1978) the iron-rich type of nodules is associated with regions of higher rates of sedimentation. The highest sedimentation rates of the study area are observed near slopes of submarine hills (Table 5). The increased accumulation of sediments apparently impedes the nodule growth and frequently stops it completely. Several authors (Margolis, 1973; Leclaire et al., 1977) have pointed out that macro nodules are concentrated near unconformities.

According to Calvert and Price (1977), the Mn-Fe ratio in the oxide fractions of the surface sediments from the northeast tropical Pacific is lower than that of the co-existing nodules. In brown clays from the Central Pacific corresponding to our sediments at the main hiatus, the Mn/Fe ratio is richer and approaches that of the surface manganese nodules (Marchig, 1978). Cronan and Tooms (1969) observed that buried nodules have a similar chemical composition to the surface nodules. Allochthony of surface nodules is a reasonable explanation for these observations.

Working on manganese nodules from a neighboring area, Sorem (1977) found that the nodule nucleus consists usually of iron-rich, isotropic delta-MnO_2; the coating, on the other hand, is mostly manganese-rich, anisotropic and consists of todorokite and/or birnessite. On his photos and on the polished sections made for the present study it can be seen that the growth is less

TABLE 5

Thickness of Facies I+II (Quaternary-Pliocene)

core	cm	core	cm	core	cm
22-KL	> 44	80-BL	2	190-BL	> 104
25-KAL	> 575	84-BL	> 101	192-BL	> 113
26-KL	98	86-BL	> 101	196-KL	>1320
27-KL	32	88-BL	> 111	198-KL	60
40-BL	> 103	90-BL	3	199-KL	288
41-BL	> 40	92-BL	10	202-BL	14
42-BL	> 108	94-BL	8	205-BL	10
43-BL	> 104	96-BL	10	207-BL	6
45-KL	8	98-BL	5	209-BL	10
46-KL	522	100-BL	6	218-BL	68
47-BL	> 107	102-BL	5	221-BL	12
48-BL	> 95	134-KL	768	225-BL	27
49-BL	> 106	135-KL	768	227-KL	847
50-BL	> 110	156-KL	784	231-KL	10
51-BL	> 109	157-KL	> 837	48-KL	34
53-BL	> 104	163-KL	6	(VA-08)	
54-BL	> 104	166-KL	501	52-KL	> 805
55-BL	> 104	169-KL	> 901	(VA-08)	
56-BL	67	172-KL	16	62-KL	300
57-BL	> 93	180-KL	452	(VA-08)	
58-BL	> 105	183-KL	700	70-KAL	850
59-KL	10	186-BL	4	(VA-08)	
72-BL	> 111	188-BL	14	73-KL	500
				(VA-08)	

directional in the center of the nodules than on the outside. For non-directional growth structures, Heye (1978) concludes high rates of growth and for layered structures lower rates. According to Calvert and Price (1977), delta-MnO_2 is precipitated from sea water, todorokite is formed mainly by diagenetic remobilization of manganese.

From this we may conclude that the growth of the inner part of the nodules took place relatively rapidly (near the unconformity) and was mostly due to precipitation from sea-water. In this initial stage the surface sediments were probably encrusted and impregnated by the precipitation of Mn-Fe oxides. Possibly, these crusts broke afterwards due to shrinkage and were overgrown discordantly by Mn-Fe oxide layers. This growth is continuing today at a much slower rate.

Observations of a positive relationship between the rate of growth and the Mn/Fe ratios are not contradicted by the discovery that the iron-rich nodule inner parts grew rapidly, because the growth rates were measured only on the outermost millimeters of the nodules, whereas the Mn/Fe ratios are averaged for the whole nodule. The variations in growth rates of both nodule types are small. The difference in rates of growth between crust and center should be considerable. They might differ even by a factor of ten, as has been determined by Harada and Nishida (1976) and Addy (1976).

Zeolites

Yellow-colored, transparent, spherical-to-irregularly shaped particles are particularly conspicuous in the coarse fraction. The spherical particles sometimes show clear impressions of radiolarian skeletons. Therefore they are considered to be internal casts. X-ray diffraction investigations of casts from three samples (45-KL, 0-2 cm; 59-KL, 0-2 cm; 180-KL, 100-102 cm) revealed phillipsite to be the main constituent after smectite. The clay minerals, of which these casts usually consist, have been almost completely replaced by phillipsite. The phillipsite casts show a radial crystal growth in the outer part. In general, the yellow casts are positively related to micro nodules and negatively to biogenous opal (Figs. 5 to 8). Just as the Mn-Fe oxides, abundant phillipsite coincides with the marked zone of opal dissolution.

Rod-like zeolite crystals up to 40 μ long, are conspicuous in smearslides of the sediment (Table 3). They are rarely found in facies IV, but are very common in facies II and III, particularly near the hiatus. Their occurrence corresponds with that of the phillipsite casts. According to X-ray refraction analyses of the fraction < 32μ of a sample from depth 815 cm of the core 183-KL, i.e., from the uppermost part of facies III, phillipsite is the main component (Table 4). The occurrence of phillipsite in casts and the positive correlation between phillipsite content and dissolution of siliceous organic remains suggest that the silica of the phillipsite originated from siliceous skeletons and not from volcanic glass.

Stonecipher (1976) concluded from Deep Sea Drilling Project data that phillipsite is associated with low sedimentation rates and sediments of younger age. Our own survey of Deep Sea Drilling Reports for the Central Pacific (Cook and Zemmels, 1971; Zemmels, 1973) revealed that phillipsite is especially concentrated near hiatus levels. This mineral is particularly frequent near the youngest sedimentation gap at the Tertiary/Quaternary boundary, but is also common at sedimentation breaks within the Miocene.

Barite

Twenty-one out of 60 smear slides showed small quantities of barite crystals (length up to 60 μ) (Table 3). The vertical distribution of this mineral is similar to that of zeolites. Barite is rare in facies I, is absent in facies IV except for 59-KL below 803 cm. It occurs in facies II and III and is particularly frequent near the unconformity. Large barite crystals have been reported near unconformities in pelagic clays by Goldberg and Arrhenius (1958) and in association with high manganese content and phillipsite (Cronan, (1974). According to Church (1970) the barium of barites in the deep-sea sediments is derived from the oxidation of organic remains.

Summary and Conclusions

Reworked microfossils preserved in the sediment cores are derived from sediments as old as Cretaceous (coccoliths) and Eocene (radiolaria and foraminifera). The oldest autochthonous material present is of early Miocene age. This sediment, a light-brown radiolarian-bearing nanno ooze with intensive and irregular bioturbation (facies IV) reflects the quiet conditions of sedimentation existing within the area affected by the equatorial bioproduction. Foraminifera have been almost completely dissolved, nannofossils are relatively well preserved, and the siliceous organic remains are well preserved.

Environmental conditions started changing in early Miocene. Due to the drifting of the Pacific plate towards the northwest, this area of the seafloor shifted out of the carbonate-rich equatorial region. Simultaneously, the water depth and carbonate dissolution increased. This resulted in the carbonate-poor facies III, a reddish-brown zeolitic mud. According to the migration tracks plotted by van Andel _et al_. (1975) this area was positioned during the early Miocene at about 4°N, 132°30'W, i.e., about 600 km closer to the equator than at present and had a water-depth of about 4500 m which is slightly above the CCD.

During middle and late Miocene the circulation of the Antarctic Bottom Water (AABW) increased producing erosion and resedimentation (van Andel _et al_., 1975). As a result effects of bioturbation in the sediments became obscured or completely destroyed. Vertical burrows penetrating deeper into the sediment are better preserved. The chemical reactivity of the bottom waters led to a complete dissolution of the calcareous fossils and to a selective dissolution of siliceous tests. Fish remains were concentrated by this process.

The dissolved silica released by this process was precipitated as cement in the radiolarian casts. In some cases the clay minerals of the casts have been completely replaced by phillipsite. Phillipsite has also been formed as fine needles in the sediment. Closely associated with opal dissolution is the occurrence of small barite crystals in the sediment. The increased formation of micro manganese nodules may have been favored by intensified reworking and resedimentation. Concentration of submarine weathering products from volcanic rocks is also attributed to the increase in the mechanical and chemical effects of the intensified AABW.

Dissolution and authigenesis are conspicuous near a particularly prominant hiatus level. Apparently, the intensity of the bottom currents increased to the extent that a break in sedimentation or erosion took place. The hiatus between facies II and III represents the period between early Miocene and late Pliocene. As the bottom current velocity decreased below a critical level, sediment accumulation was renewed, apparently under similar conditions as existed before the hiatus. Dark-brown zeolitic mud of facies II with a prevailing vertical bioturbation was deposited. Parallel to the decrease in the bottom current intensity, sediment reworking, dissolution, and formation of authigenic minerals were also diminished. They are therefore arranged symmetrically below and above the unconformity (Fig. 4).

The sedimentological evidence and bottom-current measurements from neighboring areas indicate that at present quiet water conditions prevail in the study area. Light-brown siliceous mud is typical for the surface sediments.

The grain-size distribution of the cores is determined primarily by dissolution of biogenic remains.

The thickness of the strata above the hiatus, i.e., the Pliocene-Quaternary sediments (Table 5), is influenced by bottom topography. Extreme variations in sediment thicknesses are characteristic of the hill flanks. Further, a north-south directed narrow horst in the northeast of the study area is conspicuous for its thin Pliocene-Quaternary sediments. Finer materials were stirred up on the top of the elevations and were carried down by turbid layer transportation. The major part of this material reached the basins. Considerable quantities were deposited on the flanks, especially where sediment was trapped by fault structures. Similar observations were made by Moore (1970).

The distribution of sediment thicknesses around the Eastern- and the Western-hills shows an asymmetrical configuration. Maximal sediment thicknesses were found west of W-hill (157- and 169-KL) and northwest of E-hill (166- and 196-KL). This asymmetry might be explained by directional bottom currents.

In general macro nodules occur on the sea-floor. In some cores which show high rates of sediment accumulation, they are also found in facies II above the hiatus. Structural comparison of buried nodules with surface nodules suggests that both nodule types originated at the same time, near the unconformity. During that period, increased bottom currents changed the bottom water chemistry and favored the formation of manganese nodules. Some nodules were buried by the sediments and their growth stopped. Most of the nodules were kept at the surface of the sediment by the activity of benthic organisms. This enabled the nodules to continue to grow, but at a lesser rate. Volcanism does not appear to have had any direct effect on the formation of nodules in the study area. Indirectly, it was of importance for the supply of metallic ions and the supply of nuclei material.

Thus, the surface nodules of the study area are a foreign body to the present sediments. They represent the environmental conditions of deeper horizons.

Acknowledgments

The research vessel R/V VALDIVIA was chartered by the Federal Ministry for Science and Technology from the shipping company "Reedereigemeinschaft Forschungsschiffahrt" and put at our disposal for the cruise VA-13/2. The Federal Ministry for Science and Technology supported financially the completion of the investigations.

I am obliged to Dr. Beiersdorf and Dr. Wolfart for several fruitful discussions. The biostratigraphic results, based on investigations of radiolaria (Wolfart, 1977), were fundamental to this study. Nannofossil studies by Dr. Cepek and X-ray diffraction analyses by Dr. Rösch were valuable and are gratefully acknowledged. Mr. Guenther kindly processed the sedimentological data, and Mr. Hansen compiled the bathymetric data.

I would also like to mention the technical cooperation in samples collection and preparation by Mrs. Hointza, Mrs. Stenschke and Messrs. Hointza, Karmann, Kawohl and Ostermann.

References

Addy, S.K., (1976) A study of manganese nodules and sediments in the abyssal hill province of the North West Atlantic, 25th Internatl. Geol. Congr. Sidney, Australia, Abstracts 2, 8, 337.

Baker, E.T. and Feely, R.A., (1978) Chemistry of oceanic particulate matter and sediments, Implications for bottom sediment resuspension, Science 200, 533-535.

Beiersdorf, H. und Wolfart, R., (1974) Sedimentologisch-biostratigraphische Untersuchungen an Sedimenten aus dem zentralen Pazifischen Ozean, Meerestechnik 5, 192-198.

Calvert, S.E. and Price, N.B., (1977) Geochemical variation in ferromanganese nodules and associated sediments from the Pacific Ocean, Mar. Chem. 5, 43-74.

Calvert, S.E., Price, N.B., Heath, G.R., and Moore, T.C., Jr., (1978) Relationship between ferromanganese nodule compositions and sedimentation in a small survey area of the equatorial Pacific, Jour. Mar. Res. 36, 161-183.

Chamberlain, C.K., (1975) Trace fossils in DSDP cores of the Pacific, Jour. Pal. 49, 1074-1096.

Church, T.M. (1970) Marine barite, PhD thesis, Univ. Calif. San Diego.

Cook, H.E., and Zemmels, I., (1971) X-ray mineralogy studies - Leg 8, In: Tracey, J.I., Jr. et al., Initial Reports Deep Sea Drilling Project VIII, Washington, 901-950.

Cronan, D.S. (1974) Authigenic minerals in deep-sea sediments, In: The Sea 5, Marine Chemistry, Wiley, New York, 491-525.

Cronan, D.S., and Tooms, J.S., (1969) The geochemistry of manganese nodules and associated pelagic deposits from the Pacific and Indian Oceans, Deep Sea Res. 16, 335-359.

Dürbaum, H.J. und Schlüter, H-U., (1974) Möglichkeiten der Reflexionsseismik für die Manganknollenexploration, Meerestechnik 5, 188-192.
Dugolinsky, B.K., Margolis, S.V., and Dudley, W.C., (1977) Biogenic influence on growth of manganese nodules, Jour. Sed. Pet. 47, 428-445.
Friedrich, G. und Plüger, W., (1974) Die Verteilung von Mangan, Eisen, Kobalt, Nickel, Kupfer, und Zink in Manganknollen verschiedener Felder, Meerestechnik 5, 203-206.
Friedrich, G. Plüger, W., und Kunzendorf, H., (1976) Geochemischlagerstättenjundliche Untersuchungen von Manganknollenvorkommen in einem Gebiet mit stark unterschiedlicher submariner Topograohie (Zentraler Pazifik), Erzmetall 29, 462-468.
Goldberg, E.D. and Arrhenius, G., (1958) Chemistry of Pacific pelagic sediments, Geochim. Cosmochim. Acta 13, 153-212.
Gordon, A.L. and Gerard, R.D., (1970) North Pacific bottom potential tial temperature, Geol. Soc. Am. Mem. 126, 23-39.
Greenslate, J., (1974) Manganese and biotic debris associations in some deep-sea sediments, Science 186, 529-531.
Harada, K. and Nishida, Sh., (1976) Biostratigraphy of some marine manganese nodules, Nature 260, 770-771.
Heath, G.R., (1974) Dissolved silica and deep-sea sediments, Spec. Publ. Soc. Econ. Paleont. Miner. 20, 77-93.
Heye, D., (1975) Wachstumsverhaltnisse von Manganknollen, Geol. Jahrbuch E, 5, 1-122.
Heye, D., and Marchig, V., (1977) Relationship between the growth rate of manganese nodules from the Central Pacific and their chemical constitution, Mar. Geol. 23, M 19- M 25.
Heye, D., (1978) The internal micro- structure of manganese nodules and their relationship to the growth rate, Mar. Geol. 26, M 59-M 66.
Johnson, D.A., (1972) Ocean-floor erosion in the Equatorial Pacific, Geol. Soc. Amer. Bull. 83, 3121-3144.
Leclaire, L., Clocchiatti, M., Giannesini, P.J. and Caulet, J.P., (1977) Dépôts métallifères dans l'ocean Indien austral. Données nouvelles sur la genèse et la prospection des champs de nodules, Bull. B.R.G.M., 1 Série, Sect. II, No. 1, 13-42.
Marchig, V., (1978) Brown clays from the Central Pacific-Metalliferous sediments or not?, Geol. Jahrbuch D 30, 3-25.
Margolis, St. V., (1973) Manganese deposits encountered during DSDP Leg 29 in subantarctic waters, In: The origin and distribution of manganese nodules in the Pacific and prospects for exploration, M. Morgenstein, Editor, Honolulu.
Moore, T.C., Jr., (1970) Abyssal hills in the Central Equatorial Pacific: Sedimentation and stratigraphy, Deep Sea Res. 17, 573-593.
Paul, A.Z., (1976) Deep-sea bottom photographs show that benthic organisms remove sediment cover from manganese nodules, Nature, 263, 50-51.

Piper, D.Z., and Williamson, M.E., (1977) Composition of Pacific Ocean ferromanganese nodules, Mar. Geol. 23, 285-303.

Ryan, W.B.F. and Heezen, B.C., (1976) Smothering of deep-sea benthic communities from natural disasters, Technical Report, 1-132, Lamont-Doherty Geol. Obs., Palisades.

Schnier, C., Gundlach, H., and Marchig, V., (1978) Trace elements in pore water and sea water in the radiolarian ooze area of the Central Pacific as related to the genesis of manganese nodules, In: Environmental biochemistry and geomicrobiology 3, Ann Arbor Sci., Ann Arbor, 859-867.

Sorem, R.K., (1977) Internal structure of selected manganese nodules: DOMES RP 8-OC-75, Leg 3, In: Open-file report 77-778, Menlo Park, Ca., 195-216.

Stonecipher, S.A., (1976) Origin, distribution and diagenesis of phillipsite and clinoptilolite in deep-sea sediments, Chem. Geol. 17, 307-318.

Thiel, H., (1978) The faunal environment of manganese nodules and aspects of deep sea time scales, In: Environmental biochemistry and geomicrobiology 3, Ann Arbor Sci., Ann Arbor, 887-896.

van Andel, Tj. H., Heath, G.R., and Moore, T.C., Jr., (1975) Cenozoic history and paleooceanography of the Central Equatorial Pacific Ocean, Geol. Soc. Am. Mem. 143, 1-134.

von Stackelberg, U., von Rad, U., und Zobel, B., (1976) Asymmetric distribution of displaced material in calcareous oozes around Great Meteor Seamount (North Atlantic), "Meteor"-Forsch. Ergebn. C, 25, 1-46.

Wolfart, R., (1977) Bericht über mikropaläontologisch-biostratigraphische Untersuchungen, In: Ergebnisse der Manganknollen-Wissenschaftsfahrt Va-13/2, Bundesanstalt fur Geowissenschaften und Rohstoffe, Hannover, (unpublished).

Zemmels, I., (1973) X-ray mineralogy studies - Leg 16, In: van Andel, Th.J., Heath, G.R. et al., Initial Reports of the Deep Sea Drilling Project XVI, Washington, 529-571.

EXTRACTIVE CHEMISTRY OF EQUATORIAL PACIFIC PELAGIC SEDIMENTS AND RELATIONSHIP TO NODULE FORMING PROCESSES

Carl J. Bowser and Barbara A. Mills*
Department of Geology and Geophysics, University of Wisconsin, Madison, Wisconsin 53706

E. Callender
U.S. Geological Survey, National Center
Reston, Virginia 22092

Abstract

In order to help understand early diagenetic processes in sediments potentially important to the formation of marine manganese nodules a series of selective extraction experiments were performed on sediments associated with nodule areas from the eastern equatorial Pacific Ocean. Phase selective extractants used were hydroxylamine hydrochloride (buffered and unbuffered), buffered citrate-dithionite; and hydrochloric acid. In hydroxylamine hydrochloride extractions of red-brown clay a rapid initial release of all metals within the first minute is followed by a slow, long term dissolution. The pH of the extraction solutions rises abruptly in the first minute, levels off, and drops steadily for the rest of the extraction. In sequential, batch-extraction experiments using 0.1N HCl and hydroxylamine hydrochloride, Cu is readily released in HCl, but the release of Mn, Ni, Co, and Ca is greater in hydroxylamine hydrochloride. Buffering of solutions by dissolution of calcium carbonate in an artificial carbonate/manganese nodule mixture depresses Cu release in HCl, and enhances Mn, Ni, Co, and Fe release in hydroxylamine hydrochloride.

Transition metal release in hydroxylamine hydrochloride can be modeled to a compound first order rate equation. The release rates of Fe, Cu, Ni, and Ca are evidently controlled by at least two phases in red-brown clay. The differences in Fe, Cu, and Ni behavior indicate different phases are controlling their dissolution, and is in agreement with suggested sources of Cu from

*Present address: 2321 Murel Ct., Cincinnati, OH 45219

nodules, Fe at a higher rate in sediments, and Cu accumulating at a rate an order of magnitude higher in sediments.

Introduction

Diagenetic reactions involving transition metals, Fe, Mn, Cu, Ni, Co, play an important role in manganese nodule genesis. While it may be argued whether the direct source of metals in nodules is from seawater or from diffusional transport in sediment pore-fluids (Bender, 1971, Raab, 1972, Boudreau and Scott, 1978), there is little doubt that both physical and chemical, diagenetic processes occurring at or near the sediment/water interface determine the final textural, morphological, and chemical character of nodules. Changes in the pore fluid chemistry within the top ten centimeters of sediments (Callender, et al., in press) reflect changes in the solid phases of sediments that are undergoing changes such as dissolution, solid phase transformation, and ion exchange. Metals important to the formation of manganese nodules may be released during these processes, either to be incorporated into nodules, to diagenetic products within the sediments, or lost by diffusion into the overlying water. Knowledge of the phase distribution of these elements in pelagic sediments can, thus, provide information on the mechanisms of sediment diagenesis and nodule formation.

The distribution of elements in ocean water column is controlled by a number of biological and physical processes. Particulate *biologic* matter raining on the sea floor can be divided into three general classes organic matter, carbonate and opal (Bishop, et al., 1977). Inorganic debris exists in the form of detrital silicates (aeolean or river particulates) as volcanic ejecta, or inorganic precipitates as, for example, from interaction of submarine hydrothermal fluids with normal seawater. Knowledge of the distribution of metals among the various forms of inorganic and organic particulate matter and the reactivity of the phases of absorbed metals in the water column and as sedimented matter are fundamental to questions about the origin and compositional variations of marine manganese nodules.

Metal accumulation rates in Pacific pelagic sediments measured by Bostrom, et al. (1973), show an increase in Fe and Mn accumulation rates in the area of the East Pacific Rise, and a somewhat smaller increase along the equator in the zone rich in opalline silica and carbonate; copper and nickel show similar patterns, but with a stronger association with siliceous ooze deposition north of the equatorial productivity maximum. Such patterns suggest that transport of metals to the seafloor by biologic particulates may, in part, be important controls on metal distribution in normal marine sediments. Comparing accumulation rates in ferro-manganese nodules and coatings with pelagic sediments, Kraemer and Schornick (1974) found Mn accumulating at roughly the same rate in both sediments and

organic matter (Callender, et al., in press) and Ni from siliceous tests (Sclater, et al., 1976).
Large diagenetic fluxes of Cu from the sea floor have been calculated by Boyle, et al. (1977) and corroborated by pore-fluid measurements of Callender, et al. (in press). Variations in accumulation rates for these metals and differences in their diagenetic behavior are indicative of different phase controls for copper, iron and manganese.

Total sediment analyses such as used for previously cited studies of accumulation rates and distribution between sediments and nodules do not distinguish the distribution of these metals among the various sediment phases or the reactivity of these various phase-associated metals under diagenetic conditions. Release rates of metals from sediments using selective extraction techniques can be used to characterize the reactivity and potentially the phase association of the transition metal phases.

Specific batch extraction techniques have been used previously to analyze soils and sediments. The most common extractants for transition metal oxides include hydroxylamine hydrochloride (Chester and Hughes, 1967), citrate-dithionite (Mehra and Jackson, 1960), and ammonium oxalate (Schwertmann, 1964, Heath and Dymond, 1977). These methods differ slightly in mechanism and extent of extraction.

Kinetic extraction rate studies have been conducted on natural clays including pelagic sediments. Anderson and Jenne (1970) used the change in rate of metal release in citrate-dithionite solutions to characterize the oxide phases in reference clays. Chao (1972) studied the effect of time on hydroxylamine hydrochloride extraction of Mn and Fe oxides in soils and sediments. With bulk extractions using the release of metals in ammonium oxalate, Heath and Dymond (1977) attempted a classification of East Pacific Rise sediments according to source. Chemical extraction and pore fluid data were used by Callender, et al. (in press) to characterize the geochemistry of manganese and copper in Pacific pelagic sediments.

This paper presents a study of timed chemical extractions of pelagic Pacific Ocean sediments with special emphasis on red-brown clay from the northeastern equatorial Pacific. The purpose of this phase of our investigations was an attempt to characterize the metal associations with various phases (oxide coatings, micronodules, biogenic particulates, etc.), and less to elucidate detailed mechanisms of dissolution of various phases occurring in natural sediments.

Hydroxylamine hydrochloride was chosen as the extractant because of its specificity for Mn oxides (Chao, 1972). Unbuffered hydroxylamine solutions were chosen for most of the studies presented herein. As part of this study, techniques were developed that were used to evaluate variations in phase associations of essential nodule metals with depth in sediments and nodules. Data from this study and on additional sediment types are included in the thesis research of B. Mills (M.S. unpublished).

Materials and Methods

The extractions in this study were done using a red-brown clay, a powdered manganese nodule standard and a $CaCO_3$-manganese nodule mixture (herein referred to as "artificial calcareous ooze"). The clay was sampled at a depth of 95 cm from the base of a free-fall gravity core from the northeastern equatorial Pacific Ocean, 19°40'N and 132°00'W. The core was collected in 1974 during an NSF-IDOE Manganese Nodule Project cruise (Mn 74-01).

The clay has a median size of 2.58 microns as determined by an automated Coulter Counter. Total chemical analyses of the red-brown clay and the manganese nodule standard, GRLD 126, are given in Table 1. Compositionally the clay is comparable to other deep sea red clays. Relative to average shales the sediment is enriched in Fe, Mn, Cu, Ni and Co, a result of the relatively high fraction of iron and manganese oxides in typical oxidized abyssal marine clays. The $CaCO_3$-nodule mixture was prepared by mixing 5% (by weight) of nodule standard GRLD-126 with 95% powdered reagent grade calcium carbonate.

TABLE 1

Total Analysis of Red-Brown Clay (Mn 74-01, Std 10) and Manganese Nodule Standard (GRLD-126)

Element	Red-Brown Clay		Nodule Std (GRLD-126)	
	(ppm)	(log ppm)	(ppm)	(log ppm)
Si	327,000	5.51	N.A.	N.A.
Al	93,500	4.97	N.A.	N.A.
Fe	75,000	4.88	103,700	5.02
Mn	5,460	3.74	230,900	5.36
Cu	180	2.26	670	2.83
Ni	286	2.46	988	2.99
Co	117	2.07	1,437	3.16
Ca	5,280	3.72	12,900	4.11
Mg	20,000	4.30	15,100	4.18
Zn	172	2.24	1,276	3.11
Cr	150	2.18	N.A.	N.A.

Red-clay: Analysis by J. Greenslate (room T., aqua regia/HF)
Nodule Std: Analysis by C. Eisen (120°C, aqua regia/HF)
N.A.-Not analyzed

Extracting solutions used were 0.1 N HCl, a solution of 0.05 M sodium citrate and 0.07 M sodium dithionite, and hydroxylamine hydrochloride (H/H) at various strengths ranging from 0.05 M to 1.2 M. The experiments were done in magnetically-stirred teflon reaction vessels and in 125 ml linear polyethylene flasks agitated by a gyratory water shaker bath. All experiments were thermostatted at 25°C with a refrigerated constant temperature circulator. Chao (1972) found changes in the dissolution rates of Fe and Mn oxides in H/H (hydroxylamine hydrochloride) caused by changes in solution stirring rate, thus experiments were designed to minimize the effect using constant shaking rates.

Extracting solutions were allowed to come to temperature equilibrium before a weighed portion of dry sediment was added at time $t = 0$. Samples of the extraction mixture were taken at times ranging from one minute to more than three months after solids were added. The sampling was done in a manner to quench the reaction between sediment and solution, and to maintain the sediment/solution ratio of the remaining experiment. Aliquots of 2-4 mls were taken from the suspension with plastic syringes and were then filtered through a 0.45 micron Millipore filter. The filtrate was preserved with "ultrex" nitric acid in linear polyethylene bottles until chemical analysis.

A two-part sequential extraction technique was used with the manganese nodule standard (GRLD 126) and the "artificial calcareous ooze". First the samples were leached with 0.1 N HCl for 24 hours. Then the complete samples were filtered through a 0.45 micron millipore filter, washed, dryed and weighed. The remaining sediment was extracted again with 1.2 M hydroxylamine hydrochloride. Subsamples were taken throughout both extraction periods as with the other solution extractions.

Six metals were analyzed by atomic absorption spectroscopy. Calcium, manganese and iron were determined using an air-acetylene flame on a Perkin-Elmer 303 atomic absorption spectrophotometer. Copper, nickel and cobalt were measured with a graphite furnace on a Perkin-Elmer 503 atomic absorption spectrophotometer with deuterium-arc background correction. Three to five replicates were run and averaged for each experiment. The absence of matrix effects was confirmed by comparing results with an identical analysis by standard additions. Measurements of pH were made with a combination hydrogen-ion selective electrode at either discrete times or continuously by connecting the pH meter to a strip chart recorder. Two point calibration was done for all measurements using standard buffers. Hydroxylamine hydrochloride concentration (H/H) was determined by a colorimetric method from Strickland and Parsons (1968), with minor modifications to accommodate the concentration range and spectrophotometer cell size.

Analytical precision was determined statistically from the calibration curves. The error in a concentration estimate, $s_{y.x}$, was calculated and multiplied by the appropriate dilution factors to give an error in the determined metal concentration. Table 2 gives precision ranges for the six elements analyzed. Analytical accuracy was evaluated by comparing the acid-soluble and total analyses of GRLD 126 by R. Karnauskas and C. Eisen with analyses from several other laboratories. The percent deviation ranged from 1% for Fe to 6% for Ni.

TABLE 2

Analytical Precision of Elemental Analyses by Atomic Absorption Spectroscopy

Element	$s_{y.x}$ (μg/l)	Element	$s_{y.x}$ (μg/l)
Fe	50-100	Cu	0.4-3
Mn	40-80	Ni	0.2-3
Ca	20-40	Co	0.2-3

An attempt was made to minimize the variables which would affect extraction rates. To reduce the effect of sediment heterogeneity, the samples were dried at either 70°C or 110°C to a constant weight, ground lightly and thoroughly mixed. Replicate grain-size analyses were run on a Coulter Counter to verify sample homogeneity. Most experiments were run three times and the results averaged. Other variables that were held constant were: the sediment/solution ratio (0.035 g/ml), temperature (25°C), and the stirring rate. The remaining variables that were evaluated in relation to metal release behavior were the type of extracting solution, solution strength and pH.

Results

Figs. 1 through 8 show the effect of the type of extracting solution concentration on metal release rates from the sediments studied. Because of the large changes in metal concentrations and uneven time periods of sampling, log-log plots were chosen to illustrate the changes observed. Total concentration of metals in the red-clay sample is shown in Fig. 7 and Table 1.

Metal ratios are calculated for appropriate metals and are shown in Figs. 1 through 5 and 7. Ratios were chosen because of the commonly recognized association of nickel and copper with manganese in nodules and weaker association of cobalt with iron in nodules. If, for example, both copper and manganese are associated with the same phase and homogeneous dissolution of the phases occurs, the Cu/Mn ratio should remain constant throughout the experiment.

Curves were drawn visually through the data points within the error limits of each data point. The sizes of the points plotted have no relationship to error limits, however. A computer generated spline-fit smoothing technique was used to fit some of the data, but the resultant curves differ little from the fits shown here.

Red-Brown Clay Extractions

The majority of extraction experiments were performed using unbuffered hydroxylamine hydrochloride (H/H) solutions ranging in concentration from 1.2 M (6.25%) down to 0.05 M (0.25%). For comparison additional extractions were done using acetate buffered hydroxylamine hydrochloride (Chester and Hughes, 1967) and buffered citrate-dithionite solutions (Mehra and Jackson, 1960). A summary of experimental analyses and solution strengths used is shown in Table 3.

In general all hydroxylamine extractions demonstrate a rapid, initial release of metals within the first minute of the extraction, followed by a long term slow dissolution (60-80% of manganese is released within the first minute). This initial metal release is accompanied by an increase in the pH of the solution presumably by exchange reactions. If the extracting solution is unbuffered, the pH later drops concommitant with a rise in the dissolved iron concentration.

1.2 M H/H (6.25%) Fig. 1. The dissolution of manganese is rapid and nearly complete (83.7 percent in the first minute). The nearly constant concentration for manganese suggests either equilibrium with the solution or total depletion of the dissolvable manganese phase(s). Iron continues to be released into solution throughout the extraction. Only a small percentage of the total iron present in the red-brown clay is released during the extraction (0.79 percent after the first minute). Copper, nickel, cobalt, and calcium concentrations all increase slightly with time. The slight decrease in manganese, cobalt, and calcium concentrations at greater than 10^4 minutes is probably due to analytical error.

Since the concentration of manganese in solution is nearly constant after one minute, the metal ratios against manganese simply reflect the increase in concentration of the other metals with time. Cobalt reaches steady-state concentration in relatively short time and is evidently not tied to the dissolution of the iron bearing phase. The increase in the iron to manganese ratio at the end of the experiment is not anomalous, it reflects the fact that only iron is continuing to dissolve at this time. The apparent sharp increase is an artifact of the log-log scale.

0.5 M H/H (2.60%) Fig. 2. The metal release behavior is similar to that in the 1.2 M hydroxylamine hydrochloride extraction. Again manganese release is almost complete and very rapid. The rate of release of copper, nickel, cobalt and calcium into solution is somewhat greater than that in 1.2 M hydroxylamine hydrochloride. Copper and cobalt release curves are almost the same. Again the metal ratios against manganese reflect only the change in concentration of copper, nickel, cobalt, iron and calcium, since manganese is little changed after the first minute.

TABLE 3

Summary of Experimental Conditions and Elements Analyzed for Red-Brown Clay, Nodule Standard (GRLD-126), and "Artificial Calcareous Ooze"

Sample	Extractant	Strength	Analyzed Elements									Text Figure
			Mn	Fe	Ca	Mg	Cu	Ni	Co	pH	H/H	
Red Clay	H/H	1.2 M (6.25%)	x	x	x		x	x	x	x		1
Red Clay	H/H	0.5 M (2.60%)	x	x	x		x	x	x			2
Red Clay	H/H	0.12 M (0.625%)	x	x			x	x	x	x		3
Red Clay	H/H	0.103 M (0.54%)	x	x	x	x	x	x	x	x	x	4
Red Clay	H/H	0.05 M (0.26%)	x	x	x		x	x	x			5,6
Red Clay	Buff. H/H	1.0 M (5.21%)	x	x	x	x	x	x	x	x		7
Red Clay	C/D		x	x			x	x	x			7
Nodule Std	HCl	0.1 M	x	x			x	x	x			8
Nodule Std	H/H	1.2 M	x	x			x	x	x			8
"Calc. Ooze"	HCl	0.1 M	x	x			x	x	x			8
"Calc. Ooze"	H/H	1.2 M	x	x			x	x	x			8

H/H = Hydroxylamine Hydrochloride
C/D = Citrate/Dithionite
Buff. H/H = Acetate Buffered Hydroxylamine Hydrochloride (Chester and Hughes, 1967)

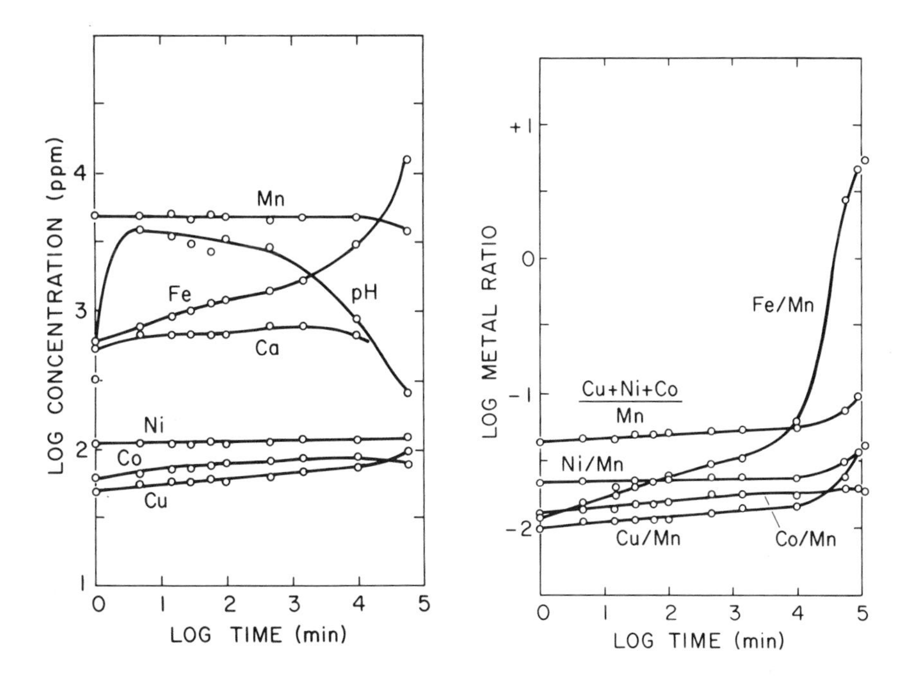

Fig. 1. Log concentration (left) and log metal ratio (right) versus time for extraction of red-brown clay in unbuffered 1.2 molar hydroxylamine solution.

It is evident that changing extracting solution strength only changes the extent toward completion of the reactions and not the relative amounts of the metals released.

0.12 M H/H (0.625%) Fig. 3. The total amount of manganese released into solution is lower in this experiment than in the previous ones (63.0 percent versus 82.2 percent and 87.9 percent in the first minute), but the change with time is still very rapid, with 89 percent of the *final value* being reached in the first minute. Iron release behavior is similar to that of the 0.5 M extraction, but the total amount in solution at any given time is always less than in higher concentration hydroxylamine hydrochloride solution. Total copper, nickel and cobalt release is also lower. The copper release, however, is somewhat different than for previously discussed experiments, giving a nearly 10 fold increase in concentration over the total time of the experiment. Evidently the hydroxylamine hydrochloride concentration

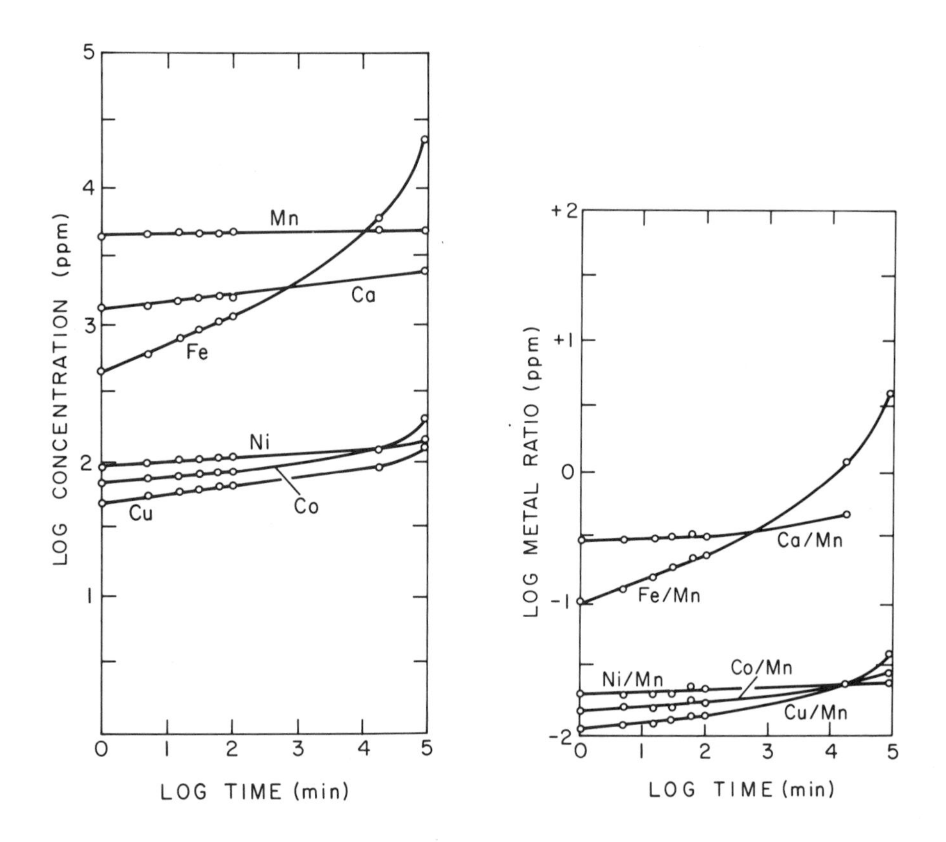

Fig. 2. Log concentration (left) and log metal ratio (right) versus time for extraction of red-brown clay in unbuffered 0.5 molar hydroxylamine solution.

was low enough to significantly change the behavior of copper without effecting the other metals nearly as much. It suggests that a significant fraction of the copper is not associated with the phase(s) that control the iron, manganese, cobalt and nickel release rates.

Cobalt release is slower, resembling release in 1.2 M rather than 0.5 M hydroxylamine hydrochloride, almost paralleling nickel release rates. The drop in pH is much more dramatic than in the first extraction, having a range of nearly 2.5 pH units, a fact which may help explain the greater range in copper concentration over the course of the experiment. Although the concentration of

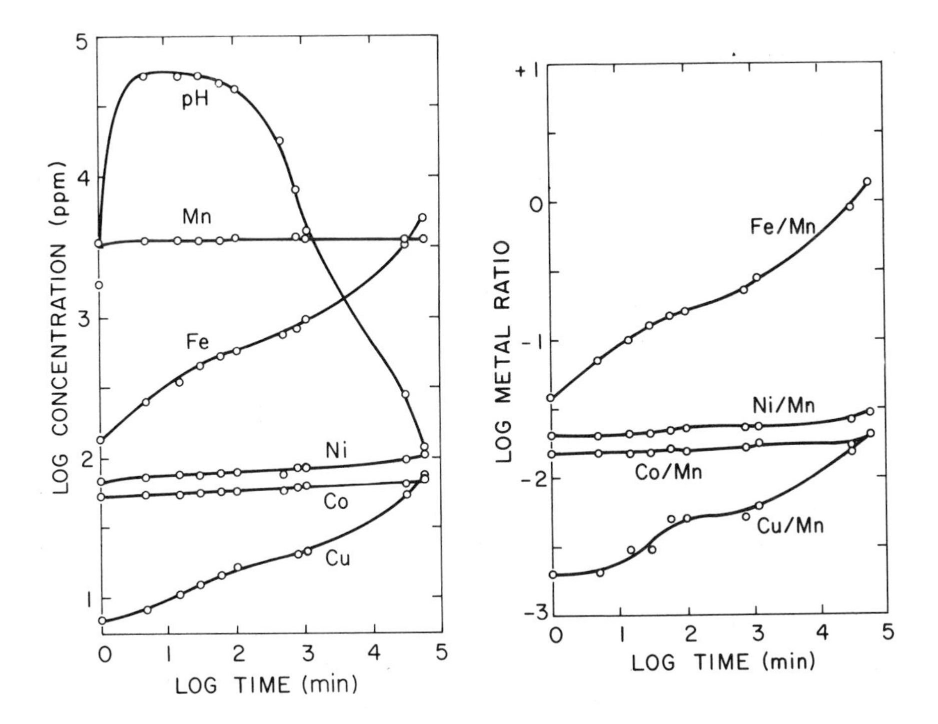

Fig. 3. Log concentration (left) and log metal ratio (right) versus time for extraction of red-brown clay in unbuffered 0.12 molar hydroxylamine solution.

manganese in solution is beginning to change slightly with time in this experiment, the metal ratios are still dominated by the changes in the concentration of the other metals. It is interesting to note that the iron to manganese and copper to manganese ratios are lower than those in the other extraction, but nickel to manganese and cobalt to manganese ratios are similar, again suggesting that the copper, manganese and iron release rates are in a large part controlled by separate phases.

0.103 M H/H (0.54%) Fig. 4. This experiment is very similar in metal release behavior to the previous extraction, an expected result since the hydroxylamine hydrochloride concentrations are very similar. However, the final concentration of manganese in solution rises slightly higher than in the 0.12 M extraction. Iron and nickel are released more slowly initially but after 100 minutes are nearly the same concentration as in the previous extractions.

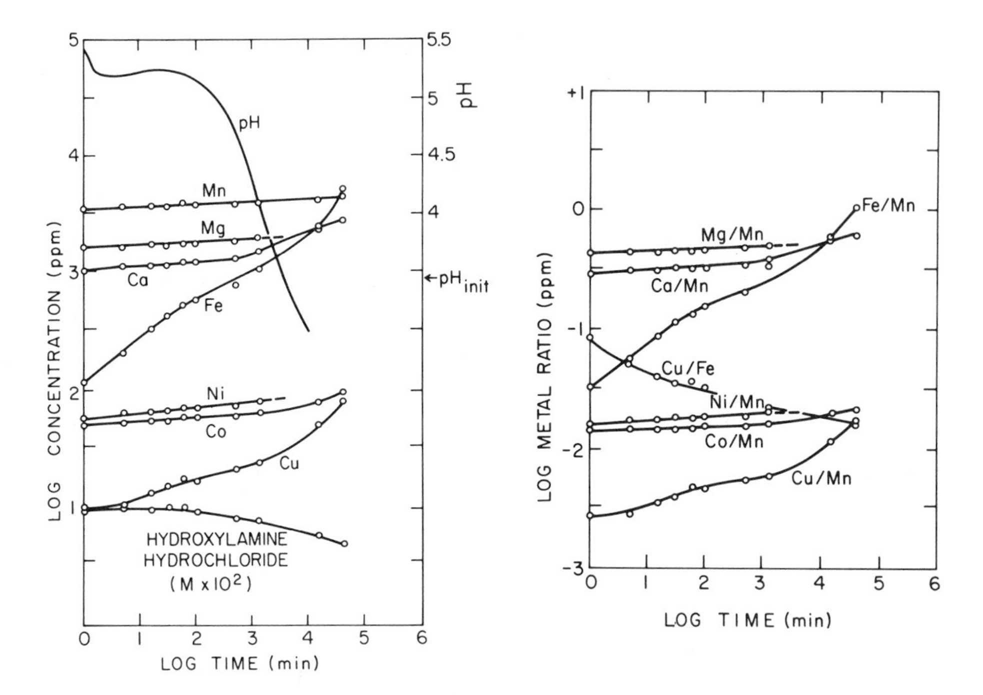

Fig. 4. Log concentration (left) and log metal ratio (right) versus time for extraction of red-brown clay in unbuffered 0.103 molar hydroxylamine solution.

The concentration of hydroxylamine hydrochloride was monitored in this experiment. It decreases in concentration slowly with almost no loss at the beginning of the experiment. The metal ratios are as expected, similar to the 0.12 M extraction results. Although the change in concentration of iron and copper in solution appears similar, a plot of their ratio (Fig. 6b) shows that copper is being released at a slower rate than iron. The dissolution kinetics of the two metals are evidently not controlled by the same phase(s).

0.05 M H/H (0.26%) Fig. 5. 0.05 M hydroxylamine hydrochloride was the weakest concentration used with the red-brown

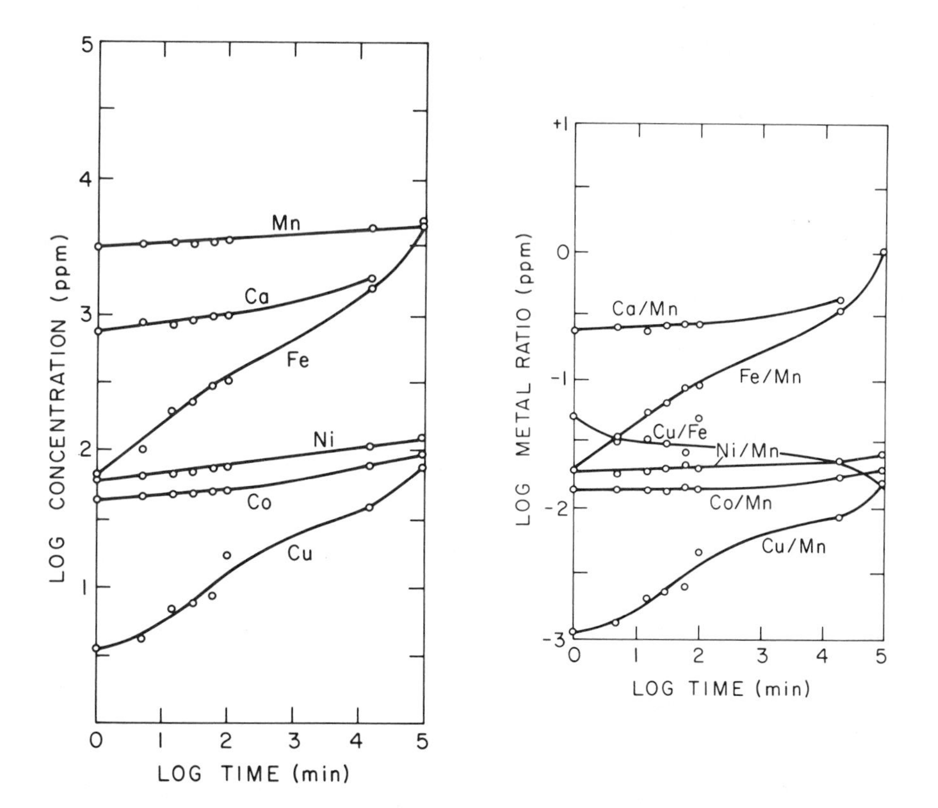

Fig. 5. Log concentration (left) and log metal ratio (right) versus time for extraction of red-brown clay in unbuffered 0.05 molar hydroxylamine solution.

clay extractions. Initial (one minute) and steady-state concentrations of manganese are considerably lower than in the other extractions. Reaction rates are significantly lowered at this extractant concentration and all metals analyzed exhibited significant increases in concentration over the 10^1 to 10^5 minute interval of the experiment. Copper and iron release show the greatest change through time of the six elements analyzed in this experiment. The change in concentration of manganese, nickel, cobalt and calcium appear almost linear on the log-log scale. In these extractions the change in metal to manganese ratios with time is determined by changes in both manganese and copper, nickel, cobalt contents. The ratios of cobalt and nickel to manganese are remarkably similar to those in the previous extractions, suggesting that cobalt and nickel are associated with the manganese phase(s). As with previous experiments, both copper and iron show release rate behavior that suggests these metals occur in different and separate phase(s) than the manganese phase(s).

To demonstrate metal dissolution as a function of the total concentration in the sediment, the percentages of total metal extracted in this experiment are plotted against log time in Fig. 6. A large percentage of the total manganese and cobalt is released over 10^5 minutes, slightly less than half of the nickel and calcium are released, and a very low percentage of iron is dissolved. Such results are expected, since hydroxylamine hydrochloride is too weak a reducing agent to dissolve much iron. In addition a large percentage of iron is tied up in the clay mineral structures in a less chemically refractive form.

Acetate buffered 1.0 M H/H (5.21%) Fig. 7. The results with buffered hydroxylamine hydrochloride extractions are significantly different and serve to point out the importance of the pH variable in controlling the release rate of metals from these sediments. In the buffered solution the continued release of all metals is fairly small after the first five minutes. The concentrations released into solution are similar to those in the 1.2 M extraction. Iron, however, shows a drastic change with buffering, its release was greatly slowed towards the end of the experiment. The metal ratios, Cu/Mn and Co/Mn, increase slightly with time, Fe/Mn and Ni/Mn are nearly constant. The ratios are reflecting the slow change in concentrations throughout the extractions, masking any indication of congruent dissolution.

Sodium Citrate-Dithionite (0.05 M citrate, 0.07 M dithionite) Fig. 7. Although buffered, this experiment resembles the unbuffered hydroxylamine hydrochloride extractions. All the metals analyzed increase in concentration rapidly in solution for the first 30 to 60 minutes, and then increase very slowly. The rate of release of iron is greater than the other four elements, but it too slows after 60 minutes. Because the extractant is less specific for manganese than hydroxylamine hydrochloride, the steady-state concentration for manganese is not reached until 50 to 60 minutes. In addition the initial iron to manganese ratios (at one minute) are higher than for comparable hydroxylamine hydrochloride extractions. Ratios over the first 60 minutes are the most interesting in this extraction because of the rapid change in all the elements during this time. The copper to manganese ratio decreases within the first five minutes, then levels out and finally increases toward the end of the extraction. The nickel to manganese ratio increases slightly with time. The cobalt to manganese ratio is nearly constant, possibly because of the reasons mentioned previously in the 0.05 M hydroxylamine hydrochloride discussion. Of all the ratios, the iron to manganese changes the greatest amount, and, again, is consistent with these metals being controlled by separate phases.

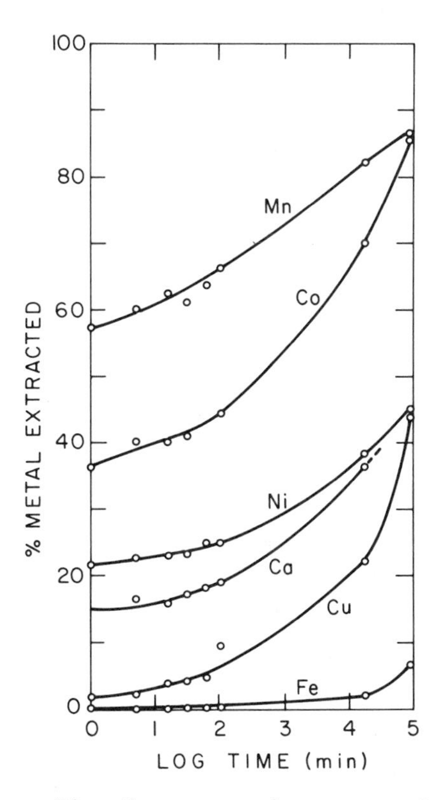

Fig. 6. Data from Fig. 5 expressed as percent of total metal extracted. Percentages based on total analysis data of Table 1.

Summary. The initial rapid metal release and increase in pH is probably due in part to ion exchange and in part to dissolution of oxides. However, simple ion exchange reactions cannot account for the high Mn release. The long-term release rate of Mn, Ni, and Co is steady or decreased through time. Fe and Cu have long term release rates that seem to reflect the drop in pH. Reduction by hydroxylamine hydrochloride appears to have a greater control on the behavior of Mn, Ni, and Co. Comparison of Cu and Fe behavior indicates that either their dissolution is controlled by a different mechanism or, more likely, that the metals are extracted from different sediment phases.

Sequential 0.1H HCl - Hydroxylamine Hydrochloride Extraction Experiments, Fig. 8

The outstanding features of these experiments are the very

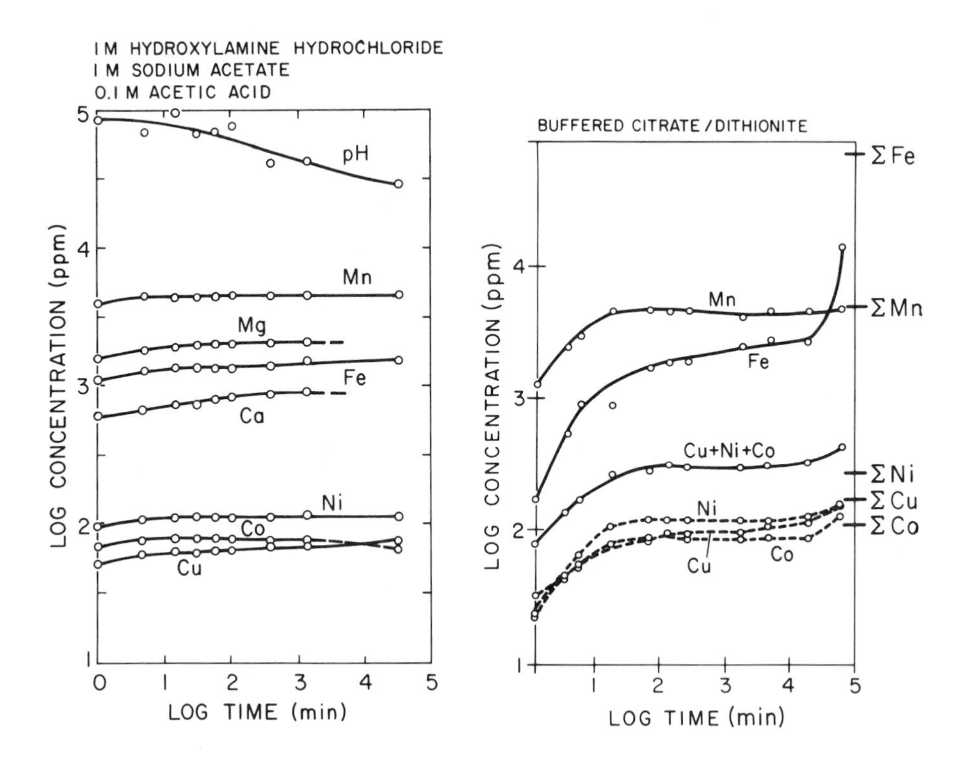

Fig. 7. Concentration-time plots for extraction of red-brown clay in acetate buffered hydroxylamine solution (left) and buffered citrate-dithionite solution (right).

rapid release of Cu in 0.1 N HCl and its subsequent slow release in H/H. The other metals exhibit minimal release in 0.1 N HCl solutions and greatly increased total metal extraction in the H/H solutions.[1] Iron demonstrates relatively low dissolution rates throughout the extractions. Similar behavior was found in red-brown clay, siliceous ooze, calcareous ooze, and hemipelagic sediments by Callender, et al. (in press). The differences in

[1]E. Glover (personal communication) has capitalized on the relative reactivity of iron and manganese oxides and carbonates in weak HCl solutions to remove calcite from manganese micronodules to simplify the interpretation of x-ray powder diffraction data. X-ray and SEM data confirm that for *short* dissolution times the manganese minerals are unaffected.

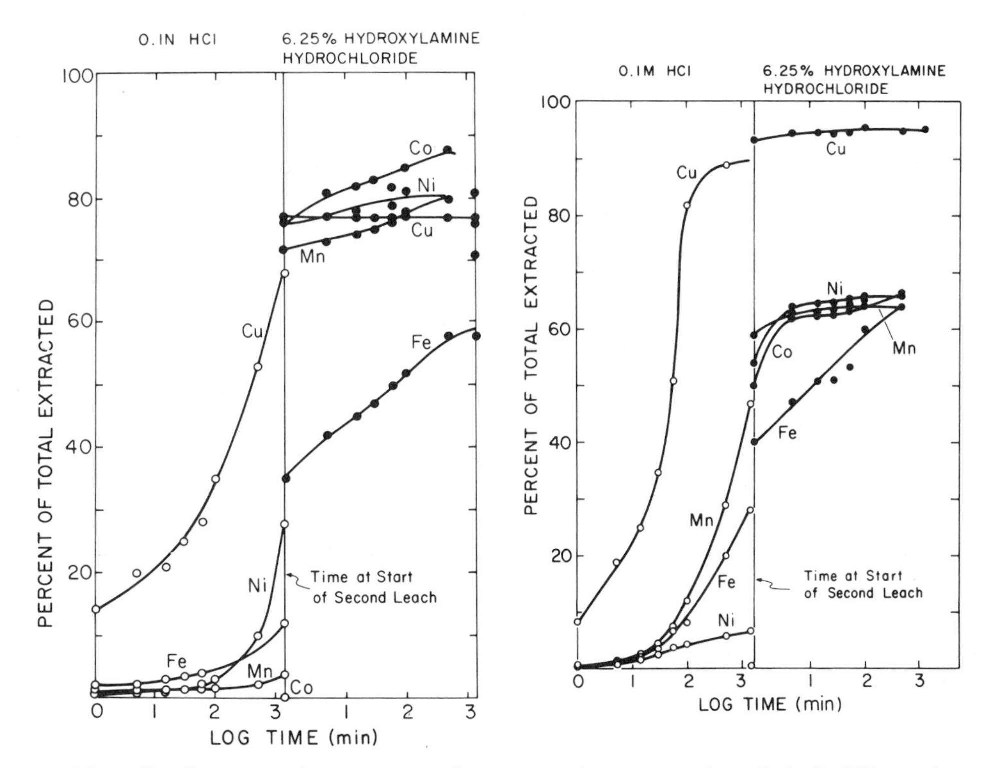

Fig. 8. Consecutive extraction experiments using 0.1 M HCl and 1.2 M hydroxylamine hydrochloride. Data expressed as percent of total metal (from bulk analysis). Open circles are data from HCl leach and closed circles are for following hydroxylamine extraction. "Artificial ooze" (left), Nodule standard (right). (See text for explanation).

percentage of metal released between GRLD-126 and the "artificial calcareous ooze" are likely due to the buffering effect of the $CaCO_3$. Buffering seems to enhance the release of Co, Ni, and Mn in H/H and depress metal release in HCl.

Discussion

Extractions of sediments involve several possible reactions; for example:

Redox

$$Fe_2O_3 + 6H^+ + 2e^- = 2Fe^{2+} + 3H_2O \quad (1)$$

Dissolution

$$Fe_2O_3 + 6H^+ = 2Fe^{3+} + 3H_2O \quad (2)$$

Ion Exchange

$$\ni\text{-Fe} + nH^{+} = Fe^{n+} + \ni\text{-}H_{n} \qquad (3)$$

The phases in which a metal occurs, the phase surface area, its degree of crystallinity, the metal's atomic charge, and the composition and temperature of the solution all determine the phase's reactivity in the extracting solution. In experiments where the relative amounts of these phases or the nature and concentration of the extracting fluids are the determining variables and other factors can be held constant, then the release rates could be tied to specific phases in which the metals are found.

To interpret these data some preliminary assumptions have been made. The reactions occurring in solution at a solid surface can be divided into three parts (Gardiner, 1972): (1) movement of the reactant to the surface - diffusion; (2) reaction at the interface; and (3) movement of the product away from the surface - diffusion. Most dissolution reactions are diffusion controlled (Berner, 1971). Dissolution reactions are sometimes incongruent, involving more than one rate determining mechanism (Pacés, 1973, Busenberg and Clemency, 1976). Other authors (Lerman, et al., 1975; Tsuzuki, et al., 1974) have cited first order dissolution reactions. For the purpose of this study the stirring rate during the extraction was assumed to be sufficient to keep the mixture homogeneous and reduce the importance of diffusion. The reaction at the sediment-solution interface should, therefore, be rate determining.

Because of the divergent extraction behavior of copper, manganese, and iron a reasonable interpretation of the data would be that each of these metals are controlled in part by separate phases and that other metals, especially nickel and cobalt, may be associated with one or another of these phases. In addition it is possible that one or more phases may control the release rate of a specific metal, as for example iron from a high surface area, reactive phase, from discrete, but slower reacting oxide phase, or from partial dissolution of iron bearing silicate phases.

The question of whether or not the metal release rates observed in this study are controlled by heterogeneous or homogeneous dissolution kinetics remains unresolved at the time of this writing. The following discussion is based on the *assumption* that metal release behavior is controlled by homogeneous dissolution.

The data in general do not fit simple first-order dissolution; however, it can be fit using a linear combination of first order rates for two or more phases. If an element is present in more than one phase, such as adsorbed on the surface of a clay or oxide, as an amorphous oxide coating, or bound with organic matter, each phase would presumably dissolve at a different rate. The observed metal behavior may be explained by combining two or three first order rate equations with different initial concentrations and rate constants.

Such order reactions have been cited for natural mineral systems (see for example: Lerman, et al., 1975, for aluminosilicates, and Anderson and Jenne, 1970, for oxides). The general form of a first order rate equation is

$$-dC/dt = kC \tag{4}$$

where: C = metal concentration in the solid phase
k = first order rate constant
t = time

After integration:

$$C = C_o \exp(-kt) \tag{5}$$

where: C_o = the initial concentration of the reactant.

For a mixture of two or more phases the equations are additive, and thus:

$$R = \frac{-d[C]}{dt} = k_1[C_1] + k_2[C_2] + \ldots k_n[C_n] = \sum_{i=1}^{n} k_i[C_i] \tag{6}$$

which, after integration, becomes:

$$C = C_{1_o}\exp(-k_1t) + C_{2_o}\exp(-k_2t) + \ldots C_{n_o}\exp(-k_nt) \tag{7}$$

$$= \sum_{i=1}^{n} C_{i_o} \exp(-k_it)$$

where: k = rate constant
t = time
C = the total concentration of the metal in the combined solids
C_{i_o} = the initial concentration of the metal in the solid phase (numbered subscripts denote separate phases).

An example of a combined reaction involving two independent first order reaction rates for a case involving dissolution is shown in Fig. 9. The ordinate, log concentration of metal *in the solid phase* is plotted against linear time, and shows the effect of various mixtures of the two phases having first order reaction rates differing by three orders of magnitude. For first order reactions, a plot of ln concentration against time will yield a straight line. Fig. 9 shows how the appearance of data plotted on a log concentration (of total solid phase) versus time scale will change if it is controlled by two rate equations. For illustration, the rate constants were arbitrarily chosen to differ

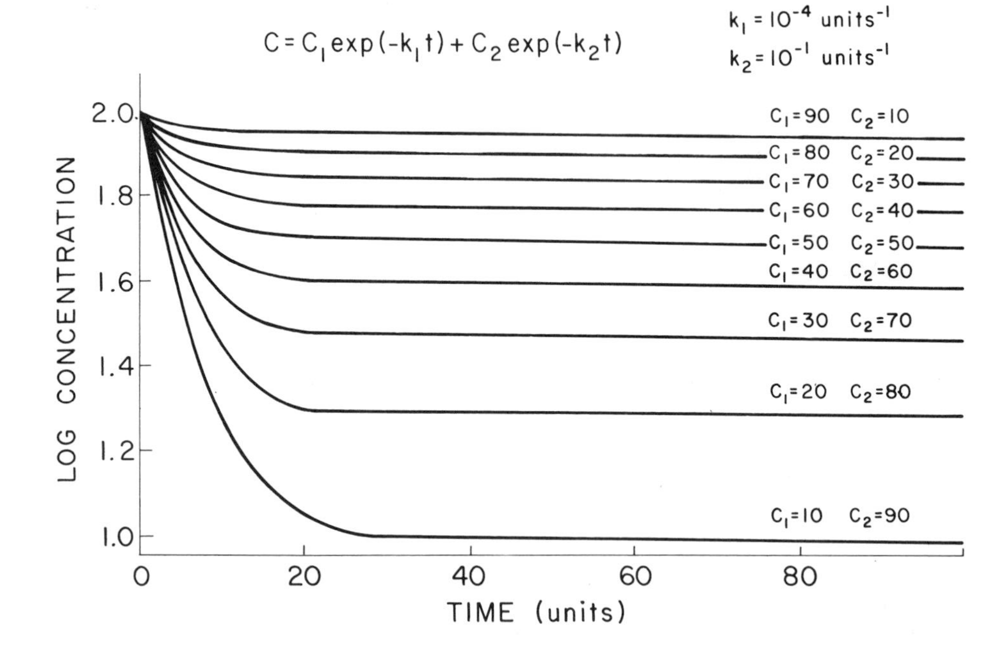

Fig. 9. Calculated plot of log total concentration in solid phase for a two phase mixture of phases with first-order rates that differ by 10^3. Curves are shown for various mixtures of C_1 and C_2 (see text for explanation).

by 10^3, and the initial concentrations of each phase in the solid mixture are varied for each line. It can be seen that the initial decrease in concentration increases as the percentage of the solid phase in the mixture having the highest rate constant increases.

Based on first order kinetics, some of the extraction data demonstrate two linear sections, implying dissolution is controlled by as many phases. Using ln concentration remaining in the sediments, one can fit the data points for the long term portion of the experiment by regression analysis to a straight line. Concentrations calculated from this line were then subtracted from the rest of the data points and the differences in concentration were used to calculate a second first order line. Figs. 10 and 11 show the original data points, the adjusted data points, and the two calculated lines for iron release in 0.5 M

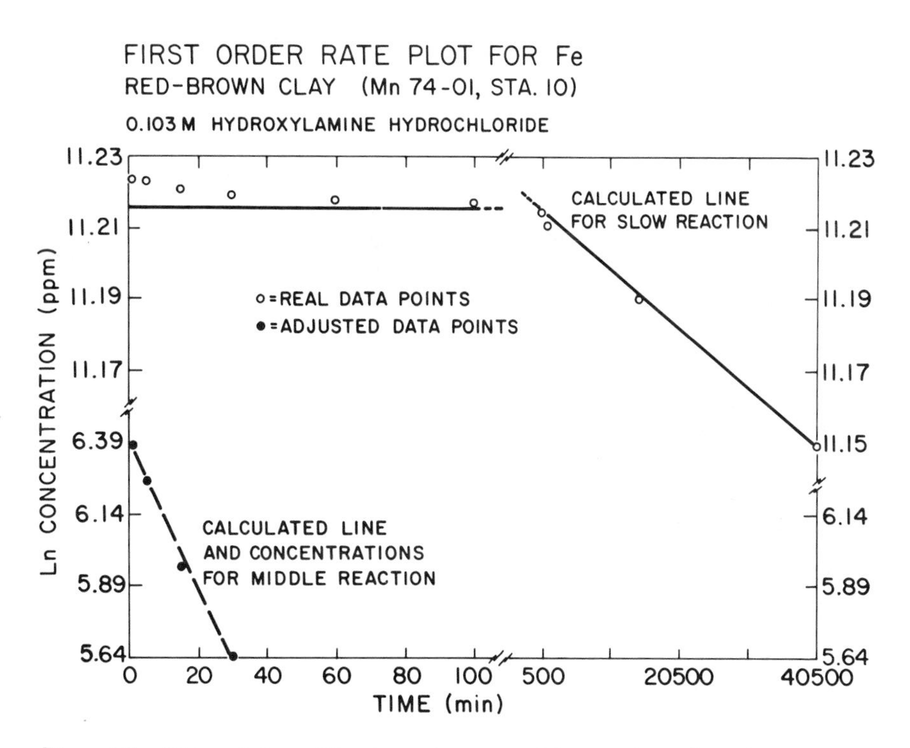

Fig. 10. Fit of two first-order reaction rates for iron dissolution in 0.103 molar H/H solution. Initial (0-1 minute) reaction neglected.

hydroxylamine hydrochloride and copper release in 0.103 M hydroxylamine hydrochloride respectively. A summary of the calculated fits, rate constants, and initial concentrations of each phase is presented in Table 4. The number of points fit for each rate constant line segment, the corresponding time range over which the fit extends, and the regression coefficient for the least squares fit of each line segment are also shown. The column labelled "initial concentration of phase" is the time zero intercept of the linear first-order segment.

In the red-brown clay, iron, copper, nickel, and calcium release can be divided into three sections: an initial rapid release in the first minute, a second intermediate release lasting for 15 to 60 minutes, and a third, slow release. Manganese can be separated into only two phases, an initial rapid release and a slow second release. Since for most experiments manganese comes

TABLE 4

Calculated First Order Parameters for Red-Brown Clay, Mn 74-01 STA10

Metal	H/H Conc. (m)	Initial Release (ppm)	Intermediate Release				
			Time Range (min)	No. of Points	Initial Conc. of Phase (ppm)	K (min^{-1})	r^2
Fe	0.05	80	1-30	4	340	$2.05x10^{-2}$	0.996
	0.10	130	1-30	4	590	$2.53x10^{-2}$	0.991
	0.12	160	1-30	4	570	$2.57x10^{-2}$	0.974
	0.50	450	1-15	3	780	$3.83x10^{-2}$	0.987
	1.20	540	1-15	3	430	$12.90x10^{-2}$	0.999
Mn	1.00 buffered	4464	1-36000	9	996	$8.44x10^{-6}$	0.998
Ca	0.05	780	1-30	4	200	$3.58x10^{-2}$	0.976
	0.50	1350	1-30	4	280	$5.29x10^{-2}$	0.993
	1.00 buffered	660	1-100	5	220	$1.73x10^{-2}$	0.975
Cu	0.05	4	1-60	5	17	$0.78x10^{-2}$	0.999
	0.10	10	1-30	4	11	$2.21x10^{-2}$	0.995
	0.12	7	1-30	4	11	$2.49x10^{-2}$	0.980
	0.50	49	1-30	4	17	$5.60x10^{-2}$	0.999
	1.20	51	1-15	3	7.9	$3.55x1C^{-2}$	0.859
Ni	0.05	64	1-30	4	20	$1.25x10^{-2}$	0.999
	0.10	60	1-30	4	14	$3.23x10^{-2}$	0.908
	0.12	70	1-30	4	11	$2.33x10^{-2}$	0.822
	0.50	94	1-15	3	10	$4.21x10^{-2}$	0.992
	1.20	100	1-60	5	4.5	$1.55x10^{-2}$	0.967

TABLE 4, continued

Metal	H/H Conc. (m)	Slow Release: Time Range (min)	No. of Points	Initial Conc. of Phase (ppm)	K (min^{-1})	r^2
Fe	0.05	60-89200	4	74580	7.24×10^{-7}	0.994
	0.10	60-40400	6	74280	1.65×10^{-6}	0.994
	0.12	60-60480	7	74270	1.05×10^{-6}	0.993
	0.50	30-89350	5	73770	3.35×10^{-6}	0.998
	1.20	30-56000	7	74030	3.92×10^{-6}	0.991
Mn	1.00 buffered					
Ca	0.05	60-17482	3	4300	1.44×10^{-5}	0.999
	0.50	60-17625	3	3650	1.46×10^{-5}	0.999
	1.00 buffered	430-36000	3	4400	4.63×10^{-6}	0.999
Cu	0.05	100-89000	3	159	5.16×10^{-6}	0.982
	0.10	60-40400	6	159	1.25×10^{-5}	0.993
	0.12	60-60480	7	162	7.11×10^{-6}	0.985
	0.50	60-89350	4	114	1.42×10^{-5}	0.999
	1.20	30-56000	7	120	6.17×10^{-6}	0.935
Ni	0.05	60-89000	4	200	3.04×10^{-6}	0.827
	0.10	60-40400	6	211	1.54×10^{-5}	0.988
	0.12	60-60480	7	204	2.61×10^{-6}	0.933
	0.50	30-89350	5	182	9.10×10^{-6}	0.994
	1.20	100-56000	5	175	7.68×10^{-7}	0.826

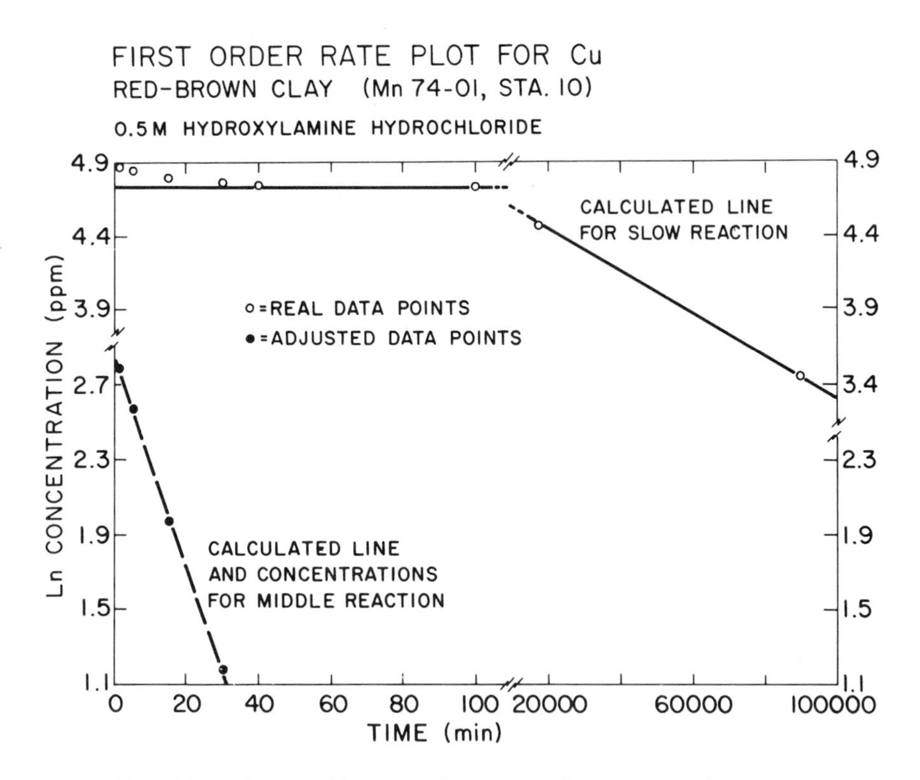

Fig. 11. Fit of two first-order reaction rates for copper dissolution in 0.5 molar H/H solution. Initial (0-1 minute) reaction neglected.

to a rapid steady-state concentration only the buffered hydroxylamine hydrochloride extraction fits well to a first order rate equation of the form:

$$[Mn] = 4464 \text{ mg/kg} + 996 \text{ mg/kg} \exp(8.44\text{x}10^{-6}\text{min}^{-1}t) \quad (8)$$

Iron release after one minute in all the unbuffered extractions agreed with a combination of two first order reactions. The rate equation corresponding to Fig. 10 (0.103 M extraction) is

$$[Fe] = 450 \text{ mg/kg} + 780 \text{ mg/kg} \exp(3.83\text{x}10^{-2}\text{min}^{-1}t) +$$

$$73770 \text{ mg/kg} \exp(3.35\text{x}10^{-6}\text{min}^{-1}t) \quad (9)$$

The dissolution of most of the iron is controlled by the slowest rate. In general, the rate constant increases with increasing hydroxylamine hydrochloride strength. The two phase control for iron dissolution is similar to the results of Anderson and Jenne (1970) for their experiments with citrate-dithionite extractions of API reference clays.

A weaker correlation to combined first order reactions was obtained for copper and nickel release. From their kinetic behavior, these metals appear similar (Table 4). An example of the rate equations for copper, and nickel in 0.103 M hydroxylamine hydrochloride follows:

$$[Cu] = 7\ mg/kg + 11\ mg/kg\ exp(2.48x10^{-2}min^{-1}t) + 162\ mg/kg\ exp(7.11x10^{-6}min^{-1}t) \quad (10)$$

$$[Ni] = 70\ mg/kg + 11\ mg/kg\ exp(2.33x10^{-2}min^{-1}t) + 162\ mg/kg\ exp(2.61x10^{-6}min^{-1}t) \quad (11)$$

The behavior of copper in the sequential extractions and in the hydroxylamine hydrochloride extractions indicates a source different from that of manganese or iron. Some copper and nickel is probably associated with the manganese phase, giving the rapid initial release in hydroxylamine hydrochloride. Some of the rapid dissolution of copper, nickel and cobalt may be due to release of these cations from interlayer sites in the structure of Mn minerals such as birnesite (Burns and Burns, 1977). However, another phase must be responsible for the long term release. Boyle, et al. (1977) conclude on the basis of dissolved copper profiles in the Pacific Ocean water column, that a strong source of copper must exist near the sediment surface. Callender, et al. (in press) suggest that rapid oxidation of organic matter present at the sediment-water interface could provide the observed copper fluxes from sediment to ocean water. This is in agreement with the observed preferred copper dissolution in HCl over dissolution in a reducing agent, hydroxylamine hydrochloride, and its anomalous behavior compared to iron and manganese. Sclater, et al. (1976) believed that ocean-water profiles of nickel in the Pacific Ocean resulted from a biogenic cycle and suggested that it may be controlled by incorporation into siliceous tests. Since nickel does not exhibit the same behavior as copper in HCl, it is unlikely that much is bound with organic matter. The long term release of nickel in hydroxylamine hydrochloride may be controlled by the dissolution of siliceous phases. This would account for the difference between iron, nickel and copper release.

At this point strict interpretation of the total number of phases controlling dissolution behavior of each element is less than satisfactory. Because pH is the main uncontrolled parameter the modelling presented herein ignores the rate dependence of solid dissolution on pH. Such is very important in the carbonate system, as shown by Berner and Morse (1974). The changes in both copper and iron release rates after 100 minutes in the 0.12 model hydroxylamine hydrochloride extraction (Fig. 4) may, in part, reflect such a pH change.

Reasons for the observed pH changes in the unbuffered hydroxylamine hydrochloride solutions are not complete at this time. Some of the initial rise in pH may be due to simple exchange reactions. Later decreases in pH are probably related to the stability of hydroxylamine hydrochloride solutions in extraction mixtures. Reactions can be written for hydroxylamine which yield four hydrogen ions per mole and nitrogen gas as breakdown products.

Additional extraction experiments have been performed using marine siliceous ooze and carbonate ooze in the same extractants (B. Mills, M.S. Thesis, unpublished). Results of these experiments are the subject of another paper (in preparation).

Relationship to Sediment Diagenesis

Of interest to the problems of deep-sea sediment diagenesis and manganese nodule formation is the association of minor elements (Cu, Ni, Co) with the "major" transition elements (Mn, Fe). In this study the release of copper, manganese and iron are shown to be quite unlike in their dissolution behavior. Cobalt and manganese are alike inasmuch as for most extractions both elements appear to come rapidly to steady-state concentrations. For that reason neither element was amenable to dissolution rate analysis, except in the cases involving very low hydroxylamine hydrochloride concentrations.

The evident association of some of the manganese and cobalt in sediment phases (presumably an oxide) can be explained by crystal field theory. McKenzie (1970) suggested that low spin Co^{3+} in the crystal structure of oxides results in a gain in crystal field stabilization energy, therefore it seems likely that the cobalt that is released is associated with the extractable manganese phase(s). The major difference between cobalt and manganese is in the percentage of total metal that is extracted (60 to 90 percent of manganese is removed in the red-brown clay hydroxylamine hydrochloride extractions and only 20 to 60 percent of cobalt is extracted). Evidently there is a large fraction of cobalt that is in phase(s) resistant to reduction by hydroxylamine hydrochloride solutions.

Manganese has been found to accumulate at roughly the same rate in nodules and in the surrounding sediments (Kraemer and Schornick, 1974). The authors argued that manganese is present as an authigenic phase in both sediments and nodules. Manganese is probably transported to and deposited in the sediments as an oxide, possibly associated with fine grained crystalline to colloidal terrigenous material (Elderfield, 1977; Callender, et al., in press). These particulates have a high surface area as well as a high negative charge, the zero point of charge being between pH 3 and 7, and the average pH of seawater at pH 8.1 (Harriss and Troup, 1970). The oxides therefore attract positively charged ions, behaving somewhat like a weak acid (Posselt, et al., 1968). The relative strength of adsorption of ions from seawater has been argued to be (Jefferies and Stumm, 1976):

$$Mn > Fe > Co > Cu = Ni > Ca > Mg$$

The preference for cobalt adsorption over copper and nickel may contribute to the association of manganese and cobalt. Once manganese is deposited in the sediment it can be diagenetically remobilized (Callender, et al., in press; Li, et al., 1969; Calvert and Price, 1972). Remobilization of the manganese oxides would also free adsorbed ions. The extraction data indicate that nearly all of the manganese in the red-brown clay is present in an easily reducible oxide form which dissolves almost within the first minute of the extraction. Of the other metals released during this time, a portion is probably from ion exchange reactions with clay minerals and other oxides and another fraction is probably associated directly with the manganese phase and/or is present as an equally easily reducible oxide. The present data do not allow for separation of these phases.

Iron release rates seem to indicate more than one phase is present in the sediments. The first order plots resolve into three reaction segments for the red-brown clay, possibly representing iron occurring as adsorbed ions and two oxide phases, one perhaps more crystalline than the other. An additional possible source is iron from structural sites within the clay minerals. Clearly the bulk of the iron is present in the chemically more refractive clay minerals, in agreement with the results of Kraemer and Schornick (1974) and Chester and Missiha-Hanna (1970) who found that up to 82 percent of the total iron in pelagic sediments is detrital.

A biological control has been proposed for copper (Bostrom, et al., 1973; Greenslate, 1975; Arrhenius, 1963; Boyle, et al., 1977). Callender, et al. (in press) calculated that rapid diagenesis of organic matter could provide observed copper fluxes. In these extractions copper behaves very differently than the other elements, especially in the sequential HCl/hydroxylamine hydrochloride solution.

First order and parabolic kinetics for copper divide into two reaction segments, indicating two phase control. Based on earlier literature, we would suggest that the copper bearing phases are organic matter and authigenic phases formed by an earlier remobilization and redeposition of copper. Presumably the latter phases are manganese oxides as micronodules and grain coatings.

The first order kinetics for nickel in the red-brown clay exhibit a behavior somewhat in between copper and cobalt. Sclater, et al., (1976) found nickel profiles in the Pacific Ocean water column suggesting that nickel is involved in the biogeochemical cycle. Since nickel and copper do not behave identically in the extractions, their phases (either as the initially deposited biogenic form or as an authigenic form such as micronodules), may well be different. It is possible that both nickel and copper are present in the sediments as diagenetically unaltered debris (organic) and part already incorporated into manganese micronodules or as oxide coatings on grains. The relative amounts of Cu and Ni proportioned between these two phases may, however, be different.

Summary of Results

In summary, the extractions by hydroxylamine hydrochloride are characterized by the following four points:
(1) very rapid release of metals, especially manganese, into solution,
(2) rapid attainment of steady-state concentrations for manganese and cobalt,
(3) long term dissolution of copper and iron, indicating that these metals are controlled by separate phases,
(4) data can be fit to kinetic models compatible with first order, with the assumption of homogeneous dissolution fundamental to such interpretation.

From interpretation of data presented herein, and consideration of metal/phase associations suggested by other workers in the field, the following transition metal bearing phases in the sediments studied are suggested:
(1) diagenetically immature - organically complexed metals,
(2) amorphous and crystalline iron and manganese oxides (and associated trace metal oxides),
(3) biogenic and inorganic debris - calcite and amorphous silica,
(4) refractory, detrital silicates - clay minerals.
The silicate, amorphous silica, and to some extent the carbonate phases show relatively little reactivity with extracting solutions. The silica and carbonate phases serve as buffers for the extractions, especially where they comprise the dominant phase in the sediment.

The proposed metal phase associations for each of the metals studied are given below:

Manganese. One or more oxide phases, partly as clastic grain coatings and discrete oxide grains (micronodules), a portion of which is probably a diagenetic alteration product.

Iron. Present in two (possibly three) phases as amorphous ferric iron oxide or crystalline ferric iron oxide, again either as clastic grain coatings or incorporated within manganese micronodules.

Copper. Present in two unrelated phases: (1) organically complexed and diagenetically immature particulate organic matter, and (2) associated with manganese oxide phase(s) (diagenetically mature). Depending on the relative amounts of organic bound and manganese oxide associated copper in the sediment, copper shows extraction behavior either different from or very similar to the manganese oxide phases. Lowering the hydroxylamine hydrochloride concentration shows pronounced effects on the rate of copper release from especially the red-brown clay, but has little effect on the very rapid release of manganese, again indicating different phase controls for both manganese and copper. Undoubtedly *some* copper in each sediment studied is associated with the manganese phase(s).

Nickel. Possibly two phases, manganese oxides and some organic matter, but in a large part associated with manganese oxide phases. Nickel to manganese ratios are nearly constant over the course of the experiments and are independent of hydroxylamine hydrochloride concentration used.

Cobalt. Evidently, cobalt is in part associated with manganese oxide phase(s) and is similar to nickel. Although commonly cited as being more often associated with iron oxides in nodules (Burns and Furstenau, 1966; Burns and Burns, 1977), it has also been claimed to be associated with manganese oxides as well (Burns and Burns, 1977; Ostwald and Frazier, 1973). The valence state of cobalt, either 2+ or 3+, is important in interpreting its manganese and iron associations. As with nickel, the slight increase in cobalt to manganese ratios of the longest time samples may be due to dissolution of more refractory phase(s) such as iron oxides or perhaps some silicate minerals.

This study is part of a continuing investigation of the dissolution behavior of modern marine sediments. Similar timed extractions on well characterized sediments or specific oxides is needed to fully understand the kinetics of hydroxylamine hydrochloride and HCl metal extractions. Known kinetic behavior of several forms of oxides could be then combined to better explain the extraction behavior of a real sediment. A series of extractions of sediments from different depths in a core to give a profile of the changes in extractable metals with depth should also be useful. This information, combined with pore fluid profiles, would yield invaluable information on the diagenetic processes at work in deep ocean sediments.

Experiments of the kind presented here probably generate more problems than are solved. Natural sediments are clearly complicated mixtures of phases having great differences in their capacity to be altered or formed in diagenetic environments. The approach to the problems of selective phases extraction using natural materials is bound to lead to ambiguous data interpretation. Ultimately one is driven to do comparable studies of simpler and better defined natural and artificial phases and phase mixtures to understand the nature of the diagenetic processes of specific phases. Some experiments of this kind have already been carried out (for example, Petrovic, 1976, and Wollast, 1967, on the dissolution of feldspars).

On the other hand, numerous empirical selective extraction schemes have been proposed to characterize bulk natural materials (Gibbs, 1977; Chester and Hughes, 1967; Mehra and Jackson, 1960; Callender et al., in press; Heath and Dymond, 1977, etc.). These schemes have variously led to definitional concepts of the "organically bound" fraction, "carbonate" fraction, "free iron oxide" fraction, "amorphous iron oxide" fraction, "sulfide" fraction, and "refractory" or silicate fraction, to name several. As hopefully is evident, from some of the studies presented here, some of these extraction schemes do not lead to unambiguous interpretations in terms of metal/phase associations. Investigations of the kind presented in this thesis are needed, however, to enhance understanding of metal/phase relationships and diagenetic reactivity of sediment phases under various conditions.

Acknowledgments

We would like to thank Dr. James Greenslate for the total analysis of the red-brown clay, and Mr. Kirk Morgen for the particle size analysis. Preliminary work on the extraction method and accuracy by Bob Karnauskas and Craig Eisen is greatly appreciated. This work was supported by funds from NSF-IDOE to the University of Wisconsin (ID075-12955, OCE77-01005).

References

Anderson, B.J., and Jenne, E.A., (1970) Free-iron and -manganese oxide content of reference clays, Soil Science 109, 163-169.

Arrhenius, G., ((1963) Pelagic sediments, In The Sea 3, M.N. Hill (ed.), Interscience Publishers, New York, 655-727.

Bender, M.L., (1971) Does upward diffusion supply the excess manganese in pelagic sediments?, J. Geophys. Res.76, 4212-4215.

Berner, R.A., (1971) Principles of Chemical Sedimentology, McGraw Hill, New York.

Berner, R.A., and Morse, J.W., (1974) Dissolution kinetics of calcium carbonate in seawater, IV, Theory of calcite dissolution, Am. Jour. Sci. 274, 108-134.

Bishop, J.K., Edmond, J.M., Ketten, D.R., Bacon, M.P., and Silker, W.B., (1977) The chemistry, biology, and vertical flux of particulate matter from the upper 400 m of the equatorial Atlantic Ocean, Deep-sea Research 24, 511-548.

Bostrom, K., Kraemer, T., and Gartner, S., (1973) Provence and accumulation rates of opaline silica, Al, Ti, Mn, Cu, Ni, and Co in Pacific pelagic sediments, Chem. Geol. 11, 123-148.

Boudreau, B.P., and Scott, M.R., (1978) A model for the diffusion controlled growth of deep-sea manganese nodules, Am. J. Sci. 278, 903-929.

Boyle, E.A., Sclater, F.R. and Edmond, J.M., (1977) The distribution of dissolved copper in the Pacific, Earth and Planet. Sci. Lett. 37, 38-54.

Burns, R.G. and Burns, V.M., (1977) Mineralogy of ferromanganese nodules, In Marine Manganese Deposits, G.P. Glasby (ed.), Elsevier, Amsterdam, 185-249.

Burns, R.G. and Fuerstenau, D.W., (1966) Electron-probe determination of inter-element relationships in manganese nodules, Am. Min. 55, 895-902.

Busenberg, E., and Clemency, C.V., (1976) The dissolution kinetics of feldspars at 25°C and one atmosphere CO_2 partial pressure, Geochimica et Cosmochimica Acta 40, 41-49.

Callender, E., Bowser, C.J., and Shedlock, R., (in press) Manganese and copper geochemistry of interstitial fluids from manganese nodule-rich pelagic sediments of the northeastern equatorial Pacific Ocean, Am. Jour. Sci.

Calvert, S.E., and Price, N.B., (1972) Diffusion and reaction profiles of dissolved manganese in the pore waters of marine sediments, Earth and Planet. Sci. Lett. 16, 245-249.

Chao, T.T., (1972) Selective dissolution of manganese oxides from soils and sediments with acidified hydroxylamine hydrochloride, Soil Science Soc. America Proceed. 36, 764-768.

Chester, R., and Hughes, M.J., (1967) A chemical technique for the separation of ferro-manganese minerals, carbonate minerals and adsorbed trace elements from pelagic sediments, Chem. Geol. 2, 249-269.

Chester, R. and Messiha-Hanna, R.G., (1970) Trace element partition patterns in north Atlantic deep-sea sediments, Geochimica et Cosmochimica Acta 34, 1121-1128.

Elderfield, H., (1977) Manganese fluxes to the oceans, Marine Chem. 4, 103-132.

Gardiner, Jr., W.C., (1972) Rates and Mechanisms of Chemical Reactions, W.A. Benjamin, Menlo Park, California.

Gibbs, R., (1977) Transport phases of transition metals in the Amazon and Yukon Rivers, Geol. Soc. Am. Bull. 88, 829-843.

Greenslate, J.L., (1975) Manganese-biota associations in northeastern equatorial Pacific sediments, Unpubl. PhD Thesis, Univ. Calif.-San Diego.

Harriss, R.C., and Troup, A.G., (1970) Chemistry and origin of fresh-water ferromanganese concretions, Limnology and Oceanography 15, 702-712.
Heath, G.R., and Dymond, J., (1977) Genesis and transformation of metalliferous sediments from the East Pacific Rise, Bauer Deep, and Central Basin, northwest Nazca Plate, Geol. Soc. Am. Bull. 88, 723-733.
Jefferies, D.S. and Stumm, W., (1976) The metal adsorption of Buserite, The Canadian Mineralogist 14, 16-22.
Kraemer, T. and Schornick, J.C., (1974) Comparison of elemental accumulation rates between ferromanganese deposits and sediments in the south Pacific Ocean, Chem. Geol. 13, 187-196.
Lerman, A., MacKenzie, F.T., and Bricker, O.P., (1975) Rates of dissolution of aluminosilicate in seawater, Earth and Planet. Sci. Lett. 25, 82-88.
Li, Y-H, Bischoff, J.L. and Mathieu, G., (1969) The migration of manganese in the Arctic Basin sediment, Earth and Planet. Sci. Lett. 7, 265-270.
McKenzie, R.M., (1970) The reaction of cobalt with manganese dioxide minerals, Australian Jour. of Soil Res. 8, 97-106.
Mehra, O.P. and Jackson, M.L., (1960) Iron oxide removal from soils and clays by a dithionite-citrate system buffered with sodium bicarbonate, Clays and Clay Min., Proceedings National Conference 7, 317-327.
Mills, B.A., (1978) Selective kinetic extractions of some northeastern equatorial Pacific pelagic sediments, Unpubl. M.S. Thesis, Dept. Geology and Geophysics, Univ. of Wisconsin, Madison.
Ostwald, J. and Frazier, F.W., (1973) Chemical and mineralogical investigations on deep-sea manganese nodules from the Southern Ocean, Miner. Deposita 8, 303-311.
Paces, Tomas, (1973) Steady state kinetics and equilibrium between groundwater and rock, Geochim. Cosmochim. Acta 37, 2641.
Petrovic, R., (1976) Rate control in feldspar dissolution - II, The protective effect of precipitates, Geochim. Cosmochim. Acta 40, 1509-1521.
Posselt, H.S., Anderson, F.J., and Weber, W.J., Jr., (1968) Cation sorption on colloidal hydrous manganese dioxide, Environmental Science and Technology 2, 1087-1093.
Rabb, W., (1967) Physical and chemical features of Pacific deep-sea manganese nodules and their implications to the genesis of nodules, in Ferromanganese Deposits on the Ocean Floor, D.R. Horn (ed.), Harriman, New York.
Schwertmann, U., (1964) Differenzierung der Eisenoxide des Bodens durch Extraktion mit Ammoniumoxalate-Losung, Z. Plansenernachr Dueng. Bodenk., 105, 194-201.
Sclater, F.R., Boyle, E., and Edmond, J.M., (1976) On the marine geochemistry of nickel, Earth and Planet. Sci. Lett. 31, 119-128.

Strickland, J.D.H., and Parsons, T.R., (1968) A Practical Handbook of Seawater Analysis, Bulletin of the Fisheries Research Board, Canada, 167 p.

Tsuzuki, Y., Mizutani, S., Shimizu, H., and Hayashi, H., (1974) Kinetics of alteration of K-feldspar and its application to the alteration zoning, Geochem. Jour. 8, 1.

Wollast, R., (1967) Kinetics of the alteration of K-feldspar in buffered solutions at low temperature, Geochim. Cosmochim. Acta 31, 635-648.

INFLUENCE OF DEEP WATER CIRCULATION AND SEA FLOOR MORPHOLOGY ON THE ABUNDANCE AND GRADE OF CENTRAL SOUTH PACIFIC MANGANESE NODULES

Guy Pautot and Marthe Melguen

CNEXO - Centre Océanologique de Bretagne
B.P. 337 - 20273 Brest Cedex, France

Abstract

Analyses of polymetallic nodules from the central south Pacific and from the underlying sediments indicate that nodule abundance is at least partly related to the degree of carbonate dissolution, which, itself, is strongly influenced by the flow of the Antarctic Bottom Water. The greatest nodule abundance is generally encountered in a 300 to 400 meter thick water layer situated between the lysocline and the calcite compensation depth levels. This range of depth is the first and main controlling factor of abundance. Inside this range, the bathymetry represents a secondary factor influencing both abundance and grade. Some correlations appear locally between nodule grade and abundance. Nodule grade seems to be generally related to the bottom morphology. Ni, Cu and Mn grades are positively correlated and are highest in topographic lows. On the contrary, Co and Cu are negatively correlated and Co presents higher grade on topographic highs.

Introduction

Recent studies undertaken by the Centre National pour l'Exploitation des Océans (Pautot and Hoffert, 1974; Pautot and Melguen, 1975; Pautot and Melguen, 1976) have stressed the importance of the environment of polymetallic nodules in the Central South Pacific.

Contribution n° 627 du Département Scientifique du Centre Océanologique de Bretagne.

Structural Framework of the Studied Area

The Tuamotu Archipelago (Fig. 1) bound in the North by the Marquesas fracture zone, has been mapped in a general way by Menard (1964), Mammerickx et al. (1975) and in more detail by Monti and Pautot (1974). The Marquesas fracture zone (Fig. 1) was described by Menard (1964) and mapped by Mammerickx et al. (1972). The Marquesas Islands Archipelago (Fig. 1), composed of a modern (Plio-Pleistocene) volcanic mass (Duncan and McDougall, 1974), is about 500 km long and is oriented NW-SE. The Society Islands Archipelago is part of a complex similar in size to the Marquesas Archipelago and with a trend which is nearly parallel.

The Tiki and the Tapu Basins (Figs. 2, 3) respectively situated to the south and west of the Marquesas Islands, correspond to two large depressions (water depth respectively greater than 4000 m and 5000 m) with respect to the average water depth in the studied area.

The Austral-Cook Archipelago (Fig. 4) extending from the Cook Islands to the Tubuai Islands, is oriented in the same direction as the Tuamotu Archipelago and is cut by Austral-Mururoa Fracture zone.

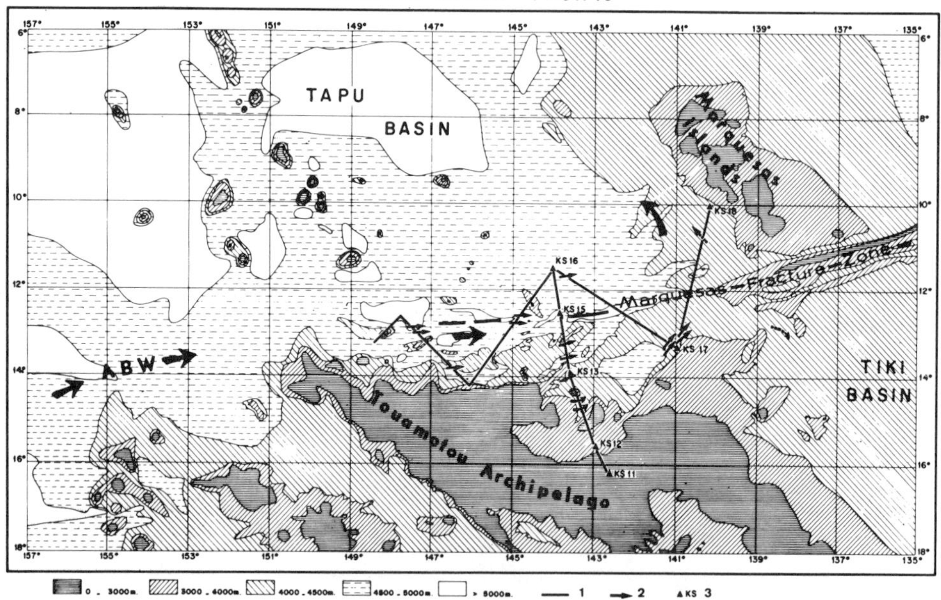

Fig. 1. Schematic map from the Tuamotu-Marquesas area after bathymetric maps from Monti and Pautot (1974). Areas with different depths levels are represented by different symbols. 1: tracks from O.V. Le Noroit during Transpac I cruise. 2: erosion features determined on seismic profiles. KS: position of gravity cores. Big arrows and dotted arrows represent hypothetical flow of AABW (Antarctic Bottom Water).

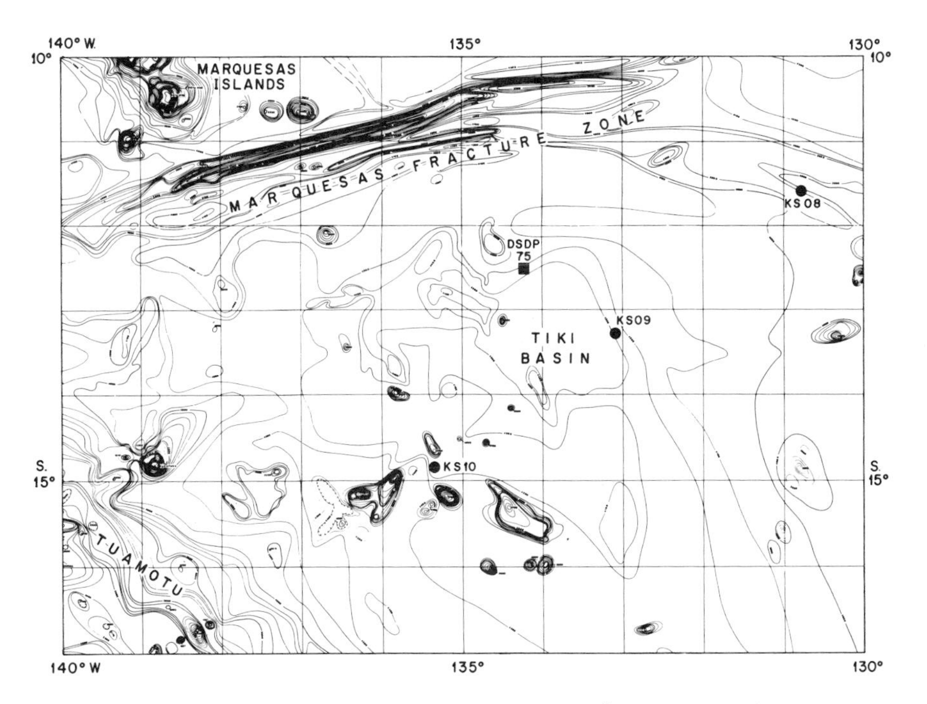

Fig. 2. Bathymetry of the Tiki Basin (after Monti and Pautot, 1974) with positions of cores (KS) taken during TRANSPAC I cruise.

The Aitutaki Area (Fig. 4, 5), situated between the western end of the Southern Cook Islands and the Manihiki Plateau, has been mapped by Mammerickx et al. (1975). A deep channel in this area was named Aitutaki Passage by Pautot and Melguen (1976) after the name of the nearest island.

The Tonga Area (Fig. 5), is located east of the Tonga trench and south of 20° S. Water depth is in general greater than 4500 m.

Objectives

The purpose of this study is to examine our previous hypothesis (Pautot and Melguen, 1976) that the nodule distribution is closely linked to the distribution of the Antarctic Bottom Water and, therefore, to the hydrographic lysocline level as defined by Peterson (1966) and later discussed by Heath and Culberson (1970), Edmond (1971) and Berger (1974).

New data concerning the environment, distribution and composition of the nodules in the Central South Pacific have now been kindly made available to us by the CNEXO/SLN group[1] and AFERNOD[2].

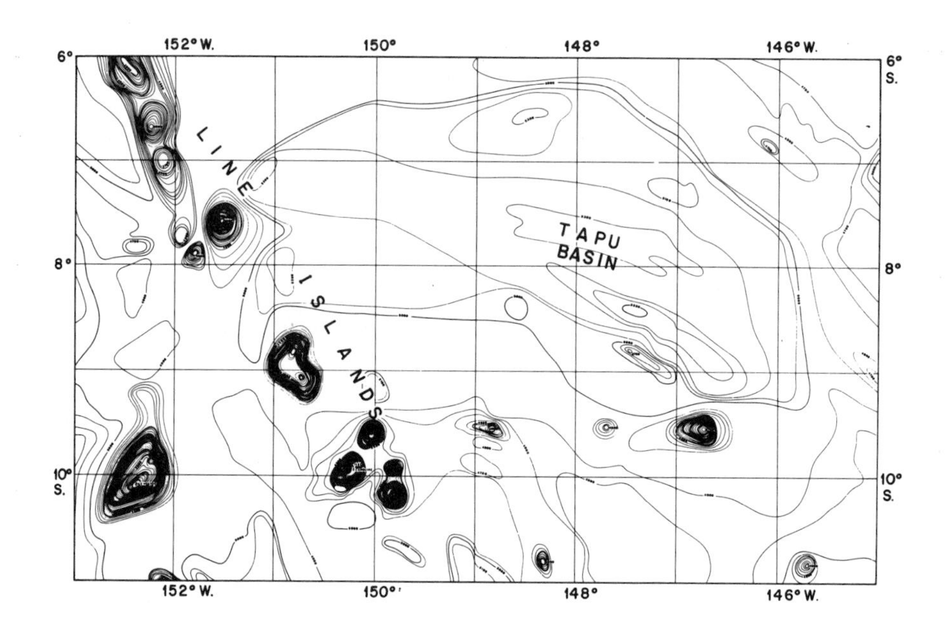

Fig. 3. Bathymetry of the Tapu Basin (after Monti and Pautot, 1974).

We also received samples (nodules and sediments) from the ORSTOM[3] Noumea team. The new data, associated with older results obtained during CNEXO/COB[4] cruises TRANSPAC I (Panama-Tahiti) and TRANSPAC II (Mururoa-Callao), have allowed us to further test our previous hypothesis and to study the importance of bottom topography on nodule abundance and grade.

Present South Pacific Patterns

Antarctic Bottom Water Circulation

The course of the Antarctic Bottom Water in the South Pacific has been described, at least partially, by Wooster and Volkman (1960) and more recently by Reid et al. (1968), Edmond et al. (1971), Johnson (1972), Hollister et al. (1974).

---

[1] Group: Centre National pour l'Exploitation des Océans/Société Le Nickel.

[2] Association Francaise pour l'Etude et la recherche des Nodules Polymétalliques.

[3] Office de la Recherche Scientifique des Territoires d'Outre-Mer.

[4] Centre Océanologique de Bretagne.

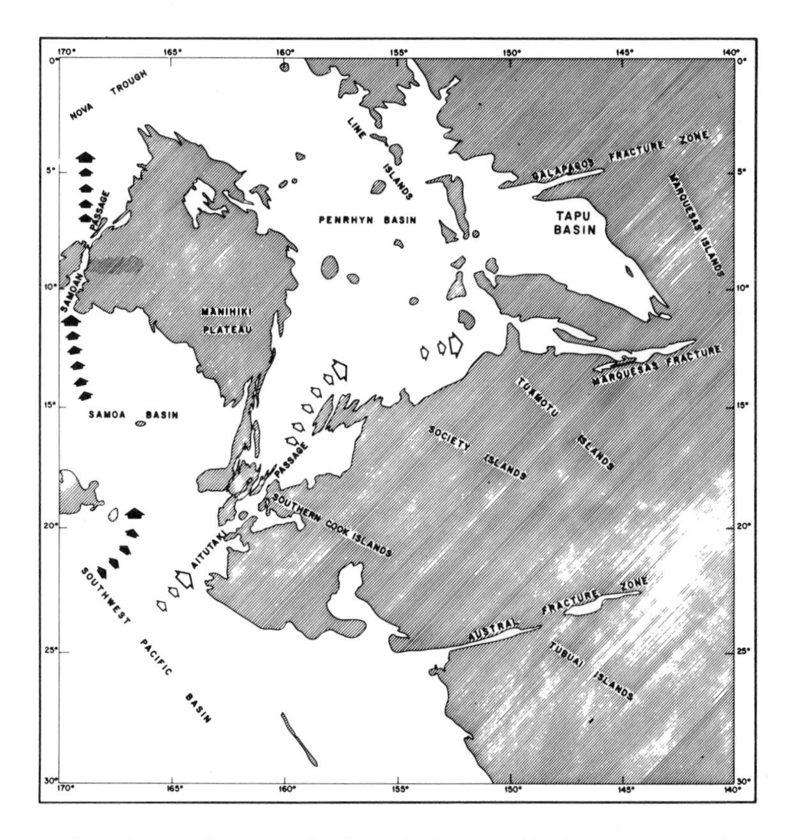

Fig. 4. General map of the whole studied area. Dashed areas are shallower than 2600 fathoms. Dark arrows are representing well-known flow of AABW. White arrows are representing the hypothetical secondary flow of AABW to the East.

The Antarctic Bottom Water AABW enters the Pacific south of Tasmania and crosses the Pacific Rise by way of fracture zones and basins at depths greater than 4000 m. From the Tonga-Kermadec trench, it flows into the Samoan Basin (Fig. 4) and then into the North Pacific through the Samoan Passage (Reid and Lonsdale, 1974). However, all the AABW reaching the Samoan Basin does not flow through the Samoan Passage. The Aitutaki Passage is a second passage (Pautot and Melguen, 1976) that may permit a branch of the Antarctic Bottom Current to reach the Penrhyn Basin and also the Marquesas fracture zone and extend into the Tiki and Tapu Basins (Fig. 4). Abundant erosional structures have been revealed by bathymetric and geophysic observations along different profiles (Fig. 6) crossing the Marquesas fracture zone area. Erosional channels, in sedimentary strata, cut by near-bottom currents, are common all along the fracture zone (Pautot and Melguen, 1976), but only in water depths greater than 4000 m.

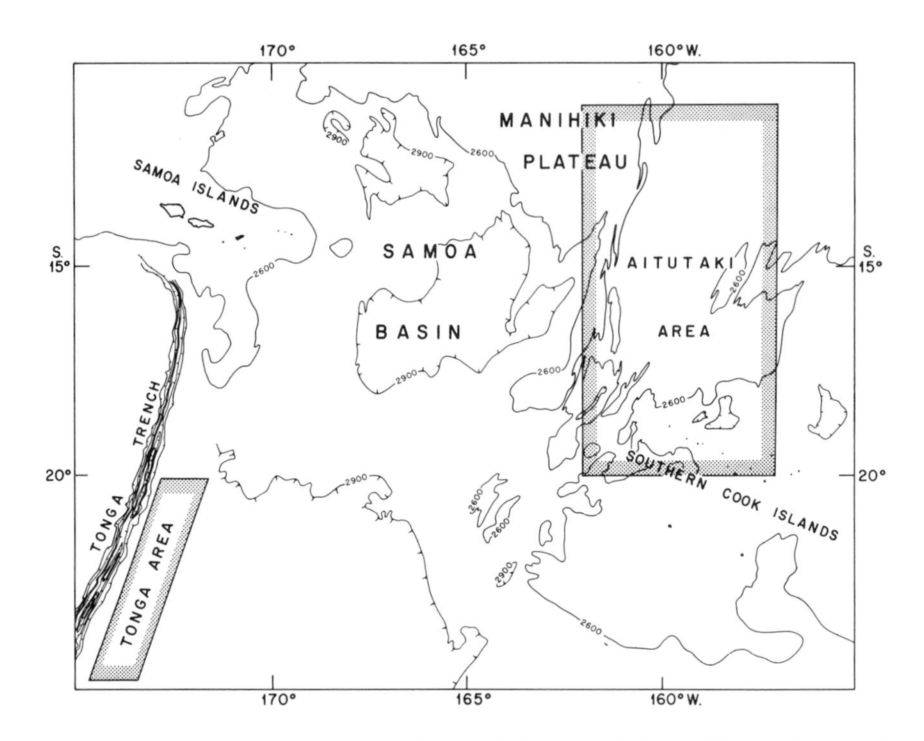

Fig. 5. Map of the Cook-Manihiki area (after Mammerickx et al., 1975). Aitutaki and Tonga areas are two of the areas studied in this paper.

No current measurements are available in the Aitutaki Passage, but a bottom water potential temperature measurement (1.08^{o} C; Wooster and Volkman, 1960) is close to the temperature (0.8^{o} C) (Craig et al., 1972) of the Antarctic Bottom Water flowing through the South West Pacific Basin. According to Craig et al. (1972), the temperature of the AABW varies between 0.6^{o} C in its deepest part to approximately 1° C at its upper limit (Fig. 7). The upper limit is not always well defined because of the presence of a transition layer (600 m thick) separating the AABW, also called Pacific Bottom Water (PBW), from the shallower Pacific Deep Water (PDW). The upper surface (3380 m) of the transition layer, which is defined by the upper inflection point (Fig. 7) is called the "benthic front" and may correspond to the "depth of no motion" between the water masses.

The AABW is one degree colder, much more oxygenated, and consequently much more aggressive with respect to calcium carbonate than the PDW. Near-bottom potential temperature of 0.83^{o} C measured at 5500 m water depth (Reid and Lonsdale, 1974) in the Penrhyn Basin must reflect the flow of PBW in that basin. This is in accordance with the observations of Wong (1972) who suggested the presence of

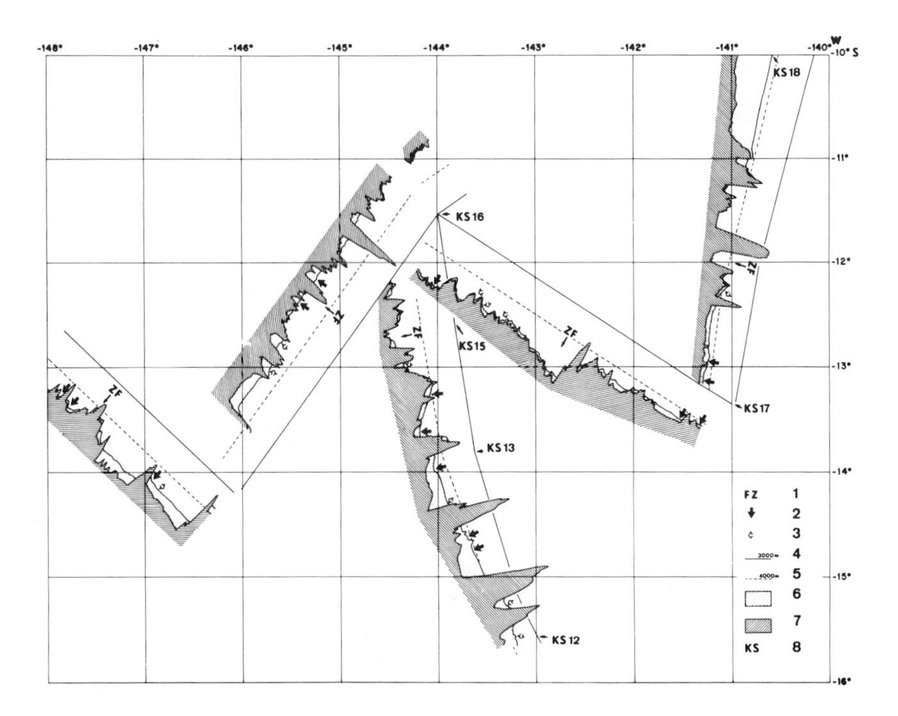

Fig. 6. Seismic profiles carried out during TRANSPAC I cruise along Marquesas Fracture Zone. Position of the tracks are drawn on Fig. 1. 1. Axis of the Fracture zone. 2. Deep erosional sedimentary features. 3. Slump sedimentary features. 4. Position of track line representing 3000 m isobath. 5. 4000 m isobath. 6. Sedimentary cover. 7. Acoustic substratum. 8. Location of gravity cores.

the AABW in the Penrhyn Basin, and between the Galapagos and the Marquesas fracture zones (from 5° to 10° S and up to 130° W). The Tapu Basin (west of the Marquesas Islands and south of the Galapagos Fracture Zone) can be reached by this flow because the Line Islands do not constitute a continuous barrier.

The Marquesas Fracture Zone is represented by a deep channel as far as the southernmost part of the Marquesas Islands. An incursion of the AABW along the Marquesas Fracture Zone and then in the Tiki Basin is thus possible.

Hydrologic measurements were made during the TRANSPAC cruises. In the Marquesas Fracture Zone we did not obtain bottom (5000 m) temperature measurements, but at 3800 m depth the potential temperature was 1.68^{o} C whereas in the neighborhood the bottom temperature is constant around 1.79^{o} C. This value is not a final argument for the presence of the AABW, but according to the observations of Craig et al. (1972), it very probably reflects the presence, at this depth, of the "transition layer" in the Marquesas Fracture Zone.

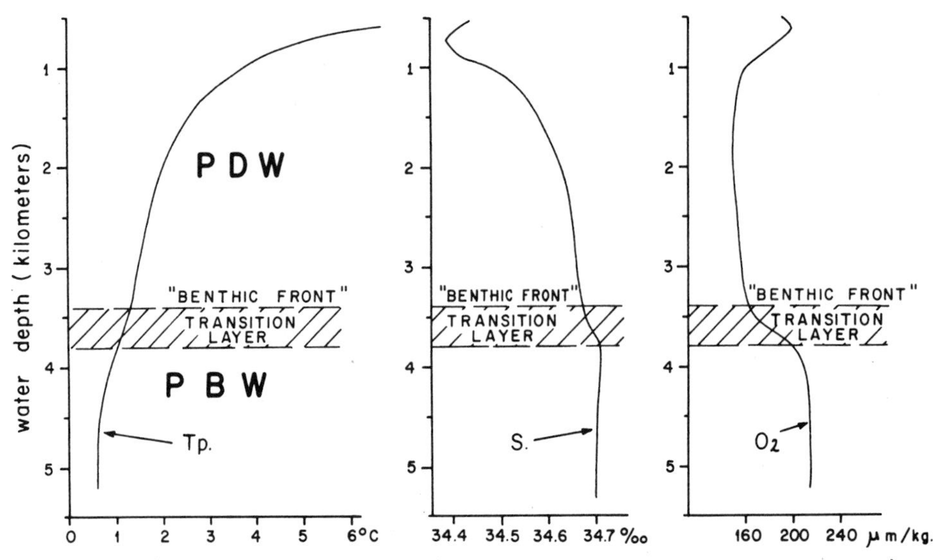

Fig. 7. Depth of the benthic front and of the transition layer separating the Pacific Bottom Water (PBW) from the Pacific Deep Water (PDW). tp: temperature; S: salinity; O_2: oxygen content. After Craig et al. (1972).

Sedimentary Facies Sampling and Analytical Methods

Surface sediments and nodules were collected by means of free-fall grabs. Subsurface sediments were sampled by piston-cores. Sample and core location and depth are given in Appendix Table 1.

A visual examination of the samples and cores was first done using the Munsell color chart. We then studied the sediment composition (including $CaCO_3$ content) and texture by means of smear slides. The Terry and Chilingar (1955) chart was used for estimating the components' frequency. Results of the microscope analysis are given in Appendix Table 2. Sedimentary facies were characterized according to the DSDP classification (Bolli et al., 1978, p. 27-28) which has been slightly modified. Four groups (< 10%, 10-30%, 30-60%, and > 60%) of components' frequencies were considered for siliceous as well as calcareous particles. According to these divisions five lithologic types were considered (see lithologic classification next page).

If nannoplankton is the dominant species in a calcareous ooze, the facies will be called a foram/nanno ooze. The second term is the dominant one. For minor, but very significant components, such as zeolites, hydroxides or volcanic glass, two specific terms were used:

- "bearing", which means containing from 5 to 10%,
- "rich", which means containing from 10 to 20%.

Lithologic classification

% biogenic siliceous particles	Lithologic type
< 10 %	Pelagic clay to siliceous debris-bearing pelagic clay.
10 - 30 %	Siliceous mud.
% biogenic calcareous particles	
< 10 %	Pelagic clay to carbonate debris-bearing pelagic clay.
10 - 30 %	Calcareous mud.
> 60 %	Calcareous ooze.

Age determinations (partly done by C. Müller) were based on calcareous nannoplankton associations, in reference to the associations described by Martini (1971). Much of the sediment was however devoid of calcareous nannofossils. A detailed analysis of the mineralogy, metal content and carbonate content of the surface sediments of the Tiki Basin was done at the Institut de Géologie de Strasbourg (Hoffert et al., internal report, 1978).

Sedimentary Facies Distribution

In the Tuamotu Archipelago-Marquesas Islands area piston-cores were taken along five profiles crossing the Marquesas fracture zone (Figs. 1, 6, 8, Appendix Table 1). Figure 8 gives an example of all the types of sedimentary facies encountered along the N-S profile going from the Tuamotu Archipelago to the North of the Marquesas Fracture Zone. Three major types of sediments were encountered: calcareous turbidites, calcareous mud/nanno ooze and pelagic clay.

Calcareous turbidites are frequent at the base of the Tuamotu Archipelago (cores KS 11 and 12; Appendix Table 2; Fig. 8) as well as at the base of the Marquesas Islands (ex. core KS 18, Appendix Table 2). They are characterized by a high carbonate content (> 60%) abundant and well preserved planktonic foraminifera associated with calcareous algae despite the relatively great (2594-3650 m) water depth. These sediments do not contain metal hydroxides.

The association calcareous mud-nanno ooze is present in core KS 13, as well as in core KS 17 (Appendix Table 2; Figs. 1, 8). The Pleistocene calcareous mud is separated from the Oligocene ooze by a hiatus. Nannoplankton (< 30%) and planktonic foraminifera (< 5%) in surface sediments of these cores, respectively taken at 4350 and 4125 m water depth, indicates that the CCD on the edge of the Tiki Basin is much deeper than 4350 m.

Pelagic clays are the most common facies throughout the studied area (cores KS 15 and KS 16; Appendix Table 2, Fig. 8). They are

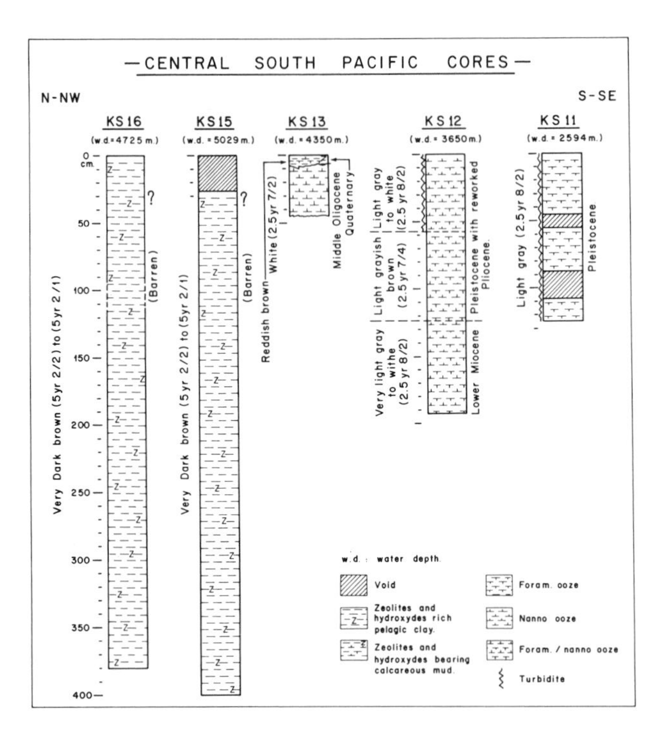

Fig. 8. Facies distribution along five piston-cores taken along a profile (Fig. 1) going from the Tuamotu Archipelago to the N-NW. The different facies types encountered throughout the studied area are present along these cores.

extremely homogeneous and rich in hydroxides (< 20%) and zeolites (< 25%). The lack of $CaCO_3$ in surface sediments of core KS 15 (taken at 5029 m water depth) and the presence of 2% of $CaCO_3$ in core KS 16 (taken at 4725 m water depth) allows us to assume that the CCD here is shallower than 5000 m but very close to 4800 m.

In the Tiki Basin (Figs. 1, 2) surface sediment characteristics (such as $CaCO_3$ content, sand fraction content, Si, Fe, Mn content) have been tentatively mapped by Hoffert et al. (1978, internal report). Calcium carbonate is abundant (> 60%) on topographic highs (water depth > 4200 m) adjacent to the Marquesas fracture zone. Even in the deepest part (> 4300 m) of the basin, sediments still contain from 10 to 30% of $CaCO_3$. Therefore, as previously mentioned, the CCD in the Tiki Basin as well as around the Tuamotu Islands is much deeper than 4300 m and probably close to 4500 m, as indicated by the cores taken in the Tiki Basin (Fig. 2, Appendix Table 1).

The deepest core (KS 09), taken at a depth of 4453 m, consists of pelagic clay containing only traces of calcareous nannoplankton (Appendix Table 2). The shallowest core (KS 08) gives us some indications about the lysocline. The core (Fig. 2, Appendix Table 1) which consists (in its upper part) of calcareous mud relatively rich (< 35%) in well preserved planktonic foraminifera, has been taken close to the lysocline level. Thus the lysocline might be close to 4000 m at that site.

In the Tapu Basin (Figs. 1, 3) no core has been taken. Surface sediments were however obtained during the CNEXO cruise Techno 8 (Appendix Table 1). The sediments sampled were associated with nodule sampling. They were taken at depths from 4690 to 4852 m. Although they are concentrated on the northeastern, they give an idea of the sedimentary facies of the Tapu Basin edge of the basin. The studied sediments are essentially pelagic clay, rich in hydroxides (< 40%) and zeolites (< 15%) and poor in calcareous particles (Appendix Table). Three samples, however, taken around 4750 m water depth contain from 5 to 30% of calcareous nannoplankton. Thus the CCD might be deeper than 4750 m. In fact, during a recent cruise in the easternmost part of the Tapu Basin, we were able to test this hypothesis and to show that the CCD there is very close to 4900 m (SONNE cruise 06.1, unpublished report).

The Aitutaki-Tonga areas (Figs. 4, 5) were surveyed by the ORSTOM Moumea group. The first one is located in the southern part of Penrhyn Basin, between Manihiki Plateau and the Cook Islands. The location of these stations was chosen to complete a preliminary survey carried out in 1976 over the same area by the CCOP/SPOAC[1] Technical Secretariat and the Cook Islands Government.

Other samples were obtained in the southern part of the Samoa Basin, between 20 and 25° S, near the Tonga trench.

The Aitutaki area, located between the Manihiki Plateau, the Society Islands and the Southern Cook Islands (Figs. 2, 5) presents an interesting sediment distribution pattern. Between 4700 and 4800 m water depth, pelagic clays are dominant, but still contain calcareous particles (foraminifera = 1-3%; nannofossils = 1-10%).

At greater depths, the pelagic clays do not contain any more calcareous particles. This shows that the CCD in this area is close to 4800 m or even deeper. Consequently, we propose a CCD level between 4800 and 5000 m.

Considering the pelagic clay composition, we may add that it is enriched in hydroxides (10-25%) and zeolites (5-15%; Appendix Tables 4, 5) around 5000 m water depth (Appendix Table 1). There are two major facies within the Aitutaki area: a pelagic clay rich in zeolites and bearing volcanic glass and siliceous debris, and a pelagic clay relatively rich in siliceous debris and bearing zeolites and volcanic glass. The first one characterizes the area extending north of 13° S. Zeolites are particularly abundant (10-25%) and of relatively large size (< 63 m) and abundance (< 3%). Furthermore,

---

[1] CCOP/SOPAC: Committee for co-ordination of joint prospecting for mineral resources in South Pacific Offshore areas.

this facies is richer than the previous one in radiolarians (5-10%; ex. GT. 204, 211, 213, Appendix Table 4) but much poorer in volcanic glass (< 1%).

Near the Tonga Trench (≃ 23° S; Fig. 5), sediments differ in an obvious way from those of the Aitutaki area by their relative richness in volcanic glass (10-25%) and siliceous debris (< 10%). This increase in siliceous debris probably reflects relative increase in the oceanic fertility.

Evolution of Nodule Distribution and Composition With Respect To Water Depth And Topography

During TRANSPAC cruises (I and II), only few nodules (≃ 3 kg/m²) were recovered from the Marquesas and Tuamotu areas (Pautot and Hoffert, 1974). However, in the light of our first observations (Pautot and Hoffert, 1974; Pautot and Melguen, 1975, 1976) we suggested that the presence of nodules seemed to be related to the following conditions:

- an old oceanic crust (up to 15 my),
- a moderate thickness of sedimentary cover (less than 100 m),
- a water depth close to the CCD,
- a smooth bathymetry.

In order to substantiate these tentative observations, as well as to define more closely the general relationship between the density and grade of nodules and their oceanic environment, we have studied much new data in each of the areas mentioned. Several hundreds of samples have been taken by the CNEXO/SLN Group and AFERNOD around the Tuamotu Archipelago, the Marquesas Archipelago, along the Marquesas fracture zone, in the Tapu and the Tiki Basins. Location of these samples, density and composition of nodules are not presented here because they are still under industrial secret. We show the evolution of nodule density (concentration on the sea floor) and grade (metal content) with increasing water depth. Nodule abundance was determined using both the weight of the nodules recovered by free-fall grabs and the abundance observed on bottom photographs.

Tuamotu-Marquesas Area (Fig. 1)

The general distribution of nodule abundance with respect to water depth (Fig. 9) in the whole Tuamotu-Marquesas area shows obvious trends.

First of all, no nodules have been found between 0 and 3900 m. Secondly, a decrease of abundance seems to appear at water depths greater than 5000 m. One may note, however, that only very few samples have been taken at such water depth in this area.

High nodule abundance occurs between 4000 m and 5000 m water depth (Fig. 9) with maximum abundance (up to 10 kg/m²) between 4350 and 4900 m water depth. As the samples are coming from an extensive

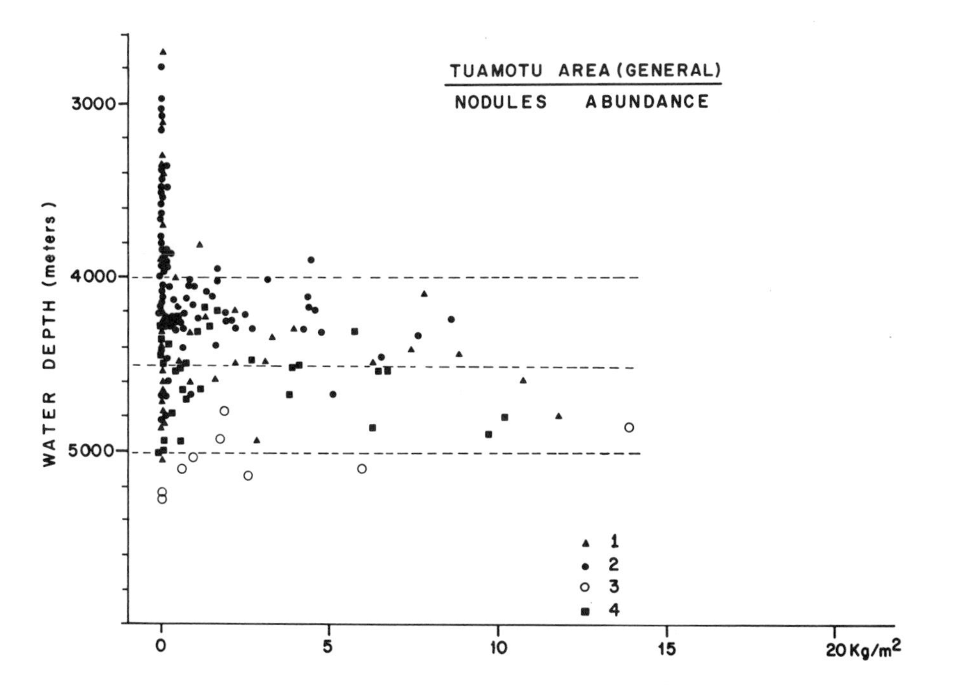

Fig. 9. Nodule abundance versus water depth in the whole Tuamotu-Marquesas area. 1, 2, 3, 4 represent values from four different CNEXO cruises.

area, it is difficult to determine in more detail distribution patterns.

Tiki-Marquesas Area (Figs. 1, 2)

As in the previous area, high nodule abundance is found between 4000 and 5000 m water depth (Fig. 10). South of the Marquesas fracture zone, the nodule distribution is, however, different from that encountered north of this zone. In the Tiki Basin, south of the Marquesas fracture zone, the highest nodule abundances have been found between 4000 m and 4350 m water depth (Fig. 1). North of the Marquesas fracture zone, highest densities are found at greater water depth (4500-4900 m). Above and below these water levels, the nodule abundance decreases drastically.

This analysis underlines some interesting observations:

- in a given morphostructural location, a preferential water depth is associated with the maximum abundance of nodules,
- the nodules are distributed through a range of water depth of about 300 or 400 meters,
- maximum values of abundance are situated in the lower part of this range.

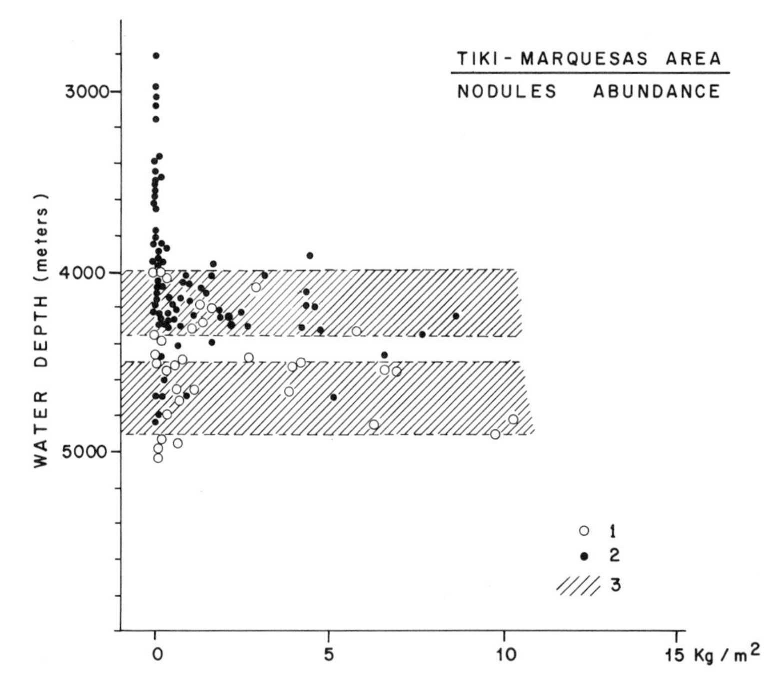

Fig. 10. Nodule abundance versus water depth in the Tiki Basin and North of the Marquesas fracture zone. 1: north of the Marquesas fracture zone; 2: south of the Marquesas fracture zone; 3: zones of highest density south and north of the fracture zone.

Ni and Cu grade do not show the same type of correlation with water depth (Fig. 11 and 12). In the depth interval (4000-4350 m) where nodules are the most abundant, Ni and Cu grade is very scattered. High Ni grade (i.e. about 1.5%) is found between 4100 and 4400 m water depth. There may be an ill-defined trend with a maximum around 4350 m. In this case, maximum abundance values and maximum Ni content are in concordance near 4350 m.

Concerning the distribution of Cu content at increasing water depth, we may also note that the highest values (about 1%) are also scattered between 4000 and 4350 m (Fig. 12), but with a swarm of values around 4300 m.

In brief, the preferential depth for nodule exploration in the Tiki Basin seems to be 4350 m. At this depth, the highest values of nodule abundance and grade converge.

Aitutaki Area (Figs. 4, 5)

Nodule abundance presents a clear maximum between 4800 and 5200 m water depth (Fig. 13) where nodule concentration varies from 5 to

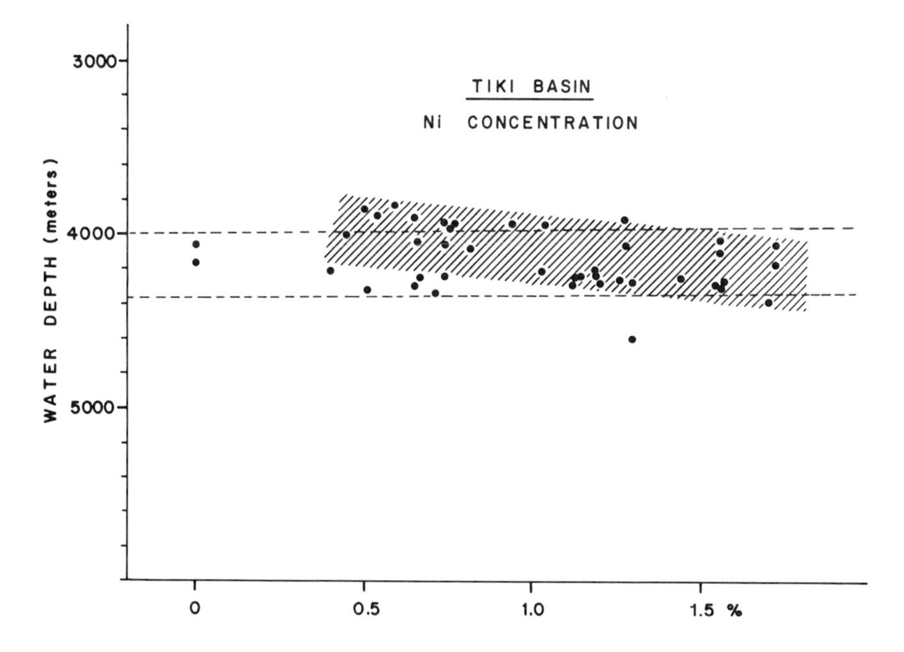

Fig. 11. Nodule composition (= Ni concentration) versus water depth in the Tiki Basin. Highest grade is comprised between 4000 and 4350 m.

25 kg/m^2. Below 5200 m, this concentration decreases markedly. Consequently, the 4800-5200 m depth range appears as being the preferential depth for nodule distribution in the Aitutaki area.

We may note (Fig. 14) that Ni content increases (0.2-1.1%) with increasing water depth, and that Cu values exhibit the same tendency to enrichment with increasing water depth but present lower grade (0.2-0.7%, Fig. 15). Co grade shows no correlation with water depth and presents its highest value at 4800 m and at 5200 m water depth (Fig. 16).

In brief, no concordance appears between maximum nodule abundance and maximum grade in the Aitutaki area.

Tapu Basin (Figs. 1, 3)

The Tapu Basin was studied first on a regional scale and then on a more detailed one.

The regional study being focused on a small area characterized by a relatively constant water depth, we were not able to show a general pattern concerning the nodule distribution in the Tapu Basin. However, it appears that nodule concentrations show great variations (0-10 kg/m^2) and are very scattered within a water depth interval of 200 m (4800-5000 m; Fig. 17). In this interval the Ni + Cu concen-

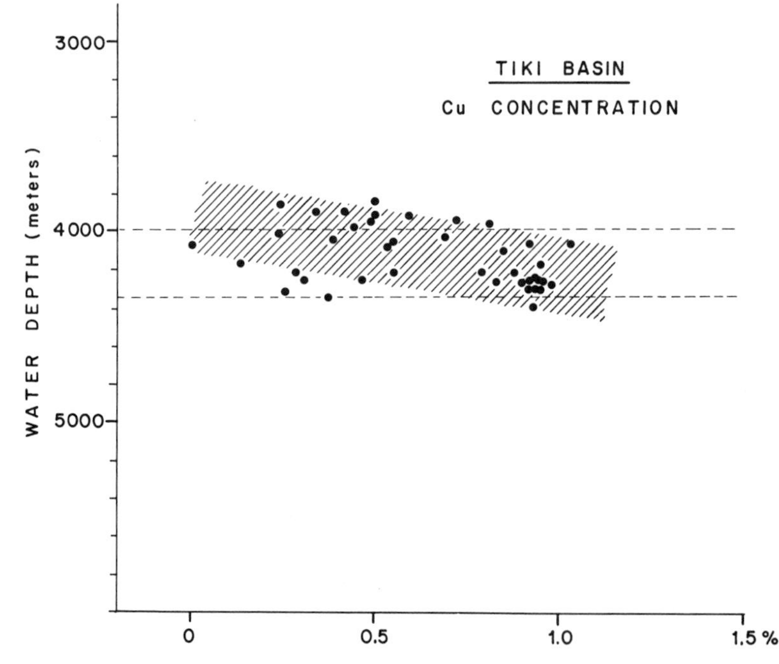

Fig. 12. Nodule composition (= Cu concentration) versus water depth in the Tiki Basin. Highest grade is comprised between 4000 and 4350 m.

tration (Fig. 18) exhibits values around 2% with maximum of 2.5%.

The detailed study was carried out around 7° S and 146° W on the flank of the Tapu Basin, along two crossing traverses. One hundred free-fall grabs were made on each traverse (NS and EW) with a mean spacing of 300 to 500 m.

The stations are regularly spaced on each traverse. In order to simplify the representation of the relationship existing between the nodule distribution and the bottom topography, only the depth of the free-fall grab stations were reported (Figs, 19, 20). Station depths were then joined to each other by a straight line. We thus obtained a schematic representation of the topography along both traverses of the "cross".

The purpose of this detailed study was to compare the nodule characteristics with two associated important factors of the oceanic environment: the depth and the topography of the ocean floor.

Depth and topography of the ocean floor. Mean depth is 4900 m with highs shallower than 4700 m and deeps around 5100 m (Figs, 19, 20). Morphology is smoother and rounder on the NS profile than on the EW profile (Figs. 19, 20). One can distinguish sharp reliefs, rounded hills, perched plateaus, rough relief, valleys.

Nodule abundance along both profiles (Figs. 19, 20A). Nodule

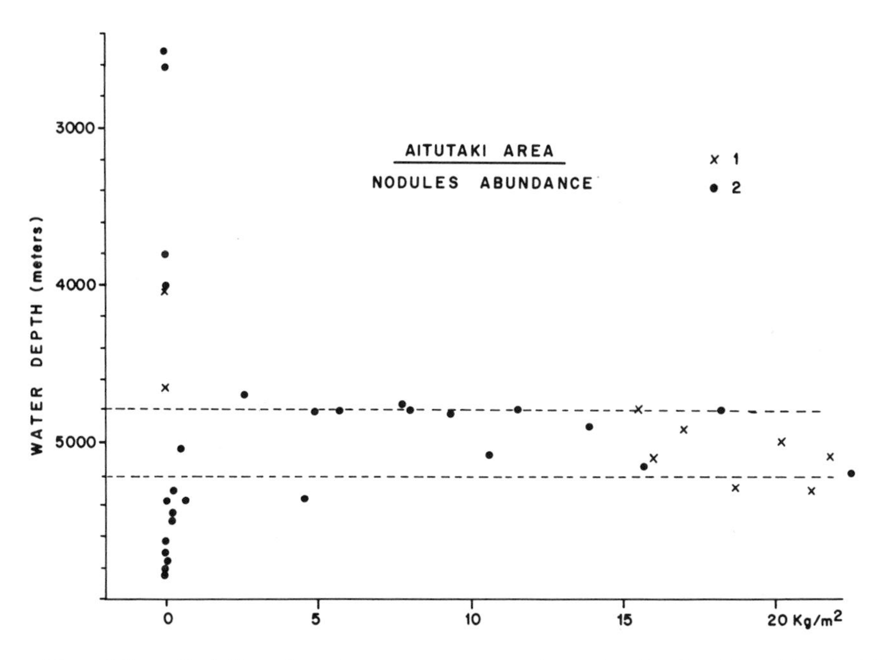

Fig. 13. Nodule abundance versus water depth in the Aitutaki area. 1: data from Landmesser et al. (1976); 2: ORSTOM/CNEXO data. Highest densities between 4800 and 5200 m.

abundance (in areas where they are present) generally varies from 2.5 kg/m^2 to 5.0 kg/m^2 (Figs. 19, 20).

Nodules are rare:

- on steep flanks of hills,
- on low and smooth areas.

Greater abundance is noted:

- on gentle slopes of hills,
- on perched plateaus and in area of rough relief.

The relationship existing between abundance and topography is much clearer along the North-South Profile than along the East-West Profile (Figs. 19, 20).

Nodule grade along both profiles (Figs, 19, 20G). Mean concentrations of Ni + Cu in this area varies from 1.8% to 2.0%. Values are scattered along both profiles, without showing a clear correlation with the topography (Figs. 19, 20). On rough topographic features, however, the nodule grade is lower than along gentle slopes. A profile of stations elsewhere in Tapu Basin (same area) was analyzed by V. Rendard (person. commun.) in order to evaluate the relationship between the bottom morphology and the grade of transition elements (Fig. 21). This approach, element by element, is more refined than the previous one, in which Ni and Cu content was taken as a whole. Figure 21 clearly shows a positive correlation between

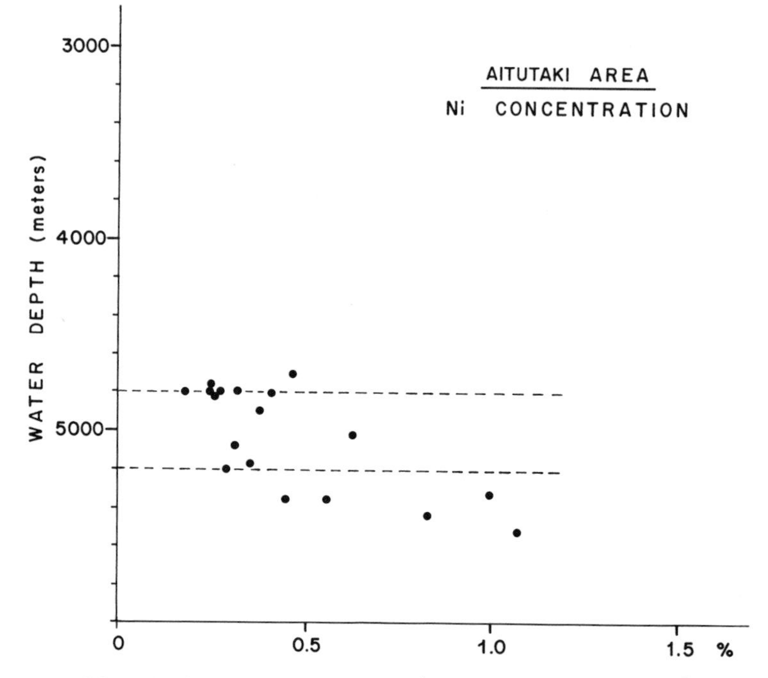

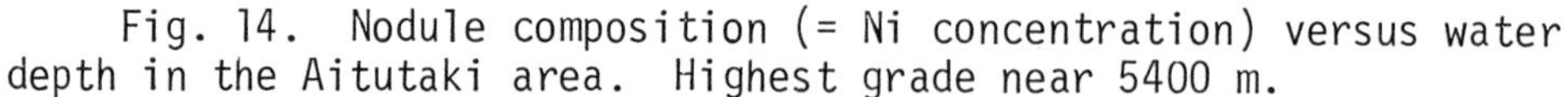
Fig. 14. Nodule composition (= Ni concentration) versus water depth in the Aitutaki area. Highest grade near 5400 m.

Cu grade and topographic lows. The Ni grade is less well correlated with depressed features. Co grade appears opposed to the Cu grade, with higher values on topographic highs. Mn grade appears to follow the same variations as Cu and Ni grade. The enrichment in Cu in topographic lows might be related either to the general enrichment of the nodules in Mn oxides, or to the presence of todorokite which incorporates selectively Ni, Cu and Mn in deep water areas (Cronan and Tooms, 1969; Cronan, 1977).

Nodule size distribution along both profiles (Figs. 19, 20S). By measuring the length of the major axis of the nodules, five classes were determined: class 1 (0 to 1 cm), class 2 (1 to 2 cm), class 3 (2 to 3 cm), class 4 (3 to 4 cm), class 5 (4 to 5 cm).

Classes 1 and 2 are dominant and often associated (Figs, 19, 20). Homogeneous populations (1 and 2) are visible on smooth reliefs or high plateaus with one class dominant (1 or 2), and presence of other nodule types in various proportions. Largest nodules are generally found in depressed areas and valleys. Along sharp reliefs, small nodules (class 1 and 2) are dominant.

Nodule morphology along both profiles (Figs. 19, 20M). The following classification is used in this paper: mononodules, polynodules and debris. Mononodules are coupled or coalescent nodules. Debris which are mainly broken, flattened ovoïd mononodules are

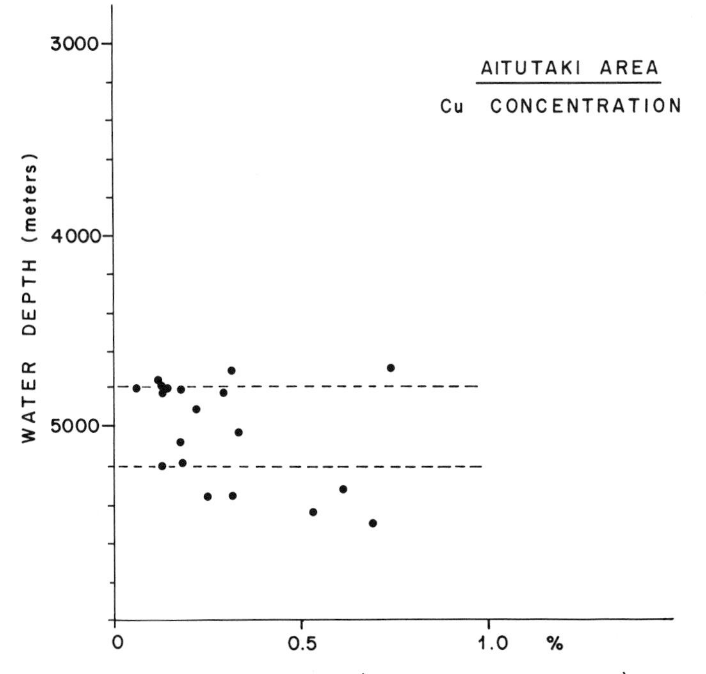

Fig. 15. Nodule composition (= Cu concentration) versus water depth in the Aitutaki area. Highest grade near 6400 m.

largely dominant in this area.

Unbroken ovoïds are present on gentle hills, and irregular nodules are preferentially found on and around topographic highs. Polynodules are seldom observed.

Nodule setting and environment (from bottom photographs). Study of bottom photographs shows that nodules are generally partially buried within the sediments. All photographs emphasize this phenomenon. Furthermore, in certain cases, one can guess the presence of nodules on the bottom only through a small mound sticking out through sediment. Some pictures even give the impression of a fossil nodule field in the process of being buried.

Discussion

In the light of the previous observations, we are now able to draw some general conclusions about the nodule abundance and grade distribution in the Central South Pacific.

First of all, nodules are generally found between 4000 m and 5200 m, and not deeper. They seem to be concentrated in a 300/400 m-thick depth interval. This interval is not everywhere the same but varies from one area to the other, as follows:

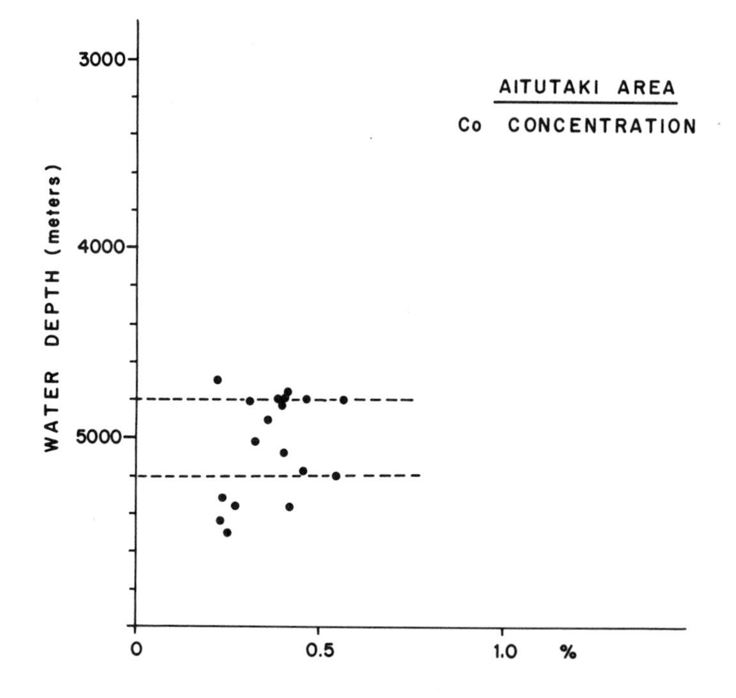

Fig. 16. Nodule composition (= Ni concentration) versus water depth in the Aitutaki area. Highest grade near 5400 m.

4000-4350 m in the Tiki Basin
4500-4800 m north of the Marquesas fracture zone,
4800-5000 m in the Tapu area,
4800-5200 m n the Aitutaki area.

Within this depth range of perferential maximum distribution exists a certain level at which abundance and grade are in good concordance. The abundance parameter is represented by a curve showing a peak in a given 300/400 m-thick interval, with maximum values at base of the peak. The grade parameter is generally represented by a straight line which exhibits a grade increasing with depth.

These two parameters may or may not be in good concordance. The abundance peak may correspond to maximum grade (like Tiki Basin around 4350 m); or the abundance peak may be situated above the maximum grade (like Aitutaki Passage around 5200 m). This conclusion contradicts the observations of Menard and Frazer (1978) for the whole Pacific Ocean.

Water depth plays obviously a major role in the distribution of manganese nodules. Topography of the sea floor appears to be, however, an important secondary factor controlling the abundance as well as the grade of nodules (Figs, 19, 20, 21).

Areas of smooth relief and gentle slopes have the highest density of nodules. Areas of sharp relief are on the contrary, unfavor-

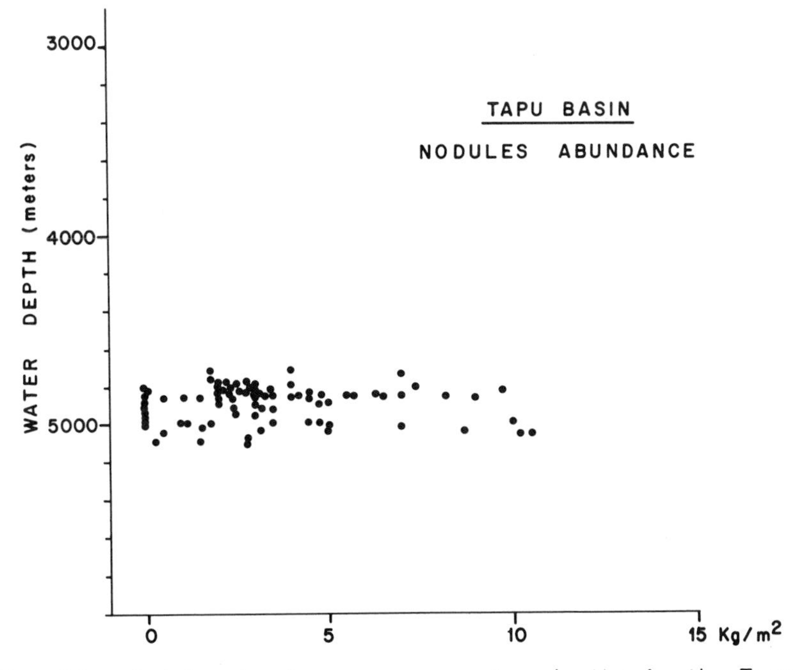

Fig. 17. Nodule abundance versus water depths in the Tapu Basin. Highest density between 4800 and 5000 m.

able in terms of abundance as well as grade. Depressed areas and valleys also have low abundances but present the highest content in Cu.

Therefore, within a 200 m thick depth interval, there are secondary fluctuations which are related to the local topography and are superimposed on major ones that are linked to more regional factors controlling the sedimentary facies, such as the bottom circulation and oceanic fertility.

We have been particularly impressed by the fact that, in most of the cases, the preferential depth of nodule distribution was not only below the Benthic Front and therefore within the Pacific Bottom Water mass (Fig. 7), but generally within the depth interval separating the calcite lysocline from the Calcite Compensation Depth (CCD). The concept of the lysocline, implicit in the findings of Peterson (1966), Berger (1967), Ruddiman and Heezen (1967), was introduced to denote a surprisingly well defined facies boundary zone between well preserved and poorly preserved foraminiferal assemblages on the floor of the Atlantic Ocean (Berger, 1968), and on that of the South Pacific (Berger, 1970a, Parker and Berger, 1971). The lysocline is generally envisaged as the level at which there is a maximum change in the composition of calcareous fossil assemblages due to differential dissolution. In areas of low oceanic fertility, such

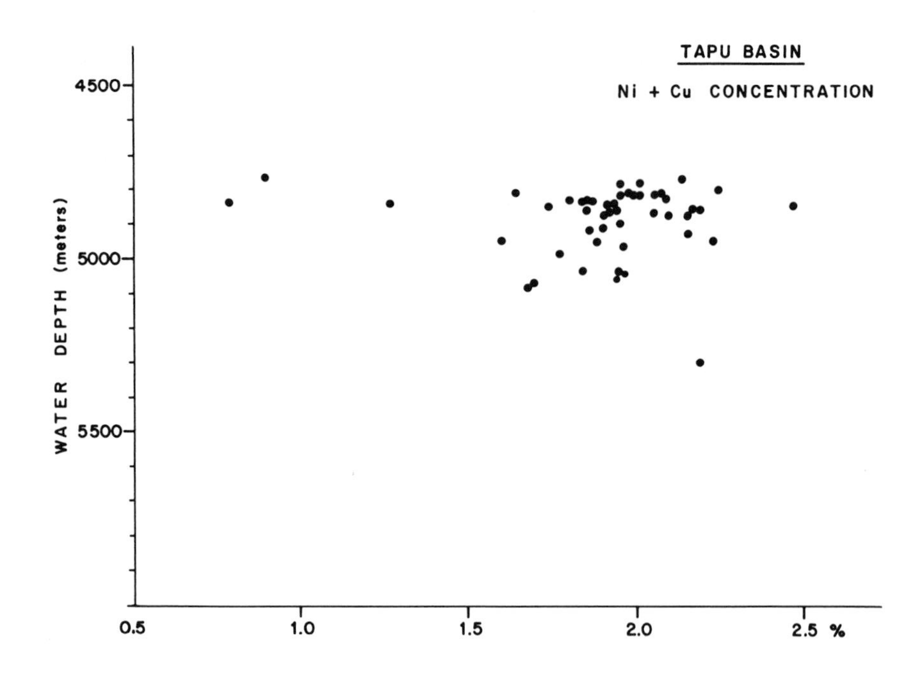

Fig. 18. Nodule composition (= Ni + Cu concentration) versus water depth in the Tapu Basin. Highest concentrations are scattered between 4700 and 5200 m.

as related to the aggressiveness of the bottom water masses, the lysocline corresponds to the "hydrographic lysocline" as defined by Peterson (1966).

In the Central South Pacific the lysocline is close to 4000 m (Berger, 1970b, 1971, 1976; Parker and Berger, 1971; Roth and Berger, 1975, Berger et al., 1976), and therefore well beneath the Benthic Front which marks the level at which the dissolution of foraminifera is first obvious (Berger, 1976). The sharpness of the transition zone from high to low values of carbonate varies from area to area, with a mean value of 300 m in the studied area (Berger et al., 1976). According to Berger et al. (1976), the CCD in the Tiki, Tapu and Aitutaki areas, varies between 4200 m and 5000 m (Fig. 22, a and b). Berger's values are not in complete agreement with our observations, according to which we propose the following CCD values:

4500 m for the Tiki Basin
4900 m for the Tapu Basin
4800-5000 m for the Atiutaki-Tonga area.

The discrepancy between the estimates of Berger et al. (1976) and ours could be explained partly by the fact that Berger et al. considered the 10% $CaCO_3$ level as being the CCD level, and partly because the density of our data in the given areas is locally greater than Berger's.

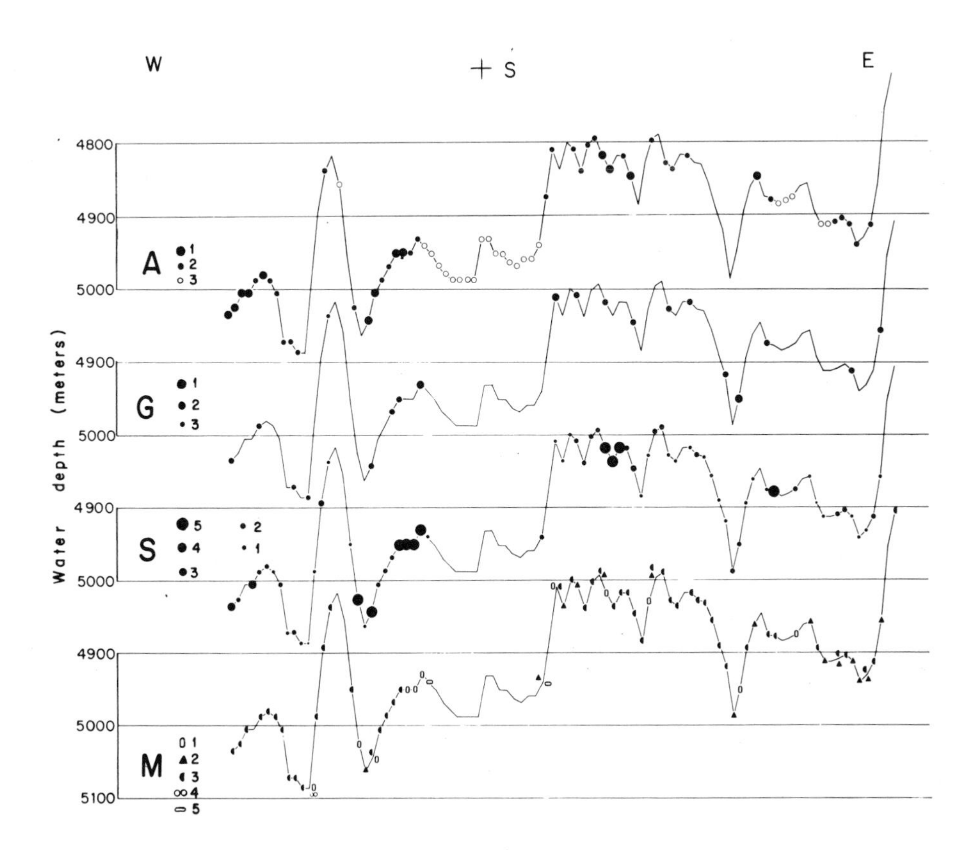

Fig. 19. Detailed study along a west-east profile in the Tapu Basin. Relationship between water depth, topography and: nodule abundance, grade, size, morphology.

A (abundance): 1. density $\geq$ 5 kg/m^2; 2. density comprised between 2.5 and 5 kg/m^2; 3. no nodules.

G (Ni + Cu grade): 1. grade $\geq$ 2.0%; 2. grade comprised between 2.0 and 1.8%; 3. grade comprised between 1.8 and 1.5%.

S (size): 1 to 5. nodules size increasing from class 1 (0-1 cm) to class 5 (4-5 cm).

M (morphology): 1. nodules of ovoïd shape; 2. nodules of irregular shape; 3. nodules debris; 4. coupled nodules or polynodules; 5. nodules of geoïd shape.

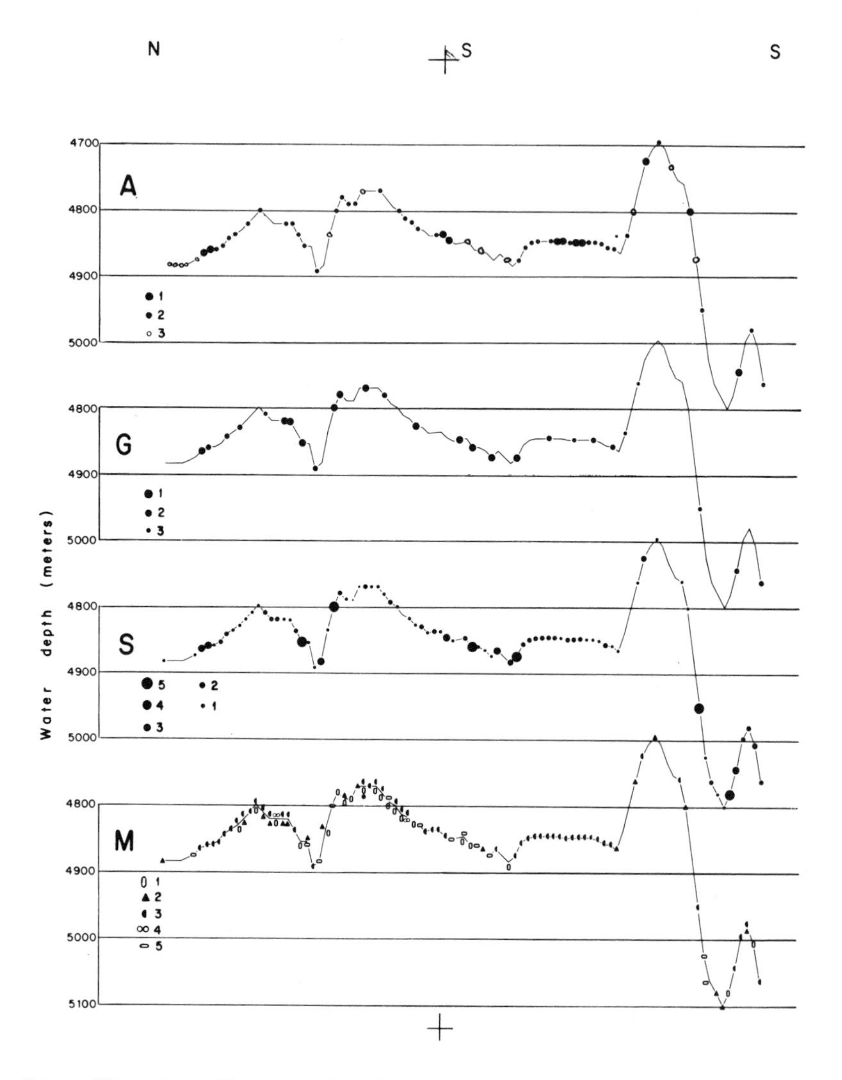

Fig. 20. Detailed study along a north-south profile in the Tapu Basin. Relationship between water depth, topography and: nodule abundance, grade, size, morphology.

A (abundance): 1. density ≥ 5 kg/m²; 2. density comprised between 2.5 and 5 kg/m²; 3. no nodules.

G (Ni + Cu grade): 1. grade ≥ 2.0%; 2. grade comprised between 2.0 and 1.2%; 3. grade comprised between 1.8 and 1.5%.

S (size): 1 to 5. nodules size increasing from class 1 (0-1 cm) to class 5 (4-5 cm).

M (morphology): 1. nodules of ovoïd shape; 2. nodules of irregular shape; 3. nodule debris; 4. coupled nodules or polynodules; 5. nodules of geoïd shape.

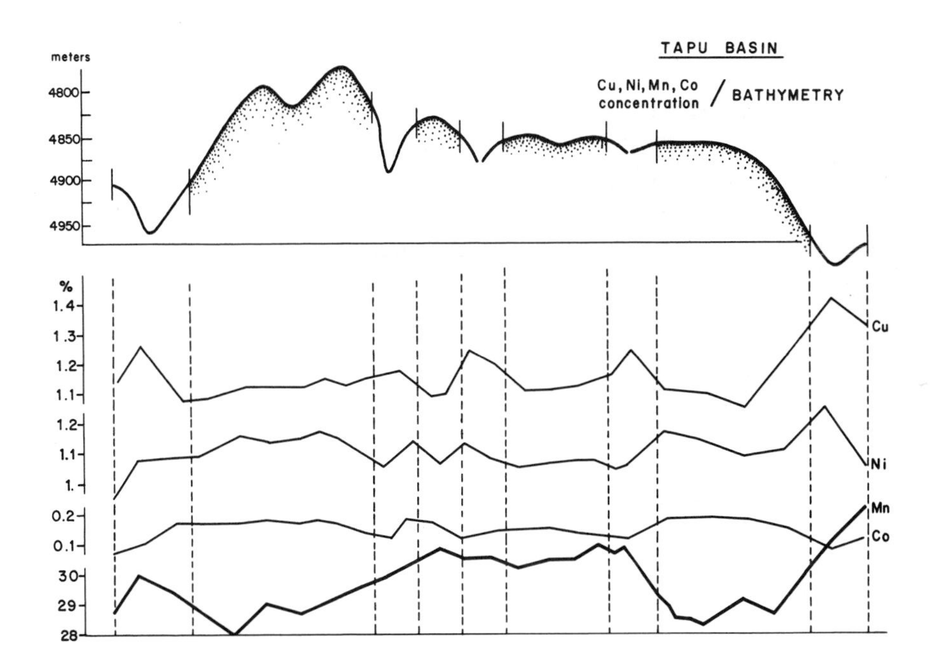

Fig. 21. Detailed study by V. Renard (pers. commun.) of the relationship between topography and nodule grade (Cu, Ni, Mn, Co content) within the Tapu Basin.

Conclusion

Our hypothesis concerning the nodule distribution is based mainly on our CCD estimates. We conclude that in the Central South Pacific, manganese nodules are found within a preferential depth interval, between the hydrographic lysocline and the calcite compensation depth, and that they are concentrated within a very specific sedimentary facies, called the "N-facies" (Fig. 22c).

This so-called "N-facies" or "nodule-rich facies" corresponds to the "R-facies" defined by Berger (1976) as being a residual facies greatly enriched in dissolution-resistant foraminifera species. Bottom Water Circulation and carbonate dissolution seem to be the major controlling factors of this distribution, while bottom topography appears to play a secondary role.

This conclusion dealing with the Marquesas-Tonga area in the South Pacific might not be applicable, for example, to the Clarion-Clipperton area in the North Pacific.

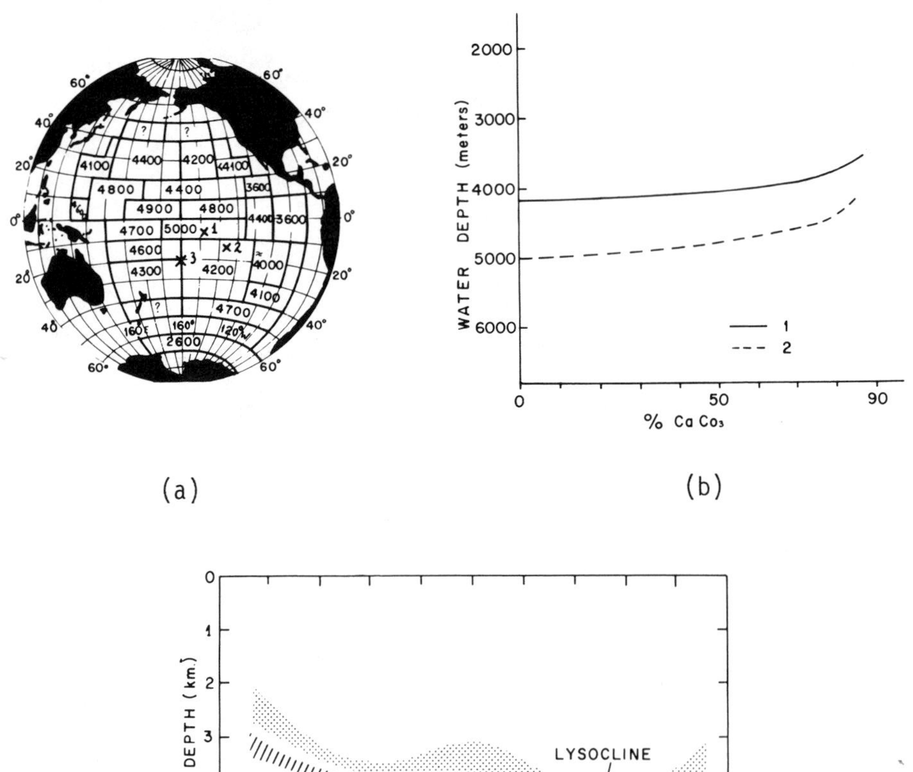

(a) (b)

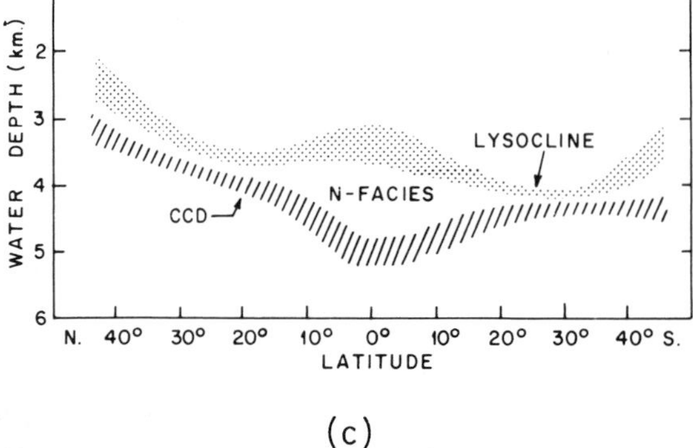

(c)

Fig. 22. a) fluctuations of the CCD levels (in meters) in the whole Pacific Ocean. 1. Tapu Basin; 2. Tiki Basin; 3. Aitutaki-Tonga area. After Berger et al. (1976).
b) distribution of $CaCO_3$ content in the sediment at increasing water depth, after Berger et al (1976); 1. Tiki Basin area 2 on Fig. 22a; 2. Tapu Basin, area 1 on Fig 22a.
c) Model of foram dissolution facies (lysocline and CCD) in a N-S profile through the central Pacific (after Berger, 1976). The "N-facies" or "nodules-rich facies" corresponds to the "R-facies" (residual facies greatly enriched in resistant foraminifera species) of Berger (1976, p. 313).

Acknowledgements

We are particularly grateful to AFERNOD and to our colleagues of CNEXO Applied Geology team for allowing us to use nodule data they gathered in the South Pacific. We also thank our colleagues of the ORSTOM Noumea team, who kindly sent sediment samples and analysis of nodules taken in the Aitutaki-Tonga area, during their cruises GEOTRANSIT II and EVA III.

We express our thanks to Dr. V. Renard for the detailed study done in the Tapu Basin during the CNEXO cruise TECHNO 08. We also thank A. Vangriesheim for the compilation done on bottom circulation in the South Pacific, R. Kerbrat, S. Marques for technical assistance, R. Thirion and A. Grotte for drawing the figures. Finally we are particularly grateful to Y. Potard and N. Guillo for so carefully typing the manuscript.

References

Berger, W. H., (1967) Foraminiferal ooze: solution at depths, Science 156, 383-385.

Berger, W. H., (1968) Planktonic foraminifera: selective solution and paleoclimatic interpretation, Deep-Sea Res. 15, 1, 31-43.

Berger, W. H., (1970a) Biogenous deep-sea sediments: fractionation by deep-sea circulation, Geol. Soc. Amer. Bull. 81, 1385-1402.

Berger, W. H., (1970b) Planktonic foraminifera, selective solution and the lysocline, Marine Geology 8, 38-111.

Berger, W. H., (1971) Sedimentation and planktonic foraminifera, Marine Geology 11, 325-358.

Berger, W. H., (1974) Deep-sea sedimentation, In Burk, C. A., and Drake, C. L., eds., The Geology of Continental Margins, Springer, New York, 213-241.

Berger, W. H., (1976) Biogenous deep-sea sediments: production, preservation and interpretation, In Riley, J. R., and Chester, R., eds., Chemical Oceanography 2nd Ed. Academic Press, London, 5, 29, 265-387.

Berger, W. H., Adelseck, C. G., Mayer, L. A., (1976) Distribution of carbonate in surface sediments of the Pacific Ocean, J. of Geophysical Res. 81, 15, 2617-2627.

Bolli, H. M., Ryan, W. F. F., et al., (1978) Initial Reports of the Deep Sea Drilling Project, Washington, D.C., U.S. Government Printing Office.

Craig, H., Chung, Y., and Fiadeiro, M., (1972) A benthic front in the South Pacific, Earth and Planetary Sci. Letters, 16, 50-65.

Cronan, D. D., (1977) Deep-sea nodules: distribution and geochemistry, In Glasby, G. P., ed., Marine Management Deposits, Elsevier, Amsterdam, 11-44.

Cronan, D. S., and Tooms, J. D., (1969) The geochemistry of manganese nodules and associated pelagic deposits from the Pacific and Indian Oceans, Deep-Sea Res. 16, 335-359.

Duncan, R. A., McDougall, I., (1974), Migration of volcanism with time in the Marquesas Islands, French Polynésie, Earth and Planetary Sci. Letters 21, 414-420.
Edmond, J. M., (1971) An interpretation of the calcite spheres experiment (abs.), EOS (Trans. Am. Geophys. Union) 52, 256 p.
Edmond, J. M., Chun , Y., and Sclater, J. G., (1971) Pacific bottom water: penetration east around Hawaii, Journ. of Geophys. Res. 76, 8089-8097.
Heath, G. R., and C. Clu erson, (1970) Calcite: degree of saturation, rate of dissolution and the compensation depth in the deep oceans, Geol. Soc. of Amer. Bull. 81, 3157-3160.
Hoffert, M., Karpoff, A. M., Schaaf, A., and Wirrmann, D., (1978) Caractéristiques sédimentologiques, minéralogiques, biostratigraphiques et géochimiques des sédiments de surface des zones prévues pour les missions Copano. Unpublished Report, Labor. de Géologie de Strasbourg.
Hollister, C. D., Johnson, D. A., Lonsdale, P. F., (1974) Current-controlled abyssal sedimentation: Samoan Passage, Equatorial west Pacific, The Journal of Geology 82, 3, 275-300.
Johnson, D. A., (1972) Ocean-floor erosion in the Equatorial Pacific, Geol. Soc. of Amer. Bull. 83, 3121-3144.
Landmesser, C. W., Kroenke, L. W., Glasby, G. P., Sawtell, G. H., Kingan, S., Utanga, A., Cowan, G., (1976) Manganese nodules from the South Penrhyn Basin, Southwest Pacific South Pacific Marine Geological Notes, Technical Secretariat, CCOP-SOPAC, Suva, 1, 3, 17-40.
Mammerickx, J., Detrick R., and Schlater, J. G., (1972) Fracture zones and magnetic lineations in the South Central Pacific, Publ. by the Institute of Marine Resources, Univ. of California, San Diego, La Jolla BP 529.
Mammerickx, J., Anderson, R. W., Menard, H. W., and Smith, S. M., (1975) Morphology and tectonic evolution of the East Central Pacific, Publ. by the Institute of Marine Resources, Univ. of California, San Diego, La Jolla, BP 529.
Martini, E., (1971) Standard Tertiary and Quaternary calcareous nannoplankton zonation, In Farinacci, A., ed., Roma Proceed. of the II Planktonic Conference Roma 1970, 739-785.
Menard, H. W., (1964) Marine Geology in the Pacific, McGraw Hill, New York, 271 p.
Menard, H. W., and Frazer, J. Z., (1978) Manganese nodules on the sea floor: inverse correlation between grade and abundance, Science 199, 969-970.
Monti, S., and Pautot, G., (1974) Cinq cartes bathymétriques au 1/1 000 000: Tahiti, Raroia, Marquises, Hao, Mururoa. Cartes éditées par le CNEXO, en vente au BRGM.
Parker, F. L., and Berger, W. H., (1971) Faunal and solution patterns of planktonic foraminifera in surface sediments of the South Pacific, Deep-Sea Res. 18, 73-107.
Pautot, G., and Hoffert, M., (1974) Pacifique Sud-Est: cadre structural, morphologique et sédimentaire. Relations avec les nodules

polymétalliques, IIème Colloque International sur l'Exploitation des Océans, Bordeaux, France, 1-4 octobre 1974, 4, BX202, 8 pp., 3 figs.
Pautot, G., and Hoffert, M., (1974) Extension du flanc nord de l' Archipel des Touamotou: analyse structurale et sédimentologique, 2ème Réun. Ann. Sc. de la Terre, Montipellier 310.
Pautot, G., and Melguen, M., Courants profonds, hiatus sédimenentaires et nodules polymétalliques, IXème Congrès International de Sédimentologie, Nice (France), 8, 57-64.
Pautot, G., and Melguen, M., (1976) Deep bottom currents, sedimentary hiatuses and polymetallic nodules, Technical Bull. CCOP-SOPAC n° 2 54-61.
Peterson, M. N. A., (1966) Calcite: rates of dissolution in a vertical profile in the central Pacific, Science 154, 1542-1544.
Reid, J., Stommel, H., Dixon, Stroup, E., Warren, B. A., (1968) Detection of a deep boundary current in the Western South Pacific Nature 217, 937.
Reid, J. L., and Lonsdale, P. F., (1974) On the flow of water through the Samoan Passage, J. of Phys. Oceanog. 4, 58-73.
Roth, P. H., and Berger, W. H., (1975) Distribution and dissolution of coccoliths in the South and Central Pacific, In Sliter, W., Bé, A. W. H., and Berger, W., eds., Dissolution of Deep-Sea Carates, Spec. Publ. Cushman Foundation for Foraminiferal Res. 13, 11-26.
Ruddiman, W. F., and Heezen, B. C., (1967) Differential solution of planktonic foraminifera, Deep-Sea Res. 14, 801-808.
Terry, R. D., and Chillinger, E. V., (1955) Charts for estimating percentage composition of rocks and sediments, J. Sediment. Petrol. 25, 229-234.
Wooster, W. S., and Volkman, G. H., (1960) Indications of Deep Pacific circulation from the properties at five kilometers, J. of Geophys. Res. 65, 4, 1239-1249.
Wong, C. S., (1972) Deep zonal water masses in the Equatorial Pacific Ocean inferred from anomalous oceanographic properties, J. of Geophys. Res. 77, 36, 7196-7202.

MINERALS, METAL CONTENTS, AND MECHANISM OF FORMATION OF MANGANESE NODULES FROM THE CENTRAL PACIFIC BASIN (GH76-1 AND GH77-1 AREAS)

Akira Usui

Department of Mineral Development Engineering
University of Tokyo
7-3-1, Hongo, Bunkyo-ku, Tokyo, Japan (113)

Abstract

Mineralogy, chemistry and microstructure of deep sea manganese nodules from the Central Pacific Basin have been studied by means of X-ray diffraction analysis, microspopy and electron microprobe analysis. The manganese nodules are composed of microscopically laminated phases of 10Å manganite and δ-MnO_2. The 10Å manganite phase is a monomineralic phase of crystalline manganese oxide with minor elements such as nickel and copper, and the δ-MnO_2 phase is a submicroscopic mixture of colloidal particles of 2 line-form δ-MnO_2, amorphous hydroxide of iron, and minute detrital silicate minerals. Regional and microscopical variations in chemical composition and inter-element relationships of manganese nodules are well explained by the variation of occurrence of the two phases in manganese nodules.

The two phases differ remarkably in mineral and chemical composition, regional distribution, and mode of occurrence on the top and bottom surfaces of individual nodules. The differences suggest that the manganese nodules have grown as a result of two material supply routes, i.e., dissolved metal ions in the interstitial water of unconsolidated sediments resulting in the formation of the 10A manganite phase, and colloidal materials in sea water resulting in the formation of the δ-MnO_2 phase. Internal structures of the phases in individual nodules show evidence of movements on the sea floor in past geologic time. In addition, the differences in morphology, mineralogy, and internal structures of the two types of nodules from the area studied, type A with internal older nodule fragments and surrounding layers and type B with small nuclei and concentric thin layers, seem to reflect the differences in growth history and mechanism of formation.

Introduction

Recent intensive and extensive research has revealed the presence of large amounts of manganese nodules on the Pacific Ocean floor (Horn et al., 1973; Meyer, 1973; Mizuno and Moritani, 1977; Frazer, 1977). A number of investigations have been carried out related to regional variations in nodule abundance, occurrence, and chemical composition. These factors are economically important particularly with respect to copper, nickel, and possibly cobalt. As more information about manganese nodules became available, hypotheses concerning the origin of manganese nodules were proposed (Bostrom et al., 1971; Morgenstein and Felsher, 1971; Greenslate et al., 1973; Crerar and Barnes, 1974). These hypotheses are generally based on geochemical features of bulk samples and regional occurrence. However, it seems that more detailed mineralogical and chemical studies are required in order to elucidate the conditions of formation of ferromanganese minerals and their relationships to the sea floor environment.

This paper presents the results of detailed chemical, mineralogical, and textural investigations of constituent phases* in manganese nodules from the Central Pacific Basin. The chemical, mineralogical and textural characteristics of these phases can help explain the variation in chemistry and morphology of manganese nodules. These characteristics can be applied to interpreting some of the geochemical features of manganese nodules that have been reported previously.

Materials and Methods

Manganese nodules used for the present study were collected from the deep sea floor of the Central Pacific Basin at a depth of 5,000-6,000 m during the GH 76-1 and GH 77-1 cruises of R/V Hakurei Maru conducted by the Geological Survey of Japan (fig. 1). Locality, description of nodules and bottom sediments, results of on-site observations, and related geological and geophysical data (Appendix) are given by Mizuno and Moritani (1977) and Moritani (in preparation). Most manganese nodules collected from the Central Pacific Basin are similar to those from the North Equatorial

* The term "phase" in the present study means that part which is homogeneous and shows invariant optical properties under the reflecting microscope.

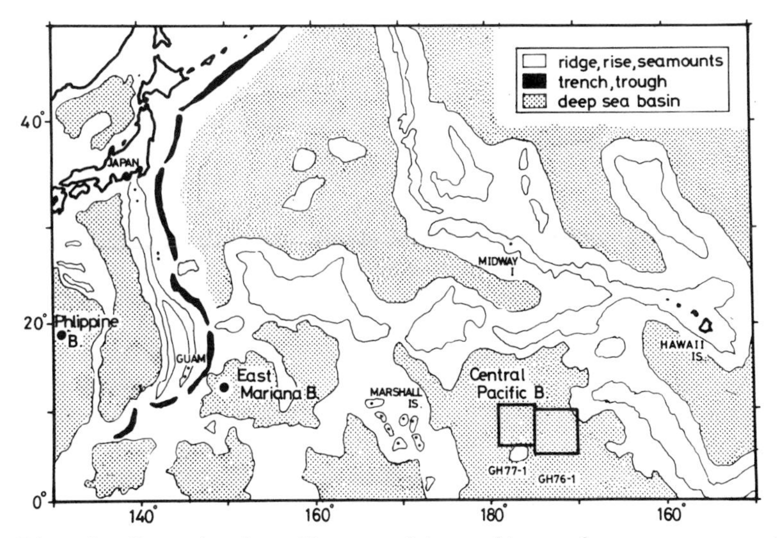

Fig. 1. Map showing the sampling sites of manganese nodules in the West and Central Pacific. Enclosed squares are the GH76-1 GH77-1 areas and solid circles are the stations of KH73-4 cruise (Usui et al., 1976)

Pacific Basin(Meyer, 1973), particularly with respect to morphology (Moritani et al., 1977) and chemistry(Fujinuki et al., 1977). Associated sediments in the area are normal pelagic sediments such as siliceous clay, siliceous ooze, and deep sea clay(Arita, 1977).

Nodule samples were washed with fresh water on the ship immediately after recovery to remove adhering sediments, and dried in air for several weeks or more at the room temperature. After description they were cut by a steel saw into two halves so that the upper and bottom surfaces and the nuclei could be observed. One half of the nodule was soaked in polyester resin in a vacuum desiccator, cured in air for several days at the room temperature, and polished with diamond paste. The other half was analyzed by X-ray diffractometer. Electron microprobe quantitative analysis of metal elements was carried out on microscopically determined phases of selected sections.

Occurrence, Morphology, and Mineralogy of Two Types of Nodules

Manganese nodules from the Central Pacific Basin occur on the surface of unconsolidated pelagic sediments. Abundance of nodules on the sea floor determined by the deep sea photography and by the measurements from grab sample ranges from trace to 30 kg/m^2 (Mizuno and Moritani, 1977). The nodules were classified into nine morphological types on the basis of size, shape, and surface structure by Moritani *et al*. (1977). The nine types are grouped into two major types on the basis of internal and external structure. Most of the A and B type nodules belong to the r and s groups of Moritani *et al*. (1977).

A type nodules contain fairly large nuclei, such as fragments of older nodules, occasionally with zeolite aggregates. They are surrounded by younger ferromanganese oxide layers of 2 to 5 mm thickness and have smooth top surfaces which are exposed to sea water and gritty bottom surfaces which are in contact with sediments. The shape is generally irregular (Figs. 2 and 3). B type nodules contain very small nuclei, such as sharks' teeth. They are composed of many concentric thin layers and the shape is nearly spherical. They are buried in surface sediments and their whole surface is gritty (Figs. 2 and 3).

These two types of nodules occur in different locales of the area surveyed. Generally, nodules from an individual grab sample are similar in morphology, showing that there is little difference in the types of nodules within any limited parts of the area.

Previous mineralogical studies of manganese nodules have shown that major constituents of marine manganese nodules are 10Å manganite (similar to todorokite in X-ray diffraction pattern), 7Å manganite (similar to birnessite or 4 line-form δ-MnO_2), poorly crystalline δ-MnO_2 (2 line form), and goethite (Burns, 1965; Barnes, 1967; Goodell *et al*., 1971; Crerar and Barnes, 1974; Glasby, 1976). Above all, 10Å manganite* and δ-MnO_2* are commonly recognized in nodules from various environments (Goodell *et al*., 1971; Glasby, 1972), whereas 7Å manganite and goethite are not common. Furthermore, the presence of 7Å manganite in nodules is very doubtful because the material which is equivalent to 7Å manganite can be easily formed through the dehydration of 10Å manganite by heating at about $100^{o}C$ or vacuum drying at the room temperature (Usui *et al*., 1977). The X-ray diffraction study of nodules collected in this study revealed that 10Å manganite, 2 line-form δ-MnO_2, quartz, plagioclase, phillipsite, clinoptilolite, montmorollonite, and carbonate-apatite are present in manganese nodules, although the relative proportions of these constituents vary considerably. Figure 4 shows typical X-ray powder diffraction patterns of nodules. Very weak reflections of 7Å manganite and 2

* In many cases, the terms "todorokite" and "birnessite" have been used for the constituents (Burns and Burns, 1977). However, few comparative studies between marine and terrestrial manganese minerals have been made on their chemical compositions, crystallographic and optical properties in detail. Thus, the writer believes that the terms "10Å manganite" and "δ-MnO_2" after the first description by Buser and Grütter should be used at the present time.

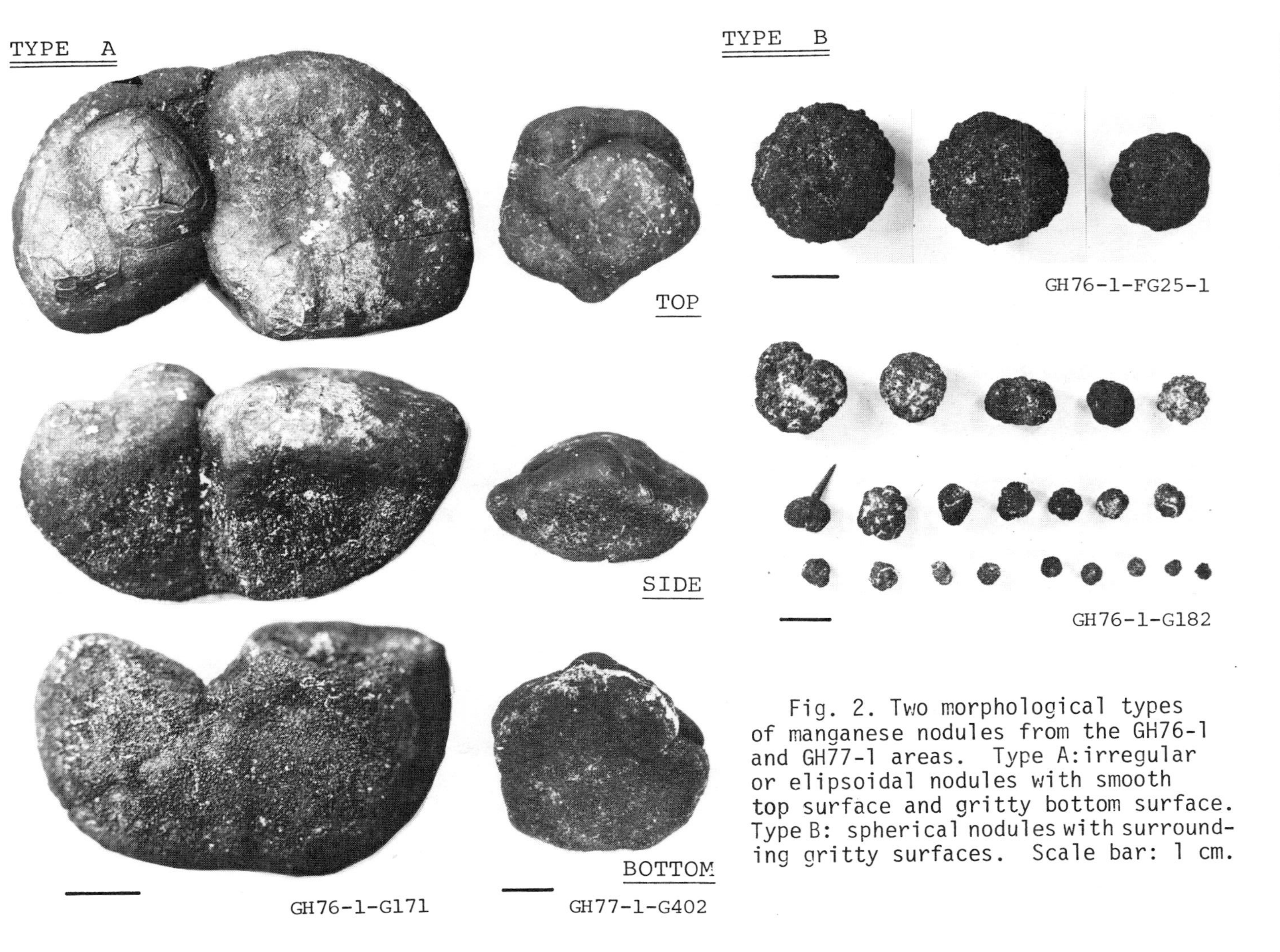

Fig. 2. Two morphological types of manganese nodules from the GH76-1 and GH77-1 areas. Type A: irregular or elipsoidal nodules with smooth top surface and gritty bottom surface. Type B: spherical nodules with surrounding gritty surfaces. Scale bar: 1 cm.

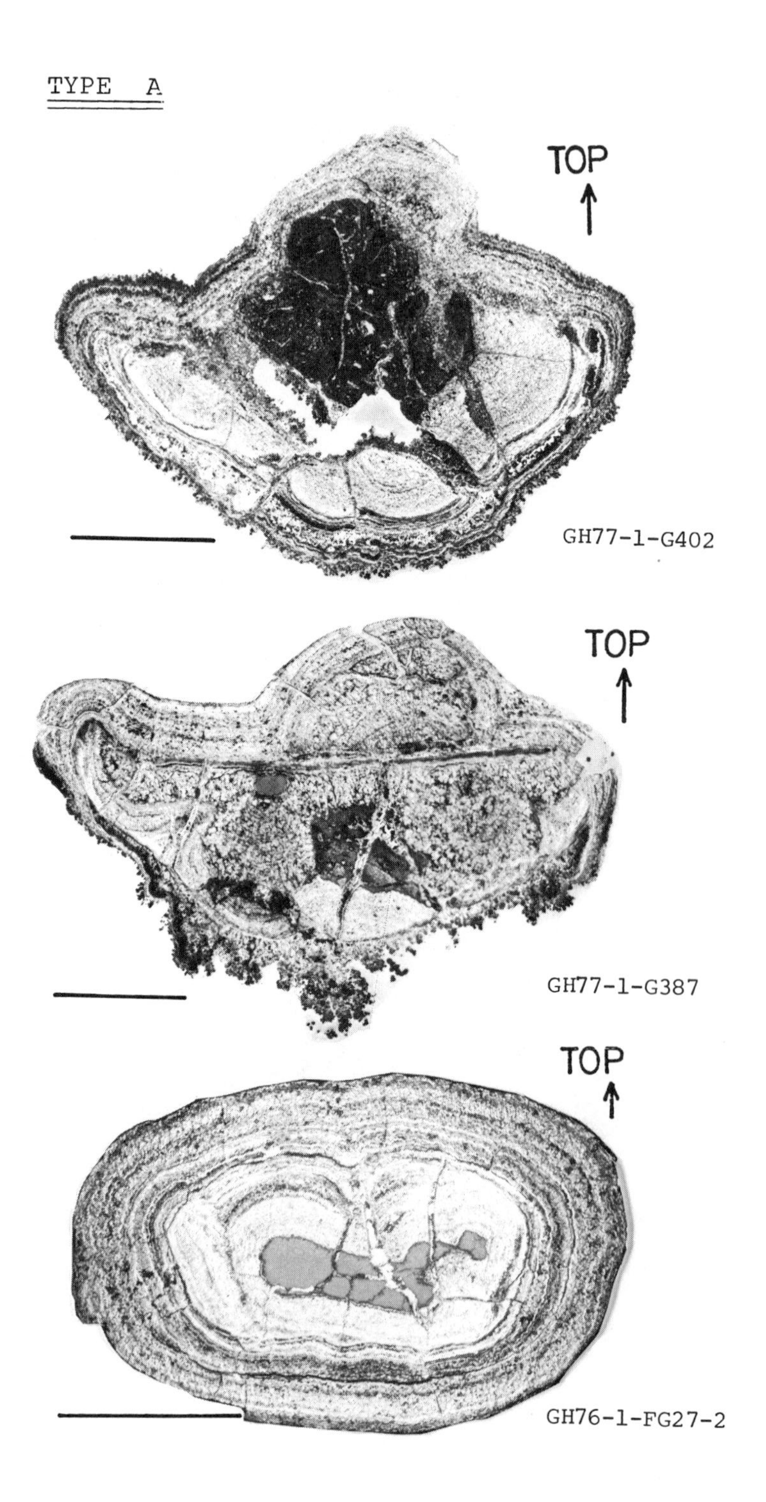
TYPE A
TOP
GH77-1-G402
TOP
GH77-1-G387
TOP
GH76-1-FG27-2

TYPE B

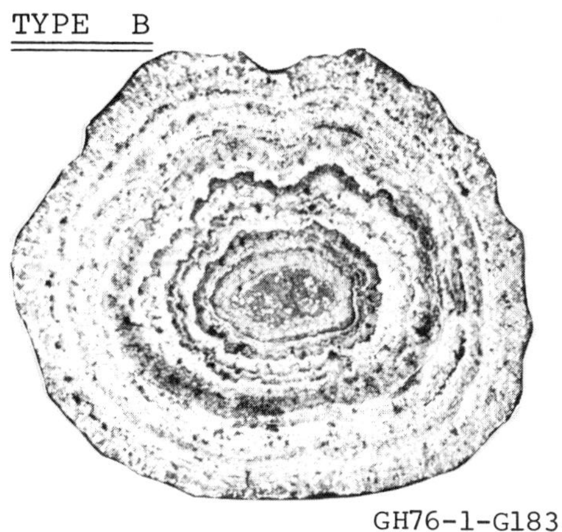

GH76-1-G183

GH76-1-G190

GH77-1-G389

Fig. 3. Internal structure of manganese nodules. Type A: irregular or elipsoidal nodules with nuclei of fragmented older nodules composed of the -MnO phase, and with surrounding 2-5 mm thick layers composed of the -MnO and 10Å manganite phases. Type B: spherical nodule with continuous concentric thin layers composed of the 10Å manganite phase. Scale bar: 1 cm.

line-form δ-MnO_2 seem to be the main ferromanganese constituent minerals of the nodules.

The two types of nodules differ from each other in mineralogy. 10Å manganite is dominant in B type nodules and surface layers of 2 to 5 mm in thickness in type A nodules, while 2 line-form δ-MnO_2 is frequently recognized in A type nodules, especially in older nodule fragments inside the nodules. Concerning the distribution of the two minerals in single nodules of type A, 10Å manganite is rarely detected in the smooth upper surface but is dominant in the gritty bottom surface, while 2 line-form δ-MnO_2 is abundant in the smooth top surface. The results of bulk X-ray diffraction analysis on manganese nodules are in accordance with the following microscopic observations which show more detailed structures of the surface and interior of nodules.

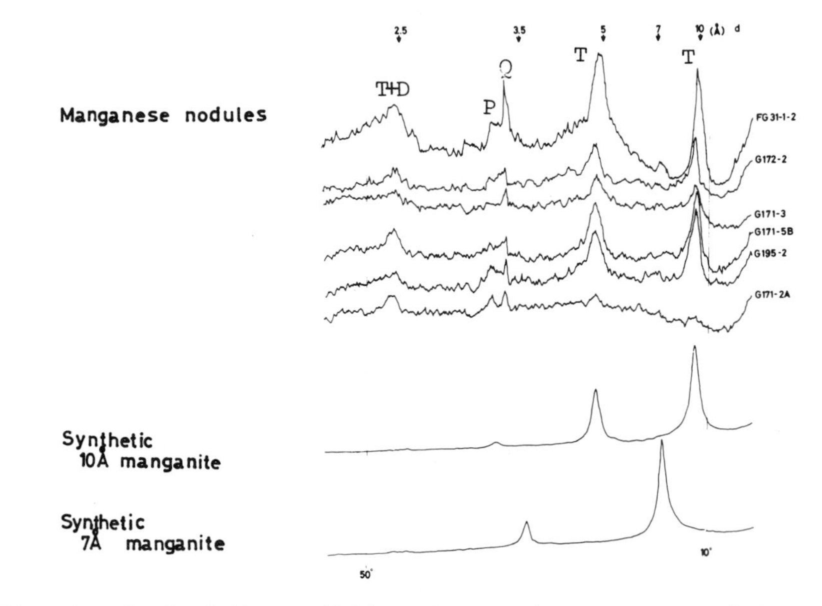

Fig. 4. Typical X-ray diffraction powder patterns of deep sea manganese nodules (Mn-filtered Fe Kα radiation). T:10Å manganite, D:δ-MnO_2, Q:quartz, P:plagioclase and/or phillipsite.

Internal Microstructures

Observations with the reflecting microscope reveal various optical characteristics of constituents and detailed internal structures of manganese nodules. Two constituents, 10Å manganite and δ-MnO_2, identified by an X-ray diffraction method, are also identified by their particular invariable optical properties under reflecting microscope. These two phases are tentatively called the 10Å manganite phase and δ-MnO_2 phase in this study. Ferromanganese phases other than the two phases were not observed. As compared to previous studies of manganese oxide minerals from land (Uyttenbogaardt and Burke, 1971), the optical properties of the 10Å manganite phase are obviously different from those of birnessite. Optical properties of the two phases are summarized in Table 1.

Each of the interlaminated thin layers in ferromanganese oxide shells of nodules, most characteristically observed under the microscope, is composed of one of these two phases. In general, each thin layer is macroscopically continuous, surrounding the nucleus of the nodules. The two phases show characteristic textures and distribution patterns in the interior of nodules, and they may reflect the formation conditions of these phases. Textures of the 10Å manganite phases are variable, showing compact conformable thin layer(several tens to several hundreds μm in thickness), dendritic structure(cauliflower-like cusps), fracture filling veinlets, networks, homogeneous massive aggregates, and a matrix of relatively large detritus such as silicate or phosphate minerals and fossils (Usui, et al., 1976). Thin layers, showing dendritic structure of the 10Å manganite phase with clayey materials in the interstices of dendrites, are predominant in the manganese nodules from the deep sea basin of this area(Fig. 5). On the other hand, textures of the δ-MnO_2 phase are relatively simple, showing stratification occasionally with radial columnar patterns(Fig. 5a and 5b). The stratification of the δ-MnO_2 phase in deep sea nodules resembles that of manganese crusts which cover exposed rocks on seamounts(Usui et al., 1976). It is believed that the characteristic textures and distribution patterns of the phases generally control both the morphology and the internal and external structure of nodules. The smooth top surface and gritty bottom surface are related to the stratification of the δ-MnO_2 phases and the dendritic structure of the 10Å manganite phase, respectively. The difference in constituents between top and bottom surfaces of A type nodules recognized by the X-ray diffraction method was also verified under reflecting microscope. Whole gritty surfaces of B type nodules are also characteristic of the dendritic structure of 10Å manganite(Fig. 5c and 5d). The nuclei of A type nodules are composed exclusively of the δ-MnO_2 phase while the surrounding interlaminated layers are composed of both phases. In B type nodules, internal concentric thin layers surround a small nucleus. The layers are composed mainly of the dendritic 10Å manganite phases and some manganese-free clayey materials(Fig. 5e and 5f).

A large number of thin layers composed of either of the two phases appear to be continuous and concentric macroscopically. However, detailed microscopic observations reveal that in many cases each thin layer is not necessarily a continuous type. Individual thin layers often fade near the equatorial zone in the case of A type nodules. This is probably due to the fact that the gritty structure develops only on the bottom surface. Even though a

TABLE 1

Characteristics of constituent phases of manganese nodules

Characteristics		10Å manganite phase	δ-MnO_2 phase
Optical Properties (under reflecting microscope)	color	light grey	dark grey
	reflectivity	high	low
	anisotropism	strong	none
	hardness(VHN)	high(52-112,ave:82)	low(10-24,ave:17)
	internal reflection	none	none
Occurrence		deep sea manganese nodules (chiefly bottom surface)	deep sea manganese nodules (chiefly top surface)
		shallow water manganese nodules	crust or coating of rocks from topographic highs
Texture		conformable thin layer dendritic (cauliflower-like cusps) network crack filling cementing of clastics massive aggregate	stratification (sometimes columnal)
Chemistry		rich in Mn	rich in Mn,Fe and Si
		Mn: 30-50 wt.% Fe: 0- 2 Ni: 1- 3 Cu: 1- 2 Co: 0-0.4 Si: 0- 1	Mn: 10-30 wt.% Fe: 11-18 Ni: 0-0.8 Cu: 0-0.8 Co:0.3-0.6 Si: 1- 8
Mineralogy		monomineralic (10A manganite)	mixture of some minerals, i.e., 2 line-form δ-MnO_2,amorphous Fe hydroxide, and minute detrital minerals, such as quartz, plagio-clase, zeolite and clay minerals.

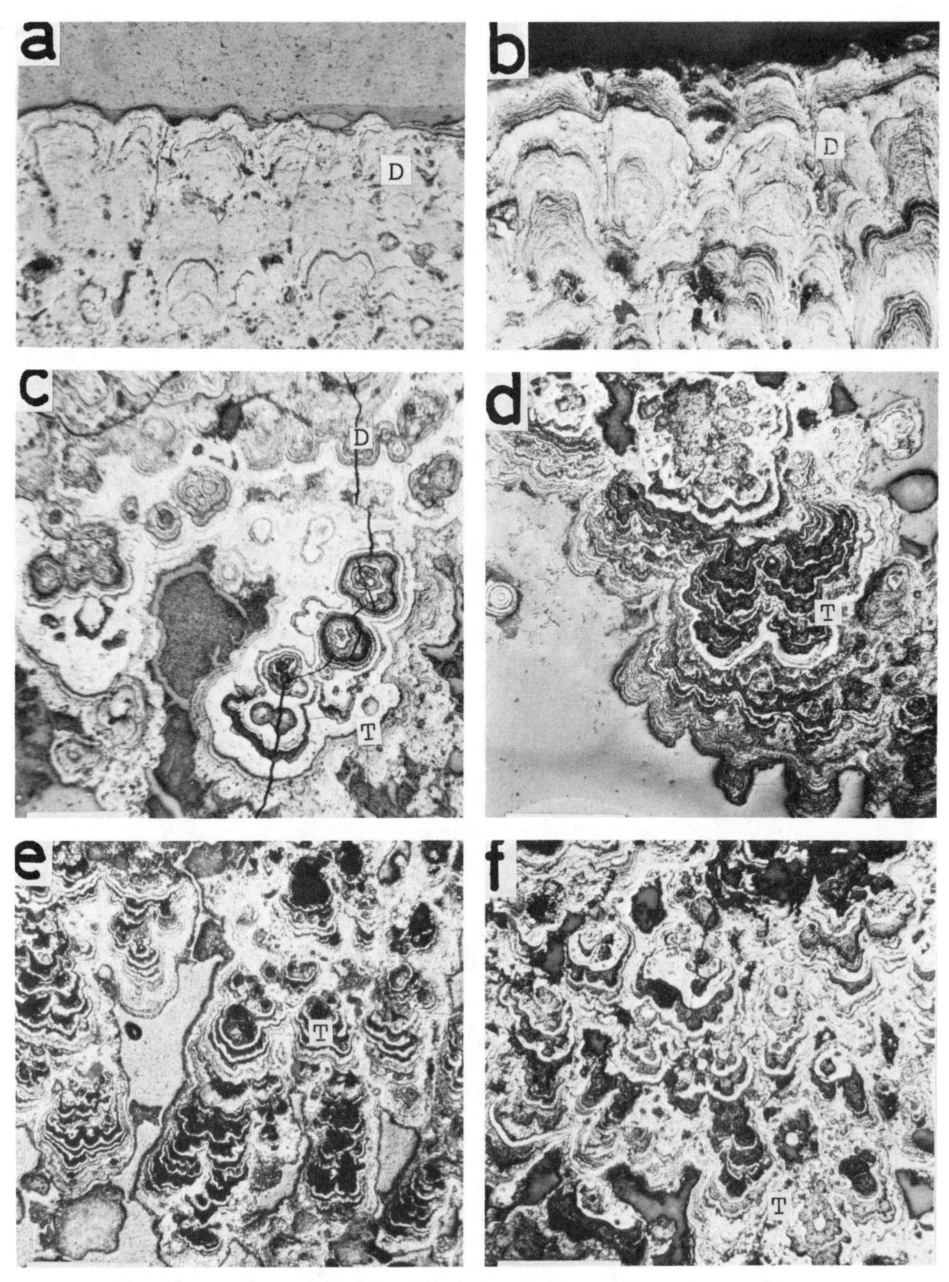

Fig. 5. Photomicrographs showing structures and mineralogy of manganese nodules. a), b) Smooth top surface of type A(stratified D). c), d) Gritty bottom surface of type A(T). e) Gritty surface of type B(T). f) Internal concentric layers of type B. This structure is similar to that of gritty surfaces(T). T: 10Å manganite phase, D: δ-MnO_2 phase, Scale bar: 20μm.

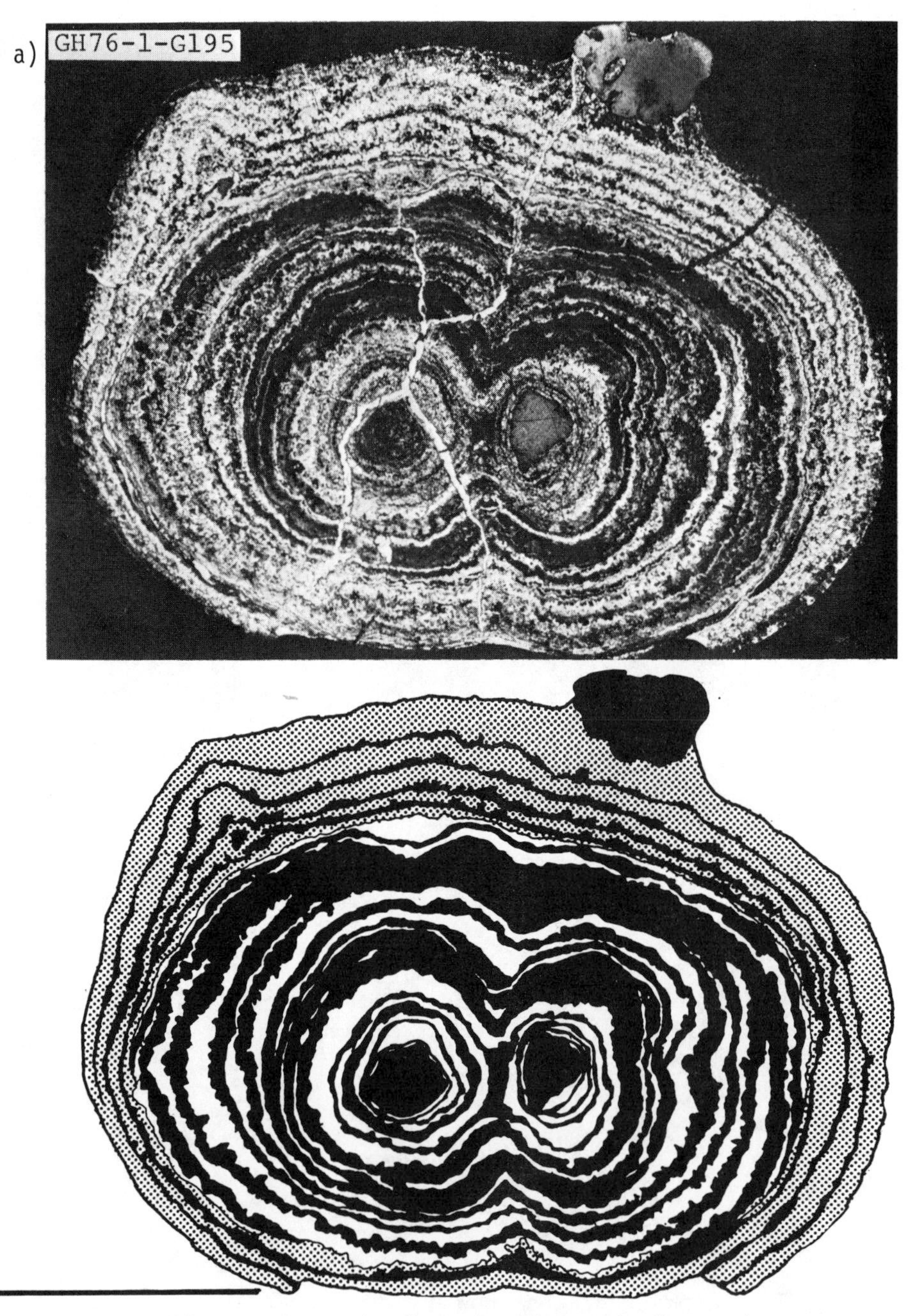

Fig. 6. Illustrations showing the discontinuity and asymmetry of thin layers in manganese nodules. White parts in the photograph are ferromanganese phases. Two phases are distinguishable under reflecting microscope. Black, white, and grey areas in the illustration are manganese-free materials, 10Å manganite phases, and δ-MnO_2 phases, respectively. Scale bar: 1 cm.

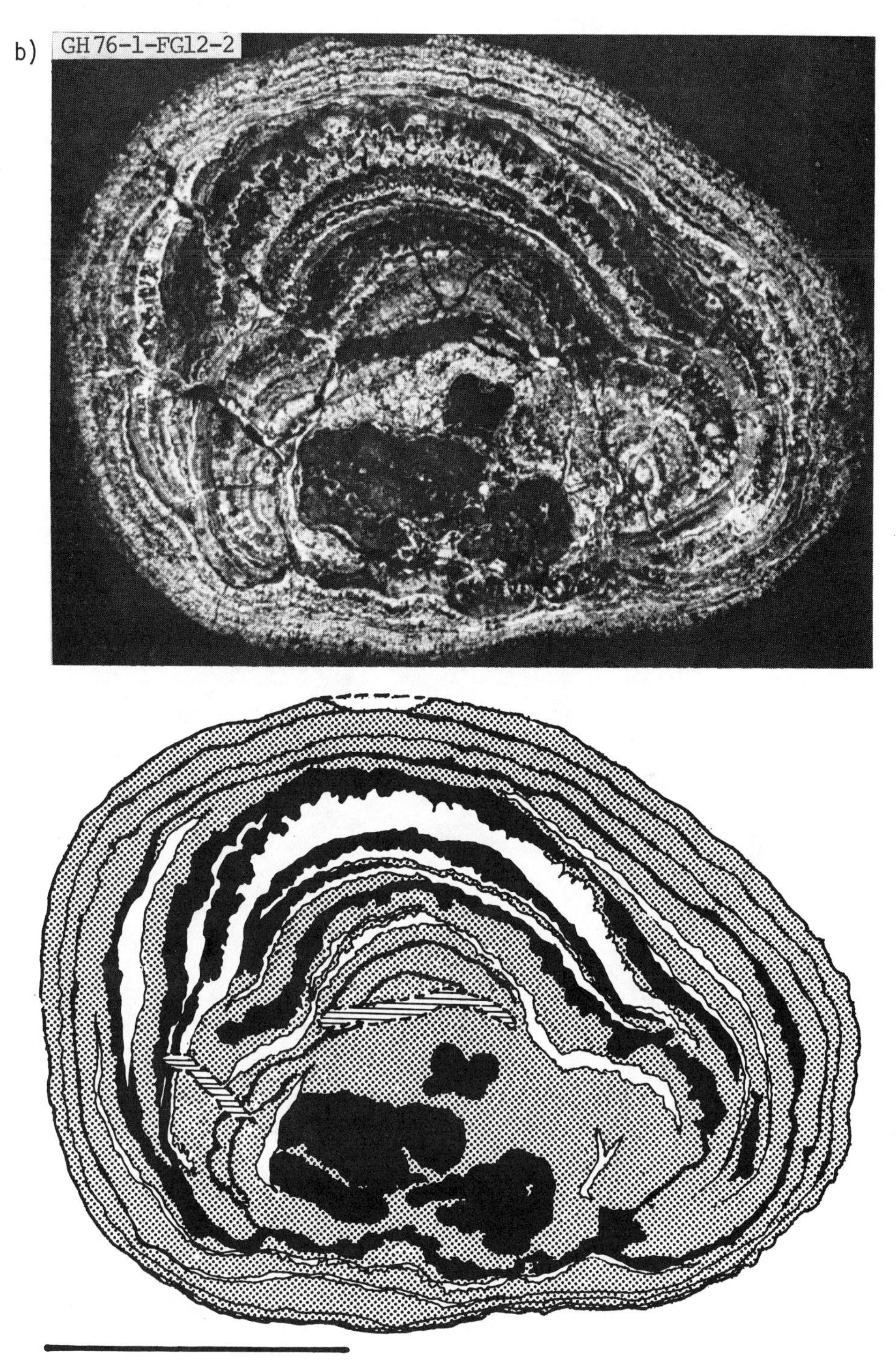
b)
GH 76-1-FG12-2

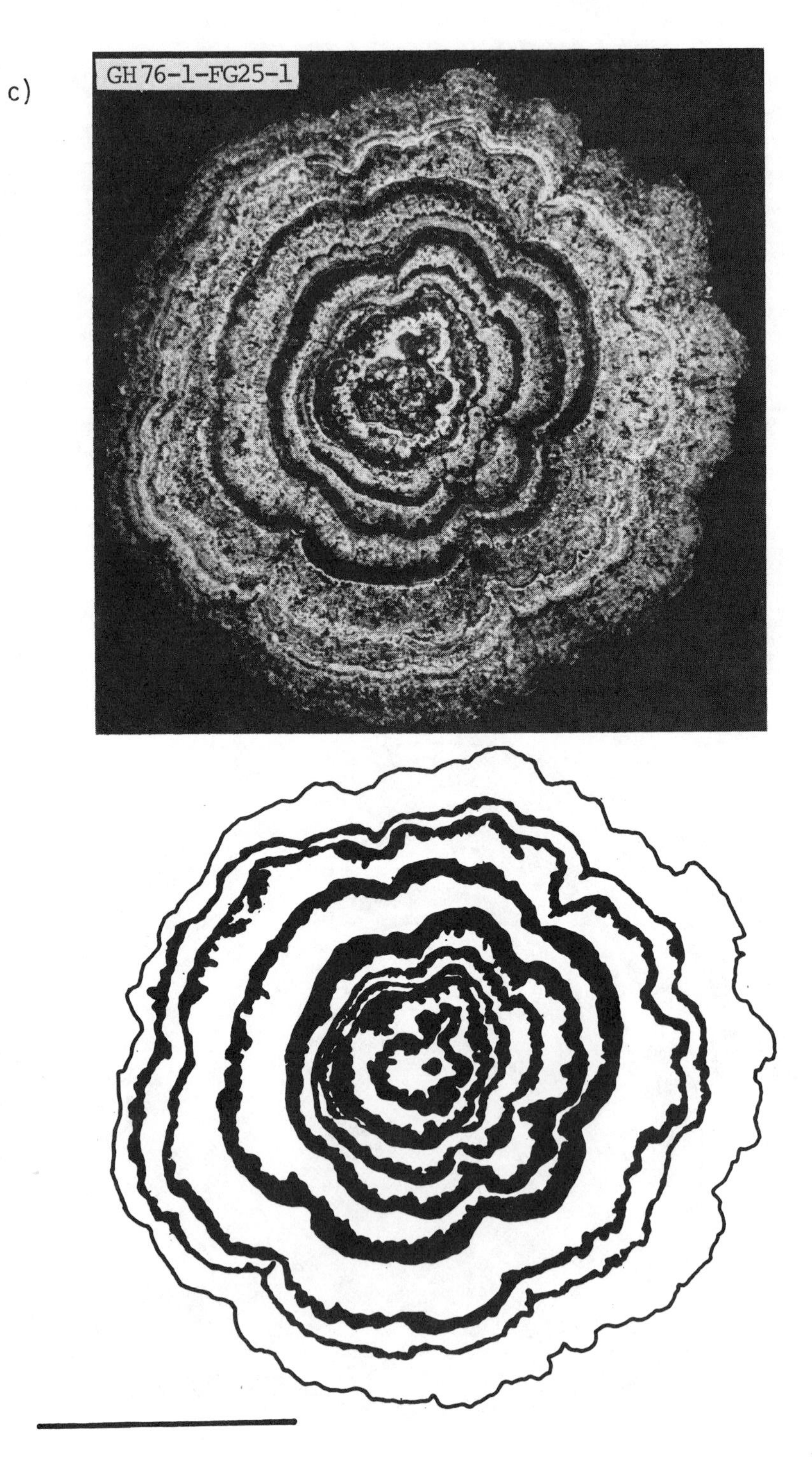
c)
GH 76-1-FG25-1

thin layer is continuous, the thickness of the layer differs appreciably between one hemisphere and the other. Photographs and illustrations of vertical cross sections of some nodules(Fig. 6a and 6b) demonstrate the discontinuity of thin layers of each phase and the asymmetry of the internal structure is more frequently observed in large or irregularly shaped nodules as compared to spherical or small nodules (fig. 6).

Chemistry of Constituent Phases

Quantitative electron microprobe analyses of the 10Å manganite phase and the δ-MnO_2 phase(Appendix) revealed a remarkable difference in chemical composition. The analyses were carried out for Mn, Fe, Cu, Ni, Co, Ca, Mg, Na, K, Si, and Al. Preferential parti-

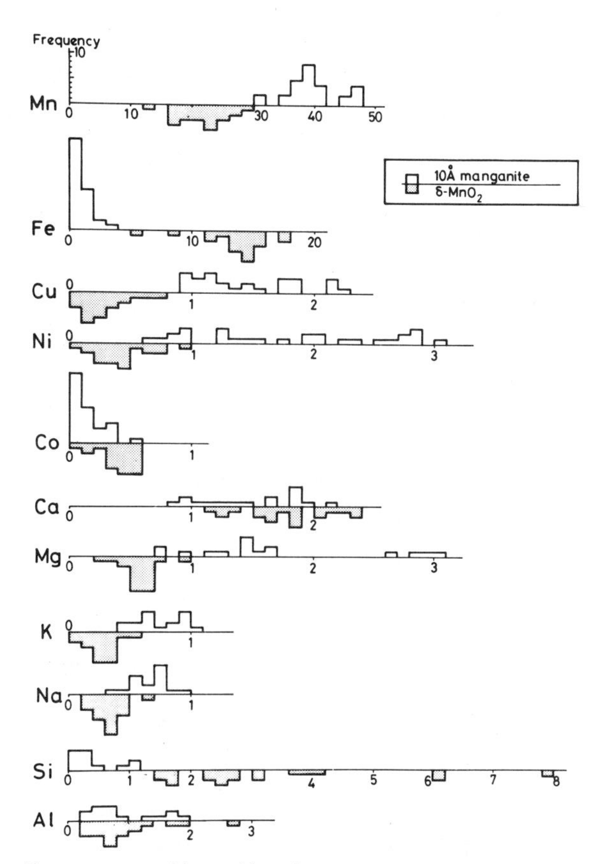

Fig. 7. Frequency distribution of metal concentration in the constituent phases of deep sea manganese noduels as determined by microprobe analysis. The number of available data sets is 39.

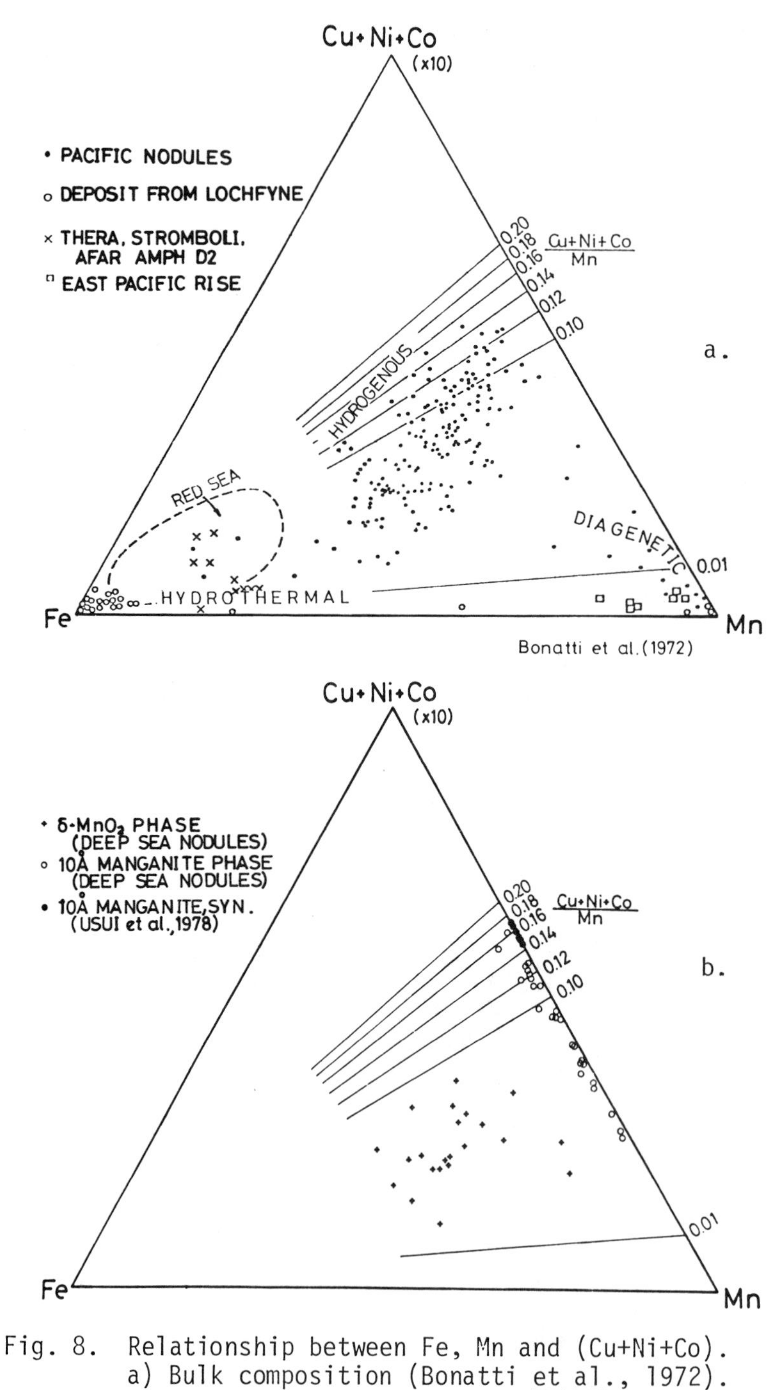

Fig. 8. Relationship between Fe, Mn and (Cu+Ni+Co).
a) Bulk composition (Bonatti et al., 1972).
b) Constituent phases in manganese nodules.

tioning of the metal elements in the two constituent phases were determined(Fig. 7). The results show significantly high concentrations of Mn, Ni, Cu, and Mg in the 10Å manganite phase, and high concentrations of Fe, Si, and Co in the δ-MnO_2 phase. The results also agree well with those of previous geochemical studies(Burns and Fuerstenau, 1966; Brooke and Prosser, 1969; Friedrich et al., 1969). Concerning the chemical composition of the 10Å manganite phase, manganese constitutes the major element and iron and silica the trace elements. It should also be emphasized that the total amount of Ni, Cu, and Co contained in the 10Å manganite phase never exceeds a certain limit, expressed as a ratio of (Cu+Ni+Co)/Mn in Fig. 8b. The upper limit of the ratio is about 0.16, which corresponds to the Ni/Mn ratio in synthetic 10Å manganite derived through the reaction of manganese oxides with nickel chloride solutions(Usui et al., 1977). The δ-MnO_2 phase shows a very different chemical composition. The major constituent elements of this phase are Mn, Fe, and Si, while the contents of the other nine elements except for Al constitute less than 1 wt.%.

Inter-element correlations found in the two phases differ markedly from those reported for the bulk composition of manganese nodules. For instance, significant positive correlation on Mn-Ni-Cu or Fe-Co, which have been previously reported, are not recognized in either phase, except for the positive correlation of Cu with Ni in the δ-MnO_2 phase. Cobalt, which is rich in the δ-MnO_2 phase, is related to manganese rather that iron (Table 2), even though the previous results of bulk chemical analyses show a positive correlation between iron and cobalt(Ahrens et al., 1967; Cronan, 1977; Fujinuki et al., 1977; and others). Inter-element correlation matrices for the eleven elements were calculated for the two phases. No significant positive correlation in the 10Å manganite phase was found, but negative correlations of Ni with Na, Ni with Ca, and Cu with K were found. For the δ-MnO_2 phase, positive correlations of Ni with Cu, Mn with Ca, and Si with Al and negative correlations of Mn with Si, Ca with Si, Co with Si, Ca with Al, and Co with Na were found. Iron is not correlated with the other metal elements. Based on their concentrations and correlations within the δ-MnO_2 phase, the metal elements are classified into three groups, i.e., 1) large amount of Mn and associated Cu, Ni, Co, Ca, and Mg, 2) Si, Al, Na, and possibly K, and 3) large amounts of Fe. The elements in the second group are presumably related to detrital silicate minerals.

Electron microprobe analyses of manganese nodules from the Daini-Komabashi Seamount, Kyushu-Palau Ridge, were previously carried out by Usui et al.(1976). When the results are compared to the present results, no significant differences in the contents of Mn, Fe, Ni, and Cu within the 10Å manganite phase between the nodules from the marginal seamount and the deep sea are recognized. There appears to be no significant difference in the contents of Mn and Fe in the δ-MnO_2 phase, but the contents of Ni and Cu in the δ-MnO_2 phase of deep sea manganese nodules are significantly higher than those of nodules from the marginal seamount.

TABLE 2

Inter-element correlation matrices for the constituent phases

10Å manganite phase

	Fe	Cu	Ni	Co	Ca	Mg	K	Na	Si	Al
Mn	0.42	-0.59	-0.07	-0.58	0.29	0.10	0.58	0.14	-0.12	-0.40
Fe	*	-0.53	0.76	0.09	-0.50	0.42	0.62	-0.76	0.18	0.29
Cu		*	0.06	0.16	-0.33	-0.02	-0.78	0.22	0.07	0.26
Ni			*	0.37	-0.88	0.60	0.16	-0.78	0.25	0.57
Co				*	-0.38	0.03	-0.42	-0.40	-0.14	0.29
Ca					*	-0.71	0.05	0.66	-0.41	-0.68
Mg						*	0.27	-0.43	0.35	0.33
K							*	-0.24	0.11	0.25
Na								*	-0.33	-0.60
Si									*	0.45
Al										*

δ-MnO_2 phase

	Fe	Cu	Ni	Co	Ca	Mg	K	Na	Si	Al
Mn	-0.45	0.35	0.48	0.60	0.74	0.17	-0.37	-0.58	-0.70	-0.55
Fe	*	0.11	0.30	-0.06	-0.44	0.37	0.46	0.20	0.15	0.18
Cu		*	0.72	0.10	-0.05	0.47	0.02	-0.26	-0.00	0.18
Ni			*	0.58	0.21	0.54	-0.31	-0.38	-0.45	-0.32
Co				*	0.58	0.23	-0.60	-0.67	-0.82	-0.77
Ca					*	-0.09	-0.37	-0.44	-0.75	-0.76
Mg						*	-0.43	-0.21	-0.20	-0.01
K							*	0.41	0.57	0.47
Na								*	0.72	0.60
Si									*	0.93
Al										*

Each correlation coefficient is calculated using 16 sets(10Å manganite phase) and 23 sets(δ-MnO_2 phase) of microprobe data. Significant correlations in the 99.9% confidence level are underlined in the table.

Discussion

Chemistry and Mineralogy of Manganese Nodules

The X-ray diffraction study and microscopic observations revealed that 10Å manganite and 2 line-form δ-MnO_2 are predominant in manganese nodules from the Central Pacific Basin. The two homogeneous phases corresponding to 10Å manganite and 2 line-form δ-MnO_2 were determined under reflecting microscope. According to the results of electron microprobe analysis, the 10Å manganite phase is composed of large amounts of manganese and minor amounts of nickel, copper, magnesium, and other alternative divalent metal ions. These divalent metal elements are regarded as essential to the crystal structure of 10Å manganite due to the close agreement between the maximum value of the total contents of copper, nickel and cobalt(ratio to manganese) in the 10Å manganite phase and the value for synthetic 10Å manganite(Usui et al., 1978). Therefore, it is concluded that the 10Å manganite phase in manganese nodules is a monomineralic phase composed of Fe- and Si-free manganese oxide. On the other hand, the δ-MnO_2 phase is rich in Fe and Si, as well as Mn, and differs appreciably from the 10Å manganite phase. A close correlation between Si and Al contents suggests that Si in the phase is related to detrital and/or authigenic silicate minerals(probably zeolites, clay minerals, plagioclase and quartz). The contents of Fe in the δ-MnO_2 phase is fairly high, ranging from 11 to 18 wt.%, and sometimes exceeds the content of Mn(Fig. 8b). Nevertheless, no crystalline iron oxide or hydroxide minerals were detected in the manganese nodules from this area. Buser and Grütter(1956) suggested that $Mn(OH)_2$ in synthetic δ-MnO_2 is isomorphically substituted by $Fe(OH)_3$, and they proposed the ideal chemical formulas for δ-MnO_2 as $4MnO_2\ Mn(OH)_2 \cdot nH_2O$ or $(Na,Mn)Mn_3O_7 \cdot nH\ O$. Other synthetic studies (Buser et al., 1954; Bricker, 1965; Giovanolli et al., 1970) showed that the ratio O/Mn in δ-MnO_2 ranges from 1.74 to1.99. This concept of isomorphic substitution, however, cannot explain the excess content of iron and the large variation of the ratio Fe/Mn in the δ-MnO_2 phase(Fig. 8b). The ratio Fe/(Fe+Mn) of the phase exceeds 0.50 in some cases, whereas the possible upper limit of substitution by iron calculated from the above formula is 0.26. Therefore, it is tentatively concluded that amorphous iron hydroxides would be present more or less in the δ-MnO_2 phase in addition to the mineral 2 line-form δ-MnO_2. That is to say, the δ-MnO_2 phase, although microscopically homogeneous, is a submicroscopical mixture of colloidal particles of more than three minerals, i.e., 2 line-form δ-MnO_2 with minor elements, non-crystalline hydroxides of iron, and detrital and/or authigenic silicate minerals.

Partition of minor metal elements shows that the contents of nickel and copper in the 10Å manganite are several times higher than those in the δ-MnO_2 phase, and that cobalt shows less preferential concentration in these two phases. Previous geochemical investigations(Arrhenius et al., 1964; Ahrens et al., 1967; Ostwald and Frazer, 1973) indicated that some of the iron in manganese nodules is substituted for by trivalent Co. This observation is based on the fact that the Fe and Co contents of nodule samples are positively correlated. If this is the case, a positive correlation between Fe and Co contents is expected in the δ-MnO_2 phase because most of the cobalt is concentrated in the δ-MnO_2 phase. However, the present results demonstrate a positive correlation between Mn and Co(+0.60) rather than between Fe and Co(-0.06)(see Table 2), indicating a closer relationship of cobalt to the mineral 2 line-form δ-MnO_2 than to the non-crystalline iron hydroxide present in the δ-MnO_2 phase. This is in good agreement with the consideration that Co^{3+} substitutes for Mn^{4+} in synthetic δ-MnO_2 (Burns, 1976).

Chemical and mineralogical characteristics and abundance of two mineralogical phases in manganese nodules are important factors in determining the chemical heterogeneity of nodules, the inter-element relationships, the regional variations of bulk chemical composition, and the relationship between chemistry and occurrence. Chemical heterogeneity, which has been characterized by the laminations of manganese-rich and iron-rich zones observed in the interior of nodules and by similar variations in the abundance of copper, nickel and zinc relative to manganese(Burns and Fuerstenau, 1966; Friedrich et al., 1969; Sano and Matsubara, 1970; Sorem and Foster, 1973; Dunham and Glasby, 1974) are related to the laminations of the 10Å manganite and the δ-MnO_2 phases. It is also suggested, from the present studies, that the strong positive correlations between Mn and Ni and Mn and Cu as determined by the bulk composition of nodules from various environments, are not due to the crystallographic property of a single mineral, but due rather to the relative abundance of 10Å manganite which contains a large amount of nickel and copper. The negative correlation between the bulk chemical compositions of Mn and Fe are similarly explained by this fact. Positive correlations between Mn and Co(Aumento et al., 1968; Sorem and Foster, 1973) may be explained by the presence of divalent cobalt in 10Å manganite, even though some of the cobalt may be present in the δ-MnO_2 phase as trivalent cobalt which is substituting for manganese in the 2 line-form δ-MnO_2.

Morphological classifications of manganese nodules have been made by some workers(Meyer, 1973; Meylan, 1974; Halbach et al., 1975; Moritani et al., 1977) and relationships between morphology and chemistry have been reported. In general, nodules with gritty surfaces contain a large amount of nickel and copper as compared to those with smooth surfaces(Halbach et al., 1975; Moritani et al., 1977). Raab(1972) also found that in some types of nodules nickel and copper are more enriched in gritty material from the bottom half than in smooth surfaces from the upper half of nodules. Manganese nodules with spherical shapes and gritty surfaces are rich in manganese, nickel and copper(Moritani et al., 1977). These relationships between the morphology and chemistry are evident when one considers the chemical characteristics and microscopic structures of the 10Å manganite and the δ-MnO_2 phases. That is, the gritty structure of the present and older surfaces of nodules consists exclusively of dendrites(cauliflower-like cusps) of the 10Å manganite phase, and the smooth structure consists of the stratification of the δ-MnO_2 phase. Therefore, some of the major factors controlling the geochemical and morphological features of manganese nodules are related to the abundance and microstructural characteristics of the two constituent phases.

A genetical classification of various ferromanganese deposits, on the basis of their bulk chemical composition, was proposed by Bonatti et al.(1972). "Hydrogenous" nodules from the Pacific Ocean show a remarkable variation in chemical composition. Taking into account the chemical characteristics of the two phases in nodules from the present work, it is suggested that most of the "hydrogenous" manganese nodules shown in Fig. 8b are an arbitrary mixture of the 10Å manganite and the δ-MnO_2 phases. Therefore, genetical discussions on manganese nodules should be based on detailed mineralogical study. Enrichment of nickel and copper in deep sea manganese nodules, low concentrations of these metals in shallow-water nodules and crusts, relatively high contents of cobalt in manganese crusts from oceanic seamounts(Price, 1967: Price and Calvert, 1970; Glasby, 1972), and consequent depth dependency of the contents of these metals(Cronan and Tooms, 1969; Cronan, 1977), are general geochemical features of marine manganese nodules. Considering the present study, these geochemical features are characteristic of the relative amounts of the two phases in manganese nodules, controlling the variation of chemical compositions of manganese nodules of various occurrences.

Formation Mechanisms and Growth History of Manganese Nodules

It has been reported that the factors controlling the formation of manganese minerals include the pH-Eh condition within the aqueous environments(Crerar and Barnes, 1974), hydrostatic pressure(Barnes, 1967), dissolved oxygen(Cronan and Tooms, 1969), and other physicochemical conditions. Cronan(1975) reported that the formation of manganese minerals is controlled mainly by redox conditions, that is , δ-MnO_2(2 line form), 7Å manganite and 10Å manganite precipitated in the order of oxidized conditions. The present results, however, show another important factor controlling the formation of manganese oxyhydroxide minerals within manganese nodules. The present detailed studies of occurrence, chemical and mineralogical characteristics, and microscopic textures of the two constituent phases in marine manganese nodules provide much information on the origin of deep sea manganese nodules. Some characteristics of these two constituent phases are summarized in Table 1.

Preferential growth of 10Å manganite in deep sea manganese nodules especially on the bottom surfaces, and the scarcity of the mineral in manganese crusts on exposed rocks of oceanic seamounts suggest that the formation of 10Å manganite is closely related to unconsolidated sediments of the deep sea. Concentrations of Mn, Cu, Ni, Co and Zn in the interstitial waters of deep sea sediments are generally high in comparison to those in normal sea water and deep sea bottom water(Presley et al., 1967; Presley et al., 1970; Raab, 1972). Lynn and Bonatti(1965), Li et al.(1969) and other geochemists(Bender, 1971; Renard et al., 1976) proposed a model of the upward migration of manganese in unconsolidated sediments, that is, Mn^{4+} in solid oxides is reduced to Mn^{2+} in the substrata under a low-Eh condition and Mn^{2+} ions diffuse upward into a high-Eh area, where they are oxidized and removed by precipitation. Iron is, however, fixed in sediments and never migrates as a soluble ion (Cheney and Vredenburgh, 1968). The geochemical behavior of iron and manganese are also supported by investigations into the stability relations in the system Mn-O_2-H_2O and Fe-O_2-H_2O(Krauskopf, 1957; Crerar and Barnes, 1974; Glasby, 1974). The present study revealed that the 10Å manganite phase contains extremely small amounts of Si and Fe, and that it preferentially occurs on the bottom surface of nodules associated with unconsolidated sediments. The present result is consistent with previous statements concerning feature deep sea sediments, suggesting that the formation of 10Å manganite is closely related to the migration of manganese in submarine sediments. On the basis of the present study and previous geochemical studies on submarine sediments, a model for the formation of the two constituent phases is proposed. Manganese supplied from the substrata, during the diagenetic process, is precipitated as 10Å manganite with nickel, copper and other divalent metal ions. Manganese oxides precipitated within surficial sediment shows characteristic growth structures such as dendrite or cauliflower-like cusps on the bottom surfaces of A type nodules in contact with the bottom sediments, and on the whole surfaces of B type nodules which are buried in the bottom sediments. Copper, nickel, and other divalent metals supplied from the interstitial waters in sediments are essential elements for the formation of crystalline 10Å manganite, and these metal ions may effectively be substituted for by additional copper and nickel ions in the crystal after the deposition of the mineral (Usui, in press). The proposed model explains the regional characteristics in the occurrence of 10Å manganite and the internal and external structures of manganese nodules.

The δ-MnO_2 phase occurs generally on the interface between sea water and solid objects, forming stratified layers on the upper surface of deep sea manganese nodules and manganese crusts on rocks from seamounts. This phase invariably contains a large amount of Fe and Si originating from detritus, and an appreciable variation in chemical composition. These facts suggest that the formation of the δ-MnO_2 phase is related to the sedimentary process in the deep sea. It is inferred that the δ-MnO_2 phase forms by the accumulation of colloidal particles of manganese and iron hydroxides precipitated from sea water and of detrital minerals, resulting in the formation of heterogeneous mixtures which vary markedly in chemical composition (Fig. 8b). This mechanism is supported by the fact that the uniform rain of manganese hydroxide particles onto the sea floor occurred throughout geological time (Bender et al., 1966 and 1970). Though a confirmative model for the accumulation of minor metal elements such as Ni and Cu in the δ-MnO_2 phase is not established, it is presumed that the surface adsorption of these minor elements on colloidal form of 2 line-form δ-MnO_2 takes place in sea water. However, the surface adsorption of metal elements is not essential in the case of 10Å manganite as stated above.

Significant information on the growth history of the nodules is obtained from the microscopic observation of the internal and external structures of nodules. Based on the present model of formation of the two constituent phases, the internal characteristic structures of nodules are interpreted in relation to the geological environments. It is presumed that each thin layer within the nodules reflects the conditions and the relation to the sediment when the thin layer itself was the nodule surface.

Therefore, when the model is applied, the asymmetry and discontinuity of each individual thin layer in a nodule (Figs. 6a and 6b) indicate the possibility of intermittent overturns and/or rotations of the nodule on the sea floor, though the major causes of these movements are still unknown in detail (Mero, 1965; Glasby and Singleton, 1975; Menard, 1976). Evidence of movement of nodules is recognized in most large nodules of type A which have irregular nuclei, whereas it is scarcely recognized in the small and spheroidal nodules of type B. A number of concentric thin layers in B type nodules are composed of the 10Å manganite phase. It is inferred with the present model that the B type nodules have been formed without movement and partially buried in the unconsolidated surface sediments.

Considering that B type nodules from the area occur generally with surfacial sediments of Quaternary siliceous clay (Arita,1977), it is inferred that the growth rate of B type nodules would be faster than average growth rates for deep sea manganese nodules (Ku and Broecker, 1969; Somayajuju et al., 1971; Heye, 1975), because the 10Å manganite-rich B type nodules are supposed to have been formed epigenetically within the sediment. Furthermore, if the surrounding 2 to 5 mm thick layers which are composed both of the 10Å manganite and δ-MnO_2 phases in A type nodules have been formed contemporaneously with the whole concentric thin layers in B type nodules which are composed exclusively of the 10Å manganite phase, it is inferred that the period of formation of the nuclei fragmented nodules in the A type nodules is considerably older. In addition, the formation conditions for the fragmented older nodules inside the A type nodules differ from that of 2 to 5 mm surface layers of A type nodules, and also from that of whole thin layers of B type nodules, since their mineralogy, occurrence, and internal structure are remarkably different. Unfortunately, detailed determination of the period of formation of the nodules is not possible because the study on geological history of this area has not been accomplished. Detailed comparative study of internal structure, mineralogy and chemistry of manganese nodules, and the geology of associated sediments is needed in terms of growth history and genesis of manganese nodules.

Summary and Conclusion

Deep sea manganese nodules from the Central Pacific Basin are composed of the 10Å manganite phase and the δ-MnO_2 phase. The 10Å manganite phase is composed solely of 10Å manganite which contains transition metals as essential elements for the formation of the crystal. The δ-MnO_2 phase is a mixture of colloidal particles of 2 line-form δ-MnO_2 and amorphous hydroxide of iron, and of minute detrital silicate minerals. Regional variation of chemical compositions, inter-element relationships on bulk compositions, and the relationship between morphology and chemistry of manganese nodules, are explained by the chemical, mineralogical, and microstructural features and relative abundance of the two phases in nodules. 10Å manganite especially has a strong correlation to nickel and copper, and controls the regional and internal distribution patterns of these elements in manganese nodules.

It has been suggested that the occurrence of manganese oxide is controlled by the physicochemical conditions in the system Mn-O_2-H_2O. However, the present study suggests that transition elements such as nickel and copper should be included in the system in considering the formation conditions of manganese minerals of nodules. The chemical and mineralogical characteristics and the mode of occurrence of the two constituent phases in and on various types of nodules are closely related to its state of contact with sediments. It is suggested with the present results and reported occurrence of manganese minerals that the 10Å manganite phase was precipitated from the interstitial waters of unconsolidated sediments, and that the δ-MnO_2 phase was formed by the accumulation of colloidal particles supplied from the sea water.

The present model agrees well with the previously proposed geochemical model of the upward migration of manganese ions through submarine sediments. Dissolved Mn^{2+} in the interstitial water of sediments supplied from substrata is reasonably precipitated as Fe-free 10Å manganite by the oxidation near the sea bottom. Addition of minor metals such as nickel and copper, which are essential to the formation of 10Å manganite would take place simultaneously and after deposition. On the other hand, manganese which is supplied from sea water as colloidal particles of 2 line-form δ-MnO_2, does not form 10Å manganite due to its poor crystallinity of particles and little capacity of incorporation of transition metal elements in the crystal structure. Colloidal particles of iron hydroxide and detrital and/or authigenic silicate minerals are simultaneously and successively trapped with 2 line-form δ-MnO_2 into the δ-MnO_2 phase. Therefore, it is concluded that deep sea manganese nodules have grown as a result of two material supply routes, from sea water and from sediment interstitial water, forming specific mineral phases, with intermittent overturns and/or rotations as suggested by the nodule internal structures.

Two types of manganese nodules from the Central Pacific Basin (GH76-1 and GH77-1 areas) are evidently different in chemical composition, mineral composition, and internal structure. It is inferred that the concentric thin layers of the 10Å manganite phase of the interior and gritty surface of spherical nodules(type B) and the surface layers of 2 to 5 mm thickness surrounding the fragmented nodules (type A) may have been formed in younger age.

Acknowledgments

The writer is grateful to Professor S. Takenouchi of Department of Mineral Development Engineering, the University of Tokyo, for helpful discussion and criticism on the manuscript.

Special thanks are due to Drs. A. Mizuno and T. Moritani of the Geological Survey of Japan, for providing nodule samples collected during the GH cruises of R/V Hakurei-Maru and for valuable information and critical reading of the manuscript.

References

Arita, M., Bottom sediments. In Cruise Report No.8, Geol. Surv. Japan(editors A. Mizuno and T. Moritani), 94-117(1977).

Ahrens, L.H., Wills, J.P. and Oothuizer, C.O., Further observations on the composition of manganese nodules, with particular reference to some of the rare earth elements, Geochim. Cosmochim. Acta 31, 2169-2180 (1967).

Arrhenius, G., Mero, G. and Korkich, J., Origin of oceanic manganese minerals, Science 144, 170-171 (1964)

Aumento, F., Lawrence, D.E. and Plant, A.G., The ferromanganese pavement on San Pablo Seamount, Geol. Surv. Pap. Can. 68-32, 30 (1968).

Barnes, S.S., Minor element composition of ferromanganese nodules, Science 157, 63-65 (1967).

Bender, M.L., Does upward diffusion supply the excess manganese in pelagic sediments?, J. Geophy. Res. 76, 4212-4215 (1971).

Bender, M.L., Ku, L. and Broecker, W.S., Manganese nodules: Their evolution, Science 151, 325-328 (1966).

Bender, M.L., Ku, L. and Broecker, W.S., Accumulation rates of manganese in pelagic sediment and nodules, Earth Planet. Sci. Lett. 8. 143-148 (1970).

Bonatti, E., Kraemer, T. and Rydell, H., Classification and genesis of submarine iron-manganese deposits. In Ferromanganese Deposits on the Ocean Floor (editor D.R. Horn), NSF, IDOE, Washington D. C., 149-166 (1972).

Boström, K., Submarine volcanism as a sources of iron, Earth Planet. Sci. Lett. 9, 348-352 (1971).

Bricker, O., Some stability relations in the system Mn-O_2-H_2O at 25° C and one atmospheric total pressure, Amer. Mineral. 51, 895-902 (1965).

Brooke, J.N. and Prosser, A.P., Manganese nodules as source of copper and nickel—mineralogical assessment and extraction, Trans. Inst. Min. Metall. 78C, 64-73 (1969).

Burns, R.G., Formation of cobalt(III) in the amorphous $FeOOH \cdot nH_2O$ phase of manganese nodules, Nature 205, 999 (1965).

Burns, R.G., The uptake of cobalt into ferromanganese nodules, soils, and manganese(IV) oxides, Geochim. Cosmochim. Acta 40, 95-102 (1976).

Burns, R.G. and Burns, V.M., Mineralogy, In Marine Manganese Deposits(editor G.P. Glasby), Elsevier, Amst. 185-248 (1977).

Burns, R.G. and Fuerstenau, D.W., Electronprobe determination of inter-element relationships in manganese nodules, Amer. Mineral. 51. 895-902 (1966).

Buser, W., Graf,P. and Feitknecht, W., Beitrag zur Kenntnis der Mangan(II)-manganit und des δ-MnO_2, Helv. Chim. Acta 37, 2322-2333 (1954).

Buser, W. and Grütter, A., Überdie Natur der Manganknollen, Schweiz. Miner. Petrogr. Mitt. 36, 49-62 (1956).

Cheney, E.S. and Vredenburgh, L.D., The role of iron sulfides in the diagenetic formation of iron poor manganese nodules, J. Sediment. Petrol. 38,1363-1365 (1968).

Crerar, D.A. and Barnes, H.L., Deposition of deep-sea manganese nodules, Geochim. Cosmochim. Acta 38 279-300 (1974).

Cronan, D.S., Manganese nodules and other ferromanganese oxide deposits from the Atlantic Ocean, J. Geophys. Res. 80, 3831-3837 (1975).

Cronan, D.S., Deep sea manganese nodules: distribution and geochemistry, In Marine Manganese Deposits(editor G.P. Glasby), Elsevier, Amst., 11-44 (1977).

Cronan, D.S. and Tooms, J.S., The geochemistry of manganese nodules and associated pelagic deposits from the Pacific and Indian Ocean, Deep Sea Res. 16, 335-359 (1969).

Dunham, A.C. and Glasby, G.P., Petrographic and electron microprobe investigation of some deep- and shallow-water manganese nodules, N.Z.J. Geol. Geophys. 17, 929-953 (1974).

Frazer, J.Z., Manganese nodule reserves: an updated estimate, Mar. Mining 1, 103-123 (1977).

Friedrich, G., Rosner, B. and Demirsoy, S., Erzmikroskopische und mikroanalytische Untersuchungen an Manganerzkonkretionen aus dem Pazifischen Ozean, Mineral. Deposita 4, 298-307 (1969).

Fujinuki, T., Mochizuki, T. and Moritani, T., Chemical composition of manganese nodules, In Cruise Report No. 8, Geol. Surv. Japan (editors A. Mizuno and T. Moritani), 162-171 (1977).

Giovanoli, R., Stähli, E. and Feitknecht, W., Über Oxidhydroxide des vier-wertigen Mangans mit Schichtengitter, 1. Mitteilung: Natriummangan(II,III)-Manganit(IV), Helv. Chim. Acta 53, 209-220 (1970).

Glasby, G.P., The mineralogy of manganese nodules from a range of environments, Mar. Geol. 13, 57-72 (1972).

Glasby. G.P., Mechanism of incorporation of manganese and associated trace elements in marine manganese nodules, Oceanogr. Mar. Biol. A. Rev. 12, 11-40 (1974).

Glasby, G.P., Manganese nodules in the South Pacific: A review, N.Z.J. Geol. Geophys. 19, 707-736 (1976).

Glasby, G.P. and Singleton, R.J., Underwater photographs of manganese nodules from the Southwestern Pacific, N.Z.J. Geol. Geophys. 18, 597-604 (1975).

Goodell, H.G., Meylan, M.A. and Grant, B., Ferromanganese deposits of the South Pacific Ocean, Drake Passage and Scotia Sea, Antarctic Research Series 15, 27-95 (1971).

Greenslate, J.L., Frazer, J.Z. and Arrhenius, G., Origin and deposition of selected transition elements in the seabed, In The Origin and Distribution of Manganese Nodules in the Pacific and Prospects for Exploitation (editor M. Morgenstein), Hawaii Inst. Geophys., 45-70 (1973).

Halbach, P., Özkara, M. and Hense, J., The influence of metal content on the physical and mineralogical properties of pelagic manganese nodules, Mineral. Deposita, 10,397-411 (1975).

Heye, D., Wachstumsverhältnisse von Manganknollen, Geol. Jb. E5, Schweiz., Hannover (1975).

Horn, D.R., Horn, B.B. and Delach, M.N., Metal content of ferromanganese deposits of the oceans, Lamont-Doherty Geol. Observ. Tech. Rept. 3, NSF/IDOE-GX33616 (1973).

Krauskopf, K.B., Separation of manganese from iron in sedimentary processes, Geochim. Cosmochim. Acta 12, 61-84 (1957).

Ku, T.L. and Broecker, W.S., Radiochemical studies on manganese of deep sea origin, Deep Sea Res. 16, 625-637 (1969).

Li, Y.H., Bischoff, J. and Mathieu, G., The migration of manganese in the Arctic Basin sediment, Earth Planet. Sci. Lett. 7, 265-270 (1969).

Lynn, D.C. and Bonatti, E., Mobility of manganese in diagenesis of deep-sea sediments, Mar. Geol. 3, 457-474 (1965).

Menard, H.W., Time, chance, and the origin of manganese nodules, Amer. Sci. 64, 519-529 (1976).

Mero, J.L., The Mineral Resources of the Sea, Elsevier, Amsterdam, (1965).

Meyer, K., Surface sediment and manganese nodule facies, encountered on R/V 'Valdivia' cruises 1972/1973, Meerestechnik 4(6), 196-199 (1973).

Meylan, M.A., Field description and classification of manganese nodules, Hawaii Inst. Geophys. Rep. HIG-74-9, 158-168 (1974).
Mizuno, A. and Moritani, T.(eds.), Deep Sea Mineral Resources investigation in the Central-eastern Part of the Central Pacific Basin, Jan-March 1976(GH76-1 cruise). Geol. Surv. Japan, Cruise Rept. No 8 (1977).
Morgenstein, M. and Felsher, M., The origin of manganese nodules: a combined theory with special reference to palagonitization, Pac. Sci. 25, 301-307 (1971).
Moritani, T.(ed), Deep Sea Mineral Resources Investigation in the Eastern Part of Central Pacific Basin, Jan.-March 1977 (GH77-1 cruise). Geol. Surv. Japan, in preparation.
Moritani, T., Maruyama, S., Nohara, M., Matsumoto, K., Ogitsu, T. and Moriwaki, H., Description, classification, and distribution of manganese nodules. In Cruise Rept. No 8, Geol. Surv. Japan (editors A. Mizuno and T. Moritani), 136-158 (1977).
Ostwald, J. and Frazer, F.W., Chemical and mineralogical investigations on deep-sea manganese nodules from the Southern Ocean, Mineral. Deposita 8, 303-311 (1973).
Presley, B.J., Brookes, R.R. and Kaplan, I.R., Manganese and related elements in the interstitial water of marine sediments, Science 158, 906-910 (1967).
Presley, B.J., Goldhaft, M.B. and Kaplan, I.R., Interstitial water chemistry, Deep Sea Drilling Project, Leg 5., Init. Rept. DSDP 5, 513-522 (1970).
Price, N.B., Some geochemical observation on manganese-iron oxide nodules from different depth environments, Mar. Geol. 5, 511-538 (1967).
Price, N.B. and Calvert, S.E., Compositional variation in Pacific ocean ferromanganese nodules and its relationship to sediment accumulation rate, Mar. Geol. 9, 145-171 (1970).
Raab, W., Physical and chemical features of Pacific deep sea manganese nodules and their implication to the genesis of nodules, In Ferromanganese Deposits on the Ocean Floor (editor D.R. Horn), NSF, IDOE, Washington, D.C. 31-49 (1972).
Renard, D., Michard, G. and Hoffert, M., Comportment geociminique du nickel st du cobalt a l'interface eau-sediment. Application a l'enrichissement en ces elements dans les formations ferromanganesiferes, Mineral. Deposita, 11. 380-393 (1976).
Sano, M, and Matsubara, H., Some aspects of the element distribution of manganese nodules and its relationship to mineral composition, Suiyokwai-Shi 17, 111-114 (1940).
Somayajuju, B.L.K., Heath, G.R., Moore, T.C. and Cronan, D.S., Rates of accumulation of manganese nodules and associated sediments from the equatorial Pacific, Geochim. Cosmochim. Acta 35, 621-624 (1971).
Sorem, R.K. and Foster, A.R., Mineralogical, chemical, and optical procedures and standards for study of growth features and economic potential of manganese nodules, In Inter-University Program of Research on Ferromanganese Deposits of the Ocean Floor—Phase I Report, NSF, IDOE, Washington D.C. 23-28 (1973).

Usui, A., Mineralogical Study of Marine Manganese Nodules, Synthesis of Hydrous Manganese Oxides, and Implications to Genesis and Geochemistry. Ph.D. Thesis, University of Tokyo, in preparation.

Usui, A., Takenouchi, S. and Shoji, Y., Distribution of metal elements in manganese nodules and formation mechanism of constituent minerals, with special reference to nodules from Komabashi-Daini Seamount, Mining Geol. 26, 371-384 (1976).

Usui, A., Takenouchi, S. and Shoji, T., Mineralogy of deep sea manganese nodules, synthesis of manganese oxides: Implications to genesis and geochemistry, Mining Geol., 28 (1978) in press.

Usui, A., Shoji, T. and Takenouchi, S., Formation of marine manganese nodules and accumulation of useful metal elements, In Preprints of Joint Meeting of Japanese Assoc. Mineral. Petrol. Econ. Geol., Mineral. Soc. Japan, and Soc. Mining Geolog., 93 [abst] (1977).

Uyttenbogaadt, W. and Burke, E.A.J., Tables for Microscopic Identification ofOre Minerals, Elsevier, Amst. (1971).

REGIONAL DISTRIBUTION OF RARE EARTH AND MINOR ELEMENTS IN MANGANESE NODULES AND ASSOCIATED SEDIMENTS IN THE SOUTHWEST PACIFIC AND OTHER LOCALITIES

P.C. Rankin and G.P. Glasby

Department of Scientific and Industrial Research
Lower Hutt, New Zealand

Abstract

Thirteen nodules and 12 sediments from the S.W. Pacific, including six nodule/sediment pairs, have been analysed by spark source mass spectrometry for 12 rare earth and 13 minor elements. The nodules appear to be relatively uniform in composition, e.g. La varies by a factor of 2.3. This finding adds weight to the contention that the S.W. Pacific constitutes a discrete nodule province distinct from that of the equatorial N. Pacific. The S.W. Pacific sediments show a much wider range of composition, e.g. La varies by a factor of 7.7. The decreasing sedimentation rate (and therefore increased oxidation state) with increasing distance from New Zealand is primarily responsible for this variation. As, Nb, REE (rare earth elements), Y, Sr, Th, Zr, Sn, Pb, U and Hf increase in concentration in the sediments with increasing distance from New Zealand, whereas Rb, Ba and Cs decrease. The former group of elements are therefore incorporated mainly in the authigenic component of the sediment, whereas the latter group are associated mainly in the silicate component. The equatorial N. Pacific nodule is lower in REE and a number of other minor elements, and higher in Rb and Cs, compared with the S.W. Pacific nodules. This finding is surprising in view of the much higher Cu and Ni contents of these nodules and may reflect the higher detrital silicate contents of the equatorial N. Pacific nodule. The REE content of the Hawaiian-Emperor Seamount Chain nodules and crusts are somewhat higher than those of the S.W. Pacific nodules. Lower concentrations of Pb in the Hawaiian-Emperor deposits suggests that this element is not always oxidized to the tetravalent state in such elevated deposits. The iron-rich

Eniwetok Atoll sample is extremely low in REE which indicates that this deposit has formed extremely rapidly as previously noted in some other near-shore deposits.

Introduction

Although a number of publications in recent years have dealt with the geochemistry of the rare earth elements (REE) in manganese nodules (e.g. Piper 1972, 1973, 1974; Glasby 1973a, b; Glasby et al. in press a), there has been no systematic attempt to deal with the regional geochemistry of the REE in nodules and their relationship to the geochemistry of the underlying sediments. The absence of studies across a major marine geologic province, such as an ocean basin or mid-ocean ridge, arises principally from the time-consuming nature of the analytical methods involved. The geochemical cycle of rare earth and minor elements in the oceans cannot be understood from the analyses of isolated samples taken at widely separated, arbitrary locations on the sea floor. A comprehensive study is therefore required to show the underlying factors controlling the incorporation of rare earth and minor elements into sediments and manganese nodules and to understand the differences in the mode of incorporation of these elements in these two particular marine phases.

The study reported here involved analysing rare earth and minor elements in manganese nodules and associated sediments from a series of locations in the S.W. Pacific. The samples were chosen to cover a large area in the Southwestern Pacific, Samoan and south Penrhyn Basins where the nodules (on the basis of size, shape, surface texture, nucleus characteristics and transition metal content) can be considered to constitute a discrete nodule province. Differences are apparent, however, within this province. The S.W. Pacific is a region of low biological productivity in the overlying surface seawater and this has a major influence on the transition metal contents of nodules (Glasby 1976). Six nodule/sediment pairs were studied from individual locations within this general area. For comparison, a limited number of samples from the equatorial N. Pacific, Hawaiian-Emperor Seamount Chain and Eniwetok Atoll were also analysed.

Nodule mineralogy was not studied here because, in a previous study, Glasby et al. (in press a) had shown significant variations in the REE contents of 5 manganese nodules from diverse localities even though these all contained δMnO_2 as the principal manganese oxide phase. A detailed study of the mineralogy of S.W. Pacific nodules has, however, been made by Meylan (1978) who showed that nodules of the Cook Island facies are predominantly δMnO_2 and those of the S.W. Pacific facies are a mixture of todorokite and δMnO_2.

Samples

The study reported here involved the analysis of 12 rare earth and 13 minor elements in 13 nodules and 12 sediments from the S.W. Pacific; 1 nodule from equatorial N. Pacific; 1 nodule, 2 crusts and 1 substrate (crust nucleus) from the Hawaiian-Emperor chain, and 1 iron-rich sample from Eniwetok Atoll. The location, depth, description and previous study of each sample are given in Appendices 1 and 2. A location map is given in Fig. 1.

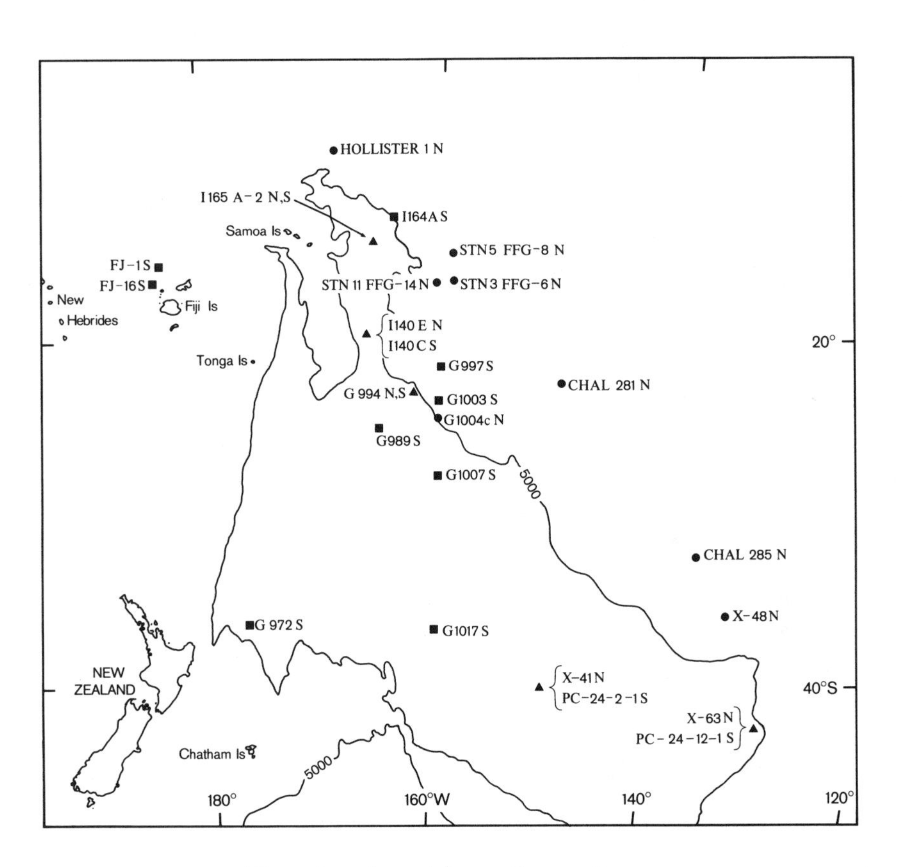

Fig. 1. Location of samples

It should be noted that the sediment sample I164A (Appendix 2) is a pure calcareous ooze derived from sieving a Samoan Basin sediment through a 240 mesh sieve to remove pelagic clay.

The two sediments from the Yasawa Trough (N.W. of Viti Levu, Fiji) were recovered during an attempt to locate metalliferous sediments in a region of subduction (K. Knedler, pers. comm.).

Sample Preparation

After collection the samples were oven dried, ground with a pestle and mortar, and passed through a 50 μm nylon sieve. The samples were prepared for spark source mass spectrometric analysis by mixing them in a ratio of 1:1, with ultra pure graphite powder. Pure Lu_2O_3 was mixed with the graphite (final concentration 500 ppm) (Rankin and Childs 1976).

Spark Source Mass Spectrometry

The analytical procedure followed closely that of Taylor (1965, 1971). Selected isotope images on the photoplate were used to determine the Q-2 emulsion response curve. The mass spectrometric data are considered to have a precision of ± 15% (based on repeat analyses of the U.S.G.S. standard BCR-1).

TABLE 1

Assessment of accuracy of spark source mass spectrometric data
All analyses in ppm.

Element	BCR-1		NOD-A-1		NOD-P-1	
	This work[2]	Flanagan (1973)[1]	This work[2]	NAA[3]	This work[2]	NAA[3]
Yb	3.6	[3.36]	13.5	14	13	13
Er	3.3	3.59	15	-	13	-
Ho	1.2	(1.2)	5.3	-	5.1	-
Dy	6.3	6.3	22	-	25	-
Tb	0.96	1.0	3.8	5.2	4.2	5.4
Gd	6.5	6.6	22	31	24	32
Eu	2.2	1.94	4.8	5.1	6.8	7.3
Sm	6.4	6.6	21	24	28	35
Nd	28	29	94	100	110	116
Pr	6.9	(7)	23	-	27	-
Ce	53	53.9	>300	724	280	306
La	24	[26]	130	122	82	115
U	1.7	(1.74)				
Th	5.6	6.0				
Pb	21	(17.6)				
Hf	3.8	[4.7]				
Ba	800	675				
Cs	1.6	0.95				
Nb	13	13.5				
Zr	220	(190)				
Y	44	[37.1]				

[1] [ ] = average; ( ) = magnitude; ___ = recommended
[2] Based on silicate matrix calibration data
[3] Average of four NAA analyses (D.Z. Piper, pers. comm.)

The accuracy of the mass spectrometric data is dependent upon the samples used to calibrate the instrument. Table 1 lists the data obtained for the U.S.G.S. standard BCR-1. The difference in major element composition of the Fe-Mn rich samples compared to the silicate-rich samples (e.g. BCR-1) raised the problem of the validity of using silicate matrix calibration data for marine manganese nodule samples. Two marine nodules (one from the Atlantic, NOD-A-1 and one from the Pacific, NOD-P-1) which had previously been analysed by neutron activation analysis for some REE (D.Z. Piper, pers. comm.) were analysed by spark source mass spectrometry using the procedure reported above. The results obtained are reported in Table 1.

The trends in the REE concentrations for NOD-A-1 and NOD-P-1 are similar for both analytical techniques. However, the absolute concentrations do differ particularly for the middle rare earths Tb-Sm (by up to 33%). It is possible that the high level of Th (estimated from spark source mass spectrometric data as 30-50 ppm) in these two standard nodule samples would cause the neutron activation data of some REE to be higher than expected.

Results and Discussion

S.W. Pacific

Nodules. Nodules from the S.W. Pacific are relatively uniform in composition e.g. La varying by a factor of 2.3 between 96 and 220 ppm (Appendix 1). Other elements show a similar range of variability. The uniformity of composition of nodules within the southwestern Pacific Basin has previously been reported by Bäcker *et al.* (1976). Table 2 (columns 1-5) lists the average data for nodules from the several sampling areas within the S.W. Pacific region and supports the observation of a relatively homogeneous nodule province.

There appear to be no well developed inter-element correlations, and element contents do not appear to vary systematically with water depth, although the depth range is fairly limited (3544-5520 m). Variations in element content within a single nodule (Hollister 1 from the Samoan Passage) are noted, but these are not systematic from the exterior to the interior of the nodule (Appendix 1) and were less than the 2.3 factor. Geographically, there appear to be no well-defined trends in composition, for example, across basins, although insufficient samples may have been analysed to examine this problem adequately. The data therefore indicate, on the basis of the rare earth and minor element data available, that S.W. Pacific nodules may indeed constitute a discrete nodule province as previously noted. The observed variability in nodule

TABLE 2

Average concentrations of elements in nodules and sediments from areas within the S.W. Pacific. All analyses in ppm.

	Nodules							Sediments				
	1*	2	3	4	5	6	7	8	9	10	11	12
U	6.1	9.4	8.0	7.7	7.3	7.9	2.4	2.4	<1.2	4.7	2.8	<0.4
Th	32	65	46	38	25	43	17	17	19	10	15	<1.2
Pb	>100	>100	>100	>100	>100	>100	>100	43	28	14	35	11
Hf	9.1	11	9.3	8.7	11	9.6	2.8	4.2	5.5	2.7	4.2	<2.0
Yb	22	25	17	20	20	21	12	7.7	-	8.2	7.3	2.3
Er	23	27	19	22	20	22	11	8.3	5.4	7.1	7.2	2.1
Ho	7.9	9.4	6.6	7.3	7.9	7.9	4.5	3.1	2.0	2.8	2.8	0.80
Dy	34	40	27	31	33	33	25	13	-	13	12	3.6
Tb	5.7	7.4	4.5	5.2	5.7	5.8	4.1	2.1	-	1.8	1.9	0.55
Gd	35	38	28	32	34	34	23	13	11	13	13	3.5
Eu	9.8	13	7.8	9.0	10	10	7.4	3.5	3.4	3.9	3.5	0.85
Sm	42	51	32	35	37	40	30	14	10	14	13	2.9
Nd	160	180	120	150	180	160	110	54	38	55	51	13
Pr	38	46	32	35	45	40	26	13	9.5	13	13	3.0
Ce	>200	>200	>200	>200	>200	>200	>200	100	150	82	106	130
La	160	140	120	150	170	150	80	47	26	41	42	13
Ba	>1000	930	>800	>1000	>1000	>1000	>1000	640	>1000	680	>750	630
Cs	0.82	<0.5	<0.7	<0.5	<0.5	<0.6	2.6	5.1	6.2	8.0	5.8	<0.5
Sn	5.7	5.2	4.6	5.4	5.3	5.2	1.6	3.0	3.5	7.7	4.0	<1.0
Nb	>40	>40	>40	>40	>40	>40	15	10	11	15	12	<1.5
Zr	>300	>440	>420	>500	>500	>430	200	170	190	160	170	74
Y	>100	>100	>90	>100	>100	>100	81	>70	38	95	60	33
Sr	>700	700	700	870	960	800	530	290	310	370	300	>700
Rb	na	11	8.8	13	8.1	11	30	>60	>60	>60	>60	9.5
As	>50	120	110	160	160	130	49	35	13	44	27	79

* 1 Average of 2 S.W. Pacific Basin nodules (G994, G1004C)
2 Average of 3 southern S.W. Pacific Basin nodules (X-41, X-48, X-63)
3 Average of 2 Samoan Basin nodules (I140E, I165A-2)
4 Average of 3 S. Penrhyn Basin nodules (FFG-6,-8,-14)
5 Average of 2 nodules from S.E. of Rarotonga (Chal 281, 285)
6 Mean analysis of S.W. Pacific nodules
7 Data for nodule RP-8-OC-76 from equatorial N. Pacific
8 Average of 7 S.W. Pacific Basin sediments (G972, G989, G994, G997, G1003, G1007, G1017)
9 Average of 2 southern S.W. Pacific Basin sediments (PC-24-2-1, PC-24-12-1)
10 Average of 2 Samoan Basin sediments (I140E, I165A-2)
11 Mean analysis of S.W. Pacific sediments
12 Average of 2 Yasawa Trough sediments (FJ-1, FJ-16)

composition may reflect in part the sampling and analytical problems involved in analysing a heterogeneous material such as manganese nodules, which may be considered to be a three component system consisting of a manganese oxide phase, an iron oxide phase and a silicate-rich phase (Glasby 1973b, Calvert et al. 1978).

The average S.W. Pacific nodule (Table 2, column 6) composition is compared to that of a nodule from the equatorial N. Pacific (column 7).

Sediments. Sediments from the S.W. Pacific vary more in composition than do nodules (Appendix 2), e.g. La varies by a factor of 7.7 between 11 and 85 ppm (ignoring the pure carbonate sediment, I164A, which is much lower in all the trace elements (cf. Glasby et al. in press b). Other elements show a similar level of variability to La. La is associated to some extent with U-Th-Nb-Zr-Y but shows no association with Pb-Hf-Ba-Cs-Sn-Sr-Rb-As. Element concentrations do not vary systematically with depth, although the two shallowest samples from the Yasawa Trough region in the vicinity of Fiji have the lowest rare earth contents, possibly because of the incorporation of terrigenous and carbonate material into these sediments. Significantly, there is no marked enrichment of rare earth or other minor elements in sample FJ-16 from the axis of the Yasawa Trough compared to sample FJ-1 taken outside the Trough, and, also, the REE content of FJ-16 is low compared to other pelagic sediments from the S.W. Pacific (Table 2) and to metalliferous sediments from other parts of the Pacific (Dymond et al. 1973, 1977). This observation suggests that the Yasawa Trough is not a centre of metalliferous sediment formation (K. Knedler, pers. comm.).

Sediment PC-24-12-1 from the southern S.W. Pacific Basin was found to have a very high level of Ba. Ba concentrations have been quoted as 'greater than' when the concentration was in excess of 1000 ppm (analytical uncertainty of spark source mass spectrometric data), however, it is still possible to obtain a qualitative estimate of Ba. Sediment PC-24-12-1 had a Ba concentration of up to 5 times that of the other S.W. Pacific sediments (i.e. of the order of 1.0%). This high Ba content resulted in interference effects on the spark spectra of some of the heavier rare earths, and data for these elements are not given. The reason for the elevated Ba concentration is not known.

Sediment samples from the S.W. Pacific and Samoan Basins and the Yasawa Trough have previously been analysed by Glasby et al. (in press a) and K. Knedler (pers. comm.) The major element data for the samples from Stns. PC-24-2-1 and PC-24-12-1 are given in Table 3.

TABLE 3

Major element data for sediments from Stns. PC-24-2-1 and PC-24-12-1. Data obtained by XRF and expressed in percentages

Element	PC-24-2-1	PC-24-12-1
Fe	5.4	4.0
Mn	0.55	0.42
Ti	0.46	0.28
Ca	0.74	0.56
K	2.5	3.0
P	0.03	0.06
Si	24.4	22.2
Al	8.7	7.5
Mg	1.7	1.2
Na	3.0	3.1
LOI*	8.8	16.5

*LOI = loss on ignition at $1000^{o}C$

In the southwestern Pacific and Samoan Basins, there appears to be evidence of a systematic increase in the rare earth content of the sediments with increasing distance from New Zealand. Lowest La concentrations are found in samples G972 and G1017 (20 ppm) and PC-24-2-1 (18 ppm) nearest New Zealand, and higher concentrations in samples G1003 (64 ppm) and G997 (85 ppm) from the dusky-brown (5YR 2/2) clay region near Rarotonga. G1003 and G997 are also enriched in U, Th, Pb, Hf, La, Nb, Zr, Y, Sr and As and depleted in Ba, Cs, and Rb relative to sediments from nearer New Zealand. In fact, an enrichment sequence can be calculated for element concentrations in sediments from Stn. G997 taken in a region of dusky-brown clay S.W. of Rarotonga to G972 at the base of the New Zealand continental slope and is as follows (enrichment ratio is in parentheses): As (8.5) > Nb (>6) > La (4.2) > Ce (3.9) > Y (>3.7) > Sr (2.8) > Th (2.0) > Zr (1.7) > Sn (1.6) > Pb (1.5) > U (1.2) > Hf (1.1) > Rb (<0.8) > Ba (0.7) > Cs (0.5). The enrichment sequence for the major elements from these two stations (based on the data of Glasby et al. (in press b) is: Mn (16.3) > P (10.2) > Ti (7.7) > Fe (4.4) > Mg (3.7) > Cr (3.0) > Al (1.1) > Si (0.6) ≃ K (0.6). Glasby et al. (in press b) have suggested that the relative contribution of terrigenous sediments from the New Zealand landmass may be a major factor in controling the trace element content of these sediments. With increasing distance from New Zealand, the influx of terrigenous sediments (and therefore the sedimentation rate) will decrease and the authigenic component of the sediment increase. This conclusion is supported by the darkening of the sediment colour, increased concentrations of Mn, Fe, Cu, Ni and Co, and increased ferric-ferrous ratios of the sediment with increasing distance from New Zealand (unpublished data). These enrichment sequences suggest that Mn, P, As, Ti, Nb, Fe, La, Ce, Y, Mg, Th, Zr, Sn, Pb, U and Hf are incorporated in the authigenic (ferromanganese) component of the sediments, whereas Al, Rb, Ba, Si, K and Cs are associated with the terrigenous (aluminosilicate) phase. Sr does not show a systematic variation across the basin. This observation is not easily explained since Sr does not appear to be incorporated in the carbonate phase of the sediment, as confirmed by the low Sr content of the carbonate-rich sample I164A. Clearly, the degree of enrichment of elements in the authigenic component of the sediment varies with element across the basin. Mn is more enriched than Fe. Those elements forming anionic species in seawater (P and As) are particularly enriched. La and Ce are enriched to the same extent, a relation suggesting that there is no preferential oxidation of Ce to the tetravalent state. The precise factors controlling the relative enrichment of elements in sediments across the basin are not known, but the increased degree of oxidation of the sediment related to the decreased sedimentation rate with increasing distance from New Zealand is undoubtedly a major factor (cf. Meylan et al. 1974; Volkov et al. 1975; Glasby 1976; Glasby et al. in press a, b).

Figure 2 illustrates the shale-normalised average data for nodules and sediments from the S.W. Pacific area. The shale data used are from Haskin and Haskin (1966). The shale-normalised curves of the nodules are similar to those observed for other deep-sea nodules from the southwest Pacific (Piper 1974; Glasby et al. in press) with a marked Ce enrichment and a slight increase in heavy rare earths relative to light rare earths. The shale-normalised curves for sediments follow a similar trend to those for the nodules, although the spread of data is much greater. The sediments exhibit a small positive Ce anomaly which is in direct contrast to the Ce data reported by Piper and Graef (1974). The sediments of the S.W. Pacific are regarded well oxidised and this characteristic is reflected in the small, positive Ce anomaly. The similarity in shape of the average shale-normalised curves for nodules and sediments from the S.W. Pacific supports the possibility of incorporation of manganese micronodules in the sediment, but other explanations, such as the well-oxidised nature of the sediment, may also be possible. In contrast to the sediments of the S.W. Pacific area, the two sediments from the Yasawa Trough have a marked negative Ce anomaly (Appendix 2) indicating the absence of metalliferous sediments. The direct incorporation of rare earth elements from the surrounding seawater into the sediments would explain the observed shale-normalised REE distribution (Piper and Graef 1974; K. Knedler, pers. comm.).

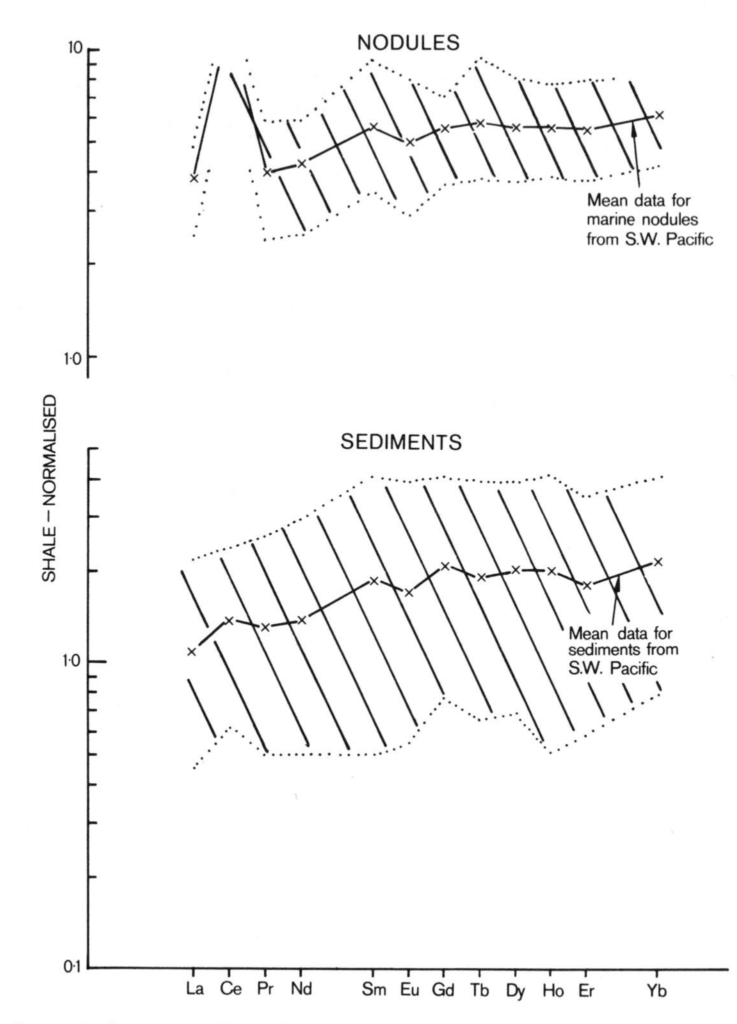

Fig. 2. Shale-normalised average data for nodules and sediments from the S.W. Pacific

Nodule-sediment relationships. Six nodule-sediment pairs from the S.W. Pacific were selected to study nodule-sediment relationships in more detail. These were G994/G994, G1003/G1004C, X-41/PC-24-2-1, X-63/PC-24-12-1, I140E/I140C and I165A-2/I165A-2 (nodule number is given on left, sediment number on right). Although the pair G1003/G1004C is not from the same site, the sediment type was the same at both stations and the stations were sufficiently close (92 km apart) to consider the samples to be from the same station.

TABLE 4

Enrichment values (nodule/sediment concentrations) for elements in 6 nodule-sediment pairs from the S.W. Pacific

	G994	G1003/G1004c	X41/ PC-24-2-1	X63/ PC-24-12-1	I140E	I165A-2
U	3.4	2.7	5.3	>25	5.8	1.0
Th	1.9	1.3	3.8	4.7	5.6	3.3
Pb	>2	1.6	>3	>5	>5	8.6
Hf	3.5	1.8	2.8	1.7	5.0	2.6
Yb	2.7	1.6	9.6	-	1.9	2.2
Er	2.6	1.5	11	-	2.3	3.0
Ho	2.5	1.2	13	-	2.1	2.6
Dy	2.4	1.4	9.3	-	2.0	2.3
Tb	2.6	1.3	10	-	2.5	2.5
Gd	2.4	1.5	8.7	2.4	2.0	2.4
Eu	3.1	1.6	13	2.9	1.8	2.2
Sm	3.3	1.5	15	3.9	2.2	2.4
Nd	2.7	1.6	9.4	3.8	1.9	2.5
Pr	3.1	1.6	8.8	4.2	2.0	2.8
Ce	-	-	>2	>2	-	-
La	2.4	2.0	7.8	4.9	2.8	3.0
Ba	-	>2	-	-	1.1	-
Cs	0.20	0.22	0.17	<0.1	<0.1	0.15
Sn	2.7	1.6	2.1	1.2	0.34	1.5
Nb	>4	>3	>4	>4	>3	>3
Zr	>2	>2	2.2	2.5	2.3	2.9
Y	-	-	>5	>2	0.79	>2
Sr	>2	>2	3.4	2.0	1.5	2.2
Rb	-	-	-	-	-	-
As	-	-	8.5	8.6	2.3	2.6

The enrichment values (nodule/sediment concentration) for the 6 nodule-sediment pairs are listed in Table 4. The considerable variation in enrichment values for individual elements is due principally to the comparison of relatively constant nodule composition to a variable sediment composition, as noted above. However, the sequence of element enrichments is similar for the nodule-sediment pairs. Of the elements reported in this study, Cs was the only one not enriched in the nodules, a finding which indicates that it is incorporated dominantly in the silicate phase of the sediments.

Equatorial N. Pacific

Only one nodule was analysed (Appendix 1) from the equatorial N. Pacific, and results must be interpreted with caution. However, the equatorial N. Pacific nodule is lower in U, Th, Hf, La, Sn, Nb, Zr, Y and Sr and higher in Cs and Rb compared to the S.W. Pacific nodules (Table 2, column 7). This is of considerable interest because the equatorial N. Pacific nodule is from the region of copper-nickel-rich nodules which are considered potentially exploitable. According to the data of Cronan (1972), these nodules are higher in Mn, Ni, Cu, Ba and Mo and lower in Fe, Co, Pb, V and Ti than those of the S.W. Pacific. From the data presented here, it appears that the equatorial N. Pacific nodules are lower in a range of rare earth and minor elements than the S.W. Pacific nodules. The exceptions, Rb and Cs, may be associated with the detrital fraction of the nodules suggesting that the equatorial N. Pacific nodule has a higher detrital silicate content that the S.W. Pacific nodules.

Hawaiian-Emperor Seamount Chain

The nodule and crusts (Appendix 1) are from a seamount province and therefore from shallower depths than the abyssal nodules considered previously. Insufficient samples are available to indicate whether element composition of the nodules and crusts varies systematically with water depth or along the chain. The analyses of these samples do, however, show a measure of similarity.

Figure 3 illustrates the shale-normalised data for nodule and crusts from the Hawaiian chain and are similar to those of nodules from the S.W. Pacific.

The seamount nodules appear to be higher in U-La-Sr, lower in Th-Pb-Hf, and similar in Ba, Cs, Sn, Nb, Zr, Y, Rb and As compared to abyssal nodules (cf. Cronan 1972, P.20, Table 1). Although these results are preliminary, they suggest that Pb is not always enriched in seamount nodules relative to abyssal nodules, casting some doubt on the proposed oxidation of Pb to a higher valency in such deposits (cf. Cronan and Tooms 1969; Cronan 1972) (see also Nohara and Nasu 1977). Cerium was not determined precisely, but the REE data suggest an enrichment of Ce due to oxidation to the Ce 4+ valency. More recent work has also shown the formation of δMnO_2 rather than birnessite as the principal manganese oxide mineral in ferromanganese deposits of the Hawaiian chain (Frank et al. 1976), again in contradiction of the findings of Cronan. The increase in the Th content of the nodules with increasing water depth (i.e. from seamount to abyssal nodules) is in agreement with the findings of Ku and Broecker (1969) and Margolis et al. (1978).

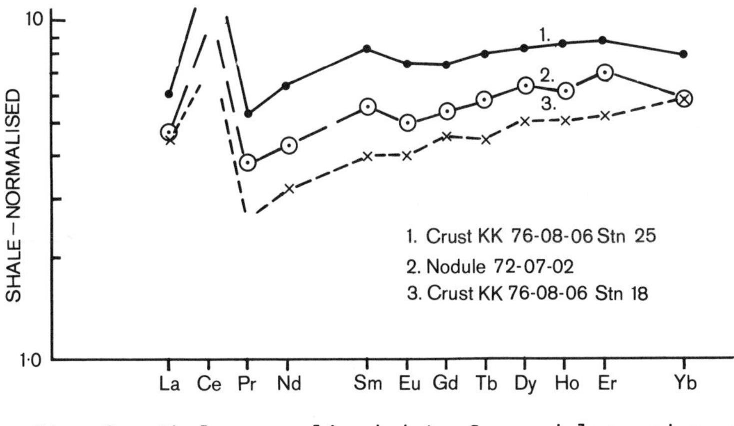

Fig. 3. Shale-normalised data for nodules and crusts from Hawaiian-Emperor Seamount Chain

The degree of enrichment of elements in the manganese nodule outer layer relative to the associated substrate (nucleus) can be derived for sample Midway 72-07-02 Dredge 32 and is in the sequence As (21) > Th (12) > La (12) > Sr (> 7.2) > Y (> 4.2) > U (3.8) > Sn (2.8) ≃ Pb (2.8) > Nb (> 2.7) > Cs (2.2) > Zr (2.0) > Ba (> 1.8) > Hf (1.1) > Rb (0.4). This compares with the sequence derived by Frank et al. (1976) for a more restricted range of elements. Rb is the only element enriched in the substrate compared with the manganese crust and indicates its association with the silicate phases of the substrate.

Eniwetok Atoll

The Eniwetok Atoll sample (Appendix 1) is a concretionary, pear-shaped deposit which may have formed around a metallic core. Mineralogical (XRD) analysis indicates magnetite (Fe_3O_4) to be the only observable mineral present. The extremely low contents of all the elements analysed suggest that this deposit may have formed rapidly (cf. Ku and Glasby 1972). Goldberg and Arrhenius (1958) have shown the rapid growth rate and low trace metal content of a near-shore deposit off California, and a similar origin is probable for the Eniwetok Atoll sample. The Eniwetok Atoll has been described by Shipek (1962).

Acknowledgements

Dr D.Z. Piper (U.S. Geological Survey, Menlo Park) kindly supplied two manganese nodules which had previously been analysed by neutron activation analysis. P. Baedecker (U.S. Geological Survey, Reston, Virginia) performed the neutron activation analyses. Other samples were provided by D.S.

Cassidy (Florida State University) (sediments from the southern sector of the S.W. Pacific Basin), Dr C.D. Hollister (Woods Hole Oceanographic Institution) (Samoan Passage nodules), Dr D.R.C. Kempe (British Museum, Natural History) (H.M.S. Challenger nodules), Dr M.A. Meylan (Hawaiian Archipelago deposits and nodules from the southern sector of the S.W. Pacific Basin), and Professor R.K. Sorem (Washington State University) (equatorial N. Pacific nodule).

References

Bäcker, H., Glasby, G.P. and Meylan, M.A., (1976) Manganese nodules from the southwestern Pacific Basin. NZOI Oceanogr. Fld Rep. 6, 88 pp.

Calvert, S.E., Price, N.B., Heath, G.R. and Moore, T.C., (1978) Relationship between ferromanganese nodule compositions and sedimentation in a small survey area of equatorial Pacific. J. Mar. Res 36, 161-183.

Cronan, D.S., (1972) Regional geochemistry of ferromanganese nodules in the world ocean. Pp. 19-30 in Horn, D.R. (Ed.) "Ferromanganese Deposits on the Ocean Floor". National Science Foundation, Washington D.C. 293 pp.

Cronan, D.S. and Tooms, J.S., (1969) The geochemistry of manganese nodules and associated pelagic deposits from the Pacific and Indian Oceans. Deep-Sea Res. 14, 239-249.

Dymond, J., Corliss, J.B., Heath, G.R., Field, C.W., Dasch, E.J. and Veeh, H.H., (1973) Origin of metalliferous sediments from the Pacific Ocean. Bull. Geol. Soc. Am. 84, 3355-3372.

Dymond, J., Corliss, J.B. and Heath, G.R., (1977) History of metalliferous sedimentation at Deep Sea Drilling Site 319, in the South Eastern Pacific. Geochim. Cosmochim. Acta 41, 741-753.

Frank, D.J., Meylan, M.A., Craig, J.D. and Glasby, G.P., (1976) Ferromanganese deposits of the Hawaiian Archipelago. Hawaiian Inst. Geophys. Rep. HIG-76-14, 71 pp.

Glasby, G.P., (1973a) Mechanisms of enrichment of the rarer elements in marine manganese nodules. Mar. Chem. 1, 105-125.

Glasby, G.P., (1973b) Manganese deposits of variable composition from north of the Indian-Antarctic Ridge. Nature, Lond. 242, 106-108.

Glasby, G.P., (1976) Manganese nodules in the South Pacific: a review. N.Z. Jl. Geol. Geophys. 19, 707-736.

Glasby, G.P., Keays, R.R. and Rankin, P.C. (In press a) Distribution of rare earth, precious metal and other trace elements in recent and fossil deep-sea manganese nodules. Geochem. J.

Glasby, G.P., Hunt, J.L., Rankin, P.C. and Darwin, J.H. (In press b) Major element analyses of marine sediments from the Southwest Pacific. N.Z. Soil Bureau Rep.

Goldberg, E.D. and Arrhenius, G.O.S., (1958) Chemistry of Pacific pelagic sediments. Geochim. Cosmochim. Acta 13, 153-212.

Haskin, M.A. and Haskin, L.A., (1966) Rare earths in European shales: a redetermination. Sci., N.Y. 154, 507-509.

Hollister, C.D., Johnson, D.A. and Lonsdale, P.F., (1974) Current-controlled abyssal sedimentation: Samoan Passage, equatorial West Pacific. J. Geol. 82, 275-300.

Ku, T.L. and Broecker, W.S., (1969) Radiochemical studies on manganese nodules of deep-sea origin. Deep-Sea Res. 16, 625-637.

Ku, T.L. and Glasby, G.P., (1972) Radiometric evidence for the rapid growth rate of shallow-water, continental margin manganese nodules. Geochim. Cosmochim. Acta 36, 699-703.

Landmesser, C.W., Kroenke, L.W., Glasby, G.P., Sawtell, G.H., Kingan, S., Utanga, E., Utanga, A., and Cowan, G., (1976) Manganese nodules from the South Penrhyn Basin, Southwest Pacific. S. Pacif. Mar. Geol. Notes 1, 17-39.

Margolis, S.V., Ku, T.L., Glasby, G.P., Fein, C.D. and Audley-Charles, M.G., (1978) Fossil manganese nodules from Timor : geochemical and radiochemical evidence for deep-sea origin. Chem. Geol. 21, 185-198.

Meylan, M.A., Bäcker, H. and Glasby, G.P., (1974) Manganese nodule investigations in the Southwestern Pacific Basin, 1974. NZOI Oceanogr. Fld Rep. 4, 24 pp.

Meylan, M.A. and Goodell, H.G., (1976) Chemical composition of manganese nodules from the Pacific-Antarctic Ocean, Drake Passage and Scotia Sea : relation to ferromanganese oxide mineralogy and nucleus type. CCOP/SOPAC Tech. Bull. 2, 99-117.

Meylan, M.A. (1978) Marine sedimentation and manganese nodule formation in the South Western Pacific Ocean. Unpubl. Ph.D. thesis. Univ. of Hawaii. 312 pp.

Meylan, M.A., Glasby, G.P., McDougall, J.C. and Singleton, R.J., (1978) Manganese nodules and associated sediments from the Samoan Basin and Passage. NZOI Oceanogr. Fld Rep. 11, 61 pp.

Murray, J. and Renard, A.F., (1891) Deep-sea deposits. Rep. Scient. Results explor. Voyage Challenger: 525 pp.

Nohara, M. and Nasu, N., (1977) Mineralogical and geochemical characteristics of manganese nodules from the Suiko Seamount, northwestern Pacific Ocean 2 - Geochemical aspects and its origin. Bull. Geol. Surv. Japan 28, 615-621.

Piper, D.Z., (1972) Rare-earth elements in manganese nodules from the Pacific Ocean. Pp. 123-130 in 'Ferromanganese Deposits on the Ocean Floor'. National Science Foundation, Washington D.C. 293 pp.

Piper, D.Z., (1973) Rare-earth elements in ferromanganese nodules and other marine phases. Pp. 7-21 in 'Inter-University Program of Research on Ferromanganese Deposits of the Ocean Floor Phase 1 Report'. National Science Foundation, Washington, D.C. 358 pp.

Piper, D.Z., (1974) Rare earth elements in ferromanganese nodules and other marine phases. Geochim. Cosmochim. Acta 38, 1007-1022.

Piper, D.Z. and Graef, P., (1974) Gold and rare-earth elements in sediments from the East Pacific Rise. Mar. Geol. 17, 287-297.

Rankin, P.C. and Childs, C.W. (1976) Rare-earth elements in iron-manganese concentrations from some New Zealand soils. Chem. Geol. 18, 55-64.

Shipek, C.J., (1962) Photographic survey of sea floor on southwest slope of Eniwetok Atoll. Bull. Geol. Soc. Am. 73, 805-12.

Taylor, S.R., (1965) Geochemical analysis by spark source mass spectrometry. Geochim. Cosmochim. Acta 29, 1243-1261.

Taylor, S.R., (1971) Geochemical applications of spark source mass spectrometry. Geochim. Cosmochim. Acta 35, 1187-1196.

Volkov, I.I., Rozanov, A.G. and Sokolov, V.S. (1975) Redox processes in diagenesis of sediments in the northwest Pacific Ocean. Soil Science 119, 28-35.

ZONAL REGULARITIES IN OCCURRENCE, MORPHOLOGY AND CHEMISTRY OF MANGANESE NODULES OF THE PACIFIC OCEAN

N. S. Skornyakova

P.P. Shirshov Institute of Oceanology Sciences

U.S.S.R.

Distribution Patterns of Nodules

Manganese nodules and crusts occur on the Pacific Ocean floor of the continental slope and shelf, the abyssal basins, and submarine rises in both the central and marginal regions. The majority of nodules, however, are concentrated in the central abyssal basins, i.e., the pelagic region of the ocean rather than the hemipelagic regions marginal to the ocean basins.

The boundary between the pelagic and hemipelagic regions is defined by the transition zone between completely oxidized sediments and sediments with subsurface reduced layers. The transition is commonly sharp (Fig. 1) and coincides with changes in the sedimentary environment such as sediment accumulation rates, content of organic matter, and abundance and trophic types of benthic fauna (Murdmaa et al., 1976).

The boundary is apparently marked also by a sharp change in the abundance of ferromanganese nodules. In the hemipelagic region where clay, siliceous mud, or tuffitic (volcanic) mud accumulates at rates in excess of 10 to 30 mm/10^3 yr, nodules are rare. Small nodules, or thin ferromanganese films covering pumice and ice-rafted pebbles are occasionally found associated with the oxidized surface layer of sediment of highly variable thickness. On submarine rises and slopes of large seamounts, where accumulation rates are extremely slow (Skornyakova and Andruschenko, 1970 and 1976), nodules and ferromanganese crusts are frequently recovered.

In the pelagic region nodules occur throughout the basins and on submarine rises. On steep slopes of abyssal hills and seamounts ferromanganese crusts tend to cover exposed rock surfaces. Nodule abundance, however, is widely variable. Highest concentrations are confined to depths below the carbonate compensation depth (CCD),

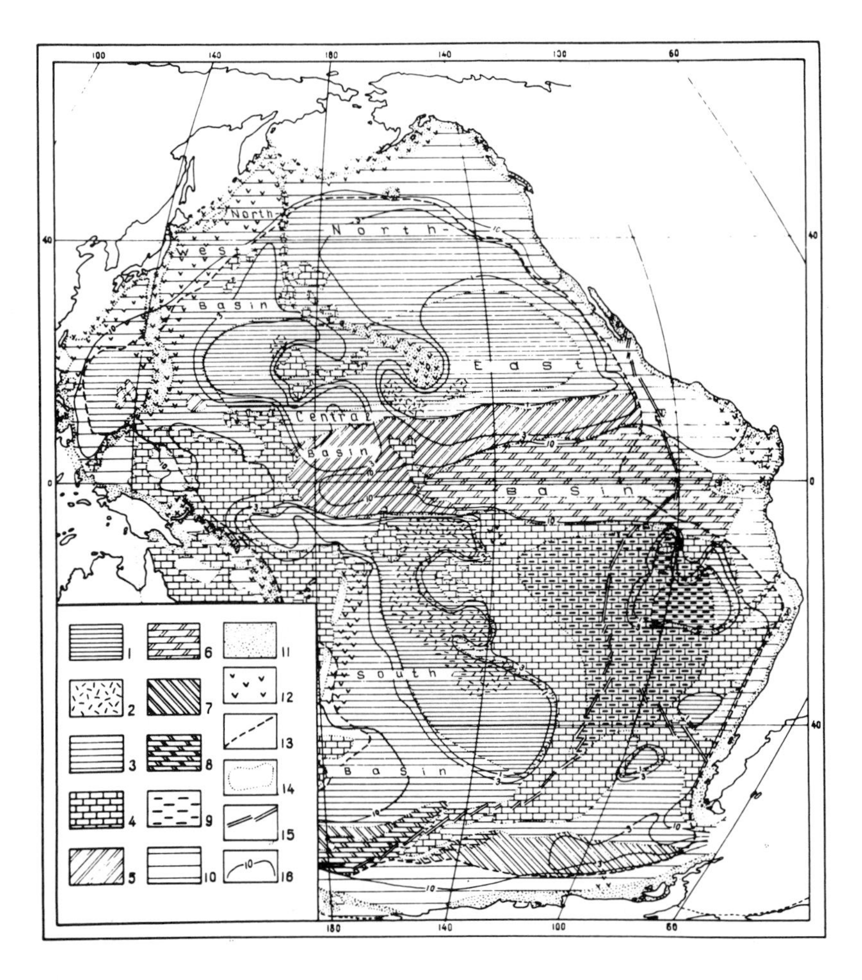

Fig. 1. Lithofacial types of sediments of the Pacific. 1 - eupelagic clays; 2 - zeolitic clays; 3 - miopelagic clay*; 4- calcareous nannoforam oozes; 5 - diatom-radiolarian clays; 6 - siliceous-calcareous (diatom-radiolarian-nannoforam oozes; 7 - diatom oozes; 8 - siliceous calcareous (diatom-nannoforam) oozes; 9 - metalliferrous pelagic sediments; 10 - hemipelagic clays (including siliceous and marly mud); 11 - terrigeneous sands and silts; 12 - sediment with abundant volcanoclastic matter; 13 - boundary between the pelagic and hemipelagic regions; 14 - boundary between lithofacial types of sediment; 15 - axis of the Mid-Pacific Ridge; 16 - sediment accumulation rates (by A. P. Lisitzin with author's modifications.

*Pelagic (red) clay is subdivided into two lithofacial types, eupelagic ("true") pelagic and miopelagic ("transitional" to hemipelagic) clays (Skornyakova and Murdmaa, 1968).

where sediment accumulation rates are low. Nodules are most abundant in association with eupelagic clay, on radiolarian mud or ooze, and less abundant on miopelagic clay and pelagic diatom ooze. In areas above the CCD, where calcareous ooze occurs, nodules are restricted mainly to areas of rugged topography.

The dependence of nodule abundance on lithofacies and sediment accumulation rates results in three latitudinal zones in the Pacific where nodule coverage of the sea floor often is greater than 25%. Two of these zones occur in the equatorial Pacific, one on either side of the equator, and the third in the subantarctic region (Fig. 2).

In the northern Pacific the nodule zone is located between 7° and 30° N latitude. Nodules are associated with eupelagic clay north of about 14° N and with siliceous ooze and siliceous mud at 7° to 14° N. Nodule coverage in this belt commonly is between 10 to 50%. Nodule abundance ranges from 0.4 kg/m^2 to 22.2 kg/m^2 and averages 7.8 kg/m^2. The highest values are usually observed on slopes and tops of abyssal hills.

The second zone of abundant nodules is located at approximately 8° S, but extends to 40° S latitude in the western Pacific. Highest concentrations are confined to eupelagic and zeolitic clay. The zone contains nodule fields which have the highest concentrations for the entire Pacific, specifically on the northeast flank of the South Pacific basin, south from Cook and Tubai Islands. Commonly nodules cover more than 25% of the bottom and the average concentration is about 50%. Weight abundances average 18 kg/m^2 with a maximum of 70 kg/m^2.

Relatively abundant nodules were found at several locations in the Bauer Deep, at approximately 8° S. This zone is separated from the zone to the west, however, by the East Pacific Rise (Lisitzin _et al._, 1976). Sea floor coverage ranges from 8% to 46% and weight abundances from 1.6 to 19 kg/m^2. Nodules have been observed frequently also in the Peru Basin (Agassiz, 1902).

The third nodule belt, which has the sharpest latitudinal definition, occurs in the subantarctic between 55° and 60° S latitude (Goodell, _et al._, 1971). Continuous ferromanganese crusts have been photographed on exposed rock surfaces, alternating with "cobblestone road-like" nodule fields. Sea floor coverage can exceed 50%. This belt apparently extends into the Tasman Sea near the junction between the South Pacific and the Antarctic-Australian Ridge (Hayes and Conolly, 1972).

Marginal to these belts, but within the pelagic environment, areas of high nodule abundance are occasionally observed where sea floor coverage may be as high as 60%. Such areas are usually restricted to large abyssal hills which extend above the sea floor approximately 400 m or more.

Calcareous ooze, in the eastern part of the equatorial belt, is virtually barren of nodules. Nodules are absent as well on accumulative aprons surrounding oceanic islands and large submarine

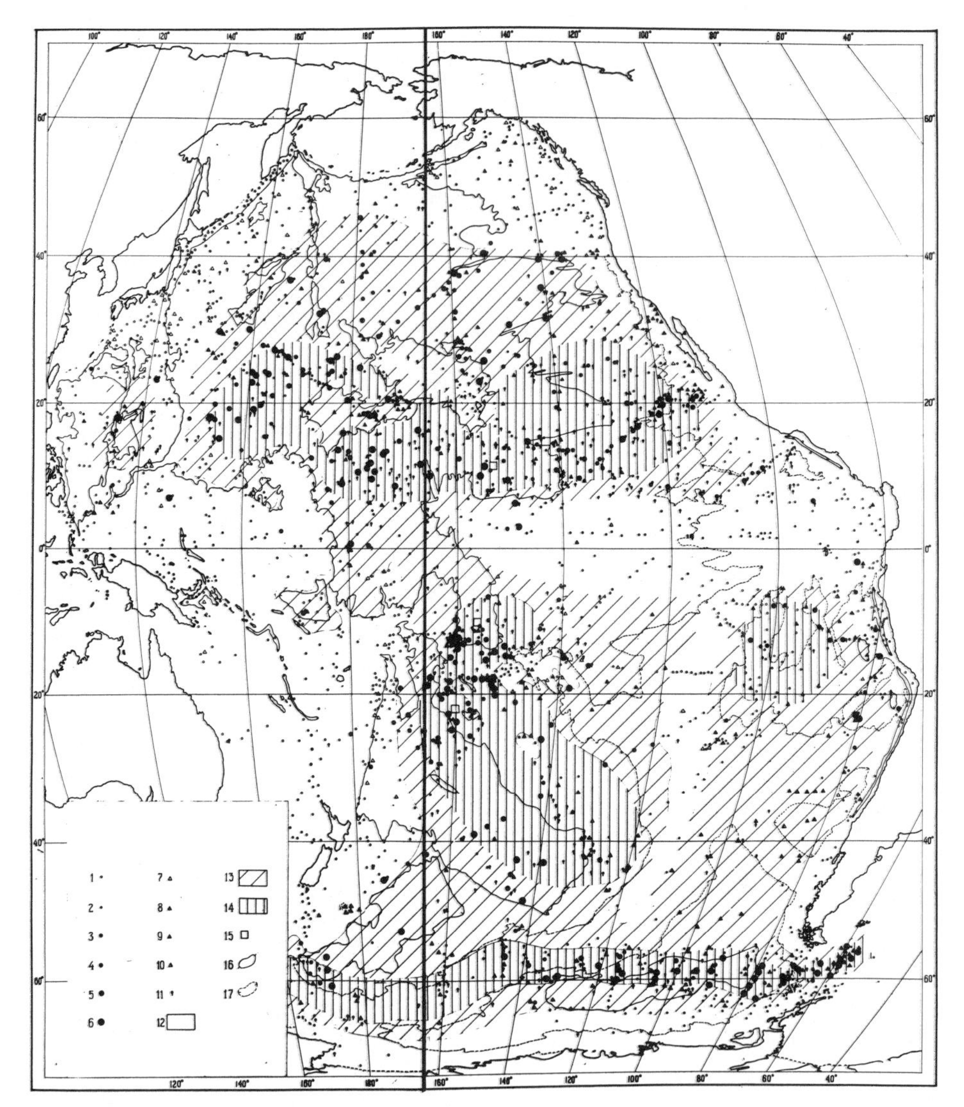

Fig. 2. Occurrence of ferromanganese nodules on the Pacific floor. 1-10 - geological stations where nodule concentrations were determined: 1-6 - by bottom photographs and grab samples: 1) no nodules; 2) < 10%; 3) 10-25%; 4) < 25% (without subdivision); 5) 25-50%; 6) > 50%; 7)-10)-by trawls: 7) no nodules; 8) few nodules; 9) abundant nodules; 10) nodules recovered but concentration not estimated; 11) nodules recovered in cores; 12) the nodules are rare; 13) the high concentrations are scanty; 14) the high concentrations are frequent; 15) polygons of detailed survey; 16) 5000 m contour; 17) 4000 m contour. The map is compiled using data collected by Soviet and foreign expeditions (Skornyakova, 1976a).

ridges or mountain chains (for example, between the Mid-Pacific Mountains).

In marginal regions nodules commonly occur as thin encrustations on rocks and other foreign debris, which constitute more than 40% of the nodule. In the pelagic region crustal nodules are found on crests, or slopes, of submarine ridges and seamounts, where angular fragments of basalt, hyaloclastites, and sedimentary rocks are covered by ferromanganese crusts. By contrast, nodules in the pelagic environment usually contain relatively small cores or even no recognizable core.

Although the shape and size of nodules can be quite different at adjacent stations, certain regional trends in these parameters are discernable (Skornyakova and Zenkevich, 1976). Towards continental margins the size of nodules generally decreases (Skornyakova, 1960). On hemipelagic clay and diatom mud, nodules are commonly small and friable and have an irregular, or flattened, shape. Nodules are small also in the vicinity of ocean islands and in the belt of calcareous ooze along the equator. Nodules associated with pelagic clay tend to be large, spheroidal, or ellipsoidal. They are commonly discoidal, botrioidal, and friable in the belt of siliceous ooze north of the equator.

Regional Variations in Chemistry

Regional variations in the chemical composition of Pacific Ocean manganese nodules were first noted by Mero (1962, 1965) and later by Skornyakova and Andruschenko (1964, 1970), Price and Calvert (1970), and Piper and Williamson (1977). However, considerable local variation is revealed from maps showing Fe, Mn, and minor elements in nodules (Figs. 3-6).

The Mn content in nodules from hemipelagic areas ranges from 0.47% to 30.5% (Fig. 3). Iron content ranges from 0.6% to 20% (Fig. 4) and the Mn/Fe ratio from 0.06 to 50.9.

We see an apparent symmetry in the Fe and Mn distribution between the western and eastern margins of the Pacific. Nodules from the western margin are enriched in Fe. They may have Fe_2O_3 concentrations as great as 70%, as in nodules from the Northwest Pacific, and have Mn/Fe ratios less than 1. The composition of Antarctic nodules falls within the limits characteristic of hemipelagic nodules of the Northwest Pacific (Goodell, *et al*., 1971). On the eastern margin, especially along California and adjacent to the Galapagos Islands, the Mn/Fe ratio of nodules is commonly more than 4 and can be as large as 50. MnO_2 comprises, in many cases, 90% of the oxides.

Concentrations of Ni and Cu in nodules from throughout the hemipelagic environment are less than 0.2%; Co is less than 0.1%; Zn is less than 0.05%; and Pb is less than 0.02%.

The Mn content of nodules from the pelagic environment ranges from 5 to 34.5%, but commonly is between 10-20% (Fig. 3). It shows

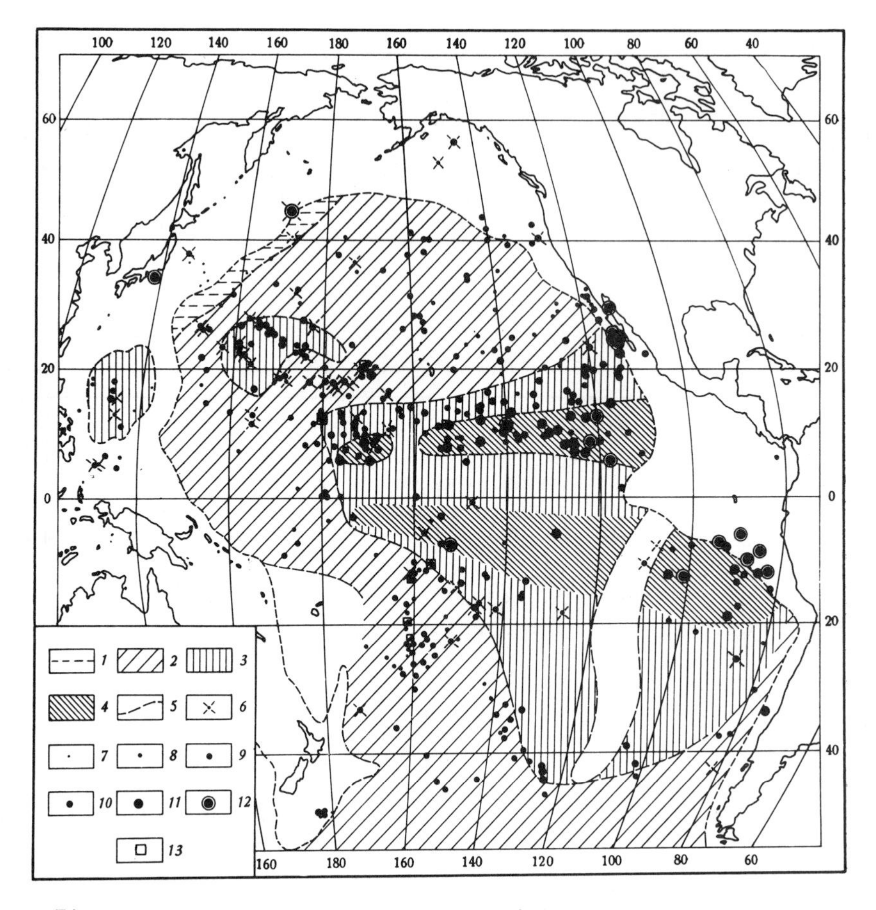

Fig. 3. Manganese concentration (%) of nodules from the Pacific Ocean: 1) < 10; 2) 10-20; 3) 15-25; 4) 20-35; 5) the boundary between the pelagic and hemipelagic regions; 6) seamounts with a height above the sea floor of more than 1000 m; 7-12) Mn concentration at sampling sites (%) - 7) < 10; 8) 10-15; 9) 15-20; 10) 20-25; 11) 25-30; 12) > 30; 13) polygons of detailed survey.

a strong latitudinal variability and a slight increase from the marginal to the central part of the ocean, with some shift toward high concentrations in the eastern Pacific. Nodules associated with radiolarian oozes and eupelagic clay north and south of the equator have relatively high Mn contents of 20-35%. Nodules from the South Pacific between 40-45° S also have somewhat higher values (Piper and Williamson, 1977).

The Fe content of pelagic nodules ranges from 3 to 26%, but usually varies from 10 to 20% (Fig. 4). High concentrations occur on submarine rises. Low values characterize nodules from topographic

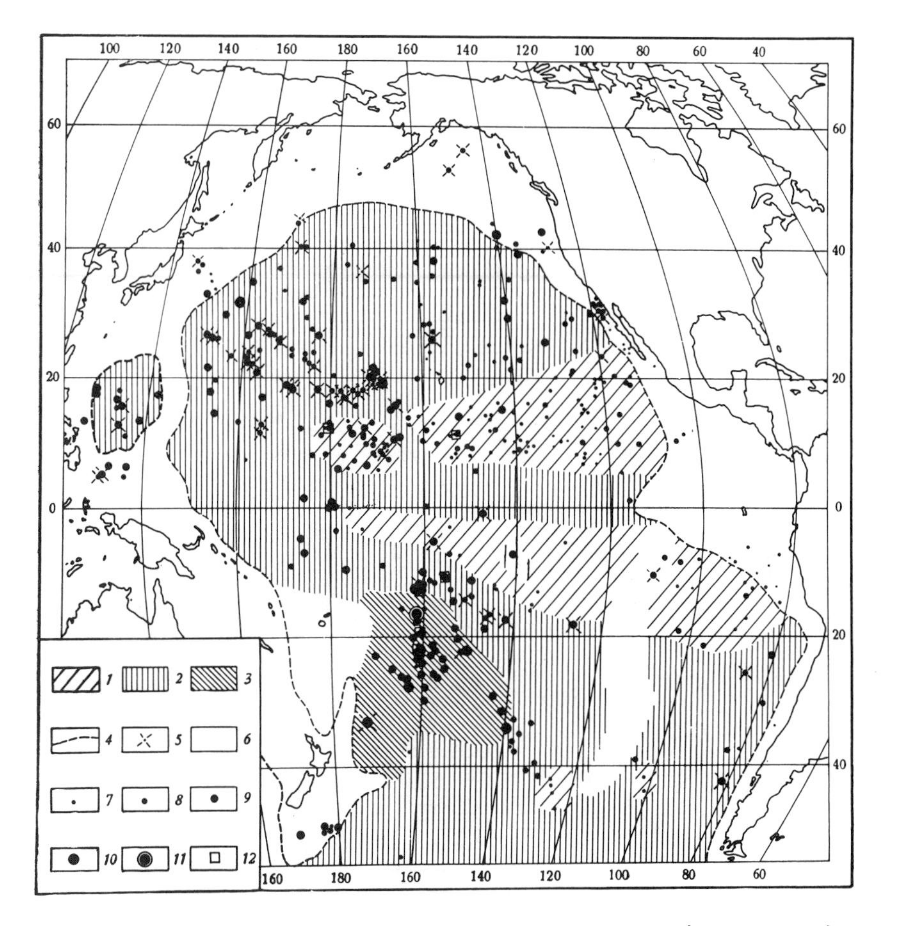

Fig. 4. Iron concentration (%) in nodules: 1) 5-15; 2) 10-20; 3) 15-25; 4) the boundary of the pelagic region; 5) seamounts with relative height more than 1000 m; 6)- 11)Fe concentration at sampling sites; 6) < 5; 7) 5-10; 8) 10-15; 9) 15-20; 10) 20-25; 11) > 25; 12) polygons of detailed survey.

depressions. Highest Fe values, however, are measured for nodules from the Southern Pacific Basin (15 to 25%), for which the Fe/Mn ratio is lower than elsewhere in the pelagic environment. These nodules also have high Ti concentration.

Within the tropical and subtropical area of the Pacific (i.e., between 25^{0} N, and 25^{0} S) regional variations in the Fe values mirror those of Mn. Relatively low values (5-15%) are found north and south of the equator (Fig. 4). West of the Lines Islands Ridge, Fe content is somewhat more (10-15%), but its concentration exhibits a strong depth dependency. Nodules from abyssal hills and seamounts have higher concentrations than those from depressions. For this latter group the Fe concentration is commonly less than 10%.

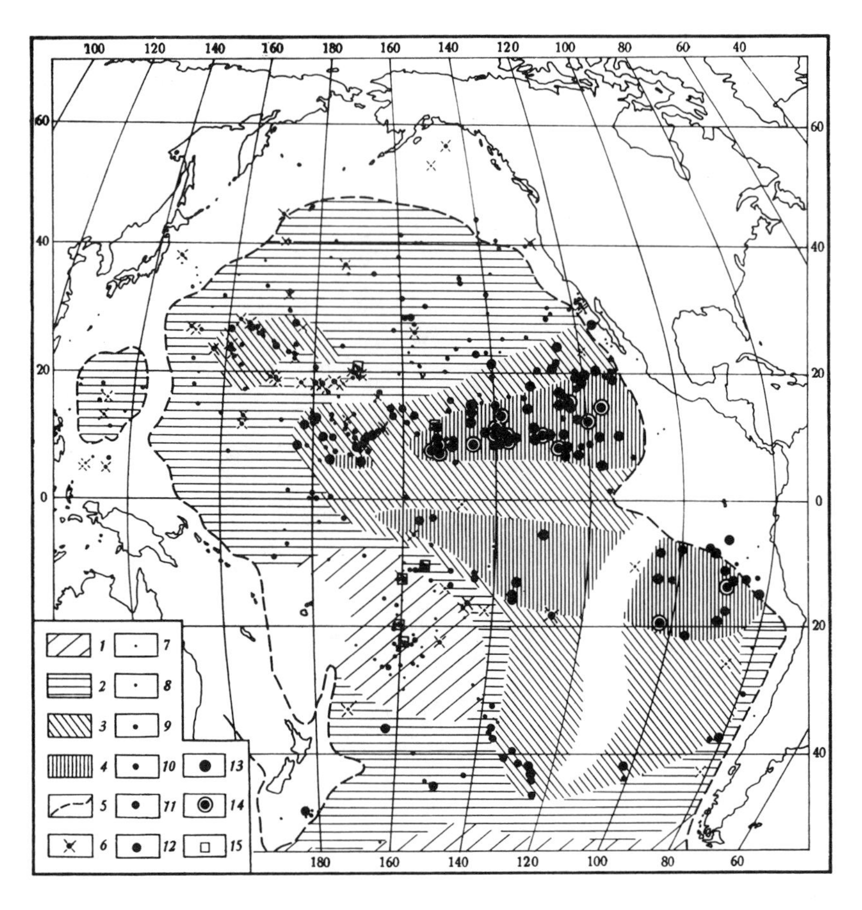

Fig. 5. Nickel concentration (%) in nodules: 1) < 0.4; 2) 0.2-0.8; 3) 0.4-1.2; 4) 0.8-2.0; 5)-6) see Figure 3; 7)-13) Cu concentration at sampling sites (%): 7) < 0.2; 8) 0.2-0.4; 9) 0.4-0.6; 10) 0.6-0.8; 11) 0.8-1.0; 12) 1.0-1.4; 13) 1.4-2.0; 14) polygons of detailed survey.

Pelagic nodules are enriched in Ni, Co, Cu, Zn, and Pb and several other minor elements relative to hemipelagic nodules. The Cu and Ni content of pelagic nodules ranges from 0.2 to 2%. In marginal regions they are approximately 0.2 to 0.8%.

The areal distribution of Cu and Ni shows the same latitudinal patterns outlined for Mn (Figs. 5 and 6). Highest concentrations show a marked eastward shift superimposed on latitudinal variations. Maximum Ni content of 0.6 to 2.0%, but commonly 1%, was measured for nodules from siliceous and siliceous-calcareous oozes to the north and south of equator. Low values of 0.1 to 0.4% were found in nodules of the South Pacific Basin.

The equatorial belts of high Cu are more narrowly delineated

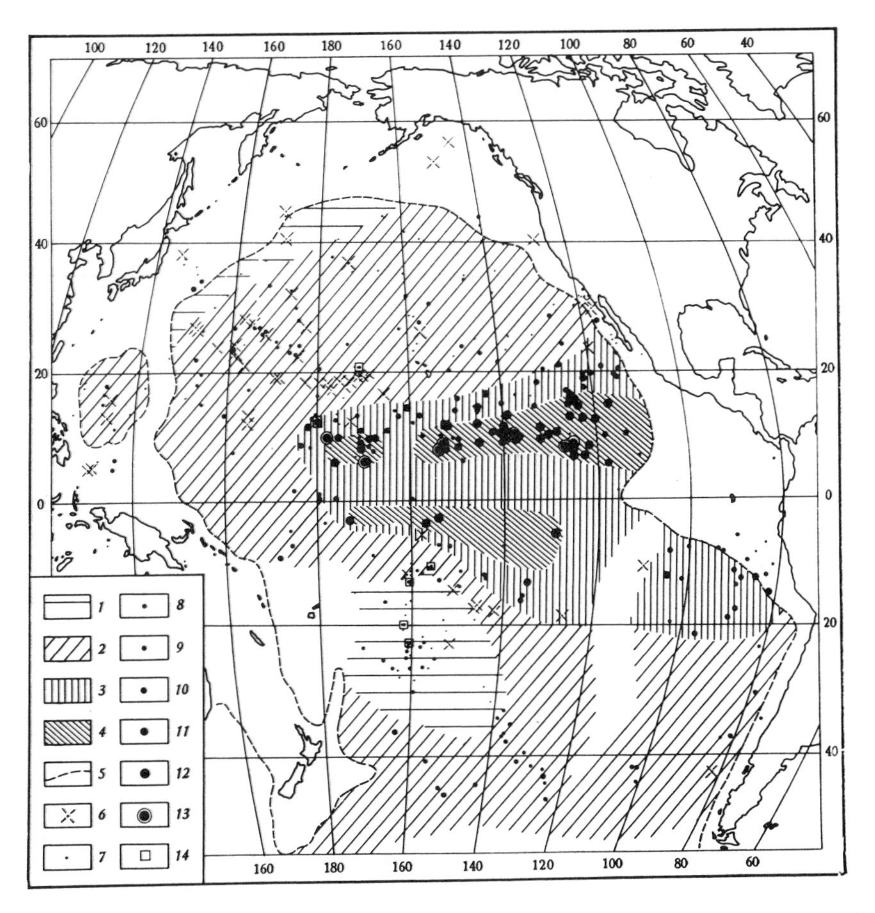

Fig. 6. Copper concentration (%) in nodules: 1) < 0.4; 2) 0.2-0.6; 3) 0.4-1.0, 4) 0.8-2.0; 5)-6) see Figure 3; 7)-13) Cu concentration at sampling sites (%): 7) < 0.2; 8) 0.2-0.4; 9) 0.4-0.6; 10) 0.6-0.8; 11) 0.8-1.0; 12) 1.0-1.4; 13) 1.4-2.0; 14) polygons of detailed survey.

than they are for Ni. The highest concentrations of Cu of 8.0%-2.0% are restricted to the diatom-radiolarian ooze of the North Pacific. In the Bauer Deep nodules had high Mn and Ni contents (0.5 to 0.8%), but relatively low Cu concentrations (0.5 to 0.8%).

Preliminary data for Mo and Zn suggest similarities in their distribution patterns with Mn.

Cobalt, lead, and vanadium show correlation with water depth, rather than showing strong regional trends similar to Mn, Ni, and Cu. Highest Co values (> 1%) were determined in nodules from seamounts (Menard, 1964; Cronan and Tooms; 1969; Skornyakova et al., 1975; Skornyakova, 1976a). Nodules recovered from basin floors have Co concentrations less than 0.4%. Somewhat higher Co concentrations are measured for nodules from the South Pacific and Central Pacific Basins and are associated with higher Ti and Pb concentrations. The lowest Co values less than 0.2% were found in nodules from the equatorial zone.

Discussion

Nodule abundance in the Pacific Ocean appears to exhibit a circumcontinental and latitudinal zonality. The circumcontinental zonation parallels the decrease in sediment accumulation rates (Bezrukov, 1962, 1970; Lisitzin, 1972, 1974), produced by a decrease in the supply of terrigenous matter from continents towards the central oceanic regions.

The latitudinal zonation in nodule abundance is observed only within the pelagic region. Two areas of maximum nodule abundance are located north and south of the equator, marginal to the equatorial current system. In these regions eupelagic clay is the most common sediment type at depths below the CCD. The accumulation rate for eupelagic and zeolitic clays of the South Pacific Basin of 0.4 mm per 1000 years. The same approximate values were obtained for marly clay from this region using benthic foraminifers (Saidova, 1974; Murdmaa _et al_., 1976).

Rates of less than 1 mm per 1000 years were estimated also for eupelagic clays of the northern nodule area. Sediment accumulation rates in this northern area increase toward its southern margin to values of 1 to 3 mm/10^3 yr, or more, as pelagic clay grades into diatom-radiolarian-bearing muds. However, the region is known to have numerous outcrops of Neogene and Paleogene sediments, or a very thin cover of Quaternary sediment (Riedel and Funnell, 1964; Jouse, 1968; Hays _et al_., 1969). Here often occur teeth of Tertiary sharks. These observations suggest extensive erosion and reworking of sediments, apparently by bottom currents.

High velocity bottom currents have actually been observed in the Antarctic belt of high nodule abundance (Heezen and Hollister, 1971; Goodell _et al_., 1971; Watkins and Kennett, 1972) and result in partial erosion and non-deposition. Thus, high nodule concentrations are related here also to slow sediment accumulation rates.

Increased sediment accumulation rates lead to decreased nodule concentrations. In the equatorial belt, where biogenic calcareous and siliceous sediments accumulate at rates of 10 to 30 mm/10^3 yr, only infrequently are small "rudimentary" nodules observed. Thus, the sediment accumulation rate, controlled by either bottom currents and/or supply of sediment, may be the main factor controlling the concentration of nodules.

The concentration of nodules may also depend on the duration of accretion. If nodule growth is continuous, the duration is simply the nodule age and its maximum age will be given by the age of the underlying sediment. In marginal regions nodules are commonly found on top of Holocene sediments. Holocene to Late Pleistocene sediments were also observed, in several cases, for pelagic nodules, principally where the sediments are miopelagic clays or calcareous oozes (Bezrukov, 1976). For example, on an abyssal hill in the eastern part of the equatorial belt (st. 5937, depth 5020, 0°19' N, 179°45' W) abundant nodules were collected from the surface of Pleistocene marly radiolarian clay (V. Mukhina, personal communication). Nodules were found with Pleistocene marly radiolarian mud at Station 5996 (10°38' N, 153°23' W), located on the northern margin of the equatorial radiolarian belt (Skornyakova _et al._, 1971). Nodules at Stations 5998 (southern part of the Central Basin, 11°04' S, 156°12' W), and 6298 (South Pacific Basin, 22°42' S, 160°51' W) occurred with Pleistocene age marly ooze. At these same locations, but below the CCD, nodules were associated with eupelagic and zeolitic clay of possibly early Pleistocene age (Murdmaa _et al._, 1976). The sediment accumulation rate at all of these stations ranged from 0.05 to 0.2 mm per 1000 years.

In many areas of the pelagic environment nodules occur with eupelagic clay or diatom-radiolarian mud of Tertiary age. Pliocene to Eocene sediments underlie nodules in the latitudinal equatorial belt of diatom-radiolarian oozes (Horn _et al._, 1973). In the southern Tasman Sea the nodule "cobblestone-road" overlies Pliocene (2.43 to 3.5 m.y.) siliceous ooze. Marginal to the "cobblestone-road", calcareous and siliceous ooze of Brunnes age (Payne and Conolly, 1972) is at the surface. In the eastern part of the Antarctic nodule belt nodules lie on sediment of Brunnes age (Meylan, 1968).

The shape of nodules, which is continuously modified during deposition of Fe and Mn oxides, appears to be related to these variations in sediment accumulation rates. In the environment of extremely slow accumulation of eupelagic clay, large spheroidal and nodules predominate. In the environment of relatively high accumulation rates, as in the area of the ocean margins and in the equatorial belt, small crustal nodules and irregular, or flattened, shaped nodules are commonly found. In the hemipelagic environment and in the areas of radiolarian ooze, the shape is influenced by the mechanism of ferromanganese oxide precipitation. This includes diagenetic migration from the underlying sediment. The characteristic collar-like rims around flat nodules on radiolarian ooze are obviously produced by the flux of Fe and Mn from the sediment.

Chemical variations in nodules also show certain relations to sedimentary environments. Nodules with the highest average Mn, Ni, Cu and Zn concentrations (Figs. 3-6) are associated with the radiolarian oozes of the equatorial belt (Table 1). The concentrations of these metals decrease successively in the miopelagic clay, eupelagic clay, and calcareous ooze. Changes in concentrations of Ni, Cu and Zn from one sediment type to the next, however, are greater than the change in Mn. For example, the Mn content in nodules from northern Pacific eupelagic clays is only 4% less than in nodules from the radiolarian belt, whereas decreases in Ni, Cu and Zn are about 30%. These trends are observed also on a local scale. Nodules from diatom-radiolarian ooze contain much more Mn, Ni, and

Table 1

Average Concentration of Metals in Nodules from the Different Ocean Areas

Area	Associated Sediment	Number of Analyses[1]	Mn/Fe Mean	Mn/Fe Range	Concentration, % Mn	Fe	Ti	Ni	Co	Cu	Zn	Pb
No. Pacific	Miopelagic Clays	38	1.17	0.5-1.8	14.55	12.38	0.61	0.47	0.24	0.29	0.074 (10)[2]	0.12(12)[2]
No. Pacific	Eupelagic Clays	72	1.74	0.64-3.9	19.14	10.98	0.72	0.65	0.33	0.54	0.084 (33)	0.12 (43)
Equatorial Pacific	Radiolarian Ooze	70	2.9	1.1-10.6	23.1	7.98	0.60	1.1	0.18	0.89	0.11 (9)	0.056 (24)
So. Pacific	Eupelagic Clays	106	0.96	0.44-1.9	15.64	16.27	1.07	0.42	0.37	0.22	0.071 (40)	0.13 (46)
Seamount of Central Pacific	-	70	1.18	0.58-4.4	18.39	15.01	0.98	0.44	0.61	0.13	0.066 (21)	0.18 (31)
Northwest Periphery of Ocean	Hemipelagic Muds	8	0.48	0.08-0.84	8.15	12.94	-	0.11	0.08	0.09	-	-
East Periphery of Ocean (near California and So. America	"	8	27.6	11.6-50.96	34.8	1.49	-	0.15	0.01	0.082	0.05 (5)	0.017 (8)

[1]Number of analyses for Fe, Mn, Ti, Ni, Co and Cu.

[2]Number of analyses for Zn and Pb.

Cu than nodules from the same approximate area, which are associated with marly ooze.

Although nodule composition responds to changes in sediment lithology, abundances of elements in nodules and surrounding sediments exhibit no apparent relationship. Correlations between sediment and nodule composition are absent at least for Mn, Ni, Co, and Cu. The Fe concentration of nodules from the pelagic region, however, commonly correlates with a high Fe content in the sediments. Absence of correlations, previously inferred from bulk sediment analyses (Skornyakova and Andruschenko, 1964, 1970; Skornyakova, 1976b; Cronan and Tooms, 1969), is even more apparent if the mobile ("hydrogenous") forms of the elements are considered.

In nodules Fe, Mn, and the minor elements are present mainly as hydrated oxides, whereas in sediments most of the Fe and essential portions of the Cu, Ni, Co and Mn are represented by stable forms, being enclosed in clay and other lithogenic minerals. Rather scant information is available about mobile forms of the elements in oceanic sediments. The most representative data have been obtained on a transoceanic section from Japan to California (Rozanov *et al.*, 1972; Glagoleva, 1972; Glagoleva *et al.*, 1975; Volkov *et al.*, 1976). Mobile Fe was defined as that fraction which was extracted by cold 3.5 M H_2SO_4. The amount of Fe extracted from the same samples by an acid-reducing solution of 1 M NH_2OH-HCl in 25% CH_3OOH (Chester and Hughes, 1967) was 25-30% less than that extracted with H_2SO_4.

Manganese was extracted by a method which leaches Mn^{4+} and not Mn^{2+} (Rozanov *et al.*, 1972; Rozanov *et al.*, 1976). In pelagic clay, however, Mn^{4+} greatly predominates, commonly comprising more than 90% of the total Mn content (Chester and Hughes, 1967). Thus, we may use the total Mn content and assume that it is essentially equal to its mobile forms.

The results of mobile Fe and total Mn determinations are plotted for three profiles, together with total Fe and Mn in nodules and with the corresponding Mn/Fe ratios for sediments and nodules (Fig. 7). Locations of the profiles are shown on Figure 8. Profile I (Fig. 7) intersects hemipelagic clay (Stations 165-6171 and 655-4265) and pelagic clay (Stations 6172-675) in the northern Pacific; Profile II crosses diatom-radiolarian ooze and clay (Stations 5934-5994, 5996-18, 5996-20) and siliceous marly ooze (Station 5996-13, 5996-17) of the equatorial belt; Profile III crosses eupelagic clay in the South and Central Pacific Basins.

Using the data from these profiles we calculated correlation coefficients between mobile Fe, Mn, and the Mn/Fe ratio in sediments and nodules (Table 2). Mn and mobile Fe in sediments have a positive correlation of $r = +0.80$, whereas in nodules $r = -0.34$. The former is indicative of a similar concentration mechanism for both elements in sediments, whereas the latter suggests separation of Fe and Mn during nodule generation.

Correlation coefficients between Fe and Mn in nodules exhibit marked regional variations. Iron and manganese in nodules from

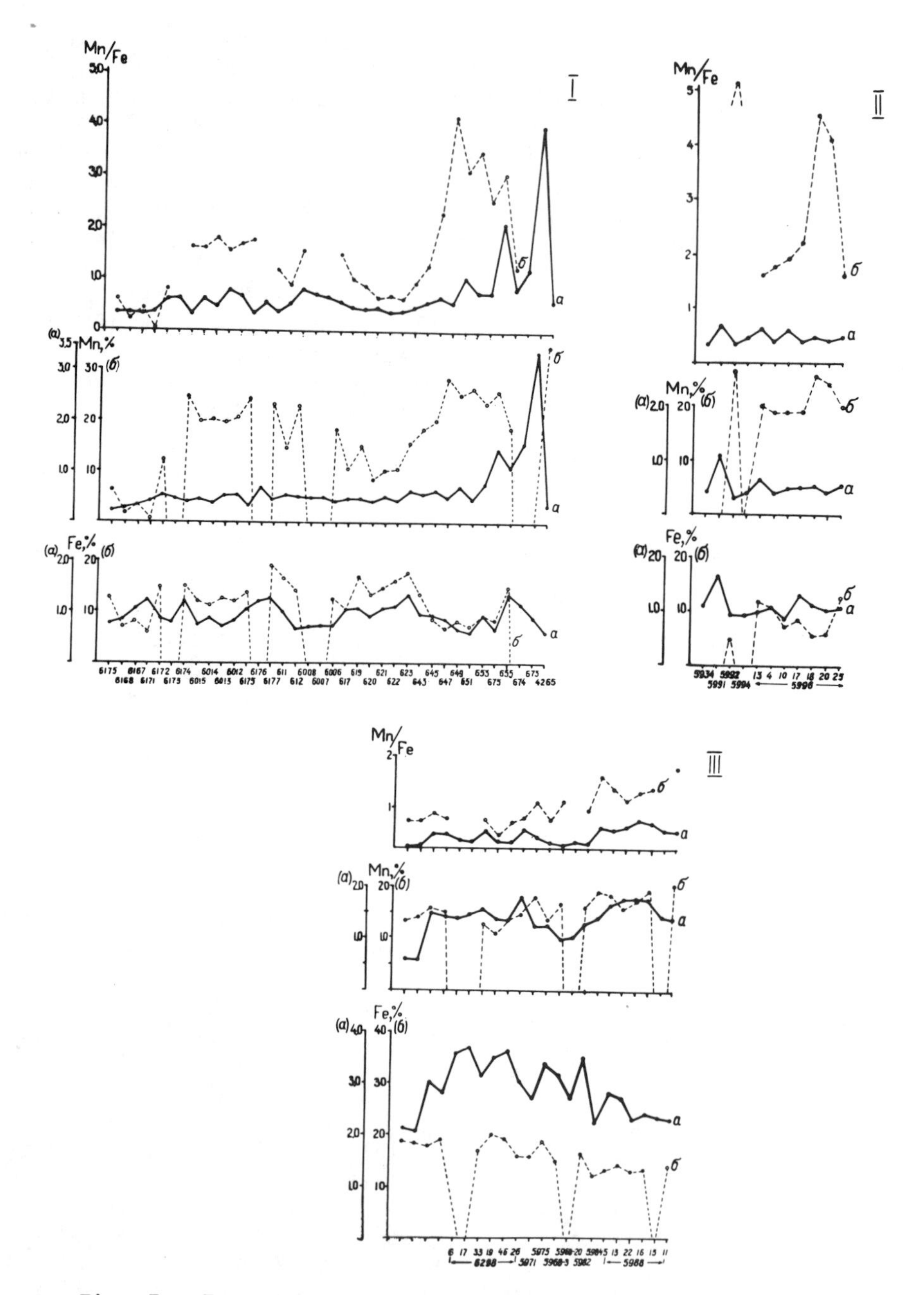

Fig. 7. Fe and Mn content and Mn/Fe ratio in nodules (b) and mobile forms of these elements and their ratios in sediments (a) in the northern Pacific (Profile I), in the equatorial Pacific (Profile II), and in the South Pacific (Profile III). The numbers below the curves refer to station numbers (Fig. 6).

TABLE 2

Correlation coefficients of Mn and Fe and their relation in nodules and associated sediments (56 pairs of analyses)

Sediments	Nodules	r
Mn_{total}/Fe_{mobile}	Mn/Fe	+0.117
Fe_{mobile}	Fe	+0.59
Mn_{total}	Mn	-0.023
	Fe to Mn	-0.34
Mn_{total} to Fe_{mobile}		-0.80

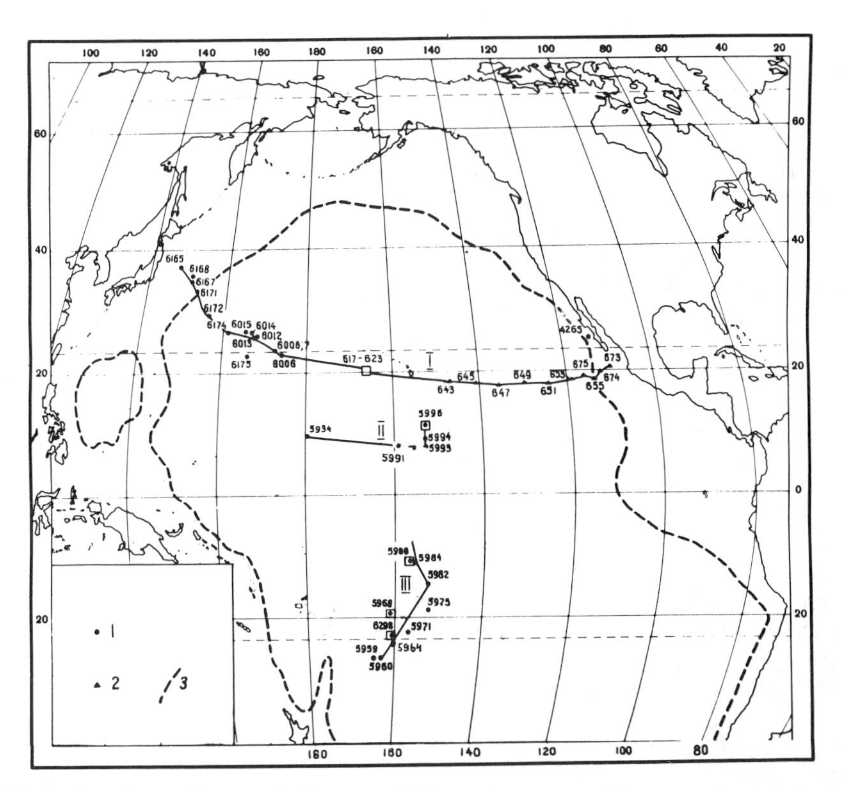

Fig. 8. Location of the profiles shown on Figures 7 and 9: 1 - stations of the R/V "Vitiaz"; 2 - stations of the R/V "D. Mendeleev"; 3 - boundary between hemipelagic and pelagic sediments.

southern Pacific eupelagic clays show essentially no correlation ($r = 0.018$ at Sta. 5908, $r = 0.008$ at Sta. 6298; Skornyakova, 1976a), whereas they show a strong negative correlation ($r = -0.81$ at Sta. 5996) in nodules from the equatorial radiolarian belt.

A positive correlation was found between mobile Fe in nodules and sediments ($r = 0.56$). The profiles in Figure 7 show that the concentration of mobile Fe is almost parallel in pelagic clay and overlying nodules (Stations 611, 612, 617-624, Profile I; 6298, 5982, Profile II). Similar interrelations were found previously (Cronan and Tooms, 1969; Skornyakova, 1976b; Volkov et al., 1976). The correlation, however, is not universal. In the Bauer Deep, where metalliferous sediments are strongly enriched in Fe, nodules are low in Fe.

The Mn content of nodules from pelagic regions apparently does not depend on the Mn concentration in the underlying sediment (Fig. 7, and Table 2). Even the presence or absence of nodules on the sediment surface does not seem to influence the Mn content of the associated clay (Fig. 7, Sta. 5960-6298-33). Mn in nodules which are associated with radiolarian ooze (Fig. 7) and hemipelagic clay (eastern end of the profile, Fig. 7, Sta. 674-4265), shows an inverse trend with the Mn content of the fine sediment fraction.

No clear correlation was found between the mobile Mn/Fe ratio in nodules of the pelagic region and the underlying sediment (Table 2). Sharp variations in nodule ratios (Fig. 7) occur with relatively constant values in pelagic sediments. In pelagic clay and radiolarian ooze mobile Mn/Fe ranges from 0.3 to 0.75, commonly 0.4 to 0.6, with only one exception (Fig. 7, Sta. 675), for which the ratio equals 2.15. The sharp increase in Mn together with several minor elements at this one station is probably caused by an exhalative influx of metals (Volkov et al., 1976). The Mn/Fe ratio in nodules from these same profiles (Fig. 7) ranges from 0.49 to 5.52, the highest values belonging to nodules from the radiolarian belt.

No correlation is observed between minor elements in pelagic nodules and sediments.

The lack of interrelations between the chemical composition of sediments and nodules suggests that there are multiple factors which control the nodule composition, including diagenesis. The degree of diagenesis can be considered in terms of the concentration factors of Fe, Mn, and minor elements. The concentration factor is defined as the ratio of the concentration of an element in nodules to its concentration in the underlying sediment (Skornyakova and Andruschenko, 1964; Skornyakova, 1965). The concentration factor for Mn ranges from 10 to 115, the factor for Fe from 2 to 20. The large difference in factors for these two elements is due to their different diagenetic mobility. The difference increases with increasing diagenetic migration. Highest concentration factors for Mn (more than 100) are found in hemipelagic clay off the California coast and adjacent to the Galapogos Islands (Mero, 1965). High values also occur in the radiolarian belt (25 to 75). Low values occur in the South Pacific Basin, where nodules are associated with eupelagic clay. The concentration factors for Fe and Mn in this latter region are almost equal (Fig. 7, Sta. 6298-26 to 6298-46).

Another indicator of the degree of diagenesis is the Mn/Fe ratio in nodules (Price and Calvert, 1970). The widest variations of the ratio appear in hemipelagic sediments (from 0.2 to 50.2). Intensity of diagenesis in this environment is limited by abundance and composition of the organic matter (Romankevich, 1977). In the marginal regions of the ocean, where sediment accumulation rates are high, much more total organic matter and more active organic compounds are buried than in pelagic regions of slow accumulation. This leads to reduction and migration to the sediment surface of the soluble forms of Mn and, to a lesser extent, Fe. The upward diffusive migration of Mn^{2+} results in the concentration of Mn^{4+} hydroxide in the oxidized surface layer. The Mn content in this oxidized layer commonly increases about tenfold relative to subsurface reduced sediments (Bonatti *et al.*, 1971).

Such diagenetic migration is the major mode of formation of nodules in the ocean margins. These "diagenetic" nodules receive their metals almost entirely from upward diffusion of the soluble reduced forms.

The asymmetry of Mn in diagenetic nodules between the eastern and western margins of the ocean, noted above, may be attributed to primary asymmetry in the Mn distribution during sedimentation. High concentrations of Mn, dispersed in oceanic sediments, show an eastward shift (Skornyakova, 1965, 1976b) as does the concentration of Mn in diagenetic nodules. The Fe-rich nodules on hemipelagic clay of the Northwest Basin (Mn/Fe ratio 0.2 to 1) were formed on sediments which were initially poor in Mn.

In hemipelagic sediments of the eastern margin, wide variations of the Mn/Fe ratio (1 to 50) reflect varying thickness of the oxidized surface layer. Highest values are found on sediments with a thin oxidized layer (several cm). For example, manganese nodules with Mn/Fe - 29.2 were found off the California coast at Sta. 4265, where the oxidized surface layer was only 3 cm thick (Fig. 7). Towards the open ocean, with increasing thickness of the oxidized layer, nodules became more Fe-rich and Mn-poor, apparantly due to a decrease in the intensity of Mn diffusion to the oxidized layer.

In the transitional zone between hemipelagic and pelagic sediments, where the thickness of the oxidized layer is 1 m or more, but where the subsurface reduced layer still exists, Mn/Fe ratios in nodules are about 1 to 1.5, similar to those in the pelagic region.

In the pelagic region, where sediment accumulation rates are slow, little organic matter is buried and reduction does not develop. Bacterial transformation of organic matter takes place only at the sediment-water interface (Romankevich, 1977).

Under extremely oxidized conditions of the pelagic environment, nodules grow only due to prolonged slow precipitation of Fe and Mn hydroxides on nucleii at the sediment-water interface. Metals in the process may, however, come from both bottom water and from sediments (Manheim, 1965). Vertical migration in pelagic sediments apparently does not take place, as in the hemipelagic environment, but may take place horizontally, involving only the surface film of the sediment (Skornyakova, 1976a; Piper and Williamson, 1977; Calvert and Price, 1977).

The Mn/Fe ratio of pelagic nodules ranges from 0.5 to 10.5 (Fig. 9). Distribution patterns of the ratio exhibit a latitudinal zonation. A belt of high values (commonly 2 to 10) occurs on each side of the equator and is associated with siliceous calcareous oozes and miopelagic clay. Also, nodules from the Bauer Deep have a high ratio. Nodules on pelagic clay have Mn/Fe ratios that range

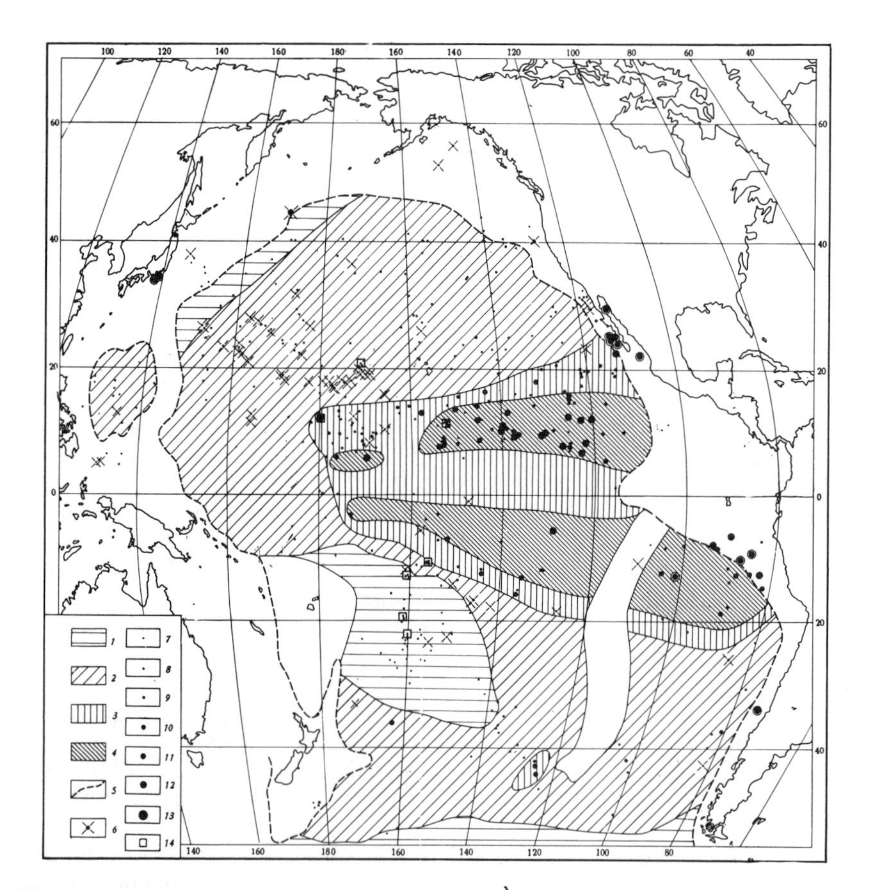

Fig. 9. Mn/Fe ratio in nodules: 1) < 1; 2) 1-2; 3) 1-4; 4) 2-10; 5)-6) see Figure 3, 7)-13) Mn/Fe ratio at sampling stations; 7) 1; 8) 1-2; 9) 2-3; 10) 3-4; 11) 4-5; 12) 5-10; 13) > 10; 14) polygons of detailed survey.

from 1 to 2. Minimum values are found in nodules from the South Pacific and Central Pacific Basins (Fig. 9).

The regular increase in the Mn/Fe ratio has been interpreted in terms of increased sedimentation rates and corresponding changes in organic matter, which controls the diagenetic migration of Mn (Price and Calvert, 1970; Calvert and Price, 1977; Skornyakova, 1976a). Piper and Williamson (1977), however, noticed that within the equatorial Pacific the highest values of the Mn/Fe ratio in nodules correspond with slow sediment accumulation rates. In this belt average accumulation rates range from less than 1 to 3 mm/10^3 yr (Lisitzin, 1972, 1974), but nodules occur only on local areas, where sediments accumulate at rates less than 1 mm/10^3 yr.

The equatorial belts of high Mn/Fe ratios correspond with the equatorial biologically productive zone. Radiolarian oozes of the nodule belts are grayish brown, mottled and extremely porous. Organic carbon is approximately twice as concentrated as in pelagic clay (Romankevich, 1977). However, only the surface sediment, approximately the uppermost cm, is enriched in organic matter (0.35 to 0.45% C_{org}). In subsurface layers it decreases to low levels (0.2% C_{org} at Sta. 5996). The surface sediment of radiolarian ooze is characterized by wide variations in Mn, Ni, Co, and Cu content (Skornyakova, 1976b). An increase in their concentration in nodules seems to correspond with a decrease in sediments. If nodules are absent, the Mn content in sediments is high (Fig. 7).

These data suggest, therefore, that the composition of nodules in the equatorial belts depends on intensity of diagenetic redeposition of Mn and minor elements. Therefore, the chemical features of nodules from the equatorial belt are related (1) to high biological productivity resulting in the diagenetic migration through organic matter supply, and (2) to slow average sediment accumulation owing to the erosional activity of high velocity bottom current.

Nodules on pelagic calcareous oozes, from the equatorial belt of high biological productivity and high carbonate accumulation rates, usually show Mn/Fe ratios less than 2. The diagenetic mobilization of Mn in this environment is apparently low, probably due to the fact that organic matter is mostly in the form of stable compounds enclosed in calcareous foram tests (Romankevich, 1977).

On eupelagic clay, in the regions of very low biological productivity and slow accumulation rates (less than 1 mm/10^3 yr), the Mn/Fe ratio is commonly less than 2 (Figs. 1 and 2).

The lowest Mn/Fe values in eupelagic clay of the South Pacific Basin correspond with minimum sediment accumulation rates. The ratio in nodules, which is commonly less than 1, is almost equal to the mobility of Mn in this environment. The nodules are produced almost entirely by direct precipitation of oxides from bottom water, with little diagenetic influx.

The high Fe content in nodules from the South Pacific Basin may be due to an additional supply of Fe by submarine weathering within this area of basaltic hyaloclastites (Skornyakova _et al_.,

1971, 1973). This interpretation is supported by the higher accumulation rate of Fe in these nodules than in those from the radiolarian belt (Table 3). These data are from three polygons of detailed investigations, one in the northern equatorial radiolarian belt and the other two in the area of eupelagic and zeolitic clay of the South Pacific and Central Pacific Basins (Skornyakova *et al.*, 1971, 1973; Murdmaa *et al.*, 1976). Sediment accumulation rates were calculated using benthic foraminifers and radiolarians. Fe accumulation rates for the South Pacific eupelagic clay are five times as high as those in the radiolarian belt.

A possible explanation for the large variation in the Mn/Fe ratio found for nodules from the Bauer Deep and the flanks of the East Pacific Rise at 40-45° S latitude (Piper and Williamson, 1977) is that there are local metal supplies from hydrothermal sources with subsequent gathering of hydroxides into nodules from the surface sediments. Such hydrothermal supply would also explain the high Mn/Fe ratio of 2 to 4.5 (Fig. 7) in nodules from pelagic clays just north of the Clarion fracture zone. There is mineralogical and chemical evidence for hydrothermal influence on the sediments both north and south of the fracture zone between 113 and 119° W longitude (Butuzova *et al.*, 1975, 1976; Lisitzin *et al.*, 1976; Volkov *et al.*, 1976). In an area of extensive hydrothermal activity in the South Pacific Basin (Sta. 6298) we observed the widest variation in the Mn/Fe ration (0.19 to 52) of both nodules and volcanic rock encrustations (Skornyakova *et al.*, 1973; Skornyakova, 1976a). Therefore, the regional variations of Mn, Fe and the Mn/Fe ratio in nodules are caused mostly by sediment supply of these elements.

Behavior of minor elements in nodules is controlled by sorption processes as well as by diagenetic migration. Low contents of minor elements in nodules from hemipelagic sediments are explained by the existence of stable metal-organic compounds in the reduced sediment layer (Price and Calvert, 1970). In the transitional zone, where the oxidized surface layer is several tens cm thick, considerable increases in Cu, Ni, and Co, together with Mn, are noted (Skornyakova, 1976b).

In pelagic nodules Ni, Cu, Zn and Mo correlate strongly with Mn. The correlation with Mn is explicable in terms of the high sorption capacity of hydrated MnO_2 for this group of minor elements.

Hydrated MnO_2, however, is also a sorbent for Fe^{3+}. If Fe^{3+} is adsorbed, the sorption capacity of MnO_2 for other elements sharply decreases (Alexeevskii, 1937). The Mn/Fe ratio should serve, to some extent, as an indicator of the capacity for minor metals. Plots of Cu versus Mn and Cu versus Mn/Fe (Fig. 10) show that Cu correlates with the Mn/Fe ratio rather than with Mn alone. Distribution patterns of Cu and Mn/Fe (Figs. 6 and 9) are almost identical. The highest Cu and Ni concentrations are found in the equatorial radiolarian belt, where the Mn/Fe ratio is highest. Minimum values of the Mn/Fe ratio, Cu, and Ni occur in South Pacific Basin nodules.

Table 3

Accumulation Rates of Pacific Nodules

Location	Sediment Type	Thickness of Pleistocene Sediments (cm)		Average rate of accum. in sediment (mm/1000 yr)	Nodule Abundance kg/m^2	% Mn	% Fe	Accumulation rates mg/1000 years+	
		Range	Mean					Mn	Fe
Northeast Basin (11° N) Station 5996	Marly diatom radiolarian ooze	9-20	16(8)*	< 0.1	9.7(4)*	19.12	10.1	0.18	0.098
Central Basin (11° S) Station 5988	Marly eupelagic clay	7-18	14(3)	< 0.1	25(2)	16.1	14.8	0.40	0.37
South Basin (22° S) Station 6298	"	7-18	15(5)	< 0.1	27(3)	14.3	16.7	0.38	0.45

* Parentheses indicated number of stations used for estimation.
\+ We assume nodule growth began 1 m.y. BP.

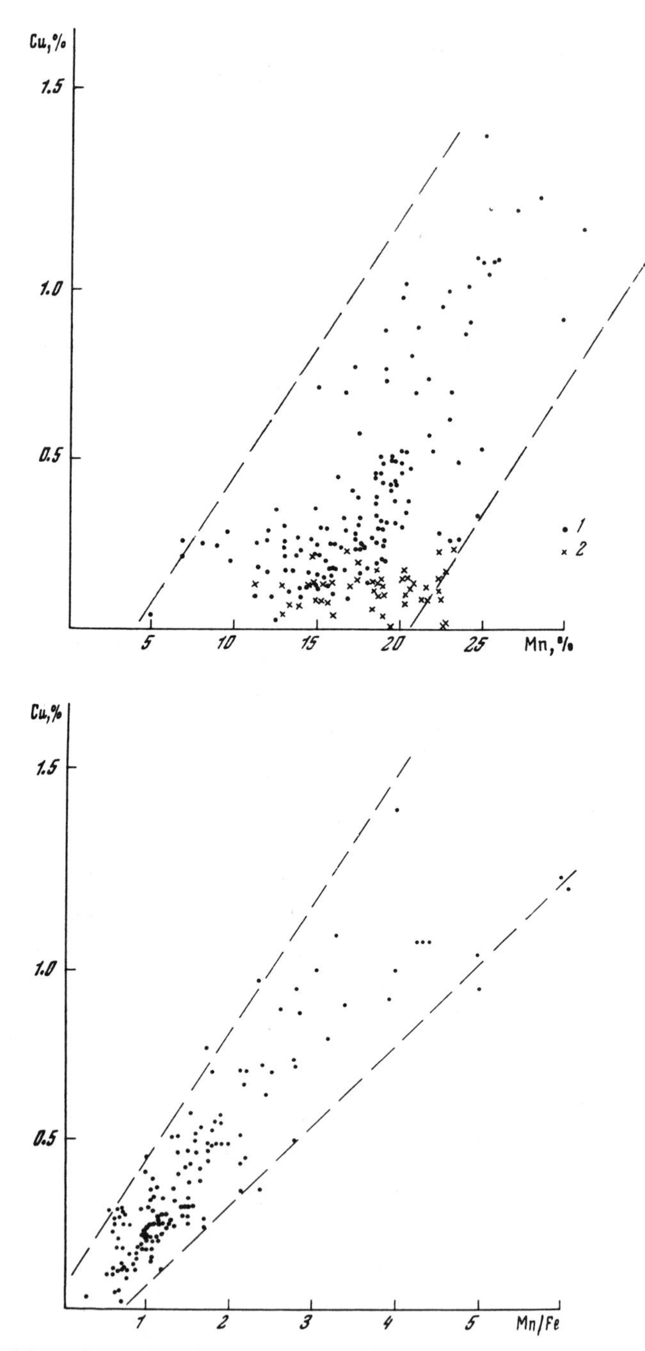

Fig. 10. Correlation between the concentrations of Cu and Mn, and between Cu and the Mn/Fe ratio. 1) nodules from ocean basins; 2) nodules from seamounts with a relative height of more than 1000 m above the sea floor.

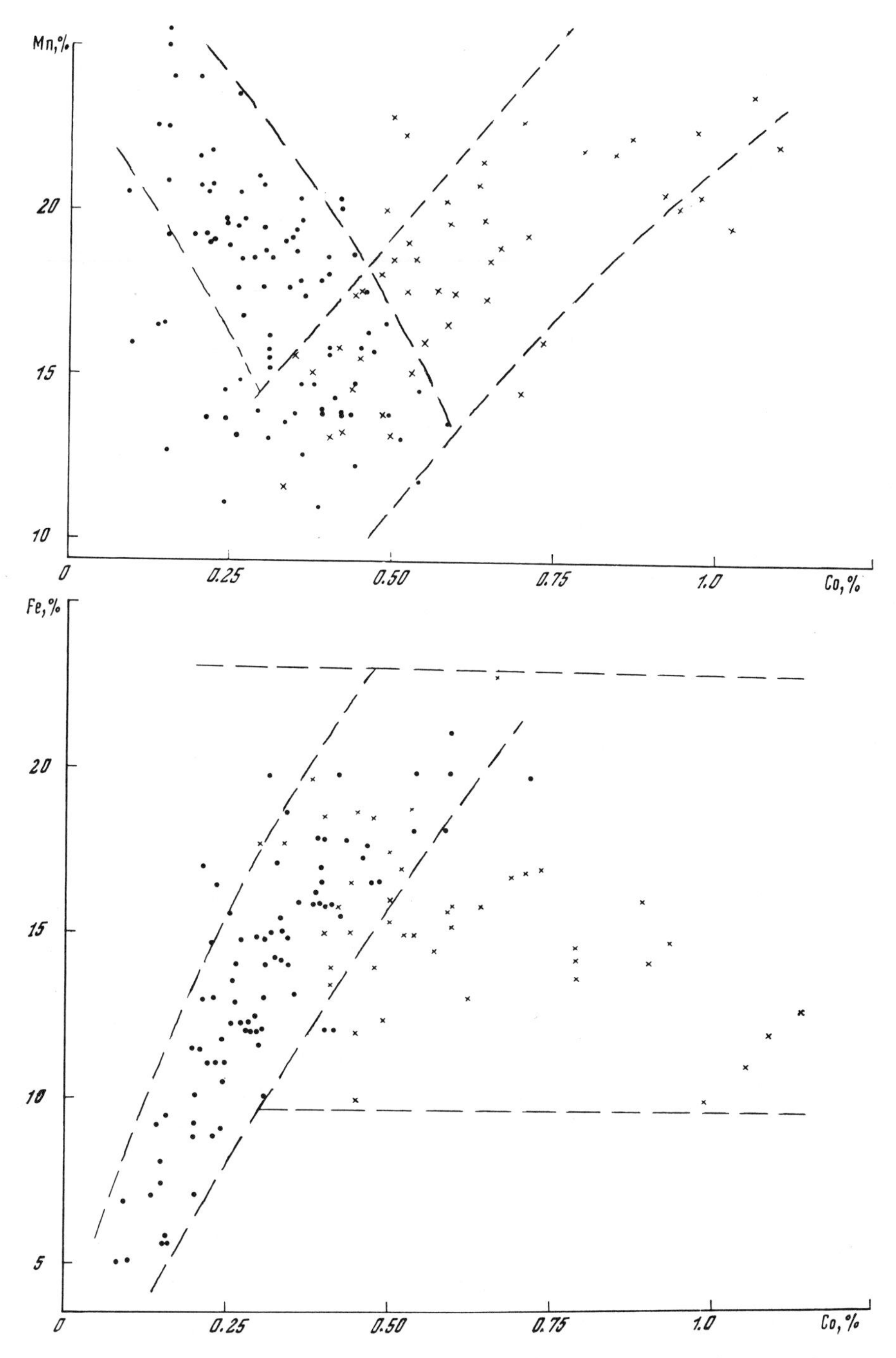

Fig. 11. Correlation between the concentration of Co and Mn, and between Co and Fe. See Figure 10 for symbols.

Table 4

Interelemental correlation coefficients for nodules from (a) Polygon 5988 from tops of abyssal hills of 200-300 m elevation (22 analyses); (b) Polygon 5996 from tops of abyssal hills of 300-500 m (20 analyses); (c) Polygon 6298 from tops of abyssal hills and seamounts 300-1800 m elevation (46 analyses); (d) mid-Pacific seamounts of 3000-5000 m elevation (56 analyses).

a. Polygon 5988

	Mn	Fe	Ti	Ni	Co	Cu	Mn/Fe
Mn	=						
Fe	-0.81	=					
Ti	-0.56	0.91	=				
Ni	0.72	-0.85	-0.90	=			
Co		0.85	0.65	(-0.40)[1)]			
Cu	0.68	-0.91	-0.89	0.84	-0.90	=	
Mn/Fe	-	-	0.90	0.89	-0.57	0.91	=
Depth	-0.58		-0.59		(-0.43)	(-0.40)	

b. Polygon 5996

	Mn	Fe	Ti	Ni	Co	Cu	Mn/Fe
Mn	=						
Fe		=					
Ti		0.61	=				
Ni	0.62	-0.65		=			
Co	0.44	0.44	0.57		=		
Cu	0.47	-0.71		0.94	(-0.43)[1)]	=	
Mn/Fe	-	-	-0.47	0.92		0.85	=
Depth					-0.56	0.50	

c. Polygon 6298

	Mn	Fe	Ti	Ni	Co	Cu	Mn/Fe
Mn	=						
Fe		-					
Ti		0.66					
Ni	0.43	-0.52	-0.48	=			
Co	0.44	0.34	0.37		=		
Cu		-0.54	-0.47	0.57	(-0.29)[1)]	=	
Mn/Fe	-	-	-0.53	0.49		0.47	=
Depth		-0.45			(-0.27)	0.43	

d. Mid-Pacific Seamounts

	Mn	Fe	Ti	Ni	Co	Cu	Mn/Fe
Mn	=						
Fe	-0.45	=					
Ti	0.28		=				
Ni	0.55	-0.71	-0.66	=			
Co	0.51				=		
Cu		-0.53	-0.55	0.68	-0.58	=	
Mn/Fe	-	-	-0.51	0.78		0.41	=
Depth		-0.43	-0.42	0.36	-0.78	0.73	

Alternatively, the interrelation between Cu and Ni content with the Mn/Fe ratio may be interpreted as evidence of diagenetic mobility of these elements during nodule formation (Calvert and Price, 1977). The assumption is supported by sharp variations in their concentrations in radiolarian ooze, as well as by decreasing trends of Cu and Ni in surface sediments in areas of increasing trends of these elements in nodules (Greenslate et al., 1973; Skornyakova, 1976b).

Variations of minor elements also are related to water depth (Menard, 1964; Cronan and Tooms, 1969), or more correctly, to the relative height of seamounts (Skornyakova et al., 1975). Co and Pb content appear to increase and Cu to decrease, with increasing height of seamounts. Maximum Co and Pb and minimum Cu concentrations were found in nodules from the tops of seamounts 3000 to 4000 m above surrounding ocean floor (water depth less than 1500 m). At the same depth (less than 1500 m) the Mn content, Mn/Fe ratio and Ni content increase (Piper and Williamson, 1977).

According to data from Stations 5996, 5988, 5968, Co is proportional to Fe in nodules from the tops of abyssal hills which are less than 1000 m above the ocean floor. At shallower depths Co is proportional to Mn. The association of Co with both Mn and Fe under different conditions is confirmed by correlation coefficients of these elements in nodules from polygons (Table 4) and by plots of Co-Mn and Co-Fe (Fig. 11).

Variations of minor elements within individual pelagic nodules have been interpreted in terms of mineralogical variations (Barnes, 1967; Cronan and Tooms, 1969). Thus, Co and Pb in seamount nodules may be associated predominantly with δ-MnO_2, whereas Ni and Cu are associated with todorokite. Investigations of numerous nodules from the Pacific and Indian Oceans, however, showed no systematic changes in mineralogy with depth (Glasby, 1972; Bezrukov, and Andruschenko, 1973). Also, considerable variations of minor elements appear in mineralogically similar nodules (Piper and Williamson, 1977).

The regular change in geochemistry of nodules with depth, i.e., vertical zonality of nodules, is rather strong, but no single explanation adequately accounts for it. Special investigations are necessary to solve this interesting problem.

References

Agassiz, A., (1902) Reports on the scientific results of the expedition to the Tropical Pacific of the steamer ALBATROSS. Mem. Museum of Compar. Zool. 26, 1-108.

Alexeevskii, E.V., (1937) Active manganese dioxide (in Russian), ONTY, Leningrad 83.

Barnes, S.S., (1967) Minor element composition of ferromanganese nodules, Science 157, 63-65.

Bezrukov, P.L., (1962) Some problems of sedimentation zonality in the World Ocean (in Russian), Trudy Okeanogr. Komissii 10, 3-9.
Bezrukov, P.L., (1970) General features of sedimentation in the Pacific Ocean (in Russian), Nauka, Moscow, 301-321.
Bezrukov, P.L., (1976) Stratigraphic position and age of ferromanganese nodules (in Russian), In Zhelyezo-Margantseviye Koncretsii Tikhovo Okeana, Nauka, Moscow, 82-90.
Bezrukov, P.L., and Andruschenko, P.F., (1973) On the geochemistry of ferromanganese nodules of Indian Ocean (in Russian), Izvestija Akad. Nauk USSR, Ser. Geol. 9-18-37.
Bonatti, E., Fisher, D.E., Joensuu, O., and Rydell, H.S., (1971) Post depositional mobility of some transition elements, phosphorus, uranium, and thorium in deep-sea sediments, Geochim. Cosmochim. Acta 25, 25-38.
Butusova, G. Yu, Lisizina, N.A., Volkov, I.I., and Lubchenko, I. Yu, (1975) The signs of exhalative activity in the bottom sediments to the south from the Gulf of California (in Russian), Litol. i Polyezn. Iskop. 6, 3-11.
Butusova, G. Yu, Lisizina, N.A., Gradusov, B.P., and Dmitrik, A. L., (1976) Authigenic montmorillonite in bottom sediments (st. 655) to the south of the Gulf of California (in Russian), Dokl. Akad. Nauk USSR 231, 192-195.
Burns, R.G., (1965) Formation of cobalt-III in the amorphous FeOOH.nH_2O phase of manganese nodules, Nature 205, 999 p.
Burns, R.G., and Fyfe, W.S., (1967) Crystal-field theory and the geochemistry of transition elements, In Researches in Geochemistry, Wiley, New York 2, 259-285.
Calvert, S. E., and Price, N.B., (1977) Geochemical variation in ferromanganese nodules and associated sediment from the Pacific Ocean, Mar. Chem. 5, 43-74.
Chester, R., and Hughes, M.S., (1967) A chemical technique for the separation of ferro-manganese minerals, carbonate minerals and adsorbed trace elements from pelagic sediments, Chem. Geol. 2, 639-654.
Cronan, D.S., and Tooms, J.S., (1969) The geochemistry of manganese nodules and associated pelagic deposit from the Pacific and Indian Oceans, Deep Sea Res. 16, 355-359.
Glagoleva, M.A., (1972) Alteration regularities in the chemical composition of iron-manganese concretions in the sediments of the Pacific Ocean (in Russian), Litol. i Polyezn. Iskop., 4, 40-49.
Glagoleva, M.A., Volkov, I.I., Sokolov, V.C., and Yagodinskaya, T., (1975) Chemical elements in Pacific sediments on the transect from Hawaiian Islands to the Mexican coast (in Russian), Litol. i Polyezn. Iskop. 9, 16-28.
Glasby, G.P., (1972) The mineralogy of manganese nodules from a range of marine environments, Mar. Geol. 2, 57-72.
Goldberg, E.D., Koide, M., (1962) Ionium-thorium chronology in deep-sea sediments of the Pacific, Science 128, 417-450.

Goldberg, E.D., (1961) Chemistry in the oceans, In Sears, M. (ed.) Oceanography, Am. Assoc. Adv. Sci. Publ. 67, 538-597.
Goodell, H.G., Meylan, M.A., and Grant, B., (1971) Ferromanganese deposits of the South Pacific Ocean, Drake Passage and Scotia Sea, In Reid, J.L. (ed.) Antarctic Oceanology I, Antarct. Res. Ser., Am. Geophys. Union, Baltimore, 27-92.
Greenslate, I.L., Frazer, I.Z., and Arrhenius, G., (1973) Origin and deposition of selected transition elements in the sea bed, In Morgenstein, M. (ed.), Origin and Distribution of Manganese Nodules in the Pacific and Prospects for Exploration, Honolulu, Hawaii, 45-60.
Hays, J.D., Saito, T., Opdyke, N.D., and Burckle, L.H., (1969) Pliocene-Pleistocene sediments of the equatorial Pacific: their paleomagnetic, biostratigraphic and climatic record, Geol. Soc. Amer. Bull. 80, 1481-1514.
Hayes, D.E., and Conolly, S.R., (1972) Morphology of the southern Indian Ocean, Antarctic Oceanology II, Amer. Geophys. Union.
Heezen, B.C., and Hollister, C.D., (1971) The Face of the Deep, Oxford Univ. Press, New York.
Horn, D.R., Horn, B.M., and Delach, M.N., (1973) Copper and nickel content of ocean ferromanganese deposits and their relation properties of the substrate, In Morgenstein, M. (ed.), papers on the Origin and Distribution of Manganese Nodules in the Pacific and Prospects for Exploration, Honolulu, Hawaii, 77-83.
Jouse, A.P., (1968) Ancient diatoms and diatomaceous rocks of the Pacific Ocean Basin (in Russian) Litol. i Polyeżn. Iskop.1, 16-32.
Lizitzin, A.P., (1972) Sedimentation in the World Ocean, Soc. of Econ. Paleont. and Mineral., Special Publ., Tulsa, Oklahoma 17, 218.
Lizitzin, A.O., (1974) Sedimentation in the Ocean (in Russian), Nauka, Moscow, 435.
Lizitzin, AJ ., Bogdanov, Yu. A., Murdmaa, I.O., Serova, V.V., Zverinskaya, I.B., Lebedev, A.I., Lukashin, V.N., and Gordeev, V.V., (1976) Metalliferous sediments of the Pacific Ocean and their genesis (in Russian), In Okeanologicheskiye Isledovaniya 29, Nauka, Moscow, 289-379.
Manheim, F.T., (1965) Manganese-iron accumulations in the shallow water environment, In Symposium on Marine Geochemistry, Occas. Publ. Univ. Rhode Island 3, 217-278.
Menard, H.W., (1964) Marine Geology of the Pacific, McGraw-Hill, 260.
Meylan, M.A., (1968) The mineralogy and geochemistry of manganese nodules from the Southern ocean, Sediment. Res. Lab., Dept. Geology, Florida State Univ., Tallahassee, Contr. 22.
Mero, J.L., (1962) Ocean-floor manganese nodules, Econ. Geol. 57, 747-767.
Mero, J.L., (1965) The Mineral Resources of the Sea, Elsevier, Amsterdam, 312.

Murdmaa, I.O., Skornyakova, N.S., and Agapova, G.V., (1976) Facial environments of ferromanganese nodule distribution in the Pacific (in Russian) In Bezrukov, P.L. (ed.) Zhelyezo-Margantseviye Koncretsii Tikhovo Okeana , Nauka, Moscow, 7-36.
Payne, R.R., and Conolly, J.R., (1972) Pleistocene manganese pavement production: it's relation to the origin of manganese in the Tasman Sea, In Horn, D. (ed.), Ferromanganese Deposits of the Ocean Floor, Washington, D.C., 81-92.
Piper, D.Z., and Williamson, M.E., (1977) Composition of Pacific Ocean ferromanganese nodules, Mar. Geol. 23, 285-303.
Price, N.B., and Calvert, S.E., (1970) Compositional variation in Pacific Ocean ferromanganese nodules and its relationship to sediment accumulation rates, Mar. Geol. 9, 145-171.
Riedel, W.R., and Funnell, B.M., (1964) Tertiary sediment cores and microfossils from the Pacific Ocean floor, Quart. J. Geol. Soc, 120, London, 305-368.
Romankevich, E.A., (1977) Geochemistry of Organic Matter in the Ocean (in Russian), Nauka, 234.
Rozanov, A.G., Sokolov, V.S., and Volkov, I.I., (1972) Iron and manganese forms in the sediments of northwestern part of the Pacific (in Russian), Litol. i Polyezn. Iskop. 4, 25-32.
Rozanov, A.G., Volkov, I.I., Sokolov, V.C., Puschkina, Z.V., and Pilipchuk, M.F., (1976) The processes of oxidation-reduction in the sediments of the Gulf of California and adjacent part of Pacific (combinations of iron and manganese) (in Russian), In Biogeokhimiya Diageneza Osadkov Okeana , Nauka, Moscow, 96-135.
Saidova, Kh. M., (1974) On large-scale facies preferment of the deep-sea bottom foraminifers (in Russian), Okeanologiya 14 (4), 665-671.
Skornyakova, N.S., (1960) Manganese concretions in sediments of the northeastern Pacific Ocean (in Russian), Dokl. Akad., Nauka, USSR 130, 653-656.
Skornyakova, N.S., (1965) Dispersed iron and manganese in Pacific sediments, Int. Geol. Rev.7, 2161-2174.
Skornyakova, N.S., (1976a) Chemical composition of ferromanganese nodules of the Pacific (in Russian), In Bezrukov, P.L. (ed.), Zhelyezo-Margantseviye Koncretsii Tikhovo Okeana , Nauka, Moscow, 190-241.
Skornyakova, N.S., (1976b) Dispersed Fe, Mn, Ti and some minor elements in the Pacific sediments (in Russian), In Bezrukov, P.L. (ed.), Zhelyezo-Margantseviye Koncretsii Tikhovo Okeana , Nauka, Moscow, 168-189.
Skornyakova, N.S., and Andruschenko, P.F., (1964) Ferromanganese nodules of the Pacific Ocean (in Russian), Litol. i Polyezn. Iskop. 5, 21-36.
Skornyakova, N.S., and Andruschenko, P.F., (1968) Ferromanganese nodules in central area of southern part of Pacific (in Russian), Okeanologiya 8 (5), 865-877.

Skornyakova, N.S., and Andruschenko, P.F., (1970) Ferromanganese nodules in the Pacific (in Russian), In Bezrukov, P.L. (ed.), Pacific Ocean, Osadkoobrazovaniye v Tikhom Okeanye , Nauka, Moscow, 202-268.

Skornyakova, N.S., and Andruschenko, P.F., (1976) Morphology and inner structure of ferromanganese nodules (in Russian), In Bezrukov, P.L. (ed.), Zhelyezo-Margantseviye Koncretsii Tikhovo Okeana , Nauka, Moscow, 91-122.

Skornyakova, N.S., Bazilevskaya, E.S., and Gordeev, V.V., (1975) Some questions of mineralogy and geochemistry of Pacific ferromanganese nodules (in Russian), Geokhimiya 7, 1064-1076.

Skornyakova, N.S., and Zenkevich, N.L., (1961) The distribution of ferromanganese concretions in the surface of sedimentary layers of the Pacific Ocean (in Russian), Okeanologiya 1 (1), 86-94.

Skornyakova, N.S., and Zenkevich, N.L., (1976) Regularities of the space distribution of ferromanganese nodules (in Russian), In Bezrukov, P.L. (ed.) Zhelyezo-Margantseviye Koncretsii Tikhovo Okeana , Nauka, Moscow, 37-81.

Skornyakova, N.S., and Murdmaa, I.O., (1968) Lithofacial types of deep-sea pelagic (red) clay Pacific Ocean (in Russian), Litol. i Polyezn Iskop. 6, 17-37.

Skornyakova, N.S., Murdmaa, I.O., Gorbunova, Z.N., and Zenkevich, N.L., (1971) On the facial variations of pelagic sediments of the Pacific Ocean (in Russian), In Istoriya Mirovovo Okeana Nauka, Moscow, 148-173.

Skornyakova, N.S., Murdmaa, I.O., Prokoptsev, N.G., and Marakuev, V.I., (1973) Bottom sediments and volcanic rocks of a polygon in the Southern Basin of the Pacific Ocean (in Russian), Litol. i Polyezn. Iskop. 1, 17-28.

Volkov, I.I., Fomina, L.S., and Yagodinskaya, T.A., (1976) Chemical composition of Pacific ferromanganese nodules of the transect from Wike atoll to the coast of Mexico (in Russian), In Biogeokhimiya Diageneza Osadkov Okeana , Nauka, Moscow, 186-204.

Watkins, N. and Kennett, J.P., (1972) Regional sedimentary disconformities and Upper Cenozoic changes in bottom water velocities between Australasia and Antarctica, Antarctic Res. Ser. 19.

CHANGES IN CHEMICAL COMPOSITION OF SOME PACIFIC MANGANESE NODULES DURING THEIR GROWTH

V. Marchig and H. Gundlach

Federal Institute for Geosciences
and Natural Resources
P.O. Box 51 01 53, D-3000 Hannover, Germany

Abstract

Manganese nodules were sampled from 133 stations in the Central Pacific during a cruise of the German research vessel Valdivia in 1976. The abundance of nodules on the sea floor of the area was calculated. A change in the abundance of nodules with water depth was found: on the slopes of the seamounts the population density varies from 0 to 7.3 kg/m^2, in the abyssal plains from 7.3 to 32 kg/m^2.

Si, Al, Mn, Fe, Ni, Cu, Co, Zn, Pb, Mg, and Ca were analysed for 76 nodules from 21 abyssal plain stations. Besides bulk analyses, analyses of different zones from the nodules were made.

The centers of the nodules are formed by the fragments of disintegrated older nodules or crusts which were recemented in dislocated positions together with some stone fragments.

There are differences in the chemical composition within the laminated layer around the center. The water side and the sediment side have similar chemical compositions. The equatorial zone, however, has elevated Ni and Cu contents and relatively low Fe and Co contents. We suggest that the reason for this enrichment is the growth of this part of the nodule layer within the sediment-water interface layer.

The loose outer layer of nodules has elevated Mn, Cu, Ni- and Zn contens. The comparison of this layer with micronodules from the surrounding sediment indicates recent deposition of diagenetically mobilized metals.

Introduction

The growth process of manganese nodules is not yet understood. In order to contribute to an understanding of this process, samples were collected during cruise 13/1 of the German research vessel Valdivia in 1976. Samples were taken from an area of 0.20' latitudinal side located at 9°05' N, 148°45'W (Fig. 1). Sediments, sea water, pore water, and manganese nodules were sampled. Only the investigations of the manganese nodules are dealt with in this paper.

Water depth within the area varies between 4310 and 5300 m. Three seamounts within the area have elevations of approximately 600 - 1000 m above an abyssal plain, which has an average water depth of 5100 - 5200 m. Seismic records show a thick sediment cover on the abyssal plain, growing thinner towards the seamount slope (Dürbaum and Schlüter 1974). Fig. 2 shows a W-E echogram across the central seamount of the area and a seismic record of sediment thickness above the basement towards the seamount. The seamounts are covered with thin sediment cover with occasional outcrops of rock (see Beiersdorf 1976 in cruise report).

The sediment in this area is a radiolarian ooze or radiolarian clay with a "peneliquid" sediment-water interface layer of few cm in the abyssal plains (Halbach et al 1976, cruise report). This "peneliquid" layer, characterized by very low shear strength values, is not present on the slopes of the seamounts.

Sampling and description of the manganese nodules

Nodules were sampled by box corer and by free fall grab at 133 stations in the investigation area. Nodule abundance, based on data from all 133 stations, varies with water depth: the deeper the water the more nodules per unit area (Fig. 3).

Stations with lower abundances (< 7.3 kg/m²) were in the area of high relief, i.e. the slopes of the central seamount ("Gundlach Höhe"). These areas have a thin sediment cover with rock outcrops. The stations with high abundances (7.3 kg/m² to 32 kg/m²) were in the plains areas, which have a thick sediment cover. The average abundance for the whole area of investigation is 9.5 kg/m².

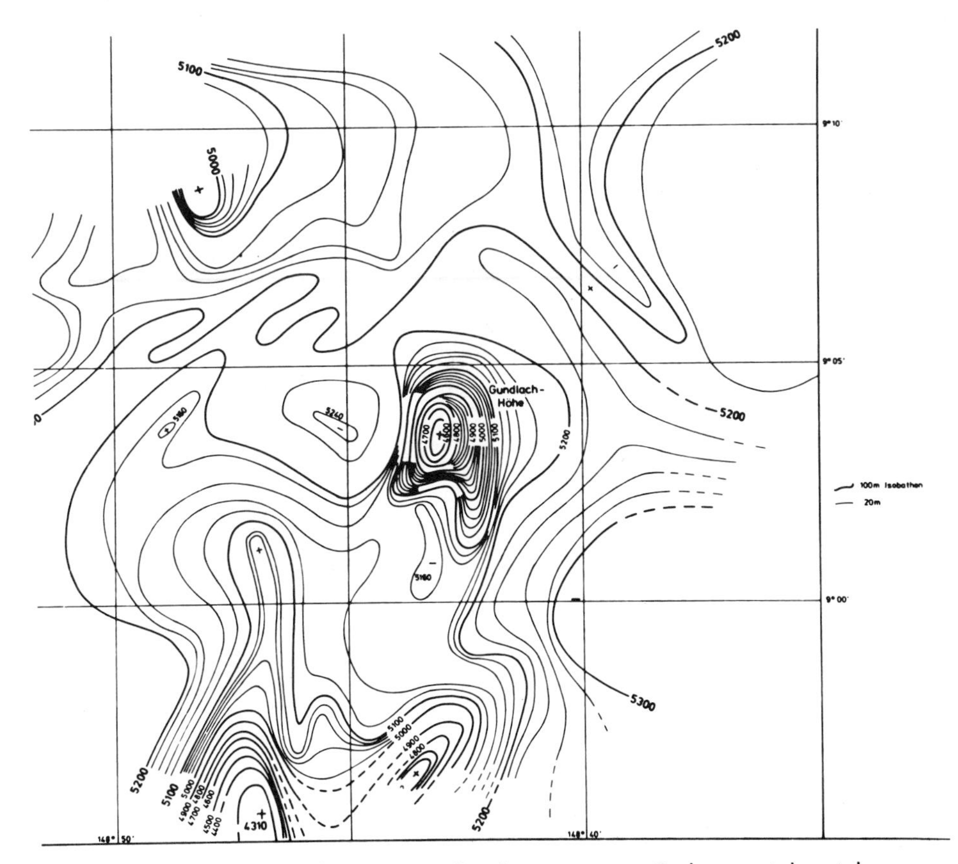

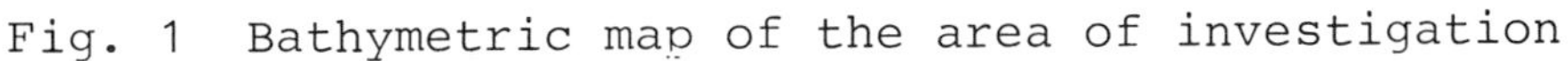
Fig. 1 Bathymetric map of the area of investigation

Nodules from the higher parts are small, have smooth surfaces, are poly-nucleated and have high Fe/Mn ratio. Nodules from the abyssal plain have a higher population density, they are usually mononodules with average diameters approximately 3 times that of the small nodule population, and have a lower Fe/Mn ratio than the polynodules from the higher areas (Halbach et al. 1976, cruise report). These two different types of nodules found in the described cruise area also show significant differences in the growth rates (Heye 1979). Nodules from the top and slopes of the seamounts have grown uniformly slowly with rates of a few millimeters in a million years, whereas nodules from the surrounding abyssal plain exibit phases of rapid growth. The latter grow at an average of three times as fast as those from the seamounts.

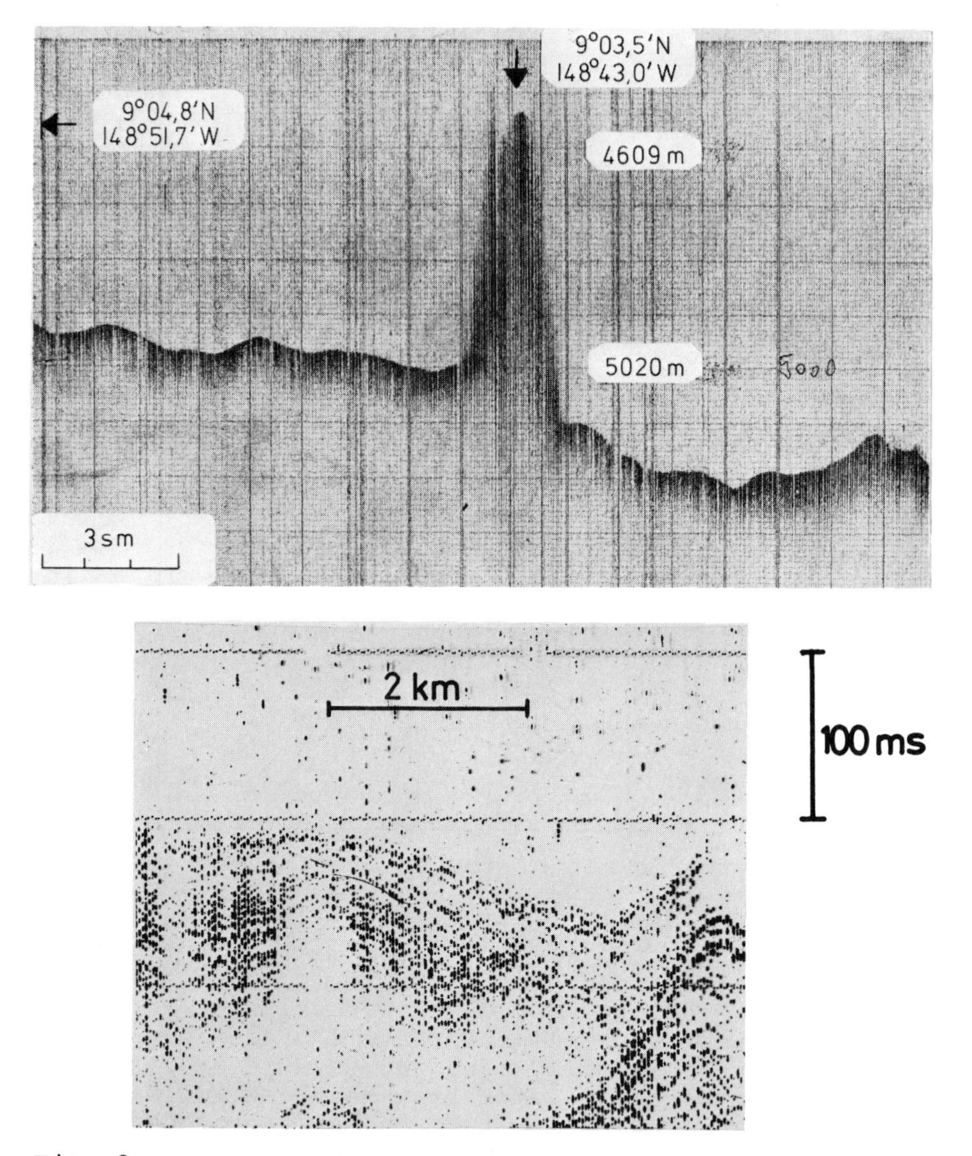

Fig. 2

W-E bathymetric profile across the central seamount in the investigation area (from Beiersdorf 1976, cruise report) and the seismic record of part of the same profile from the W to the seamount). The sediment above the basement grows thinner and vanishes toward the seamount (figure: courtesy of H. Schlüter).

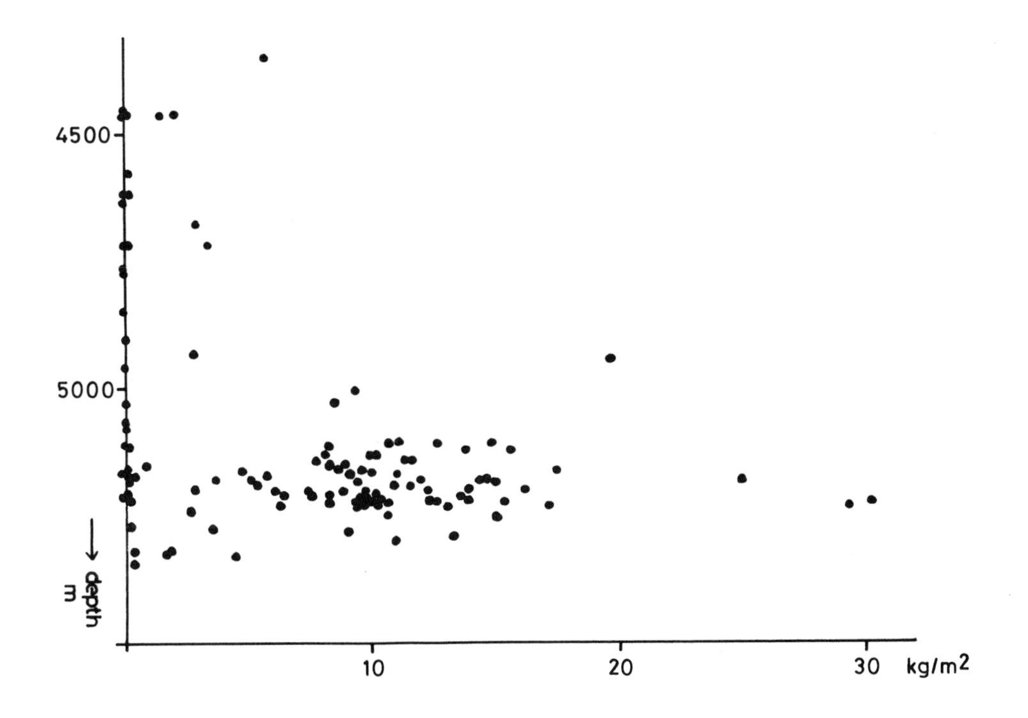

Fig. 3

The population density of manganese nodules on the sediment surface in the area of investigation plotted against the water depth.

Box corers could be used only in the area of abyssal plain; in most cases short box corers brought on board ship a sediment column of about 40 cm and a very well preserved sediment surface of 30 x 40 cm with the nodules in their original position. Thus, only nodules from the plains were analysed for this study, i.e. large mono-nodules with Mn content and higher growth rates.

The upper surface of these nodules is smooth where they were in contact with sea water. On the lower part, where they were covered with sediment (bottom and equatorial part), they have a rough, botryoidal surface. This rough surface is covered with small growth structures, which are more frequent around the equatorial part than on the bottom of the nodules. In this work we call these small growth structures "loose outer layer".

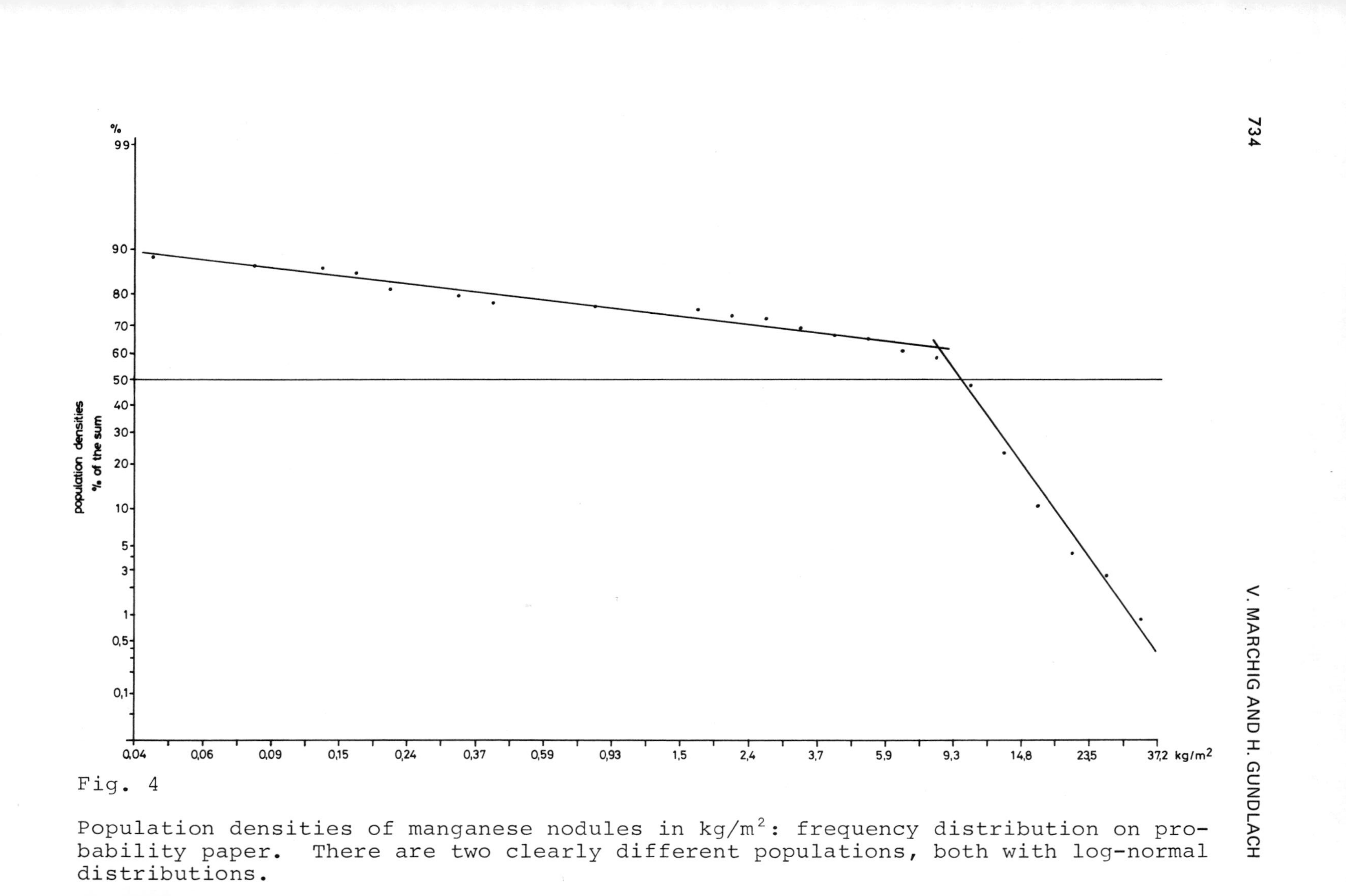

Fig. 4

Population densities of manganese nodules in kg/m^2: frequency distribution on probability paper. There are two clearly different populations, both with log-normal distributions.

Three distinct growth zones could be distinguished in all nodules from the sediment surface (see Fig. 5):

1. the center - fragments of manganese nodules or crusts recemented in dislocated positions together with some stone fragments;

2. a laminated zone - the nodule material surrounding the inner part is laminated. This laminated zone has nearly the same thickness around the entire nodule; it is only about 20 % thicker around the equatorial part than on the upper and lower parts;

3. a loose outer layer - the small growth structures growing only on those parts of the laminated zone in contact with sediment.
   This zone is the last growth phase of manganese nodules.

Nodules were also found at depths of 25 to 40 cm under the sediment surface in two box cores. These buried manganese nodules had an additional zone of highly variable thickness around the laminated zone. This zone was extremely porous. The presence of this additional zone in buried nodules indicated that they can continue growing after being buried (Heye et al. in press).

Experimental procedures

The small growth structures forming a loose outer layer on the lower and equatorial surface of the nodules were abraded with a wooden spatula for separate analysis. The abraded nodules were soaked in araldite before being sawed; nodules without araldite treatment crumble during sawing. A thin slice was sawed from the middle of each manganese nodule vertical to its position on the sea floor (see Fig. 5). These discs were used for bulk chemical analysis of the nodule.

Two nearly complete halves remained from each nodule. From one of these halves, four samples were bored from the laminated zone: one from the upper part of the nodule, one from the lower part of the nodule, and two from equatorial part (see Fig. 5). The upper part of the nodule is in contact with sea water, the equatorial part and the lower part of the nodule are in contact with sediment. Assuming that the nodules had not changed their position, the upper part of laminated zone was in contact with sea water and the equatorial and lower part of the laminated zone were in contact with sediment during

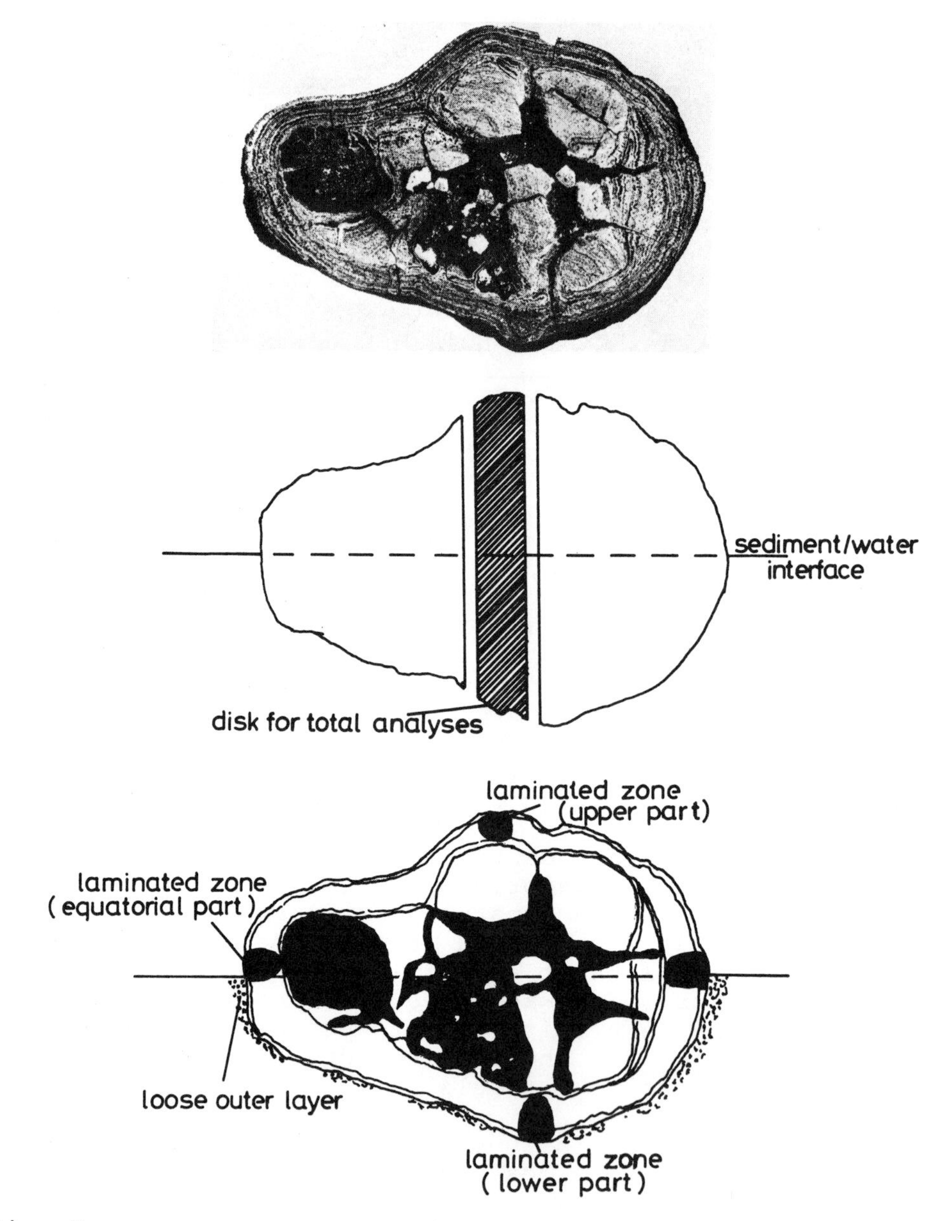

Fig. 5
Polished cross-section of a typical nodule. The nodule is cut vertically to its original position on the sea floor. (2 x magnification). The drawing of a manganese nodule shows the places where samples were taken for analysis.

the time of growth of laminated zone.

The material obtained this way (6 samples from each nodule) was dissolved in 1 : 1 diluted HCl and analysed by atomic adsorption for the elements Mn, Ni, Cu, Co, Zn, Pb, Ca, Mg, Fe, and Al. Si was determined gravimetrically on the larger samples.

Results

The results of the chemical analyses are shown in the appendix. The average values and standard deviations for the analyses for different parts of the nodules are shown in Fig. 6. The analyses were grouped as follows:

1. micronodules separated from the top 5 cm of the sediment (from Marchig and Gundlach, 1978);

2. the loose outer layer in contact with sediment (material abraded with a wooden spatula);

3. material from the equatorial part of the laminated zone of the nodules (contact with sediment during the growth);

4. material from the upper part of the laminated zone of the nodules (contact with sea water);

5. material from the bottom part of the laminated zone of the nodules (contact with sediment during the growth);

6. the thin sections from the middle of each nodule from which total analyses for the entire nodule were obtained.

The concentrations of Mn, Ni, Cu, Zn, Mg, Co, and Ca in the total analyses lie between the values for the various parts of the nodules. The concentrations of Fe and Al in these analyses, however, are higher than in the different parts of the nodules due to the presence of stone fragments in the central part of the nodules. These stone fragments have higher Fe and Al contents than the rest of the nodule.

The near-surface micronodules and the loose outer layer of the nodules show great similarities in their chemical composition. Their chemical composition is also considerably different from the laminated material of the nodules. They have elevated concentrations of Mn, Ni, Cu, Mg, and

Zn, and relatively low concentrations of Fe.

There are, however, some chemical differences between the loose outer layer of nodules and the near-surface micronodules: the micronodules are even more enriched in Mn, Cu, Mg, and Zn.

Within the laminated zone of the nodules, the upper and lower sides have the same chemical composition. The equatorial part shows differences in the concentration of a few elements: Ni and Cu have elevated concentrations and Fe and Co have lower concentrations compared with the upper and lower sides of the laminated zone. Differences in Mn, Zn, Mg, Al, and Ca within the laminated zone are not statistically significant.

Discussion

Nodules were found buried at 25, 33, and 40 cm beneath the sediment surface in two box cores from the investigation area (Heye et al. in press). These buried nodules differed from surface nodules by having an additional outer zone of high porosity containing frequent sediment inclusions. This additional zone had high α-activity, showing that it has grown recently. The chemical composition of this zone shows enrichment in Cu and Mn and depletion in Fe and Co. These observations suggest that mobilisation within the sediment can supply sufficient metals, especially diagenetically mobile metals such as Mn, to produce the precipitation of manganese nodule matter from the pore water.

Our analysis of manganese micronodules (Marchig and Gundlach, 1978) brought us to the same conclusion. The population of micronodules decreases with depth in the sediment within the first 35 cm. Only 5 % of the amount of micronodules at the sediment surface was found at 35 cm. In addition the Mn, Ni, Cu, and Zn contents of the micronodules decrease with depth and the Fe content increases.

These observations indicate partial dissolution and remobilization of microndoules buried under the "sediment-water interface boundary layer" as Sorem et al. (1978) defined it. Mn, as the more soluble component, and the ions incorporated in or adsorbed on manganese hydroxide are preferentially remobilized; Fe, Si, and Al, being less soluble than Mn, become more concentrated in remaining micronodules.

The reprecipitation of the mobilized material can take place in the sediment-water interface boundary layer in two ways:

1. precipitation as micronodules

2. precipitation as loose outer layer on that part of nodule surface which is in contact with the sediment.

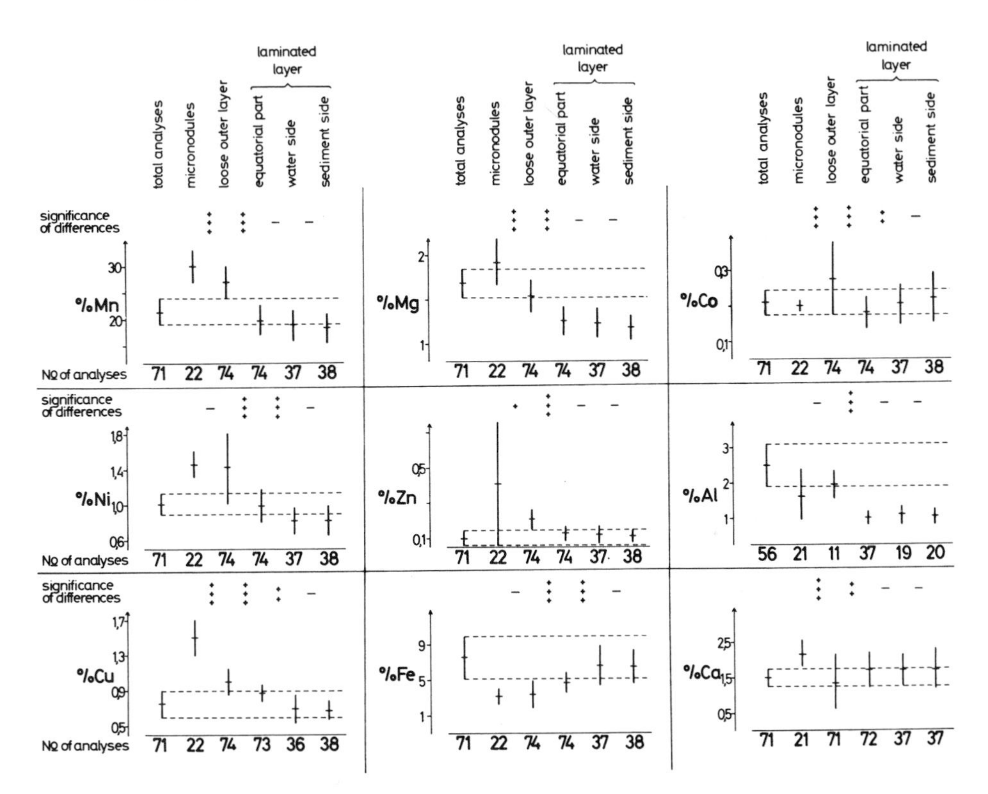

Fig. 6

Bulk chemical composition of nodules, of different parts of the nodules, and of associated micronodules. The lenghts of the vertical lines give standard deviation. Arithmetic mean is marked by a short horizontal line. The significance of differences between average values is indicated as follows:

-- difference is not significant (> 5 %)
* difference is weakly significant (P < 5 %)
** difference is significant (P < 1 %)
*** difference is highly significant (P < 0.1 %)

This mechanism would account for the similarity in composition of the near-surface micronodules and the loose outer layer abraded from the lower surface of the macronodules (Fig. 6).

Table 1

Correlation matrix for the chemical composition of total manganese nodules. Only the correlation coefficients are shown which are with more than 99 % probability different from zero. See Fig. 5 for the number of chemical analyses used for the calculation of correlation coefficients.

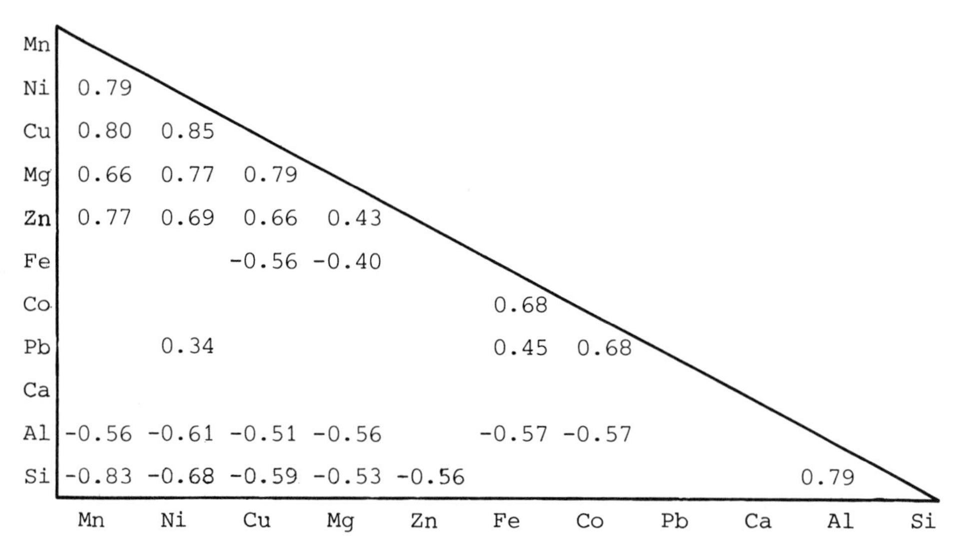

	Mn	Ni	Cu	Mg	Zn	Fe	Co	Pb	Ca	Al	Si
Mn											
Ni	0.79										
Cu	0.80	0.85									
Mg	0.66	0.77	0.79								
Zn	0.77	0.69	0.66	0.43							
Fe			-0.56	-0.40							
Co						0.68					
Pb		0.34				0.45	0.68				
Ca											
Al	-0.56	-0.61	-0.51	-0.56		-0.57	-0.57				
Si	-0.83	-0.68	-0.59	-0.53	-0.56					0.79	

Manganese nodules in this work, therefore, seem to be made from two different sources:

from pore water -- in this case the diagenetically mobile elements such as Mn, Ni, Cu, Mg, and Zn are enriched in the precipitate;

from sea water -- in this case the most insoluble components such as Fe should have elevated concentrations.

Manganese nodules growing on the basalt of seamounts are suggested to be products of precipitation only from sea water. They have higher Fe and Co contents and lower Mn, Ni, and Cu contents than the average nodules from the area (Halbach et al. 1976, cruise report).

The recently deposited "loose outer layer" of nodules as discovered in this work is suggested to be a purely diagenetical product, formed by precipitation from pore water. This accounts for the similarity of composition with micronodules from sediment-water interface boundary layer, which were formed on the same way. This also accounts for the difference in the chemical composition to that of the laminated layer. This laminated layer of the nodules also has differences in chemical composition within the layer itself. The chemical composition is the same in the parts of the layer on the upper and lower sides of the nodules (see Fig. 6). In contrast, the parts of this layer on the equatorial side of the nodules have higher concentrations of Ni and Cu and lower concentrations of Fe and Co.

Raab (1972) has investigated nodules from approximately the same area. He found differences between upper and lower sides of nodules, i.e. the water and sediment sides, respectively. These results are in accordance with the observation that the supply of matter for the growth of manganese nodules comes from two sources, sea water and pore water, and that these sources have different relative concentrations of metal ions. However, we could not observe these differences in the laminated zone of our nodules. This laminated zone must have had the same growth environment on both sides, only the equatorial part may have had a slightly different growth environment.

The appearance of the outer zone of the nodules (laminated and equally thick on all sides) suggests that this zone must have been formed under conditions radically different from those of recent growth. The composition of the material supplied through pore water and sea water should have been identical during the growth of the laminated zone. Another possibility, which is hardly imaginable for nodules of these dimensions, is that during the growth of the laminated zone the nodules were frequently turned upside down. (All of these possibilities are valid only under the assumption that the nodules stayed on the sediment/water interface during the entire time of their growth).

The different chemical composition of the equatorial part of the laminated zone can be explained by the different environment within the sediment-water interface boundary layer in which only the equatorial part of nodules is si-

Table 2

Correlation matrix for the chemical composition of micro-nodules from the sediment-water interface boundary layer (0 - 5 cm). Only the correlation coefficients are shown which are with more than 99 % probability different from zero. See Fig. 5 for the number of chemical analyses used for the calculation of correlation coefficients.

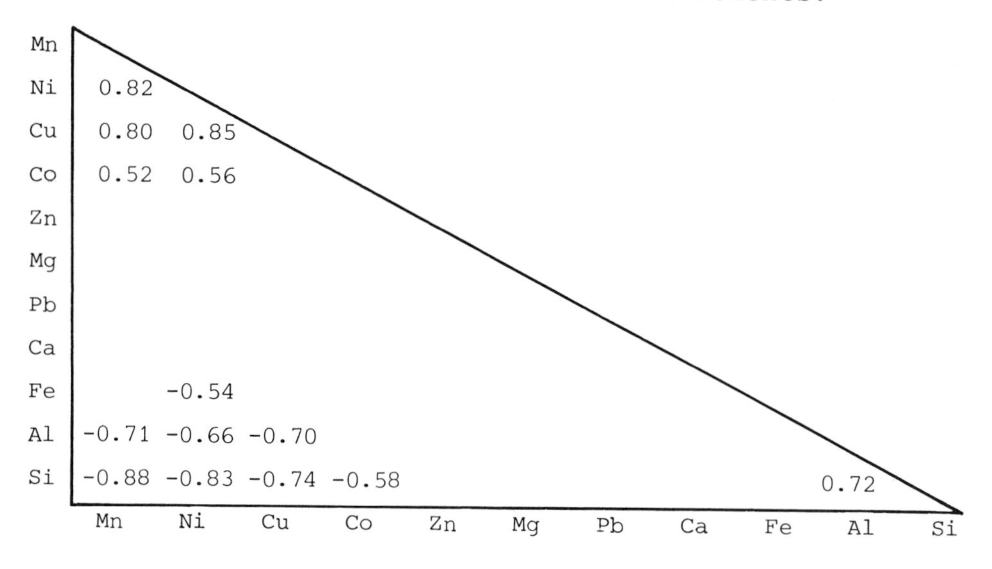

	Mn	Ni	Cu	Co	Zn	Mg	Pb	Ca	Fe	Al	Si
Mn											
Ni	0.82										
Cu	0.80	0.85									
Co	0.52	0.56									
Zn											
Mg											
Pb											
Ca											
Fe		-0.54									
Al	-0.71	-0.66	-0.70								
Si	-0.88	-0.83	-0.74	-0.58						0.72	

tuated. This boundary layer could be detected on our cruise in all the cores from the abyssal plain, due to its characteristically low shear strength values ("peneliquid" layer after Halbach et al. 1976, cruise report). It was absent on the slopes of seamounts, as expected, because of its "peneliquid" consistence. This sediment-water interface boundary layer can influence oxide deposition, promoting precipitation of manganese oxides (Sorem et al. 1978) and in this way creates optimal conditions for the growth of manganese nodules.

A possible turning of the nodules would not change the position of the equatorial zone because the flattened shape of the nodules allow them to lie on the ocean floor in only two positions.

The total analyses of manganese nodules show, as already mentioned, higher concentrations of Fe and Al than the

Table 3

Correlation matrix for the chemical composition of the "loose outer layer" of manganese nodules. Only the correlation coefficients are shown which are with more than 99 % probability different from zero. See Fig. 5 for the number of chemical analyses used for the calculation of correlation coefficients.

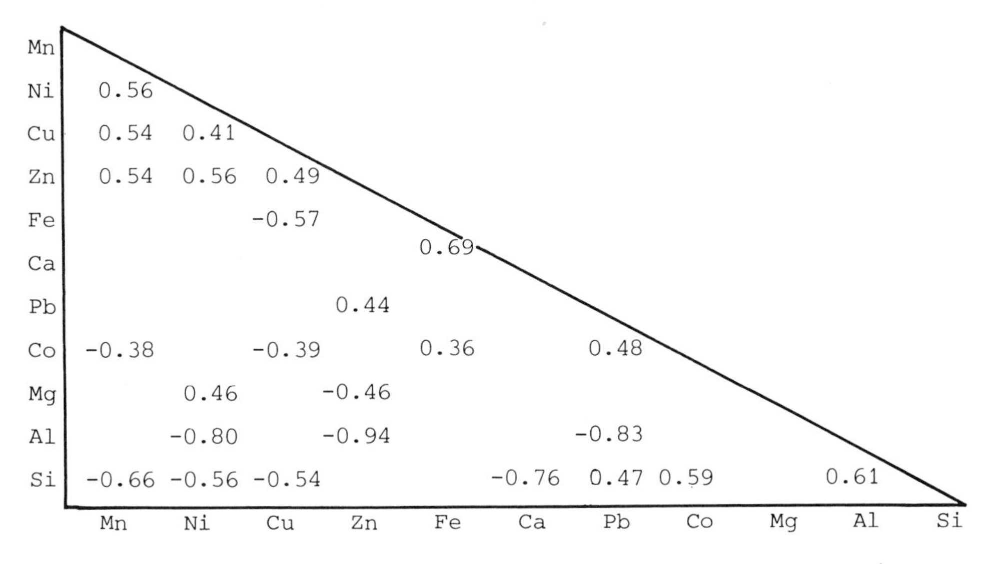

	Mn	Ni	Cu	Zn	Fe	Ca	Pb	Co	Mg	Al	Si
Mn											
Ni	0.56										
Cu	0.54	0.41									
Zn	0.54	0.56	0.49								
Fe			-0.57								
Ca					0.69						
Pb				0.44							
Co	-0.38		-0.39		0.36		0.48				
Mg		0.46		-0.46							
Al		-0.80		-0.94			-0.83				
Si	-0.66	-0.56	-0.54			-0.76	0.47	0.59		0.61	

laminated or "loose outer" layers, owing to the presence of stony fragments in the central parts of the nodules. The concentrations of Mn, Ni, Cu, Zn, Mg, Co, and Ca are, as expected, between the values for the various parts of the nodules.

The correlation matrix of total analyses shows three well defined groups (Table 1):

1. Mn - Ni - Cu - Mg - Zn
2. Fe - Co - Pb
3. Al - Si

In contrast the correlation matrixes of micronodules and of loose outer layer of the nodules show only two groups of elements (Tables 2 and 3): the first contains Mn, Ni, and Cu, and the second contains Si and Al. Co correlates with the Mn - Ni - Cu group in micronodules and Zn correlates with this group in the loose outer layer of nodules. Fe is not clearly identified with either group.

The Fe - Co - Pb group, common in manganese nodules, was not found in the micronodules and the loose outer layer of manganese nodules. We suggest two possible reasons for the absence of this group:

1. The content of Fe in this material is relatively low. Therefore, the Fe content from silicate material, which is always present in manganese nodules, is high compared with total Fe content. Thus the Fe from silicates masks any correlation of the Fe with other elements.

2. The correlation between Fe, Co, and Pb could be the result of diagenesis during the aging of the nodules and not of coprecipitation as it is the case with Mn, Cu, Ni, Mg, and Zn.

Summary

The growth of manganese nodules from the area of investigation comprises three different phases.

Little can be said about the first phase, because it ended with disintegration of earlier manganese nodules.

The dislocated pieces were recemented with manganese nodule material, together with stone fragments. The reason for the disintegration could be fracturing with aging as Heye (1975) suggested.

The second phase of growth produced a laminated zone of nearly equal thickness around the entire nodule. This zone shows enrichment of diagenetically mobile metals in the equatorial part of the nodules and not, as expected, on the sediment side of the nodules.

The third phase of growth began relatively recently and produced a loose outer layer on the lower, sediment side of the nodules. The material for this growth comes from the diagenetic remobilization which can be observed in the top 35 cm of the sediment. The recently formed loose outer layer has elevated concentrations of diagenetically mobile elements such as Mn, Ni, and Cu.

We were not able to conclude from these investigations whether diagenesis changes the chemical composition of manganese nodules or whether the nodules have retained their original chemical composition. The proportion of Mn to Cu and Ni seems to have been determined at the time of precipitation of the oxides. A correlation of Fe to Co and Pb could be observed only in older material. These proportions could, but need not be the result of diagenesis of oxide material.

References

Beiersdorf, H. (1976): Überblick über die morphologischen und geologischen Verhältnisse in den Arbeitsgebieten der Forschungsfahrt VA 13/1, in: Ergebnisse der Manganknollen - Wissenschaftsfahrt VA 13/1, Fachlicher Bericht.

Dürbaum, H.J. and H.U. Schlüter (1974): Möglichkeiten der Reflexionsseismik für Manganknollen-Exploration, Meerestechnik 5, 188 - 192.

Halbach, P., M. Özkara, E. Rehm, and J Craig (1976): Untersuchungen von Manganknollen und Sedimentproben der VA 13/1, in: Ergebnisse der Manganknollen - Wissenschaftsfahrt VA 13/1, Fachlicher Bericht.

Heye, D. (1975): Wachstumsverhältnisse von Manganknollen, Geol. Jb., E, 5, 3 - 122, Hannover.

Heye, D. (1975): Comparison of growth rates of manganese nodules from different locations of the Pacific. Tenth international congress on sedimentology Jerusalem, Abstracts, P 305 - 307.

Heye, D., V. Marchig and H. Meyer: The growth of buried manganese nodules. Deep Sea Res. (In press).

Marchig, V. and H. Gundlach (1978): Diagenetic changes in the radiolarian oozes of the Central Pacific and their influence on the growth of manganese nodules. International colloquium on the genesis of manganese nodules; Gif sur Yvette, 25 - 30 September.

Raab, W (1972): Physical and chemical features of Pacific deep sea manganese nodules and their implications to the genesis of nodules, in: Ferromanganese deposits on the ocean floor, ed: D.R. HORN, p. 31 - 50.

Sorem, R.K., R.H. Fewkes, W.D. McFarland, and W.G. Reinhart (1978): Physical aspects of the growth environment of manganese nodules in the "Horn Region", east equatorial Pacific Ocean, Proc. colloque international du C.N.R.S. No 289 sur la genese des nodules de manganese, Paris, 1978.

MANGANESE NODULE RESOURCES IN THE NORTHEASTERN EQUATORIAL PACIFIC

V.E. McKelvey, Nancy A. Wright, and Robert W. Rowland

U.S. Geological Survey

Reston, Virginia 22092

Abstract

Recent publication of maps at scale 1:1,000,000 of the northeastern equatorial Pacific region showing publicly available information on the nickel plus copper content of manganese nodules has made it possible to outline the prime area between the Clarion and Clipperton fracture zones which has been the focus of several recent scientific and commercial studies. The area, defined as that in which the nodules contain more than 1.8 percent nickel plus copper, is about 2.5 million km^2. The available evidence suggests that about half of it contains nodules in concentration (reported in wet weight units) greater than 5 kg/m^2 and averaging 11.9 kg/m^2. If we assume that 20 percent of the nodules in this area of 1.25 million km^2 are recoverable, its potential recoverable resources are about 2.1 billion dry metric tons of nodules averaging about 25 percent Mn, 1.3 percent Ni, 1.0 percent Cu, 0.22 percent Co, and 0.05 percent Mo--enough to support about 27 mining operations each producing an average of 75 million metric tons of nodules over their lifetimes. Estimates based on other plausible assumptions would be higher or lower, but of the same order of magnitude. Thus it seems probable that the magnitude of the potentially recoverable nodule resources of the Clarion-Clipperton prime area--the most promising now known--is at most in the range of several tens of the average-size operations postulated.

Introduction

Shown on Fig. 1 are the approximate limits of areas in the northeastern equatorial Pacific within which publicly available analyses indicate manganese nodules contain more than 1.8 percent nickel plus copper. The existence of high-grade nodules in this region has been known for many years. Mero (1960, p. 226) included this region as part of a broader one in which the nodules are high in copper; Menard (1964, p. 186) identified part of the region as containing nodules high in nickel. Subsequently Horn and others (1972, 1973) referred to the area between the Clarion and Clipperton fracture zones and 120° and 155° W. longitude as being unquestionably the one having the best economic potential for nodules rich in nickel and copper and showed that the rich nodules correspond in their distribution to that of siliceous sediments of Miocene age. Cronan (1972), Skornyakova (1976, p. 206-207), Margolis and Burns (1976), Bezrukov (1976, p. 252) and Piper and Williamson (1977) also roughly delineated the area as one containing nodules high in nickel and copper, and several authors (Archer, 1976, 1978; Holser, 1976; Pasho and McIntosh, 1976; Frazer, 1977; and Bastien-Thiry and others, 1977) have recently attempted to estimate the magnitude of the resources in the general region and the number of potential mine sites that it might contain. The region has been the focus of studies sponsored by the National Science Foundation in support of the International Decade of Ocean Exploration (Frazer and Arrhenius, 1972; Horn and others, 1972, 1973), by the Federal Republic of Germany's ship "Valdivia" (Schultze-Westrum, 1973; Friedrich and others, 1973), a French group (Bastien-Thiry and others 1977), and the Deep Ocean Mining Environmental Study (DOMES) sponsored by the National Oceanic and Atmospheric Administration (see Piper and others, 1978 and Sorem and others, 1978). The Deepsea Ventures Inc. "claim" is within the high-grade area between the Clarion and Clipperton fracture zones, as are other sites selected by deep sea mining consortia for prospecting and experimental mining. Nodules containing more than 1.8 percent nickel plus copper are known in other parts of the world (see especially the world map of Rawson and Ryan, 1978), but the area between the Clarion and Clipperton fracture zones (hereafter referred to as the C-C prime area) shown on Fig. 1 is clearly the one of greatest current interest in regard to the possible recovery of manganese nodules.

Our purpose here is to explain the basis for delineating the C-C prime area, to summarize the analytical data on the metal content of the nodules in the area on the basis of publicly available data, and to make an independent estimate of the area's potentially recoverable resources. For the latter purposes, we have drawn on the kind cooperation of Jane Z. Frazer who furnished us with all the pertinent data available on the metal content and concentration of the nodules in the Scripps Institution of Oceanography Sediment Data Bank, and of Ronald Sorem and others (1978) and David Piper and others (1978) who permitted us to use analyses of samples collected as a part of the DOMES project that were not yet in the Scripps Data Bank.

Basis for Delineating the High-Grade Nodule Areas

Because of the interest in the region, the Ocean Mining Administration sponsored, through a grant from the U.S. Bureau of Mines,

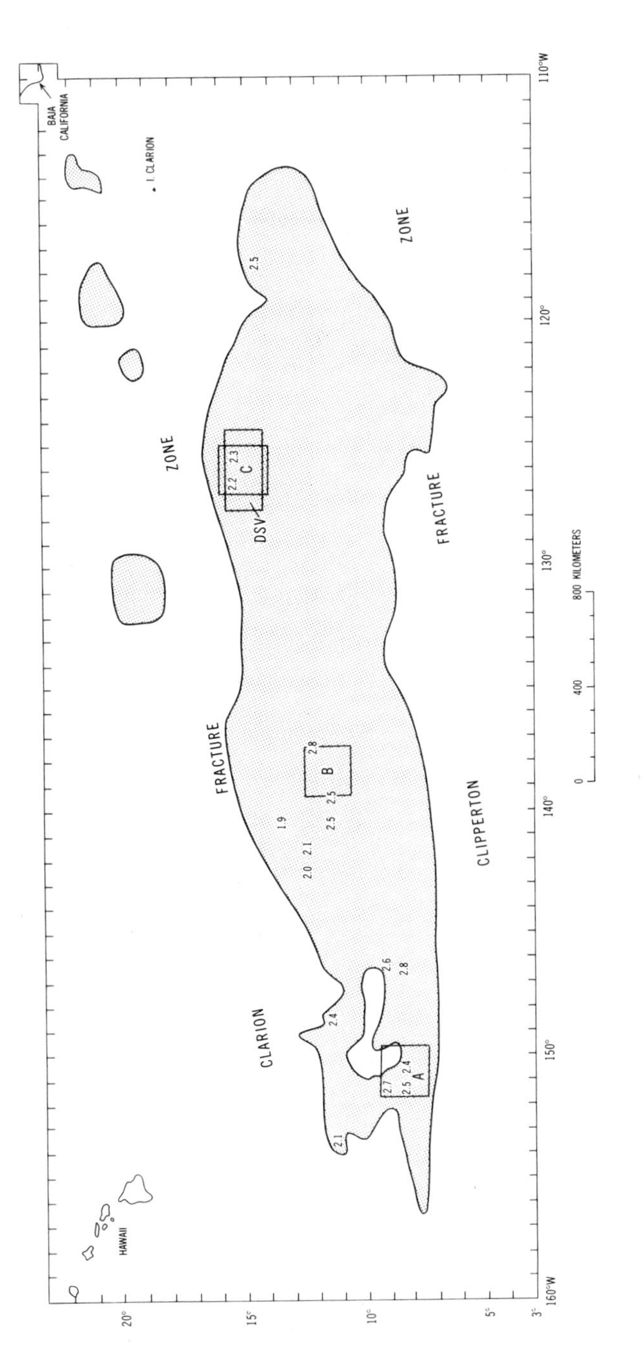

Fig. 1. Areas in the northeastern equatorial Pacific in which manganese nodules contain more than 1.8 percent nickel plus copper. The numbers represent the average percent nickel plus copper of samples in one-degree squares where analyses are available at five or more station (See Table 1). The rectangles outline the Deepsea Ventures, Incorporated (DSV) "claim" and the DOMES sites (A, B, and C).

the preparation of a bathymetric map of the northeast equatorial Pacific Ocean under the direction of Heezen and Tharp covering the area from 110° to 160° W. latitude and 3° to 23° N. longitude at scale 1:5,000,000, and of 15 bathymetric maps at scale 1:1,000,000 covering the same area. To the 1:1,000,000 maps were added the data compiled by Jane Z. Frazer and Mary B. Fisk from the Scripps Institution of Oceanography Data Bank showing the location of publicly available observation points related to manganese nodules, and the nickel plus copper content of the nodules where analyses were available. The maps were published by the U.S. Geological Survey in March, 1978 (I-1094 A-O and I-1095).

Examination of these maps suggested that it might be possible to delineate more closely than had been done previously in the public literature the areas in the northeast equatorial Pacific in which the nodules are richer in nickel and copper. This was done on the 1:1,000,000 maps, using 1.8 percent nickel plus copper as the minimum grade. The result is shown in Fig. 1.

The maps contain approximately 650 observation points within the higher grade areas, nearly one-fourth of which indicate that nodules are absent or, if present, cover less than 20 percent of the seabottom. The higher grade areas include a few stations where the analyses show less than 1.8 percent nickel plus copper, but such stations are included only where the average with closely adjacent samples is more than 1.8 percent; isolated samples containing less than 1.8 percent are excluded and the boundaries of the higher grade areas are the result of interpolation between samples above and below the minimum grade. Because of the paucity of samples, we have little confidence in the limits ascribed to the three high-grade areas shown on Fig. 1 north of the Clarion fracture zone, and they are not considered further here. The existence of the lower grade area centered by 10° N. latitude between 146° and 151° 30'W. longitude is more certain, for it is defined by 29 stations where the grade is less than 1.8 percent nickel plus copper.

Table 1 lists the average nickel plus copper content for 16 one-degree squares in the C-C prime area from which analyses are available at 5 or more stations within each square; rounded averages for these values are shown on Fig. 1. Two sets of contiguous one-degree squares in the western part of the area with analyses averaging 2.5 percent or more nickel plus copper may reflect relatively continuous higher grade deposits. Such deposits may also be present in other parts of the area, but the data are not sufficient to delineate them now.

The C-C prime area outline on Fig. 1 totals approximately 2.5 million km^2. As previously indicated, parts of it do not contain nodules and other parts may not contain them in minable quantities, but the data available are not sufficient to delineate areas of low concentration.

TABLE 1

One degree squares in the C-C prime area from which analyses for copper and nickel are available at five or more stations within each square

Location*	Number of Stations	Average Ni and Cu (percent)
8°-146°	8	2.78
8°-150°	15	2.41
8°-151°	7	2.54
9°-146°	7	2.62
9°-151°	7	2.74
11°-139°	5	2.49
11°-140°	5	2.53
11°-148°	5	2.41
11°-153°	6	2.06
12°-137°	6	2.77
12°-141°	23	2.12
12°-142°	7	1.99
13°-140°	43	1.92
14°-117°	6	2.47
15°-125°	24	2.27
15°-126°	18	2.23
	190	Average 2.40

*Southeastern corner of a one degree square

Metal Content and Concentration

The statistical data on metal content and nodule concentration in the C-C prime area are summarized in four groups: 1) all localities in the C-C prime area for which information is available on both grade and nodule concentration (Table 2); 2) all stations in the C-C prime area, including those for which only grade is available (Table 3); 3) all DOMES stations in the C-C prime area (Table 4); and 4) all stations where concentration of nodules is 5 kg/m^2 or more (Table 5). For some stations, DOMES' in particular, analyses are available for more than one sample; for the purpose of arriving at the statistics in these tables, such analyses were averaged to yield one value per station. The values for all stations are listed in the appendix.

The reasons for assembling the statistics in these groups are as follows: 1) Menard and Frazer (1978) found that for the 182 sites in the Pacific for which data on both nickel plus copper content and nodule concentration were available in the Scripps Data Bank there was an inverse correlation between the two, with the correlation coefficient -0.40 at the 99.9 percent confidence level, and that for the 120 samples available from the northeast equatorial Pacific the correlation coefficient was -0.28 at the 99.9 percent confidence level. The data assembled here for the C-C prime area (Table 2) show a correlation coefficient between nickel plus copper content and concentration of -0.24 at the 98 percent confidence level--only slightly less negative than Menard and Frazer found for the northeast equatorial Pacific as a whole and weak enough to suggest that in this area only a slight positive bias would result from treating grade and abundance as independent variables; 2) on the latter assumption,

TABLE 2

Metal content (percent), nodule concentration (kg/m^2), and metal ratios of manganese nodules in the C-C prime area at stations where information on both is available

	Number of Stations	Mean	Range	Standard Deviation	Correlation Coefficient
Mn	90	27.15	7.30 -35.50	4.92	
Fe	90	7.25	3.31 -12.51	2.17	
Ni	90	1.40	0.52- 1.91	0.31	
Cu	90	1.06	0.26 - 1.64	0.30	
Ni+Cu	90	2.45	0.78 - 3.50	0.58	
Co	60	0.24	0.09 - 0.39	0.05	
Mo	6	0.06	0.03 - 0.07	0.02	
Concentration	90	8.45	0.20 -27.25	6.14	-0.24*
Mn/Fe	90	4.24	1.30 - 9.70	1.84	-0.64**
Ni/Mn	90	0.052	0.026- 0.148	0.013	0.77**
Cu/Mn	90	0.039	0.013- 0.114	0.011	0.80**
(Ni+Cu)/Mn	90	0.091	0.039- 0.261	0.022	0.82**
Co/Mn	60	0.009	0.003- 0.027	0.004	0.009**
Mo/Mn	6	0.002	0.002- 0.002	0.000	***
Cu/Ni	90	0.75	0.49 - 1.04	0.12	0.83**

*Between nickel plus copper and concentration.
**Between the respective metals.
***Insufficient number of stations for meaningful values.

the grade of all available samples has been averaged in Table 3, with results only slightly different from those shown in Table 2; and 3) the DOMES stations (Table 4), which make up 15 percent of the data on nickel and copper content and 48 percent of those on concentration, are concentrated in three relatively small parts of the C-C prime area and might have been thought to give the overall results an undue local bias. As may be seen from a comparison of Tables 2, 3, and 4, both the DOMES samples and others for which concentration data are available, are higher in metal content and do raise the average for the area slightly. But the average metal contents of the three groups differ only slightly and probably not significantly except for

TABLE 3

Metal content (percent) and metal ratios of manganese nodules in the C-C prime area.

	Number of Stations	Mean	Range	Standard Deviation	Correlation Coefficient
Mn	362	25.43	2.50 -36.43	4.98	
Fe	359	6.66	0.50 -15.31	1.99	
Ni	366	1.27	0.11 - 1.91	0.27	
Cu	366	1.02	0.11 - 1.66	0.28	
Ni+Cu	366	2.29	0.22 - 3.51	0.52	
Co	323	0.22	0.05 - 0.56	0.07	
Mo	93	0.05	0.02 - 0.08	0.01	
Mn/Fe	359	4.38	0.30 -57.00	3.41	-0.30*
Ni/Mn	362	0.05	0.02 - 0.15	0.01	0.75*
Cu/Mn	362	0.04	0.01 - 0.11	0.01	0.70*
(Ni+Cu)/Mn	362	0.09	0.04 - 0.26	0.02	0.77*
Co/Mn	319	0.009	0.003- 0.027	0.004	0.028*
Mo/Mn	93	0.002	0.001- 0.004	0.000	0.59*
Cu/Ni	366	0.80	0.44 - 1.34	0.14	0.79*

*Between the respective metals.

molybdenum. Because only six molybdenum analyses are available for the stations represented in Table 2, compared with 93 for those in Table 3, the average shown in Table 3 must be considered more reliable, although the difference is slight. Table 5 has been added on the assumption that it may be possible to restrict operations to areas containing higher concentrations of nodules.

Also shown in the tables are metal ratios and their correlation coefficients. Noteworthy are the high correlation coefficients for the ratios of nickel and copper to manganese and of nickel and copper to each other. Molybdenum shows the same trend, but the values may be misleading due to the paucity of data.

The analyses represented in the tables are drawn from many sources, and may include some of poor quality. As Frazer (1978) has pointed out, however, the fact that they are drawn from several sources may improve the accuracy of the averages, for the errors are probably random and tend to cancel each other. Because the averages for manganese, iron, nickel, copper, and cobalt in the C-C prime area are based on several hundred samples, they are probably close to the true value--possibly, as Frazer (1978) points out, slightly on the low side as some of the analyses may have been made on air-dried samples still containing 10-15 percent water. The confidence that the large number of nickel plus copper analyses engenders must be tempered, however, by the facts that it represents an average of only one per 6,800 km^2 and the sample locations are unevenly distributed. An indication of the latter is that the 16 one-degree squares listed in Table 1 make up 7 percent of the C-C prime area but contain 52 percent of the localities where nickel and copper analyses are available.

TABLE 4

Metal content (percent), nodule concentration (kg/m^2) and metal ratios in the C-C prime area at stations sampled in the Deep Ocean Mining Environmental Study

	Number of Stations	Mean	Range	Standard Deviation	Correlation Coefficient
Mn	55	27.15	7.30 -35.50	4.89	
Fe	55	7.32	3.56 -11.70	2.15	
Ni	55	1.35	0.82 - 1.87	0.22	
Cu	55	1.10	0.53 - 1.64	0.27	
Ni+Cu	55	2.45	1.35 - 3.41	0.48	
Co	55	0.24	0.15 - 0.39	0.05	
Mo	1	0.06			
Concentration	43	8.04	0.20 -20.00	5.96	-0.39*
Mn/Fe	55	4.16	1.30 - 8.37	1.77	-0.50**
Ni/Mn	55	0.051	0.040- 0.148	0.014	0.75**
Cu/Mn	55	0.041	0.024- 0.114	0.012	0.69**
(Ni+Cu)/Mn	55	0.092	0.067- 0.262	0.026	0.74**
Co/Mn	55	0.009	0.005- 0.027	0.004	-0.17**
Mo/Mn	1	0.002	***	***	***
Cu/Ni	55	0.80	0.56 - 1.04	0.11	0.88

*Between nickel plus copper and concentration.
**Between the respective metals
***Insufficient number of stations for meaningful values.

The average concentration of nodules in the C-C prime area is much more uncertain than the average metal content, partly because the determinations are fewer, but also because the measurement itself is less certain. Bastien-Thiry and others (1977) consider that the error in estimates of concentration ranges from 20 to 50 percent.

The weak inverse relation between grade and nodule concentration in the C-C prime area may be seen from a comparison of Tables 2 and 5. These tables show a decrease of 5 percent or 0.15 weight percent in the average content of nickel plus copper compared to a 41 percent increase in average concentration from 8.45 to 11.94 kg/m^2. The low-grade nodules in the high concentrations shown in Table 5, however, contain more nickel plus copper per unit area (0.27 kg/m^2) than the high-grade but low concentrations shown in Table 2 (0.21 kg/m^2).

TABLE 5

Metal content (percent), concentration (kg/m^2), and metal ratios of manganese nodules in the C-C prime area at stations where the nodule concentration is 5 kg/m^2 or more

	Number of Stations	Mean	Range	Standard Deviation	Correlation Coefficient
Mn	58	26.68	16.20 -33.50	4.28	
Fe	58	7.85	3.31 -12.51	2.17	
Ni	58	1.34	0.52 - 1.88	0.33	
Cu	58	0.96	0.26 - 1.43	0.28	
Ni+Cu	58	2.30	0.78 - 3.21	0.59	
Co	38	0.25	0.15 - 0.39	0.05	
Mo	4	0.07	0.07 - 0.07	0.00	
Concentration	58	11.94	5.50 -27.25	4.80	0.06*
Mn/Fe	58	3.82	1.58 - 9.70	1.69	-0.76**
Ni/Mn	58	0.050	0.026- 0.068	0.007	0.88**
Cu/Mn	58	0.035	0.013- 0.047	0.007	0.91**
(Ni+Cu)/Mn	58	0.085	0.039- 0.109	0.012	0.92**
Co/Mn	38	0.010	0.007- 0.019	0.003	-0.05**
Mo/Mn	4	0.002	0.002- 0.002	0.000	***
Cu/Ni	58	0.71	0.49 - 0.92	0.11	0.88**

*Between nickel plus copper and concentration.
**Between the respective metals.
***Insufficient number of stations for meaningful values.

Resources of the C-C Prime Area

Previous estimates of the potentially recoverable resources in the C-C prime area were for the region as a whole, reduced to an area considered minable on the basis, for example, of the percentages of samples considered to be above some "cut-off" grade and concentration. Archer (1976) estimated that the area contains 23-27 potential mine sites each capable of producing an average of 3 million metric tons of dry nodules per year for 25 years. Holser (1976) showed on his Fig. 4 approximately 20 "first generation" mine sites for the Clarion-Clipperton area each averaging 75 million metric tons for their lifetime production. For the northeast equatorial Pacific, bounded by latitudes 0° N. and longitudes 100° W. and 170° E., and excluding the Mid-Pacific Mountains in the region north of 20° N. and west of 160° W., Pasho and McIntosh (1976) estimated that there is a 95 percent chance that it contains at least 5 mine sites capable of producing 3 million metric tons a year for 20 years, a 50 percent chance that there are more than 30 such sites and a 5 percent chance that there are more than 140 sites that may become commercially exploitable before the year 2000; Pasho (1977) later refined this estimate to indicate a probable range of 10-95 mine sites. In the north equatorial Pacific Frazer (1977) identified an area of 3.3 million km^2 in 11 five degree squares between 5° and 20° N. and 110° and 155° W., 38 percent of which she predicts has sufficient nodule abundance for first generation mining. For this area she estimated nodule resources comparable to 26 mine sites each capable of producing 75 million dry metric tons over their lifetimes. She identified two additional mine site equivalents in the South Pacific and allowing for error by a factor of 2, she expressed her final total estimate as 14-56 mine site equivalents.

On the basis of extensive surveys between the Clarion and Clipperton fracture zones by the French group, Bastien-Thiry and others (1977) estimated that a 2.25 million km^2 area (the size actually given is "2.25-million-sq.-m" but from other parts of the paper "m" is believed to be a typographical error)--presumably essentially the same as that outlined on Fig. 1--contains 7.75 billion wet metric tons of nodules in place or 5-10.6 billion wet metric tons averaging 2.25 percent nickel plus copper and 0.24 percent cobalt measured as dry weight. At a "cutoff" nodule concentration of 5 kg/m^2 and an assumed 20 percent recovery they estimated that the recoverable tonnage would be 670 million wet metric tons averaging 2.5 percent nickel plus copper--enough to support 8 to 11 mining operations each producing 3-4 million wet metric tons for 20 years. The concentration of nodules is reported to average 3.45 kg/m^2 (including sites where they are absent), significantly less than the concentration shown in Tables 2, 4, and 5. Bastien-Thiry and others report that their results are based on grid surveys involving 262 localities where 1,844 sampling stations were established. If concentration measurements were made at all of these sites, their data base would be much more reliable than that formed by the publicly available information analyzed here.

Because of the paucity of the publicly available data, particularly on nodule concentration, not much accuracy can be claimed for estimates of nodule resources in the C-C prime area, but a rough estimate confirms the general order of magnitude indicated by the previous estimates. Of the nodule concentrations reported in Table 1, 36 percent are less than 5 kg/m^2, which may be taken as the minimum average concentration for mining. As previously indicated, nodules are reported to be absent at 22 percent of the stations shown on the maps, or, if present, to cover less than 20 percent of the bottom. On the assumption that these two percentages may be combined to indicate how much of the area does not contain minable resources, the minable area would be 1.25 million square km^2 containing an average (rounded from the Tables) or 11.9 kg/m^2 of nodules containing about 1.3 percent nickel, 1 percent copper, 25 percent manganese, 0.22 percent cobalt, and 0.05 percent molybdenum. If we assume that only 20 percent of the nodules would be recoverable and that the moisture content is 30 percent, the C-C prime area would contain about 2.1 billion dry metric tons of potentially recoverable nodules (in the resource classification system recently adopted by the U.S. Bureau of

Mines and the U.S. Geological Survey (U.S.G.S. Bulletin 1450-A) the nodules in this area would be classed as identified paramarginal resources) -- enough to support about 27 mining operations, each producing an average of 75 million metric tons of nodules over their lifetimes.

Estimates based on other plausible assumptions would be higher or lower but of the same order of magnitude. When we consider the paucity and uncertainty of the concentration data and the fact that data stations are not randomly distributed it is readily conceivable that the size of the minable area could range from one half to twice the size estimated, with proportionate effects on the estimates. If the distribution of the nodules is found to be locally irregular, a minimum average concentration may not be followed in mining, possibly increasing somewhat the potentially recoverable resource but reducing initially the percentage of the resources recoverable. If the percentage of recoverable nodules is increased, as a result of technological advance over time, then potentially recoverable tonnage and the number of average-sized operations, such tonnage would support, would increase proportionately. It seems clear, however, from this and from the previous estimates cited that this most promising area now known, contains potentially recoverable manganese nodule resources adequate to support not more than several tens of the average-sized operations postulated.

Perspective

On the basis of the publicly available data, the Clarion-Clipperton prime area of the northeastern equatorial Pacific appears to contain extensive deposits of metal-rich manganese nodules. Clearly, the deposits are not of unlimited extent, but they are large enough to justify the commercial interest the area has been receiving in recent years. Also, indications of high-grade deposits in other regions are sufficient to suggest the possibility that additional search may turn up other promising areas. And there is the prospect that if the industry is able to get underway and improve its recovery technology, deposits lower in grade than are being considered for mining now may become minable in the more distant future. Clearly it will take extensive prospecting, exploration, and technologic development to test, much less realize, any of these possibilities.

Addendum

In Feburary 1979, after camera copy of this paper had been prepared, we obtained from Jane Fraser all of the analyses of samples of nodules in the Scripps Data Bank lying within the C-C prime area outlined in Fig. 1. These included additional analyses of the six metals listed in Tables 2-5 that had been added to the bank since April 1978, when the data for those tables were obtained, as well as analyses of other constituents. The statistical characteristics of these analyses are listed in Table 6. As may be seen from that Table, the number of analyses available for other constituents are insufficient to serve as the basis for conclusions as to their average content in the nodules in the C-C prime area, but they suggest that several other elements--for example, zinc, lead, vanadium, and some of the rare earths--are present in the nodules in amounts considerably greater than their average in the earth's crust, and deserve more attention in further analytical work.

TABLE 6
Composition (percent) of manganese nodules in the C-C prime area

	Number of samples	Minimum	Maximum	Mean	Standard Deviation
Mn	402	2.50	37.50	28.80	4.79
Fe	402	.18	15.31	6.64	2.09
Co	397	.05	.91	.23	.09
Ni	409	.11	1.95	1.22	.26
Cu	407	.11	1.66	.99	.28
Zn	226	.04	.25	.13	.04
Pb	161	.015	.17	.048	.02
Al	84	.44	5.18	2.84	.78
Si	96	2.20	20.43	7.81	2.91
Ca	195	.02	4.19	1.47	.60
H_2O	55	9.10	47.15	23.91	7.87
Be	1	-	-	.0002	-
B	7	.01	.027	.016	.006
C	14	.07	.40	.22	.10
Na	51	1.12	4.22	1.87	.67
Mg	82	.30	2.83	1.80	.45
P	64	.02	2.2	.23	.29
S	39	.05	1.00	.22	.19
K	80	.26	2.05	.82	.30
Sc	5	.0009	.27	.05	.12
Ti	165	.12	18.90	.61	1.45
V	43	.002	.13	.03	.026
Cr	49	.0002	.07	.02	.027
Ga	4	.0007	.003	.002	.001
As	20	.005	.01	.008	.002
Rb	19	.0007	.002	.002	.0003
Sr	43	.01	.17	.066	.026
Y	26	.004	.015	.01	.002
Zr	26	.019	.08	.036	.012
Nb	1	-	-	.003	-
Mo	168	.004	.13	.048	.02
Cd	17	.0004	.003	.001	.0007
I	1	-	-	.25	-
Ba	65	.0001	1.52	.32	.23
La	2	.01	0.15	.013	.004
Ce	20	.016	.07	.035	.016
Nd	1	-	-	.03	-
Yb	6	.0016	.0026	.002	.0004
Tl	17	.006	.03	.016	.008
Bi	1	-	-	.0006	-
Th	25	.001	.003	.002	.0006
U	6	.0002	.0035	.0013	.0015

References

Archer, A.A., (1976) Prospects for the exploitation of manganese nodules: the main technical, economic, and legal problems In Papers Presented at the I.D.O.E. Workshop, Suva, Fiji, 1-6 September, 1975 (editors G.P. Glasby, and H.R. Katz), Tech. Bull. 2, U.N., Economic and Social Commission for Asia and the Pacific, CCOP/SOPAC, 21-38.

Archer, A.A., (1978) Resources and potential reserves of nickel and copper in manganese nodules: United Nations, in press.

Bastien-Thiry, Hubert, Lenoble, Jean-Pierre, and Rogel, Pierre, (1977) French exploration seeks to define minable nodule tonnages on Pacific floor. Eng. and Min. J. July 1977, 86-87, 171.

Bezrukov, P.L., (1976) Geological background of the exploitation of ferromanganese deposits. Academy of Sciences of the U.S.S.R., Trans. P.P. Shirshov Institute of Oceanology, 109, 250-257.

Cronan, D.S., (1972) Regional geochemistry of ferromanganese nodules in the world oceans. In Papers from a conference on ferromanganese deposits of the ocean floor, (editor, D.R. Horn) Nat. Sci. Foundation, 19-30.

Frazer, Jane Z., (1977) Manganese nodule reserves: An updated estimate: Mar. Min. 1, 103-123.

Frazer, Jane Z., (1978) The reliability of available data on element concentrations in seafloor manganese nodules. United Nations, in press.

Frazer, J.Z., and Arrhenius, G., (1972) Worldwide distribution of ferromanganese nodules and element concentration in selected Pacific Ocean nodules. Nat. Sci. Foundation Tech. Rept. 2.

Friedrich, G.H., Kunzendorf, H., and Pluger, W.L. (1973) Geochemical investigation of deep sea manganese nodules from the Pacific on board R/V Valdivia--an application of the EDX-technique, In Papers on the origin and distribution of manganese nodules in the Pacific, and prospects for exploration (editor M. Morgenstein), Valdivia Manganese Exploration Group and the Hawaii Institute of Geophysics, 31-44.

Heezen, Bruce C., and Tharp, Marie, (1978) Bathymetric map of the northeast equatorial Pacific Ocean. U.S. Geol. Survey Misc. Invest. Ser. Map I-1095.

Heezen, Bruce C., Tharp, Marie, Frazer, Jane Z., and Fisk, Mary B., (1978) Bathymetric and nodule assessment maps, northeast equatorial Pacific Ocean: U.S. Geol. Survey Misc. Invest. Ser. I-1094A-).

Holser, A.F., (1976) Manganese nodule resources and mine site availability. Ocean Mining Administration, U.S. Dept. of Interior, Prof. Staff Study, 1-12.

Horn, D.R., Horn, B.M., and Delach, M.N., (1972) Ferromanganese deposits of the North Pacific: Nat. Sci. Foundation Tech. Rept. 1.

Horn, D.R., Delach, M.N., and Horn, B.M., (1973) Metal content of ferromanganese deposits of the oceans: Nat. Sci. Foundation Tech. Rept. 3.

Horn, D.R., Delach, M.N., and Horn, B.M., (1973) Ocean manganese nodules, metal values, and mining sites: Nat. Sci. Foundation Tech. Rept. 4.

Horn, D.R., Horn, B.M., and Delach, M.N., (1973) Factors which control the distribution of ferromanganese nodules and proposed research vessel's track, North Pacific. Nat. Sci. Foundation Tech. Rept. 8.

Margolis, S.V., and Burns, R.G., (1976) Pacific deep-sea manganese nodules: their distribution, composition, and origin. Ann. Rev. Earth and Planet. Sci., 4, 229-263.

Menard, H.W., (1964) Marine geology of the Pacific. McGraw-Hill.

Menard, H.W., and Frazer, J.Z., (1978) Manganese nodules on the sea floor: inverse correlation between grade and abundance. Science, 199, 969-971.

Mero, John L., (1960) The mineral resources of the sea. Elsevier.

Pasho, D.W., (1977) Review of the development of deep seabed manganese nodules. The Northern Miner, April 14, 1977, B6, B9, B16.

Pasho, D.W., and McIntosh, J.A., (1976) Recoverable nickel and copper from manganese nodules in the northeast equatorial Pacific--preliminary results. Canadian Inst. of Min. and Metal. Bull. 69, 15-16.

Piper, David Z., Cannon, W.F., and Leong, K., (1978) Manganese nodules and surface sediment composition, DOMES sites A, B, and C. Written communication.

Piper, David Z., and Williamson, Michael E., (1977) Composition of Pacific Ocean ferromanganese nodules. Mar. Geol., 23, 285-303.

Rawson, Martine Dreyfus, and Ryan, William B.F., (1978) Ocean floor sediment and polymetallic nodules. Lamont-Doherty Geological Observatory, Columbia University.

Skornyakova, N.S., (1976) Chemical composition of ferromanganese nodules of the Pacific. Academy of Sciences of the U.S.S.R., Trans. P.P. Shirshov Institute of Oceanology, 109, 190-240.

Schultze-Westrum, H.H., (1973) The station and cruise pattern of the R/V Valdivia in relation to the variability of manganese nodule occurrences. In Papers on the origin and distribution of manganese nodules in the Pacific, and prospects for exploration (editor M. Morgenstein), Valdivia Manganese Exploration Group and the Hawaii Institute of Geophysics, 145-149.

Sorem, Ronald K., Reinhart, William R., Fewkes, R.H., and McFarland, W.D., (1978) Occurrence and character of manganese nodules in the "Horn region". Written communication.

AMINO ACID DATING OF BONE NUCLEI IN MANGANESE NODULES FROM THE NORTH PACIFIC OCEAN

Keith A. Kvenvolden and David J. Blunt

U.S. Geological Survey
345 Middlefield Road
Menlo Park, California 94025

Abstract

The extent of racemization of isoleucine and aspartic acid has been measured on cetacean ear bones from the center of eight manganese nodules in order to estimate the ages of the nuclei of the nodules and the rates of accretion of ferromanganese oxide layers. Age estimates obtained range from about 0.4 to 6 m.y., and rates of growth from less than 1 to about 10 mm/m.y. These rates are consistent with those obtained by other nucleus dating methods (K-Ar and fission track) and by layer dating methods (^{230}Th, ^{231}Pd, ^{10}Be and ^{26}Al).

Introduction

Manganese nodules are not distributed uniformly on the floors of the deep oceans. One factor apparently controlling the variable distribution is the availability of suitable nuclei around which ferromanganese oxides can deposit (Cronan, 1977). Almost every nodule has a recognizable nucleus such as volcanic rock or glass, solid organic remains (mainly teeth from sharks and bones from whales), or fragments of older nodules (Raad and Meylan, 1977). Manganese nodules were generally considered scientific curiosities before their potential as a source for valuable elements, chiefly manganese, nickel, copper, cobalt, and lead, was recognized in the early 1950s.

Growth or accretion rates of manganese nodules may provide possible clues to some of the processes contributing to their formation (Ku, 1977). There are two approaches to determining rates of accretion: (1) dating the nuclei around which the ferromanganese oxide layers accumulate; (2) measuring the age differences among successive layers. In this paper we estimate the growth rates of eight manganese nodules by using the first approach and determining the extent of racemization of amino acids in their bone nuclei. Racemization is a chemical reaction involving the interconversion of amino acid enantiomers (optical isomers). Only L-amino acids are present in the protein of living organisms. When an organism dies, the L-amino acids slowly interconvert to D-amino acids until equilibrium mixtures of the two isomers remain. Racemization of amino acids in bone apparently follows reversible first-order kinetics essentially to equilibrium (Bada, 1972). By measuring the extent of racemization of amino acid (D/L ratios), it is possible to calculate time or age if the rate constants of the reaction are known. This method was first used to date manganese nodules by Bada (1972) who measured the amino acids in shark vertebra found in the core of a nodule from Horizon Guyot in the North Pacific Ocean.

The extent of racemization of seven amino acids was measured in the bone nuclei of eight manganese nodules from the North Pacific Ocean. Because of the uncertainties in using amino acid racemization for geochronology, this work was not designed to yield absolute ages of the bone nuclei but rather to try to set limits on the age of the nuclei. The limits set are independent of radiometric ages. The results, therefore, contribute to a resolution of the controversy concerned with whether or not rates of growth of manganese nodules are very rapid as proposed by Lalou and Brichet (1972) or very slow as proposed by Ku (1977). The nodules examined were selected by J.L. Bischoff, U.S. Geological Survey, from a collection obtained from the Summa Corporation. They came from stations at water depths ranging from 4396 to 5124 meters (Table 1); four came from a single sample station. All samples are from an area bounded by 7° and 28°N latitude and 148° and 169°W longitude. The nucleus of each of the nodules was identified by Charles A. Repenning, U.S. Geological Survey, as a cetacean ear bone.

Methods

The manganese nodules were cut approximately in half along their largest dimension by diamond saw with double distilled water as the coolant. (Double-distilled water was used to avoid contamination of the samples by amino acids during the cutting process.) At the core of each nodule was bone. The thickness of the ferromanganese accumulation was measured (Table 1) and the maximum thickness deposited on the nuclei found to range from about 1 to 8 mm, averaging about 3 mm.

Half of the sample was prepared for amino acid analysis by removing ferromanganese oxide deposits with a dental tool. A sample

TABLE 1

D/L Ratios of Amino Acids in Bone Nuclei of Manganese Nodules from the North Pacific Ocean

Sample No.	Depth (m)	Lat. Long.	Thickness (mm)	VAL	ALA[1]	ILEU[2]	LEU[1]	PRO	ASP	GLU
Mn 48	5124	$10^{o}58.7$'N $148^{o}19.5$W	2-3	0.13	0.25	0.13	0.13	0.17	0.37	0.18
Mn 53a	5003	$11^{o}58.2$'N $150^{o}18.5$'W	1-2	.12	.22	.12	.13	.13	.40	.17
Mn 53b	5003	$11^{o}58.2$'N $150^{o}18.5$'W	6-8	nd	.62	.28	.40	.71	.61	.38
Mn 53c	5003	$11^{o}58.2$'N $150^{o}18.5$'W	0-1	.20	.41	.21	.23	.29	.51	.30
Mn 53d	5003	$11^{o}58.2$'N $150^{o}18.5$'W	3-5	.30	.57	.24	.34	.54	.46	.37
Mn 57	4849	$25^{o}03.7$'N $163^{o}38.8$'W	1-3	nd	nd	nd	.46	.57	.69	.52
Mn 58	4396	$28^{o}23.8$'N $169^{o}03.0$'W	2-3	nd	nd	nd	nd	nd	.70	.56
Mn 86	4908	$7^{o}49.8$'N $154^{o}49.3$'W	3-4	.43	.75	.40	.46	.50	.73	.61

[1]Average of values obtained with two different chromatographic columns

[2]D-alloisoleucine/L-isoleucine

was sonicated for one minute each in double distilled water, then in 1 N HCl, and again in double distilled water. The clean sample was dried for 45 minutes in a vacuum oven at 40°C and weighed (approximately one gram of sample). The sample was acidified with 10 N HCl. After dissolution, the sample was dried on a rotary evaporator at less than 40°C. The dried residue was taken up in 6 N HCl, and this solution hydrolyzed for 20 hours at 110°C. The resulting hydrolysate was evaporated to dryness, taken up in distilled water and desalted on Dowex 50W-X8 (H^+). The amino acids were eluted with 2N NH_4OH; this solution was dried on a rotary evaporator at less than 40°C, then placed in a vacuum oven for one hour at 40°C to insure the removal of ammonia. The samples were taken up in distilled water and divided into a portion for analyses by automated ion exchange chromatography (IEC) and a portion for analysis by gas chromatography (GC). The portion for IEC was evaporated to dryness and taken up in dilution buffer before analysis. For gas chromatography N-pentafluorpropionyl-(+)-2-butyl esters of the amino acid were prepared following the general procedures given by Kvenvolden et al., (1972). Gas chromatographic analyses were carried out on two chromatographic columns each coated with a different phase (UCON 75H-90,000 and Carbowax 20M).

Results

The extent of racemization of seven amino acids, valine (VAL), alanine (ALA), isoleucine (ILEU), leucine (LEU), proline (PRO), aspartic acid (ASP), and glutamic acid (GLU) is shown in Table 2. The D/L ratios were obtained by measuring peak heights on gas chromatograms (for example chromatogram, see fig. 1). Ratios for VAL, PRO, ASP, and GLU were obtained with the capillary column coated with UCON 75H-90,000. Values for ALA and LEU are averages obtained by use of both capillary columns (Carbowax 20 M and UCON 75H-90,000). Measurement of the ratio of isoleucine (ILEU) to alloisoleucine (ALILEU) was to be made on chromatograms from the automated ion exchange amino acid analyzer (IEC). With this system, however, an unknown compound interfered with ALILEU making determination of the ratio impossible. Bada (personal communication, 1978) has observed this kind of interference in unpublished analyses of bones from manganese nodules. The interfering compound has not yet been identified. Because the ratios of L-ILEU/D-ALILEU could not be obtained by IEC, GC with a capillary column caoted with Carbowax 20M was employed. The region of the chromatograms was free of interference and D- and L-ALA are well resolved from D-ALILEU and L-ILEU.

Discussion

The integrated rate equation derived by Bada and Schroeder

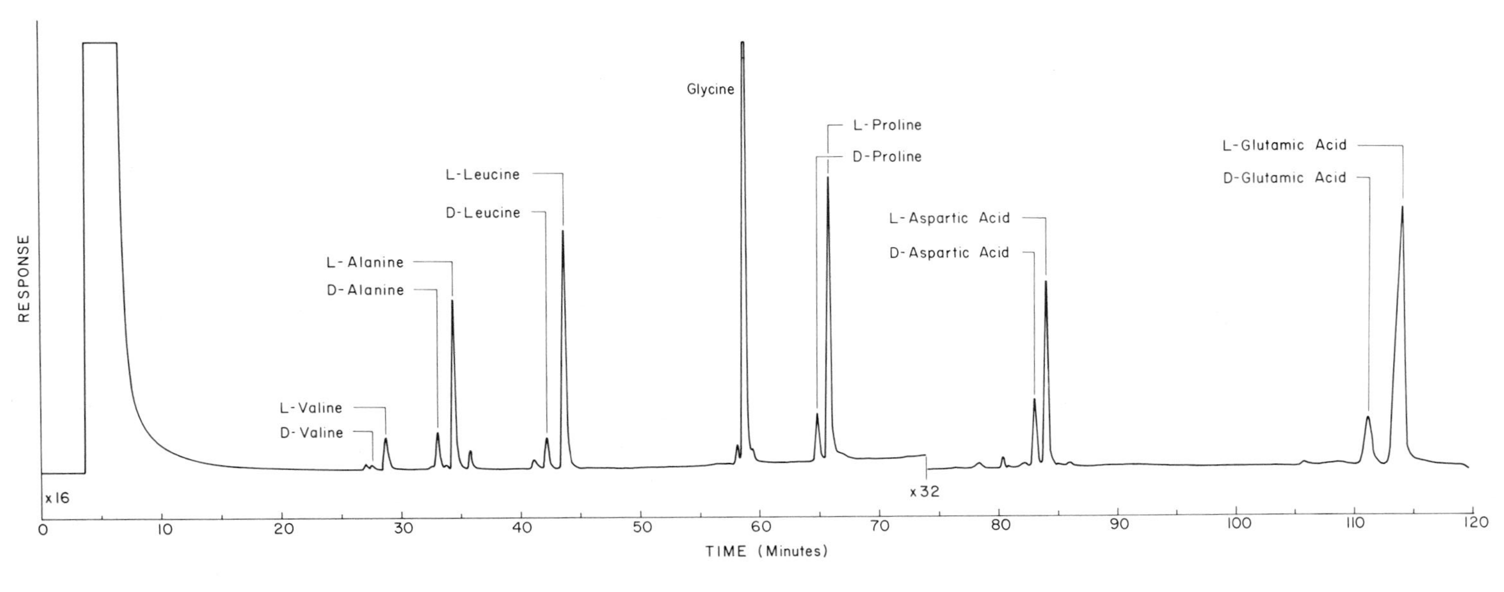

Fig. 1 Gas chromatogram showing the separation and resolution of diastereomeric derivatives of amino acids recovered from the bone nucleus of Sample Mn-48. Chromatographic conditions: Column, 200' x 0.02", stainless steel capillary coated with UCON 75 H-90,000; He flow rate, 6 ml/min; program rate, initial temp. 80°C for 8 min., programmed to 145° @ 1/min., isothermal at 145° for 35 min.; instrument, Perkin-Elmer 900.

(1972) for the first-order reversible reaction of racemization is:

$$\ln\left[\frac{1+\frac{D}{L}}{1-\frac{k_2}{k_1}\frac{D}{L}}\right]_t - \ln\left[\frac{1+\frac{D}{L}}{1-\frac{k_2}{k_1}\frac{D}{L}}\right]_{t=0} = \left(1+\frac{k_2}{k_1}\right) kt \qquad (1)$$

where k_1 and k_2 are the rate constants for the forward and reverse reactions, D/L the ratio of the enantiomers of a given amino acid, and t time. The second term is the constant of integration evaluated at t=o. The relation between the rate constant and temperature is expressed by the Arrhenius equation:

$$k = Ae^{-Ea/RT} \qquad (2)$$

where A is the pre-exponential factor, Ea the activation energy, R the gas constant, and T the absolute temperature.

Two amino acid pairs were used to attempt to calculate ages of bone nuclei and rates of ferromanganese oxide accumulation. These amino acid pairs, L-ILEU: D-ALILEU and L-ASP: D-ASP have been employed previously in dating bones (Bada, 1972; Bada and Protsch, 1973, Dungworth et al., 1974, Bada and Helfman, 1975), and information from some of these studies applies to the dating of bone nuclei of manganese nodules.

Bada (1972) dated a shark vertebra from a manganese nodule by means of the ratio ALILEU/ILEU. By kinetic experiments at high temperature on modern bone, he obtained an Arrhenius equation that was used to calculate the k_{ILEU} at 3^oC, the bottom water temperature at Horizon Guyot.

$$\log k_{ILEU} = 19.41 - 7304.0/T \qquad (3)$$
$$\text{where } Ea = 33.4 \text{ kcal/mole}$$

In addition, for equation (1) the values of $k_2/k_1 = 0.725$ and the integration constant t = o = 0.028 were determined. In Bada's sample, the ALILEU/ILEU ratio was 0.78; from this, the age was calculated to be 8.7 m.y. and the accumulation rate of manganese to be about 0.6 to 1.2 mm/m.y.

The average bottom temperature in the area where the eight nodules of this study were collected is 1.5^oC (Wooster and Volkman, 1960). We have substituted this temperature (274.65^o) into equation (3) to obtain $k_{ILEU} = 6.55 \times 10^{-8} yr^{-1}$. This value was substituted into equation (1) along with the measured D/L ratios to obtain an estimate of the ILEU ages of bone nuclei (Table 2). The age estimates range from 1.6 to 5.8 m.y., the minimum rate of accumulation from 0.4 to 2.0 mm/m.y.

Age estimates have also been made using the degree of racemization of ASP. For these estimates, two calibration sites were chosen for which D/L ratios, temperatures, and ages from independent methods are known (Bada and Helfman, 1975). From this information, it is possible to calculate first order rate constants using equation (1) and to adjust these rates to 1.5°C by means of equation (2) following methods of Bada and Protsch (1973) and Bada, Schroeder and Carter (1974). In equation (1) for ASP, $k_1/k_2 = 1$ and the value of the t=o term is 0.14 (Bada and Protsch, 1973).

	^{14}C Age (Yr)	D/L ASP	k_{ASP} Yr^{-1}	T (mean annual)
Muleta Cave, Mallorca	8570	0.214	1.7×10^{-5}	17°C
Olduvai Gorge, Tanzania	1390	0.165	7.0×10^{-5}	23°C

From this information and Ea = 33.4 kcal/mole (Bada and Helfman, 1975), two Arrhenius equations can be derived by which k_{ASP} can be adjusted to 1.5°C:

Muleta Cave, Mallorca $\log k_{ASP} = 20.386 - 7298.9/T$, k_{ASP} @ 1.5°C = $6.47 \times 10^{-7}\ yr^{-1}$

Olduvai Gorge, Tanzania $\log k_{ASP} = 20.491 - 7298.9/T$, k_{ASP} @ 1.5°C = $8.24 \times 10^{-7}\ yr^{-1}$

With these rate constants and D/L ASP ratios, the ASP ages of bone nuclei were calculated (Table 2). These age estimates are lower than the ILEU ages by as much as a factor of 5 and the minimum rates of accumulation are greater by the same amount. The rates of accumulation range from a low of 1.3 to a high of 10.3 mm/m.y. depending on choice of k_{ASP}. The ASP ages and the ILEU ages all contain uncertainties in the validity of the calibration temperatures and in the adjustments of the rate constants via the Arrhenius equation. Nevertheless, our results show that the minimum rates of accumulation of ferromanganese oxides on bone nuclei may range from less than 1 mm/m.y. to as much as 10 mm/m.y. In addition to the discrepancy in ILEU and ASP ages, the relations among D/L ratios (Table 1) have some ambiguities. The order of the rates of racemization of amino acids in uncontaminated bone was postulated to be ASP > ALA=GLU > ILEU=LEU (Bada et al., 1973). The data of Table 1 follow this pattern only broadly and there are many exceptions. This observation suggests either that this criterion for contamination is not strictly valid or that the nodules may be contaminated to varying degrees. Despite the ambiguities, however, we believe that the amino acid data show that the growth rates of manganese nodules are slow and of the order of less than 10 mm/m.y.

TABLE 2

Estimated Ages and Rates of Accretion of Manganese Nodules from the North Pacific Ocean

Sample No.	Max. Thickness (mm)	D-ALLOILEU / L-ILEU	Age (m.y.)	Rate[1] (mm/m.y.)	D/L ASP	Age[2] (m.y.)	Rate[2] (mm/m.y.)	Age[3] (m.y.)	Rate[3] (mm/m.y.)
Mn 48	3	0.13	1.7	1.8	0.37	0.49	6.1	0.39	7.6
Mn 53a	2	.12	1.6	1.3	.40	.55	3.6	.43	4.7
Mn 53b	8	.28	4.0	2.0	.61	.99	8.1	.78	10.3
Mn 53c	1	.21	2.9	0.4	.51	.76	1.3	.60	1.6
Mn 53d	5	.24	3.4	1.5	.46	.66	7.6	.52	9.6
Mn 57	3	-	-	-	.69	1.2	2.5	.94	3.2
Mn 58	3	-	-	-	.70	1.2	2.5	.97	3.1
Mn 86	4	.40	5.8	0.7	.73	1.3	3.1	1.04	3.9

[1] $k_{ILEU} = 6.55 \times 10^{-8}\ yr^{-1}$

[2] $k_{ASP} = 6.47 \times 10^{-7}\ yr^{-1}$

[3] $k_{ASP} = 8.24 \times 10^{-7}\ yr^{-1}$

Two other dating methods, K-Ar and fission track, have been applied to the nuclei of manganese nodules to ascertain the growth rates of ferromanganese layers. Potassium-argon measurements by Barnes and Dymond (1967) on volcanic minerals and glass shards from the nuclei of nodules from the Pacific Ocean showed a minimum growth rate of the order of 1 to 4 mm/m.y. Aumento (1969) measured fission track and K-Ar ages for basalts from nodules of the Mid-Atlantic Ridge and therefrom estimated the accumulation rates of ferromanganese layers on the basalts to be 1.6 to 4.1 mm/m.y. The three "nucleus dating" methods (K-Ar, fission track, and amino acid racemization) thus all yield nodule growth rates in the same range. These methods, however, provide no information on whether the growth has been continuous or episodic or whether the nodules are growing at the present time. Layer dating methods (^{230}Th, ^{231}Pa, ^{10}Be, and ^{26}Al) provide some insight to these problems. These methods depend on the assumption that radioactive nuclides are incorporated into nodule layers at a constant rate as the layers are deposited in seawater. Ku (1977) in reviewing much of the available information on rates of accretion of manganese nodules, notes that the rates of accretion, based on the majority of radiometric measurements of nuclides in layers of nodules, fall in the relatively narrow range of 2 to 8 mm/m.y. The ^{10}Be method applied to three nodules shows rates of accumulation of 0.8, 1.8, and 3.8 mm/m.y. (Ku, 1977). Guichard et al. (1978) recently reported the growth rates of manganese nodules to be 2.8 mm/m.y. by ^{10}Be and 2.3 mm/m.y. by ^{26}Al methods. The fact that nucleus dating and layer dating methods consistently yield the same range of rates of ferromanganese accumulation suggests that manganese nodules on the deep sea floor grow slowly and that the growth has probably been continuous. Our amino acid dates support the premise that the rate of growth of manganese nodules in the deep sea is slow, as summarized by Ku (1977), and not rapid, as proposed by Lalou and Brichet (1972).

Summary

Amino acid dating methods have been applied to the cetacean ear bones that have served as nuclei of eight manganese nodules from the North Pacific Ocean. From the extent of racemization of isoleucine and aspartic acid, the ages and rates of ferromanganese oxide accretion were estimated. Age estimates range from about 0.4 to 6 m.y., rates of growth from less than 1 to about 10 mm/m.y. These growth rates compare very favorably with rates obtained by other nucleus dating methods (K-Ar and fission track) and by layer dating methods (^{230}Th, ^{231}Pa, ^{10}Be, ^{26}Al). The growth rates determined by amino acid racemization support the general consensus that manganese nodules on the floor of the deep sea have a slow rate of accretion of ferromanganese oxides.

References

Aumento, F., (1969) The Mid-Atlantic Ridge near 45 N; V: fission track and ferromanganese chronology, Can. Jour. Earth Sci., 6, 1431-1440.

Bada, J.L., (1972) The dating of fossil bones using the racemization of isoleucine, Earth and Planetary Science Letters, 15, 223-231.

Bada, J.L. and Schroeder, R.A. (1972) Racemization of isoleucine in calcareous marine sediments: kinetics and mechanisms. Earth and Planetary Science Letters, 15, 1-11.

Bada, J.L., Kvenvolden, K.A. and Peterson, E., (1973) Racemization of amino acids in bones, Nature, 245, 308-310.

Bada, J.L. and Protsch, R., (1973) Racemization reaction of aspartic acid and its use in dating fossil bones. Proc. Natl. Acad. Sci. USA, 70, 1331-1334.

Bada, J.L., Schroeder, R.A. and Carter, G.F., (1974) New evidence for the antiquity of man in North America deduced from aspartic acid racemization, Science, 184, 791-793.

Bada, J.L. and Helfman, P.M., (1975) Amino acid racemization dating of fossil bones, World Archaeology, 7, 160-173.

Barnes, S.S. and Dymond, J.R., (1967) Rates of accumulation of ferromanganese nodules, Nature, 213, 1218-1219.

Cronan, D.S., (1977) Deep-sea nodules: distribution and geochemistry, In: Glasby, G.P., ed., Marine Manganese Deposits, 11-44, Elsevier, Amsterdam.

Dungworth, G., Vincken, N.J. and Schwartz, A.W., (1974) Racemization of aliphatic amino acids in fossil collagen of Pleistocene, Pliocene and Miocene ages, In: Tissot, B., and Biener, eds., Advances in Organic Geochemistry 1973, 689-700. Editions Technip, Paris.

Guichard, F., Reyss, J.L., and Yokoyama, Y., (1978) Growth rate of manganese nodules measured with ^{10}Be and ^{26}Al, Nature, 272, 155-156.

Ku, T.L., (1977) Rates of accretions, In: Glasby, G.P., ed., Marine Manganese Deposits, 249-267, Elsevier, Amsterdam.

Kvenvolden, K.A., Peterson, E., and Pollock, G.E., (1972) Geochemistry of amino acid enantiomers: gas chromatography of their diastereomeric derivatives, In: V. Gaertner, H.R. and Wehner, H., Advances in Organic Geochemistry, 1971, 387-401, Pergamon Press, Braunschweig.

Lalou, C. and Brichet, E., (1972) Signification des mesures radiochimiques dans l'evaluation de la vitesse de croissance des nodules de manganese. Comptes Rendus Acad. Sci., Paris, 275D, 815-818.

Raad, W.J. and Meylan, M.A., (1977) Morphology, In: Glasby, G.P., ed., Marine Manganese Deposits, 109-146, Elsevier, Amsterdam.

Wooster, W.S. and Volkman, G.H., (1960) Indications of deep Pacific circulation from the distribution properties at five kilometers, Jour. Geophys. Res. 65, 1239-1250.

THE DISTRIBUTION OF TOTAL ALPHA RADIOACTIVITY IN SELECTED MANGANESE NODULES FROM THE NORTH PACIFIC: IMPLICATIONS FOR GROWTH PROCESSES

J. D. Macdougall

Scripps Institution of Oceanography A-020

La Jolla, California 92093

Abstract

Analyses of the total alpha radioactivity distribution in 31 North Pacific manganese nodules yield an average growth rate of 6.8 (±2) mm/m.y. Samples from a single box core show almost identical total activity vs. depth profiles, regardless of nodule size or shape. A characteristic feature of many nodules examined is an apparent change (steepening) in slope of the alpha activity profile between depths of 0.5-1.5 mm. This feature may reflect a fundamental feature of nodule growth, possibly mobility of ^{230}Th or its daughters in a regular way, interrupted growth, or nodule turnover. Near-surface decreases in total alpha activity probably result from ^{226}Ra loss.

Introduction

Knowledge of the rate at which manganese nodules accumulate is important for a number of reasons. Coupled with chemical analyses, growth rate studies permit calculation of the rate of incorporation of individual chemical constituents into nodules. Thus knowledge of growth rates may be directly useful for comparing hypotheses of manganese nodule formation, and for elucidating the source of nodule components. In addition, accurate knowledge of the rate of incorporation of some elements (e.g. Mn, Ni, Cu, etc.) is important in understanding their overall oceanic budgets.

Several methods have been used to obtain accumulation rates for manganese nodules (see Ku, 1977, for a recent review). The most common is the excess ^{230}Th method, sometimes referred to as the ionium or ionium-thorium technique. In this approach, the decay rate of excess ^{230}Th--which is produced by radioactive decay of dissolved seawater uranium, rapidly removed from the water column, and incorporated into the surface layers of nodules and other sediments--is measured as a function of depth below the nodule surface. For such analyses, material is scraped successively from thin layers of the nodule, uranium and thorium are separated chemically, and isotopic measurements are made by alpha spectrometry. In this report a somewhat different technique, total alpha counting, is employed. Since excess ^{230}Th ($^{230}Th_{ex}$) and its daughter products in the ^{238}U decay series dominate the alpha activity near the surface of most manganese nodules, profiles of total alpha activity vs. depth can be expected to approximate $^{230}Th_{ex}$ profiles. Ku and Broecker (1969) found that accumulation rates determined by thick source total alpha counting agreed to within 30% with those obtained for the same nodules by chemical separation and alpha counting of ^{230}Th. In the present work cellulose nitrate sheets, which record tracks of alpha particles, were used to make autoradiographs of the distribution of alpha activity in slices cut through the centers of nodules. Alpha activity vs. depth profiles were then measured nondestructively by optical microscopy on the plastic sheets. This technique, discussed by Andersen and Macdougall (1977), has the advantages over radiochemical procedures that it is simple, rapid, nondestructive and permits high spatial resolution. It also has the disadvantage that mobility of any ^{230}Th daughter products will produce spurious results. However, empirically, such effects do not appear to be overly serious in most cases (Andersen and Macdougall, 1977). A conceptually similar technique employing nuclear emulsion detectors has been used by Heye (1975).

Experimental

The experimental procedures used in this work are described in detail by Andersen (1978), and briefly summarized here. A sheet of cellulose nitrate (Kodak CA80-15) was fastened to a section cut through the center of each nodule. Exposure time varied, but was typically 3-6 months. After etching the cellulose nitrate in a sodium hydroxide solution, alpha track measurements were made by optical microscopy at a magnification of $\sim$600 X. Typically the profiles were measured along a 1- to 2- mm-wide radial strip. Locations at which the profiles were measured were carefully chosen. Unless otherwise noted, all measurements were made for the top (seawater-facing) nodule side if it was identifiable. Additionally, only regions where the outermost surface was intact were chosen for measurement.

Only alpha particles with energies $\lesssim$ 4 MeV are detected under the experimental conditions of this work. Because ^{230}Th and its daughter isotopes produce alpha particles spanning a range of energies, particles which form tracks at a particular location on the detector do not necessarily all originate at the same position in the nodule. Because of this variability in alpha particle range, the spatial resolution of the alpha track method is of the order of 30 μm. Figure 1 illustrates the appearance of alpha particle tracks in a cellulose nitrate detector.

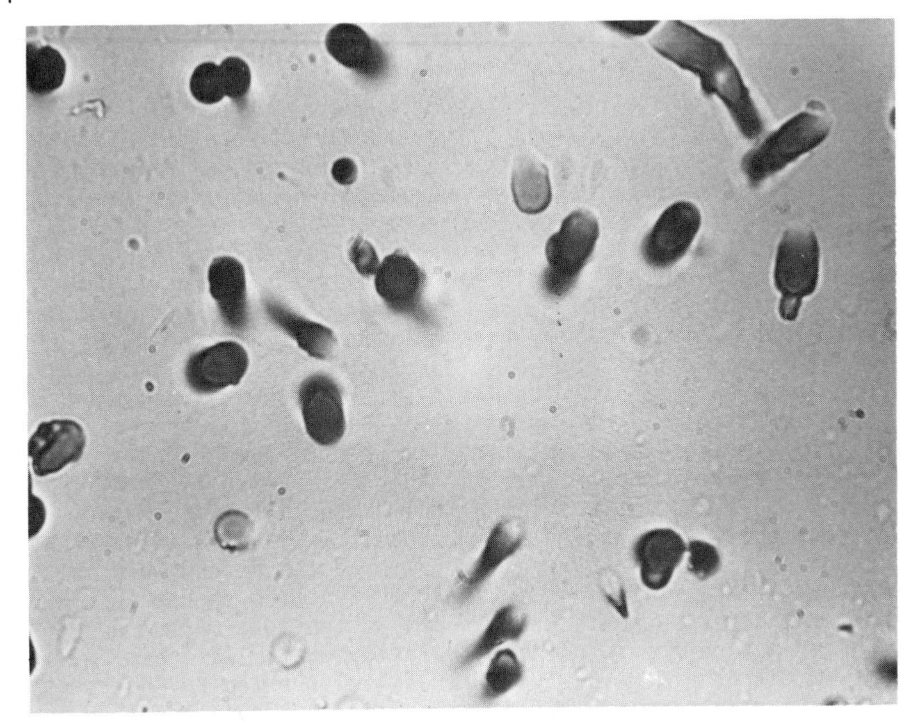

Fig. 1. Photomicrograph of alpha particle tracks in cellulose nitrate detector from DOMES nodule P2. Width of field is 93 μm.

The raw data collected in this work are track densities, normally measured for 100 μm depth intervals. The final results are accumulation rates, calculated from background and ^{231}Pa corrected data. Background corrections (for the contribution from alpha emitters such as ^{238}U, ^{232}Th, and their daughters) were based on the observed alpha activity in each nodule at depths where $^{230}Th_{ex}$ was expected to be negligible and the alpha track profiles became flat, i.e. at depths of a few mm. For each nodule, several background values were measured and averaged. Maximum point-to-point variations in background activity were less than a factor of two, and generally 20-30%. Typically, background track densities were 1-2 orders of magnitude less than those at the nodule surfaces. The ^{231}Pa correction was calculated during an iterative, weighted least squares routine used to determine growth rates. A surface $^{230}Th_{ex}/^{231}Pa_{ex} = 5$ was assumed. For the growth rates reported here, the actual magnitude of the ^{231}Pa correction is $\lesssim$ 10% if the real $^{230}Th/^{231}Pa$ ratio is >2. Radiochemical data indicate that such a ratio is likely (e.g. Ku and Broecker, 1969).

Results

All nodules for which growth rate data are reported here are from the North Pacific (0°-40°N, 110°-180°W) and were collected at depths greater than 2500 m. In Table 1, locations, water depths and calculated growth rates are summarized for individual samples. Some of these results were previously reported by Andersen (1978). With a few exceptions, noted in later sections, all of these samples exhibit "Type 1" (simple exponential) alpha activity profiles (Heye, 1975). Following the scheme introduced by Andersen (1978), only profiles for which no more than four consecutive data points (i.e. counts over a total depth interval of ~400 μm) fall more than 2σ away from the least squares line fit to the data were considered to be Type 1 profiles. Although this is a useful criterion and preserves objectivity, it undoubtedly conceals some interesting variations from strictly exponential profiles. For example, Heye (1975), in a somewhat more subjective examination of alpha track profiles measured using nuclear emulsions, interpreted many such deviations as changes in growth rate. Recent work by Krishnaswami et al. (1978) on several nodules from the northeast equatorial Pacific suggests that breaks in the total alpha activity curves may be the result of nodule turnover. In addition, close examination of the 65 Type 1 profiles reported by Andersen (1978) shows that many exhibit a common type of deviation from linearity on a plot of log track density vs. depth: a sharp decrease in total alpha activity at a depth of ~1 (±0.5) mm, followed by a region of relatively constant, or in some cases even slightly increased, activity. A similar feature is present in many of DOMES and PUMICE Nodule samples reported here. One such profile is shown in Fig. 2a (Domes, Station 25, Nodule F). In this case, four of the 22 data points fall away from the least squares fit just outside 2σ error limits (note that errors plotted in Fig. 2a are ±1σ). Only two of these four are consecutive data points (at 0.45 and 0.55 mm). Thus, following the criteria described earlier, this is a "Type 1" activity profile. Averaging over intervals of either 300 μm or 400 μm--a typical depth interval used for radiochemical $^{230}Th_{ex}$ studies in the outer portions of nodules--makes no difference in the calculated growth rate, but it does obscure the apparent fine structure evident from the 100 μm points (Fig. 2b). If the assumption of the total alpha activity method that measured profiles primarily reflect $^{230}Th_{ex}$

TABLE 1

Samples discussed in this paper

Sample	Location	Water depth (m)	Growth rate (mm/m.y.)
Vitiaz A17863	29.97°N 125.92°W	4324	4.1 (±0.5)
MR A17875	22.50°N 113.00°W	3603	2.9 (±0.4)
Ris 4G	22.28°N 117.40°W	3890	7.6 (±1.2)
Styxx 48	21.45°N 126.72°W	4300	8.9 (±0.8)
DWD BD1	21.45°N 126.72°W	4300	6.6 (±1.4)
MR 55	20.75°N 114.25°W	3740	10.9 (±1.5)
Explore A17865	19.33°N 114.20°W	3400	7.8 (±0.9)
DSV 53P-8	14.97°N 125.00°W	5000	5.4 (±0.4)
Mn7402 FFG8	14.97°N 125.00°W	4460	9.2 (±0.9)
Mn7402 D1	14.97°N 125.00°W	4465	5.0 (±0.4)
Mn7402 D1	14.97°N 125.00°W	4500	5.9 (±0.4)
Expo 60-4	14.47°N 107.8 °W	3678	5.9 (±1.7)
Mid Pac 5-1	14.37°N 133.12°W	4990	6.9 (±0.9)
Mid Pac 16-1	13.75°N 149.25°W	4564	2.8 (±2.3)
Mn7402 FFG 37	11.10°N 140.00°W	4800	4.6 (±0.3)
Mn7601 20-B2	11.07°N 140.05°W	4722	6.7 (±0.4)
MSN 150-G	10.98°N 143.37°W	4978	3.8 (±0.3)
DOMES F2	14.25°N 124.98°W	4560	5.9 (±0.4)
DOMES H2	14.25°N 124.98°W	4560	8.2 (±0.6)
DOMES V2	14.25°N 124.98°W	4560	6.6 (±0.6)
DOMES E3	14.25°N 124.98°W	4560	6.9 (±0.6)
DOMES R3	14.25°N 124.98°W	4560	6.0 (±0.9)
DOMES P2	14.25°N 124.98°W	4560	7.0 (±0.5)
DOMES U	14.25°N 124.98°W	4560	6.4 (±0.5)
DOMES M	14.25°N 124.98°W	4560	6.9 (±0.5)
PUMICE Mn68-1	40.34°N 175.63°W	5307	10.7 (±1.5)
PUMICE Mn68-2	40.34°N 175.63°W	5307	6.8 (±1.0)
PUMICE Mn67-1	40.30°N 173.77°E	4399	10.7 (±1.9)
PUMICE Mn67-2	40.30°N 173.77°E	4399	6.7 (±1.2)
PUMICE Mn61	31.29°N 174.03°E	5362	6.7 (±0.5)
PUMICE 70-1	38.94°N 178.62°E	5398	11.2 (±1.1)

decay is correct, these results imply that the typically exponential decay profiles observed in radiochemical $^{230}Th_{ex}$ studies of nodules may be artifacts, and that finer scale sampling will be necessary to determine the true shape of the profiles.

The cause of the peculiar shape of many of the alpha track profiles cannot be determined from the total alpha data alone and

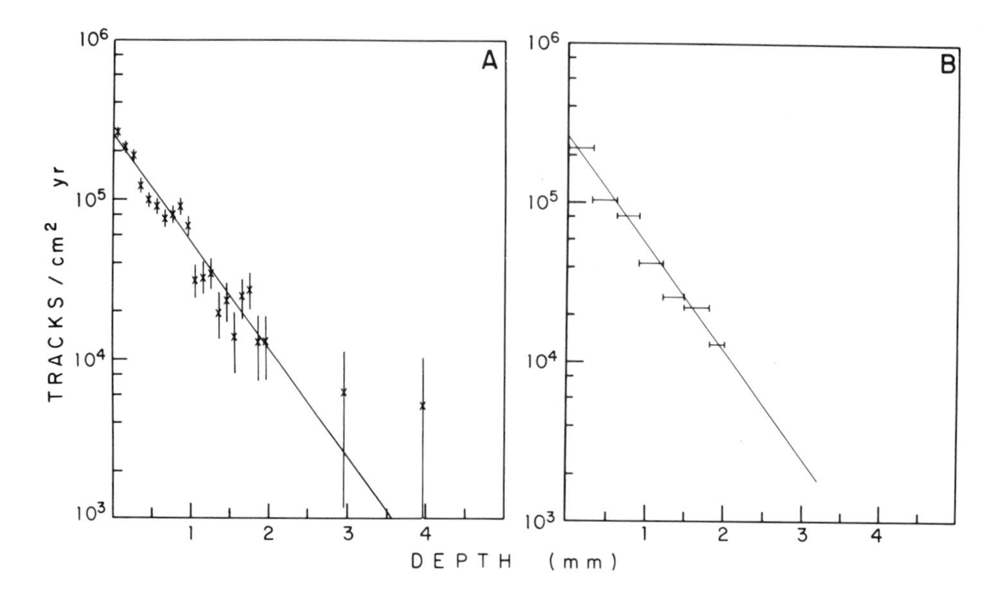

Fig. 2. Alpha track data for DOMES nodule F2, corrected for ^{231}Pa and background contributions. a) Data as measured in 100 μm intervals for an exposure of 183 days. Errors shown are $\pm 1\sigma$ counting errors. b) Same data averaged over 300 μm intervals (no uncertainty indicated). The regression line is the fit to data from (a), which yields a growth rate of 5.9 ($\pm$0.4) mm/10^6 yr.

will obviously require detailed sampling for radiochemical measurements. In addition to causes related to the distribution of ^{230}Th itself, migration of daughters in the decay series also may be involved. For example, Ku and Broecker (1969) have observed ^{226}Ra/^{230}Th disequilibrium in a number of nodules. The repetition of the same general shape in the total alpha track profiles for nodules from widely varying locations suggests that a general process is involved rather than a growth feature peculiar to individual nodules or to specific nodule locations.

Andersen and Macdougall (1977) showed that, in general, growth rates determined from total alpha track profiles are consistent with those based on radiochemical $^{230}Th_{ex}$ measurements. In addition, Ku (1977) has shown that the $^{230}Th_{ex}$ method gives results consistent with other types of radiochemical measurements. Therefore, in this report the calculated growth rates based on alpha track profiles are considered to be true nodule growth rates, except in the cases where profiles are "anomalous" (Andersen, 1978).

As discussed by Andersen and Macdougall (1977), diffusion of ^{226}Ra from nodule surfaces (e.g. see Cochran and Krishnaswami, 1977) may have the effect of flattening the slope of the alpha track profiles, thus increasing slightly the accumulation rates calculated from the total alpha data. For the 31 nodules listed in Table 1, the mean growth rate is 6.8 mm/m.y. The distribution of growth rates about the mean is reasonably symmetrical (Fig. 3), with a standard deviation of 2.0 mm/m.y. In contrast, growth rate measurements for North Pacific nodules made by the radiochemical

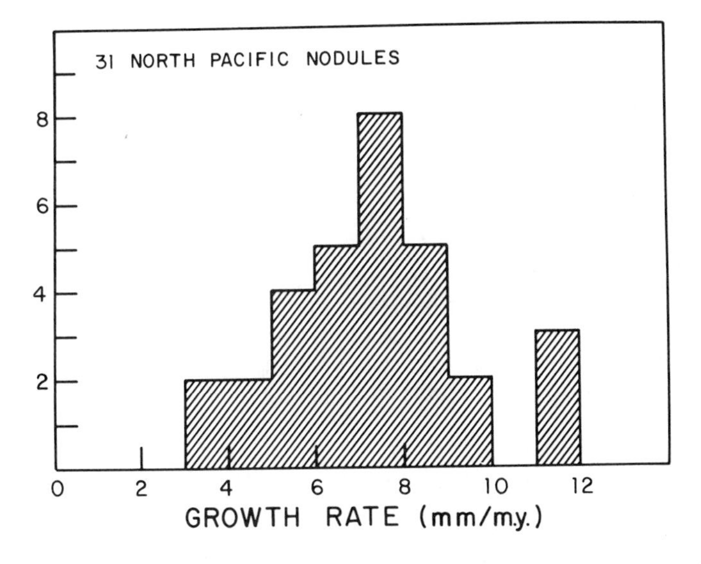

Fig. 3. Histogram showing the distribution of growth rates for nodules discussed in this paper.

^{230}Th technique (tabulated by Ku, 1977) range from 2.1 to 40 mm/m.y. The mean of the 13 measurements compiled by Ku is 10.7 mm/m.y., with a standard deviation of 12.3 mm/m.y. However, the mean for this set of values is heavily influenced by the high growth rates measured for two nodules by Nikolayev and Yefimova (1963). Omitting these two samples, and a third nodule collected from a shallow depth of 1464 m which is not strictly comparable with the samples examined in this study, gives a mean growth rate for the remaining ten samples of 5.6 mm/m.y., with a standard deviation of 4.7 mm/m.y. Thus, although the distribution of measured rates about the mean is broader in the case of the radiochemical measurements, the average growth rate calculated for North Pacific nodules agrees well with that calculated from the alpha track data. In the following sections, some individual samples and groups of samples are discussed in greater detail.

DOMES Station 25

Several nodules from a box core taken by NOAA personnel on board the R/V *Oceanographer* (RP-6-OC-76) were analyzed in order to examine both the relationship among a number of nodules from the same location and the distribution of total alpha activity as a function of size. Surface nodules from this core, arranged according to size, are shown in Fig. 4. The eight samples measured for alpha activity are identified on this photograph using designations given by Greenslate (1977) (nodules M,U,F2,H2,P2,V2,E3, and R3), and the alpha activity profile for one (F2) is shown in Fig. 2. Growth rates calculated from the measured alpha track profiles range between 5.9 and 8.2 mm/m.y. (Table 1), with a mean of 6.7 mm/m.y. With two exceptions (nodules U and R3), the alpha track profiles are the Type 1 variety. The two exceptions are both cases in which the type of fine structure evident in the alpha track profile for nodule F2 (see Fig. 2) is more pronounced. Overall, the shapes of the profiles are remarkably similar from nodule to nodule. There is no correlation between nodule size and either profile shape or calculated growth rate. The calculated growth rates cluster quite closely, with only two of the eight falling more than 1σ away from the mean.

For this set of nodules, then, it appears that growth rates have been relatively constant and approximately equal over about the past 300,000 years. In spite of this, the surface total alpha activity, derived from the least squares fit to the data, varies by a factor of 3 among the eight nodules (Table 2). The variation in the measured alpha track density for the outermost (0-100 μm) nodule interval, uncorrected for background or ^{231}Pa contribution, is almost as great (Table 2). The reason for these differences is not clear. Because a rather small portion of the nodule is involved in these measurements, the variations may be due to the differences in surface geometry (and hence ^{230}Th supply) at the location chosen for measurement. Alternatively, variable surface abrasion, turnover, or periods of no growth for some nodules may be responsible. Measurements of a large number of adjacent profiles for a single nodule slice may help to clarify this problem.

Pumice Nodules

A series of six manganese nodules developed as crusts on nuclei of pumice (now highly altered) from the North Pacific between 31° and 38°N and 174°W to 178°E were analyzed. The crusts are very

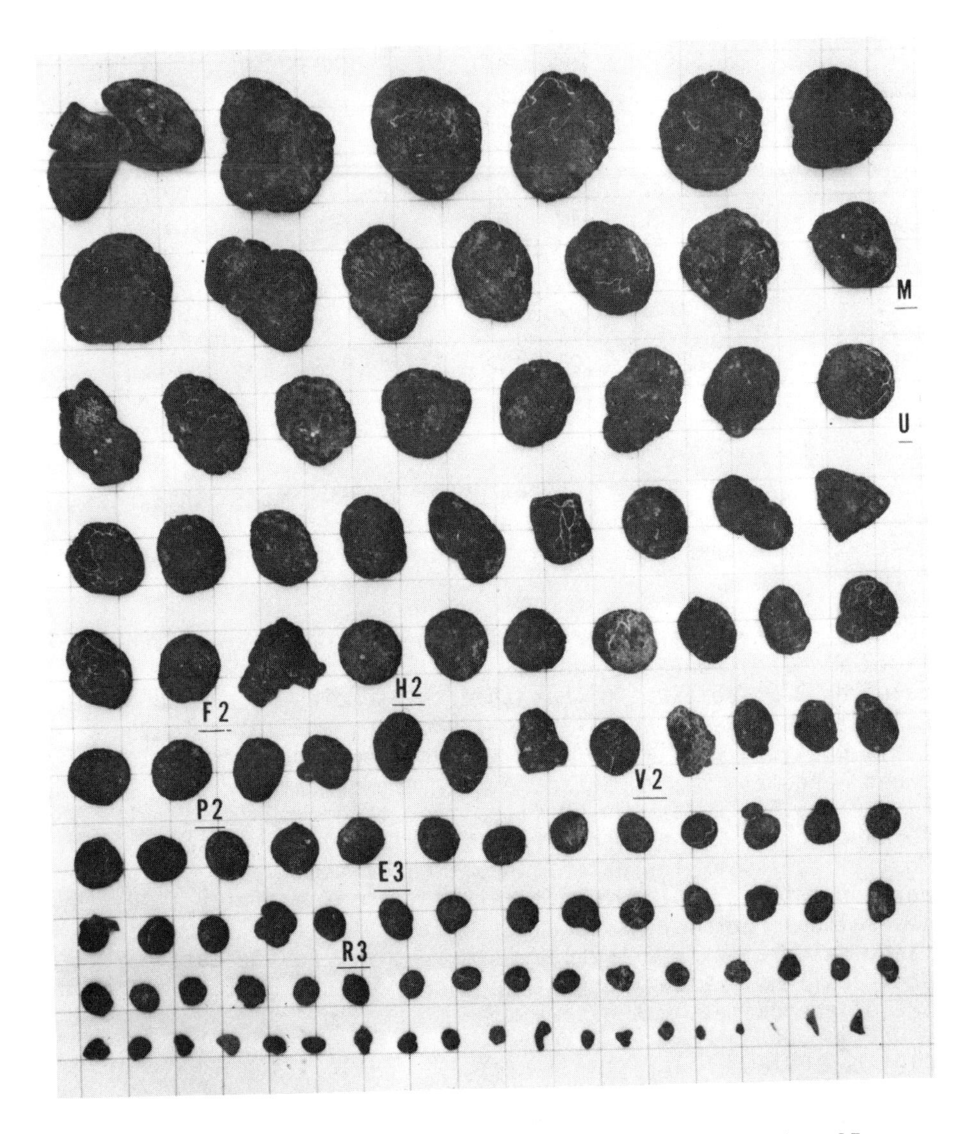

Fig. 4. Surface nodules recovered from DOMES Station 25 Box Core #2. Nodules analyzed in this work are designated by symbols at the lower right side of each sample. Grid lines are 1.5 cm apart.

TABLE 2

Surface activities

Sample	Surface intercept from least squares fit (tracks/cm^2·yr)	Measured track density, 0-100 μm interval (tracks/cm^2)
M	1.40(±0.09)x10^5	0.87(±0.07)x10^5
U	2.28(±0.17)x10^5	1.27(±0.08)x10^5
F2	2.56(±0.16)x10^5	1.61(±0.09)x10^5
H2	1.47(±0.08)x10^5	0.91(±0.07)x10^5
P2	0.94(±0.07)x10^5	0.61(±0.06)x10^5
V2	1.57(±0.11)x10^5	1.03(±0.07)x10^5
E3	2.79(±0.21)x10^5	1.55(±0.09)x10^5
R3	1.90(±0.21)x10^5	1.13(±0.07)x10^5

finely laminated, range between about 1 and 8 mm in thickness, and are quite uniformly developed around the volcanic nuclei. In most cases the manganese coating is rather thin compared to the pumice core.

The alpha track growth rates for these nodules are given in Table 1. They range between 4.8 and 11.2 mm/m.y. with a mean value of 8.5, somewhat higher than the samples discussed above. All but one (68-2) exhibit Type 1 profiles according to the criteria outlined earlier. There appears to be no correlation between manganese thickness and either growth rate or profile shape. Figure 5 illustrates two profiles from this group of samples, one for a thin (67-2) and one for a thick (70-1) manganese coating. Most of these samples exhibit profiles of the type described earlier, with a rapid dropoff in total alpha activity at depths of several hundred microns to a millimeter. Surface alpha activities, calculated from the least squares fit to the data points, show a slightly wider range than those for the DOMES samples, but are of approximately the same magnitude.

Pleiades MN7601: detailed profiles

As part of a larger interdisciplinary study of a single nodule, the alpha activity distribution in a slice of an oriented nodule from a box core taken on the Pleiades expedition (See Spiess and Greenslate, 1976) was measured. A sketch showing the shape of the slice cut for alpha track work and locations of the measured profiles is given in Fig. 6. In Fig. 7 are shown the four profiles measured to date. For this study measurements were made in 50-μm- depth intervals over the first half millimeter, and

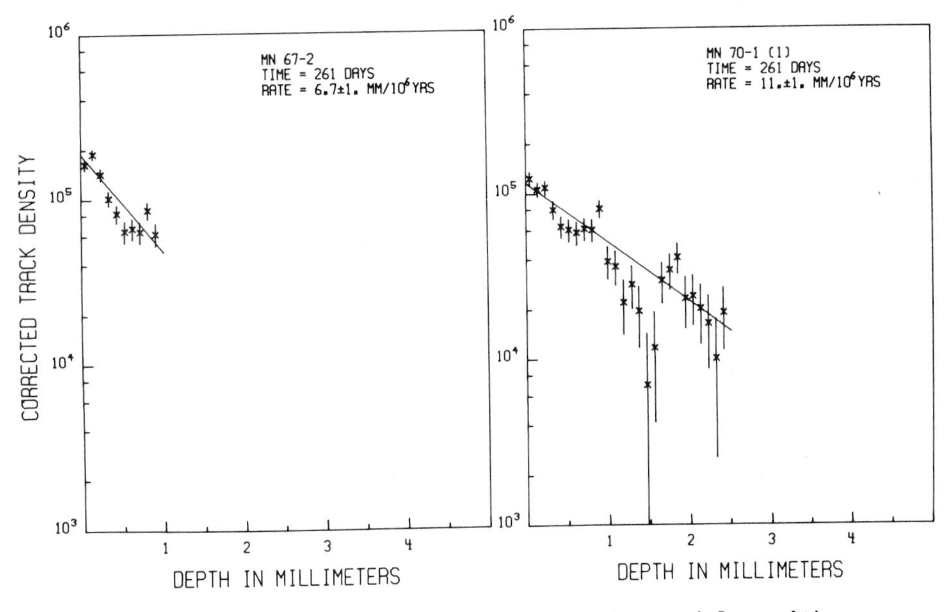

Fig. 5. Alpha track profiles for pumice nodules with thick (Mn 70-1) and thin (Mn 67-2) manganese crusts.

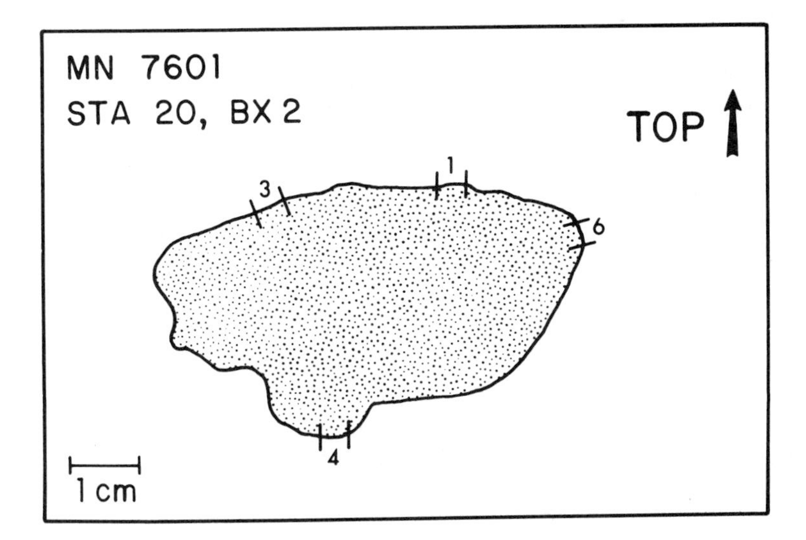

Fig. 6. Sketch showing locations of the profiles presented in Fig. 7.

thereafter every 500 μm.

The weighted least squares fits to the data points for profiles 1 and 3, both on the "top" side of the nodule, give growth rates of 7.6 (±1.0) and 8.2 (±1.0) mm/m.y. respectively. Thus, ignoring

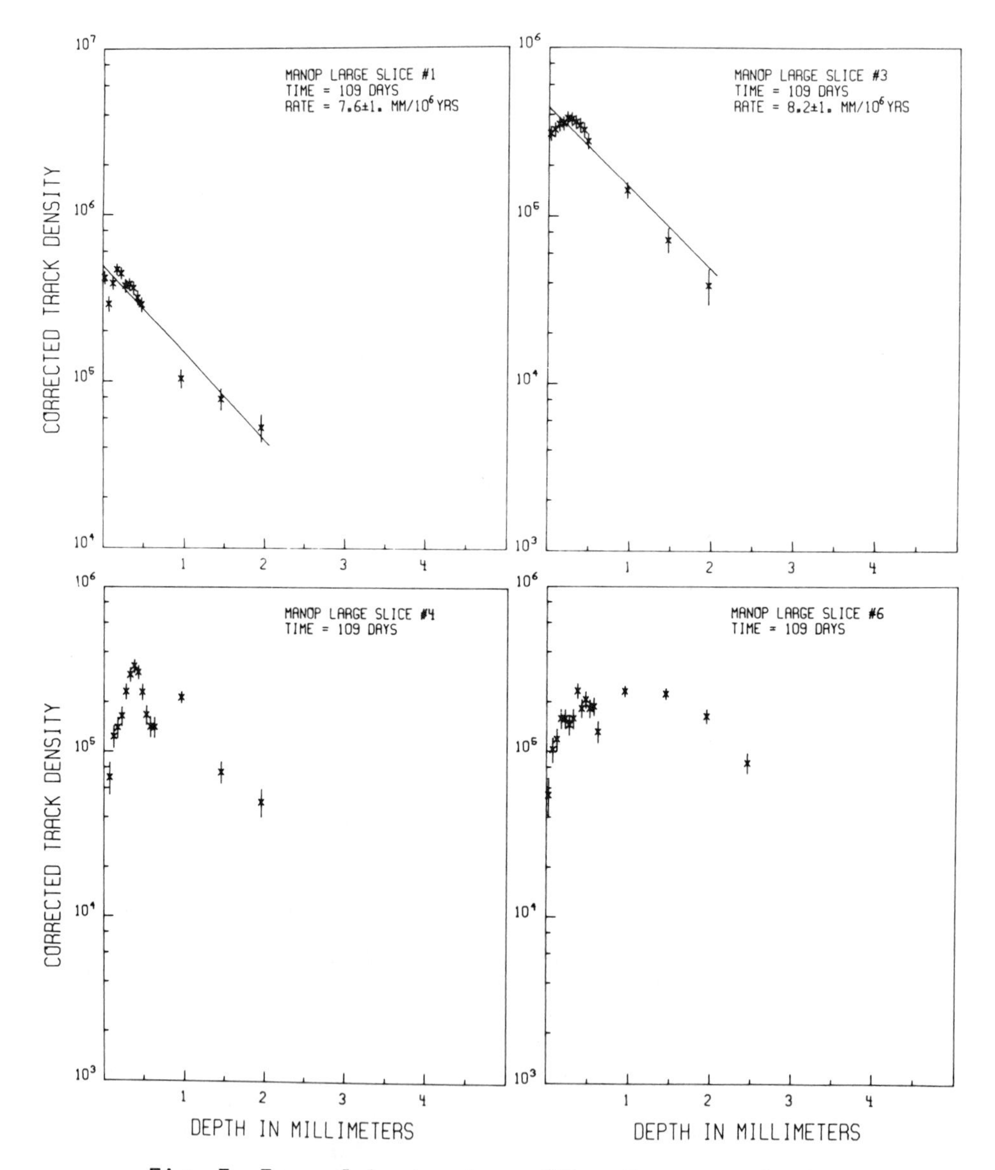

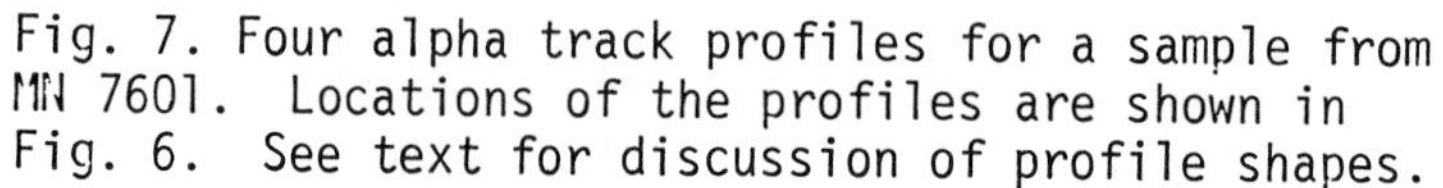

Fig. 7. Four alpha track profiles for a sample from MN 7601. Locations of the profiles are shown in Fig. 6. See text for discussion of profile shapes.

the fine structure which is evident, two profiles about a centimeter apart correspond to accumulation rates which are the same within experimental error. However, both of these profiles also show activities over the first 150-200 μm which are lower than that at the 250 μm point. Especially from the data for profile #3, it is evident that the low near-surface points are not due to experimental error, but exhibit a regular behavior, with the track density increasing from the surface to a depth of 250 μm. This is likely due to loss of radium and/or radon and its daughters from the outermost nodule layers. If this is the case, the profiles suggest that some process, perhaps diagenesis, acts to inhibit significant loss at deeper positions in the nodule. Alternatively, simultaneous inward and outward radium diffusion from the near-surface radium maximum might explain the observed profile shapes. Loss of radium and radon from nodule surface regions, but retention at depths below a few hundred microns, would explain the generally exponential total alpha activity profiles observed by Andersen (1978) and the slightly higher growth rates calculated by this method in comparison with those determined from radiochemical $^{230}Th_{ex}$ studies. Some evidence for such a process exists from radiochemical data from a single North Pacific nodule (Cochran and Krishnaswami 1977) which shows a deficiency of ^{226}Ra to a depth of ∿0.4 mm.

If the first three and four points respectively are omitted from profiles number 1 and 3, the calculated growth rates become 6.6 (±0.5) and 6.7 (±0.3) mm/m.y. The latter value is listed in Table 1 for this nodule.

Profiles 4 and 6, from the bottom and side of the nodule respectively, are quite different from those observed for the upper surface. Both exhibit low activities at the very surface, and a rather rapid increase in activity to depths of about 400 μm. Total activities at these depths are roughly similar to those at the same depth in the "top" profiles. Neither the shape of these two profiles nor the magnitude of the total alpha activity can be understood solely on the basis of the alpha track data. Radiochemical analyses for ^{230}Th and its daughters, particularly ^{226}Ra, will be necessary. However, it is clear that at least with respect to alpha emitting nuclides, quite different processes of incorporation and possibly migration occur on the "up and "down" sides.

Conclusions

Analyses of total alpha activity in a large number of North Pacific manganese nodules indicate that growth rates average 6.8 mm/m.y. Most of the measured total alpha activity profiles are approximately exponential with depth, although deviations occur.

For the most part, these deviations do not appear to be due to changes in growth rate unless they reflect a Pacific-wide change. This is considered unlikely because the widespread presence of a sharp dropoff in total alpha activity occurs at different depths in different nodules. In general it is observed at depths between 0.5 and 1.5 mm, and occasionally even outside these limits. Its depth of occurrence does not correlate with growth rate. For a growth rate of 6.8 mm/m.y., 1 mm corresponds to a time interval of $\sim$150,000 years.

To the extent that the sample is representative, analyses of a group of nodules from a single box core show that nodules from a single location grow at about the same rate. The average growth rate for these samples is close to the average for all North Pacific samples measured in this work, and the variation is almost within experimental error at the 2σ level.

For a single North Pacific nodule, detailed alpha track profiles along four traverses at locations on the top, side and bottom surfaces reveal quite different alpha activity distributions. Profiles from the seawater-facing surface are approximately exponential and indicate growth rates consistent with measurements on other nodules. However, the high spatial resolution employed for these profiles (50 μm) also reveals low activity near the surface, possibly due to radium and/or radon diffusion. Diagenetic changes may decrease mobility at greater depths. Bottom and side profiles also show low surface values, increasing inward over the first several hundred microns, but their overall shape is quite different. Presumably this difference results from different incorporation processes on the two sides of the nodule.

Acknowledgements

I thank J. Carlson and K. Garret for assistance. M. Andersen kindly allowed use of data from his M.S. thesis, and J. Bischoff and J. Greenslate provided some of the samples analyzed. I am grateful to J. Greenslate for permission to use Fig. 4. Work supported by NSF-IDOE (OCE77-01158).

References

Andersen, M. E. (1978) Accumulation rates of manganese nodules and sediments: An alpha track method. Unpublished M.S. thesis, Univ. Calif. San Diego, 113 p.

Andersen, M.E. and Macdougall, J.D. (1977) Accumulation rates of manganese nodules and sediments: An alpha track method, Geophys. Res. Lett. 4, 351-353.

Cochran, J. K. and Krishnaswami, S. (1977) U/Th series nuclides in sediment, pore water and manganese nodules of the North Equatorial Pacific, Trans. Am. Geophys. Union 58, 420.

Greenslate, J. (1977) Manganese concretion wet density: a marine geochemical constant, Marine Mining 1, 125-148.

Heye, D. (1975) Wachstumverhaltnisse von Manganknollen, Geologisches Jahrbuck, Reihe E, Heft 5.

Krishnaswami, S., Cochran, J. K., Turekian, K. K. and Sarin, M. M. (1978) Time scales of deep-sea ferromanganese nodule growth based on ^{10}Be and alpha track distributions and their relation to uranium decay series measurements. Preprint.

Ku, T. L. (1977) Rates of manganese accretion. In Marine Manganese Deposits (editor G. P. Glasby) Elsevier.

Ku, T. L. and Broecker, W. S. (1969) Radiochemical studies on manganese nodules of deep-sea origin, Deep-Sea Res. 16, 625-637.

Nikolayev, P. S. and Yefimova, E. I. (1963) On the age of iron-manganese concretions from the Indian and Pacific Oceans, Geokhimiya 7, 678-688.

Spiess, F. N. and Greenslate, J. (1976) Preliminary cruise report, Pleiades Expedition Leg 4, Mn76-01. Seabed Assessment Program, IDOE-NSF, Technical Report No. 15, 87 p.

Be^{10} AND U-SERIES ISOTOPES IN MANGANESE NODULES FROM THE CENTRAL NORTH PACIFIC

T. L. Ku, A. Omura* and P. S. Chen

Department of Geological Sciences
University of Southern California
Los Angeles, California 90007

Abstract

The exponential decreases with depth of Be^{10}, Th^{230} and Pa^{231} in two manganese nodules from the central North Pacific is interpreted in terms of slow growth of the nodules: at average rates of millimeters per million years. We observe in one specimen (Mn 139): Be^{10}-based rate < Th^{230}-based rate < Pa^{231}-based rate. This suggests that diffusion-mixing may have modified the depth gradients of these nuclides, such that the true rate would be close to or even lower than the observed Be^{10} rate of 1.3 mm/10^6yr. Quantitative assessment using a diffusion-decay model indicates that in Mn 139, Th^{230} and Pa^{231} could have been subjected to diffusion-like transport with an effective diffusion coefficient of $\sim 1 \times 10^{-8}$cm^2/yr. After correction for this effect, the Th^{230} and Pa^{231} data give a concordant rate of 1.9 mm/10^6yr, about 2-3 times lower than their uncorrected values. The small and variable amounts of integrated dpm/cm^2 of Be^{10}, Th^{230}_{ex} and Pa^{231}_{ex} found in nodules are best explained by their incorporating only a fraction of the nuclides supplied due to frequent coverage by sediments. The alternative explanation of young "exposure" ages is refutable on several grounds. Among these, we show that diffusion of nuclides required by the exposure-age concept yields depth profiles that are not of the commonly observed exponential form.

---

*Presently at Department of Earth Sciences, Kanazawa University, Kanazawa 920, Japan.

Introduction

The slow growth rates of deep-sea manganese nodules, by and large, have been documented by radiometric measurements. Their long-term, average rates are of the order of millimeters per million years (see Ku, 1977). This is compared with the accumulation rates of millimeters per thousand years for sediments in the nodule fields. The disparity raises the question of how nodules can escape burial by the fast depositing sediments, often leading to arguments against the validity of radiometric dating. One such argument states that the rapid exponential decrease with depth of unsupported Th^{230} and Pa^{231} found in nodule surface layers could be due to surface adsorption and diffusion of these nuclides and/or to sampling artifact resulting from the highly crenulated nodule morphology (Arrhenius, 1967; Lalou and Brichet, 1972), rather than due to radioactive decay. This reasoning is clearly unacceptable if concordant rates are obtained from the distribution of radioactive nuclides with vastly different half-lives. Such has been the case for the simultaneous measurements of Th^{230} and Be^{10} on two nodules: Zetes-3D and Tripod-2D (Bhat et al., 1973). However, the data base for this concordancy check is still considered limited. For Zetes-3D, as an example, the comparison was made based on a total of five analyses of Th^{230} and Be^{10}. The Th^{230} data were from two layers down to a depth of 0.14 mm, whereas the three analyses of Be^{10} covered the depth range of 0-13 mm; no Pa^{231} measurements were made. Recently published results of Guichard et al. (1978) show more detailed Be^{10} profiles obtained for a Mn-encrustation from the south Pacific. Comparative studies of the U and Th decay series nuclides on this same specimen are in progress (Lalou et al., 1978). We now report a similar study carried out on two manganese nodules from the central North Pacific. It was hoped that, with the more detailed measurements than previously available, we would be able to evaluate in a more rigorous way the possibility and implications of the "open system" brought about by the hypothesis of rapid growth of nodules.

Materials Studied

The specimens taken for study are ARIES 13D and Mn 139. ARIES 13D was dredged from a terrace of a seamount in an area with rugged topography near 20°45'N, 173°40'E. Water depths recorded during the dredging operation were 3816-2955 m. The sample is discoidal in shape and about 11 cm at its longest side. As shown in Fig. 1A, the Mn-Fe oxide coating is very asymmetrical around a nucleus of altered basaltic rock. Three layers, L1, L2, and L3, scraped parallel to the surface, are sampled from the thick side (2.2 cm) of the oxide coating. The thickness of each layer is measured at several points using calipers read to 0.05

mm. The opposite side of the coating measures only about 0.2-0.3 cm thick.

The Mn 139 nodule, recovered from a depth of 3916 m at 20°01'N, 136°36'W, is also discoid-shaped with a slightly convexed (upward) bottom (see Fig. 1B, also for dimensions). It has a fairly unusual morphological appearance in that it can be divided into two parts according to the external as well as internal structures. The surface of one part is relatively smooth whereas the other part exhibits rough and knobby surfaces. The "knobs" are surface expressions of small concretions ranging from mm to 2 cm in diameter. Nodule material beneath the smooth surface, similar to that of ARIES 13D, is uniform and almost structureless. The nucleus material is inconspicuous. As sketched in Fig. 1B, layers of various thickness, L1 through L5, are sampled from the "smooth" side. The thickness of the layers is also measured using the calipers. Before sampling of L1-L5, six thin layers (T1 through T6) have been successively scraped

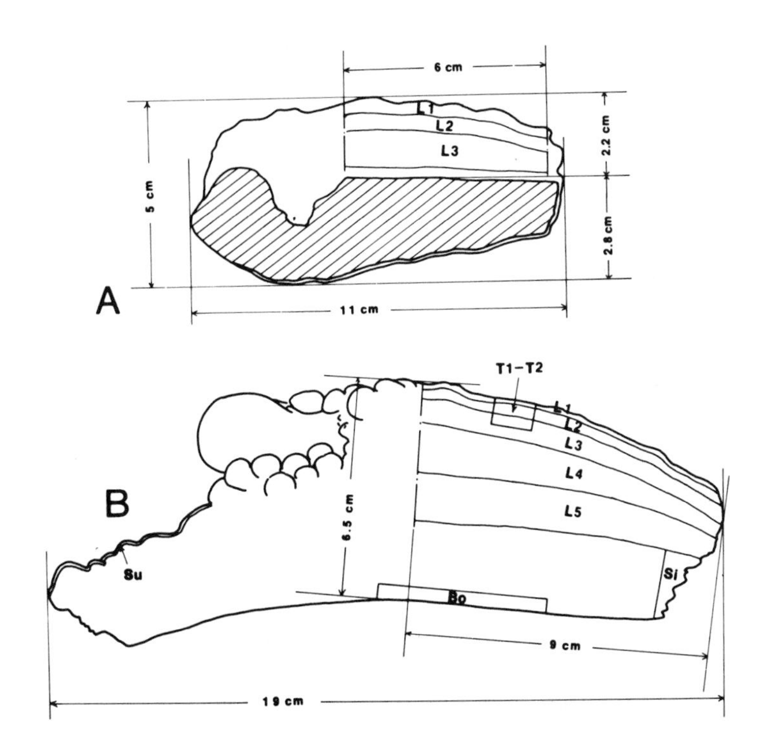

Fig. 1. Sketches of ARIES 13D (A) and a side-view of Mn 139 (B) showing the dimensions and sampled areas. Note the morphological difference in two parts of Mn 139.

from an area of 2.25 cm^2 from the top surface down to a depth of about 7 mm (Fig. 1B). The thickness of each of these thin layers is estimated by the method of Bender et al. (1966), taking the density of nodule material as 2 gr/cm^3.

Three more samples, designated as Bo, Su, and Si, are taken from Mn 139 for analysis. Their approximate position and thickness are shown in Fig. 1B and Table 1, respectively.

Experimental

The sampled nodule material is pulverized to powder form, homogenized, and dried at 110°C before analysis.

Figure 2 gives a flow diagram for Be^{10} analysis. The sample powder is first dissolved in a HCl-H_2O_2 mixture. The solution is centrifuged. The residue is washed with water and the washings added to the solution, to which a $BeSO_4$ carrier-yield tracer (30-60 mg BeO) is added. After heating to decompose H_2O_2, the solution is made to 8N HCl from which iron is removed by extraction with isopropyl ether. Removal of manganese is accomplished by boiling the solution with NaOH. The NaOH washings of the precipitate and the supernatant are combined, from which hydroxides of Al and Be are precipitated with the addition of first HCl, and then NH_4OH. The hydroxides are redissolved in HCl to a pH of 2-3. At this point, a 20% EDTA·2Na solution is added as a masking agent for Al. The solution is heated and its pH raised to 8.5-9 with NaOH. The $Be(OH)_2$ precipitate thus formed is dissolved in HCl to a strength of 8N.

Anion- and cation- exchange techniques are performed afterwards for Be purification. The 8N HCl sample solution is passed through an anion column (Dowex 1x8, 100-200 mesh) prepared with 1 column volume (cv) of 0.1N HCl and 2 cv of 8N HCl. The column is washed with 3 cv of 8N HCl; the washing and the effluent are combined and evaporated to dryness. Beryllium is taken up in 0.1N HCl and the solution is then passed through a cation column (Dowex 50x8, 200-400 mesh) prepared successively with 1 cv of 0.5N HCl, 4 cv of 4N HCl and 2 cv of H_2O. After the column is washed with 2 cv of H_2O and 15 cv of 0.5M oxalic acid, Be is desorbed by elution with 10 cv of a mixture of 0.5N NH_4OH and 0.1N HCl.

For counting of the purified Be, it is precipitated as $Be(OH)_2$ and then ignited to form BeO. The latter is powdered and wetted before being transferred onto the lucite source holder of the beta detector, dried under an infrared lamp, and covered with mylar (0.84 mg/cm^2). The source holder provides an area of 3.9 cm^2 for counting. For each sample mount, the beta activity of Be^{10} is counted twice, usually about 1½ months apart, and using two different counters. These counters are of the design of Lal and Schink (1960). "Q" gas (98.7% helium and 1.3% butane

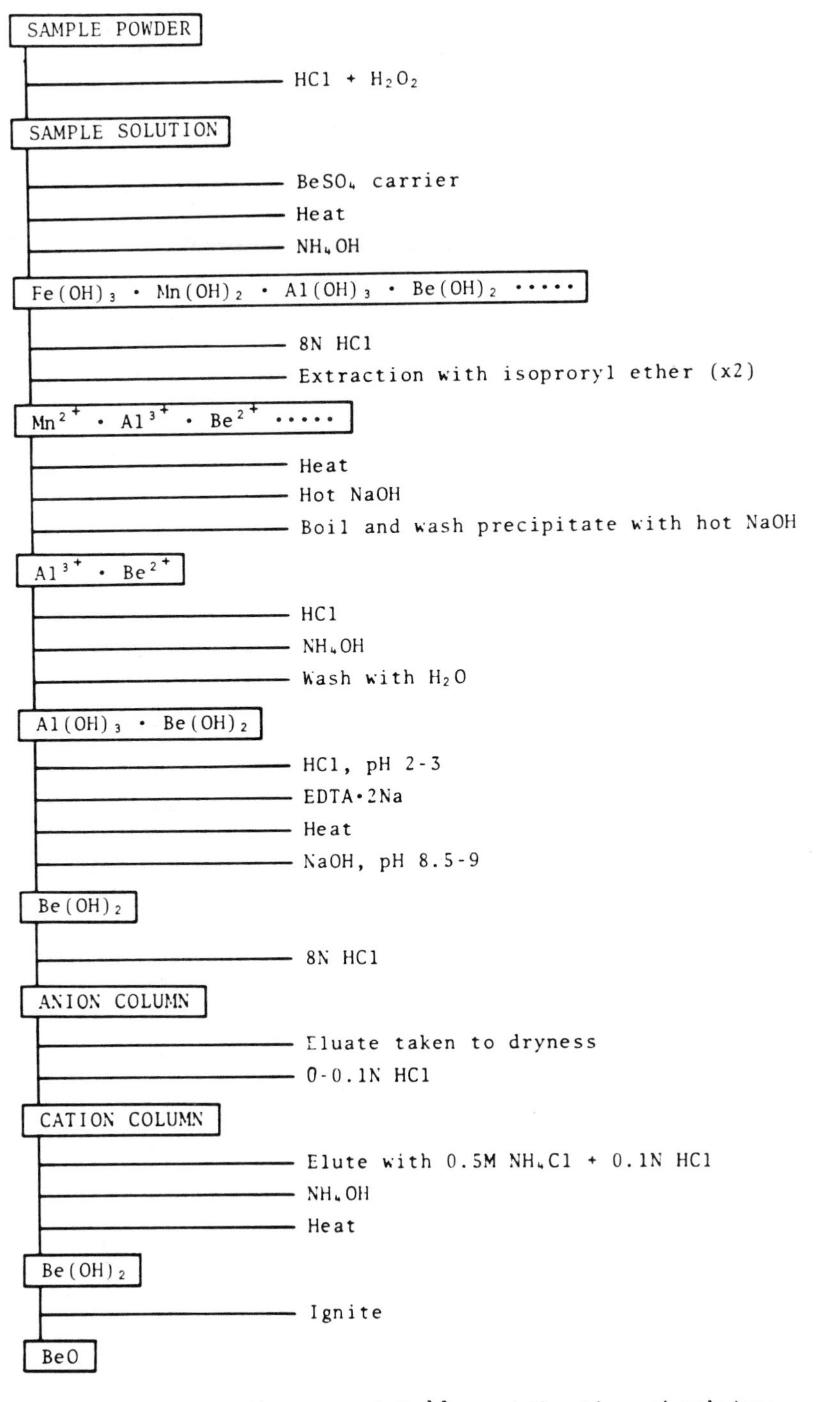

Fig. 2. Flow diagram of Be^{10} purification chemistry.

mixture) is used as the flowing counting gas. The backgrounds of the two counters have remained stable at 10.4 ± .3 and 13.1 ± .6 counts per hour respectively, throughout the 10-month period of the experiment. The counting efficiency, as checked with the four Pb^{210} standards supplied by Dr. Y. C. Chung of Scripps Institution of Oceanography, is 43% and 44% for the two counters. Correction factors for external and self absorption as well as counting efficiency are applied to the observed counting rates for obtaining the absolute disintegration rates.

A blank for Be^{10} has been measured on all the chemicals and apparatus used as in sample runs, following the procedures described above.

Methods for the measurements of uranium, thorium, and protactinium isotopes are those of Ku and Broecker (1969) with minor modifications. The common beryllium (Be^9) is measured using atomic absorption spectrophotometry (Perkin-Elmer 305B, equipped with HGA-2100 graphite furnace).

Analytical Results

Results of the beryllium measurements are presented in Tables 1 and 2; those of the decay series isotopes, in Tables 3 and 4. The quoted errors are standard deviations (1σ) derived from counting statistics. The estimated error in the Be^9 analyses is about ±5%.

As the number of Be^{10} analyses reported herein almost equals those which have appeared in literature to date, an appraisal of the Table 1 data is in order. It is apparent that for the very low abundance of Be^{10} to be detected, purification of Be^{10} against the numerous possible beta-emitting contaminants is exceedingly important. The purity can be assessed via counting techniques: by checking the energy of the beta radiation using absorbers or a two-element counter. But, perhaps a more effective and straight-forward means of checking is to recount the samples after chemical repurification. The results of this test for samples of Mn 139 and the blank are listed in Table 1. Clearly, the high reproducibility of the two purification results attest to the soundness of the analytical scheme. Sample Mn 139-L1, analyzed at the very beginning of the experiment, is one exception. A third purification of the sample has been carried out and found to agree with the second. It should be noted that in repurifying the already prepared BeO, much smaller quantities of reagents are used as compared to those required during the initial Be concentration and purification steps. Hence almost negligible blank corrections (as also reflected by the "Blank" data in Table 1) are applied to the repurification results. The repurifications usually yield high chemical recovery, about 70-90%; the listed chemical yields in Table 1 refer to % recovery of

TABLE 1

Be^{10} Analyses on Manganese Nodules ARIES 13D and Mn 139

Sample No.	Depth (mm)	Weight (g)	Residue (wt.%)	Chemical[1] yield (%)	Net count rate[2] (cph)	Be^{10} Conc.[3] (dpm/kg nodule)
ARIES 13D						
L1	0-4.3	16.7	6.8	24.0	3.56±.47	48.9±6.5
L2	4.3-8.8	29.6	11.8	33.	3.08±.56	17.7±3.2
L3	8.8-18.2	71.1	6.3	44.0	3.21±.57	5.48±.97
Mn 139						
L1	0-0.8	9.22	12.2	(1) 33.4	8.64±.63	171±13
				(2) 21.4	3.80±.48	114±14
				(3) 17.5	3.34±.44	121±16
L2	0.8-3.6	21.0	18.0	(1) 42.9	6.18±.62	46.0±4.7
				(2) 31.1	4.63±.42	46.3±4.1
L3	3.6-9.4	35.5	18.2	(1) 45.3	3.89±.44	16.4±1.9
				(2) 37.8	3.32±.41	16.3±2.1
L4	9.4-24.1	64.0	18.4	(1) 48.3	2.75±.43	5.13±.80
				(2) 38.6	2.18±.40	4.97±.91
L5	24.1-38.4	102	17.3	(1) 45.8	2.06±.44	2.93±.62
				(2) 39.9	1.64±.40	2.65±.65
Bo	59.1-65.0	23.3	9.7	(1) 47.1	0.13±.78	0.75±4.50
				(2) 41.2	0.18±.31	1.16±2.00
Su	1.0	6.6	22.7	(1) 51.7	2.05±.39	42.8±8.3
				(2) 43.2	1.49±.43	37.7±10.8
Si	0-<1.5 (irregular)	15.8	6.9	(1) 44.6	3.95±.41	33.4±3.5
				(2) 39.2	2.95±.41	27.8±3.7
Blank	----	----	---	(1) 57.0	0.89±.37	--------
				(2) 44.4	0.96±.38	--------

1. Number in parenthesis denotes 1st purification, 2nd purification, etc.
2. Beta count rate (for samples) after correction of counter background and blank (yield-normalized).
3. The weight of nodule material is defined as that leachable by $HCl-H_2O_2$.

TABLE 2

Be^{10} Concentration and Be^{10}/Be^{9} in ARIES 13D and Mn 139

Sample No.	Be^{10} conc.[1] (dpm/kg nodule)	Be^{9} conc. (mg/kg nodule)	Be^{10}/Be^{9} (10^{3} dpm/g)
ARIES 13D			
L1	48.9±6.5	9.8	5.0±.7
L2	17.7±3.2	8.8	2.0±.4
L3	5.48±.97	9.0	0.62±.11
Mn 139			
L1	118±11	8.5	13.9±1.5
L2	46.2±3.1	8.3	5.6±.5
L3	16.4±1.4	8.4	2.0±.2
L4	5.05±.61	9.2	0.55±.07
L5	2.79±.45	9.5	0.29±.05
Bo	0.96±2.46	8.9	0.10±.25
Su	40.3±6.8	9.1	4.4±.8
Si	30.6±2.5	9.4	3.3±.3

1. For Mn 139, weighted averages of the 1st and 2nd purification results are reported here (except for layer L1, in which case the average of 2nd and 3rd purification results is used).

the originally added BeO carrier.

One can safely assume from the reproducibility that virtually all beta emitters except Be^{10} are eliminated after the single cycle of purification steps shown in Fig. 2. This assumption has been applied to the ARIES 13D samples for which no repurifications are performed. The Be^{10} data summarized in Table 2 for Mn 139 samples are, accordingly, weighted averages (based on counting statistics) of the two purification results.

The counters used have good stability in background and efficiency. Alternate countings of each sample mount in the two counters as mentioned earlier have given consistent results.

However, the counter backgrounds clearly constitute the dominant signal in the present analyses, and they are the limiting factor for precision as well as sensitivity.

Discussion

Nodule Growth Rate

We assume that within the datable interval the newly formed nodule layers contain a fixed amount of ^{10}Be which is not subjected to post-depositional migration. Then the slope of the plot of $\ln Be^{10}$ concentration versus depth gives a measure of the nodule growth rate, according to the relationship:

TABLE 3

U, Th, Pa in ARIES 13D and Mn 139
(on sample splits measured for Be)

Sample No.	Depth (mm)	U^{238} (dpm/g)	Th^{232} (dpm/g)	$\frac{U^{234}}{U^{238}}$ [1]	Th^{230}_{ex} [2] (dpm/g)	Pa^{231}_{ex} [2] (dpm/g)
ARIES 13D						
L1	0-4.3	9.19±.42	3.42±.17	0.97±.05	12.4±.8	4.55±.42
L2	4.3-8.8	9.30±.38	1.80±.14	0.99±.04	0.26±.54	-0.04±.03
L3	8.8-18.2	6.56±.25	0.75±.06	0.96±.04	0.18±.33	0.12±.08
Mn 139						
L1	0-0.8	8.62±.60	28.7±3.2	1.14±.10	388±12	22.1±1.9
L2	0.8-3.6	10.3±.44	18.4±2.3	0.96±.04	55.4±7.9	4.37±.30
L3	3.6-9.4	7.42±.44	11.0±.8	0.98±.06	10.3±.9	0.70±.10
L4	9.4-24.1	7.35±.34	8.81±.54	1.02±.06	4.62±.69	0.43±.15
L5	24.1-38.4	7.15± 43	7.94±.46	1.01±.07	1.13±.64	0.67±.23
Bo	59.1-65.0	8.94±.37	3.31±.22	1.06±.05	-0.47±.56	-0.03±.10
Su	< 1.0	7.67±.46	13.1±.9	1.03±.07	18.0±1.7	0.95±.14
Si	0- < 1.5 (irregular)	9.43±.57	4.55±.42	1.09±.07	6.10±1.34	3.73±.39

1. Activity ratio
2. Th^{230}_{ex} = dpm/g(Th^{230}) - dpm/g(U^{234}); Pa^{231}_{ex} = dpm/g(Pa^{231}) - dpm/g(U^{235}).

$$\frac{d\ln C}{dz} = \frac{-\lambda}{S} \quad (1)$$

where C = Be^{10} concentration
λ = decay constant of Be^{10}
S = growth rate
z = space (depth) coordinate, origin fixed at nodule surface, positive axis into nodule.

Figures 3 and 4 show plots of the Be^{10} concentration data for ARIES 13D and Mn 139, respectively. Similar plots can be made using the Be^{10}/Be^{9} ratio or excess Th^{230} (Th_{ex}^{230}) and Pa^{231} (Pa_{ex}^{231}) concentrations. Because of the uniformity of Be^{9} in the two nodules, depth plots of the Be^{10}/Be^{9} ratio give essentially the same results as the Be^{10} concentration shown in Figs. 3 and 4. Figure 5 is the equation (1) plot of Th_{ex}^{230} and Pa_{ex}^{231} for Mn 139, based on the data of Table 4.

TABLE 4

U, Th and Pa Data in Top Layers of Mn 139

Sample No.	Depth (mm)	U^{238} (dpm/g)	Th^{232} (dpm/g)	U^{234}/U^{238} [1]	Th_{ex}^{230} [2] (dpm/g)	Pa_{ex}^{231} [2] (dpm/g)
T1	0-0.3	9.11±.44	33.2±2.1	1.18±.07	2210±117	93.2±3.7
T2	0.3-1.2	8.29±.42	25.6±2.4	1.11±.06	270±25	6.25±.31
T3	1.2-2.5	7.60±.39	18.6±1.3	1.10±.06	18.2±1.9	0.49±.03
T4	2.5-3.8	7.24±.45	12.3±.7	1.03±.07	8.56±.99	-0.01±.02
T5	3.8-4.9	6.51±.32	10.6±.8	1.00±.06	--------	-0.01±.02
T6	4.0-7.0	6.09±.91	9.83±.45	0.95±.04	0.12±.09	0.06±.04

1. Activity ratio
2. Th_{ex}^{230} = dpm/g(Th^{230}) - dpm/g(U^{234}); Pa_{ex}^{231} = dpm/g(Pa^{231}) - dpm/g(U^{235}).

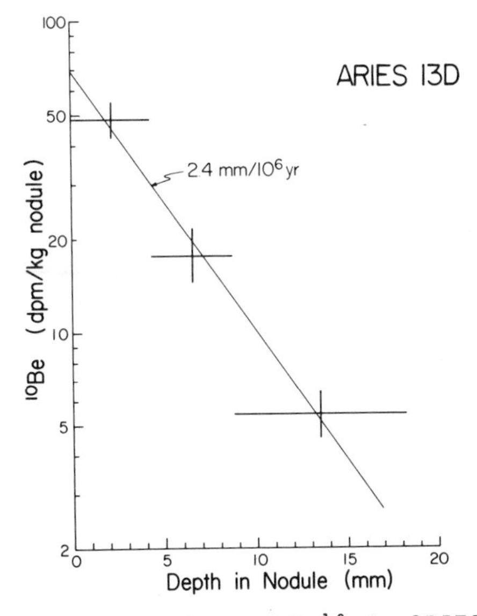

Fig. 3. Semilog depth plot of Be^{10} in ARIES 13D. The slope of the best-fitting line gives an apparent growth rate of 2.4 mm/10^6yr.

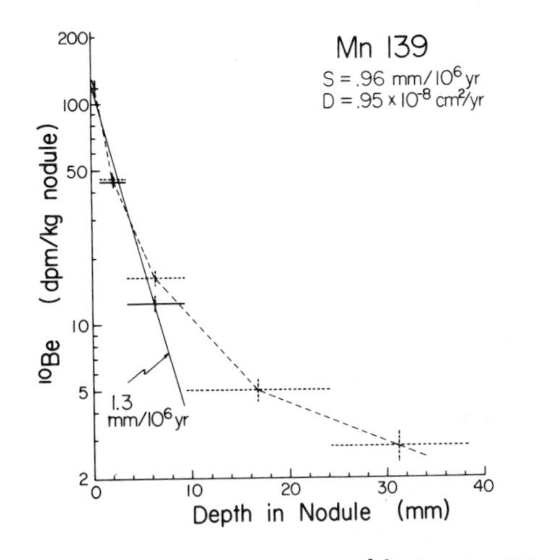

Fig. 4. Semilog depth plot of Be^{10} in Mn 139. Measured data are shown by dashed lines. Correction for activities from side of the nodule (see text) yields a rate of 1.3 mm/10^6yr (based on solid lines). The growth rate would be 0.96 mm/10^6yr if Be^{10} in nodule had an effective diffusion coefficient of 0.95 x 10^{-8} cm^2/yr.

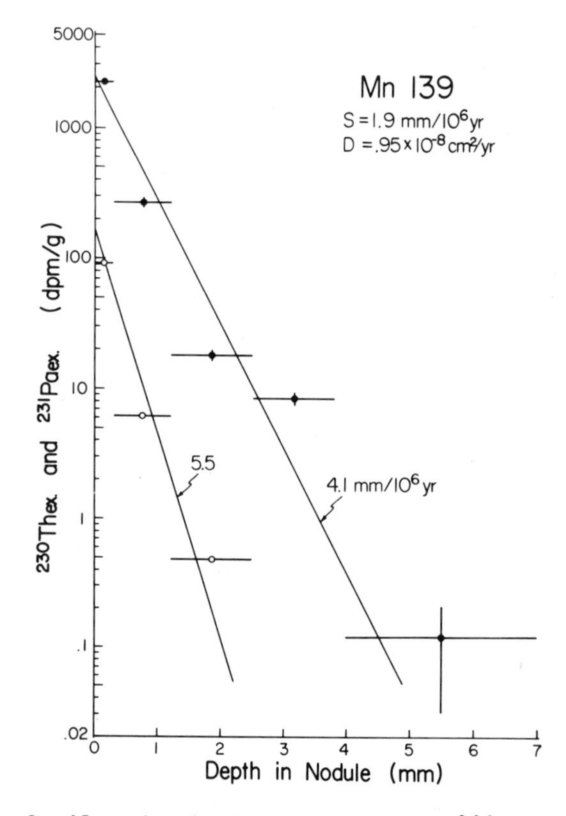

Fig. 5. Semilog depth plot of excess Th^{230} and Pa^{231} in Mn 139. The 5.5 and 4.1 mm/10^6yr figures are estimates assuming no diffusion of the nuclides (equation (1) plots). Calculated effective diffusion coefficient for Th^{230} and Pa^{231} is 0.95×10^{-8} cm^2/yr, which results in a growth rate of 1.9 mm/10^6yr.

A growth rate of 2.4 mm/10^6yr averaged over the top ∿ 15 mm of ARIES 13D is obtained from Be^{10} (Fig. 3). This information would predict absence of measurable Th_{ex}^{230} and Pa_{ex}^{231} in the deeper two layers of the nodule, L2 and L3. Data in Table 3 agree with this prediction.

Before examining the Be^{10} results of Mn 139, it should be pointed out that the apparent increase in growth rate for the deeper layers, as indicated by the decrease in Be^{10} gradient with depth (Fig. 4), could well be a sampling artifact. In peeling off different growth layers for analysis, the possibility exists that the scraped surfaces are not always parallel to the growth surfaces. This situation is likely to be more aggravated for deeper layers. Furthermore, in the present case where a large area has been scraped, we have overlooked the "bending" of the growth layers at the marginal part of the nodule specimen while sampling layers L1 through L5 (Figs. 1B and 6). The "bending" is a result of the concentric nature of accretion, leading to the effect that material on the side of Mn 139 can be much younger than that of the interior at the same "depth level" from top. This is evidenced by the amount of Be^{10}, Th_{ex}^{230}, and Pa_{ex}^{231} in sample Si being significantly higher than the layers "above" (Tables 2 and 3; cf. Fig. 6). Contamination of younger material into deeper layers from the side could, therefore, cause the curvature in the Fig. 4 plot.

We can estimate the magnitude of the contamination. As Fig. 6 illustrates, the sampled area is approximated by a quadrant with a radius of 90 mm. Assuming that the thickness of growth layers remains the same on all sides of the nodule, we estimate that the % volumes (or weights) of nodule material contaminated into underlying layers are about 1.8% of L1, 6.0% of L2, 12% of L3, and 27% of L4. Based on these estimates and the measured Be^{10} concentrations for each layer, corrections can be made of the contaminations. The corrected Be^{10} values are listed in Table 5. In the same table are also presented the corrected values of excess Th^{230} and excess Pa^{231} obtained in the same manner as for Be^{10}. The negative values assigned to layers L4 and L5 (Table 5) suggest that the corrections may be overly made so that the corrected Be^{10} concentration gradient becomes an upper bound. However, as shown in Fig. 4, for the first 3 data points, the effect of corrections is minor.

Iterative least-squares fit to an exponential function (Wolberg, 1967) gives the following slope values (in cm^{-1}) for Be^{10}, Th_{ex}^{230} and Pa_{ex}^{231} in the plots of Figs. 4 and 5:

$$\left.\frac{d\ln C}{dz}\right|_{Be} = -3.58, \quad \left.\frac{d\ln C}{dz}\right|_{Th} = -22.5, \quad \left.\frac{d\ln C}{dz}\right|_{Pa} = -36.9.$$

From (1), the Be^{10}-derived rate of growth for the top 10 mm of

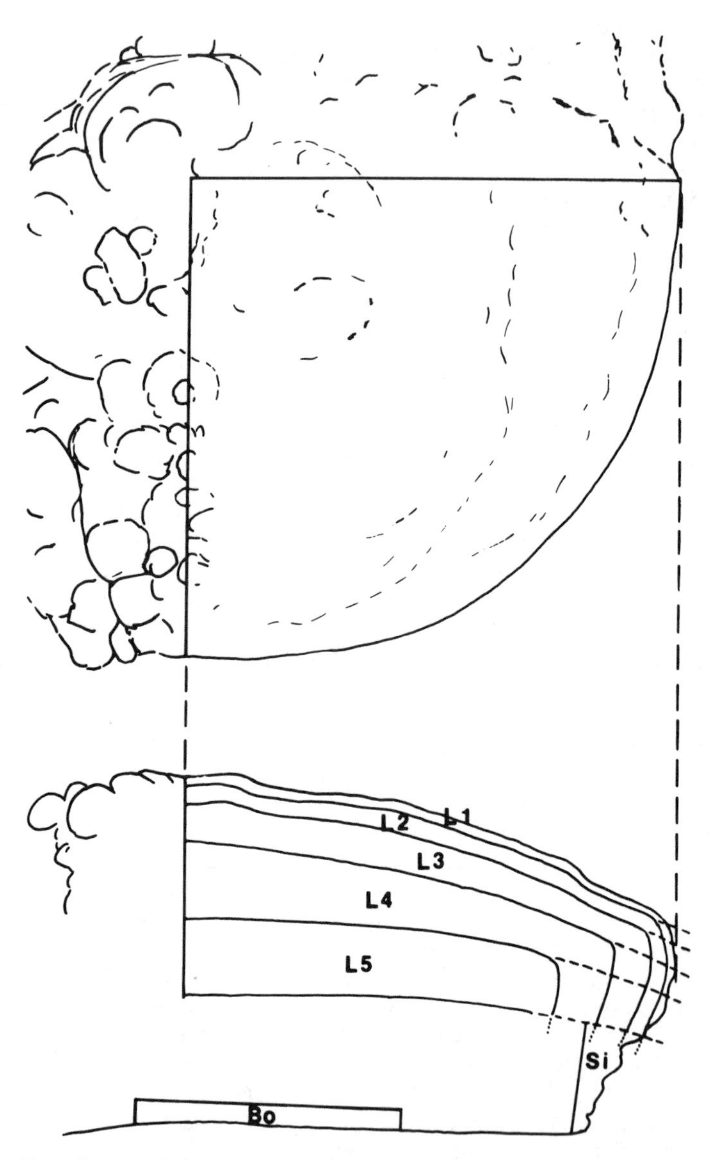

Fig. 6. Top is a plan view of part of Mn 139, showing the quadrant-shaped sampled area. The side view (bottom) shows that the boundaries of the sampled layers are assumed to follow those of the growth layers, except near the margin where the latter could be "bended" around (cf. Fig. 1B). Extent of contamination of material from upper to lower layers, as given in the text, is estimated assuming that the thickness of growth layer remains nearly constant around the nodule.

Mn 139 is 1.3 mm/10^6yr. This compares with the Th^{230}_{ex}-based rate of 4.1 mm/10^6yr and Pa^{231}_{ex}-based rate of 5.5 mm/10^6yr. Although these values have uncertainties of at least ±20% and are averages over different depth ranges (hence time scales), a three- to four-fold increase in the Th^{230} and Pa^{231} rates in relation to the Be^{10} rate merits discussion.

TABLE 5

Corrected Be^{10}, Th^{230}_{ex} and Pa^{231}_{ex} in Layers L1-L5 of Mn 139

Layer	%* Contamination	Corrected Be^{10} (dpm/kg)	Corrected Th^{230}_{ex} (dpm/g)	Corrected Pa^{231}_{ex} (dpm/g)
L1	0	118±11	388±12	22.1±1.9
L2	1.8	44.9±3.1	49.3±7.9	4.05±.30
L3	7.8	12.6±1.4	0.39±.9	0.064±.10
L4	19.8	-1.60±.61	-6.76±.69	-0.27±.15
L5	46.8	-5.87±.45	-13.3±.64	0.12±.23

* L2 is assumed to be contaminated by 1.8% (L1), L3 by 1.8% (L1) + 6.0% (L2), etc.

It is possible that the Pa^{231} as well as the Th^{230} rates are too high. This stems from the consideration that any homogenization processes such as diffusion and mixing (due largely to sampling on crenulated growth surfaces) would tend to reduce the concentration gradients of shorter-lived radioisotopes more than those of longer-lived ones. To a first approximation, the true growth rate could be close to the 1.3 mm/10^6yr figure depicted by Be^{10} decay.

If diffusion-mixing does occur, its effect on growth rate estimates from equation (1) can be evaluated as the following.

Diffusion-Decay Model

It is assumed that any mixing or transport mechanism for radio-nuclides in Mn 139 along its growth direction is likened to a diffusional process, characterized by a constant "effective" diffusion coefficient. The steady state distribution of the nuclide as a function of depth (normal to growth surface) in the nodule can be written as:

$$D\frac{d^2C}{dz^2} - S\frac{dC}{dz} - \lambda C = 0 \qquad (2)$$

where D = effective diffusion coefficient (cm^2yr^{-1})
C = concentration (dpm cm^{-3})
λ = decay constant (yr^{-1})
S = growth rate (cm yr^{-1})
z = depth in nodule, origin at top surface, positive downward (cm).

For the boundary conditions: $C = C_o$ at $z = 0$, $C = 0$ at $z = \infty$, the solution is:

$$C = C_o e^{Az} \qquad (3)$$

$$\text{where } A = \frac{S}{2D} - \left(\frac{S^2}{4D^2} + \frac{\lambda}{D}\right)^{\frac{1}{2}} \qquad (4)$$

Note $A = \frac{d\ln C}{dz}$; its numerical values for Be^{10}, Th^{230} and Pa^{231} in Mn 139 have been given earlier.

To estimate the magnitude of S and D, we proceed with two approaches. First, taking the aforementioned approximation of S = 1.3 mm/10^6yr and applying it to the intervals measured for Th^{230} and Pa^{231}, we obtain from equation (4) D values of 1.2 x 10^{-8} cm^2/yr and 1.1 x 10^{-8} cm^2/yr for Th^{230} and Pa^{231} respectively. Another approach is to suppose that the data for Th^{230} and Pa^{231} would yield common S and D, in view of the two nuclides' rather similar geochemical behavior and applicable depth range. Therefore, with A values known for Th^{230} and Pa^{231}, we solve equation (4) simultaneously for S and D: S = 1.9 mm/10^6yr and D = 0.95 x 10^{-8} cm^2/yr. Assuming Be^{10} to have a similar D of 0.95 x 10^{-8} cm^2/yr, then the Be^{10}-based growth rate would be reduced from 1.3 to 0.96 mm/10^6yr.

The model calculations thus show that, because of the slow growth nature of nodules, diffusion-mixing transport of radio-isotopes, if occurring, could significantly affect growth rate determination. As expected, given the same effective diffusivity, the shorter-lived isotopes would be more vulnerable to giving high apparent rates than the longer-lived ones. In the present case, with a D value of $\sim 10^{-8}$ cm^2/yr, the Th^{230}-based and Pa^{231}-based apparent rates obtained for the surface 2.5-5 mm of Mn 139 are shown to be too large by about 2 to 3 times. The same D applied to Be^{10} would result in an apparent rate about 30% too high (cf. Figs. 4 and 5).

Amounts of Be^{10}, Th_{ex}^{230}, and Pa_{ex}^{231} in Nodules

Ku and Broecker (1969) first pointed out that only a fraction of Th^{230} and Pa^{231} produced in the water column entered manganese nodules. Bhat et al. (1973) showed the same for Be^{10}. These observations are also indicated by the present data. Let p be the production rate, and C, z, and λ as defined before, the fraction of radionuclides incorporated in nodule is:

$$f = \frac{\lambda}{p}\left[\int_0^\infty C(z)dz\right] \quad (5)$$

Taking a water depth of 3430 m, a uranium concentration of 3.3 µg/l and $U^{234}/U^{238} = 1.14$, we estimated at the sampling sites of ARIES 13D and Mn 139 the production rates (in dpm cm^{-2} yr^{-1}) of Th^{230} and Pa^{231} to be: $p(Th^{230}) = 8.77 \times 10^{-3}$ and $p(Pa^{231}) = 7.73 \times 10^{-4}$. The production rate of Be^{10} is about 4.7×10^{-7} dpm $cm^{-2}yr^{-1}$ (Amin et al., 1975).

Values of f in the two nodules are thus calculated from (5) as:

ARIES 13D: $f_{Be} = 0.07$, $f_{Th} = 0.01$, $f_{Pa} = 0.10$; (assuming no activities on the bottom side).

Mn 139: $f_{Be} = 0.06$, $f_{Th} = 0.20$, $f_{Pa} = 0.18$.

One notes that these values are generally small and highly variable. Published data tend to reinforce this notion (Ku and Broecker, 1969; Krishnaswami, et al., 1972). The variability may occur even in a single specimen. As in the case of Mn 139, the depth integrated Be^{10}, Th_{ex}^{230}, and Pa_{ex}^{231} in the "smooth" half of the nodule appear to be much more than those in the "knobby" half (Fig. 1B). This refers to the data of sample L1 versus sample Su (Tables 3 and 5). Even in the smooth half, ΣTh_{ex}^{230} under Area T (Fig. 1) is more than that under Area L (Tables 4 and 5). It is tempting to suggest the variation as being due to the sediment veneer on nodule surface often revealed by the bottom photographs. Knobby surfaces tend to trap sediment particles more readily which then shield the nodule from isotopes scavenged from above. That the parameter f reflects the fraction of a nodule's lifetime uncovered by sediments (Bhat et al., 1973) is a plausible proposition. In a sense, the process of nodule growth is not a continuous one.

An Evaluation of the Concept of "Exposure-Diffusion"

It has been hypothesized (Arrhenius, 1967; Lalou and Brichet, 1972) that nodules could form very rapidly without inclusion of any unsupported Th^{230}. After formation they receive the rain of radionuclides but experience almost no growth on the sea floor. The observed rapid decrease with depth of the nuclides could result from inward diffusion-mixing processes. Accordingly, the age of the nodule can be derived from the observed integrated amount of a nuclide, knowing its production rate. That is, the length of time during which nodules are "exposed" to the radioactivity flux is calculated from (with notations defined as above):

$$t = -\frac{1}{\lambda}\ln[1-\frac{\lambda}{P}(\int_0^\infty Cdz)] \tag{6}$$

Or from equation (5):

$$t = -\frac{1}{\lambda}\ln(1-f) \tag{7}$$

This "exposure-diffusion" hypothesis has several serious flaws as viewed from the radiochemical consequences it entails.

(1) Commonly observed values for f are small; this leads to much younger ages for the nodules than implied in our previous discussion. The "exposure" ages of Mn 139 are calculated to be 127,000 yr, 24,200 yr, and 9,800 yr, based on the observed f's for Be^{10}, Th^{230}, and Pa^{231} respectively. For ARIES 13D, the disparity in these ages is even larger, and it remains to be explained.

(2) Implicit in the age calculation is the assumption that nodules collect with 100% efficiency the Be^{10}, Th^{230} and Pa^{231} produced above them, on a unit area basis. The fact that, due to their reactivity, a significant portion of these nuclides could be scavenged by particulate phases into the adjacent sediments rather than incorporated into nodules, puts strain on this assumption.

(3) As has been pointed out (Ku, 1977), the absence of deficiencies of Th^{230} and Pa^{231} relative to their uranium parents in the inner parts of both nodules studied is unexpected in view of the young "exposure" ages and the rapid formation requirement.

(4) Recent works (Ku et al. 1975; Ku and Knauss, 1978) have shown that the internal fractures of manganese nodules may contain filling material > 300,000 years old. Clearly, this age cannot be reconciled with the young exposure ages.

(5) Given the assumptions required by the exposure-diffusion (with no growth) concept, one can model as follows the distribution of radionuclides in a nodule, giving further insight into the reality of the concept.

Exposure-diffusion model. The concept can be approximated by considering the Fick's second-law equation (radio-decay is omitted owing to $t \ll \lambda^{-1}$):

$$\frac{\partial C}{\partial t} = D\frac{\partial^2 C}{\partial z^2} \tag{8}$$

The initial condition is t=0, C=0. The upper boundary

receives a constant flux of radionuclides (with hardly any volume) equivalent to their production rates, p, namely:

$t>0$, $p = -D\frac{\partial C}{\partial z}\Big|_{z=0}$ = constant. At the lower boundary: $z \rightarrow \infty$, $C \rightarrow 0$.

The solution can be found in Carslaw and Jaeger (1959, p. 75) and in Duursma and Hoede (1967):

$$C(z,t) = \frac{2p}{D}\left[\left(\frac{Dt}{\pi}\right)^{\frac{1}{2}} \exp\left(\frac{-z^2}{4Dt}\right) - \frac{z}{2}\,\mathrm{erfc}\,\frac{z}{2(Dt)^{\frac{1}{2}}}\right] \qquad (9)$$

In the first place, we see that, according to the model, lnC varies with z^2, rather than with z as depicted by equations (1) and (3). Also, the dependency of C on z is rather strong due to the error function. This results in the convex-upward curvature in the lnC vs. z plots of equation (9), as shown in Figs. 7 and 8 for the cases of Be^{10} and Th^{230}_{ex} in Mn 139. In these figures, using the exposure ages (t) and the P values given earlier, C(z) is plotted for different D's.

The upward convexity in the depth plot of lnC have rarely been observed. In fact, measured profiles not uncommonly show opposite (concave-upward) curvature, such as the cases in Figs. 7 and 8. This cannot always be the result of sampling artifact, since the Be^{10} data of Fig. 7 have already been somewhat overly corrected for this effect, and since such concave characteristics have also been exhibited by alpha-track profiles (e.g., Lalou *et al.*, 1978).

Magnitude of Parameter D

The effective diffusion coefficients computed from the diffusion-decay model for Mn 139 are of the order of 10^{-8} cm^2/yr. Values significantly higher than this, in effect, are not allowed by the model. This is seen from (4):

$$S = AD - \lambda/A \qquad (10)$$

Since S must be > 0 and $A < 0$, it follows:

$$D < \lambda/A^2 \qquad (11)$$

For Mn 139, equation (11) depicts that $D_{Be} < 3.6 \times 10^{-8}$ cm^2/yr, $D_{Th} < 1.8 \times 10^{-8}$ cm^2/yr and $D_{Pa} < 1.5 \times 10^{-8}$ cm^2/yr.

From Figs. 7 and 8, the best D values according to the exposure-diffusion model are of the order of 5×10^{-7} cm^2/yr, although the curve fitting for the model is far from ideal. In these cases, smaller D values such as 10^{-8} cm^2/yr depicted by the diffusion-decay model are clearly not permissible. One might thus suggest that an evaluation of D for Th^{230} and Pa^{231} in nodules would be useful to further appraise the two models.

The parameter D could characterize: (1) solid-state diffusion, (2) diffusion through the nodule pore water in the presence of instantaneous exchange with the solid, (3) physical mixing (*in*

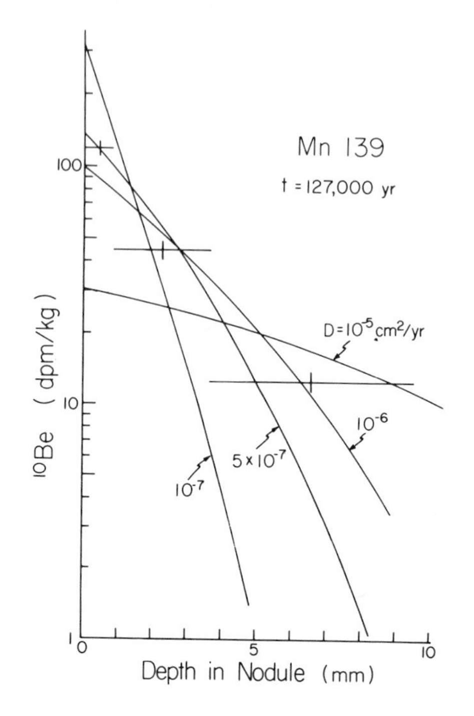

Fig. 7. Curves representing the Be^{10} distribution in Mn 139 as a function of D, as predicted by the "exposure-diffusion" model. Measured data points are shown for comparison.

situ, or artifact of sampling), or (4) combination of all three.

Krishnaswami and Cochran (1978) pointed out that pure solid-state diffusion is possible but unlikely. These workers discussed the Ra^{226} distribution in nodules in terms of diffusion of Ra^{226} through nodule pore water. They obtained a model diffusion coefficient $D_{Ra} \simeq 10^{-6}$ cm^2/yr, and attributed this low value (compared to $\sim 10^2$ cm^2/yr for molecular diffusivity, D_m, in free solution: Li and Gregory, 1974) to rapid equilibrium adsorption of Ra^{226} on solids and the high tortuosity of diffusion paths in the nodule. This arises from the relationship:

$$D \simeq D_m/(\theta^2 \cdot K) \qquad (12)$$

where θ^2 denotes tortuosity and K the ion exchange or adsorption equilibrium constant ($= C_{nodule}/C_{pore\ water}$). The value of K for Ra^{226} was given as $\sim 10^5$ (Krishnaswami and Cochran, 1978). The activity ratio of Th^{230} to Ra^{226} in nodules is close to unity, whereas that in the pore water, taken to be similar to that of

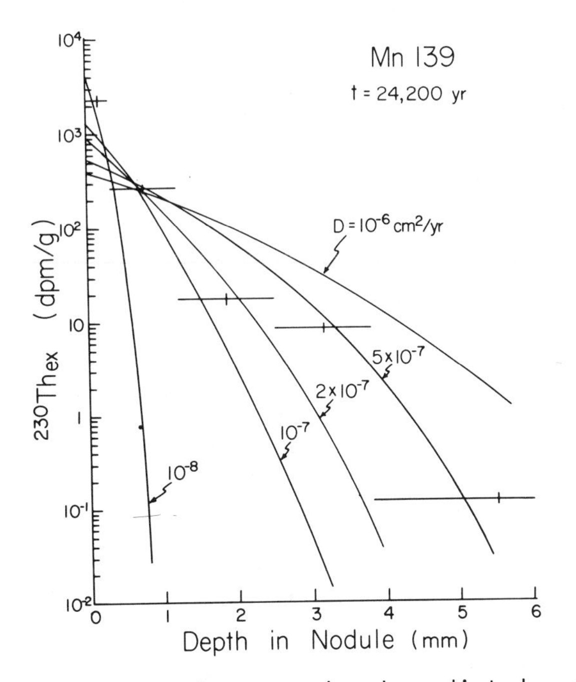

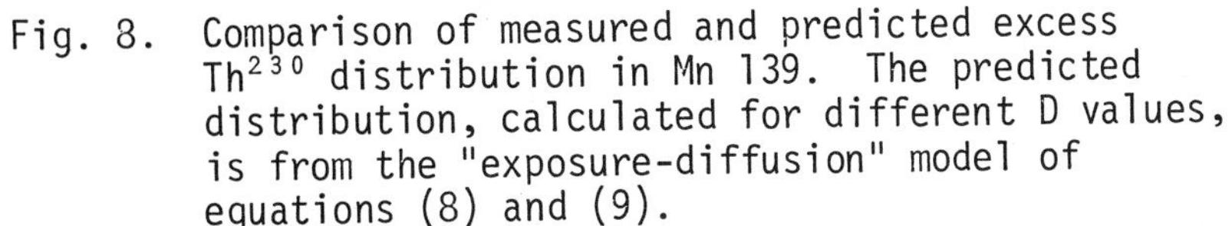
Fig. 8. Comparison of measured and predicted excess Th^{230} distribution in Mn 139. The predicted distribution, calculated for different D values, is from the "exposure-diffusion" model of equations (8) and (9).

bottom water, is about 0.01 - 0.05 (e.g., Ku, 1972). Therefore, K for Th^{230} should be $(0.2\text{-}1) \times 10^7$. Since $D_m(Th) \simeq 0.2 D_m(Ra)$ (Li and Gregory, 1974), from (12) we calculate:

$$D_{Th} \simeq \frac{D_m(Th)/(\theta^2 \cdot 10^7)}{D_m(Ra)/(\theta^2 \cdot 10^5)} \times D_{Ra} \simeq (0.2\text{-}1) \times 10^{-8} \ cm^2/yr \quad (13)$$

In view of the strong chemisorption property of Pa^{231}, we expect that $D_{Pa} < D_{Th}$. Hence, the D values of $\sim 10^{-8}$ cm^2/yr deduced earlier from the diffusion-decay model appear to be quite acceptable to our limited evidence about the possible magnitude of D.

Apparently, the extent of physical mixing must be small (as constrained by the rather small D values) in order for the diffusion-decay model to be applicable. Such could well be generally the case under careful sampling conditions, as shown by the Pb^{210} measurements of Krishnaswami and Cochran (1978).

Summary

Various radioactivities--Th^{230}, Pa^{231}, Be^{10}, Al^{26}--present within the Mn-Fe oxide layers provide the most direct and useful clues to the ages of manganese nodules. A long established observation is their logarithmic decrease with depth in nodules. The decrease is such that it is more gradual for the longer-lived radionuclides than for the shorter-lived ones. The gradients, taken to reflect radioactive decay, generally yield concordant (to within error limits) growth-rate estimates of mm's/10^6yr for the nodules.

A question arises as to what extent the gradients could be affected by processes other than radiodecay, i.e., by diffusion and/or diffusion-like mixing. A manifestation of these latter processes would be the apparent discordant rates based on Be^{10}, Th^{230}, and Pa^{231}, such as found in Mn 139. Quantitative evaluation using a diffusion-decay model shows that the apparent discordancy can be explained by mixing processes characterized by "effective" diffusion coefficients of the order of 10^{-8} cm^2/yr. The true growth rate, obviously, would be slower than the apparent rates.

The depth integrated dpm/cm^2 of Be^{10}, excess Th^{230} and excess Pa^{231} in nodules in general is variable and small in amount, compared to their potential source of supply and their inventory in sediments. In Mn 139 the variability occurs even in different parts of its upper surface. The observations are best explained by the suggestion that on a unit area basis nodules receive a small share of the isotopes relative to that received by other sedimentary phases.

Critiques of the proposed catastrophic formation and young "exposure" ages for nodules are presented. Mathematical modeling of such concept of nodule formation leads to depth distribution of the radioisotopes rather different from the simple exponential form that is commonly observed.

Acknowledgments

D. Z. Piper and J. L. Bischoff of the U. S. Geological Survey and T. Walsh of the Scripps Institution of Oceanography (SIO) kindly provided the nodule specimens from collections of their respective institutions. We are indebted to F. Guichard of Centre des Faibles Radioactivites for his knowledge in Be^{10} analytical chemistry, to Y. C. Chung of SIO for providing the beta counting standards, and to A. Sycip of the U.S.C. Environmental Engineering Program for assistance in the Be^9 measurements. Mr. S. Y. Meng ably aided in the U-series isotope analyses.

Financial support from the National Science Foundation through grant OCE77-04061 administered by the Seabed Assessment

Program of I.D.O.E. is gratefully acknowledged.
This is contribution #385 from the Department of Geological Sciences, University of Southern California.

References

Amin, B. S., Lal, D. and Somayajulu, B. L. K. (1975) Chronology of marine sediments using the Be^{10} method: intercomparison with other methods. Geochim. Cosmochim. Acta 39, 1187-1192.

Arrhenius, G. (1967) Deep-sea sedimentation: a critical review of U.S. work. Trans. Am. Geophys. Union, 48, 604-631.

Bender, M. L., Ku, T. L. and Broecker, W. S. (1966) Manganese nodules: Their evolution. Science 151, 325-328.

Bhat, S. G., Krishnaswami, S., Lal, D., Rama and Somayajulu, B. L. K. (1973) Radiometric and trace elemental studies of ferromanganese nodules. Proc. Symp. Hydrogeochem. Biogeochem., I, 443-462, Clarke, Washington, D.C.

Carslaw, H. S. and Jaeger, J. C. (1959) Conduction of Heat in Solids (2nd edition) Oxford Univ. Press.

Duursma, E. K., and Hodge, C. (1967) Theoretical, experimental and field studies concerning molecular diffusion of radioisotopes in sediments and suspended solid particles of the sea, Part A: theoretical and mathematical calculation. Netherland Jour. Sea Res. 3, 423-457.

Guichard, F., Reyss, J. L. and Yokoyama, Y. (1978) Growth rates of manganese nodules measured with ^{10}Be and ^{26}Al. Nature 272, 155-156.

Krishnaswami, S. and Cochran, J. K. (1978) Uranium and thorium series nuclides in oriented ferromanganese nodules: growth rates, turnover times and nuclide behavior. Earth Planet. Sci. Lett. 40, 45-62.

Krishnaswami, S., Somayajulu, B. L. K. and Moore, W. S. (1972) Dating of manganese nodules using beryllium-10. In Papers from a Conference on Ferromanganese Deposits on the Ocean Floor (ed. D. R. Horn) 117-122. NSF, Washington, D.C.

Ku, T. L. (1972) Radium in the ocean. In Encyclopedia of Geochemistry and Environmental Sciences (ed. R. W. Fairbridge), 1008-1014, Van Nostrand Reinhold, N.Y.

Ku, T. L. (1977) Rates of accretion. In Marine Manganese Deposits (ed. G. P. Glasby) 249-267, Elsevier, N.Y.

Ku, T. L. and Broecker, W. S. (1969) Radiochemical studies on manganese nodules of deep-sea origin. Deep-Sea Res. 16, 625-637.

Ku, T. L. and Knauss, K. G. (1978) Radiocative disequilibrium in fissure-filling material and its implication in dating of manganese nodules. Colloque International du C.N.R.S. No. 289 sur la Genèse des Nodules de Manganèse, Gif-sur-Yvette, 25-30 September.

Ku, T. L., Knauss, K. G. and Lin, M. C. (1975) An evaluation of dating nodules by the uranium series isotopes. EOS, Trans. Am. Geophys. Un. 56, 999.

Lal, D. and Schink, D. R. (1960) Low background thin-wall flow counters for measuring the beta activity of solids. Rev. Sci. Instr. 31, 395-398.

Lalou, C. and Brichet, E. (1972) Signification des mesures radiochimiques dans l'évaluation de la vitesse de croissance des nodules de manganèse. C. R. Acad. Sc. Paris. Série D, 275, 815-818.

Lalou, C., Ku, T. L., Brichet, E., Poupeau, G. and Romary, P. (1978) TECHNO encrustation I: radiometric studies. Colloque International du C.N.R.S. No. 289 sur la Genèse des Nodules de Manganèse, Gif-sur-Yvette, 25-30 September.

Li, Y. H. and Gregory, S. (1974) Diffusion of ions in sea water and in deep-sea sediments. Geochim. Cosmochim. Acta 38, 703-714.

Wolberg, J. R. (1967) Prediction Analysis. Van Nostrand, Princeton, New Jersey.

GROWTH RATES AND POSSIBLE AGE OF A NORTH PACIFIC MANGANESE NODULE

C. Lalou, E. Brichet, G. Poupeau, P. Romary,
and C. Jehanno

Centre des Faibles Radioactivites
Laboratoire Mixte, CNRS-CEA
91190 - Gif-sur-Yvette, France

Abstract

The rate of accumulation of a ferromanganese coating on a fragment of pillow basalt was estimated using a variety of techniques. Unsupported ^{230}Th activity decrease in the oxide layer, K/A dating of the basalt, fission tracks dating of the glassy layer around the basalt, thickness of the palagonitization rind, and integrated ^{230}Th activity give ages from approximately 3×10^6 years to 5×10^3 years.

Data suggest that the ferromanganese material formed rapidly (333 mm/10^6 years) and by hydrothermal or volcanic processes.

Introduction

A 3cm fragment of fresh pillow lava, coated with Fe-Mn oxide deposit (courtesy of J. Greenslate, Scripps Institution of Oceanography, La Jolla) was recovered during DPSN Leg 2, in rock dredge 3D on 23 February 1976, at 9°08 65'N; 105°12 94' W from between 3214 and 3265 meters depth.

Figure 1 shows the location between the Siqueiros and Clipperton fracture zones, near the East Pacific Rise, on magnetic anomaly 3. The maximum age of the basement is $5 \pm 1 \times 10^6$ years. Based on the empirical relationship of Sclater *et al.*, (1971) between ridge elevation and age of the oceanic crust the depth at which the nodule has been dredged would suggest an age of about 4×10^6 years.

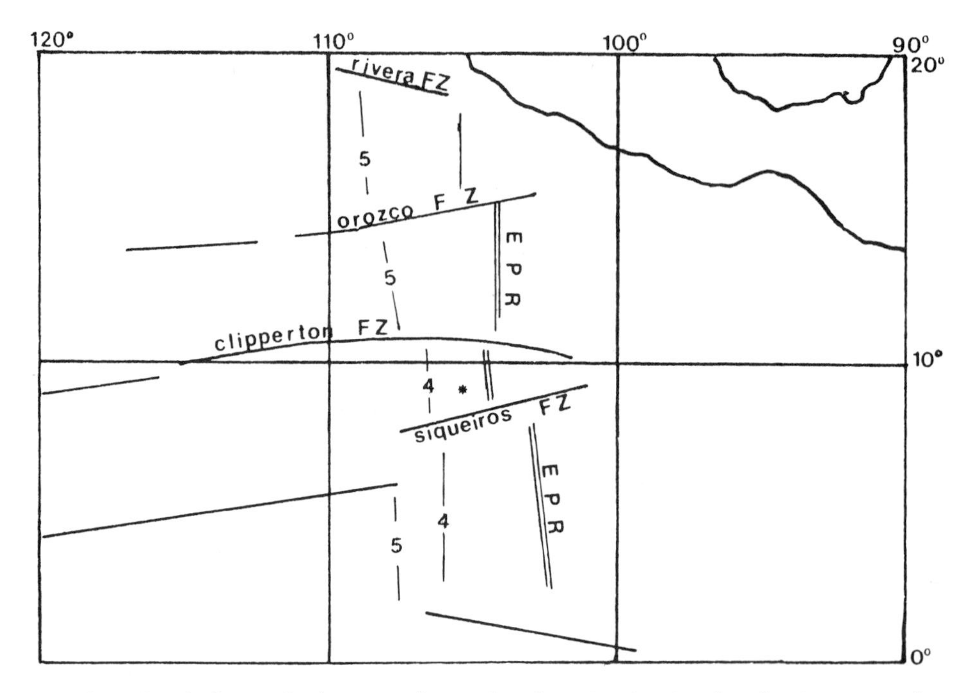

Fig. 1. Position of the sample relative to tectonic features and magnetic anomalies. (The map is due to G. Pautot, personal communication.)

The sample has an ovoid shape (fig. 2) and appears to have been recently broken from a rock outcrop. The left portion has no Fe-Mn deposit, while the upper and right sides, as shown on the sketch, are covered with Fe-Mn oxide averaging 40 mm thick. This material is less than 10 mm thick on the lower portion.

At the boundary between the Fe-Mn oxide and the basalt, the basalt exhibits a thin layer of alteration. On the right side there is a complex structure consisting of alternating basalt and oxide.

The general structure of the oxide is cauliflower in form, made of neighboring cones with their tips oriented towards the core.

This sample seemed particularly amenable to several dating methods, and we were interested to determine if the various methods might give consistent results.

The growth rate of the manganese oxide deposit has been evaluated in a conventional manner from the unsupported ^{230}Th activity decrease. The age of the core has been measured by K/A dating, fission track, and by hydration rind dating. A structural and chemical study has also been made.

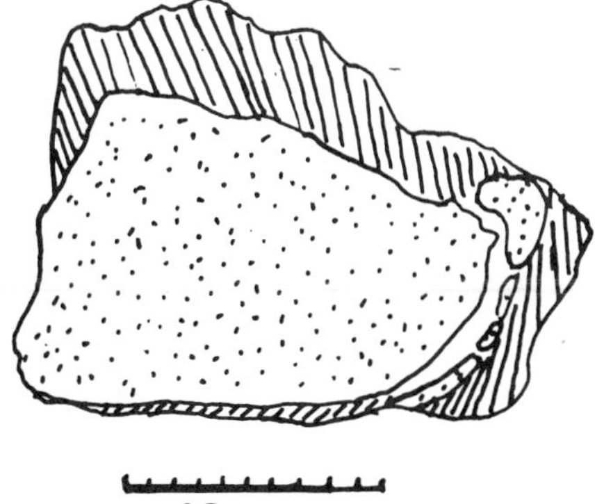

Fig. 2. Schematic cross section of the sample

Growth Rate of the Oxide Layer

Minimum Growth Rate of the Oxide Deposit

A minimum growth rate for the oxide deposit may be obtained, using the age of the basement, and considering that the basaltic core is part of this basement. Using the mean thickness of oxide (40 mm) and the theoretical age of the basement (5 Mys), a growth rate of 8 mm/10^6 years is obtained. Using the same assumption, and using the age deduced from the relationship of Sclater et al. (1971), a minimum growth rate of 10 mm/10^6 years is obtained.

Growth Rate by Unsupported ^{230}Th Activity Decrease

a) Sampling procedure and radiochemical separation.

The decrease with depth of unsupported ^{230}Th activity in the oxide of manganese nodules can be used to calculate a growth rate (Bender et al. (1966), Barnes and Dymond (1967), Ku and Broecker (1967, 1969), Bhat et al. (1973). In this method, it is assumed that the unsupported ^{230}Th activity coming from the dissolved uranium in the overlaying water column is regularly incorporated in the nodule along with the Fe-Mn oxides.

For this study, we have sampled twelve successive layers having areas varying from 3.5 cm^2 for the outermost to 1 cm^2 for the deepest layer.

The thickness of the layers has been calculated assuming a dry density of 1.7 g/cm^3. The methods of analysis are similar to those described by Ku and Broecker (1969) for uranium and thorium-

TABLE 1

Uranium and thorium series analysis in twelve successive layers of the oxide deposit. All the concentration values are given for oxide dried for 3 hours at 110°C, while the density for calculating the thickness of the layer is wet density.

Layer N°	Depth in oxide (mm)	^{238}U (ppm)	$^{234}U/^{238}U$ (act. ratio)	^{234}U (dpm/g)	^{230}Th (dpm/g)	$^{230}Th_{ex}$ (dpm/g)	^{230}Th dpm/cm^2	^{232}Th (ppm)
1	0 - 0.44	39.5 ± 1.5	1.14 ± 0.04	33.3 ± 1.2	88.8 ± 3.8	55 ± 5 82[1]	3.5 5.2	15.2 ± 1.6
2	0.44- 0.89	8.4 ± 0.6	1.090 ± 0.039	6.8 ± 0.5	94 ± 3	87.2 ± 3.5	5.6	15.9 ± 1.2
3	0.89- 1.33	8.3 ± 0.4	1.05 ± 0.06	6.6 ± 0.3	79.8 ± 2.3	73.1 ± 2.6	4.6	10.1 ± 0.8
4	1.33- 1.79	8.3 ± 0.4	1.14 ± 0.07	7.0 ± 0.3	55.1 ± 2.9	48.1 ± 3.2	3.2	23.3 ± 1.4
5	1.79- 2.30	6.9 ± 0.5	1.252 ± 0.095	6.3 ± 0.4	39.2 ± 1.9	32.9 ± 2.3	2.4	8.0 ± 0.8
6	2.30- 2.85	10.2 ± 0.5	1.14 ± 0.06	8.6 ± 0.4	32.4 ± 1.7	23.8 ± 2.1	2	9.0 ± 1.0
7	2.85- 3.50	25.0 ± 1.7	1.220 ± 0.075	22.4 ± 1.5	28.4 ± 1.1	6.0 ± 2.6 21.4[1]	0.6 3.3	8.1 ± 0.7
8	3.50- 4.75	7.2 ± 0.3	1.050 ± 0.047	5.5 ± 0.2	24.0 ± 0.9	18.5 ± 1.1	3.3	8.1 ± 0.6
9	4.75- 6.22	5.9 ± 0.8	0.91 ± 0.16	3.9 ± 0.6	7.8 ± 0.6	3.9 ± 1.2	0.8	4.1 ± 0.7
10	6.22- 9.58	7.1 ± 0.4	1.35 ± 0.07	7.0 ± 0.3	16.5 ± 1.4	9.5 ± 1.7	4.7	8.7 ± 1.1
11	9.58-11.25	8.9 ± 0.7	0.95 ± 0.08	6.3 ± 0.5	13.1 ± 0.8	6.8 ± 1.3	1.7	8.1 ± 0.9
12	11.25-13.70	12.1 ± 0.8	1.114 ± 0.073	9.9 ± 0.6	18.8 ± 1.3	8.9 ± 1.9	3.2	19.2 ± 1.7

[1] values corrected for a ^{234}U activity of 7 dpm/g

isotope determination. Uranium and thorium isotopes are measured with reference to a ^{232}U + ^{228}Th spike (activity ratio = 1). After separation and purification, uranium and thorium are extracted in a TTA solution and deposited by evaporation on a stainless steel disc. In general, the sources prepared in this manner have been counted twice: on in a gas flow grid chamber and secondly with a surface barrier detector (ORTEC 16/194 A).

b) Results and discussion.

The results of these analyses are given in Table 1 and the decrease of activity of unsupported thorium v.s. depth in Fig. 3.

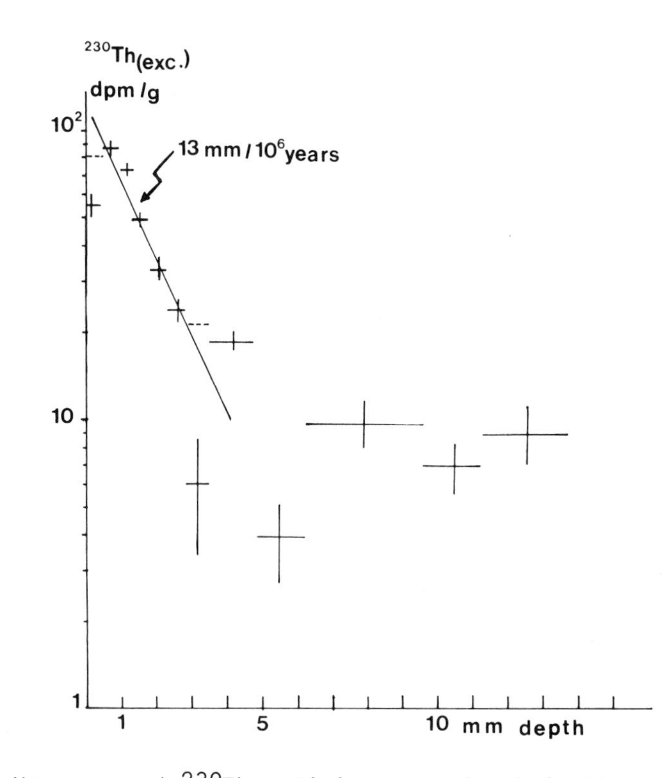

Fig. 3. Unsupported ^{230}Th activity v.s. depth in the oxide layer

Except for two points (layer 1 and layer 7) a reasonably regular decrease of unsupported ^{230}Th activity is seen as deep as 3 mm. If interpreted as a radioactive decrease this permits the calculation of a mean growth rate of 13 mm/10^6 years. Two characteristics have to be noted, however:

1) The anomalously low values of unsupported ^{230}Th relative to the surrounding layers 1 and 7, corresponding to an anomalous ^{238}U content. While the overall mean value of uranium in the other layers is about 8 ppm, values of 39.5 and 25.0 ppm have been respectively measured for layers 1 and 7. If instead of subtracting the measured ^{234}U activity from total measured ^{230}Th activity in these two layers, we subtract the mean ^{234}U activity found in the other layers to obtain the unsupported ^{230}Th activity, the new values (dashed lines on Fig. 3) agree fairly well with the normal decrease. This seems to indicate that for some reason, a part of the uranium in these two layers is not in equilibrium with its daughter products.

The extreme outer layer (layer 1) is highly enriched in uranium, which might be attributed to a concentration of sea water uranium by the oxide. $^{234}U/^{238}U$ ratio of 1.14 ± 0.04 in this layer supports this hypothesis. Layer 7 could therefore be considered contaminated during sampling by the very active superficial layer. In this case we would expect to have an excess of ^{230}Th accompanying this uranium. Moreover, the activity ratio of uranium in layer 7 is definitely different from that of sea water (1.22 ± 0.075). If, as deduced from the measured growth rate, this layer 7 is about 250 x 10^3 years, extra-uranium must have been introduced recently without daughter products. If this uranium had time to produce some ^{230}Th, the ^{230}Th activity value would not fall exactly on the line of ^{230}Th decrease when subtracting the mean uranium value.

2) The presence of a significant amount of unsupported ^{230}Th deeper than 5 mm in the nodule does not fit the exponential decrease observed until 3 mm depth. This phenomenon is frequently observed when the measurements are made on a portion of nodules deeper than the outermost 3 or 4 mm (Lalou et al., 1978).

Assuming a constant rate of formation of the oxide layer, the mean growth rate calculated at 13 mm/10^6 years (without taking into account the values for layers 1 and 7 discussed above) leads to an age for the beginning of deposition of oxide on this sample of 3 x 10^6 years, which is compatible with the age of the basement.

Dating of the Basalt Core

The age of a core of a manganese nodule yields a minimum growth rate, assuming the oxide formation began immediately after the formation of the core and has operated continuously.

K/A dating

We have performed K/A dating on a central part of the core, using the "whole rock". An apparent age of 1.7 ± 10^6 years has been obtained, leading to a minimum growth rate of approximately 23 mm/10^6 years. Potassium-argon dating of nodule cores has been performed by Barnes and Dymond (1967) who calculated growth rates for Fe-Mn deposits. They have assumed that the age obtained in this manner is representative of the time when the basalt has extruded on the sea floor, that the precipitation of oxide took place as soon as the basaltic core reached the sea floor, and that the oxide precipitation is a continuous phenomenon. This method can give only a maximum age as the degassing of argon in submarine basalts is not completely achieved and some residual radiogenic argon results in calculation of ages which are too great (Funkhouser _et al_., 1968).

Fission track dating

Fission track dating (Fleischer _et al_., 1975) has been successfully applied to deep-seated basalts, taking advantage of the track etching facility offered by the glassy crust of pillow-lavas (Fleischer _et al_., 1968; Aumento, 1969). It was later shown (Mac Dougall, 1976) that very young samples ($<10^6$ years) should be little affected by fission track annealing at normal deep sea temperature of a few degrees centigrades.

The basaltic fragment forming the core of the studied nodule does not present any evidence of alteration, as shown from petrographic observations. The basaltic pillow lava core has a well preserved glassy crust, up to 5 mm thick. This glass shows no evidence of devitrification, thus allowing use of the fission track dating. A first survey having shown that the density of fossil tracks, due to the spontaneous fission of ^{238}U was very low, a glassy crust fragment was exposed to thermal neutrons in a nuclear reactor (EL 3 reactor in Saclay, France) without previous thermal annealing treatment in order to induce the fission of ^{235}U. The thermal neutron flux was measured within ±3% using an activated cobalt disc. The irradiated sample was then mounted in epoxy resin together with an unirradiated piece of basaltic glass, polished and etched in 20% HF for 20 sec. at 20°C. Both pieces of glass were then scanned on different surface areas with an optical magnification of x 1000 by two different observers. The observations were made under reflective light with an interferential contrast system. The results are reported in Table 2.

In the unirradiated sample, over a total surface area of about 0.4 cm^2, only two features could be considered as possible fission tracks. In the irradiated sample, the induced fission tracks are homogeneously distributed and the track density measurements in good agreement between the two observers. The total surface area counted for induced fission tracks is much smaller than the one scanned for fossil tracks. In order to compensate this effect, the induced tracks were measured as follows: each field of view counted (measuring $\simeq 2 \times 10^4$ μm^2) was randomly chosen along a grid with a constant spacing of $\simeq$ μm between two successive countings, allowing coverage of a total surface of $\simeq 0.5$ cm^2. The samples were repolished and etched between observer n°1 and n°2 measurements. The counts were therefore made on different areas. Observer 1 did not see any fossil tracks, whereas observer 2 considered

TABLE 2

Results of fossil and induced fission tracks counting

Observer	Surface scanned (mm^2)	Fossil tracks ρs (cm^{-2})	N	Induced tracks * ρi ($x10^{-4} cm^{-2}$)	N	T + (years)
1	12.4	< 8	0	1.96	448	< 440,000
2	20.3	< 10	⩽ 2	1.90	218	< 570,000
1 + 2	32.7	< 6	⩽ 2			⩽ 330,000

ρs and ρi, respectively fossil and induced track density ; N, number of tracks counted ; * thermal neutron flux : 1.78×10^{16} n. cm^{-2} ; +$\lambda_F = 7.00 \times 10^{-17}$ $years^{-1}$.

conservatively two features as possible fission tracks. Application of Poisson statistics (which would lead to a combined age of 330,000 ± 240,000 yr (2σ), may lack validity as regards the small number of fossil tracks. Taking into account the conservative criteria used in track counting, we would consider the ages in Table 2 as upper limits to the time elapsed since the last cooling of this material. If no thermal event occurred since the basalt extrusion, this age dates the time of formation of this pillow lava fragment.

From the individual results of each observer, an upper limit of about 600,000 years can be set for the age of the basalt. Taking into account that only two possible fission tracks have been observed in the total surface area scanned, one can restrict this upper limit to < 330,000 years.

Thus, a growth rate for the Fe-Mn deposit of 67 to 114 mm/10^6 years is calculated.

Hydration rinds dating

Assuming a constant rate of 2.91 microns/10^3 years for formation of the palagonitized rind on basaltic glass (Morgenstein, 1969, 1972) has measured the growth rates for Fe-Mn deposits of the Waho shelf in Hawaii. According to this author, during hydration of basaltic glass, CaO is depleted and K_2O enriched in the palagonite alteration product.

In our sample, at the limit between the fresh basaltic glass and the innermost layer of Fe-Mn oxide precipitation, there is a layer of palagonite. A polished section of the sample was studied with the SEM/EDAX system. On Figure 4 is shown (a) an enlargement

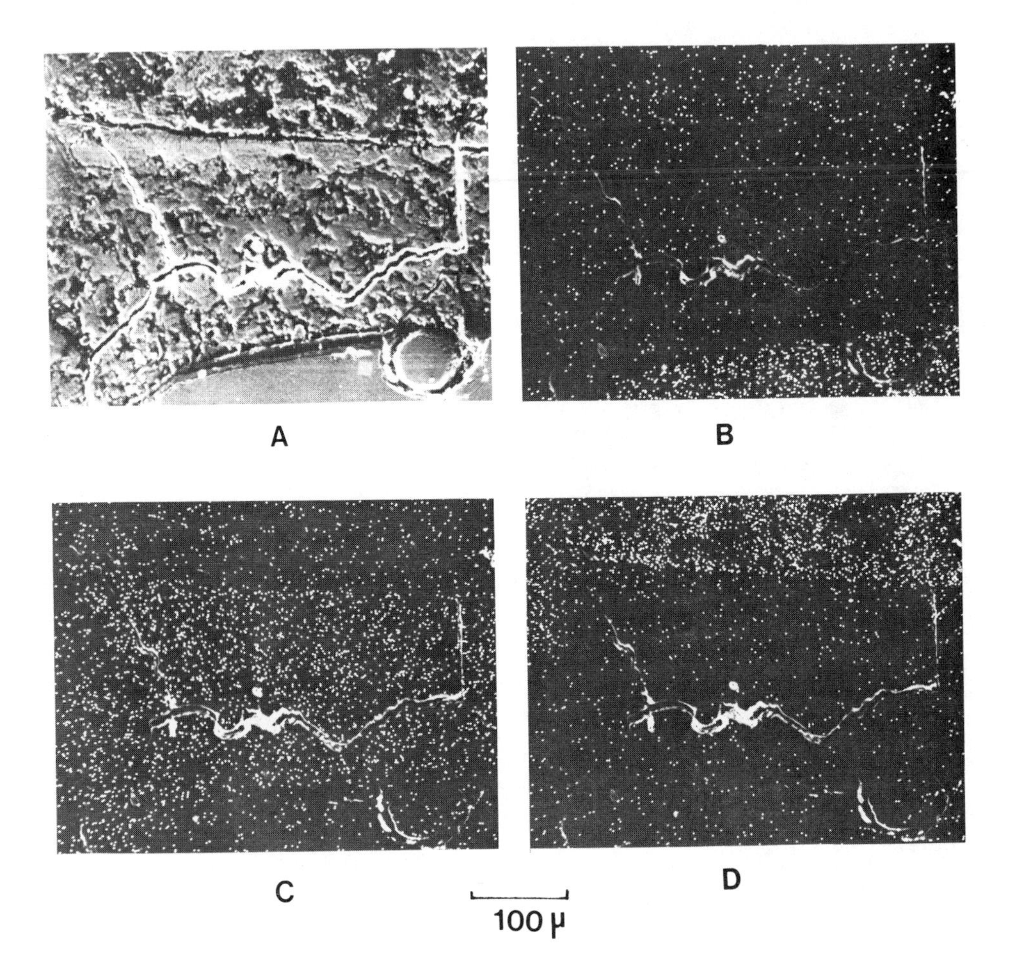

Fig. 4. SEM/EDAX observation of the palagonitized rind of the basaltic core

(a) secondary electrons image of the contact between basalt and oxide
(b) Ca image showing the limit of the basaltic glass rich in Ca
(c) K image, showing the concentration of K in the palagonite
(d) Mn image, showing the sharp contact between Fe-Mn oxide layer and the core.

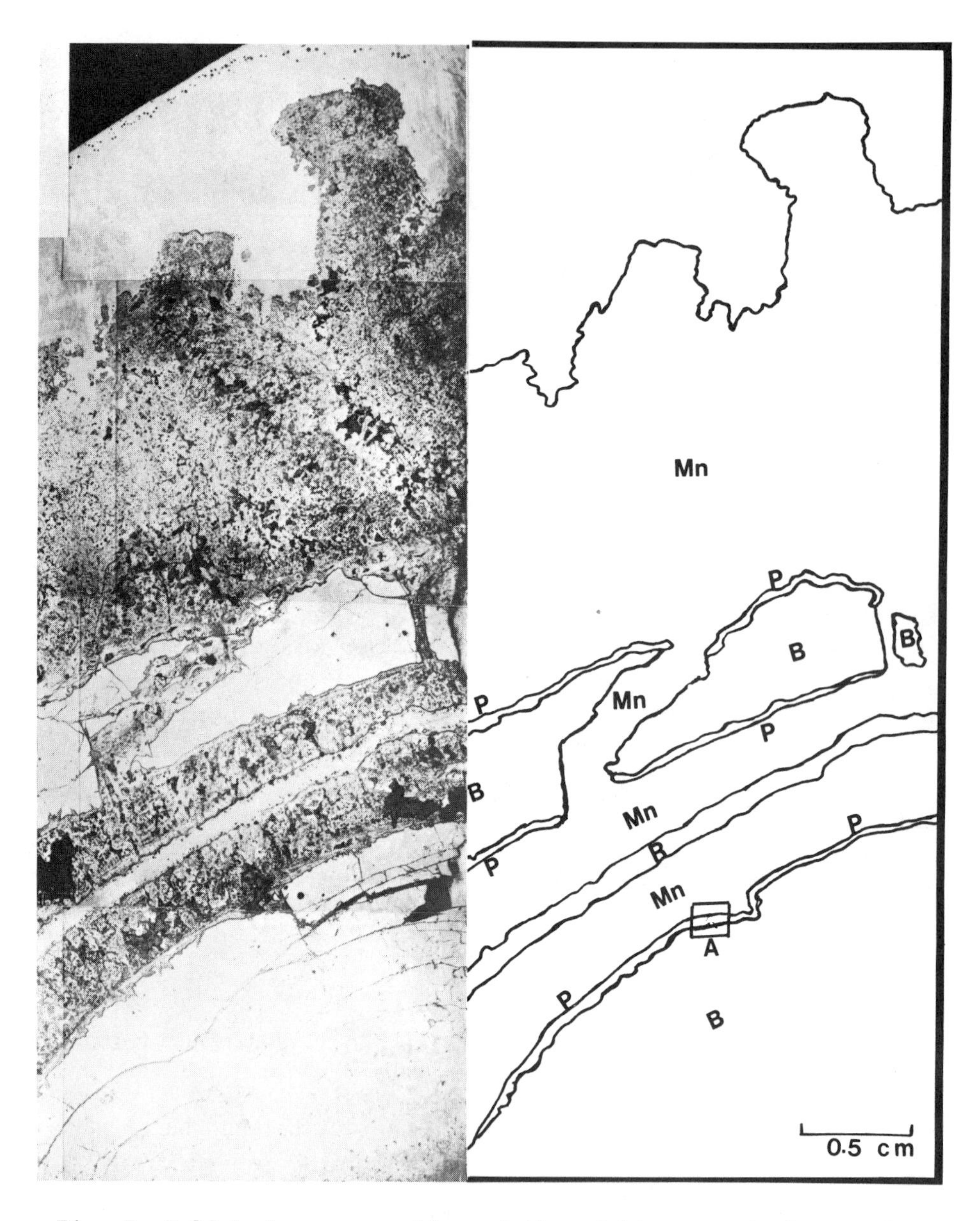

Fig. 5. Polished cross section of the right part of the sample
Mn = Fe-Mn deposit ; B = basaltic glass of the core ; B_2 = second basaltic glass layer ; P = palagonitized layer ; R = epoxy resin in a crack ; A = limit of the area represented in Fig. 4 .

of the contact between the fresh basaltic glass and the Fe-Mn deposit; in (b) the Ca image, showing clearly the limit of the basalt, calcium being depleted in the palagonitized band; in (c) the K image showing the enrichment of the palagonite and in (d) the Mn image, showing the sharp contrast between the palagonitized band and the Fe-Mn deposit. The entire cross section of the sample is presented on Fig. 5, together with a schematic diagram showing the different layers as determined by EDAX analysis. This shows that the palagonitized thickness is quite constant with a mean value of 350 microns, leading to an hydration-rind age of about 120,000 years. Using this "age" to calculate the Fe-Mn oxide accretion rate, a value of 333 mm/10^6 years is obtained.

$\Sigma^{230}Th/cm^2$ dating

According to Bonatti and Nayudu (1965), Arrhenius (1967) and Cherdyntsev (1971), there is another hypothesis for nodule formation. After a rapid formation without inclusion of unsupported ^{230}Th, the nodule, staying at the surface of sediment before being buried will receive a rain of unsupported ^{230}Th. In effect, different authors (Ku and Broecker, 1969, Krishnaswami et al, 1972) have noted that the integrated activity of unsupported ^{230}Th in nodules is deficient compared to the theoretical value that would be found if nodules were older than 0.5×10^6 years.

In this nodule, the integrated $^{230}Th/cm^2$ is 35.6 dpm if we take the measured values of layer 1 and 7, and 40 dpm/cm^2 if we use the corrected values for these layers. Using a ^{230}Th flux of 7.1 dpm/cm^2/1000 years at 3,200 meters depth, this leads to an "exposition age" of 5,000 to 5,600 years.

Discussion of dating results

The results obtained with the different methods, as well as the significance of these results are summarized in Table 3.

The different methods applied here to calculate growth rates give very scattered values by a factor of at least 50.

The first three methods in Table 3 give growth rates which are consistant but which are certainly too slow. K/A dating of the core gives a somewhat faster growth rate which must be taken as the lower limit, even if we consider that total degassing of previous argon has occurred.

However, there is a complete disagreement between these and the other methods. Of all the methods used to date the core, the one which seems least affected by environmental factors is fission track dating. Even this method gives only an upper limit for the age, and is corroborated by hydration-rind dating. The value found by this last method may be influenced by environmental conditions as in fact the rate of hydration used is calculated for normal deep-sea temperature conditions (around 1°C). If for any reason the temperature has been higher, the rate would have been faster and consequently the true age would be younger.

TABLE 3

Summary of the results of the different methods of dating

Method	Mn oxide growth rate mm/10^6 years	Observations
Dating basement by magnetic anomaly	8	Minimum values, assuming core contemporaneous of basement and Mn deposit beginning immediately
by sea-floor depth	10	
$^{230}Th_{exc}$	13	Assuming ^{230}Th activity decrease is a radioactive decrease. Problem of excess U and high $^{234}U/^{238}U$ ratio and of ^{230}Th excess too deep. This value implies an age of 3 x 10^6 years for the core.
K/A	23	Minimum value, assuming K/A dating of the core being not a too old apparent age (due to incomplete degassing) and Mn deposit beginning immediately.
Fission track	67 to 114	Minimum value, as fission track dating gives only upper limit to the age of the core (less than 600,000 or 330,000 years)
Hydration rind	333	Minimum value. Using a hydration rate of 2.91 μm/10^3 years, the age of the core would be 120,000 years
$\Sigma^{230}Th$	?	Assuming a rapid formation of oxide layer and that ^{230}Th is the amount of this isotope accumulated on the nodule after formation about 5,000 years ago.

Those considerations lead us to question the validity of the only method which theoretically gives a measure, and not a calculation, of the growth rate: the unsupported ^{230}Th activity decrease. In effect, we have seen that different anomalies exist in the uranium series distribution: (1) an excess of uranium in two layers, (2) ^{234}U/^{238}U ratio too high deep in the oxide, (3) unsupported ^{230}Th in layers as deep as 10 mm, incompatible with the growth rate of 13 mm/10^6 years as determined considering that unsupported ^{230}Th activity decrease is a radioactive decrease. We hypothesize that for this nodule the best approach is the last. The method using ^{230}Th/cm^2 to calculate an "exposition age" probably gives a somewhat lower limit. Two principal reasons may be invoked for this:

a) during dredging operations, the surface of the sample may have had a part of its ^{230}Th washed out,

b) the ^{230}Th flux calculated is theoretical and, although Fe-Mn oxides are known as good scavengers (Goldberg, 1954), the effective flux on the ocean bottom may be lower than the theoretical value (Ku, 1976).

With such an hypothesis, we accept the idea of Bonatti and Nayudu (1965), Arrhenius (1967) and Cherdintsev (1971) of an abrupt formation of the oxide layer, and, as this hypothesis would imply hydrothermal or volcanic source for the metals, we need to look to the structure and/or chemical composition of this nodule to test this hypothesis.

SEM/EDAX Study

The SEM/EDAX study was carried out on different parts of the sample, after embedding in epoxy resin and polishing:

a) a cauliflower structure sliced through the middle and taken from the top surface.

b) a piece in the complex structure of the right side (fig.2), which is shown in Figure 5.

c) eleven successive slices about 1 mm thick have been sawed perpendicularly to the axis of a cauliflower structure in an effort to illustrate whether or not the sampling for uranium series measurements are truly representative.

When observing samples "a" and "b", it appears that Fe-Mn deposit has no well developed structures as it is made of a mixture of iron and manganese hydroxides, very poorly crystallized, enclosing patches of sedimentary material rich in foraminifera, coccoliths, some quartz and feldspar. Some of the foraminifera retain their calcitic structures, show evidence of dissolution; others, like those in Figure 6 (a) have their external structure transformed into silica and iron, with some calcite in their central part. Coccoliths are either transformed into silica and iron or into barium sulfate or, as shown in Figure 6, show a nearly complete alteration by dissolution. The different minerals are not

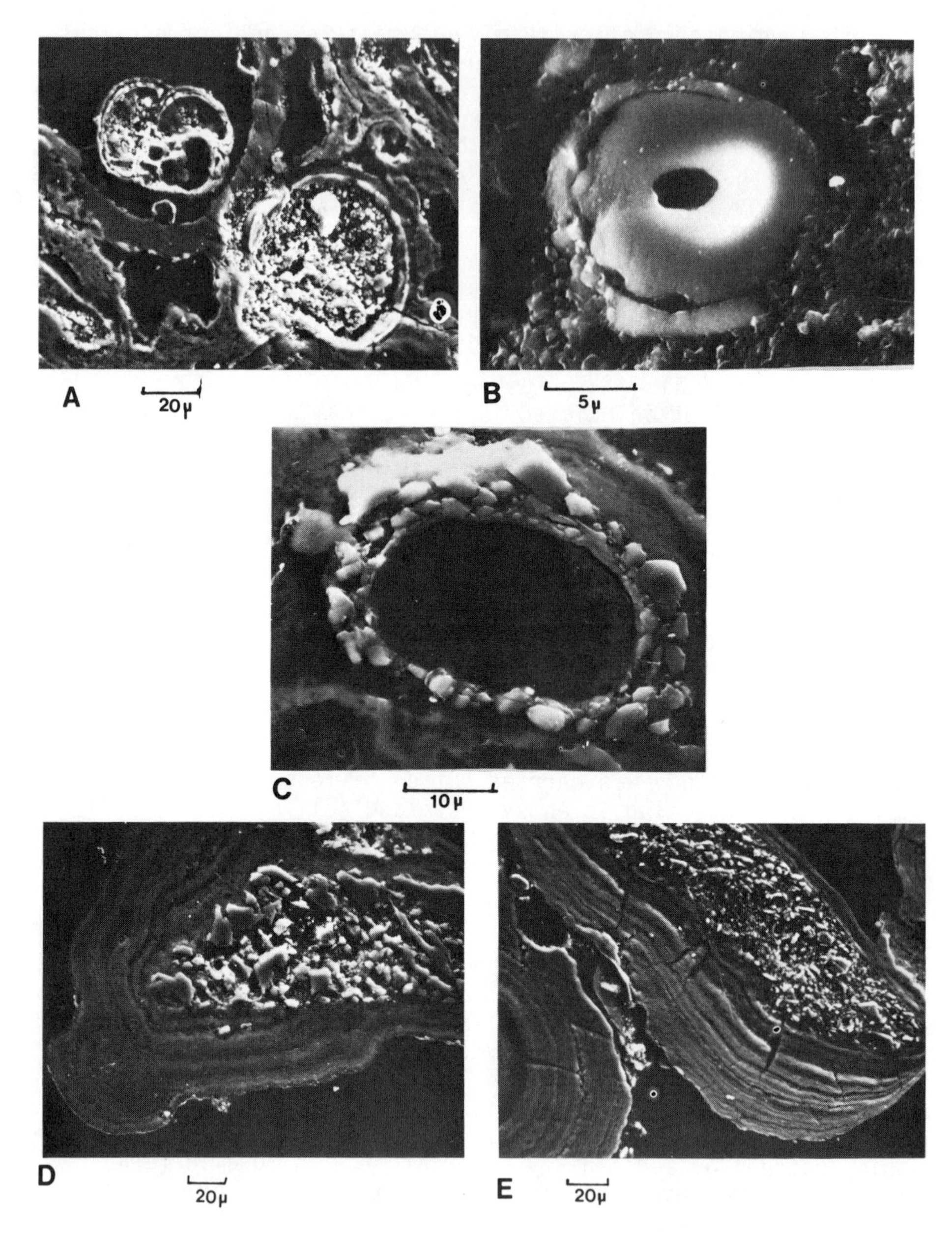

Fig. 6. Secondary electron image of (a) foraminiferal test partially transformed to Si and Fe; (b) coccolith showing alteration by dissolution; (c) quartz and feldspar around void; (d) Fe-Mn botrvoid with calcitic precipitation; (e) Fe-Mn botryoid with pure Si precipitation

randomly distributed in the sedimentary patches, but generally grouped around voids (Fig. 6 (c)).

Another morphological characteristic of this sample is the presence of precipitates of calcite (Fig. 6 (d) or of silica (Fig. 6 (e) in nearly identical Fe-Mn botryoides. In contrast, sections (samples c) made perpendicular to the direction of samples "a" and "b" show substructures developed across the direction of the sampling (Fig. 7) showing that the sampling for radiochemistry cannot represent a chronological succession. This sample is characterized by the presence, in all the oxide, of sulfur and of barium sulfate particles. Figure 8 shows in (a) the secondary electron image made in a deep part of the oxide; in (b) it can be seen, by a "barium image" that the two curved structures are composed entirely of barium sulfate, while (c) shows that the feature between the two is composed of silica. Figure 8 (d) shows that Mn is completely segregated from these structures of barite and silica. Figure 8 is a greater enlargement of a part of the barium structure showing that it is made with an accumulation of microorganisms and coccoliths transformed into barite as shown by their EDAX spectrum (Fig. 8 (f)).

Such accumulation of barite particles has been seen generally well localized in different parts of the oxide layer. Even though we have not been able to separate these particles for accurate chemical analysis, due to their small size, we think that they may be responsible for the anomalous uranium content. We have verified that there are also such concentrations of barite particles at the surface of the sample (and also that the very high concentration of uranium is not found everywhere on the surface of the sample, which indicates that uranium excess is also well localized).

The unusual quantity of barite, the high uranium content of some layers, and the high $^{234}U/^{238}U$ activity ratio strongly suggests a hydrothermal origin for this precipitate (Bonatti et al., 1972 a and b).

Another interesting feature may be seen in Figure 5, in the second basaltic glass layer (b_2). This layer is crossed by a fracture which is filled with Fe-Mn oxide. The edges of the basaltic glass show a hydration rind, except in the fracture itself. This may give a more precise possible "age" of this sample. In effect, this indicates that the filling of this fracture has occurred at a time sufficiently recent so that no hydration rind can be detected and gives support to a very rapid formation of the oxide layer about 5,000 years ago, as indicated by the "exposition age".

Summary and Conclusions

The sample we have studied comes from a relatively young area of the North Pacific Ocean. It is located between the East Pacific Rise, the Siqueiros fracture zone and Clipperton fracture zone, consequently, in an area where recent volcanic or hydrothermal

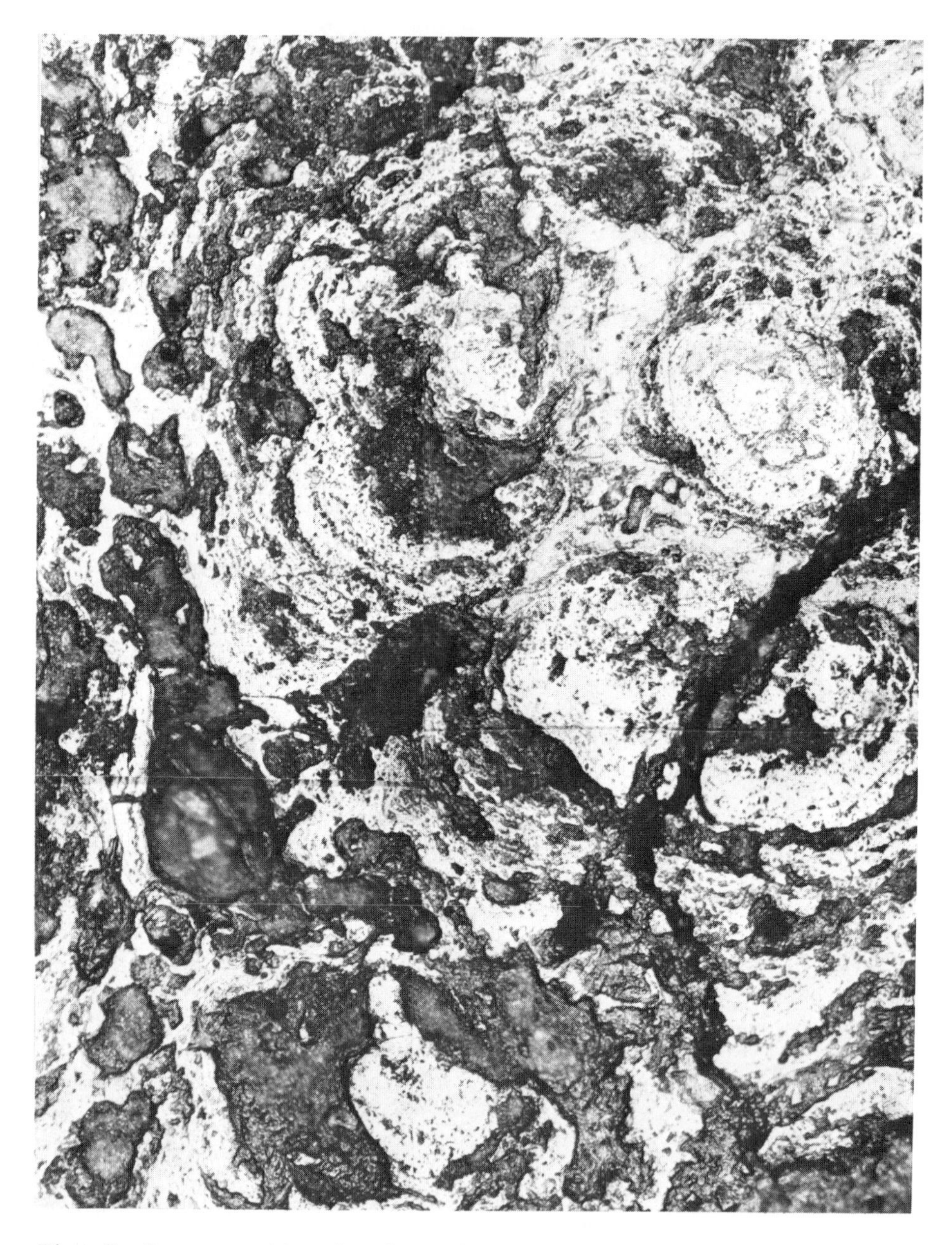

Fig. 7. Cross section in the oxide, perpendicular to the section shown in Fig. 5, showing the substrucures of the oxide which are crossed during radiochemical sampling. This layer is about 10 mm depth in the oxide.

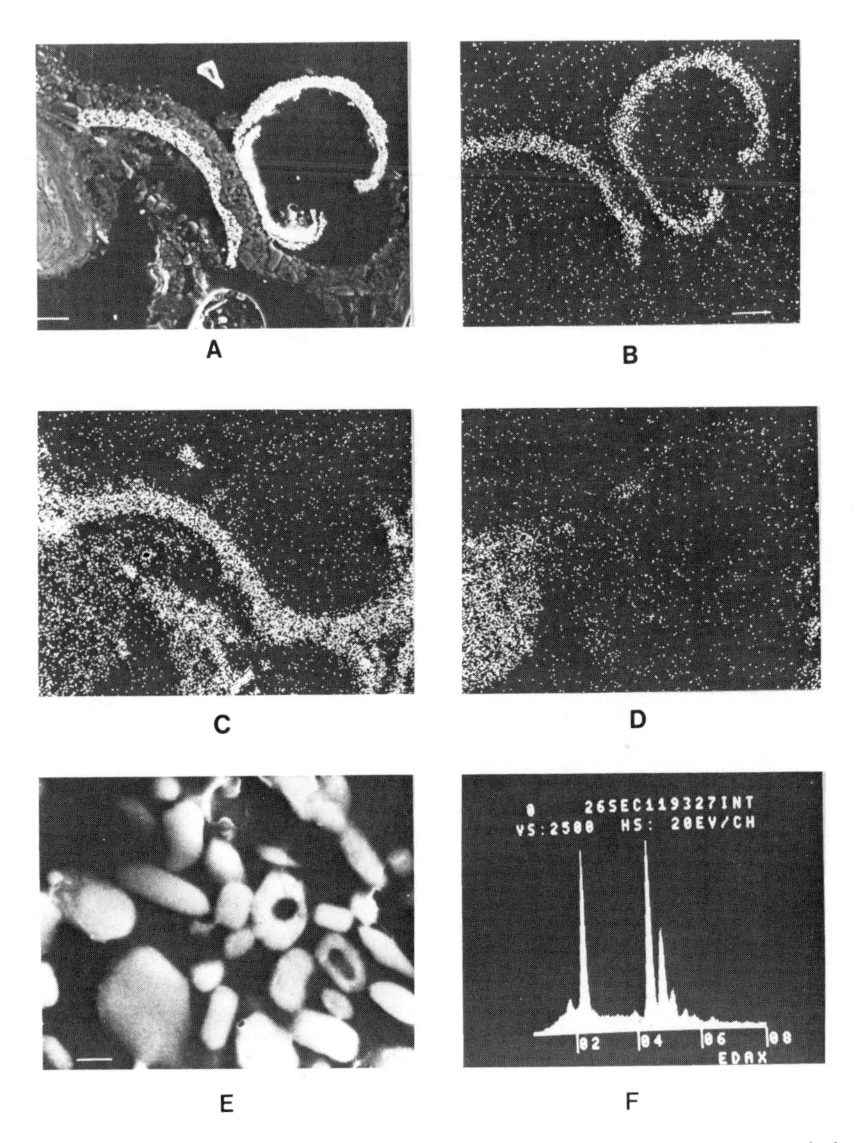

Fig. 8. Barite particles in the Fe-Mn oxide of the nodule. (a) Secondary electron images (white bar is 20 microns); (b) barium; (c) silicon; (d) manganese; (e) larger magnification of a part of barite crown showing the presence of micro-organisms and coccoliths epigenized into barite (white bar is 1 micron); (f) EDAX spectrum of the particles shown in (e) typical of $BaSO_4$.

activity may have taken place. The fission track age, hydration rind age and exposition age are consistent with a recent origin.

As the SEM/EDAX study shows characteristics that may be attributed to hydrothermal activity (high content of barite, transformation of micro-organisms into Si-Fe material, high level of dissolution of calcium carbonate, purely chemical deposition of silica and high uranium content with high isotopic activity ratio), we may propose for this sample the following model:

The basaltic block which acts as a core, reached the sea floor not earlier than about 120,000 years BP; 5,000 years ago a hydrothermal process formed the oxide precipitate which then received the ^{230}Th rain from the ocean. This ^{230}Th partially penetrates the very porous material of the Fe-Mn oxide deposit giving an apparently decreasing gradient.

Acknowledgements

We thank Dr. J. Greenslate, from Scripps Institution of Oceanography, for providing the sample. Dr. W. Dudley helped us in manuscript preparation. Dr. C. Cassignol and P.Y. Gillot performed the K/A dating.

The authors are particularly indebted to Dr. J. Labeyrie, J.C. Duplessy and L. Labeyrie, Jr. for comments and constructive criticisms.

This study was supported by DGRST grant 77-7-0018.

REFERENCES

Arrhenius, G., Deep sea sedimentation, Trans. Amer. Geophys. Union U.S. National report (1963-1967), 604 (1967).

Aumento, F., The Mid-Atlantic Ridge near 45°N. Fission track and manganese chronology, Can. J. Earth Sc. 6, 1431 (1969).

Barnes, S.S. and Dymond, J.R., Rate of accumulation of ferromanganese nodules, Nature 213, 1218 (1967)

Bender, M.L., Ku, T.L. and Broecker, W.S., Manganese nodules, their evolution, Science 151, 325 (1966).

Bhat, S.G., Krishnaswami, S., Lal, D.R. and Somayajulu, B.L.K., Radiometric and trace elemental studies of ferromanganese nodules, Proc. Intern. Symp. Hydrochem. and Biochem., Tokyo, Sept. 1970, The Clark Co. Wash.DC, I, 443 (1973).

Bonatti, E., Fisher, D.E., Joensuu, O., Rydell, H.S. and Beyth, M. Iron manganese barium deposit from the Afar rift (Ethiopia), Econ. Geol. 67, 717 (1972a).

Bonatti, E., Honnorez, J., Joensuu, O., and Rydell, H., Submarine iron deposits from the Mediterranean sea, In The Mediterranean sea (editor D.J. Stanley) Hutchinson and Ross, Strondsburg, Pa. 701 (1972b).

Bonatti, E. and Nayudu R., The origin of manganese nodules on the ocean floor, Amer. J. Sci. 263, 17 (1965).

Cherdyntsev, V.V., Kadyrova, N.B. and Novichkova, N., Origin of manganese nodules of the Pacific ocean from radioisotope data, Trans. from Geokhimiya 3, 339 (1971).

Fleischer, R.L., Price, P.B. and Walker, R.M., Nuclear tracks in solids, University California Press, San Francisco, (1975).

Fleischer, R.L., Viertl, J.R.M., Price, P.B. and Aumento, F., Mid-Atlantic Ridge : age and spreading rate, Science 161, 1339 (1968).

Funkhouser, J.G., Fisher, D.E. and Bonatti, E., Excess argon in deep sea rocks, Earth Planet. Sci. Lett. 5, 95 (1968).

Goldberg, E.D., Marine geochemistry I-Chemical scavengers of the sea, J. Geol. 62, 249 (1954).

Krishnaswami, S., Somayajulu, B.L.K. and Moore, W.S., Dating of manganese nodules using beryllium 10, In Ferromanganese deposits on the ocean floor, (editor D.R. Horn) IDOE-NSF 117 (1972).

Ku, T.L., The uranium series methods of age determination, Ann.Rev. Earth Planet. Sci. 4, 347 (1976).

Ku, T.L. and Broecker, W.S., Uranium, thorium, protactinium in a manganese nodule, Earth Planet. Sci. Lett. 2, 317 (1967).

Ku, T.L. and Broecker, W.S., Radiochemical study of manganese nodules of deep sea origin, Deep Sea Res. 16, 625 (1969).

Lalou, C., Brichet, E., Ku, T.L., Poupeau, G. and Romary, P., Techno incrustation : I. Radiochemical study and possible growth rate. In Colloque Intern. CNRS n°289, September 1978.

Mac Dougall, J.D., Fission track annealing and correction procedures for oceanic basalt glasses, Earth Planet. Sci. Lett. 30, 19 (1976).

Morgenstein, M., Composition and development of palagonite in deep sea sediments from the Atlantic and Pacific oceans, Master's thesis, Syracuse Univ. (1969).

Morgenstein, M., Sedimentary diagenesis and rate of manganese accretion on the Waho shelf, Kauai channel, Hawaii, In Investigations of ferromanganese deposits from the central Pacific (HIG 72-23, Hawaii Inst. Geophys.) 1 (1972).

Sclater, J.G., Anderson, R.N. and Lee Bell, M., Elevation of ridges and evolution of the central eastern Pacific, J. Geophys. Res. 76, 7888 (1971).

APPENDIX

The appendix of this volume consists of 6 microfiche cards (in a pocket attached to the inside of the back cover) which provide the raw data for the various papers included in the volume. The data for each article start on the card and page indicated.

INDEX